Lecture Notes in Computer Science　　9474

Commenced Publication in 1973
Founding and Former Series Editors:
Gerhard Goos, Juris Hartmanis, and Jan van Leeuwen

More information about this series at http://www.springer.com/series/7412

George Bebis · Richard Boyle
Bahram Parvin · Darko Koracin
Ioannis Pavlidis · Rogerio Feris
Tim McGraw · Mark Elendt
Regis Kopper · Eric Ragan
Zhao Ye · Gunther Weber (Eds.)

Advances in Visual Computing

11th International Symposium, ISVC 2015
Las Vegas, NV, USA, December 14–16, 2015
Proceedings, Part I

 Springer

Editors

George Bebis
University of Nevada
Reno, NV, USA

Tim McGraw
Purdue University
West Lafayette, IN, USA

Richard Boyle
NASA Ames Research Center
Moffett Field, CA, USA

Mark Elendt
Side Effects Software
Santa Monica, CA, USA

Bahram Parvin
Lawrence Berkeley National Laboratory
Berkeley, CA, USA

Regis Kopper
The DiVE
Durham, NC, USA

Darko Koracin
Desert Research Institute
Reno, NV, USA

Eric Ragan
Texas A&M University
College Station, TX, USA

Ioannis Pavlidis
University of Houston
Houston, TX, USA

Zhao Ye
Kent State University
Kent, OH, USA

Rogerio Feris
IBM T.J. Watson Research Center
Yorktown Heights, NY, USA

Gunther Weber
Lawrence Berkeley National Laboratory
Berkeley, CA, USA

ISSN 0302-9743   ISSN 1611-3349   (electronic)
Lecture Notes in Computer Science
ISBN 978-3-319-27856-8   ISBN 978-3-319-27857-5   (eBook)
DOI 10.1007/978-3-319-27857-5

Library of Congress Control Number: 2015957779

LNCS Sublibrary: SL6 – Image Processing, Computer Vision, Pattern Recognition, and Graphics

Printed on acid-free paper

This Springer imprint is published by SpringerNature
The registered company is Springer International Publishing AG Switzerland

Preface

It is with great pleasure that we welcome you to the proceedings of the 11th International Symposium on Visual Computing (ISVC 2015), which was held in Las Vegas, Nevada, USA. ISVC provides a common umbrella for the four main areas of visual computing including vision, graphics, visualization, and virtual reality. The goal is to provide a forum for researchers, scientists, engineers, and practitioners throughout the world to present their latest research findings, ideas, developments, and applications in the broader area of visual computing.

This year, the program consisted of 16 oral sessions, one poster session, eight special tracks, and six keynote presentations. The response to the call for papers was very good; we received over 260 submissions for the main symposium from which we accepted 80 papers for oral presentation and 35 papers for poster presentation. Special track papers were solicited separately through the Organizing and Program Committees of each track. A total of 43 papers were accepted for oral presentation in the special tracks.

All papers were reviewed with an emphasis on the potential to contribute to the state of the art in the field. Selection criteria included accuracy and originality of ideas, clarity and significance of results, and presentation quality. The review process was quite rigorous, involving two to three independent blind reviews followed by several days of discussion. During the discussion period we tried to correct anomalies and errors that might have existed in the initial reviews. Despite our efforts, we recognize that some papers worthy of inclusion may have not been included in the program. We offer our sincere apologies to authors whose contributions might have been overlooked.

We wish to thank everybody who submitted their work to ISVC 2015 for review. It was because of their contributions that we succeeded in having a technical program of high scientific quality. In particular, we would like to thank the ISVC 2015 area chairs, the organizing institutions (UNR, DRI, LBNL, and NASA Ames), the industrial sponsors (BAE Systems, Intel, Ford, Hewlett Packard, Mitsubishi Electric Research Labs, Toyota, General Electric), the international Program Committee, the special track organizers and their Program Committees, the keynote speakers, the reviewers, and especially the authors who contributed their work to the symposium. In particular, we would like to express our appreciation to MERL and Drs. Jay Thornton and Mike Jones for sponsoring the "best" paper award this year.

We sincerely hope that ISVC 2015 offered participants opportunities for professional growth.

October 2015

George Bebis
ISVC'15 Steering Committee and Area Chairs

Organization

Steering Committee

Bebis George	University of Nevada, Reno, USA
Boyle Richard	NASA Ames Research Center, USA
Parvin Bahram	Lawrence Berkeley National Laboratory, USA
Koracin Darko	Desert Research Institute, USA

Area Chairs

Computer Vision

Pavlidis Ioannis	University of Houston, USA
Feris Rogerio	IBM, USA

Computer Graphics

McGraw Tim	Purdue University, USA
Elendt Mark	Side Effects Software Inc., USA

Virtual Reality

Kopper Regis	Duke University, USA
Ragan Eric	Texas A&M University, USA

Visualization

Ye Zhao	Kent State University, USA
Weber Gunther	Lawrence Berkeley National Laboratory, USA

Publicity

Erol Ali	Eksperta Software, Turkey

Local Arrangements

Morris Brendan	University of Nevada, Las Vegas, USA

Special Tracks

Wang Junxian	Microsoft, USA

Keynote Speakers

Ravi Ramamoorthi	University of California, San Diego, USA
Benjamin Kimia	Brown University, USA
Claudio Silva	New York University, USA
Oncel Tuzel	Mitsubishi Electric Research Laboratories, USA
Evan Suma	University of Southern California, USA
Luc Vincent	Google, USA

International Program Committee

(Area 1) Computer Vision

Abidi Besma	University of Tennessee at Knoxville, USA
Abou-Nasr Mahmoud	Ford Motor Company, USA
Aboutajdine Driss	National Center for Scientific and Technical Research, Morocco
Aggarwal J.K.	University of Texas, Austin, USA
Albu Branzan Alexandra	University of Victoria, Canada
Amayeh Gholamreza	Foveon, USA
Ambardekar Amol	Microsoft, USA
Angelopoulou Elli	University of Erlangen-Nuremberg, Germany
Agouris Peggy	George Mason University, USA
Argyros Antonis	University of Crete, Greece
Asari Vijayan	University of Dayton, USA
Athitsos Vassilis	University of Texas at Arlington, USA
Basu Anup	University of Alberta, Canada
Bekris Kostas	Rutgers University, USA
Bhatia Sanjiv	University of Missouri – St. Louis, USA
Bimber Oliver	Johannes Kepler University Linz, Austria
Bourbakis Nikolaos	Wright State University, USA
Brimkov Valentin	State University of New York, USA
Cavallaro Andrea	Queen Mary, University of London, UK
Charalampidis Dimitrios	University of New Orleans, USA
Chellappa Rama	University of Maryland, USA
Chen Yang	HRL Laboratories, USA
Cheng Hui	Sarnoff Corporation, USA
Cheng Shinko	HRL Labs, USA
Cui Jinshi	Peking University, China
Dagher Issam	University of Balamand, Lebanon
Darbon Jerome	CNRS-Ecole Normale Superieure de Cachan, France
Demirdjian David	Vecna Robotics, USA
Diamantas Sotirios	Ecole Nationale Superieure de Mecanique et des Microtechniques, France
Duan Ye	University of Missouri – Columbia, USA

Doulamis Anastasios	Technical University of Crete, Greece
Dowdall Jonathan	Google, USA
El-Ansari Mohamed	Ibn Zohr University, Morocco
El-Gammal Ahmed	University of New Jersey, USA
Eng How Lung	Institute for Infocomm Research, Singapore
Erol Ali	ASELSAN, Turkey
Fan Guoliang	Oklahoma State University, USA
Fan Jialue	Northwestern University, USA
Ferri Francesc	Universitat de Valencia, Spain
Ferzli Rony	Intel, USA
Ferryman James	University of Reading, UK
Foresti GianLuca	University of Udine, Italy
Fowlkes Charless	University of California, Irvine, USA
Fukui Kazuhiro	The University of Tsukuba, Japan
Galata Aphrodite	The University of Manchester, UK
Georgescu Bogdan	Siemens, USA
Goh Wooi-Boon	Nanyang Technological University, Singapore
Ghouzali Sanna	King Saud University, Saudi Arabia
Guerra-Filho Gutemberg	Intel, USA
Guevara Angel Miguel	University of Porto, Portugal
Gustafson David	Kansas State University, USA
Hammoud Riad	BAE Systems, USA
Harville Michael	Hewlett Packard Labs, USA
He Xiangjian	University of Technology, Sydney, Australia
Heikkil Janne	University of Oulu, Finland
Hongbin Zha	Peking University, China
Hou Zujun	Institute for Infocomm Research, Singapore
Hua Gang	IBM T.J. Watson Research Center, USA and Stevens Institute, USA
Huang Yongzhen	Chinese Academy of Sciences, China
Imiya Atsushi	Chiba University, Japan
Kamberov George	Stevens Institute of Technology, USA
Kampel Martin	Vienna University of Technology, Austria
Kamberova Gerda	Hofstra University, USA
Kakadiaris Ioannis	University of Houston, USA
Kettebekov Sanzhar	Keane Inc., USA
Kimia Benjamin	Brown University, USA
Kisacanin Branislav	Texas Instruments, USA
Klette Reinhard	Auckland University of Technology, New Zealand
Kokkinos Iasonas	Ecole Centrale de Paris, France
Kollias Stefanos	National Technical University of Athens, Greece
Komodakis Nikos	Ecole Centrale de Paris, France
Kosmopoulos Dimitrios	Technical Educational Institute of Crete, Greece
Kozintsev Igor	Intel, USA
Kuno Yoshinori	Saitama University, Japan
Kim Kyungnam	HRL Laboratories, USA

Latecki Longin Jan	Temple University, USA
Lee D.J.	Brigham Young University, USA
Levine Martin	McGill University, Canada
Li Baoxin	Arizona State University, USA
Li Chunming	Vanderbilt University, USA
Li Xiaowei	Google Inc., USA
Lim Ser N.	GE Research, USA
Lisin Dima	VidoeIQ, USA
Lee Hwee Kuan	Bioinformatics Institute, A*STAR, Singapore
Lee Seong-Whan	Korea University, Korea
Li Shuo	GE Healthcare, Canada
Lourakis Manolis	ICS-FORTH, Greece
Loss Leandro	Lawrence Berkeley National Laboratory, USA
Luo Gang	Harvard University, USA
Ma Yunqian	Honyewell Labs, USA
Maeder Anthony	University of Western Sydney, Australia
Makrogiannis Sokratis	Delaware State University, USA
Maltoni Davide	University of Bologna, Italy
Maroulis Dimitris	National University of Athens, Greece
Maybank Steve	Birkbeck College, UK
Medioni Gerard	University of Southern California, USA
Melenchn Javier	Universitat Oberta de Catalunya, Spain
Metaxas Dimitris	Rutgers University, USA
Ming Wei	Konica Minolta Laboratory, USA
Mirmehdi Majid	Bristol University, UK
Morris Brendan	University of Nevada, Las Vegas, USA
Mueller Klaus	Stony Brook University, USA
Muhammad Ghulam	King Saud University, Saudi Arabia
Mulligan Jeff	NASA Ames Research Center, USA
Murray Don	Point Grey Research, Canada
Nait-Charif Hammadi	Bournemouth University, UK
Nefian Ara	NASA Ames Research Center, USA
Nguyen Quang Vinh	University of Western Sydney, Australia
Nicolescu Mircea	University of Nevada, Reno, USA
Nixon Mark	University of Southampton, UK
Nolle Lars	The Nottingham Trent University, UK
Ntalianis Klimis	National Technical University of Athens, Greece
Or Siu Hang	The Chinese University of Hong Kong, Hong Kong, SAR China
Papadourakis George	Technological Education Institute, Greece
Papanikolopoulos Nikolaos	University of Minnesota, USA
Pati Peeta Basa	CoreLogic, India
Patras Ioannis	Queen Mary University, London, UK
Pavlidis Ioannis	University of Houston, USA
Petrakis Euripides	Technical University of Crete, Greece
Peyronnet Sylvain	LRI, University of Paris-Sud, France

Pinhanez Claudio	IBM Research, Brazil
Piccardi Massimo	University of Technology, Australia
Pietikainen Matti	LRDE/University of Oulu, Finland
Pitas Ioannis	Aristotle University of Thessaloniki, Greece
Porikli Fatih	Australian National University, Australia
Prabhakar Salil	DigitalPersona Inc., USA
Prokhorov Danil	Toyota Research Institute, USA
Qian Gang	Arizona State University, USA
Raftopoulos Kostas	National Technical University of Athens, Greece
Regazzoni Carlo	University of Genoa, Italy
Regentova Emma	University of Nevada, Las Vegas, USA
Remagnino Paolo	Kingston University, UK
Ribeiro Eraldo	Florida Institute of Technology, USA
Robles-Kelly Antonio	National ICT Australia (NICTA), Australia
Ross Arun	Michigan State University, USA
Rziza Mohammed	Agdal Mohammed-V University, Morocco
Samal Ashok	University of Nebraska, USA
Samir Tamer	Allegion, USA
Sandberg Kristian	Computational Solutions, USA
Sarti Augusto	DEI Politecnico di Milano, Italy
Savakis Andreas	Rochester Institute of Technology, USA
Schaefer Gerald	Loughborough University, UK
Scalzo Fabien	University of California at Los Angeles, USA
Scharcanski Jacob	UFRGS, Brazil
Shah Mubarak	University of Central Florida, USA
Shi Pengcheng	Rochester Institute of Technology, USA
Shimada Nobutaka	Ritsumeikan University, Japan
Singh Rahul	San Francisco State University, USA
Skodras Athanassios	University of Patras, Greece
Skurikhin Alexei	Los Alamos National Laboratory, USA
Souvenir Richard	University of North Carolina – Charlotte, USA
Su Chung-Yen	National Taiwan Normal University, Taiwan (R.O.C.)
Sugihara Kokichi	University of Tokyo, Japan
Sun Zehang	Apple, USA
Syeda-Mahmood Tanveer	IBM Almaden, USA
Tan Kar Han	Hewlett Packard, USA
Tavakkoli Alireza	University of Houston – Victoria, USA
Tavares Joao	Universidade do Porto, Portugal
Teoh Eam Khwang	Nanyang Technological University, Singapore
Thiran Jean-Philippe	Swiss Federal Institute of Technology Lausanne (EPFL), Switzerland
Tistarelli Massimo	University of Sassari, Italy
Tong Yan	University of South Carolina, USA

Tsui T.J.	Chinese University of Hong Kong, Hong Kong, SAR China
Trucco Emanuele	University of Dundee, UK
Tubaro Stefano	Politecnico di Milano, Italy
Uhl Andreas	Salzburg University, Austria
Velastin Sergio	Kingston University London, UK
Veropoulos Kostantinos	GE Healthcare, Greece
Verri Alessandro	Università di Genova, Italy
Wang Junxian	Microsoft, USA
Wang Song	University of South Carolina, USA
Wang Yunhong	Beihang University, China
Webster Michael	University of Nevada, Reno, USA
Wolff Larry	Equinox Corporation, USA
Wong Kenneth	The University of Hong Kong, Hong Kong, SAR China
Xiang Tao	Queen Mary, University of London, UK
Xu Meihe	University of California at Los Angeles, USA
Yang Ming-Hsuan	University of California at Merced, USA
Yang Ruigang	University of Kentucky, USA
Yin Lijun	SUNY at Binghampton, USA
Yu Ting	GE Global Research, USA
Yu Zeyun	University of Wisconsin-Milwaukee, USA
Yuan Chunrong	University of Tübingen, Germany
Zabulis Xenophon	ICS-FORTH, Greece
Zervakis Michalis	Technical University of Crete, Greece
Zhang Jian	Wake Forest University, USA
Zheng Yuanjie	University of Pennsylvania, USA
Zhang Yan	Delphi Corporation, USA
Ziou Djemel	University of Sherbrooke, Canada

(Area 2) Computer Graphics

Abd Rahni Mt. Piah	Universiti Sains Malaysia, Malaysia
Abram Greg	Texas Advanced Computing Center, USA
Adamo-Villani Nicoletta	Purdue University, USA
Agu Emmanuel	Worcester Polytechnic Institute, USA
Andres Eric	Laboratory XLIM-SIC, University of Poitiers, France
Artusi Alessandro	GiLab, Universitat de Girona, Spain
Baciu George	Hong Kong Poly, Hong Kong, SAR China
Balcisoy Selim Saffet	Sabanci University, Turkey
Barneva Reneta	State University of New York, USA
Belyaev Alexander	Heriot-Watt University, UK
Benes Bedrich	Purdue University, USA
Berberich Eric	Max Planck Institute, Germany
Bilalis Nicholas	Technical University of Crete, Greece
Bimber Oliver	Johannes Kepler University Linz, Austria

Bouatouch Kadi	University of Rennes I, IRISA, France
Brimkov Valentin	State University of New York, USA
Brown Ross	Queensland University of Technology, Australia
Bruckner Stefan	Vienna University of Technology, Austria
Callahan Steven	University of Utah, USA
Capin Tolga	Bilkent University, Turkey
Chaudhuri Parag	Indian Institute of Technology Bombay, India
Chen Min	University of Oxford, UK
Cheng Irene	University of Alberta, Canada
Chiang Yi-Jen	New York University, USA
Choi Min-Hyung	University of Colorado at Denver, USA
Comba Joao	Universidade Federal do Rio Grande do Sul, Brazil
Cremer Jim	University of Iowa, USA
Culbertson Bruce	HP Labs, USA
Dana Kristin	Rutgers University, USA
Debattista Kurt	University of Warwick, UK
Deng Zhigang	University of Houston, USA
Dick Christian	Technical University of Munich, Germany
Dingliana John	Trinity College, Ireland
El-Sana Jihad	Ben-Gurion University of The Negev, Israel
Entezari Alireza	University of Florida, USA
Fabian Nathan	Sandia National Laboratories, USA
De Floriani Leila	University of Genoa, Italy
Fu Hongbo	City University of Hong Kong, Hong Kong, SAR China
Fuhrmann Anton	VRVis Research Center, Austria
Gaither Kelly	University of Texas at Austin, USA
Gao Chunyu	Epson Research and Development, USA
Geist Robert	Clemson University, USA
Gelb Dan	Hewlett Packard Labs, USA
Gotz David	University of North Carolina at Chapel Hill, USA
Gooch Amy	University of Victoria, Canada
Gu David	Stony Brook University, USA
Guerra-Filho Gutemberg	Intel, USA
Habib Zulfiqar	COMSATS Institute of Information Technology, Lahore, Pakistan
Hadwiger Markus	KAUST, Saudi Arabia
Haller Michael	Upper Austria University of Applied Sciences, Austria
Hamza-Lup Felix	Armstrong Atlantic State University, USA
Han JungHyun	Korea University, Korea
Hand Randall	Lockheed Martin Corporation, USA
Hao Xuejun	Columbia University and NYSPI, USA
Hernandez Jose Tiberio	Universidad de los Andes, Colombia
Hou Tingbo	Google Inc., USA
Huang Jian	University of Tennessee at Knoxville, USA
Huang Mao Lin	University of Technology, Australia

Huang Zhiyong	Institute for Infocomm Research, Singapore
Hussain Muhammad	King Saud University, Saudi Arabia
Jeschke Stefan	IST Austria, Austria
Jones Michael	Brigham Young University, USA
Julier Simon J.	University College London, UK
Kamberov George	Stevens Institute of Technology, USA
Klosowski James	AT&T Research Labs, USA
Ko Hyeong-Seok	Seoul National University, Korea
Lai Shuhua	Virginia State University, USA
Le Binh	Virginia Disney Research, USA
Lewis R. Robert	Washington State University, USA
Li Bo	Samsung, USA
Li Frederick	University of Durham, UK
Li Xin	Louisiana State University, USA
Lindstrom Peter	Lawrence Livermore National Laboratory, USA
Linsen Lars	Jacobs University, Germany
Liu Feng	Portland State University, USA
Loviscach Joern	Fachhochschule Bielefeld (University of Applied Sciences), Germany
Magnor Marcus	TU Braunschweig, Germany
Martin Ralph	Cardiff University, UK
McGraw Tim	Purdue University, USA
Min Jianyuan	Google, USA
Meenakshisundaram Gopi	University of California – Irvine, USA
Mendoza Cesar	NaturalMotion Ltd., USA
Metaxas Dimitris	Rutgers University, USA
Mudur Sudhir	Concordia University, Canada
Musuvathy Suraj	Siemens, USA
Nait-Charif Hammadi	University of Dundee, UK
Nasri Ahmad	American University of Beirut, Lebanon
Noh Junyong	KAIST, Korea
Noma Tsukasa	Kyushu Institute of Technology, Japan
Okada Yoshihiro	Kyushu University, Japan
Olague Gustavo	CICESE Research Center, Mexico
Oliveira Manuel M.	Universidade Federal do Rio Grande do Sul, Brazil
Owen Charles	Michigan State University, USA
Ostromoukhov Victor M.	University of Montreal, Canada
Pascucci Valerio	University of Utah, USA
Patchett John	Los Alamos National Lab, USA
Peters Jorg	University of Florida, USA
Pronost Nicolas	Utrecht University, The Netherlands
Qin Hong	Stony Brook University, USA
Rautek Peter	Vienna University of Technology, Austria
Razdan Anshuman	Arizona State University, USA
Rosen Paul	University of Utah, USA
Rosenbaum Rene	University of California at Davis, USA

Rudomin Isaac	Barcelona Supercomputing Center, Spain
Rushmeier Holly	Yale University, USA
Sander Pedro	The Hong Kong University of Science and Technology, Hong Kong, SAR China
Sapidis Nickolas	University of Western Macedonia, Greece
Sarfraz Muhammad	Kuwait University, Kuwait
Scateni Riccardo	University of Cagliari, Italy
Sequin Carlo	University of California – Berkeley, USA
Shead Timothy	Sandia National Laboratories, USA
Sourin Alexei	Nanyang Technological University, Singapore
Stamminger Marc	REVES/Inria, France
Su Wen-Poh	Griffith University, Australia
Szumilas Lech	Research Institute for Automation and Measurements, Poland
Tan Kar Han	Hewlett Packard, USA
Tarini Marco	University of Insubria (Varese), Italy
Teschner Matthias	University of Freiburg, Germany
Tong Yiying	Michigan State University, USA
Torchelsen Rafael Piccin	Universidade Federal da Fronteira Sul, Brazil
Umlauf Georg	HTWG Constance, Germany
Vanegas Carlos	University of California at Berkeley, USA
Wald Ingo	University of Utah, USA
Walter Marcelo	UFRGS, Brazil
Wimmer Michael	Technical University of Vienna, Austria
Wylie Brian	Sandia National Laboratory, USA
Wyman Chris	University of Calgary, Canada
Wyvill Brian	University of Iowa, USA
Yang Qing-Xiong	University of Illinois at Urbana, Champaign, USA
Yang Ruigang	University of Kentucky, USA
Ye Duan	University of Missouri – Columbia, USA
Yi Beifang	Salem State University, USA
Yin Lijun	Binghamton University, USA
Yoo Terry	National Institutes of Health, USA
Yuan Xiaoru	Peking University, China
Zhang Jian Jun	Bournemouth University, UK
Zeng Jianmin	Nanyang Technological University, Singapore
Zara Jiri	Czech Technical University in Prague, Czech Republic
Zeng Wei	Florida Institute of Technology, USA
Zordan Victor	University of California at Riverside, USA

(Area 3) Virtual Reality

Alcaniz Mariano	Technical University of Valencia, Spain
Arns Laura	Purdue University, USA
Bacim Felipe	Virginia Tech, USA
Balcisoy Selim	Sabanci University, Turkey

Behringer Reinhold	Leeds Metropolitan University, UK
Benes Bedrich	Purdue University, USA
Bilalis Nicholas	Technical University of Crete, Greece
Billinghurst Mark	HIT Lab, New Zealand
Blach Roland	Fraunhofer Institute for Industrial Engineering, Germany
Blom Kristopher	University of Barcelona, Spain
Bogdanovych Anton	University of Western Sydney, Australia
Brady Rachael	Duke University, USA
Brega Jose Remo Ferreira	Universidade Estadual Paulista, Brazil
Brown Ross	Queensland University of Technology, Australia
Bues Matthias	Fraunhofer IAO in Stuttgart, Germany
Capin Tolga	Bilkent University, Turkey
Chen Jian	Brown University, USA
Cooper Matthew	University of Linkoping, Sweden
Coquillart Sabine	Inria, France
Craig Alan	NCSA University of Illinois at Urbana-Champaign, USA
Cremer Jim	University of Iowa, USA
Edmunds Timothy	University of British Columbia, Canada
Egges Arjan	Universiteit Utrecht, The Netherlands
Encarnaio L. Miguel	ACT Inc., USA
Figueroa Pablo	Universidad de los Andes, Colombia
Friedman Doron	IDC, Israel
Fuhrmann Anton	VRVis Research Center, Austria
Gregory Michelle	Pacific Northwest National Lab, USA
Gupta Satyandra K.	University of Maryland, USA
Haller Michael	FH Hagenberg, Austria
Hamza-Lup Felix	Armstrong Atlantic State University, USA
Herbelin Bruno	EPFL, Switzerland
Hinkenjann Andre	Bonn-Rhein-Sieg University of Applied Sciences, Germany
Hollerer Tobias	University of California at Santa Barbara, USA
Huang Jian	University of Tennessee at Knoxville, USA
Huang Zhiyong	Institute for Infocomm Research (I2R), Singapore
Julier Simon J.	University College London, UK
Johnsen Kyle	University of Georgia, USA
Jones Adam	Clemson University, USA
Kiyokawa Kiyoshi	Osaka University, Japan
Klosowski James	AT&T Labs, USA
Kohli Luv	InnerOptic, USA
Kopper Regis	Duke University, USA
Kozintsev Igor	Samsung, USA
Kuhlen Torsten	RWTH Aachen University, Germany
Laha Bireswar	Stony Brook University, USA
Lee Cha	University of California, Santa Barbara, USA

Liere Robert van	CWI, The Netherlands
Livingston A. Mark	Naval Research Laboratory, USA
Luo Xun	Qualcomm Research, USA
Malzbender Tom	Hewlett Packard Labs, USA
MacDonald Brendan	National Institute for Occupational Safety and Health, USA
Molineros Jose	Teledyne Scientific and Imaging, USA
Muller Stefan	University of Koblenz, Germany
Owen Charles	Michigan State University, USA
Paelke Volker	University of Ostwestfalen-Lippe, Germany
Peli Eli	Harvard University, USA
Pettifer Steve	The University of Manchester, UK
Pronost Nicolas	Utrecht University, The Netherlands
Pugmire Dave	Los Alamos National Lab, USA
Qian Gang	Arizona State University, USA
Raffin Bruno	Inria, France
Ragan Eric	Oak Ridge National Laboratory, USA
Rodello Ildeberto	University of Sao Paulo, Brazil
Roth Thorsten	Bonn-Rhein-Sieg University of Applied Sciences, Germany
Sandor Christian	Nara Institute of Science and Technology, Japan
Sapidis Nickolas	University of Western Macedonia, Greece
Schulze Jurgen	University of California – San Diego, USA
Sherman Bill	Indiana University, USA
Singh Gurjot	Virginia Tech, USA
Slavik Pavel	Czech Technical University in Prague, Czech Republic
Sourin Alexei	Nanyang Technological University, Singapore
Steinicke Frank	University of Würzburg, Germany
Suma Evan	University of Southern California, USA
Stamminger Marc	REVES/Inria, France
Srikanth Manohar	Indian Institute of Science, India
Wald Ingo	University of Utah, USA
Whitted Turner	TWI Research, UK
Wong Kin Hong	The Chinese University of Hong Kong, Hong Kong, SAR China
Yu Ka Chun	Denver Museum of Nature and Science, USA
Yuan Chunrong	University of Tübingen, Germany
Zachmann Gabriel	Clausthal University, Germany
Zara Jiri	Czech Technical University in Prague, Czech Republic
Zhang Hui	Indiana University, USA
Zhao Ye	Kent State University, USA

(Area 4) Visualization

AAndrienko Gennady	Fraunhofer Institute IAIS, Germany
Avila Lisa	Kitware, USA

Apperley Mark	University of Waikato, New Zealand
Balizs Csibfalvi	Budapest University of Technology and Economics, Hungary
Brady Rachael	Duke University, USA
Benes Bedrich	Purdue University, USA
Bilalis Nicholas	Technical University of Crete, Greece
Bonneau Georges-Pierre	Grenoble University, France
Bruckner Stefan	Vienna University of Technology, Austria
Brown Ross	Queensland University of Technology, Australia
Bihler Katja	VRVis Research Center, Austria
Burch Michael	University of Stuttgart, Germany
Callahan Steven	University of Utah, USA
Chen Jian	Brown University, USA
Chen Min	University of Oxford, UK
Chevalier Fanny	Inria, France
Chiang Yi-Jen	New York University, USA
Cooper Matthew	University of Linkoping, Sweden
Chourasia Amit	University of California – San Diego, USA
Crossno Patricia	Sandia National Laboratories, USA
Daniels Joel	University of Utah, USA
Dick Christian	Technical University of Munich, Germany
Duan Ye	University of Missouri-Columbia, USA
Dwyer Tim	Monash University, Australia
Entezari Alireza	University of Florida, USA
Ertl Thomas	University of Stuttgart, Germany
De Floriani Leila	University of Maryland, USA
Geist Robert	Clemson University, USA
Gotz David	University of North Carolina at Chapel Hill, USA
Grinstein Georges	University of Massachusetts Lowell, USA
Goebel Randy	University of Alberta, Canada
Gregory Michelle	Pacific Northwest National Lab, USA
Hadwiger Helmut Markus	KAUST, Saudi Arabia
Hagen Hans	Technical University of Kaiserslautern, Germany
Hamza-Lup Felix	Armstrong Atlantic State University, USA
Healey Christopher	North Carolina State University at Raleigh, USA
Hochheiser Harry	University of Pittsburgh, USA
Hollerer Tobias	University of California at Santa Barbara, USA
Hong Lichan	University of Sydney, Australia
Hong Seokhee	Palo Alto Research Center, USA
Hotz Ingrid	Zuse Institute Berlin, Germany
Huang Zhiyong	Institute for Infocomm Research (I2R), Singapore
Jiang Ming	Lawrence Livermore National Laboratory, USA
Joshi Alark	Yale University, USA
Julier Simon J.	University College London, UK
Koch Steffen	University of Stuttgart, Germany
Laramee Robert	Swansea University, UK

Lewis R. Robert	Washington State University, USA
Liere Robert van	CWI, The Netherlands
Lim Ik Soo	Bangor University, UK
Linsen Lars	Jacobs University, Germany
Liu Zhanping	Kentucky State University, USA
Lohmann Steffen	University of Stuttgart, Germany
Maeder Anthony	University of Western Sydney, Australia
Malpica Jose	Alcala University, Spain
Masutani Yoshitaka	The Hiroshima City University, Japan
Matkovic Kresimir	VRVis Research Center, Austria
McCaffrey James	Microsoft Research/Volt VTE, USA
Melancon Guy	CNRS UMR 5800 LaBRI and Inria Bordeaux Sud-Ouest, France
Miksch Silvia	Vienna University of Technology, Austria
Monroe Laura	Los Alamos National Labs, USA
Morie Jacki	University of Southern California, USA
Moreland Kenneth	Sandia National Laboratories, USA
Mudur Sudhir	Concordia University, Canada
Museth Ken	Linköping University, Sweden
Paelke Volker	University of Ostwestfalen-Lippe, Germany
Papka Michael	Argonne National Laboratory, USA
Peikert Ronald	Swiss Federal Institute of Technology Zurich, Switzerland
Pettifer Steve	The University of Manchester, UK
Pugmire Dave	Los Alamos National Lab, USA
Rabin Robert	University of Wisconsin at Madison, USA
Raffin Bruno	Inria, France
Razdan Anshuman	Arizona State University, USA
Reina Guido	University of Stuttgart, Germany
Rhyne Theresa-Marie	North Carolina State University, USA
Rosenbaum Rene	University of California at Davis, USA
Sadana Samik	Georgia Tech, USA
Sadlo Filip	University of Stuttgart, Germany
Scheuermann Gerik	University of Leipzig, Germany
Shead Timothy	Sandia National Laboratories, USA
Sips Mike	Stanford University, USA
Slavik Pavel	Czech Technical University in Prague, Czech Republic
Sourin XavierAlexei	Nanyang Technological University, Singapore
Thakur Sidharth	Renaissance Computing Institute (RENCI), USA
Theisel Holger	University of Magdeburg, Germany
Thiele Olaf	University of Mannheim, Germany
Xavier Tricoche	Purdue University, USA
Umlauf Georg	HTWG Constance, Germany
Viegas Fernanda	IBM, USA
Wald Ingo	University of Utah, USA
Wan Ming	Boeing Phantom Works, USA

Weinkauf Tino	Max-Planck-Institut für Informatik, Germany
Weiskopf Daniel	University of Stuttgart, Germany
Wischgoll Thomas	Wright State University, USA
Wongsuphasawat Krist	Twitter Inc., USA
Wylie Brian	Sandia National Laboratory, USA
Wu Yin	Indiana University, USA
Xu Wei	Brookhaven National Lab, USA
Yeasin Mohammed	Memphis University, USA
Yuan Xiaoru	Peking University, China
Zachmann Gabriel	Clausthal University, Germany
Zhang Hui	Indiana University, USA
Zhao Jian	University of Toronto, USA
Zhao Ye	Kent State University, USA
Zheng Ziyi	Stony Brook University, USA
Zhukov Leonid	Caltech, USA

Special Tracks

1. Computational Bioimaging Organizers

Tavares João Manuel R.S.	University of Porto, Portugal
Natal Jorge Renato	University of Porto, Portugal

2. 3D Surface Reconstruction, Mapping, and Visualization Organizers

Nefian Ara	Carnegie Mellon University/NASA Ames Research Center, USA
Edwards Laurence	NASA Ames Research Center, USA
Huertas Andres	NASA Jet Propulsion Lab, USA

3. Observing Humans Organizers

Savakis Andreas	Rochester Institute of Technology, USA
Argyros Antonis	University of Crete, Greece
Asari Vijay	University of Dayton, USA

4. Advancing Autonomy for Aerial Robotics Organizers

Alexis Kostas	ETH Zurich, Switzerland
Chli Margarita	University of Edinburgh, UK
Achtelik Markus	ETH Zurich, Switzerland
Kottas Dimitrios	University of Minnesota, USA
Bebis George	University of Nevada, Reno, USA

5. Spectral Imaging Processing and Analysis for Environmental, Engineering and Industrial Applications Organizers

Doulamis Anastasions (Tasos)	National Technical University of Athens, Greece
Loupos Konstantinos	Institute of Communications and Computer Systems, Greece

6. Unconstrained Biometrics: Challenges and Applications Organizers

Proena Hugo   University of Beira Interior, Portugal
Ross Arun   Michigan State University, USA

7. Intelligent Transportation Systems Organizers

Ambardekar   Amol, Microsoft, USA
Morris Brendan   University of Nevada, Las Vegas, USA

8. Visual Perception and Robotic Systems Organizers

La Hung   University of Nevada, Reno, USA
Sheng Weihua   Oklahoma State University, USA
Fan Guoliang   Oklahoma State University, USA
Kuno Yoshinori   Saitama University, Japan
Ha Quang   University of Technology Sydney, Australia
Tran Anthony (Tri)   Nanyang Technological University, Singapore
Dinh Kien   Rutgers University, USA

Organizing Institutions and Sponsors

Contents – Part I

ST: Computational Bioimaging

Graph-Based Visualization of Neuronal Connectivity Using Matrix
Block Partitioning and Edge Bundling   3
   Tim McGraw

Fuzzy Skeletonization Improves the Performance of Characterizing
Trabecular Bone Micro-architecture   14
   Cheng Chen, Dakai Jin, and Punam K. Saha

Thermal Infrared Image Processing to Assess Heat Generated
by Magnetic Nanoparticles for Hyperthermia Applications   25
   Raquel O. Rodrigues, Helder T. Gomes, Rui Lima, Adrián M.T. Silva,
   Pedro J.S. Rodrigues, Pedro B. Tavares, and João Manuel R.S. Tavares

Visualization Techniques for the Developing Chicken Heart   35
   Ly Phan, Cindy Grimm, and Sandra Rugonyi

InVesalius: An Interactive Rendering Framework for Health Care Support . . .   45
   Paulo Amorim, Thiago Moraes, Jorge Silva, and Helio Pedrini

Computer Graphics

As-Rigid-As-Possible Character Deformation Using Point Handles   57
   Zhiping Luo, Remco C. Veltkamp, and Arjan Egges

Image Annotation Incorporating Low-Rankness, Tag
and Visual Correlation and Inhomogeneous Errors   71
   Yuqing Hou

Extracting Surface Geometry from Particle-Based Fracture Simulations   82
   Chakrit Watcharopas, Yash Sapra, Robert Geist, and Joshua A. Levine

Time-Varying Surface Reconstruction of an Actor's Performance   92
   Ludovic Blache, Mathieu Desbrun, Céline Loscos, and Laurent Lucas

Interactive Procedural Building Generation Using Kaleidoscopic
Iterated Function Systems   102
   Tim McGraw

Motion and Tracking

Motion Priors Estimation for Robust Matching Initialization
in Automotive Applications   115
    Nolang Fanani, Marc Barnada, and Rudolf Mester

Multi-target Tracking Using Sample-Based Data Association
for Mixed Images   127
    Ting-hao Zhang, Hsiao-Tzu Chen, and Chih-Wei Tang

A Hierarchical Frame-by-Frame Association Method Based on Graph
Matching for Multi-object Tracking.   138
    Sourav Garg, Ehtesham Hassan, Swagat Kumar, and Prithwijit Guha

Experimental Evaluation of Rigid Registration Using Phase Correlation
Under Illumination Changes   151
    Alfonso Alba and Edgar Arce-Santana

Multi-modal Computer Vision for the Detection of Multi-scale
Crowd Physical Motions and Behavior in Confined Spaces   162
    Zoheir Sabeur, Nikolaos Doulamis, Lee Middleton,
    Banafshe Arbab-Zavar, Gianluca Correndo, and Aggelos Amditis

HMM Based Evaluation of Physical Therapy Movements
Using Kinect Tracking.   174
    Carlos Palma, Augusto Salazar, and Francisco Vargas

Segmentation

Segmentation of Partially Overlapping Nanoparticles Using Concave Points   187
    Sahar Zafari, Tuomas Eerola, Jouni Sampo, Heikki Kälviäinen,
    and Heikki Haario

Temporally Object-Based Video Co-segmentation   198
    Michael Ying Yang, Matthias Reso, Jun Tang, Wentong Liao,
    and Bodo Rosenhahn

An Efficient Non-parametric Background Modeling Technique
with CUDA Heterogeneous Parallel Architecture.   210
    Brandon Wilson and Alireza Tavakkoli

Finding the N-cuts of Watershed Partitions for Image Segmentation   221
    Chao Zhang and Sokratis Makrogiannis

A Novel Word Segmentation Method Based on Object Detection
and Deep Learning   231
    Tomas Wilkinson and Anders Brun

Recognition

Estimating the Dominant Orientation of an Object Using Image
Segmentation and Principal Component Analysis   243
Sravan Bhagavatula and Nashlie Sephus

Label Propagation for Large Scale 3D Indoor Scenes   253
Keke Tang, Zhe Zhao, and Xiaoping Chen

Symmetry Similarity of Human Perception to Computer Vision Operators . . .   265
Peter M. Forrest and Mark S. Nixon

UT-MARO: Unscented Transformation and Matrix Rank Optimization
for Moving Objects Detection in Aerial Imagery.   275
Agwad ElTantawy and Mohamed S. Shehata

Architectural Style Classification of Building Facade Towers   285
Gayane Shalunts

Visualization

Visualizing Document Image Collections Using Image-Based Word Clouds . . .   297
Tomas Wilkinson and Anders Brun

Guided Structure-Aligned Segmentation of Volumetric Data.   307
*Michelle Holloway, Anahita Sanandaji, Deniece Yates, Amali Krigger,
Ross Sowell, Ruth West, and Cindy Grimm*

Examining Classic Color Harmony Versus Translucency Color Guidelines
for Layered Surface Visualization   318
Sussan Einakian and Timothy S. Newman

Guidance on the Selection of Central Difference Method Accuracy
in Volume Rendering   328
Kazuhiro Nagai and Paul Rosen

Deep Learning of Neuromuscular Control for Biomechanical
Human Animation.   339
Masaki Nakada and Demetri Terzopoulos

NEURONAV: A Tool for Image-Guided Surgery - Application
to Parkinson's Disease.   349
*José Bestier Padilla, Ramiro Arango, Hernán F. García,
Hernán Darío Vargas Cardona, Álvaro A. Orozco,
Mauricio A. Álvarez, and Enrique Guijarro*

ST: 3D Mapping, Modeling and Surface Reconstruction

Generation of 3D/4D Photorealistic Building Models. The Testbed Area
for *4D Cultural Heritage World* Project: The Historical Center
of Calw (Germany)   361
  José Balsa-Barreiro and Dieter Fritsch

Visual Autonomy via 2D Matching in Rendered 3D Models   373
  D. Tenorio, V. Rivera, J. Medina, A. Leondar, M. Gaumer, and Z. Dodds

Reconstruction of Face Texture Based on the Fusion of Texture Patches   386
  Jérôme Manceau, Renaud Séguier, and Catherine Soladié

Human Body Volume Recovery from Single Depth Image   396
  Jaeho Yi, Seungkyu Lee, Sujung Bae, and Moonsik Jeong

Dense Correspondence and Optical Flow Estimation Using Gabor,
Schmid and Steerable Descriptors   406
  Ahmadreza Baghaie, Roshan M. D'Souza, and Zeyun Yu

ST: Advancing Autonomy for Aerial Robotics

Efficient Algorithms for Indoor MAV Flight Using Vision
and Sonar Sensors   419
  Kyungnam Kim, David J. Huber, Jiejun Xu, and Deepak Khosla

Victim Detection from a Fixed-Wing UAV: Experimental Results   432
  *Anurag Sai Vempati, Gabriel Agamennoni, Thomas Stastny,
  and Roland Siegwart*

Autonomous Robotic Aerial Tracking, Avoidance, and Seeking
of a Mobile Human Subject   444
  *Christos Papachristos, Dimos Tzoumanikas, Kostas Alexis,
  and Anthony Tzes*

Inspection Operations Using an Aerial Robot Powered-over-Tether
by a Ground Vehicle   455
  Lida Zikou, Christos Papachristos, Kostas Alexis, and Anthony Tzes

Autonomous Guidance for a UAS Along a Staircase   466
  *Olivier De Meyst, Thijs Goethals, Haris Balta, Geert De Cubber,
  and Rob Haelterman*

Nonlinear Controller of Quadcopters for Agricultural Monitoring   476
  *Víctor H. Andaluz, Edison López, David Manobanda,
  Franklin Guamushig, Fernando Chicaiza, Jorge S. Sánchez,
  David Rivas, Fabricio Pérez, Carlos Sánchez, and Vicente Morales*

Medical Imaging

Groupwise Shape Correspondences on 3D Brain Structures
Using Probabilistic Latent Variable Models   491
Hernán F. García, Mauricio A. Álvarez, and Álvaro Orozco

Automatic Segmentation of Extraocular Muscles Using Superpixel
and Normalized Cuts   501
Qi Xing, Yifan Li, Brendan Wiggins, Joseph L. Demer, and Qi Wei

More Usable V-EGI for Volumetric Dataset Registration   511
Chun Dong and Timothy S. Newman

A Robust Energy Minimization Algorithm for MS-Lesion Segmentation   521
*Zhaoxuan Gong, Dazhe Zhao, Chunming Li, Wenjun Tan,
and Christos Davatzikos*

Impact of the Number of Atlases in a Level Set Formulation
of Multi-atlas Segmentation   531
*Yihua Song, Zhaoxuan Gong, Dazhe Zhao, Chaolu Feng,
and Chunming Li*

Probabilistic Labeling of Cerebral Vasculature on MR Angiography   538
*Benjamin Quachtran, Sunil Sheth, Jeffrey L. Saver, David S. Liebeskind,
and Fabien Scalzo*

Virtual Reality

Lateral Touch Detection and Localization for Interactive,
Augmented Planar Surfaces   551
A. Ntelidakis, X. Zabulis, D. Grammenos, and P. Koutlemanis

A Hybrid Real-Time Visual Tracking Using Compressive RGB-D Features . . .   561
*Mengyuan Zhao, Heng Luo, Ahmad P. Tafti, Yuanchang Lin,
and Guotian He*

High-Quality Consistent Illumination in Mobile Augmented Reality
by Radiance Convolution on the GPU   574
Peter Kán, Johannes Unterguggenberger, and Hannes Kaufmann

Efficient Hand Articulations Tracking Using Adaptive Hand Model
and Depth Map   586
Byeongkeun Kang, Yeejin Lee, and Truong Q. Nguyen

Eye Gaze Correction with a Single Webcam Based on Eye-Replacement   599
Yalun Qin, Kuo-Chin Lien, Matthew Turk, and Tobias Höllerer

ST: Observing Humans

Gradient Local Auto-Correlations and Extreme Learning Machine
for Depth-Based Activity Recognition   613
  Chen Chen, Zhenjie Hou, Baochang Zhang, Junjun Jiang, and Yun Yang

An RGB-D Camera Based Walking Pattern Detection Method
for Smart Rollators   624
  He Zhang and Cang Ye

Evaluation of Vision-Based Human Activity Recognition
in Dense Trajectory Framework   634
  Hirokatsu Kataoka, Yoshimitsu Aoki, Kenji Iwata, and Yutaka Satoh

Analyzing Activities in Videos Using Latent Dirichlet Allocation
and Granger Causality   647
  Dalwinder Kular and Eraldo Ribeiro

Statistical Adaptive Metric Learning for Action Feature Set Recognition
in the Wild.   657
  Shuanglu Dai and Hong Man

ST: Spectral Imaging Processing

Learning Discriminative Spectral Bands for Material Classification   671
  Chao Liu, Sandra Skaff, and Manuel Martinello

A Deep Belief Network for Classifying Remotely-Sensed
Hyperspectral Data   682
  Justin H. Le, Ali Pour Yazdanpanah, Emma E. Regentova,
  and Venkatesan Muthukumar

Variational Inference for Background Subtraction in Infrared Imagery   693
  Konstantinos Makantasis, Anastasios Doulamis,
  and Konstantinos Loupos

Image Based Approaches for Tunnels' Defects Recognition
via Robotic Inspectors   706
  Eftychios Protopapadakis and Nikolaos Doulamis

Deep Learning-Based Man-Made Object Detection
from Hyperspectral Data   717
  Konstantinos Makantasis, Konstantinos Karantzalos,
  Anastasios Doulamis, and Konstantinos Loupos

Hyperspectral Scene Analysis via Structure from Motion   728
  Corey A. Miller and Thomas J. Walls

ST: Intelligent Transportation Systems

Detecting Road Users at Intersections Through Changing Weather
Using RGB-Thermal Video 741
   Chris Bahnsen and Thomas B. Moeslund

Safety Quantification of Intersections Using Computer Vision Techniques . . . 752
   Mohammad Shokrolah Shirazi and Brendan Morris

Vehicles Detection in Stereo Vision Based on Disparity Map Segmentation
and Objects Classification 762
   Djamila Dekkiche, Bastien Vincke, and Alain Mérigot

Traffic Light Detection at Night: Comparison of a Learning-Based Detector
and Three Model-Based Detectors. 774
   Morten B. Jensen, Mark P. Philipsen, Chris Bahnsen,
   Andreas Møgelmose, Thomas B. Moeslund, and Mohan M. Trivedi

Modelling and Experimental Study for Automated Congestion Driving 784
   Joseph A. Urhahne, Patrick Piastowski, and Mascha C. van der Voort

Visualization

Aperio: A System for Visualizing 3D Anatomy Data
Using Virtual Mechanical Tools 797
   T. McInerney and D. Tran

Quasi-Conformal Hybrid Multi-modality Image Registration
and its Application to Medical Image Fusion 809
   Ka Chun Lam and Lok Ming Lui

CINAPACT-Splines: A Family of Infinitely Smooth, Accurate
and Compactly Supported Splines. 819
   Bita Akram, Usman R. Alim, and Faramarz F. Samavati

Vis3D+: An Integrated System for GPU-Accelerated Volume Image
Processing and Rendering 830
   I. Nisar and T. McInerney

Ontology-Based Visual Query Formulation: An Industry Experience 842
   Ahmet Soylu, Evgeny Kharlamov, Dmitriy Zheleznyakov,
   Ernesto Jimenez-Ruiz, Martin Giese, and Ian Horrocks

ST: Visual Perception and Robotic Systems

Dynamic Target Tracking and Obstacle Avoidance using a Drone. 857
   Alexander C. Woods and Hung M. La

An Interactive Node-Link Visualization of Convolutional Neural Networks . . .   867
  Adam W. Harley

DPN-LRF: A Local Reference Frame for Robustly Handling Density
Differences and Partial Occlusions   878
  Shuichi Akizuki and Manabu Hashimoto

3D Perception for Autonomous Robot Exploration   888
  Jiejun Xu, Kyungnam Kim, Lei Zhang, and Deepak Khosla

Group Based Asymmetry–A Fast Saliency Algorithm   901
  Puneet Sharma and Oddmar Eiksund

Prototype of Super-Resolution Camera Array System   911
  Daiki Hirao and Hitoshi Iyatomi

Author Index   921

Contents – Part II

Applications

Hybrid Example-Based Single Image Super-Resolution   3
  Yang Xian, Xiaodong Yang, and Yingli Tian

Automated Habit Detection System: A Feasibility Study   16
  Hiroki Misawa, Takashi Obara, and Hitoshi Iyatomi

Conductor Tutoring Using the Microsoft Kinect   24
  Andrea Salgian, Leighanne Hsu, Nathaniel Milkosky,
  and David Vickerman

Lens Distortion Rectification Using Triangulation Based Interpolation   35
  Burak Benligiray and Cihan Topal

A Computer Vision System for Automatic Classification of Most
Consumed Brazilian Beans.   45
  S.A. Araújo, W.A.L. Alves, P.A. Belan, and K.P. Anselmo

3D Computer Vision

Stereo-Matching in the Context of Vision-Augmented Vehicles   57
  Waqar Khan and Reinhard Klette

A Real-Time Depth Estimation Approach for a Focused Plenoptic Camera . . .   70
  Ross Vasko, Niclas Zeller, Franz Quint, and Uwe Stilla

Range Image Processing for Real Time Hospital-Room Monitoring.   81
  Alessandro Mecocci, Francesco Micheli, and Claudia Zoppetti

Real–Time 3-D Surface Reconstruction from Multiple Cameras   93
  Yongchun Liu, Huajun Gong, and Zhaoxing Zhang

Stereo Correspondence Evaluation Methods: A Systematic Review   102
  Camilo Vargas, Ivan Cabezas, and John W. Branch

Computer Graphics

Guided High-Quality Rendering   115
  Thorsten Roth, Martin Weier, Jens Maiero, André Hinkenjann,
  and Yongmin Li

User-Assisted Inverse Procedural Facade Modeling and Compressed
Image Rendering. 126
   Huilong Zhuo, Shengchuan Zhou, Bedrich Benes,
   and David Whittinghill

Facial Fattening and Slimming Simulation Based on Skull Structure 137
   Masahiro Fujisaki and Shigeo Morishima

Many-Lights Real Time Global Illumination Using Sparse Voxel Octree 150
   Che Sun and Emmanuel Agu

WebPhysics: A Parallel Rigid Body Simulation Framework
for Web Applications 160
   Robert (Bo) Li, Tasneem Brutch, Guodong Rong, Yi Shen,
   and Chang Shu

Segmentation

A Markov Random Field and Active Contour Image Segmentation Model
for Animal Spots Patterns 173
   Alexander Gómez, German Díez, Jhony Giraldo, Augusto Salazar,
   and Juan M. Daza

Segmentation of Building Facade Towers. 185
   Gayane Shalunts

Effective Information and Contrast Based Saliency Detection 195
   Aditi Kapoor, K.K. Biswas, and M. Hanmandlu

Edge Based Segmentation of Left and Right Ventricles Using Two Distance
Regularized Level Sets 205
   Yu Liu, Yue Zhao, Shuxu Guo, Shaoxiang Zhang, and Chunming Li

Automatic Crater Detection Using Convex Grouping and Convolutional
Neural Networks. 213
   Ebrahim Emami, George Bebis, Ara Nefian, and Terry Fong

ST: Biometrics

Segmentation of Saimaa Ringed Seals for Identification Purposes 227
   Artem Zhelezniakov, Tuomas Eerola, Meeri Koivuniemi, Miina Auttila,
   Riikka Levänen, Marja Niemi, Mervi Kunnasranta,
   and Heikki Kälviäinen

Fingerprint Matching with Optical Coherence Tomography 237
   Yaseen Moolla, Ann Singh, Ebrahim Saith, and Sharat Akhoury

Improve Non-graph Matching Feature-Based Face Recognition
Performance by Using a Multi-stage Matching Strategy   248
  Xianming Chen, Wenyin Zhang, Chaoyang Zhang, and Zhaoxian Zhou

Neighbors Based Discriminative Feature Difference Learning
for Kinship Verification   258
  Xiaodong Duan and Zheng-Hua Tan

A Comparative Analysis of Two Approaches to Periocular Recognition
in Mobile Scenarios.   268
  João C. Monteiro, Rui Esteves, Gil Santos, Paulo Torrão Fiadeiro,
  Joana Lobo, and Jaime S. Cardoso

Applications

Visual Perception and Analysis as First Steps Toward Human–Robot
Chess Playing.   283
  Andreas Schwenk and Chunrong Yuan

A Gaussian Mixture Representation of Gesture Kinematics for On-Line
Sign Language Video Annotation   293
  Fabio Martínez, Antoine Manzanera, Michèle Gouiffès,
  and Annelies Braffort

Automatic Affect Analysis: From Children to Adults.   304
  Rizwan Ahmed Khan, Alexandre Meyer, and Saida Bouakaz

A Study of Hand Motion/Posture Recognition in Two-Camera Views   314
  Jingya Wang and Shahram Payandeh

Pattern Recognition

Automatic Verification of Properly Signed Multi-page Document Images. . . .   327
  Marçal Rusiñol, Dimosthenis Karatzas, and Josep Lladós

CRFs and HCRFs Based Recognition for Off-Line Arabic Handwriting.   337
  Moftah Elzobi, Ayoub Al-Hamadi, Laslo Dings, and Sherif El-etriby

Classifying Frog Calls Using Gaussian Mixture Models.   347
  Dalwinderjeet Kular, Kathryn Hollowood, Olatide Ommojaro,
  Katrina Smart, Mark Bush, and Eraldo Ribeiro

Ice Detection on Electrical Power Cables   355
  Binglin Li, Gabriel Thomas, and Dexter Williams

Facial Landmark Localization Using Robust Relationship Priors
and Approximative Gibbs Sampling 365
  Karsten Vogt, Oliver Müller, and Jörn Ostermann

Recognition

Off-the-Shelf CNN Features for Fine-Grained Classification of Vessels
in a Maritime Environment.................................... 379
  Fouad Bousetouane and Brendan Morris

Joint Visual Phrase Detection to Boost Scene Parsing 389
  Keke Tang, Zhe Zhao, and Xiaoping Chen

If We Did Not Have ImageNet: Comparison of Fisher Encodings
and Convolutional Neural Networks on Limited Training Data 400
  Christian Hentschel, Timur Pratama Wiradarma, and Harald Sack

Investigating Pill Recognition Methods for a New National Library
of Medicine Image Dataset.................................... 410
  *Daniela Ushizima, Allan Carneiro, Marcelo Souza,
  and Fatima Medeiros*

Realtime Face Verification with Lightweight Convolutional
Neural Networks.. 420
  *Nhan Dam, Vinh-Tiep Nguyen, Minh N. Do, Anh-Duc Duong,
  and Minh-Triet Tran*

Virtual Reality

Relighting for an Arbitrary Shape Object Under Unknown Illumination
Environment.. 433
  Yohei Ogura and Hideo Saito

Evaluation of Fatigue Measurement Using Human Motor Coordination
for Gesture-Based Interaction in 3D Environments 443
  Neera Pradhan, Angela Benavides, Qin Zhu, and Amy Ulinski Banic

JackVR: A Virtual Reality Training System for Landing Oil Rigs......... 453
  *Ahmed E. Mostafa, Kazuki Takashima, Mario Costa Sousa,
  and Ehud Sharlin*

DAcImPro: A Novel Database of Acquired Image Projections
and Its Application to Object Recognition 463
  *Aleksandr Setkov, Fabio Martinez Carillo, Michèle Gouiffès,
  Christian Jacquemin, Maria Vanrell, and Ramon Baldrich*

Deformable Object Behavior Reconstruction Derived Through
Simultaneous Geometric and Material Property Estimation   474
  Shane Transue and Min-Hyung Choi

Poster

Accidental Fall Detection Based on Skeleton Joint Correlation
and Activity Boundary   489
  *Martha Magali Flores-Barranco, Mario-Alberto Ibarra-Mazano,
  and Irene Cheng*

Generalized Wishart Processes for Interpolation Over Diffusion
Tensor Fields   499
  *Hernán Darío Vargas Cardona, Mauricio A. Álvarez,
  and Álvaro A. Orozco*

Spatio-Temporal Fusion for Learning of Regions of Interests Over
Multiple Video Streams   509
  *Samaneh Khoshrou, Jaime S. Cardoso, Eric Granger,
  and Luís F. Teixeira*

Patch Selection for Single Image Deblurring Based on a Coalitional Game . . .   521
  Jung-Hsuan Lin, Rong-Sheng Wang, and Jing-wei Wang

A Robust Real-Time Road Detection Algorithm Using Color
and Edge Information   532
  Jae-Hyun Nam, Seung-Hoon Yang, Woong Hu, and Byung-Gyu Kim

SeLibCV: A Service Library for Computer Vision Researchers   542
  Ahmad P. Tafti, Hamid Hassannia, Dee Piziak, and Zeyun Yu

Bicycle Detection Using HOG, HSC and MLBP   554
  *Farideh Foroozandeh Shahraki, Ali Pour Yazdanpanah,
  Emma E. Regentova, and Venkatesan Muthukumar*

On Calibration and Alignment of Point Clouds in a Network
of RGB-D Sensors for Tracking   563
  George Xu and Shahram Payandeh

Semantic Web Technologies for Object Tracking and Video Analytics   574
  *Benoit Gaüzère, Claudia Greco, Pierluigi Ritrovato, Alessia Saggese,
  and Mario Vento*

Home Oriented Virtual e-Rehabilitation   586
  Yogendra Patil, Iara Brandão, Guilherme Siqueira, and Fei Hu

WHAT2PRINT: Learning Image Evaluation   597
  Bohao She and Clark F. Olson

Use of a Large Image Repository to Enhance Domain Dataset
for Flyer Classification.   609
   Payam Pourashraf and Noriko Tomuro

Illumination Invariant Robust Likelihood Estimator for Particle Filtering
Based Target Tracking.   618
   Buti Al Delail, Harish Bhaskar, M. Jamal Zemerly,
   and Mohammed Al-Mualla

Adaptive Flocking Control of Multiple Unmanned Ground Vehicles
by Using a UAV.   628
   Mohammad Jafari, Shamik Sengupta, and Hung Manh La

Basic Study of Automated Diagnosis of Viral Plant Diseases
Using Convolutional Neural Networks.   638
   Yusuke Kawasaki, Hiroyuki Uga, Satoshi Kagiwada,
   and Hitoshi Iyatomi

Efficient Training of Evolution-Constructed Features.   646
   Meng Zhang and Dah-Jye Lee

Ground Extraction from Terrestrial LiDAR Scans Using 2D-3D
Neighborhood Graphs   655
   Yassine Belkhouche, Prakash Duraisamy, and Bill Buckles

Mass Segmentation in Mammograms Based on the Combination
of the Spiking Cortical Model (SCM) and the Improved CV Model   664
   Xiaoli Gao, Keju Wang, Yanan Guo, Zhen Yang, and Yide Ma

High Performance and Efficient Facial Recognition Using Norm
of ICA/Multiwavelet Features.   672
   Ahmed Aldhahab, George Atia, and Wasfy B. Mikhael

Dynamic Hand Gesture Recognition Using Generalized Time Warping
and Deep Belief Networks   682
   Cristian A. Torres-Valencia, Hernán F. García, Germán A. Holguín,
   Mauricio A. Álvarez, and Álvaro Orozco

Gaussian Processes for Slice-Based Super-Resolution MR Images.   692
   Hernán Darío Vargas Cardona, Andrés F. López-Lopera,
   Álvaro A. Orozco, Mauricio A. Álvarez, Juan Antonio Hernández Tamames,
   and Norberto Malpica

Congestion-Aware Warehouse Flow Analysis and Optimization   702
   Sawsan AlHalawani and Niloy J. Mitra

Building of Readable Decision Trees for Automated Melanoma
Discrimination   712
   Keiichi Ohki, M. Emre Celebi, Gerald Schaefer, and Hitoshi Iyatomi

A Novel Infrastructure for Supporting Display Ecologies   722
   Christian Eichner, Martin Nyolt, and Heidrun Schumann

Visualizing Software Metrics in a Software System Hierarchy   733
   Michael Burch

Region Growing Selection Technique for Dense Volume Visualization   745
   Lionel B. Sakou, Daniel Wilches, and Amy Banic

Computing Voronoi Diagrams of Line Segments in $\mathbb{R}^K$ in $O(n \log n)$ Time   755
   Jeffrey W. Holcomb and Jorge A. Cobb

Visualizing Aldo Giorgini's Ideal Flow   767
   Esteban Garcia Bravo and Tim McGraw

Restoration of Blurred-Noisy Images Through the Concept
of Bilevel Programming.   776
   Jessica Soo Mee Wong and Chee Seng Chan

Free-Form Tetrahedron Deformation   787
   Ben Kenwright

Innovative Virtual Reality Application for Road Safety Education
of Children in Urban Areas   797
   Taha Ridene, Laure Leroy, and Safwan Chendeb

Vision-Based Vehicle Counting with High Accuracy for Highways
with Perspective View   809
   Mohammad Shokrolah Shirazi and Brendan Morris

Automatic Motion Classification for Advanced Driver Assistance Systems . . .   819
   Alok Desai, Dah-Jye Lee, and Shreeya Mody

Shared Autonomy Perception and Manipulation of Physical Device
Controls.   830
   Matthew Rueben and William D. Smart

Condition Monitoring for Image-Based Visual Servoing Using Kalman
Filter   842
   Mien Van, Denglu Wu, Shuzi Sam Ge, and Hongliang Ren

Author Index   851

ST: Computational Bioimaging

Graph-Based Visualization of Neuronal Connectivity Using Matrix Block Partitioning and Edge Bundling

Tim McGraw$^{(\boxtimes)}$

Purdue University, West Lafayette, USA
tmcgraw@purdue.edu

Abstract. Neuronal connectivity matrices contain information vital to the understanding of brain structure and function. In this work we present graph-based visualization techniques for macroscale connectivity matrices that retain anatomical context while reducing the clutter and occlusion problems that plague 2D and 3D node-link diagrams. By partitioning the connectivity matrix into blocks corresponding to brain hemispheres and bundling graph edges we are able to generate intuitive visualizations that permit investigation at multiple scales (hemisphere, lobe, anatomical region). We demonstrate our approach on connectivity matrices computed using tractography of high angular resolution diffusion images acquired as part of a Parkinson's disease study.

1   Introduction

The central nervous system (CNS) is an intricate network of interconnected cells with structural features at multiple scales. Visualizing and understanding this network is critical to investigating the mysteries of cognition, consciousness, memory and diseases of the brain. Advances in medical imaging have made it possible to map parts of the CNS "wiring diagram" in vivo. The term "connectome" [1] was coined to convey the importance of mapping the neuronal connections in the brain by comparing it to the human genome project. Just as mapping the genome will lead to improved diagnosis and treatment of inherited diseases, so will understanding the connectome lead to advances in diagnosis and treatment of neurological disorders.

Functional MRI permits measurement of patterns of brain activation during cognitive tasks which suggest functional connectivity [2]. Microscopy [3] and diffusion MRI modalities permit micro- and macroscale structural connectivity, respectively, to be estimated. Random molecular motion causes transport of water at a microscopic scale within biological systems. The structure of the surrounding tissue can affect the diffusion process, sometimes making it anisotropic (or directionally dependent). Within an oriented structure, such as a bundle of axonal fibers in the brain, the diffusion direction tends to be parallel to the fiber direction. In the 1990s diffusion tensor imaging (DTI) [4] enabled white matter architecture to be assessed indirectly by observing the attenuation of

© Springer International Publishing Switzerland 2015
G. Bebis et al. (Eds.): ISVC 2015, Part I, LNCS 9474, pp. 3–13, 2015.
DOI: 10.1007/978-3-319-27857-5_1

the MR signal due to diffusion of water molecules. Soon after that, *tractography* (the process of tracing white matter pathways) at the scale of axonal fiber bundles was developed [5]. Over time, refinement of the imaging technique and improved fiber tractography techniques have permitted more detailed mapping of structural connectivity in the brain [6].

In this work we focus on macroscale visualization of structural connectivity computed from diffusion imaging, and apply our methods to a Parkinson's disease study. Our approach uses a bundled edge graph layout technique to present an intuitive and uncluttered view to the user. Many previous approaches to visualizing diffusion MRI and the underlying diffusion process lead to cluttered images with occluded features.

2 Related Work

Visualization of DT-MRI and neuronal connectivity has taken many forms in the scientific visualization community. In this section we give an overview of some common approaches, but the reader is directed to the surveys by Margulies et al. [7] and Pfister et al. [8] for a more thorough treatment. Many approaches from vector- and tensor-field visualization, such as glyphs and streamlines, have been applied to the problem of connectivity visualization. Early tractography methods [5] were similar to streamline computations in fluid mechanics. Fibers are traced from a starting point by repeatedly stepping in the local direction of principal diffusion. However, visualization of the connectome by displaying streamlines or streamtubes is not practical due to the large number of streamlines required to represent whole brain connectivity. Fibers near the cortical surface of the brain occlude the inner structure, and the resulting image appears very cluttered, as seen in Fig. 1. Volume rendered connectivity maps [9,10] only display connectivity from a single seed point or region of interest as shown in Fig. 1. Although GPU implementations permit exploration of the full dataset at interactive rates, it cannot give a comprehensive visualization of whole brain connectivity.

Connectivity matrices are square matrices $C_{i,j}$ where the row and column indices correspond to either individual voxels or groups of voxels. A macroscale anatomical connectivity matrix can be assembled by registering an expert-labeled anatomical atlas to the diffusion weighted images. For each fiber path the endpoints can be mapped to the regions-of-interest (ROIs) defined in the atlas. Each element of $C_{i,j}$ then represents the connectivity of region i to region j.

A **connectivity graph** corresponding to a connectivity matrix can be assembled by creating a node for each ROI, and computing node adjacency from $C_{i,j}$. Existing tools from network analysis and graph theory have been applied to this graph to determine its characteristics. As with other complex graphs, such as social networks, neuronal connectivity graphs are characterized by highly connected hub nodes and a small world topology (meaning there is a short path between every pair of nodes) [11].

Connectivity matrices are often visualized as an image with a colormap applied to the matrix elements. Permuting matrix rows and columns so that

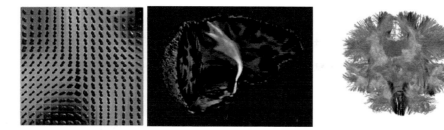

Fig. 1. Glyph-based visualization of white matter fiber orientation probabilities (left), volume rendered connectivity (middle), and neuronal fiber tracts (right) colored according to local direction. The glyphs convey the local structure of fiber orientations within a voxel but the global connectivity cannot be discerned. The volume rendering of connectivity shows only a small portion of the full connectivity matrix. The fiber tract visualization does convey global structure, but the clutter makes it difficult to visually follow individual fibers from end to end.

the ROIs are sorted from left to right results in a matrix that can readily yield inter- and intrahemispheric connectivity information. Sorted and unsorted connectivity matrices are shown in Fig. 2. The sorted matrix can be split roughly (neglecting the brainstem) into quadrants: the upper-left is connectivity within the left hemisphere of the brain, the lower-right is connectivity within the right hemisphere, and the other two quadrants represent interhemispheric connectivity. Many approaches to improving or augmenting adjacency matrix visualization have been proposed [12–14], but the drawback shared by these methods is that the nodes lose locational meaning, which we wish to preserve.

Node-link diagrams are a popular approach to visualizing **graphs and networks** which may represent such diverse data as transportation routes and social

Fig. 2. Connectivity matrix with rows and columns ordered arbitrarily (left), and with rows and columns sorted left to right by ROI position (middle). A 3D node-link diagram of the neuronal connectivity with anatomically embedded nodes (right). Note the occlusion and clutter due to the large number of edges, even though there are only 116 nodes.

relations. But even graphs of a moderate size can result in an unusable "hairball" visualization, such as in Fig. 2, right. Nodes in graph based visualization may correspond to physical locations which should, in some applications, be preserved in the visualization because the relative positions of nodes provide valuable contextual information. A major problem with node-link visualizations is the visual clutter that occurs when many edges overlap. This can be avoided by using edge clustering and force-based edge bundling [15] to group edges and bend them around nodes. Previous approaches to connectivity visualization suffer from problems of visual clutter [16,17] or loss of anatomical context [18,19]. Our goal in this work is to strike a balance between maintaining the spatial relations and anatomical meaning of the nodes of the connectivity graph while minimizing clutter due to the large number of edges.

3   Methods

In this section we describe our visualization application which was implemented in Matlab. The input to our method is a connectivity matrix computed from diffusion MRI. Guided by psychological principles, we aim to simplify use of our application by reducing visual clutter and exploiting the user's existing mental models. Overcrowded and disorganized displays have been shown to lead to decreased performance on visual search and recognition tasks [20]. We minimize visual clutter by separating inter- and intrahemispheric displays, and bundling edges. The existing mental models we build upon are the radiological conventions, such as standard imaging planes, and human anatomical knowledge. In the design of our application we eschewed a 3D approach to visualization since the degree of visual clutter and edge overlap is view dependent and therefore difficult to control.

Node Layout. Node locations were computed from the automated anatomical labeling (AAL) brain atlas which consists of 116 gray matter structures which were manually segmented from a human subject registered into a standard coordinate system [21]. A slice of the atlas is shown in Fig. 3. We have further grouped the 116 labels into brain hemispheres (left, right) and regions roughly corresponding to brain lobes (shown in Fig. 4).

Anatomists and clinicians are trained to analyze medical images and identify structures by looking at images in 3 standard planes: axial, coronal and sagittal. We laid out our nodes in these standard planes by positioning each one at the centroid of its AAL atlas region and discarding one of the 3 coordinates of each node. Node positions were adjusted to prevent overlap using the method described by Misue et al. [22]. Nodes are drawn as color-coded circles with radius proportional to the degree (number of incident edges) of the node. Color is determined by which of the 7 lobes the node belongs to. The results we present in the next section use the coronal imaging plane for node layout.

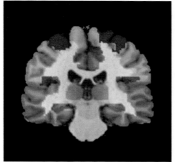

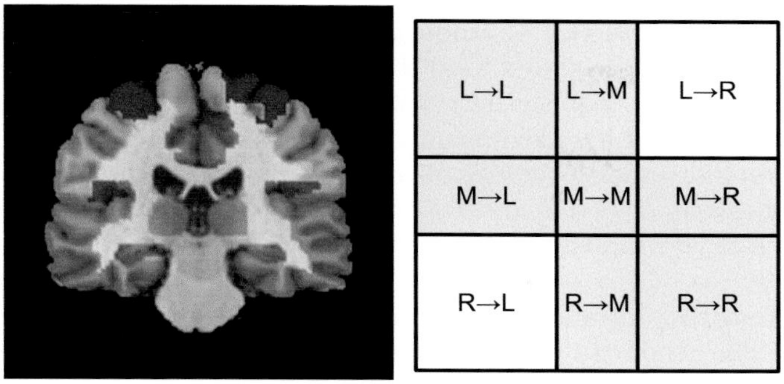

Fig. 3. A representative coronal slice of the AAL atlas with colored regions overlaid on T1-weighted MRI (left), and partitioning of the sorted connectivity matrix (right). The brain is partitioned into left (L) and right (R) hemisphere, and medial (M) structures. The shaded blocks and unshaded blocks are visualized separately.

Edge Routing. The connectivity matrix was thresholded to discard the lowest 1 % of connectivity values, then edges consisting of line segments were created for each nonzero value in the connectivity matrix. Initially the edges consist of few segments (2–10, depending on length), but as the iterative process of bundling proceeds we refine the edges using midpoint subdivision. Edge bundling is performed in a manner similar to the process described by Holten and Van Wijk [15] with a few application specific differences. Edge bundling criteria are based on length compatibility, orientation, and visibility as in [15] but we add an anatomical compatibility criterion. We only bundle edges together if one or more of their endpoints is in the same lobe. We also add a small repulsion force between incompatible edges to try to separate them and avoid clutter.

Temporal Lobe	Hippocampus, Parahippocampus, Amygdala, Fusiform gyrus, Heschl gyrus, Superior temporal gyrus, Temporal pole: superior temporal gyrus, Middle temporal gyrus, Temporal pole: middle temporal gyrus, Inferior temporal gyrus
Posterior Fossa	Cerebellum, Vermis, Medulla, Midbrain, Pons
Insula and Cingulate Gyri	Insula, Cingulate gyrus (ant., mid, post.)
Frontal Lobe	Precentral gyrus, Superior frontal gyrus, Middle frontal gyrus, Inferior frontal gyrus, Rolandic operculum, Supplementary motor area, Olfactory cortex, Gyrus rectus, Paracentral lobule
Occipital Lobe	Calcarine fissure and surrounding cortex, Cuneus, Lingual gyrus, Occipital lobe (sup., mid. and inf.)
Parietal Lobe	Postcentral gyrus, Superior parietal gyrus, Inferior parietal gyrus, Supramarginal gyrus, Angular gyrus, Precuneus
Central Structures	Caudate nucleus, Putamen, Pallidum, Thalamus

Fig. 4. Lobes and their constituent AAL labels. Most of the 116 AAL regions consist of left-right pairs.

Block Partitioning of the Connectivity Matrix. To achieve a similar connectivity grouping as the sorted matrix visualization in Fig. 2, we display inter- and intrahemispherical connectivity separately. Interhemispheric fibers include some long pathways which are difficult to avoid when routing edges. By separating these fibers into their own visualization we reduce the problem to bipartite graph visualization.

The connectivity matrix was sorted by node location from left to right and partitioned by the scheme shown in Fig. 3. The shaded blocks of the connectivity matrix, representing intrahemispheric connectivity, are visualized together in a single view. The other two blocks which represent interhemispheric connections are shown in a separate view. In our experiments, even with edge bundling, the long-range edges that cross the midline of the brain resulted in too much visual clutter. So interhemispheric connectivity is shown with a local scaling applied to node positions to shift them out of this region. The scaling applied to each hemisphere is given by the matrix product $T_c^{-1}ST_c$, where T_c translates the centroid of the hemisphere to the origin and S is a nonuniform scaling by 0.25 in the x-direction.

Our application also supports interactive selection and highlighting of multiple lobes and atlas labels. For each selection the constituent nodes and all edges between them are shown in color. All other nodes and edges are drawn in a light gray color for context.

4   Results

To generate the results shown in this section we used images from a publicly available dataset from the Neuroimaging Informatics Tools and Resources Clearinghouse (NITRC). Data for a set of 53 subjects in a cross-sectional Parkinson's disease (PD) study was acquired. The dataset contains diffusion-weighted images (DWI) of 27 PD patients and 26 age, sex, and education-matched control subjects. The diffusion-weighted images were acquired with 120 unique gradient directions, b=1000 and b=2500 s/mm^2, and isotropic 2.4 mm^3 voxels. The acquisition used a twice-refocused spin echo sequence in order to avoid distortions induced by eddy currents. The data were postprocessed to compensate for patient motion. Tractography and subsequent connectivity computation and visualization were performed on a Dell Optiplex workstation with a 3.4 GHz Intel Core i7-3770 CPU, and 8 GB RAM.

Connectivity Matrix Computation. From the diffusion weighted images we computed 4th order fiber orientation distribution tensors using the methods described by Weldeselassie et al. [23]. One million fiber tracts were generated by randomly seeding within the white matter and deterministically tracking until the fiber terminated. We use the AAL atlas to define 116 anatomical ROIs, so we initialize a 116 × 116 connectivity matrix to all zeroes ($C_{i,j} = 0$). The nearest AAL atlas region to each endpoint of the fiber was found and then the corresponding element of the connectivity matrix was incremented

Fig. 5. Visualization of the control group mean connectivity without matrix partitioning and edge bundling.

$(C_{i,j} = C_{i,j} + 1)$. Since we have no basis on which to assume fiber direction-
ality we make the connectivity matrix symmetric $(C = C + C^T)$, resulting in an
undirected connectivity graph.

Control Group and Parkinson's Group Visualization. After computing
C for each subject, the mean and variance of connectivity values for subjects
in each group (PD and control) were computed. A node-link diagram of the
resulting graph of 1842 edges with no partitioning or edge bundling is shown in
Fig. 5. Note the visual clutter near the midline where many edges cross as they
pass between hemispheres.

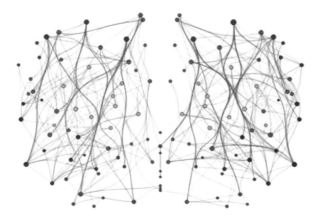

Fig. 6. Visualization of the control group mean intrahemispherical connectivity. Lobes
are color-coded and connectivity weights are represented by edge thickness.

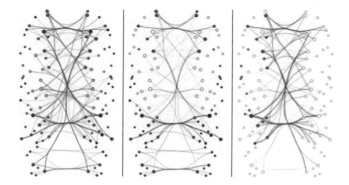

Fig. 7. Visualization of control group average interhemispherical connectivity for the whole brain (left), 3 selected lobes (middle) and a few selected nodes (right).

The mean control group intrahemispheric connectivity is shown using our proposed method in Fig. 6. Since the interhemispheric connectivity is displayed separately the clutter near the midline in greatly reduced, and the edge bundling within each hemisphere makes the connectivity patterns more easily discerned. Large bundles of edges can be seen connecting to high degree nodes (represented with larger markers), emphasizing their importance as connectivity hubs. Left-right hemisphere asymmetry is clearly visible in this view, and is expected. This is due to many factors, including image noise, hard thresholding of connectivity values and lateralization of brain function. Functions such as speech and language are known to be controlled by the left cerebral hemisphere, especially the temporal and parietal lobes. Note that bilateral symmetry is not related to

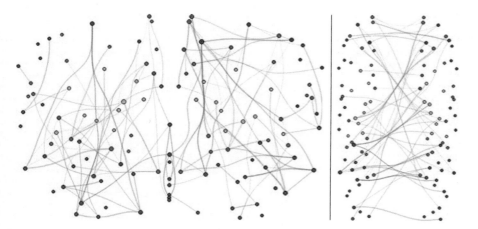

Fig. 8. Visualization of PD and control group connectivity differences. Statistically insignificant differences in connectivity are not shown.

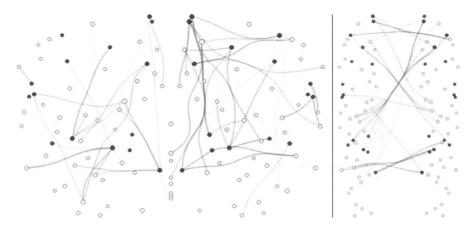

Fig. 9. Visualization of PD and control group connectivity differences in the nodes belonging to the frontal lobe. The left frontal lobe, shown in the right side of the image by radiological convention, is seen to be the most affected lobe.

symmetry of the connectivity matrix ($C = C^T$). Control group interhemispheric connectivity visualization results are shown in Fig. 7. Label and lobe selection by the user results in visual highlighting of connectivity. Although the node locations have been locally scaled, it is still possible to differentiate the nodes near the midline from the more lateral nodes.

Parkinson's disease is associated with impaired motor function and reduced cognitive performance. The disease is associated with dysfunction of the frontal lobe of the brain, and changes in functional connectivity have been observed in PD patients [24]. Connectivity differences between the PD and control groups were visualized by computing the absolute difference between connectivity matrices $C_{\text{diff}} = |C_{\text{PD}} - C_{\text{control}}|$. An unpaired two-sample t-test at the 5 % confidence level was performed for each pair of nodes (i, j). If the test did not reject the null hypothesis (that the mean connectivity in the groups is equal) then we set $C_{\text{diff}(i,j)} = 0$. Visualization of this difference is shown in Fig. 8. By selecting the nodes of the frontal lobe we can clearly see in Fig. 9 that many of the connectivity differences between the PD and control group have connections to the frontal lobe.

5   Conclusion and Future Work

In this work we have presented an approach to visualizing macroscale neuronal connectivity graphs which is designed to reduce user cognitive load by maintaining aqnatomical context. Visual clutter is reduced by partitioning the connectivity matrix into hemispheric blocks which are separately visualized using node-link diagrams with bundled edges. Hierarchical relations between brain hemispheres, lobes and regions-of-interest can be explored interactively, supporting coarse-to-fine investigation of connectivity. Examples of our technique were presented using data from a Parkinson's disease study. Edge bundling revealed long-range connectivity patterns that are not visible in the naive node-link diagram or matrix visualization. In future work we plan to conduct a user study to assess which views and interactions permit experts to recognize meaningful connectivity patterns in the connectome.

References

1. Sporns, O., Tononi, G., Kötter, R.: The human connectome: a structural description of the human brain. PLoS Comput. Biol. **1**, e42 (2005)
2. Behrens, T.E., Sporns, O.: Human connectomics. Curr. Opin. Neurobiol. **22**, 144–153 (2012)
3. Micheva, K.D., Smith, S.J.: Array tomography: a new tool for imaging the molecular architecture and ultrastructure of neural circuits. Neuron **55**, 25–36 (2007)
4. Basser, P.J., Mattiello, J., LeBihan, D.: MR diffusion tensor spectroscopy and imaging. Biophys. J. **66**, 259 (1994)
5. Basser, P.J., Pajevic, S., Pierpaoli, C., Duda, J., Aldroubi, A.: In vivo fiber tractography using DT-MRI data. Magn. Reson. Med. **44**, 625–632 (2000)
6. Cammoun, L., Gigandet, X., Meskaldji, D., Thiran, J.P., Sporns, O., Do, K.Q., Maeder, P., Meuli, R., Hagmann, P.: Mapping the human connectome at multiple scales with diffusion spectrum MRI. J. Neurosci. Methods **203**, 386–397 (2012)
7. Margulies, D.S., Böttger, J., Watanabe, A., Gorgolewski, K.J.: Visualizing the human connectome. NeuroImage **80**, 445–461 (2013)
8. Pfister, H., Kaynig, V., Botha, C.P., Bruckner, S., Dercksen, V.J., Hege, H.C., Roerdink, J.B.: Visualization in connectomics. In: Hansen, C.D., Chen, M., Johnson, C.R., Kaufman, A.E., Hagen, H. (eds.) Scientific Visualization: Uncertainty, Multifield, Biomedical, and Scalable Visualization, pp. 221–245. Springer, London (2014)
9. McGraw, T., Nadar, M.: Stochastic DT-MRI connectivity mapping on the GPU. IEEE Trans. Visual. Comput. Graphics **13**, 1504–1511 (2007)
10. McGraw, T., Herring, D.: High-order diffusion tensor connectivity mapping on the GPU. In: Bebis, G., Boyle, R., Parvin, B., Koracin, D., McMahan, R., Jerald, J., Zhang, H., Drucker, S.M., Kambhamettu, C., El Choubassi, M., Deng, Z., Carlson, M. (eds.) ISVC 2014, Part II. LNCS, vol. 8888, pp. 396–405. Springer, Heidelberg (2014)
11. Sporns, O.: The human connectome: a complex network. Ann. NY Acad. Sci. **1224**, 109–125 (2011)
12. Dinkla, K., Westenberg, M.A., van Wijk, J.J.: Compressed adjacency matrices: untangling gene regulatory networks. IEEE Trans. Visual. Comput. Graphics **18**, 2457–2466 (2012)
13. Sheny, Z., Maz, K.L.: Path visualization for adjacency matrices. In: Proceedings of the 9th Joint Eurographics/IEEE VGTC Conference on Visualization, Eurographics Association, pp. 83–90 (2007)
14. Henry, N., Fekete, J.D., McGuffin, M.J.: Nodetrix: a hybrid visualization of social networks. IEEE Trans. Visual. Comput. Graphics **13**, 1302–1309 (2007)
15. Holten, D., Van Wijk, J.J.: Force-directed edge bundling for graph visualization. Comput. Graphics Forum **28**, 983–990 (2009)
16. Bottger, J., Schafer, A., Lohmann, G., Villringer, A., Margulies, D.S.: Three-dimensional mean-shift edge bundling for the visualization of functional connectivity in the brain. IEEE Trans. Visual. Comput. Graphics **20**, 471–480 (2014)

17. Li, K., Guo, L., Faraco, C., Zhu, D., Chen, H., Yuan, Y., Lv, J., Deng, F., Jiang, X., Zhang, T., et al.: Visual analytics of brain networks. NeuroImage **61**, 82–97 (2012)
18. Al-Awami, A., Beyer, J., Strobelt, H., Kasthuri, N., Lichtman, J., Pfister, H., Hadwiger, M.: Neurolines: a subway map metaphor for visualizing nanoscale neuronal connectivity. IEEE Trans. Visual. Comput. Graphics (Proceedings IEEE InfoVis) **2014**(20), 2369–2378 (2014)
19. Irimia, A., Chambers, M.C., Torgerson, C.M., Van Horn, J.D.: Circular representation of human cortical networks for subject and population-level connectomic visualization. NeuroImage **60**, 1340–1351 (2012)
20. Rosenholtz, R., Li, Y., Nakano, L.: Measuring visual clutter. J. Vis. **7**, 1–22 (2007)
21. Tzourio-Mazoyer, N., Landeau, B., Papathanassiou, D., Crivello, F., Etard, O., Delcroix, N., Mazoyer, B., Joliot, M.: Automated anatomical labeling of activations in SPM using a macroscopic anatomical parcellation of the MNI MRI single-subject brain. NeuroImage **15**, 273–289 (2002)
22. Misue, K., Eades, P., Lai, W., Sugiyama, K.: Layout adjustment and the mental map. J. Vis. Lang. Comput. **6**, 183–210 (1995)
23. Weldeselassie, Y.T., Barmpoutis, A., Atkins, M.S.: Symmetric positive semidefinite Cartesian tensor fiber orientation distributions (CT-FOD). Med. Image Anal. **16**, 1121–1129 (2012)
24. Baggio, H.C., Sala-Llonch, R., Segura, B., Marti, M.J., Valldeoriola, F., Compta, Y., Tolosa, E., Junqué, C.: Functional brain networks and cognitive deficits in Parkinson's disease. Hum. Brain Mapp. **35**, 4620–4634 (2014)

Fuzzy Skeletonization Improves the Performance of Characterizing Trabecular Bone Micro-architecture

Cheng Chen[1]([✉]), Dakai Jin[1], and Punam K. Saha[1,2]

[1] Department of Electrical and Computer Engineering,
University of Iowa, Iowa City, USA
`cheng-chen@uiowa.edu`
[2] Department of Radiology, University of Iowa, Iowa City, USA

Abstract. Skeletonization provides a compact, yet effective representation of an object. Despite limited resolution, most medical imaging applications till date use binary skeletonization which is always associated with thresholding related data loss. A recently-developed fuzzy skeletonization algorithm directly operates on fuzzy objects in the presence of partially volumed voxels and alleviates this data loss. In this paper, the performance of fuzzy skeletonization is examined in a popular biomedical application of characterizing human trabecular bone (TB) plate/rod micro-architecture under limited resolution and compared with a binary method. Experimental results have shown that, using the volumetric topological analysis, fuzzy skeletonization leads to more accurate and reproducible measure of TB plate-width than the binary method. Also, fuzzy skeletonization-based plate-width measure showed a stronger linear association ($R^2 = 0.92$) with the actual bone strength than the binary skeletonization-based measure.

Keywords: Binary and fuzzy skeletonization · Trabecular bone · Volumetric topological analysis

1 Introduction

Skeletonization provides a compact yet effective representation of an object while preserving important topological and geometrical features; see [1,2] for through surveys on applications of skeletonization. Various implementations of skeletonization following the basic principle of Blum's grassfire propagation are available in literature [1,3]. Traditional skeletonization algorithms are defined on binary objects. Recently, Jin and Saha [4] have presented a comprehensive solution for skeletonization of fuzzy object using the theory of fuzzy grassfire propagation and collision impact. However, the influence of fuzzy skeletonization in different applications has not been studied.

Image resolution is a major bottleneck in medical imaging. Often, anatomic structures are acquired in the presence of partial voxel voluming. Despite this

© Springer International Publishing Switzerland 2015
G. Bebis et al. (Eds.): ISVC 2015, Part I, LNCS 9474, pp. 14–24, 2015.
DOI: 10.1007/978-3-319-27857-5_2

bottleneck, most medical imaging applications [1,2] use the binary skeletonization, which is associated with thresholding-induced data-loss and the effects are magnified at low image resolutions. On the other hand, fuzzy skeletonization may be directly applied on the fuzzy representation of an object in the presence of partially volumed voxels without requiring the binarization step. In this paper, we examine the role of fuzzy skeletonization in measuring trabecular bone (TB) structure-width or plate-width with limited resolution.

The significance of TB plate/rod distribution in assessing osteoporosis and low-trauma fracture-risk has long been recognized in histologic studies [5,6], which have confirmed the relationship between erosion of trabeculae from plates to rods and higher fracture risk. Various approaches have been reported to distinguish between rod-like and plate-like trabeculae. The volumetric topological analysis (VTA) [7] is an effective and powerful method to quantitatively characterize TB plate/rod [8] by computing local TB plate-width based on topological classification, geodesic distance analysis, and feature propagation on the skeleton of an object. The performance of VTA is highly dependent on the accuracy of the skeleton used for plate-width analysis. The original VTA algorithm [7] was developed for binary objects requiring thresholding on TB images, which is a sensitive and undesired step at *in vivo* image resolution [9].

The purpose of this work is to determine the influence of fuzzy skeletonization on the performance of VTA at *in vivo* image resolution, where most bone voxels are partially volumed. More specifically, we examine the accuracy and reproducibility of plate-widths computed using the fuzzy and binary skeletonization-based VTA and analyze their abilities to predict the actual bone strength.

2   Methods and Algorithms

In this section, a few definitions and notations and brief outlines of the fuzzy and binary skeletonization algorithms are presented followed by a short description of the VTA [7] method used for TB plate-width computation. The *3-D cubic grid*, denoted as $\mathbb{Z}^3$, where $\mathbb{Z}$ is the set of integers, is used for image representation; each grid element is referred to as a *voxel*. Conventional definitions of 8- and 4-adjacencies in 2-D and 6-, 18-, and 26-adjacencies in 3-D are followed. This paper starts with the assumption that the target object is fuzzily segmented using a suitable segmentation algorithm [10,11]. A *fuzzy object* $\mathcal{O} = (O, f_\mathcal{O})$ is a fuzzy set of $\mathbb{Z}^3$, where $f_\mathcal{O} : \mathbb{Z}^3 \rightarrow [0,1]$ is the *membership function* and $O = \{p \in \mathbb{Z}^3 | f_\mathcal{O}(p) > 0\}$ is its *support*. In this paper, 26-adjacency is used for object voxels, i.e., voxels in O, while 6-adjacency is used for background voxels, i.e., voxels in $\bar{O} = \mathbb{Z}^3 - O$.

2.1   Skeletonization Algorithms

The fuzzy and binary skeletonization algorithms using morphological erosion under certain topologic and geometric constraints [4,12–15] were chosen for our comparative study. This simple yet effective approach, outlined in the following, has become popular [1].

Primary Skeletonization

- Locate all *quench* or *skeletal voxels* in the input object.
- Filter noisy quench voxels and mark *significant skeletal voxels*.
- Delete unmarked voxels in the increasing order of distance values while preserving the *object topology* and the *continuity of skeletal surfaces*.

Final Skeletonization

- Convert *two-voxel thick* structures into *single-voxel thin* surfaces and curves.
- Remove voxels with *conflicting topologic* and *geometric* properties.

Skeleton Pruning

- Remove *noisy skeletal branches* with low *global significance*.

Binary Skeletization. In binary skeletonization, a fuzzy object $\mathcal{O} = (O, f_\mathcal{O})$ is first binarized into $O_{\text{bin}} = \{p|p \in O \wedge f_\mathcal{O}(p) > \text{thr}\}$, which is used as input. A constant threshold value of '0.5' is used for 'thr' for all experiments presented in this paper. During primary skeletonization, the centers of maximal balls (CMBs) [16] are located as *binary quench voxels* using 3-4-5 weighted distance transform (DT) [17,18]. The filtering algorithm by Saha *et al.* [15] is applied to remove noisy quench voxels and select significant surface and curve quench voxels. The four-condition constraint of the (26,6) simple point/voxel characterization by Saha *et al.* [13] is applied for topology preservation. Primary skeletonization produces a "thin set" [19] that has at least one background neighbor except at very busy intersections. During the final skeletonization step, two-voxel thick structures are eroded under topology preservation and some additional geometric constraints [14,15,20] to generate a one-voxel thin skeleton while maintaining its overall shape. Finally, noisy skeletal branches are removed during the pruning step using a *global significance measure* [7].

Fuzzy Skeletization. The fuzzy skeletonization algorithm is directly applied on the fuzzy object $\mathcal{O} = (O, f_\mathcal{O})$ without requiring the thresholding step. Here, fuzzy distance transform (FDT) [21] is used instead of binary DT. A *fuzzy quench voxel* [4,22] is located in a fuzzy object by detecting singularity voxels on the FDT map that holds the following inequality for each 26-neighbor q:

$$\text{FDT}(q) - \text{FDT}(p) < \tfrac{1}{2}(f_\mathcal{O}(p) + f_\mathcal{O}(q))|p - q|, \tag{1}$$

where $|p - q|$ is the Euclidean distance between p, q. During the filtering step in primary skeletonization, the measure of *collision impact* ξ_D [4] is used to define the local significance of individual fuzzy quench voxels.

$$\xi_\text{D}(p) = 1 - \max_{q \in N_{26}^*(p)} \frac{f_+(\text{FDT}(q) - \text{FDT}(p))}{\tfrac{1}{2}(f_\mathcal{O}(p) + f_\mathcal{O}(q))|p - q|}. \tag{2}$$

Collision impact at a quench voxel relates to the angle where independent fire-fronts collide. The value is high when fire-fronts make a head-on collision. Quench voxels on a core skeletal structure have high values of collision impact, while those on a skeletal branch emanating from a noisy protrusion have low values [4]. Jin and Saha [4] locally characterized the surface- and curve-like fuzzy quench voxels and argued the use of different filtering kernels for them.

For topology preservation, the constraints of 3-D simple pints/voxels [13] are applied on the support of the fuzzy object. Moreover, it is possible to find examples in fuzzy skeletonization, where the criteria of quench voxels and 3-D simple voxels fail to maintain the continuity of skeletal surfaces. Thus, to ensure the continuity of skeletal surfaces, additional constraints of (8,4) 2-D simple points are applied on each of the three coordinate planes through the candidate voxel [4]. Similar to binary skeletonization, two-voxel thick structures are converted into a single-voxel thin skeleton during final skeletonization. A few thick-voxels survive the first step of final skeletonization, which are eliminated using the conflict between their topologic and geometric properties. Finally, a collision impact-weighted path length of individual skeletal branches is used as their global significance during the pruning step [4].

Volumetric Topological Analysis. VTA calculates the plate-width of individual trabeculae in the unit of microns and uses this measure to locally classify individual trabecular type on the continuum between a perfect plate (green) and a perfect rod (red) (Fig. 1a). Note that the VTA-measured plate-width (Fig. 1b) is different from either thickness or skeleton-width. Figure 2 presents the results of intermediate steps of VTA. First, a surface representation of an object is computed using a suitable skeletonization algorithm [3]. Different topological entities, including surface-interior, surface-edge, curve, and junction voxels, are identified on the surface skeleton using digital topological analysis [23] (Fig. 2b). The geodesic distance transform (GDT) (Fig. 2c) is applied on the surface skeleton, which computes the geodesic distance of individual skeletal voxels from the edge; GDT at surface-edge as well as curve voxels are initialized as their FDT values, which enables the computation of plate-width instead of skeleton-width (see Fig. 1b). Using this convention for initialization, the GDT value at an axial voxel, i.e., a voxel on the arc skeleton, gives the value of half plate-width. However, the same is not true at non-arc skeletal voxels, where the plate-width measure is derived from the nearest arc skeletal voxel using a feature-propagation algorithm [24] (Fig. 2d). Finally, another level of feature propagation is applied to propagate plate-width measures from the surface-skeleton to the entire object volume, i.e., from Fig. 2d to 2e. Both feature propagation steps are performed using the principle established by Liu et al. [24] that is independent of scan or processing order. See [7] for detail and accurate description of different steps.

Let $\mathcal{O}_{\mathrm{BVF}} = (O_{\mathrm{BVF}}, f_{\mathrm{BVF}})$ denote the bone volume fraction (BVF) image of TB, where O_{BVF} is the support with non-zero BVF . The method presented in [24] was used to compute the BVF map in a CT image. Let $\mathrm{VTA}_{\mathrm{FSK}}(p)$ denote the TB plate-width at p using fuzzy skeletonization based VTA algorithm.

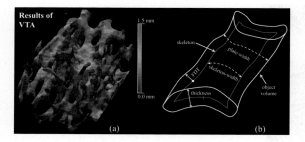

Fig. 1. (a) Trabecular bone plate/rod classification using volumetric topological analysis (VTA). (b) Various measures on a schematic drawing of a plate-like structure (color figure online).

The fuzzy skeletonization-based average plate-width (PW_{FSK}) characterizing the overall plate/rod micro-architecture of $\mathcal{O}_{\text{BVF}}$ is defined as follows:

$$\text{PW}_{\text{FSK}} = \frac{\sum_{p \in \mathcal{O}_{\text{BVF}}} \text{VTA}_{\text{FSK}}(p) f_{\text{BVF}}(p)}{\sum_{p \in \mathcal{O}_{\text{BVF}}} f_{\text{BVF}}(p)} \tag{3}$$

The average plate-width measure PW_{BSK} using binary skeletonization was similarly computed using $\text{VTA}_{\text{BSK}}(p)$.

3   Experiments and Results

Results of surface skeleton and local plate-width computation from VTA are presented in Fig. 3. In general, binary skeletonization over erodes TB surfaces, which makes the computed plate-width lower than true plate-width and results in more rod-like (reddish) trabeculae. Also, binary skeletonization created a few noisy branches (indicated by red arrows) and less-smooth skeletal surfaces (indicated by blue arrows). On the other hand, fuzzy skeletonization generated smoother surfaces and yielded expected measures of plate-width.

Three quantitative experiments were designed to evaluate the following — (1) accuracy, (2) repeat scan reproducibility, and (3) ability to predict bone strength. Computer-generated phantoms were used for Experiment 1. Multi-row detector CT (MD-CT) imaging of cadaveric TB specimens were used for Experiments 2 and 3. For Experiment 3, bone strength of cadaveric specimens was determined by mechanical testing.

3.1   Experimental Methods

Computer-Generated Phantoms. Computerized phantoms with known plate-widths were generated to examine the accuracy of fuzzy and binary skeletonization-based VTA algorithms. First, 3-D binary objects with boundary-noise and their true skeletons were generated at a high resolution over an array

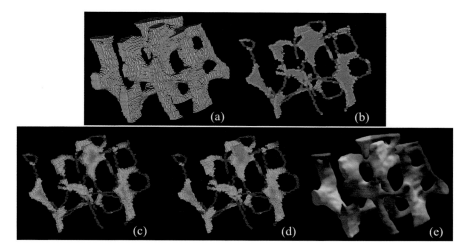

Fig. 2. Intermediate steps of the VTA algorithm. (a) A region from a micro-CT image of a TB specimen. (b) Classified topological entities including surface (green), curves (red), edges (light), and junctions (blue) on the surface skeleton. (c-e) Color coded display of geodesic distance transform (c), plate-width (d) and surface rendition of VTA over the bone volume (e) (color figure online).

of $512 \times 512 \times 512$. Test phantoms were generated from binary objects by down-sampling. The process starts by sampling ideal sinusoids of known skeleton-width, say w_S; let S denote the set of sampled points. A 3-D object V was computed from S by computing a distance transform from S, thresholding it at a value, say w_d, and, finally, adding random noisy protrusion of three-voxel diameter on the object boundary. The plate-width of the binary object V was recorded as $w_V = w_S + 2 \times w_d$. The test phantom V_{test} was generated by down-sampling V using a $3 \times 3 \times 3$ window to simulate fuzziness. Ten test phantoms were generated with their plate-width $w_V = 3, 5, \cdots, 21$ in the voxel unit of the down-sampled image. For fuzzy skeletonization-based VTA, the test phantom V_{test} was directly used as the input. For binary skeletonization, a threshold of 0.5 was applied on V_{test}.

Cadaveric Specimens and MD-CT Imaging. Fifteen fresh-frozen human cadaveric ankle specimens were obtained from 11 body donors (age: 55 to 91 years) under the Deeded Bodies Program, The University of Iowa. The ankle specimens were removed at the mid-tibia region. Exclusion criteria for this study were evidence of previous fracture or knowledge of bone tumor or bone metasta-sis. These specimens were kept frozen until the performance of MD-CT imaging. High resolution MD-CT scans of the distal tibia were acquired on a 128-slice SOMATOM Definition Flash scanner (Siemens, Munich, Germany) using the following CT parameters: single tube spiral acquisition at 120 kV, 200 effective mAs, 1 sec rotation speed, pitch factor: 1.0, nominal collimation: 16×0.3 mm, scan length: 10 cm beginning at the distal tibia end-plateau, and total effective

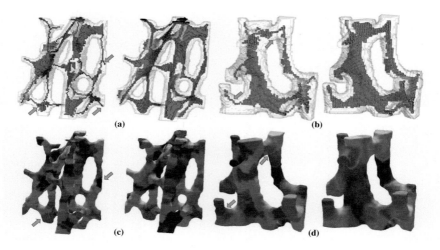

Fig. 3. Results of binary and fuzzy skeletonization on two small regions from TB images are shown in (a) and (b), respectively. Results of local plate-width on the same image regions using binary and fuzzy skeletonization-based VTA are shown in (c) and (d) respectively. The same color coding bar of Fig. 1 are used here (color figure online).

dose equivalent to 17 mrem $\approx$ 20 days of environmental radiation in the USA. Images were reconstructed at 0.2 mm slice thickness and 0.2×0.2 in-plane resolution using a special U70u kernel achieving high structural resolution. Three MD-CT repeat scans were acquired for each specimen with repositioning the phantom between scans.

Mechanical Testing for Bone Strength. To determine TB strength, a cylindrical TB core 8 mm in diameter and 20.9±3.3 mm in length was cored from the distal tibia *in situ* along the proximal-distal direction. Each TB core was mechanically tested for compression using an electromechanical materials testing machine. To minimize specimen end effects, strain was measured with a 6 mm gauge length extensometer attached directly to the midsection of the bone. A compressive preload of 10 N was applied and strains were set to zero. At a strain rate of 0.005 sec^{-1}, each specimen was preconditioned to a low strain with at least ten cycles and then loaded to failure. Yield stress was determined as the intersection of the stress-strain curve and a 0.2 % strain offset of the modulus.

3.2   Results and Discussion

Accuracy. To examine the accuracy of computed plate-width, an error was defined as the mean absolute difference between computed and true plate-widths; let Error$_{FSK}$ and Error$_{BSK}$ denote errors of fuzzy and binary skeletonization-based plate-width computation. The mean and standard deviation of Error$_{BSK}$ over ten test phantoms were 1.50±0.15 in the down-sampled voxel unit. The observed mean and standard deviation of $\text{Error}_{\text{FSK}}$ was 0.40±0.02, a 73 % reduction compared to $\text{Error}_{\text{BSK}}$. A paired t-test result ($p < 0.001$) confirmed the significance of the error reduction. Also, it is encouraging to note that the mean error of $\text{Error}_{\text{FSK}}$ is around 0.4 voxel unit, while the digital localization error in skeletonization is 0.38 [25]. Thus, it may be inferred that the primary source of error in $\text{Error}_{\text{BSK}}$ is digital localization.

Reproducibility. Two different analyses were performed using the three repeat-scan MD-CT images of fifteen cadaveric TB specimens to examine the reproducibility. First, the mean absolute mutual difference (MAMD) of PW_{FSK} (or PW_{BSK}) was computed over matching regions in post-registered repeat MD-CT scans of each TB specimen. The results of MAMD analysis are presented in Fig. 4(a). It is observed that MAMD of PW_{FSK} is consistently lower than that of PW_{BSK} for each TB specimen and a paired t-test confirmed that the reduction in MAMD using fuzzy skeletonization is statistically significant ($p < 0.001$)

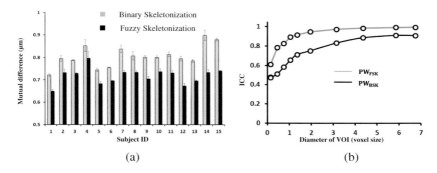

(a)   (b)

Fig. 4. (a) Mutual difference between fuzzy skeletons and binary skeletons of three repeat-scan CT images of fifteen cadaveric trabecular bone specimens. (b) Reproducibility of plate-width computed from binary and fuzzy skeleton in three repeat scans of CT images.

The intraclass correlation (ICC) of PW_{FSK} (or PW_{BSK}) was computed over matching spherical volume of interests (VOIs) in post-registered TB repeat MD-CT scans. Ten spherical VOIs were randomly selected in the first scan of each TB specimen (a total of 150 VOIs). Each VOI was located at least 8 mm proximal to the distal endplate. A post-registration algorithm was used to locate the matching VOIs in the second and third repeat scans. The VOI size was varied, and the ICC values were presented in Fig. 4(b) as a function of VOI diameter. The fuzzy skeletonization-based measure PW_{FSK} achieves an ICC of 0.95 at a VOI diameter of 1.05 mm or greater and it converges to the value of 0.98. In contrast, the binary skeletonization-based measure PW_{BSK} achieves the highest ICC value of 0.91 over the range of VOI diameters of our experiment.

Ability to Predict Bone Strength. Results of correlation analysis between TB yield stress and the average plate-width computed from binary and fuzzy

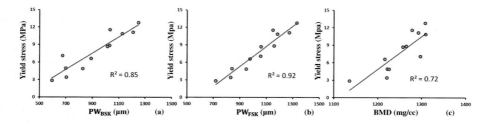

Fig. 5. Ability of different TB measures to predict experimental bone strength: (a) PW_{BSK} (b) PW_{FSK} (c) BMD. The ability is computed in terms of the R^2 of linear correlation between bone strength and respective measures.

skeletonization are presented in Fig. 5. PW_{FSK} achieves the value of 0.92 for R^2 or the coefficient of determination from linear regression analysis. On the other hand, R^2 of PW_{BSK} was 0.85. Note that both measures achieves higher linear correlation with yield stress as compared to the simple measure of average bone mineral density (BMD), in which the R^2 was 0.74. This observed results reaffirm the importance of TB plate/rod mirco-architecture as well as the superiority of fuzzy skeletonization over the binary method in capturing TB micro-structural properties at *in vivo* resolution.

4   Conclusion

This paper has evaluated the role of fuzzy skeletonization in characterizing TB micro-architecture at *in vivo* image resolution. The experimental results have demonstrated that fuzzy skeletonization effectively eliminates the binarization step which is always associated with data loss, especially, at regions with limited resolution. It is experimentally confirmed that fuzzy skeletonization-based TB plate-width is significantly more accurate and reproducible as compared to the binary skeletonization-based measure. Further, it was found in a cadaveric study that the fuzzy skeletonization-based TB plate-width measure has a stronger association with actual bone strength than the binary method. In the context of the specific application, the quality of surface skeleton influences the performance of volumetric topological analysis (VTA) where surface skeleton heavily determines the accuracy of plate-width. The improvement of plate-width using fuzzy skeletonization also indicates that fuzzy skeletonization generates more accurate skeleton than the binary method, possibly, by removing over erosion and false branches.

References

1. Saha, P.K., Borgefors, G., di Baja, G.S.: A survey on skeletonization algorithms and their applications. Pattern Recogn. Lett. (2015). http://www.sciencedirect.com/science/article/pii/S0167865515001233

2. Saha, P., Strand, R., Borgefors, G.: Digital topology and geometry in medical imaging: a survey. IEEE Trans. Med. Imag. **34**(9), 1940–1964 (2015)
3. Siddiqi, K., Pizer, S.M.: Medial Representations: Mathematics, Algorithms and Applications, vol. 37. Springer, The Netherlands (2008)
4. Jin, D., Saha, P.K.: A new fuzzy skeletonization algorithm and its applications to medical imaging. In: Petrosino, A. (ed.) ICIAP 2013, Part I. LNCS, vol. 8156, pp. 662–671. Springer, Heidelberg (2013)
5. Kleerekoper, M., Villanueva, A.R., Stanciu, J., Rao, D.S., Parfitt, A.M.: The role of three-dimensional trabecular microstructure in the pathogenesis of vertebral compression fractures. Calc. Tiss. Int. **37**, 594–597 (1985)
6. Recker, R.R.: Architecture and vertebral fracture. Calc. Tiss. Int. **53**(Suppl 1), S139–142 (1993)
7. Saha, P.K., Xu, Y., Duan, H., Heiner, A., Liang, G.: Volumetric topological analysis: a novel approach for trabecular bone classification on the continuum between plates and rods. IEEE Trans. Med. Imag. **29**(11), 1821–1838 (2010)
8. Chen, C., Jin, D., Liu, Y., Wehrli, F.W., Chang, G., Snyder, P.J., Regatte, R.R., Saha, P.K.: Volumetric topological analysis on in vivo trabecular bone magnetic resonance imaging. In: Bebis, G., Boyle, R., Parvin, B., Koracin, D., McMahan, R., Jerald, J., Zhang, H., Drucker, S.M., Kambhamettu, C., Choubassi, M., Deng, Z., Carlson, M. (eds.) ISVC 2014, Part I. LNCS, vol. 8887, pp. 501–510. Springer, Heidelberg (2014)
9. Krug, R., Burghardt, A.J., Majumdar, S., Link, T.M.: High-resolution imaging techniques for the assessment of osteoporosis. Radiol. Clin. North Am. **48**(3), 601–621 (2010)
10. Saha, P.K., Udupa, J.K., Odhner, D.: Scale-based fuzzy connected image segmentation: theory, algorithms, and validation. Comp. Vis. Imag. Und. **77**, 145–174 (2000)
11. Sonka, M., Hlavac, V., Boyle, R.: Image Processing, Analysis, and Machine Vision, 3rd edn. Thomson Engineering, Toronto, Canada (2007)
12. Jin, D., Iyer, K.S., Chen, C., Hoffman, E.A., Saha, P.K.: A robust and efficient curve skeletonization algorithm for tree-like objects using minimum cost paths. Pattern Recogn. Lett. (2015). http://www.sciencedirect.com/science/article/pii/S0167865515001063
13. Saha, P.K., Chaudhuri, B.B.: Detection of 3-D simple points for topology preserving transformations with application to thinning. IEEE Trans. Patt. Anal. Mach. Intell. **16**, 1028–1032 (1994)
14. Sanniti di Baja, G.: Well-shaped, stable, and reversible skeletons from the (3,4)-distance transform. J. Vis. Commun. Image Represent. **5**, 107–115 (1994)
15. Saha, P.K., Chaudhuri, B.B., Majumder, D.D.: A new shape preserving parallel thinning algorithm for 3D digital images. Pat. Recog. **30**, 1939–1955 (1997)
16. Arcelli, C., Sanniti di Baja, G.: Finding local maxima in a pseudo- euclidean distance transform. Comp. Vis. Graph Imag. Proc. **43**, 361–367 (1988)
17. Borgefors, G.: Distance transformations in digital images. Comp. Vis. Graph Imag. Proc. **34**, 344–371 (1986)
18. Borgefors, G.: Distance transform in arbitrary dimensions. Comp. Vis. Graph Imag. Proc. **27**, 321–345 (1984)
19. Arcelli, C., Sanniti di Baja, G.: A width-independent fast thinning algorithm. IEEE Trans. Patt. Anal. Mach. Intell. **7**(4), 463–474 (1985)
20. Arcelli, C., Sanniti di Baja, G., Serino, L.: Distance-driven skeletonization in voxel images. IEEE Trans. Patt. Anal. Mach. Intell. **33**(4), 709–720 (2011)

21. Saha, P.K., Wehrli, F.W., Gomberg, B.R.: Fuzzy distance transform: theory, algorithms, and applications. Comp. Vis. Imag. Und. **86**, 171–190 (2002)
22. Svensson, S.: Aspects on the reverse fuzzy distance transform. Pat. Recog. Lett. **29**, 888–896 (2008)
23. Saha, P.K., Chaudhuri, B.B.: 3D digital topology under binary transformation with applications. Comp. Vis. Imag. Und. **63**, 418–429 (1996)
24. Liu, Y., Jin, D., Li, C., Janz, K.F., Burns, T.L., Torner, J.C., Levy, S.M., Saha, P.K.: A robust algorithm for thickness computation at low resolution and its application to in vivo trabecular bone CT imaging. IEEE Trans. Biomed. Eng. **61**(7), 2057–2069 (2014)
25. Saha, P.K., Wehrli, F.W.: Measurement of trabecular bone thickness in the limited resolution regime of in vivo MRI by fuzzy distance transform. IEEE Trans. Med. Imag. **23**, 53–62 (2004)

Thermal Infrared Image Processing to Assess Heat Generated by Magnetic Nanoparticles for Hyperthermia Applications

Raquel O. Rodrigues[1], Helder T. Gomes[1], Rui Lima[2,3,4], Adrián M.T. Silva[5], Pedro J.S. Rodrigues[2], Pedro B. Tavares[6], and João Manuel R.S. Tavares[7(✉)]

[1] LCM – Laboratory of Catalysis and Materials – Associate Laboratory LSRE-LCM, Polytechnic Institute of Bragança, Bragança, Portugal
[2] Polytechnic Institute of Bragança, Bragança, Portugal
[3] Mechanical Engineering Department, University of Minho, Guimarães, Portugal
[4] CEFT, Faculdade de Engenharia, Universidade do Porto, Porto, Portugal
[5] LCM – Laboratory of Catalysis and Materials – Associate Laboratory LSRE-LCM, Faculdade de Engenharia, Universidade do Porto, Porto, Portugal
[6] CQVR, Chemical Department, Universidade de Trás-os-Montes e Alto Douro, Vila Real, Portugal
[7] Departamento de Engenharia Mecânica, Faculdade de Engenharia, Instituto de Ciência e Inovação em Engenharia Mecânica e Engenharia Industrial, Universidade do Porto, Porto, Portugal
tavares@fe.up.pt

Abstract. Magnetic fluid hyperthermia (MFH) is considered a promising therapeutic technique for the treatment of cancer cells, in which magnetic nanoparticles (MNPs) with superparamagnetic behavior generate mild-temperatures under an AC magnetic field to selectively destroy the abnormal cancer cells, in detriment of the healthy ones. However, the poor heating efficiency of most NMPs and the imprecise experimental determination of the temperature field during the treatment, are two of the majors drawbacks for its clinical advance. Thus, in this work, different MNPs were developed and tested under an AC magnetic field (~1.10 kA/m and 200 kHz), and the heat generated by them was assessed by an infrared camera. The resulting thermal images were processed in MATLAB after the thermographic calibration of the infrared camera. The results show the potential to use this thermal technique for the improvement and advance of MFH as a clinical therapy.

Keywords: Thermal imaging · Image calibration · Image processing

1 Introduction

Magnetic fluid hyperthermia (MFH) is considered a promising therapeutic technique for cancer treatment that arise from the emerging of nanotechnology and the clinical urgency to develop new medicines, treatments and detection strategies for this disease [1]. This technique is based on the use of magnetic nanoparticles

© Springer International Publishing Switzerland 2015
G. Bebis et al. (Eds.): ISVC 2015, Part I, LNCS 9474, pp. 25–34, 2015.
DOI: 10.1007/978-3-319-27857-5_3

(MNPs), usually with superparamagnetic properties, to generate heat under an alternating magnetic field by the Brownian and Néel relaxation mechanisms [2], and also in the physiological intolerance of the abnormal cancer cells to temperatures in the range between 41 and 46 °C [3]. In fact, MFH has several advantages over the common used therapeutic strategies, such as the capability of MNPs to be bio-functionalized, allowing them to be used as a non-invasive imaging agent in magnetic resonance imaging (MRI) simultaneously with MFH, offering the possibility to develop an image-guided therapy system. Thus, this therapeutic technique allows the combination of two functionalities, diagnostic and therapy (also defined as theragnostic) [4]. Furthermore, although hyperthermia has been shown to cause direct damage to cancer cells when applied as single therapy, this technique can also be applied as an adjuvant to other treatments, such as radiotherapy, chemotherapy, gene therapy and immunotherapy, with minimal or no injury at all of the normal tissues [1, 4]. However, the majors drawbacks of MFH are still the poor heating efficiency of most MNPs [5], and the imprecise experimental determination of the temperature field during the treatment, which may result in insufficient heating or in over-heating with collateral damage of the healthy tissues [6]. Usually, the temperature field is measured during the treatment using fiber optic probes. Nevertheless, these captors cannot be precisely implanted in the tumor tissues and only give local temperature information [6]. Infrared thermography, based on the physical phenomenon that all bodies at temperatures above absolute zero generate heat by radiation in the infrared portion of the electromagnetic spectrum [7], can overcome the drawbacks of the fiber optic probes, through the creation of a thermal map with the temperature range of the detected infrared radiation emitted from the body (both healthy and tumor tissues), as well as from the agglomeration of MNPs.

Thus, the aim of this work was to use infrared thermography and its potentialities to assess the heat released by MNPs when subjected to an AC magnetic field, which will serve as a screening tool to evaluate the potentiality of novel synthetized MNPs to be further applied in hyperthermia. For this purpose, the thermal images acquired by the infrared camera were calibrated and processed in MATLAB (Mathworks, USA).

2   Materials and Methods

2.1   Synthesis of Magnetic Nanoparticles (MNPs)

Before the synthesis of the MNPs, solutions of $FeCl_3.6H_2O$ (134 mM), $FeCl_2.4H_2O$ (67 mM) and NH_4OH (1 M) were prepared in distilled and deionized water at room temperature. Samples S1 and S2 were synthetized by co-precipitation, mixing the iron precursors, under vigorous magnetic stirring, in the stoichiometric $Fe^{2+}:Fe^{3+}$ ratio of 1:2 and heating the resultant mixture until the desired temperature (30 and 75 °C, respectively for S1 and S2). Then, the basic NH_4OH (1 M) solution was added dropwise until the pH ~10 was reached to promote the co-precipitation of the precursors. Sample S3 was synthetized in a similar way at room temperature. However, after the MNPs co-precipitation, the black solution was transferred to a teflon-lined stainless steel high pressure batch reactor (125 mL) and maintained at 180 °C under its own atmosphere during 24 h. Sample S3 was further coated with graphene oxide (GO) by using a simple electrostatic self-assembly process (S3@rGO), with a similar method as the one described by Wei et al. [8].

2.2 Physical and Chemical Characterization of MNPs

The crystallite phases and core sizes of the synthesized MNPs (S1, S2, S3 and S3@rGO) were analyzed by X-ray powder diffraction (XRD, PAN'alytical X'Pert PRO, Netherlands) by using the Bragg's law.

2.3 Experimental Setup

The adequate heat release promoted by MNPs when subjected to an AC magnetic field is one of the most important key-points for the success of MFH, and thus, directly dependent on the crystalline microstructure of the MNPs, the intensity of the AC magnetic field and the method for temperature acquisition. In order to study these three points, the experimental work was divided into two main experimental setups: (i) *infrared camera calibration* and (ii) *thermal imaging acquisition of MNPs under an AC magnetic field*.

Infrared Camera Calibration. In this infrared camera calibration test, the influence of the surface material of the observed body (beakers) and solutions used as working fluids for the dispersion of MNPs was assessed. Two beakers of 100 mL fabricated in borosilicate glass and polypropylene (PP) polymer, with the same dimensions, 50 mm × 70 mm (Outer Diameter × Height) were used, each one containing 100 mL of the working fluid, deionized water and glycerine 70 %, respectively. The objects and working fluids were further heated, using a magnetic heating stirrer plate with an integrated thermocouple (C-Mag HS7, IKA, Germany), at a heating increment of 5 °C, between 20 and 50 °C. In addition, the temperature measured by the integrated thermocouple, was confirmed with a second individual thermocouple. The thermal images were acquired with the infrared camera (E30, FLIR, USA) placed at 50 cm of the experimental setup, and transferred to a PC for posterior analysis using MATLAB (R2015a). The experimental tests and the parameters considered in the infrared camera calibration are shown in Table 1.

Table 1. Calibration experimental tests and the parameters considered in the infrared camera.

Calibration test	Object material	Working fluid	Temperature range
1	Glass	Deionized water	20–50 °C
2	PP	Deionized water	20–50 °C
3	PP	Glycerine 70 %	20–50 °C

Thermal Imaging Acquisition of MNPs Under AC Magnetic Field. After the calibration of the thermography method, the experimental setup for the assessment of the heat released by the synthetized MNPs was assembled as shown in Fig. 1. The AC magnetic field apparatus used in this work is a non-adiabatic in-house-made device, designed to work at a maximum intensity of 1.10 kA/m and with a frequency of 200 kHz.

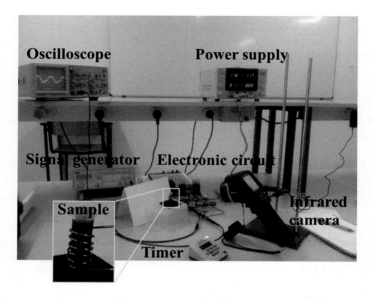

Fig. 1. Experimental setup used for the assessment of heat generated by MNPs under an AC magnetic field, monitored by an infrared camera (E30, FLIR, USA)

To investigate the influence of the working fluids in the heat released by the MNPs under an AC magnetic field, different fluids were considered to perform the infrared camera calibration (deionized water and glycerine 70 %). The infrared camera was placed at 50 cm of the samples, which were involved by the coil-shaped heating station (detail shown in Fig. 1), generating the AC magnetic field. The experimental assays and the parameters considered to assess the heat released by the different synthetized MNPs (S1, S2 and S3@rGO) are shown in Table 2.

Table 2. Experimental assays and the parameters considered to assess the heat generated by the different MNPs under an AC magnetic field

Assay	Carrier fluid	Density of the fluid (g/cm^3)	MNPs Concentration (mg/mL)	RMS magnetic field (H) (kA/m @ kHz)
1	Deionized water	1.00	5.00	1.10 @ 200
2	Glycerine 70 %	1.18	5.00	1.10 @ 200
3	Glycerine 70 %	1.18	5.00	0.80 @ 250

The thermal imaging assays shown in Table 2 were repeated for all the synthetized MNPs (S1, S2 and S3@rGO), as well as for blank tests (working without MNPs). These blank tests allowed to check and remove the heat generated by the coil (Joule heating) from the global heat generated with the MNPs during the experimental tests. Periodic infrared images were taken in all thermal image assays at 0, 60, 120, 300 and 600 s, stored in the format *.JPEG* and further processed in MATLAB.

2.4   Thermal Infrared Image Processing

The thermal infrared images acquired in the calibration and experimental tests were recorded in the format *.JPEG* with 160×120 pixels, transferred to a PC, and further converted into MATLAB readable files, with the extension *.MAT*, using the software *ThermaCAM Researcher* (trial version 2.10 Pro, FLIR, USA). This new format enables the use of the original information acquired by the infrared camera, where each pixel corresponds to the "real" captured temperature in Kelvin (K).

Infrared Camera Calibration. The thermal images acquired in the calibration tests, with extension *.MAT* as a matrix 120×160 corresponding to temperatures in K, were converted to degrees Celsius (°C) using the following code [9]:

```
name='Ir_0323'; %Name of the thermal image
t=sprintf('%s.mat', name); %Format data into string
O=load(t); %Load variables from file into workspace
I1=O.(name);
a=I1-273.15; %Kelvin into Celsius
```

After the creation of the new matrix (*a*), with the temperature in °C for each pixel, the APP *Image Viewer* of MATLAB was used to open it with the following command: *File > Import from workspace > a > ok*. Then, by selecting the icon *contrast*, the image was displayed, zoomed to 200 % and the desired part of the image containing the warmed fluid was selected by using the icon *crop image*. The region of interest was further transferred to the workspace of MATLAB using the menu *File > Export To Workspace > Image variable name > I*. Finally, calling the new matrix (*I*) it was possible to calculate the mean temperature of this matrix, as shown in the following Matlab code line:

```
>> mean(mean (I))
```

Heat Released by the MNPs Under an AC Magnetic Field. In a similar way to the image processing described above for the infrared camera calibration, the infrared images obtained in the thermal imaging assays (assays 1, 2 and 3, as shown on Table 2), were transferred to MATLAB with the extension *.MAT* and the matrix 120×160 corresponding to temperatures in K. The temperature was converted to a matrix in degrees °C, using the same code above described. The new matrix (*a*), in degrees °C, was further open in the APP *Image Viewer* of MATLAB, using the following command: *File > Import from workspace > a > ok*. Then, selecting the icon *contrast* the image was displayed, zoomed to 800 % and the desired parts of the image (R_1, R_2, R_3 and R_4), containing the fluid heated by the MNPs under the AC magnetic field within the loops of the coil, were selected with the icon *crop image*, as shown in Fig. 2.

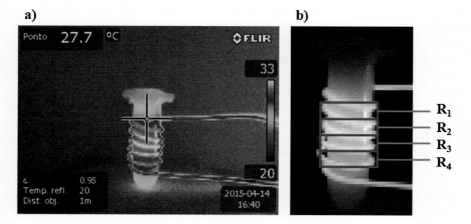

Fig. 2. Representation of an infrared image acquired in the MNPs thermal imaging assays under an AC magnetic field: (a) Thermal image obtained according or the.*JPEG* format by the infrared camera; (b) Thermal imaged transformed to the.*MAT* format with the identification of the four regions of interest (R_1, R_2, R_3 and R_4)

3   Results and Discussion

3.1   Physical and Chemical Characterization of MNPs

Table 3 shows a summary of the physical and chemical characterization parameters of the synthetized MNPs (crystallite size (d_{XRD}), lattice constant (a) and core phases) obtained from XRD analysis.

Table 3. Physical and chemical characterization parameters obtained by XRD analysis of the synthetized MNPs.

Sample	Synthesis Temp. (°C)	d_{XRD} (nm)	Lattice constant, a (Å)	Core phases (v/v)
S1	30 °C	12.80 ± 0.20	8.3652	100 % Magnetite
S2	75 °C	32.20 ± 0.30	8.3770	100 % Magnetite
S3@rGO	180 °C	63.05 ± 1.00	5.0345	49.9 % Hematite
		30.04 ± 0.30	8.3683	50.1 % Magnetite

Table 3 puts in evidence the influence of the synthesis temperature on the crystallite sizes of the MNPs. As observed for S1 (30 °C), S2 (75 °C) and S3 (180 °C), the increase of the synthesis temperature directly influences the growth of the magnetic core, higher crystallite sizes being obtained with higher synthesis temperatures. It is also concluded that the MNPs synthetized at 180 °C /24 h, under its own atmosphere (sample S3) not only lead to higher crystallite sizes on the magnetic core, but also results in the conversion of part of the magnetite composition into hematite (as observed in the lattice constant, a). It is worth mentioning that hematite is a form of iron oxide with less magnetic saturation (Ms) [10] and therefore, with lower heating power than magnetite under the same AC magnetic field.

3.2 Infrared Camera Calibration

As previously shown in Table 1, the infrared camera calibration was performed with three tests, where the material of the containers and the working were taken into account to determine their influence in the acquisition of the thermal images, as well as to obtain the correlation between inner and outer temperatures, Table 4.

Table 4. Temperature results obtained by the infrared camera calibration tests regarding the temperature measurements in TC (Thermocouple temperature), FLIR (Infrared camera pointer temperature) and MAT (mean temperature of the region of interest).

Calibration test 1			Calibration test 2			Calibration test 3		
TC (°C)	FLIR (°C)	MAT (°C)	TC (°C)	FLIR (°C)	MAT (°C)	TC (°C)	FLIR (°C)	MAT (°C)
19.3	18.4	18.7	20.0	19.6	20.1	21.7	21.7	21.7
25.0	24.0	24.1	25.0	24.2	24.4	25.0	24.8	24.9
30.0	28.3	28.5	30.0	29.1	29.6	30.0	29.5	29.5
35.0	32.9	32.8	35.0	34.7	34.7	35.0	33.8	33.8
40.0	37.2	37.1	40.0	39.3	39.4	40.0	38.9	38.7
45.0	41.8	41.7	45.0	46.1	46.2	45.0	43.3	43.3
50.0	46.3	46.1	50.0	49.8	50.1	50.0	48.8	48.8

In general, the local temperature given by the pointer of the infrared camera are in good agreement with the mean temperature obtained from the region of interest acquired as thermal map and processed in MATLAB. The good agreement between the local and mean temperatures of the region of interest can be explained by the magnetic stirring of the liquid samples that allowed the homogenization of solutions, and thus the homogenization of the heating transfer along the thermal tests. However, Table 4 also shows a discrepancy between the temperatures recorded by the thermocouple (TC) (placed inside the fluids) and the FLIR temperatures (obtained at the outer surface of the beakers). Therefore, calibration curves and linear fits (data not shown) between the temperatures acquired by the TC and the MALTAB analysis for the three calibration tests were determined to be further used in the evaluation of the heat release caused by the MNPs under AC magnetic field tests. In addition, the results presented in Table 4 allowed the calculation of the relative errors according to:

$$\text{Relative error} = \frac{\sum_{i=0}^{n}(\text{Inlet temperature (Thermocouple)} - \text{Outlet temperature (MATLAB)})}{n} \qquad (1)$$

where, n is the total number of measures and i the measurement number.

As a result, the relative error was found to be 2.19 °C for the calibration test 1, 0.23 °C for the calibration test 2, while in the calibration test 3 an error of 0.86 °C was obtained. These results show that the use of polypropylene polymer (PP) as material of the container brings a more accurate thermal measurement between the inner and the outer surface temperatures of the observed object. Therefore, PP was found to be a better choice of material to be used in this kind of experiments. Nevertheless, the difference in the accuracy of the results obtained between the use of PP and of glass in the containers cannot be explained by the thermal conductivity of the materials. In fact, the borosilicate glass has a higher thermal conductivity than PP, being expected *a priori* that the difference between the inner and outer temperatures for the glass material was lower than in the PP material. Thus, a possible explanation for the results obtained, could be the interference of glass reflection in the thermal images acquisition, which in fact was observed in the experimental tests performed with this material.

3.3 Heat Released by MNPs Under AC Magnetic Field

One of the crucial steps to perform hyperthermia tests is the colloidal stabilization of the MNPs within the working fluid, also known as ferrofluid. Therefore, in order to compare the influence of two working fluids with different densities (deionized water and glycerin 70 %) on the stabilization parameters and heat dispersion of the MNPs samples, two thermal imaging assays (thermal assays 1, 2) were performed at 1.10 kA/m and 200 kHz, during 600 s (cf. Table 2). In addition, thermal assay 3 was performed to check the influence of the frequency increase (250 kHz) in the heat generated by the MNPs, Fig. 3.

As shown in Fig. 3, the thermal assay 2, performed with glycerine 70 % as working fluid, with 1.10 kA/m and 200 kHz, was conducted under conditions that allowed the higher heat released by synthetized MNPs, of the sample considered. Comparing the heating results of the thermal assays 1 and 2, it is also concluded that the stabilization of the MNPs in the proper working fluid is a crucial step to the performance of hyperthermia tests. The importance of this stabilization is related to the homogenization of the ferrofluid in contact with the AC magnetic field, which reach a maximum magnetic intensity within the coil. Without this stabilization, the MNPs tend to sediment in the bottom of the vials. This fact is especially critical when the AC magnetic field apparatus generate a maximum magnetic field intensity that is 20 times lower (~1.10 kA/m, like in our case) than a commercial apparatus, which normally perform these hyperthermia tests around 20 kA/m. Nevertheless, the developed apparatus used in this work allowed, with a very low AC magnetic field, a maximum temperature increase of 3.45 °C ± 0.54 for sample S2 (co-precipitation at 75 °C), after 600 s under the AC magnetic field. In fact, this result is in good agreement with the literature, which points to the fact that MNPs under the mono-domain size and superparamagnetic properties increased their heating power as they grow in size.

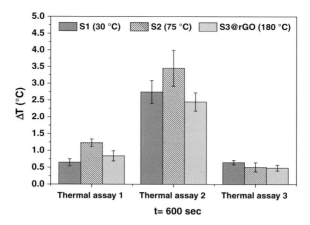

Fig. 3. Comparison of the heat released by the suspended MNPs (5 mg/mL), measured as the increment of the mean temperature of the suspensions, in different working fluids under an AC magnetic field. The mean temperatures presented are corrected by the respective calibration curves calculated from the infrared calibration tests and normalized according to the heat generated by the coil (blank tests). Error bars correspond to the standard deviation between the four regions of interest (R_1–R_4).

For the case of the MNPs synthetized at 180 °C and coated with GO (S3@rGO), the presence of almost 50 % of hematite (very lower Ms and thus, low heating power), puts these MNPs into the ferromagnetic materials. In fact, the ferromagnetic materials are considered in literature inappropriate materials to be used in biomedical application, due to the fact that these MNPs remain magnetized for a long period of time, even after the applied magnetic field is removed (high coercivity). Once more, this screening results obtained with the developed AC magnetic field are in good agreement with the characterization results and the literature in the field, resulting by this way, in a good methodology to perform screening hyperthermia characterization of MNPs.

4   Conclusions

Hyperthermia is a promising therapeutic technique for cancer treatment, but the clinical successful application is intrinsically linked to the development of new nanomaterials and the improvement of new methodologies to determinate the temperature fields during treatment. Therefore, in this study, MNPs with different sizes were synthetized at different synthesis temperature (30, 75 and 180 °C), and its heating efficiency assessed through the use of a developed AC magnetic field apparatus and an infrared camera. The acquired thermal images were transferred to a PC and the images processed in MATLAB. In order to correlate the inner and outer temperatures of the containers and working fluids, several calibration tests were performed with deionized water and glycerine 70 % as well as with different container materials (glass and PP). These calibration were further used in the experimental acquired data, allowing the effectively assessment of the inner temperatures achieved by the different MNPs in the different thermal assays.

The screening hyperthermia assays revealed that the MNPs synthetized at 75 °C (S2) are those presenting higher heating power for hyperthermia applications, which is related to the fact that MNPs under the mono-domain size and superparamagnetic properties increase their heating power as they grow in size. Therefore, the main conclusion of this study is that the use of infrared cameras are very effective in the assessment of the heat released by MNPs for hyperthermia applications, even when using an AC magnetic system able to create low magnetic fields.

As final remark, the good screening results obtained in this work also demonstrated the potentiality to use some of the synthetized MNPs in real hyperthermia applications. Therefore, other hyperthermia studies with commercial hyperthermia apparatus and higher magnetic intensities (~20 kA/m) are now being considered in a near future using the same thermal image acquisition and processing framework (infrared camera and MATLAB), allowing the calculations of the specific absorption rate (SAR) of the MNPs, a crucial parameter in the characterization of MNPs for real magnetic hyperthermia applications.

Acknowledgments. R.O.R acknowledges the PhD scholarship SFRH/BD/97658/2013 granted by FCT – Fundação para a Ciência e a Tecnologia, in Portugal. A.M.T.S acknowledges the FCT Investigator 2013 Programme (IF/01501/2013), with financing from the European Social Fund and the Human Potential Operational Programme.

References

1. Chicheł, A., Skowronek, J., Kubaszewska, M., Kanikowski, M.: Hyperthermia – description of a method and a review of clinical applications. Rep. Pract. Oncol. Radiother. **12**, 267–275 (2007)
2. Laurent, S., Dutz, S., Häfeli, U.O., Mahmoudi, M.: Magnetic fluid hyperthermia: Focus on superparamagnetic iron oxide nanoparticles. Adv. Colloid Interface Sci. **166**, 8–23 (2011)
3. Deatsch, A.E., Evans, B.A.: Heating efficiency in magnetic nanoparticle hyperthermia. J. Magn. Magn. Mater. **354**, 163–172 (2014)
4. Ito, A., Shinkai, M., Honda, H., Kobayashi, T.: Medical application of functionalized magnetic nanoparticles. J. Biosci. Bioeng. **100**, 1–11 (2005)
5. Di Corato, R., Espinosa, A., Lartigue, L., Tharaud, M., Chat, S., Pellegrino, T., Ménager, C., Gazeau, F., Wilhelm, C.: Magnetic hyperthermia efficiency in the cellular environment for different nanoparticle designs. Biomaterials **35**, 6400–6411 (2014)
6. COST: Action TD1402 - Multifunctional Nanoparticles for Magnetic Hyperthermia and Indirect Radiation Therapy (RADIOMAG) 054/14 Brussels (2014)
7. Calin, M.A., Mologhianu, G., Savastru, R., Calin, M.R., Brailescu, C.M.: A review of the effectiveness of thermal infrared imaging in the diagnosis and monitoring of knee diseases. Infrared Phys. Technol. **69**, 19–25 (2015)
8. Wei, H., Yang, W., Xi, Q., Chen, X.: Preparation of Fe3O4@graphene oxide core–shell magnetic particles for use in protein adsorption. Mater. Lett. **82**, 224–226 (2012)
9. Bento, D.: Modelação matemática da variação da temperatura no pé. Master thesis, Polytechnic Institute of Bragança, Bragança (2011) (in Portuguese)
10. Liu, Q., Barrón, V., Torrent, J., Qin, H., Yu, Y.: The magnetism of micro-sized hematite explained. Phys. Earth Planet. Inter. **183**, 387–397 (2010)

Visualization Techniques for the Developing Chicken Heart

Ly Phan[1], Cindy Grimm[2]($\boxtimes$), and Sandra Rugonyi[3]

[1] Intel Corp., Portland, USA
[2] Oregon State University, Corvallis, OR, USA
cindy.grimm@oregonstate.edu
[3] Oregon Health and Science Institute, Portland, OR, USA

Abstract. We present a geometric surface parameterization algorithm and several visualization techniques adapted to the problem of understanding the 4D peristaltic-like motion of the outflow tract (OFT) in an embryonic chick heart. We illustrated the techniques using data from hearts under normal conditions (four embryos), and hearts in which blood flow conditions are altered through OFT banding (four embryos). The overall goal is to create quantitative measures of the temporal heart-shape change both within a single subject and between multiple subjects. These measures will help elucidate how altering hemodynamic conditions changes the shape and motion of the OFT walls, which in turn influence the stresses and strains on the developing heart, causing it to develop differently. We take advantage of the tubular shape and periodic motion of the OFT to produce successively lower dimensional visualizations and quantifications of the cardiac motion.

1 Introduction

Cardiac development depends on genetic programs that are modulated by hemodynamic forces and environmental factors. Since the heart starts pumping blood early during embryonic development [10,20] cardiogenesis mainly occurs under blood flow conditions. Blood flow is essential for proper cardiac development, with altered flow or absence of flow leading to cardiac malformations [13,24,26].

At early stages of embryonic development the heart has a tubular structure that soon begins pumping blood, and then bends and loops to form a looping tubular heart, and eventually the four-chambered heart. Although it is known that altering blood flow causes heart defects [9,21,23] exactly how this happens is not well understood. The focus of the analysis techniques in this paper is understanding how the pumping tubular structure's motion, specifically the Out Flow Tract (see Fig. 1), is altered by changing blood flow conditions, for example, by comparing normal embryos to ones with bands placed around them.

The motion exhibited by the OF is both peristaltic-like (successive cross-sectional contraction) and longitudinal (the tube lengthens and shrinks). The aortic sac end of the OFT is relatively fixed in space, but the ventricular end moves — and, unfortunately, it moves in and out of the segmented volume. This

© Springer International Publishing Switzerland 2015
G. Bebis et al. (Eds.): ISVC 2015, Part I, LNCS 9474, pp. 35–44, 2015.
DOI: 10.1007/978-3-319-27857-5_4

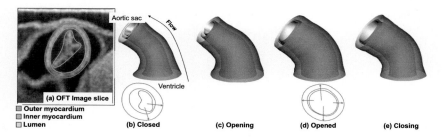

Fig. 1. Heart outflow tract (OFT) surfaces reconstructed from OCT data: (a) Original OCT data with contours to illustrate the myocardial walls (external ring-like structure, with the outer and inner myocardium surfaces delineating the myocardium) and lumen-wall interface (bluish line). OFT surfaces depicted when the ventricular end of the OFT is (b) fully closed, (c) opening, (d) fully opened, and (e) closing (Color figure online).

motion makes consistent parameterization of the tube over time non-trivial. For this reason, we first present an algorithm (Sect. 4) that defines the parameterizations of the surfaces over time using domain-specific knowledge about how the tissue is deforming *and* the desired down-stream analysis.

Next, we extend the algorithm to support temporal alignment of multiple chick heart data sets. Once data sets are consistently parameterized, we apply data reduction and analysis techniques to examine OFT motion. Specifically, we employ area and volume analysis to produce both qualitative and quantitative measurements of the OFT contraction rate and how contraction travels along the OFT tube and derivative analysis to generate qualitative visualizations of *how* the OFT contraction and expansion occur.

Contributions: We present a consistent parameterization algorithm suited for analysis of cylindrical-like biological surfaces. Using these parameterizations, we develop analysis techniques which map complex surface data to 2D images and video streams, and then use standard image processing techniques to generate quantitative plots capturing global behavior.

2   Related Work

Previous work has been performed in the areas of parameterization techniques, and visualization approaches for biological data. Here we summarize previous works and their relevance to our study.

2.1   Parameterization Approaches

There are a wide variety of techniques for creating consistent parameterizations of a related collection of surfaces. The simplest approach is to map all the surfaces to a common domain or base mesh [12]. Some parameterization techniques are ad-hoc, but most are based on the minimization of a 2D surface metric, for example, conformal or area measures [2, 4–7, 17, 25], geodesics on a shape-space manifold [3, 16], or strain [22]. Alternatively, one can create a deformation of one shape to the other and minimize the 3D deformation energy [14, 28] or a statistical measure [1]. The techniques that come from the computer graphics or vision literature primarily deal with surfaces that have distinctive features, and aim to map those features on each surface [15]; unfortunately, we have no such features.

2.2   Visualization Approaches

For visualization purposes, the embryonic heart OFT is essentially a deforming tube. Visualizing characteristics of the OFT tube (e.g. curvature or some measure of texture) can be done by color-coding the tube surface and rotating the tube in order to view the tube from all sides. An obvious approach for visualizing the whole tube in a single image is to cut the tube surface open and lay it out flat in the plane, resulting in a 2D image that is easy to inspect and understand. This technique has been used, for example, in virtual colonoscopies [11, 29], in which the primary challenge is to effectively recreate the relatively complex geometry of the interior of the colon (e.g., polyps) while avoiding topology inaccuracies that arise from the scanning process. To better visualize the OFT tube, we employ the same unfolding technique.

Because our data has a temporal element we need to extend this approach A simple technique is to select several cross-sectional planes along the heart tube and plot either cross-sectional contours or areas within the contours versus time [8, 19]. This allows examination of the wall motion along the heart tube and how this motion changes over time and space. We extend this approach to visualize the entire temporal tube.

3   Data Acquisition

The data sets we use come from chick embryonic OFTs imaged using *in-vivo* OCT imaging (fully described in [18, 19, 27]). The final output from these procedures, which is used as input for this study, is a set of three 3D surface meshes (outer and inner myocardium surfaces and the lumen surface) at 195 time points over the entire cardiac cycle, which gives a total of $3 \times 195 = 585$ surface meshes per embryo. At the embryonic stage studied here (approximately 3 days of incubation, HH18), the period of the cardiac cycle is about 400 ms, and thus the time span between consecutive meshes is about 2 ms. Note that these meshes are *not* in correspondence. We have eight data sets (four normal and four banded) which we use to validate our algorithms; we do not attempt to make strong biological claims from this small sample size.

4   Consistent Parameterization

In this section we describe how we generate a consistent mesh parameterization of the heart surfaces both temporally within one chicken embryonic OFT, and across multiple chicken embryos. We are interested in tracking the motion of the OFT over time (Fig. 2).

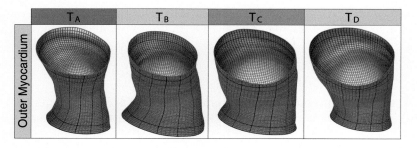

Fig. 2. Producing a consistent parameterization of the OFT tube at four time points (T_A - minimum, T_C - maximum, T_B and T_D are mid-way between the two). The geodesic constraint results in a relatively uniform geodesic spacing of the grid lines. The inner curve of the tube (darkest line) is used to align the grid radially between time points and to prevent "twisting" of the parameterization.

Motivations and observations: While during the cardiac cycle the OFT tube moves in space, it remains roughly C-shaped. The longitudinal geodesics (traced from one end of the tube to the other) have a distinct minimum length along the inside of the C-shape (see Fig. 1 for side view). Further, the myocardium can be assumed to be incompressible, which enables estimation of tissue motion given that the aortic sac side of the OFT is fixed in space. Let M_t represent the set of surfaces extracted from the 4D images by the segmentation algorithm. Informally, we want (A) a geodesic parameterization of these surfaces, (B) to use the curve on the inside of the tube (shortest geodesic length) as a stable feature for alignment, and (C) cross sections along the tube that do not intersect.

A Let $u, v \in [0,1]X[0,1]$ be the parameterization of the surface at each time, and let G_t be the mapping from the domain $[0,1]X[0,1]$ to the surface M_t at time $t \in [0,1]$. Specifically, u is the circumferential parameter (around the tube) and v is the longitudinal parameter (along the tube). Then even increments in u and v result in even (geodesic) increments along the surface.

B For all time t, the curve $G_t(0, v)$, $v \in [0,1]$, corresponds to the shortest longitudinal geodesic on the surface G_t. By shortest longitudinal geodesic we mean the shortest geodesic from the class of geodesics that connect a point on the inlet boundary to a point on the outlet boundary.

C Define a cross section at $V(v), v \in [0,1]$, by intersecting G_t with the plane defined by the three points $G_t(0, v)$, $G_t(1/3, v)$, and $G_t(2/3, v)$ (essentially three equally-spaced points on the cross-section contour at v). Further we impose that the family of cross sections V for all v do not intersect inside of G_t (see Fig. 3).

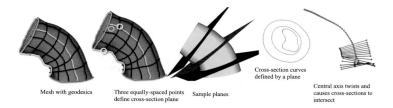

Mesh with geodesics   Three equally-spaced points   Sample planes   Central axis twists and
                        define cross-section plane   causes cross-sections to
                                                                           intersect

Cross-section curves
defined by a plane

Fig. 3. Finding consistent cross-sectional planes that do not intersect. (a) Sketch of the OFT mesh with the geodesics highlighted. (b) Three equally spaced points along a circumferential geodesic, which define a cross-sectional plane. (c) Intersection of cross-sectional planes with the surface mesh at three locations along the OFT. (d) Cross-sectional contours for the three surfaces in a cross-sectional plane. (e) Using the centerline of the mesh results in planes that self-intersect within the OFT.

Recall that, at each time t, we have three surface meshes representing the OFT wall: the outer and inner myocardial surfaces, and the lumen surface. Our algorithm creates a single, deforming mesh G for each of those three surfaces such that the *shape* of G_t matches the original surface M_t for each time step, but the *parameterization* of G has the properties given above.

We represent G using a mesh that has a grid structure, with $n \times m$ mesh nodes, with n nodes aligned around cross-sectional contours, and m nodes in the longitudinal direction, so that m is also the number of cross-sectional intersecting planes selected for generating the mesh. Note that cross-sectional planes are not necessarily perpendicular to the heart medial axis. The mesh is created by minimizing differences in geodesic distances among neighboring nodes, while ensuring that cross-sectional planes V do not intersect within G_t (see Fig. 3).

The minimization leading to the mesh parameterization should be done for each of the three OFT surfaces so that the outer and inner myocardial surfaces, and lumen surface, remain in correspondence. In practice, we parameterize the outer myocardial surface and then project the parameterization to the inner myocardial surface (for each vertex we found the closest point on the inner surface). Because the lumen surface folds, lumen parameterization must be done separately (the projection is not unique); we align the end points of the shortest lumen geodesic to the projection of the myocardial end points.

In what follows, we describe in more detail the algorithms for creating a deforming mesh for a single chick heart OFT over time, and for aligning deforming meshes between two (or more) chick heart OFTs. The parameterization itself provides geometric correspondence.

4.1 Single OFT Alignment Algorithm

Our input is a set of 195 surfaces, M_t, consisting of outer myocardial meshes that are not consistently parameterized. For each M_t, we find the shortest geodesic connecting any two end points on the tube boundary (see (B) above).

We cut the mesh along the shortest geodesic and map it to the unit square using Desbrun's parameterization [2] with area weighting. We place the cut line so that it aligns with $u = 0$, satisfying (B) above. This mapping defines the parameterization $G_t(u, v)$. It approximately satisfies the desired geodesic constraint (A), in that the mapping attempts to preserve area. (C) is satisfied by the construction of the planes from three equally-spaced points on the contour with constant v, provided the original surface M_t does not fold back on itself longitudinally in space. Informally, if the curves $G_t(0, v)$, $G_t(1/3, v)$, $G_t(2/3, v)$ do not fold back on themselves then the sequence of triangles defined by those points will not intersect with each other.

To construct our deforming mesh from the parameterization we then place a uniformly-spaced grid (80 nodes in the circumferential direction, 50 in the longitudinal) on the flattened mesh and re-sample the mapped geometry to get the locations for the mesh at time t.

Following this procedure, however, the shortest geodesic can "slide" on the surface between time frames because of slight changes or noise in the tube geometry. To address this, we stabilize the *end-points* of the $v = 0$ cut line from frame to frame.

The re-parameterization of the existing meshes is then used to create a single, deforming surface that roughly maintains equal geodesic spacing and is temporally smooth. We next discuss finding consistent temporal and spatial alignment between multiple chick OFTs.

4.2   Temporal Alignment (Multiple Hearts)

Our goal is to define a starting point $t = 0$ that is at the same time in the cardiac cycle for all chicks. The heart-cycle has distinct points of maximum contraction (minimum cross-sectional surface area) and maximum expansion (maximum surface area). For each chick heart we find the point of maximal contraction by finding the outer myocardial surface with the minimum volume. We set this to be the $t = 0$ time point. While not perfect, this alignment is enough for our purposes.

4.3   Between Chick Spatial Alignment

Exact structural alignment between multiple chicks is, in general, not possible. This is because imaging varies from chick to chick. Although every effort has been made to consistently image the same portion of the heart, there are no features to align the images. This is compounded by imaging quality, which degrades as the OFT descends into the embryo. Even though we were careful to measure embryos at the same stage of development (HH18), the embryos do differ in size, which also influences how far into the embryo the OFT descends.

Within these constraints our parameterization can be used to align the OFTs in the radial direction, and provide roughly similar spacing in the longitudinal direction, even if the starting and ending points are not the same. Specifically, we can compare the overall change along and around the tube but we cannot guarantee that these plots image the exact same portion of the tube.

5   Cardiac Motion Visualization and Analysis

In this section we describe several visualization and quantification techniques for analyzing the motion of the heart OFT. OFT wall motion can be described as a peristaltic-like contraction wave that starts at the ventricular end and travels towards the aortic sac. Central to analyzing wall motion along the OFT is the ability to extract cross-sections of the OFT tube as outlined in Sect. 4.

Specific questions we are interested in answering are: Does the contraction wave travel at a uniform speed? Do contraction and later expansion happen in a similar fashion along the tube, or do they vary from one end to the other? How does the band affect contraction waves?

Using the parameterized meshes, we calculate the internal area of each cross-sectional contour (50 cross-sections at each of the 195 time points, using contours from the three OFT surfaces; see Fig. 3). Because for each surface the cross-sectional areas change both in time (t) and space (v), areas can be mapped into a 2D image of v (vertical direction) versus t (horizontal direction), mapping areas to colors (blue is smallest, red is largest) and plotting each contour's area as a pixel in the 2D image (see Fig. 4A). From these images, we can readily visualize that areas in the ventricle end (bottom) are larger than at the aortic sac end (top), clearly demonstrating that the OFT is a tapered tube. Further, these images enable visualization of the contraction wave that travels through the OFT.

The 2D area images described above can be analyzed to extract characteristics of the OFT cardiac motion, such as peristaltic wave speed, contraction and expansion rates. Our approach to visualize and quantify cardiac motion is described next.

Peristaltic Wave Visualization: We extracted, for each OFT cross-section, the point at which the area reaches a maximum (the middle black line of Fig. 4A). These data can be further summarized as 2D plots showing the maximal area along the OFT medial axis (Fig. 4B). These plots are useful for comparing OFT sizes and how cross-sectional area changes along the OFT.

Contraction and Expansion Motion: Our next measure examines how long the OFT expansion/contraction lasts, as a percentage of the cardiac cycle. For this measure we tried two approaches, which yielded quantitatively similar results. Our first approach was to use the cross-sectional area data over time (for each position along the OFT), and fit a Gaussian curve centered on the maximum expansion time for that cross-section. We verified that the data was Gaussian-shaped by visually examining a plot of the data with the fitted Gaussian (see supplemental materials). Then, we found the time at which the area was one standard deviation from the maximum area (see Fig. 4A, stars). Our second approach was to use the same data but directly find the time at

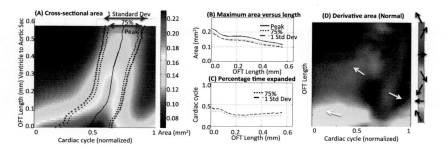

Fig. 4. Two-dimensional cross-sectional area images and derived quantifications. (A) Cross-sectional area (color coded) as it changes over the cardiac cycle (horizontal direction) and along the OFT (vertical direction). We overlay on the 2D area image the point of maximum area (maximum OFT expansion) along with the points at which the cross-sectional area reaches 75 % of the maximum (crosses) or one standard deviation from the maximum (stars). (B) 2D plot showing, for each cross-section, maximum area along the OFT. The plot also shows the area that is one standard deviation from the maximum in comparison with 75 % maximal area along the OFT length. (C) 2D plot showing the percentage of time the OFT is expanded, that is OFT cross-sectional area is above 75 % of the maximal area or above one standard deviation from the maximum. (D) The derivative of the cross-sectional area image, with angle mapped to hue and magnitude mapped to intensity.

which the area is 75 % of the maximum (see Fig. 4A, crosses). These data were plotted together with maximum area along the OFT (Fig. 4B). Further, we generated 2D plots showing the time span over which area is expanded, by plotting the percentage of time the area is either above 75 % of maximum or above one standard deviation from the maximum (e.g. Fig. 4C). Of course, we can also chose to plot the percentage time in which the OFT area is contracted. These data are useful for comparing whether contraction/expansion timings change from heart to heart and along the OFT length.

Contraction and Expansion Rates: As one final visualization we can show the *change* in area by taking the image derivative of the 2D area images. The image derivative calculates the gradient of the area, which results in a vector field (vector components are the derivative of area with respect to time, t, and the derivative of area with respect to the circumferential direction, u). We map the angle of the gradient vector to hue and the magnitude of the vector to intensity (Fig. 4D). From these data, maximum cardiac expansion and contraction rates can also be extracted.

6   Conclusion

In this paper we have presented an algorithm for producing consistent spatial correspondence for the deforming heart OFT and extending this correspondence (both temporally and spatially) to multiple embryos, within the limitations of the imaging data. This algorithm is suitable for aligning other tubal surfaces undergoing deformation.

The primary purpose of the correspondence algorithm is to enable comparison within the cardiac cycle (single embryo) and between embryos. We have presented several visualization and analysis approaches that can be used to reduce the dimensionality of the data and quantify some aspects of the overall motion (e.g., peristaltic wave speed and cross sectional shape change).

Acknowledgment. This research was funded in part by NSF grants DBI-1052688 and IIS-1302142 and NIH grant R01-HL094570.

References

1. Datar, M., Cates, J., Fletcher, P.T., Gouttard, S., Gerig, G., Whitaker, R.: Particle based shape regression of open surfaces with applications to developmental neuroimaging. In: Yang, G.-Z., Hawkes, D., Rueckert, D., Noble, A., Taylor, C. (eds.) MICCAI 2009, Part II. LNCS, vol. 5762, pp. 167–174. Springer, Heidelberg (2009)
2. Desbrun, M., Meyer, M., Alliez, P.: Intrinsic parameterizations of surface meshes. CGF **21**(3), 209–218 (2002)
3. Drury, H.A., van Essen, D.C., Joshi, S.C., Miller, M.I.: Analysis and comparison of areal partitioning schemes using two-dimensional fluid deformations. NeuroImage **3**, S130 (1996)
4. Drury, H., Van Essen, D.: Functional specializations in human cerebral cortex analyzed using the visible man surface-based atlas. Hum. Brain Mapp. **5**(4), 233–237 (1997)
5. Eck, M., DeRose, T., Duchamp, T., Hoppe, H., Lounsbery, M., Stuetzle, W.: Multiresolution analysis of arbitrary meshes. In: SIGGRAPH 1995. pp. 173–182. ACM, New York, NY, USA (1995)
6. Essen, D.C.V., Drury, H.A., Dickson, J., Harwell, J., Hanlon, D., Anderson, C.H.: An integrated software suite for surface-based analyses of cerebral cortex. J. Am. Med. Inform. Assoc. **8**(5), 443–459 (2001)
7. Fischl, B., Sereno, M.I., Tootell, R.B., Dale, A.M.: High-resolution intersubject averaging and a coordinate system for the cortical surface. Hum. Brain Mapp. **8**(4), 272–284 (1999)
8. Garita, B., Jenkins, M.W., Han, M., Zhou, C., Vanauker, M., Rollins, A.M., Watanabe, M., Fujimoto, J.G., Linask, K.K.: Blood flow dynamics of one cardiac cycle and relationship to mechanotransduction and trabeculation during heart looping. Am. J. Physiol. Heart Circulatory physiol. **300**(3), H879–H891 (2011)
9. Goodwin, R.L., Biechler, S.V., Junor, L., Evans, A.N., Eberth, J.F., Price, R.L., Potts, J.D., Yost, M.J.: The impact of flow-induced forces on the morphogenesis of the outflow tract. Front. Physiol. **5**(225), 1–9 (2014)
10. Hamburger, V., Hamilton, H.L.: A series of normal stages in the development of the chick embryo. J. Morphol. **88**(1), 49–92 (1951)
11. Hong, W., Gu, X., Qiu, F., Jin, M., Kaufman, A.: Conformal virtual colon flattening. In: Proceedings of the 2006 ACM Symposium on Solid and Physical Modeling, SPM 2006, pp. 85–93. ACM, New York, NY, USA (2006)

12. Hormann, K., Polthier, K., Sheffer, A.: Mesh parameterization: theory and practice. In: ACM SIGGRAPH ASIA 2008 courses, pp. 47:1–47:87. SIGGRAPH Asia 2008 (2008)
13. Hove, J.R., Koster, R.W., Forouhar, A.S., Acevedo-Bolton, G., Fraser, S.E., Gharib, M.: Intracardiac fluid forces are an essential epigenetic factor for embryonic cardiogenesis. Nature **421**, 172–7 (2003)
14. Huang, Q.X., Adams, B., Wicke, M., Guibas, L.J.: Non-rigid registration under isometric deformations. In: SGP, pp. 1449–1457. Eurographics Association (2008)
15. van Kaick, O., Zhang, H., Hamarneh, G., Cohen-Or, D.: A survey on shape correspondence. In: Proceedings of Eurographics State-of-the-art Report, pp. 1–22 (2010)
16. Kurtek, S., Srivastava, A., Klassen, E., Laga, H.: Landmark-guided elastic shape analysis of spherically-parameterized surfaces. CGF **32**(2.4), 429–438 (2013)
17. Lévy, B., Petitjean, S., Ray, N., Maillot, J.: Least squares conformal maps for automatic texture atlas generation. ACM TOG **21**, 362–371 (2002)
18. Liu, A., Wang, R., Thornburg, K.L., Rugonyi, S.: Efficient postacquisition synchronization of 4-d nongated cardiac images obtained from optical coherence tomography: application to 4-d reconstruction of the chick embryonic heart. J. Biomed. Opt. **14**(4), 044020 (2009)
19. Liu, A., Yin, X., Shi, L., Li, P., Thornburg, K.L., Wang, R., Rugonyi, S.: Biomechanics of the chick embryonic heart outflow tract at HH18 using 4D optical coherence tomography imaging and computational modeling. PLoS ONE **7**(7), e40869 (2012)
20. Martinsen, B.J.: Reference guide to the stages of chick heart embryology. Dev. Dyn. **233**(4), 1217–1237 (2005)
21. Midgett, M., Rugonyi, S.: Congenital heart malformations induced by hemodynamic altering surgical interventions. Front. Physiol. **5**, 287 (2014)
22. Phan, L., Knutsen, A.K., Bayly, P.V., Rugonyi, S., Grimm, C.: Refining shape correspondence for similar objects using strain. In: EG 3D OR, pp. 17–24. Eurographics Association (2011)
23. Rothenberg, F., Fisher, S.A., Watanabe, M.: Sculpting the cardiac outflow tract. Birth Defects Res. C **69**(1), 38–45 (2003)
24. Sedmera, D., Pexieder, T., Rychterova, V., Hu, N., Clark, E.B.: Remodeling of chick embryonic ventricular myoarchitecture under experimentally changed loading conditions. Anat. Rec. **254**(2), 238–252 (1999)
25. Thompson, P., Mega, M., Woods, R., Blanton, R., Moussai, J., Zoumalan, C., Aron, J., Cummings, J., Toga, A.: A probabilistic atlas of the human brain in alzheimer's disease: emerging patterns of variability, asymmetry and degeneration. NeuroImage **9**, S597 (1999)
26. Tobita, K., Garrison, J.B., Liu, L.J., Tinney, J.P., Keller, B.B.: Three-dimensional myofiber architecture of the embryonic left ventricle during normal development and altered mechanical loads. Anat. Rec. Part A Discov. Mol. Cell. Evol. Biol. **283A**(1), 193–201 (2005)
27. Yin, X., Liu, A., Thornburg, K.L., Wang, R.K., Rugonyi, S.: Extracting cardiac shapes and motion of the chick embryo heart outflow tract from four-dimensional optical coherence tomography images. J. Biomed. Opt. **17**(9), 1–10 (2012)
28. Zhang, H., Sheffer, A., Cohen-Or, D., Zhou, Q., van Kaick, O., Tagliasacchi, A.: Deformation-driven shape correspondence. In: SGP, pp. 1431–1439. Eurographics Association (2008)
29. Zhao, J., Cao, L., Zhuang, T., Wang, G.: Digital eversion of a hollow structure: an application in virtual colonography. J. Biomed. Imag. **2008**, 1–6 (2008)

InVesalius: An Interactive Rendering Framework for Health Care Support

Paulo Amorim[1], Thiago Moraes[1], Jorge Silva[1], and Helio Pedrini[2]([✉])

[1] Tridimensional Technology Division,
Center for Information Technology Renato Archer,
Campinas, SP 13069-901, Brazil
[2] Institute of Computing, University of Campinas,
Campinas, SP 13083-852, Brazil
helio@ic.unicamp.br

Abstract. This work presents InVesalius, an open-source software for analysis and visualization of medical images. The tool has supported several surgeries in hospitals and has been downloaded from more than a hundred countries around the world. Its main characteristics, aspects of implementation, and applications in areas such as image segmentation, mesh generation, volume rendering, and 3D printing of anatomic models are described.

Keywords: Medical image software · Health care · 3D printing

1 Introduction

Medical diagnostic imaging procedures have been used since the early twentieth century with the emergence of X-ray technique. There are nowadays a variety of medical imaging modalities, such as computed tomography (CT), magnetic resonance imaging (MRI), ultrasonography (US), among others. With the continuous advance of technology and the popularity of these medical imaging techniques, it is now common to find facilities for diagnostic imaging in major cities, even in developing countries. The imaging techniques are highly dependent on software tools for processing, analysis and visualization; however, they are not always available to interested users due to cost issues or support for their end-user computing platform.

In the late 90's, one of the first 3D printers of Latin America was installed at the Renato Archer Information Technology Center in Campinas - CTI, São Paulo State, Brazil, whose main purpose was the 3D printing of anatomical models to assist Brazilian surgeons in the surgical planning process and design of patient-specific prostheses. At that time, only a few proprietary and expensive software packages for medical image processing were available worldwide becoming unaffordable for the majority of Brazilian public hospitals. In such scenario, it arose the need to develop a free software for processing, analysis, visualization and 3D printing support for medical images. The developed software was called

G. Bebis et al. (Eds.): ISVC 2015, Part I, LNCS 9474, pp. 45–54, 2015.
DOI: 10.1007/978-3-319-27857-5_5

InVesalius, as a tribute to the Belgian anatomist Andreas Vesalius (1514–1564), considered the "father of modern anatomy".

This paper presents the InVesalius, an open-source medical framework, including its main applications, relevant concepts used in its development, potential applications, as well as some statistics of its worldwide use.

The text is organized as follows: Sect. 2 briefly presents some concepts and works related to the topic under investigation. Section 3 describes some computer graphics and image processing techniques used to develop the tool. Section 4 illustrates some applications of InVesalius and its use around the world. Finally, Sect. 5 outlines new directions for the development of InVesalius software.

2   Background

This section presents some of the key concepts and related works on medical imaging, 3D printing, image processing and computer graphics techniques associated with InVesalius software development.

2.1   Medical Images

Medical imaging is used in the diagnosis of several diseases. There are currently various types of medical images, such that each modality or a combination of them is appropriate for a certain kind of tissue, organ or, in some cases, for a particular type of disease.

The intrinsic physical principles used in a set of image acquisition differentiate the imaging modalities. Computed tomography using X-rays measures the attenuation of X-rays in various directions in the tissue study, forming a three-dimensional image using reconstruction mathematical algorithms. Denser materials attenuate a larger amount of X-ray and therefore appear more prominently in the resulting image, which makes this modality more suitable to visualize hard tissues. Magnetic resonance imaging measures the spin relaxation time of the water molecule hydrogen atoms after undergoing radio frequency pulses and a high magnetic field highlights soft tissues more clearly. Ultrasonography, on the other hand, measures the sound turnaround time (echo) produced by the sound emitted by focusing on a tissue. Distinct types of tissues produce different echoes and consequently differences in the images.

These devices store images in DICOM (Digital Imaging and Communication in Medicine) [1] format. DICOM is an international standard protocol for file storage and network transmission of a set of medical images. This format allows the interoperability among medical imaging equipments and tools from different manufacturers. Due to its advantageous features and to be compliant with international standardization, DICOM is the image format supported by InVesalius.

2.2   Medical Image Software

There are several software packages available for the processing, analysis and visualization of medical images, some of them with support for exporting files for 3D printing anatomical parts, such as InVesalius. In this section, we give a brief description of some tools with similar functionalities to InVesalius.

The Slicer 3D is an open-source software available under BSD (Berkeley Source Distribution) licensing modality and has several tools for processing, analysis and visualization of images. However, it has a complex graphical user interface (GUI) for users in the health care field, consequently mainly used in the academic environment. OsiriX software is widely used by surgeons and radiologists around the world, has a friendly interface and is an open-source code licensed under the GNU GPL 3. It has only a version for OS X operating system and is dependent on Apple Inc. hardware, which makes it expensive when compared to hardware from other manufacturers with similar processing power. Moreover, its version for 64-bit architecture, which is required to process high-resolution images, is not for free. The FreeSurfer software, designed for segmentation and analysis of brain images, runs on Linux and OS X platforms and has a complex graphical interface, requiring the use of Unix command lines to perform some tasks. Despite having open-source-code, it has a proprietary license. The Volview software (Kitware Inc.) is an open-source, BSD license, with a user-friendly graphical interface that runs on Windows, Linux and OS X platforms; however, its main purpose is for visualizing of DICOM images and does not support triangular mesh generation for 3D printing.

Among the main proprietary software packages, Analyze (Mayo Clinic) has various tools, but a complex graphical user interface. Mimics software (Materialise Corp.) and 3D Doctor (Able Software Corp.) have friendly interfaces with a large number of tools. In Mimics software, some modules need to be purchased separately. Amira 3D, developed by FEI, contains many features but has a complex graphical user interface. ScanIP, developed by Simpleware, has a simple interface with a smaller number of tools compared with the previously mentioned software packages. Vitrea (Vital Images) and Vizua are widely used mainly in radiology field. These tools are normally quite expensive for public hospitals in developing countries, since it is often necessary to have a separate license for each computer intended to run the application. Table 1 summarizes the main characteristics of the previously mentioned software tools.

2.3   3D Printing

3D printing [2], also known as rapid prototyping or, more recently, standardized (ASTM F2792) as additive manufacturing (AM), is nowadays considered a groundbreaking technology for the production of high added-value products. 3D printing began as a way to produce higher quality prototypes more quickly, with minimum human intervention. However, what happened during the almost 30 years during which this technology has existed is that, beyond a myriad of

Table 1. Main characteristics of similar tools to InVesalius.

Name	License	Operating system	Country
3D Doctor	Proprietary	Windows	USA
3D Slicer	BSD	Windows/Linux/OS X	USA
Amira 3D	Proprietary	Windows/Linux	USA
Analyze	Proprietary	Windows/Linux	USA
FreeSurfer	Other/Open Source	Linux/OS X	USA
InVesalius	GNU GPL 2	Windows/Linux/OS X	Brazil
Mimics	Proprietary	Windows	Belgium
OsIrix	GNU GPL 3	OS X	Switzerland
ScanIP	Proprietary	Windows	UK
Vitrea	Proprietary	Windows	USA
Vizua	Proprietary	Windows/Linux/OS X	France
Volview	BSD	Windows/Linux/OS X	USA

available processes today, they largely migrate to various sectors of the industry, thanks to the evolution of the AM processes and their associated materials. Therefore, one of the more promising area is the health care sector, where customization is a quantum leap beyond traditional way to produce prostheses and medical devices.

The 3D printing process starts with a 3D virtual model, normally represented by means of a simple triangle mesh called STL (Stereolithography) file format. Every single triangle in a STL file is composed of a Cartesian three spatial coordinates (x, y, z) and a normal vector pointing out of each triangle as a way to inform the 3D printing machines where the structures material to be printed are. The STL format is one of the simplest ways to represent a solid and is used as a 3D printing *de facto* standard since the first machines available in the market; however, STL is not accurate and carries several redundancies. Today, there is a better representation called AMF (Additive Manufacturing File Format) or ASTM F2915 but still not widely used. The STL file can be originated from many sources such as a CAD modeling system, a 3D laser or light scanner, or even medical scanners. The former is the way a medical imaging software integrates to a 3D printing machine. The 3D STL file is then sliced by the 3D printing machine that, by means of a layer-by-layer paradigm, can print objects in many different materials depending upon the 3D printing process used.

3   InVesalius Resources

The development of InVesalius aimed at including the following features: a multi-language user-friendly graphical interface, cross-platform, and open-source coding. The code was implemented using Python and C++ programming languages, VTK library (Visualization Toolkit from Kitware) for 2D and 3D visualization,

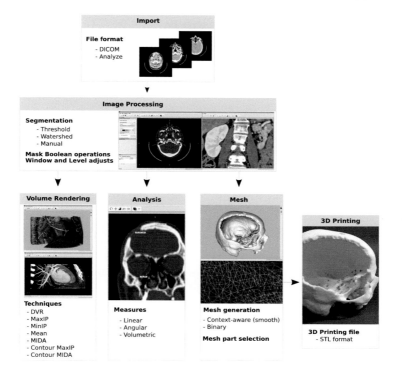

Fig. 1. Main components of InVesalius.

Numpy (Numeric Python), Scipy (Scientific Python), PIL (Python Imaging Library) and GDCM (Grassroots DICOM). The graphical user interface was built through the wxPython library, which keeps the same look and feel on any operating systems. Figure 1 illustrates the main components of Invesalius. The following sections briefly describe the main software modules and theoretical aspects associated with InVesalius.

3.1   DICOM Import

InVesalius imports DICOM files using GDCM (Grassroots DICOM), which supports JPEG 2000 compression. In addition, the library is composed of methods for verifying the orientation of image volumes (axial, coronal, sagittal or oblique) and sorting the slices.

A class has been implemented to classify DICOM files taking into account patient and series information. It is common for an examination of a unique patient to contain several series. For instance, in the case of computed tomography, acquisition can be used for displaying more clearly bones and other acquisition using contrast agents for better visualization of the vascular system.

After performing a sequence of DICOM file importing, InVesalius stacks the images and applies an interpolation algorithm according to the spacing that is indicated in the appropriate field of DICOM file, considering the spacing in x and y axes. It is important to keep the real dimensions when making measurements or exporting the model as an STL mesh file for 3D printing. The volume is saved on hard disk and is used as a file mapping technique of memory for accessing it. The advantage is the lower use of memory (RAM), enabling 32-bit architecture operating systems to work with larger sets of images.

With the importation of DICOM files, a multiplanar reconstruction (MPR) is performed. This kind of reconstruction provides the visualization of structures under different anatomical orientations, allowing more precise image segmentation process and anatomical structure measurement and visualization.

3.2 Image Segmentation

Segmentation techniques allows for the "digital dissection" of different tissues or region of interest. Among those techniques, the manual segmentation, thresholding and watershed techniques are available in InVesalius. A mask is employed to select and represent the region of interest. Furthermore, each mask has a level of transparency, preserving the original image in the background. During manual edition, the user can delineate the region of interest in each image (slice) of the dataset using a brush tool or erase the selections performed by other manual editions tools.

In the thresholding-based segmentation, the user can select an initial and final grayscale level to be kept in the segmentation mask. Additionally, there are some presets for computed tomography images based on the Hounsfield scale.

The watershed segmentation method requires for the user to enter markers to indicate portions of the image that represent objects and background. This method considers the image as a drainage basin, where the graylevel intensities correspond to altitude values, forming valleys and mountains, whereas background and object markers correspond to water sources. These water sources will flood the relief and construct barriers when different water sources meet together, resulting the image segmented into background and objects. The image segmentation methods were implemented with Numpy and Scipy libraries. The interaction and visualization of the mask were implemented with VTK.

3.3 Triangle Mesh

After the segmentation process, it is possible to generate a triangle mesh (Fig. 2) by means of Marching Cubes [3]. This technique considers the segmented volume as an input, analyzes each voxel and its neighborhood, then searches a 256-triangle-layout table in order to determine the most suitable triangle.

Binary segmented volumes can generate triangle meshes with certain staircase artifacts (Fig. 3(b)), mainly in regions with high curvature. These artifacts are not natural to the patient anatomy, such that the resulting model is not suitable for 3D printing. Thus, a mesh smoothing is required. The selection of the

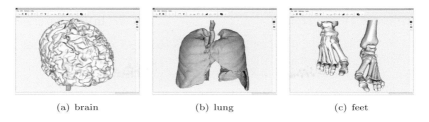

(a) brain   (b) lung   (c) feet

Fig. 2. Examples of triangle meshes.

smoothing algorithm is crucial since it may oversmooth the resulting mesh and cause the loss of fine details. In Fig. 3(c), for instance, it is possible to observe some holes in the eyeball area due to the mesh extraction from the grayscale image (Fig. 3(a)). To address this problem, a context-aware smoothing algorithm [4] was implemented in InVesalius, where higher weights are assigned to artifact regions and lower weights to non-artifact regions. These weights determine the degree of smoothing to be applied to the images with respect to their original dimensions. A sample of a mesh smoothed through this algorithm is shown in Fig. 3(d). By using the Hausdorff distance and taking the grayscale mesh as reference, the difference from the context-aware mesh to the grayscale mesh is on average 0.000239 in units of the bounding box diagonal, whereas the difference from the Gaussian smoothing mesh to the grayscale mesh is on average 0.000431 in units of the bounding box diagonal. A lower Hausdorff distance indicates a more reliable method.

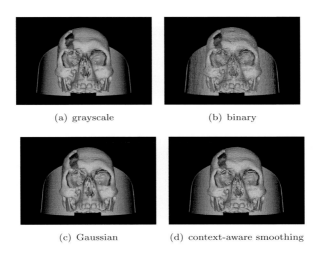

(a) grayscale   (b) binary

(c) Gaussian   (d) context-aware smoothing

Fig. 3. Comparison among meshes extracted from a (a) grayscale image, (b) binary image, (c) binary image smoothed through a Gaussian filter and (d) binary extracted mesh smoothed through context-aware smoothing.

3.4   Volume Rendering

Volume rendering is a type of scientific visualization of medical imaging data. Direct Volume Rendering (DVR) [5] employs presets that highlight certain anatomic structures and is widely used by radiologists. From each pixel that belongs to the view window, rays are emitted in the direction of the volume. The voxels intersected by these rays have their graylevel intensities represented by a color and transparency value. A transfer function is employed to assign color to the voxels. This module of the InVesalius tool is implemented through VTK library.

InVesalius also offers projection techniques, such as MIP (Maximum Intensity Projection) and MIDA (Maximum Intensity Difference Accumulation). Similarly to direct volume rendering, MIP and MIDA emit rays to the direction of the volume. In MIP, each ray intersects the voxel with highest value, which is useful to highlight high contrast regions, such as bone regions and nodules; however, it presents some visual drawbacks. In MIDA, on the other hand, rays capture and accumulate changes from low to high values. Thus, regions with large variations of intensity are highlighted in the resulting image. MIDA provides visual information of depth, allowing for the user to better comprehend the image. Both techniques, MIP and MIDA, are less sensitive to occlusion, because they focus on high contrast areas at the expense of low contrast.

4   Distribution and Applications

The InVesalius software was a milestone for the use of three-dimensional technologies (virtual and physical through 3D printing) in Brazilian public hospitals. In this section, we present some usage statistics and applications of InVesalius in the health care field.

InVesalius checks on the server if there is a new version of the software every time a user initializes it and, according to the IP (Internet Protocol), it is possible to identify the country that requests the license and count as an installation. To avoid counting the same installation twice, a unique identifier is sent to the InVesalius server. Since it is an open-source software, new features can be incorporated into the tool as needed.

From March 2013 to April 2015, there were 7945 installations, distributed in 115 countries. In Brazil, 1794 installations were downloaded from our servers. We can see a map with the InVesalius distribution in Fig. 4. Although developed countries have the highest number of installations along with Brazil, countries in development, BRICS and some of the Asian continent have a significant amount of installations, which demonstrates the popularization of computed tomography and magnetic resonance imaging for medical diagnoses. Unfortunately, use of InVesalius in Africa is low. One of the reasons for high dissemination of InVesalius software is its support for seven languages (English, Portuguese, German, Spanish, French, Greek and Korean).

Since its first version started to be developed in January 2001, as the pioneer open-source solution integrating medical scanners and 3D printing, until

Fig. 4. Installation distribution of InVesalius software between March 2013 to April 2015. A legend is shown on the top right of the figure.

April 2015, InVesalius enabled 3D printing (Fig. 5) of 3808 anatomical models, in the CTI's ProMED Program - 3D Technologies for Healthcare - that were used for surgical planning. In the context of ProMED, approximately 284 Brazilian hospitals were supported, between the years 2011 and 2014. Surgeons from all over Brazil use models to simulate surgical procedures, fixation plates modeling, and patient-specific prosthesis design and production. A few Latin America hospitals outside Brazil were also supported in countries such as Argentina, Chile, Paraguay, Ecuador, Peru, Uruguay, Colombia and Mexico.

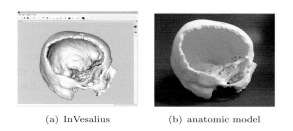

(a) InVesalius   (b) anatomic model

Fig. 5. Virtual and physical model for surgical planning.

InVesalius has also been a basis for new research and developments, such as a free laparoscopic surgery simulator [6] under development by the Tecnológico of Monterrey, in Mexico, and a neuronavigator [7] for applying transcranial magnetic stimulation (TMS) that supports different types of spatial trackers.

In addition to improvements on surgery planning, as mentioned previously, InVesalius has contributed for Brazil to be on par with developed countries in regard to the development of software for processing, analysis and visualization of medical images.

5   Conclusions

This paper described the open-source software InVesalius, its main resources and some applications related to health care. The software has contributed to the scientific and technological development of Brazil in terms of infrastructure and medical facilities, promoting more efficient procedures in health care with cost-effectiveness in public hospitals.

Several new functionalities such as generation of triangle meshes using graphics processing units (GPUs), the integration with mobile devices, touchless interfaces for using in operating rooms, among several other resources already tested as stand alone tools are under development to be incorporated into InVesalius.

Acknowledgements. The authors are grateful to FAPESP for the Brazilian Research Institute for Neuroscience and Neurotechnology - BRAINN (CEPID process 2013/07559-3) and for the Thematic Project (Grant 2011/22749-8) for the financial support.

References

1. DICOM: Digital Imaging and Communications in Medicine (2015). http://dicom.nema.org/
2. Schubert, C., van Langeveld, M., Donoso, L.: Innovations in 3D printing: a 3D overview from optics to organs. Br. J. Ophthalmol. **98**, 159–161 (2014)
3. Lorensen, W.E., Cline, H.E.: Marching cubes: a high resolution 3D surface construction algorithm. ACM SIGGRAPH Comput. Graph. **21**, 163–169 (1987). ACM
4. Moench, T., Gasteiger, R., Janiga, G., Theisel, H., Preim, B.: Context-aware mesh smoothing for biomedical applications. Comput. Graph. **35**, 755–767 (2011)
5. Rogers, D.F., Earnshaw, R.: State of the Art in Computer Graphics: Aspects of Visualization, 1st edn. Springer, New York (1994)
6. Guerra, M.R.M., Amorim, P.H.J., Moraes, T.F., Silva, J.V.L., Quinones, K.L.B., Villalba, E.F., Zuniga, A.E., Rodriguez, C.A.: Soft tissue modeling for virtual surgery simulation. In: XXIV Brazilian Congress of Biomedical Engineering, pp. 813–816 (2014)
7. Rondinoni, C., Souza, V., Matsuda, R., Salles, A., Santos, M., Baffa Filho, O., dos Santos, A., Machado, H., Noritomi, P., Silva, J.: Inter-institutional protocol describing the use of three-dimensional printing for surgical planning in a patient with childhood epilepsy: from 3D modeling to neuronavigation. In: IEEE 16th International Conference on e-Health Networking, Applications and Services (Healthcom), pp. 347–349 (2014)

Computer Graphics

As-Rigid-As-Possible Character Deformation Using Point Handles

Zhiping Luo$^{(\boxtimes)}$, Remco C. Veltkamp, and Arjan Egges

Utrecht University, Utrecht, The Netherlands
{Z.Luo,R.C.Veltkamp,J.Egges}@uu.nl

Abstract. In this paper, we present a versatile, point handles based character skinning scheme. Point handles are easier to design and fit into an object's volume than a skeleton. Moreover, with a conventional blending technique such as linear blending, point handles have been successfully demonstrated to handle stretching, twisting, and supple deformation, which are difficult to achieve with rigid bones. In the context of only blending, limbs, however, are not bent rigidly with point handles, limiting the space of possible deformations. To address this, we propose a method that automatically recovers the local rigidities of limbs via minimizing a surface-based, nonlinear rigidity energy. The minimization problem is subjected to the positions of a set of point handles' proximal vertices. The positions fitting point transformations are computed by linear blend skinning, leading to speedups of the minimization in particular for large deformations. The use of nonlinear energy also allows versatile posing by intuitively selecting which point handles provide their proximal vertices on-the-fly. The degrees of freedom in modeling user constraints are reduced, and the skinning process is automated by relevant functionalities included in our scheme. The effectiveness of our scheme is demonstrated by a variety of experimental results, showing that the scheme could be an alternative to skeletal skinning.

1 Introduction

To date, skeletal skinning has dominated the practical usages in deformable character animation due to its simplicity and efficiency. In this skinning scheme, the skeleton is a hierarchy of rigid bones referred to as *handles* inside the character, and the skin is moved as an explicit function of handle transformations.

A manual pipeline of skeleton-based character rigging mainly consists of three sequential phrases: first, designing and fitting a skeleton into the character body, and second, painting the influence weights per bone, and finally specifying the transformations of a set of bones. Before painting, usually a method that automatically computes the weights is applied in order to reduce the time, as in this context artists only need to adjust some of the weights. There are several cases where the number of corrections is reduced. First, the computed weights are good enough. Second, artists can easily determine the influencing bone(s) of a body segment so that they will intuitively repaint the regions assigned with

© Springer International Publishing Switzerland 2015
G. Bebis et al. (Eds.): ISVC 2015, Part I, LNCS 9474, pp. 57–70, 2015.
DOI: 10.1007/978-3-319-27857-5_6

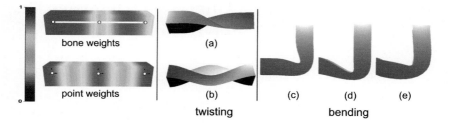

Fig. 1. The weights are visualized using a jet colormap. The white chain denotes the skeleton and the coordinate frames denote point handles. Point weights vary over the region between two point handles, while bone weights only vary near a joint. **Twisting**: twisting the middle joint (point handle), skeleton-based LBS twists the shape only near the middle joint (a), while LBS with point handles produces interesting twisting spread over the entire shape (b). **Bending**: skeleton-based LBS produces a rigid bending (c), while LBS with point handles fails (d). Alternatively, for a rigid bending point handles can be placed at bone centers if one already knew the characteristics of point weights. However, the intuitiveness of specifying transformations is lost. Our scheme aims to approximately recover the local rigidities of a region between two point handles (e).

improper weights. At last but not least, the rotation center of each bone is correct enough, as artists will transform the bones for animation previewing, which is necessary to detect problematic regions and would be repeated. A common solution is to fit a skeleton to the character body and anatomically collocate it with the skin. However, this solution is cumbersome, as transforming a joint often shifts its incident bone(s) outside the body due to the skeleton hierarchy.

The hierarchical structure also complicates skeleton-based automatic skinning. On the one hand, automatic skeleton extraction [1] does not guarantee that the resulting line segments lie inside the object's volume. On the other hand, embedding an existing skeleton inside a shape is a difficult problem. Although well embedded skeletons have been produced, they were archived by solving a continuous optimization of joint locations specific to humanoid shapes [2], or a nonlinear optimization where users provide joint location hints [3].

In contrast to the skeleton, point handles are easy to place inside an object's volume, and each point handle is independently positioned without influencing the positioning of other point handles. Moreover, point handles with associated weight functions (e.g. [4]) are highly suitable for handling stretching, twisting [5], and supple skin regions [4], which are difficult to achieve by linear blend skinning (LBS) with only a scalar weight function per bone. However, LBS with point handles are unsuitable to bend limbs rigidly if the point handles are placed at joint locations so that they are rigged as intuitively as joints. In this paper, the influence weights computed for point handles are referred to as *point weights*, and the weights computed for bones are referred to as *bone weights*. As Fig. 1 illustrates, interesting twisting is spread over the entire shape by LBS with point weights, while LBS with bone weights bends a shape rigidly. This is because point weights by construction vary over the shape more than bone weights which only vary around joints.

Favoring the aforementioned flexibility of point handles, we seek a fully point handles based skinning scheme that also produces rigid bending desired for articulated deformable characters. As a result, this novel scheme will richly expand the space of deformations possible with LBS with only per point handle scalar weight functions.

To address the limitation of bending, we propose a method that automatically recovers the local rigidities of limbs via minimizing a surface-based, as-rigid-as-possible deformation energy. The minimization problem is subjected to deformed positions of a set of point handles' proximal vertices, computed by LBS with point weights. As point handles are placed exactly at joints which are located where they would be in a real world skeleton, a region between two point handles deforms as-rigidly-as-possible as a rigid limb by solving the constrained minimization problem. Point handles are not in a hierarchical structure as rigid bones, hence inverse kinematics or motion capture techniques cannot be applied to facilitate the input of transformations. Instead, their transformations have to be manually specified through a two-dimensional interface common to animation authoring. To reduce the degrees of freedom, a functionality inferring rotations from translations is developed. So only translations, easily and intuitively prescribed by moving the mouse, are required to users. Compared to automatic skeletal skinning, automatic skinning with point handles is easily achieved based on a mesh-segmentation method.

Our scheme has shown advantages by a variety of experimental results:

i. It provides a cheap and robust way of producing rigid bending, and maintaining shape details, in terms of user interactions and computational time.
ii. It allows versatile posing by adjusting the positional constraints of the as-rigid-as-possible deformation energy on-the-fly.
iii. It retains the intuitiveness of rigging a skeletal character by placing point handles at joint locations, and the simplicity of LBS.

2   Previous Work

Linear blend skinning (LBS) with a hierarchy of rigid bones gives rise to *joint-collapse* artifacts near joints such as shoulders, wrists, or even elbows if they exhibit large rotations. To cope with such artifacts, one possibility is to expand the expressive power of LBS by either supplying additional weights per bone [6,7], or adding virtual bones expressed as their transformations and weights [8,9]. The new data has to be automatically computed using example poses [5], which often do not exist. Another solution is to linearly blend unit dual quaternions instead of transformation matrices, forming the method called dual quaternion skinning (DQS) [10] which is more computationally expensive than LBS.

It is also possible to reduce the number of LBS artifacts by replacing rigid bones with other types of handles. Works such as [11,12] use curve skeletons to fix joint-collapse. Curved skeletons need new rigging controls that are inconsistent with the existing rigging pipeline [13]. A lattice of control points, used in free-form deformation [14], is easy to design, but it is unsuitable for the control of concave objects due to its regular structure. A cage [15], as a general control polyhedron loosely enclosing the target surface, has a better match of degrees of freedom to the enclosed geometry, thereby allowing precise area control such as bulging and thinning in regions of interest. However, a control mesh is tedious to establish and manipulate.

Points are perhaps the easiest handles to establish. Existing automatic bone weights methods such as [2] may be trivially adapted to compute the influence weights for point handles. Jacobson et al. [4] proposed a method well suited for computing point weights. Although the method is slow to compute and requires a volume discretization, its resulting weights produce the highest quality deformations [5]. Once good point weights are available, the transformation of a vertex at runtime is computed by linearly blending point transformations. LBS with point handles is suitable for handling stretching, twisting [5], and supple deformation which requires a skeleton with many bones in skeletal skinning. However, it is much less effective to bend limbs rigidly than skeleton-based LBS. Furthermore, LBS, as a closed-form blending technique, has no shape-preserving property. Variational methods [16,17] compute high-quality shape preserving deformations for arbitrary handles at points, but at a greater runtime cost.

Since each of the aforementioned handle types shows its own advantages, a hybrid of two or more handle types has been demonstrated to expand the space of deformations possible with LBS: a hybrid of point and bones [5], and a hybrid of points, cages and bones [4]. Although the flexibility of point handles has been demonstrated, until now a fully point handles based LBS remains to be developed.

3   Skinning with Point Weights

To achieve rigid bending with point handles is difficult. On the one hand, this is because it is not so clear where the deformation point handles should be placed for rigid bending. On the other hand, rigging the point handles for a rigid bending is not as intuitive as rigging a skeleton, in case intuitively and anatomically the point handles are placed between conjoint near-rigid segments. In contrast, tediously and carefully rotating the point handles is required, as point weights vary over the region between two point handles.

Assuming that good point weights are available, it is no doubt that associated weights of the vertices in some distances to each point handle fall into a rather small range close to 1. These vertices form a small region in the vicinity of each point handle. Here, good weights should at least satisfy the properties of locality and smoothness. So in the context of blend skinning, on the one hand the deformation of the proximal region per point handle is nearly rigid. In other words, the weight map per point handle captures the local rigidities of such a region. On the other hand, the region's deformation accurately and intuitively fits the transformation of point handle since the associated weights are large enough.

Our solution is to impose a rigidness regularization to the vertices between two point handles, by solving an energy minimization problem. The objective function is subjected to the positions of proximal vertices per point handle, which are computed in the spirit of LBS. Thus, our framework keeps the simplicity of modeling the user constraints as blend skinning, where users specify the handle transformations rather than to directly prescribe the vertex positions.

3.1 Modeling Framework

Given a skinning model consisting of H point handles, a skin mesh S containing N vertices $\{v_1, ..., v_N\}$ and M triangles $\{t_1, ..., t_M\}$ at rest pose, each vertex v_i is bound to the handles by influence weights expressed as a vector $\mathbf{w}_i = (w_{i,1}, ..., w_{i,H})$. Given the transformations of point handles $(C_1^f, ..., C_H^f)$ at frame f, optimal deformation is computed by solving the minimization problem

$$\min \quad E_s(S') = \sum_{i=1}^{N} \sum_{j \in \mathcal{N}_1(v_i)} \boldsymbol{\omega}_{i,j} \parallel (v_i' - v_j') - R_i(v_i - v_j) \parallel^2$$

$$\text{s.t.} \quad v_k' = v_k \sum_{j=1}^{H} w_{k,j} C_j^f, \quad k \in \mathscr{H},$$

(1)

where $\mathscr{H}$ is the set of indices of the proximal vertices, $\boldsymbol{\omega}_{i,j}$ are per-edge weights between v_i and each one-ring neighbor $v_j \in \mathcal{N}_1(v_i)$. R_i of vertex v_i is derived from the singular value decomposition (SVD) of the covariance matrix $S_i = \sum_{j \in \mathcal{N}_1(v_i)} (\boldsymbol{\omega}_{i,j} e_{ij} e_{ij}'^T) = U_i \sum_i V_i^T$, and $R_i = V_i U_i^T$ [16], where $e_{ij} = v_i - v_j$ and e_{ij}' is the edge after reconstruction.

The solution consists of two steps. First, compute an initial guess as $v_i' = v_i \sum_{j=1}^{H} w_{i,j} C_j^f$, and approximate the local rotations R_i based on v_i'. Second, new positions v_i' are obtained by minimizing Eq. (1). We follow the same procedures as [16] to compute the partial derivatives w.r.t. v_i'. By setting the partial derivatives to zero w.r.t. each v_i', a sparse linear system of equations is achieved, which can be solved efficiently by the sparse Cholesky factorization. The minimization is iteratively performed by alternately re-computing local rotations and solving the equation until it reaches the number of iterations specified by users.

Closely similar to the surface modeling framework [16], our framework also minimizes the changes of edge lengths to obtain as-rigid-as possible (ARAP) deformations. However, we compute positional constraints in the sprit of linear blend skinning. Thus, users only need to move the point handles without manually selecting a set of vertices on the surface. Also, an initial guess is computed by LBS rather than using the Laplacian surface editing (LSE) [18] method. Since LBS, having a closed-form formula, is much faster than LSE, and can produce a reasonable degrees of local rigidities in bending with point weights. Thus, the number of iterations required to reach a desired rigid bending will be reduced.

Our method is depicted in Fig. 2. Since the objective function is subjected to hard constraints that are computed positions using LBS, discontinuities appear

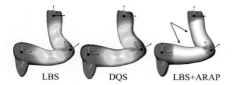

Fig. 2. Point weights are displayed as colors in the two images on the left. Both LBS and DQS with point handles are unsuitable to bend a shape rigidly. Our scheme (LBS + ARAP) automatically refines the region between two point handles to be as-rigid-as-possible via minimizing a rigidity energy constructed over the shape. The minimization problem is subjected to the positions of a set of point handles' proximal vertices (reddish parts of the rightmost image).

Fig. 3. The torus is deformed. (a) Discontinuities appear around the point handles. (b) With a Laplacian smoothness regularization, such artifacts are suppressed.

in the handle proximity. Such artifacts can be easily suppressed, see Fig. 3, by moving the vertices in the direction of the Laplacian

$$v'_{i\_opt} = v'_i + \frac{1}{\sum_{j \in \mathcal{N}(i)} \omega_{i,j}} \sum_{j \in \mathcal{N}(i)} \omega_{i,j}(v'_j - v'_i), \tag{2}$$

In fact, only the positions of proximal vertices will be re-optimized. For the regions between two point handles, optimal results have been reached due to the rigidness regularization resulting from a solution of Eq. 1.

3.2 Degrees of Freedom

The transformations of point handles are manually specified through a 2D user interface common to animation authoring. Through such an interface, rotations that are necessary to LBS (see Fig. 4), however, are non-trivial and less intuitive to specify relative to translations that are directly proportional to mouse movements. We present methods that automatically infer rotations from user-specified translations.

Given P point handles denoted by $p_i \in \mathbb{R}^3, i \in \{1, ..., P\}$, each point handle has E user-defined incident pseudo-edges, defined as three-dimensional vectors $p_{i,k} = p_i - p_j, k \in \{1, ..., E\}$ at rest pose, and $p'_{i,k} = p'_i - p'_j$ after translations. The rotation between two vectors $p_{i,k}$ and $p'_{i,k}$ is represented using quaternion denoted by $q_{i,k}$. The final rotation denoted by q_i of point handle p_i is an average of all $q_{i,k}, k \in \{1, ..., E\}$.

Notice that, unlike the skeleton, the pseudo-edges do not necessarily lie inside the volume and conform to a skeleton structure. In fact, there is a large degree of freedom on connecting the point handles as long as no one is isolated, such that, for each point handle, the rotation axis is the cross product of the two vectors $p_{i,k}$ and $p'_{i,k}$ defined by the pseudo-edges.

The half-quaternion method is used to calculate the rotation about an axis between two vectors, simply because it saves some square-root calculations and ensures a "shortest arc" solution. For an average of only two quaternions, the well-known spherical linear interpolation (Slerp) can be applied. In the presence of more quaternions, a simple and fast way is to average by a recursion sum algorithm

$$avg(q_{i,E}) = \begin{cases} \frac{1}{E}(q_{i,E} + avg(q_{i,E-1})) & E \geq 2 \\ q_{i,1} & E = 1. \end{cases} \tag{3}$$

This method yields a valid average only if the quaternions are relatively close to each other.

A more accurate method comes at the price of reduced speed. However, because the number of incident edges per point handle is small, the overload is minimal. This method first converts the quaternions $q_{i,k}(w, x, y, z)$ as four-dimensional column vectors $q_{i,k}(x, y, z, w)$ and then build a matrix as

$$\mathbf{M} = \frac{1}{E} \sum_{k=1}^{E} q_{i,k} q_{i,k}{}^{T}. \tag{4}$$

The final quaternion is the (convert back) eigenvector corresponding to the largest eigenvalue of the matrix. Figure 4 depicts a visual comparison between two methods.

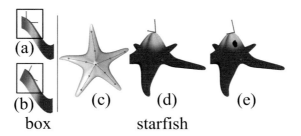

Fig. 4. Box: With the presence of only translation of the transforming point handle (coordinate frame), LBS gives rise to shape distortion (a). Additionally providing a rotation removes the distortion artifact (b). **Starfish**: given the pseudo-edges (dash lines) (c), accompanying rotations are inferred from user-prescribed translations. For the middle point handle, the fast recursion sum method taking about 6 ms yields self-intersection, rendered into a black hole (e). The method based on computing eigenvector of a matrix addresses this limitation (d), though it takes about 16.6 ms.

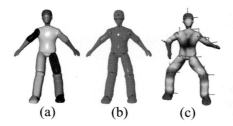

Fig. 5. Segments with different thicknesses, indicated by different colors, are obtained based on SDF values (a). Point handles are automatically placed between conjoint segments (b), and their influence weights, displayed as colors in (c), are computed. The character is posed by translating point handles (coordinate frames). Note that rotations are inferred.

3.3   Automatic Skinning

Point handles are easily placed within the character volume with the help of ray-triangle intersection for instance, and their positions can be anatomically guided to users. For example, the positions of point handles tend to be joint locations in a humanoid character. Targeting a fully automatic skinning scheme, a method is developed to automatically determine the placements of deformation point handles.

In articulated deformable characters, the thickness of a body segment is often different to its conjoint segments, and the skin of such a segment deforms as-rigidly-as-possible. Recall that in our method the region between two point handles is as-rigid-as-possible, thus, for a polygonal mesh, it is natural to place the point handles between conjoint segments with different thickness. The shape-diameter-function (SDF) values [1] measure the diameter of an object's volume in the neighborhood of each vertex on the surface, providing a direct way to detect near-rigid segments. Our method is based on SDF values.

First, a hard-clustering is performed on the raw SDF values to obtain pairwise facet and cluster-id. Second, candidate vertices occupying multiple clusters are picked out if their hosting triangles have a different cluster-id. For each candidate: a ray along the inverse normal is cast from the vertex position, and then a point as an intermediate result is picked on the line of the ray according to half of the associated SDF value. This ensures that all picked points lie inside the volume. Finally, the position of a point handle is the center of a spatially grouped intermediate points, guaranteeing that all point handles are inside the volume.

Once the skin is bound to deformation point handles by influence weights computed using an automatic weights method such as [4], the character is ready for animation. Figure 5 depicts our method applied to a humanoid character.

4   Results and Discussion

Our method was implemented in C++ on a laptop with an Intel Core i7 2.4 Ghz processor and 6 GB memory. The sparse Cholesky solver and two-sided Jacobi

SVD decomposition, provided by the Eigen library [19], were used to solve the minimization formula and to compute rotations R_i, respectively. The cotangent-weight formula [20] instead of uniform weights is used to compute per-edge weights $\omega_{i,j}$, avoiding deformation artifacts [16].

Bone weights are much more intuitive to paint by artists than point weights. However, seeking an automatic scheme is a primary goal in skinning. Thus, to obtain fair comparisons, the influence weights of bones and point handles are computed. Bounded biharmonic weights (BBW) [4] are computed as influence weights for point handles and bones, respectively, because they are smooth, local and shape-aware, and have experimentally produced high-quality deformations [4,5]. Since BBW requires a priori volume discretization, we obtain it by voxelization instead of tetrahedral meshing, avoiding dealing with self-intersection and non-manifoldness. The computation then could be largely increased as usually a voxel grid is denser than a tetrahedral representation, but it is a tolerable, one-time precomputation taking dozens of seconds for a polygonal mesh with reasonable geometric complexity.

Two animation models mainly containing deformations of rigid bending are used to demonstrate the effectiveness of our method. They include a hand model with 33 frames and 1,926 vertices, a human model with 54 frame and 3,557 vertices. Deformations resulting from skeleton-based LBS (LBS + bones) are considered as the ground-truth, as bones are naturally effective for rigid bending. Joint locations are directly used as point handle positions. For rendering of the deformed mesh, vertex normals are recomputed on CPU.

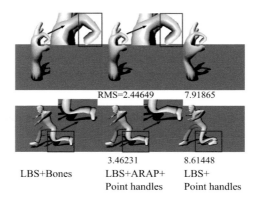

Fig. 6. $l = 0.98$ for the handle model, $l = 0.95$ for the human model, and the number of iterations is 3, giving rise to the results. The numbers below are RMS errors. LBS with point handles fails to bend limbs rigidly. Our scheme gives rise to comparable, visual and quantitative results w.r.t. skeleton-based LBS.

For each point handle, our method needs to determine $v_k, k \in \mathcal{H}$ (recall Sect. 3.1) as positional constraints. A vertex is selected if its largest weight is larger than a threshold l. Note that the associated weights of a vertex consist of a set of per-point weights, as mostly a vertex is influenced by multiple point handles. Due to the locality offered by BBW, such vertices are well located in the vicinity of point handles.

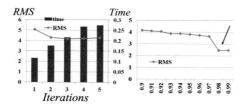

Fig. 7. Left ($l = 0.9$): Computational time (in seconds) increases along with the number of ARAP iterations. The shape is bent more rigidly by increasing iterations, hence mainly the error is reduced along with iterations. By specifying 3 iterations, the lowest error is obtained. **Right** (*iterations* $= 3$): The pattern of the discrete curve shows that mainly the error is reduced by increasing l. In the handle model, mostly, the largest weights of the vertices in the vicinity of a point handle lie in the range from 0.97 to 0.98. Specifying l as a value in this range yields positional constraints, in turn deformations, different to the results generated by other values also in that range. As a result, a sudden drop happens.

An error metric w.r.t. the ground truth is formulated to quantitatively evaluate the methods. The lower the computed value is, the more rigidly the limbs are bent. We first resize the models so that their rest poses are tightly enclosed by a unit cube. Then, the metric is calculated as

$$E_{RMS} = \frac{\sum_{f=1}^{F} \sum_{i=1}^{N} \| bv_i^f - pv_i^f \|^2}{\sqrt{3NF}}, \tag{5}$$

where F denotes the number of frames, N is the number of vertices, bv^f and pv^f are deformed vertices at frame f resulting from a method with bone weights and point weights, respectively. This form of metric has been reported to be less sensitive to global motions [21].

As Fig. 6 shows, LBS with point handles fails in producing rigid bending. Our method, called $LBS + ARAP + Point \; handles$, produces better visual and quantitative results.

Relationships between computational time and the number of iterations used to solving the minimization problem, called $ARAP$ *iterations* for short, between E_{RMS} and the number of ARAP iterations, between E_{RMS} and threshold l, are tested on the hand model. Figure 7 reports the results. Though the computational time increases along with the number of iterations, our energy minimization runs three iterations to yield the lowest error. In cases of larger iterations, the shape is bent more rigidly by our scheme than by skeleton-based LBS. So, the error then is slightly increased. This happened as the bone lengths of the hand model are small. Since BBW yields local and shape-aware weights [4,5], the threshold l is decided intuitively, reducing user errors. As a result, no sharp line is observed in the right subplot.

Discussion. The use of skeleton is known to facilitate the input of bone motions by either motion capture techniques or inverse kinematics. Currently, point handles' counterparts are joints of a corresponding skeletal model in order to maintain an anatomical embedding. Due to this setting, bone transformations were directly reused as point handle transformations in our experiments. As a consequence, the aforementioned techniques facilitating motion input have been investigated in skinned models based on the point handle metaphor, and encouraging results have been obtained.

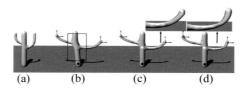

     (a)   (b)   (c)   (d)

Fig. 8. The cactus at rest pose (a) is skinned with four point handles. A pose is achieved by moving two point handles in branch (coordinate frames) and fixing the one at the base (b). In this case, only these three point handles are used to define positional constraints of ARAP. If one desires locally controlled deformation of a branch, our scheme allows it by additionally enabling the middle point handle (white) to provide its proximal vertices as new constraints on-the-fly (c). LBS also produces such a deformation style (d), but the branches are over-stretched, compared to our scheme, see close ups.

Compelling deformations can be obtained by ARAP (as-rigid-as-possible deformation energy) with few user edits. This has been demonstrated in ARAP-based surface modeling [16] for instance. This advantage can be augmented to reduce the degrees of freedom for certain character postures in our skinning scheme based on point handles. Given a skinned model with many point handles, user specifies only a subset of the degrees of freedom, and the rest deformations are automatically determined using the nonlinear ARAP. This is similar to the method [22] that uses nonlinear, rigidity energies to infer the rest skinning transformations. It is also allowed to adjust the positional constraints of ARAP on-the-fly. As Fig. 8 shows, an extra point handle is enabled to provide its proximal vertices as positional constraints. As a result, resulting deformations are locally controlled if it is desired. In this case, the newly enabled point handle actually serves to occlude the rotation propagations from the moving point handle to the regions far away. As Figs. 8 and 9 show, closed-form blending techniques, such as LBS and DQS, do not support shape-reserving property, but our scheme better maintains the shape locality.

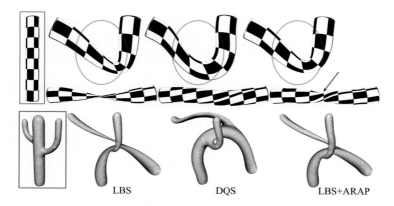

Fig. 9. LBS and DQS do not maintain surface details during deformations, while our method (LBS + ARAP) does. DQS aims at volume preservation, but without considering the realism of posture. For example, the interesting twisting of the cylinder is lost when using DQS. Models in rectangle are in their rest poses.

5   Conclusions

In this paper, we have presented a novel skinning scheme based on point handles, which are much easier to design and embed into the character body than a skeleton. Our method addressed the limitation of bending, when using a conventional blend skinning with point weights, by optimizing the vertex positions using a nonlinear, rigidity energy. We have demonstrated the effectiveness of our method by experimental results of two animation models mainly containing rigid bending. Automatic skinning has been explored in order to reduce the manual work. Also, our method allows for versatile posing and it is robust to large deformations.

Currently, point handles are considered as counterparts of joints. In cases of skinned models with long bones, the mesh part wrapping around a long bone might not be rigid enough when bending by our scheme. This is because we only defines constraints on edge lengths in order to obtain substantial speedups. To overcome this, we plan to develop a method that automatically selects better placements of point handles based on *Schelling points* [23]. To further obtain speedups, we will incorporate existing weight reduction techniques [24,25] to reduce the computational cost of a dense-weight skinning model.

References

1. Shapira, L., Shamir, A., Cohen-Or, D.: Consistent mesh partitioning and skeletonisation using the shape diameter function. Vis. Comput. **24**, 249–259 (2008)
2. Baran, I., Popović, J.: Automatic rigging and animation of 3d characters. ACM Trans. Graph. **26**, 72 (2007)

3. Jacobson, A., Panozzo, D., Glauser, O., Pradalier, C., Hilliges, O., Sorkine-Hornung, O.: Tangible and modular input device for character articulation. ACM Trans. Graph. **33**, 82:1–82:12 (2014)

4. Jacobson, A., Baran, I., Popović, J., Sorkine, O.: Bounded biharmonic weights for real-time deformation. ACM Trans. Graph. **30**, 78:1–78:8 (2011)

5. Jacobson, A., Sorkine, O.: Stretchable and twistable bones for skeletal shape deformation. ACM Trans. Graph. **30**, 165:1–165:8 (2011)

6. Wang, X.C., Phillips, C.: Multi-weight enveloping: least-squares approximation techniques for skin animation. In: Proceedings of the 2002 ACM SIGGRAPH/Eurographics Symposium on Computer Animation, SCA 2002, pp. 129–138 (2002)

7. Merry, B., Marais, P., Gain, J.: Animation space: a truly linear framework for character animation. ACM Trans. Graph. **25**, 1400–1423 (2006)

8. Wang, R.Y., Pulli, K., Popović, J.: Real-time enveloping with rotational regression. ACM Trans. Graph. **26**, 73 (2007)

9. Kavan, L., Collins, S., O'Sullivan, C.: Automatic linearization of nonlinear skinning. In: Proceedings of the 2009 Symposium on Interactive 3D Graphics and Games, I3D 2009, pp. 49–56 (2009)

10. Kavan, L., Collins, S., Žára, J., O'Sullivan, C.: Skinning with dual quaternions. In: Proceedings of the 2007 Symposium on Interactive 3D Graphics and Games, I3D 2007, pp. 39–46 (2007)

11. Yang, X., Somasekharan, A., Zhang, J.J.: Curve skeleton skinning for human and creature characters: research articles. Comput. Animat. Virtual Worlds **17**, 281–292 (2006)

12. Forstmann, S., Ohya, J., Krohn-Grimberghe, A., McDougall, R.: Deformation styles for spline-based skeletal animation. In: Proceedings of the 2007 ACM SIGGRAPH/Eurographics Symposium on Computer Animation, SCA 2007, pp. 141–150 (2007)

13. Kavan, L., Collins, S., Žára, J., O'Sullivan, C.: Geometric skinning with approximate dual quaternion blending. ACM Trans. Graph. **27**, 105:1–105:23 (2008)

14. Sederberg, T.W., Parry, S.R.: Free-form deformation of solid geometric models. In: Proceedings of the 13th Annual Conference on Computer Graphics and Interactive Techniques, SIGGRAPH 1986, pp. 151–160 (1986)

15. Ju, T., Schaefer, S., Warren, J.: Mean value coordinates for closed triangular meshes. ACM Trans. Graph. **24**, 561–566 (2005)

16. Sorkine, O., Alexa, M.: As-rigid-as-possible surface modeling. In: Proceedings of the Fifth Eurographics Symposium on Geometry Processing, SGP 2007, pp. 109–116 (2007)

17. Botsch, M., Sorkine, O.: On linear variational surface deformation methods. IEEE Trans. Vis. Comput. Graph. **14**, 213–230 (2008)

18. Sorkine, O., Cohen-Or, D., Lipman, Y., Alexa, M., Rössl, C., Seidel, H.P.: Laplacian surface editing. In: Proceedings of the 2004 Eurographics/ACM SIGGRAPH Symposium on Geometry Processing, SGP 2004, pp. 175–184 (2004)

19. Guennebaud, G.: Eigen: a c++ linear algebra library, version 3.0 (2011). http://eigen.tuxfamily.org/ (2011). Accessed 22 November 2014

20. Meyer, M., Desbrun, M., Schröder, P., Barr, A.H.: Discrete differential-geometry operators for triangulated 2-manifolds. In: Hege, H.-C., Polthier, K. (eds.) Visualization and mathematics III, pp. 35–57. Springer, Heidelberg (2003)

21. Kavan, L., Sloan, P.P., O'Sullivan, C.: Fast and efficient skinning of animated meshes. Comput. Graph. Forum **29**, 327–336 (2010)

22. Jacobson, A., Baran, I., Kavan, L., Popović, J., Sorkine, O.: Fast automatic skinning transformations. ACM Trans. Graph. **31**, 77:1–77:10 (2012)
23. Chen, X., Saparov, A., Pang, B., Funkhouser, T.: Schelling points on 3d surface meshes. ACM Trans. Graph. **31**, 29:1–29:12 (2012)
24. Landreneau, E., Schaefer, S.: Poisson-based weight reduction of animated meshes. Comput. Graph. Forum **29**, 1945–1954 (2010)
25. Le, B.H., Deng, Z.: Two-layer sparse compression of dense-weight blend skinning. ACM Trans. Graph. **32**, 124:1–124:10 (2013)

Image Annotation Incorporating Low-Rankness, Tag and Visual Correlation and Inhomogeneous Errors

Yuqing Hou$^{(\boxtimes)}$

Key Laboratory of Machine Perception (MOE), School of EECS,
Peking University, Beijing 100871, China
houyuqing1988@gmail.com

Abstract. Tag-based image retrieval (TBIR) has drawn much attention in recent years due to the explosive amount of digital images and crowdsourcing tags. However, TBIR is still suffering from the incomplete and inaccurate tags provided by users, posing a great challenge for tag-based image management applications. In this work, we propose a novel method for image annotation, incorporating several priors: Low-Rankness, Tag and Visual Correlation and Inhomogeneous Errors. Highly representative CNN feature vectors are adopted to model the tag-visual correlation and narrow the semantic gap. And we extract word vectors for tags to measure similarity between tags in the semantic level, which is more accurate than traditional frequency-based or graph-based methods. We utilize the Accelerated Proximal Gradient (APG) method to solve our model efficiently. Extensive experiments conducted on multiple benchmark datasets demonstrate the effectiveness and robustness of the proposed method.

1 Introduction

The prevalence of social network and digital photography in recent years makes image retrieval an urgent need. Image retrieval methods can be classified into two categories: content-based image retrieval (CBIR) and tag-based image retrieval (TBIR). The performance of CBIR algorithms are limited due to the semantic gap between the low-level visual features used to represent images and the high-level semantic meaning behind images. Tags can represent the semantics of images more precisely than low-level visual features, giving rise to research on TBIR.

However, tags are usually noisy and incomplete due to the arbitrariness of user tagging behaviors, leading to performance degradations of TBIR systems [1]. What's more, manual annotation is laborious, error prone, and subjective, making automatic image annotation an attractive research task.

Many machine learning methods have been developed for image annotation. They can be roughly grouped into three categories: supervised methods, unsupervised methods and semi-supervised methods.

© Springer International Publishing Switzerland 2015
G. Bebis et al. (Eds.): ISVC 2015, Part I, LNCS 9474, pp. 71–81, 2015.
DOI: 10.1007/978-3-319-27857-5_7

Supervised methods use the tagged images to train a dictionary of concept models and formulate image annotation as a supervised learning problem. They annotate images using the likelihood between images and tags. [2] formulates the annotation problem in a probabilistic framework and images are represented as bags of localized feature vectors. [3] learns a two-dimensional Multi-resolution Hidden Markov Model (2D-MHMM) on a fixed-grid segmentation of all category examples. [4] models image annotation procedure as a translation problem between image blobs and tags.

Unsupervised methods, e.g. search based-methods, learn the distribution of images and tags and annotate tags among clusters. Search-based methods always search in the feature space to find the most relevant images to the query image, and transfer tags to it using various tag transfer algorithms [5–7]. JEC [6] demonstrates that simple baseline algorithm can achieve high performance. TagProp [5] applies metric learning in the neighborhood of the feature space to annotate query images.

In recent years, semi-supervised approaches have been proposed in this field [8–10]. Semi-supervised algorithms can exploit the unlabelled information to improve the learning procedure and achieve satisfactory performance. [9] models the annotation task as a matrix completion problem, assuming the low-rankness property of the underlying matrix. [11] combined language model with matrix completion by assuming the independency of tags. Kernel trick and metric learning are exploited in [12] to capture the nonlinear relationships between visual features and semantics of the images. Semi-supervised relational topic model (ssRTM) is exploited to explicitly model the image content and their relations [13].

To utilize the large amount of unlabeled dataset for removing noisy tags and completing the missing ones, we propose a semi-supervised method. We formulate the annotation task as a transduction matrix completion problem, taking the following four priors into consideration:

1. **Low-rankness.** Many methods formulated the image annotation problem in a matrix completion framework by constructing and refining the image-tag matrix [8, 10–12]. Existing works have demonstrated that semantic space spanned by tags can be approximated by a much smaller subset of words derived from the original space [14]. As text information, tags are consequently subjected to such subset property [8]. According to the subset property, we assume that the image-tag matrix is a low rank matrix. Thus we can exploit the low rank matrix completion techniques to complete the matrix, thereby annotating the images.

2. **Tag Correlation.** Tags have high level semantic meanings and often appear correlatively at the semantic level. However, traditional methods treat tags merely as labels, reducing the annotation task as a multi-label classification problem. In recent years, researchers have explored the relation among tags. Graph-based methods calculated the semantic correlation among tags using the WordNet distance [15]. Frequency-based methods [10, 16] estimated tag correlations using Jaccard coefficient or co-occurrence in text search results. Jensen-Shannon divergence is introduced in the Flickr Distance to make the algorithm

more precise and reasonable [17]. However, these methods are still imprecise and inefficient. In this work, we utilize the vector representations for tags instead of labels. Word vectors [18], which are seldom used in this field, can present a much higher level semantic meanings than labels, thus we can measure the tag similarity much more easily and precisely.

3. **Tag-Visual Correlation.** Tag-Visual Correlation describes the correlation between the content level and the semantic level. Visually similar images often belong to similar themes and thus are annotated with similar tags. This prior has been widely explored in the image classification field [19,20]. However, there still exists a semantic gap between the content level and the semantic level. Traditional methods usually adopt low level image features, such as color, texture or shape descriptors, to represent the images, which are not so correlated with the semantic level. To narrow the semantic gap and make full use of the correlation property, we utilize high level visual features in our model, such as $DeCAF_6$, which demonstrate much stronger tag-visual correlation than low level visual features.

4. **Inhomogeneous Errors.** Tagging errors come from two aspects: missing tags and noisy tags. Since human-beings are relatively reasonable, we should assume that the tagging results are reasonably accurate. We can observe from the datasets that one image usually has relation with only a few tags, but we have to calculate its association with hundreds or even thousands of tags. For example, images from the MIRFlickr-$25K$ have about 12.7 tags on average [21], but the dataset has $1,386$ unique tags, which means that each image should only be annotated with less than 1% of all the tags. Hence users are more likely to adding noisy tags than missing noisy tags since there are too many unrelated tags. And the errors are mainly composed of noisy tags rather than missing tags. Thus we should treat these two errors with different strategies. We should put more emphasis on denoising rather than completing, paying more attention to the annotated tags rather than the unannotated ones. In other words, if an image is not originally annotated with a tag, it is more likely that they really have no relation at all.

Existing methods never model these two kinds of errors separately. They simply model the errors as Laplacian noise [8] or Gaussian noise [10]. To our knowledge, our model is the first to model the missing errors and noisy errors separately. The model can further adapt to different datasets according to their noise levels.

The novelties and main contributions of this paper are summarized as follows:

- We propose a new image annotation model that incorporates four priors: Low-rankness, Tag Correlation, Tag-Visual Correlation, and Inhomogeneous Errors.
- We utilize the word vectors and CNN features for the tag and the visual features, respectively. These high level features can narrow the semantic gap effectively. It is the first time to utilize both the features for image annotation.
- We model tag correlation and tag-visual correlation in different ways according to their semantic levels.

– We model two kinds of errors separately, the model can adapt to different datasets according to the noise level.
– We utilize the APG to solve our model efficiently.

The most related work to our model is LRES [8]. In their work, the authors formulated the image annotation task as a Robust PCA [22] framework, decomposing the original tag matrix into a refined tag matrix and a sparse error matrix. LRES also takes the tag correlation and tag-visual correlation into consideration and achieves good performance. However, our model is different from LRES in several aspects. First, our model measures tag correlation and tag-visual correlation using different models according to their different semantic levels, rather than using the same Graph Laplacian model in LRES. Second, we adopt more representative features such as CNN features and word vectors to narrow the semantic gap. Third, we do not model the error matrix simply as a sparse matrix, since thee errors are inhomogeneous and the distribution varies across different datasets.

2   Our Image Annotation Model

2.1   Low-Rankness

Denote the image collection $I = \{i_1, i_2, \ldots, i_m\}$, where m is the size of the image set. All original tags appearing in the set form a tag set $W = \{w_1, w_2, \ldots, w_n\}$, where n denotes the total number of unique tags. We can construct a binary matrix $\hat{T} \in \{0,1\}^{m \times n}$ whose element $\hat{T}_{i,j}$ indicates the relation between image i_i and tag w_j, i.e. if i_i is annotated with tag w_j, $\hat{T}_{i,j} = 1$, otherwise $\hat{T}_{i,j} = 0$. We use T to represent the refined tag matrix, where $T_{i,j} \in (0,1)$ the confidence score of assigning w_j to i_i. As mentioned above, we want the refined matrix T to be low rank. Since the low-rankness constraint on T is NP-hard to solve, we replace it with the standard relaxation, the trace norm, i.e. sum of singular values : $\|T\|_*$.

2.2   Tag Correlation Using Word Vectors

To narrow the semantic gap, we extract 300-dimensional word vectors [18] for each tag rather than treating tags merely as labels. Word vectors contain rich semantic information, e.g. semantic similarity. We denote the word vectors as $WV = \{wv_1, wv_2, \ldots, wv_n\}$. Given the completed tag matrix, T^i and T^j are the ith and jth columns of the tag matrix T. Thus we can measure the correlation between tag i and tag j in two ways: (1) similarity between word vectors wv_1 and wv_2, (2) similarity between tag vectors T^i and T^j.

The tag correlation prior can be enforced by solving the following optimization

$$\min_{T} \sum_{i=1}^{n} \sum_{j=1}^{n} \|T^i - T^j\|^2 S_{i,j}, \tag{1}$$

where $\|T^i - T^j\|^2$ measures the similarity between tag vectors T^i and T^j and $S_{i,j}$ measures the similarity between word vectors wv_i and wv_j. The formulation forces tag vectors with large similarities also have large similarity in their corresponding word vectors and vice versa, which essentially embodies the tag correlation prior.

The formulation can be rewritten as $Tr(T^T LT)$, where $L = diag(S^T 1) - S$ is the Graph Laplacian [23]. In our formulation, we define $S_{i,j} = \cos(wv_i, wv_j)$.

2.3 Tag-Visual Correlation Using CNN Features

The tag-visual correlation is not as strong as the correlation between tags owing to the semantic gap. Thus we just formulate the problem in a widely used model [10], which is much more simple and intuitive compared with the Graph Laplacian framework. Denote the image visual features as matrix V, where each visual image is represented as a row vector in $V \in R^{m \times f_v}$. Given the visual feature matrix, we can compute the visual similarity between i_i and i_j as $V_i^T V_j$, where V_i^T and V_j^T are the ith and jth rows of matrix V. Given the completed tag matrix, we can compute the similarity between i_i and i_j basing on the overlap between their corresponding tags, i.e., $T_i^T T_j$, where T_i^T and T_j^T are the ith and jth rows of the tag matrix T [10]. To model the aforementioned tag-visual correlation, we expect $|T_i^T T_j - V_i^T V_j|^2$ to be as small as possible. Thus we can model the tag-visual correlation using the Frobenius norm as $\sum_{i,j}^n |T_i^T T_j - V_i^T V_j|^2 = \|TT^T - VV^T\|_F^2$.

2.4 Inhomogeneous Errors

To model the inhomogeneous errors, we set different weight to the annotated positions and unannotated positions separately: $\lambda_0 \|P_\Omega(\hat{T} - T)\|_F^2 + \lambda_1 \|P_{\Omega^\perp}(\hat{T} - T)\|_F^2$. Ω represents the positions where the images are annotated with tags, P_Ω and $P_{\Omega^\perp}$ are projection operators, λ_0 and λ_1 are positive weighting parameters. λ_0 and λ_1 will change adaptively in different datasets according to their noisy levels. Different from the assumption of sparse errors [8], we model the errors using the Frobenius norm since we observe that large scale noisy datasets tend to be contaminated with dense Gaussian noises rather than Laplacian noises. Experiments on noisy datasets have confirmed our assumption.

2.5 Object Function Formulation: The Four Priors Model

Based on the terms regarding low-rankness, tag correlation, tag-visual correlation and inhomogeneous errors, we formulate the objective function as follows:

$$\min_T F(T) = \|T\|_* + \lambda_0 \|P_\Omega(\hat{T} - T)\|_F^2 + \lambda_1 \|P_{\Omega^\perp}(\hat{T} - T)\|_F^2 + \\ \lambda_2 Tr(T^T LT) + \lambda_3 \|TT^T - VV^T\|_F^2. \tag{2}$$

λ_2 and λ_3 are also weighting parameters.

The proposed Four Priors method belongs to transductive learning category, which means it reasons from both labeled and unlabeled data. We can further turn it into a inductive model using traditional machine learning approaches [24].

3   Solving the Four Priors Model

We set $\lambda_0 = 1$ for computational efficiency, and denote the nuclear norm as $g(T)$ and the other terms together as $f(T)$. And $F(T) = g(T) + f(T)$, where $g(\cdot)$ is nonsmooth and $f(\cdot)$ is smooth. We pursuit an effective iterative procedure to solve this optimization based on Accelerated Proximal Gradient method (APG) [25].

Given the following unconstrained problem

$$\min_X F(X) = \mu g(X) + f(X). \tag{3}$$

where $g(\cdot)$ is nonsmooth, $f(\cdot)$ is smooth and its gradient is Lipschitz continuous. To avoid the computation of subgradient, proximal gradient algorithms minimize a sequence of separable quadratic approximations to $F(X)$, denoted as $Q(X, Y)$, formed at specially chosen points Y

$$Q(X, Y) \triangleq \mu g(X) + f(Y) + \langle \nabla f(Y), X - Y \rangle + \frac{L_f}{2} \|X - Y\|^2. \tag{4}$$

Let $M = Y - \frac{1}{L_f} \nabla f(Y)$, we get

$$X = \underset{X}{argmin}\, Q(X, Y) = \underset{X}{argmin}\{\mu g(X) + \frac{L_f}{2} \|X - M\|^2\}. \tag{5}$$

APG set $Y_k = X_k + \frac{b_{k-1}-1}{b_k}(X_k - X_{k-1})$ for a sequence $\{b_k\}$ satisfying $b_{k+1}^2 - b_{k+1} \leq b_k^2$ to get an $O(k^{-2})$ convergence rate. The APG method is described in Algorithm 1.

Algorithm 1. APG Method

Require:
1: **while** not converged **do**
2:     $Y_k = X_k + \frac{b_{k-1}-1}{b_k}(X_k - X_{k-1})$
3:     $M_k = Y_k - \frac{1}{L_f}\nabla f(Y_k)$
4:     $X_{k+1} = \underset{X}{argmin}\{\mu g(X) + \frac{L_f}{2}\|X - M_k\|^2\}$
5:     $b_{k+1} = \frac{1+\sqrt{4b_k^2+1}}{2}$
6:     $k = k + 1$
7: **end while**
Ensure:

The main advantage of the APG method is that the minimizer X_{k+1} has a simple or even closed-form solution when the $g(\cdot)$ is ℓ_1 norm or nuclear norm [8].

It is obvious that the APG method naturally fits for the Four Priors model. We estimate the L_f using backtracking method and calculate the $\nabla f(T)$:

$$\nabla f(T) = 2[P_\Omega^* P_\Omega(\hat{T} - T) + \lambda_1 P_{\Omega^\perp}^* P_{\Omega^\perp}(\hat{T} - T) + \lambda_2 LT + \lambda_3(TT^T - VV^T)T] \quad (6)$$

where P_Ω^* and $P_{\Omega^\perp}^*$ are the adjoint operators of P_Ω and $P_{\Omega^\perp}$, respectively.

Basing on Eqs. (5) and (6) we can obtain the subproblem (Step 4 in Algorithm 1) for our model.

$$T_{k+1} = \underset{T}{argmin}\left\{ \|T\|_* + \frac{L_f}{2}\|T - M_k\|^2 \right\}, \quad (7)$$

where $M_k = T_k + \frac{b_{k-1}-1}{b_k}(T_k - T_{k-1}) - \frac{1}{L_f}\nabla f[T_k + \frac{b_{k-1}-1}{b_k}(T_k - T_{k-1})]$. The solution to (7) is:

$$T_{k+1} = U S_{\frac{1}{L_f}}(\Sigma)V^T \quad (8)$$

where $U\Sigma V^T$ is the singular value decomposition (SVD) of M_k and $S_\tau(\cdot)$ is the singular value thresholding operator [26].

4 Experimental Evaluation

4.1 Datasets and Experimental Setup

The proposed algorithm is denoted as Four Priors and is evaluated on two well known benchmark datasets: MIRFlickr-25K , Corel5K and Labelme. MIRFlickr-25K is collected from Flickr. Compared to the Corel5K, tags in Labelme and MIRFlickr-25K are rather noisy and many of them are misspelled or meaningless words. Hence, a pre-processing is performed. We match each tag with entries in a Wikipedia thesaurus and only retain the tags in accordance with Wikipedia. We use the pre-trained word and phrase vectors [18] to extract tag vectors from the tags in these two datasets. To narrow the semantic gap, we utilized DeCAF [27] to extract the DeCAF$_6$ features, which have high level semantic meanings (Table 1).

We compare the proposed Four Priors model with the state-of-the-art methods, including matrix completion-based model LRES [8], TCMR [11], RKML [12], search-based algorithms (i.e. JEC [6], TagProp [5], and TagRelevance

Table 1. Statistics of 3 datasets

Statistics	Corel5K	Labelme	MIRFlickr-25K
No. of images	4,918	2,900	25,000
Vocabulary size	260	495	1,386
Tags per Image (mean/max)	3.4/5	10.5/48	12.7/76
Images per Tag (mean/max)	65.3/1,120	67.1/379	416.5/76,890

[7]), mixture models (i.e. CMRM [28] and MBRM [29]), tag recommendation approaches (i.e. Vote+ [30] and Folk [31]), co-regularized learning model Fast-Tag [32] and Bayesian network model InfNet [33]. Note that the parameters of adopted baselines are also carefully tuned on the validation set of Corel5K with corresponding proposed tuning strategy.

We measure all the algorithms in terms of *average precision@N* (i.e. *AP@N*), *average recall@N* (i.e. *AR@N*) and *coverage@N* (i.e. *C@N*). In the top N completed tags, *precision@N* is to measure the ratio of correct tags in the top N competed tags and *recall@N* is to measure the ratio of missing ground-truth tags, both averaged over all test images. *Coverage@N* is to measure the ratio of test images with at least one correctly completed tag.

4.2   Evaluation of Tag Completion on Corel5K

We adopt the tuning strategy used in [10] to set $\lambda_1 = 0.6, \lambda_2 = 1$, and $\lambda_3 = 0.8$. Table 2 demonstrates the performance comparisons. Due to the space limit, we only report results when $N = 2, 3, 5, 10$.

4.3   Evaluation of Tag Completion on MIRFlickr-25K and Labelme

We tuned $\lambda_1 = 0.2, \lambda_2 = 1.0$, and $\lambda_3 = 0.5$ using cross validation on MIRFlickr-25K. The two datasets use the same parameters since they are both noisy. Note that as the datasets become large or noisy, the semantic gap expands, leading to the decrease of λ_3. And λ_1 varies according to different noisy level.

Table 2. Performance comparison on Corel5K dataset

| | Corel5K | | | | | | | | | | | |
| | $N = 2$ | | | $N = 3$ | | | $N = 5$ | | | $N = 10$ | | |
	AP	AR	C	AP	AR	C	AP	AR	C	AP	AR	C
Four Priors	**0.58**	**0.42**	**0.52**	**0.50**	**0.50**	**0.62**	**0.42**	**0.60**	**0.66**	0.37	**0.65**	**0.92**
LRES [8]	0.56	0.39	0.47	0.48	0.48	0.57	0.41	0.53	0.62	0.37	0.62	0.85
TCMR [11]	0.57	0.39	0.49	0.48	0.47	0.58	0.44	0.55	0.64	**0.38**	0.62	0.88
RKML [12]	0.29	0.21	0.24	0.25	0.24	0.29	0.23	0.25	0.34	0.19	0.29	0.67
JEC [6]	0.36	0.34	0.39	0.31	0.40	0.47	0.27	0.32	0.59	0.20	0.33	0.76
TagProp [5]	0.46	0.40	0.50	0.38	0.48	0.57	0.33	0.51	0.63	0.26	0.54	0.86
TagRel [7]	0.43	0.41	0.48	0.37	0.47	0.57	0.31	0.50	0.60	0.26	0.53	0.90
CMRM [28]	0.29	0.20	0.23	0.24	0.24	0.27	0.21	0.25	0.35	0.16	0.27	0.63
MBRM [29]	0.35	0.29	0.35	0.28	0.34	0.42	0.24	0.24	0.39	0.17	0.28	0.70
FastTag [32]	0.54	0.31	0.45	0.46	0.44	0.51	0.40	0.52	0.63	0.36	0.63	0.82
Vote+ [30]	0.41	0.34	0.40	0.35	0.40	0.48	0.29	0.35	0.56	0.24	0.37	0.81
Folk [31]	0.29	0.29	0.34	0.22	0.34	0.41	0.20	0.24	0.41	0.18	0.30	0.61
InfNet [33]	0.26	0.19	0.24	0.20	0.22	0.29	0.17	0.24	0.30	0.12	0.19	0.64

Table 3. Performance comparison on Labelme dataset

	Labelme											
	N = 2			N = 3			N = 5			N = 10		
	AP	AR	C	AP	AR	C	AP	AR	C	AP	AR	C
Four Priors	**0.53**	**0.36**	**0.44**	**0.50**	**0.39**	**0.53**	**0.46**	**0.50**	**0.65**	**0.35**	**0.62**	**0.79**
LRES [8]	0.42	0.32	0.39	0.40	0.36	0.50	0.35	0.45	0.55	0.27	0.56	0.69
TCMR [11]	0.44	0.32	0.42	0.41	0.36	0.51	0.37	0.45	0.60	0.29	0.55	0.75
RKML [12]	0.21	0.14	0.20	0.20	0.16	0.21	0.19	0.20	0.23	0.14	0.22	0.28
JEC [6]	0.33	0.29	0.31	0.30	0.32	0.37	0.27	0.38	0.45	0.20	0.48	0.58
TagProp [5]	0.39	0.31	0.36	0.35	0.37	0.45	0.33	0.45	0.52	0.25	0.56	0.64
TagRel [7]	0.43	0.32	0.36	0.37	0.35	0.44	0.34	0.45	0.51	0.11	0.55	0.62
CMRM [28]	0.20	0.14	0.18	0.18	0.15	0.20	0.18	0.19	0.25	0.12	0.22	0.29
MBRM [29]	0.23	0.14	0.18	0.21	0.16	0.21	0.18	0.20	0.25	0.12	0.27	0.37
FastTag [32]	0.43	0.34	0.40	0.48	0.36	0.44	0.37	0.44	0.53	0.28	0.57	0.70
Vote+ [30]	0.32	0.28	0.32	0.31	0.30	0.38	0.28	0.38	0.47	0.20	0.50	0.60
Folk [31]	0.25	0.24	0.30	0.19	0.30	0.36	0.17	0.20	0.39	0.14	0.45	0.51
InfNet [33]	0.22	0.19	0.20	0.16	0.20	0.24	0.14	0.24	0.26	0.09	0.16	0.49

Table 4. Performance comparison on MIRFlickr-25K dataset

	MIRFlickr-25K											
	N = 2			N = 3			N = 5			N = 10		
	AP	AR	C	AP	AR	C	AP	AR	C	AP	AR	C
Four Priors	**0.55**	**0.39**	**0.46**	**0.50**	**0.43**	**0.55**	**0.40**	**0.47**	**0.61**	**0.31**	**0.61**	**0.80**
LRES [8]	0.43	0.35	0.40	0.40	0.39	0.53	0.32	0.40	0.57	0.26	0.45	0.73
TCMR [11]	0.45	0.35	0.44	0.43	0.38	0.54	0.35	0.41	0.60	0.28	0.48	0.77
RKML [12]	0.21	0.15	0.15	0.23	0.22	0.25	0.13	0.23	0.31	0.13	0.22	0.55
JEC [6]	0.33	0.30	0.32	0.31	0.38	0.45	0.25	0.34	0.55	0.19	0.35	0.66
TagProp [5]	0.39	0.35	0.39	0.36	0.42	0.51	0.28	0.37	0.59	0.20	0.41	0.73
TagRel [7]	0.42	0.34	0.37	0.37	**0.43**	0.52	0.30	0.37	0.57	0.20	0.40	0.78
CMRM [28]	0.20	0.15	0.16	0.18	0.21	0.24	0.13	0.18	0.30	0.11	0.20	0.50
MBRM [29]	0.22	0.16	0.18	0.17	0.30	0.35	0.13	0.18	0.33	0.10	0.22	0.55
FastTag [32]	0.43	0.35	0.38	0.39	0.48	0.51	0.30	0.41	0.57	0.27	0.42	0.75
Vote+ [30]	0.34	0.29	0.33	0.28	0.33	0.40	0.23	0.33	0.52	0.21	0.37	0.70
Folk [31]	-	-	-	-	-	-	-	-	-	-	-	-
InfNet [33]	-	-	-	-	-	-	-	-	-	-	-	-

Tables 3 and 4 demonstrate the performance comparisons. Note that Folk and InfNet is unable to run on the large dataset MIRFlickr-25K. Besides, search-based baselines (JEC, TagProp, and TagRel) cost a lot of time to run on the dataset.

4.4   Observations on Experimental Results

We observe that: (1) Generally algorithms achieve better performance on Corel5K, since tags in MIRFlickr-25K are more noisy. (2) Matrix completion-based methods, such as Four Priors, LRES and TCMR, usually achieve the best performances. (3) Four Priors shows increasing advantage to LRES as the data become more and more noisy, justifying our assumption and model of the noises. (4) Four Priors nearly outperforms all the other algorithms in all cases. (5) Performance on MIRFlickr-25K in some sense provides an evidence for the robustness of Four Priors.

5   Conclusions and Future Work

We have proposed an effective method for image annotation. The model takes four priors into consideration: Low-Rankness, Tag Correlation, Tag-Visual Correlation and Inhomogeneous Errors. This is the first work to model inhomogeneous errors in the image annotation field. We utilize word vectors to calculate tag correlation and CNN features to measure tag-visual correlation. It achieves the state-of-the-art performance in extensive experiments conducted on benchmark datasets for image annotation.

References

1. Ntalianis, K., Tsapatsoulis, N., Doulamis, A., Matsatsinis, N.: Automatic annotation of image databases based on implicit crowdsourcing, visual concept modeling and evolution. Multimedia Tools Appl. **69**, 397–421 (2014)
2. Carneiro, G., Chan, A.B., Moreno, P.J., Vasconcelos, N.: Supervised learning of semantic classes for image annotation and retrieval. IEEE Trans. Pattern Anal. Mach. Intell. **29**, 394–410 (2007)
3. Li, J., Wang, J.Z.: Automatic linguistic indexing of pictures by a statistical modeling approach. IEEE Trans. Pattern Anal. Mach. Intell. **25**, 1075–1088 (2003)
4. Duygulu, P., Barnard, K., de Freitas, J.F.G., Forsyth, D.: Object recognition as machine translation: learning a lexicon for a fixed image vocabulary. In: Heyden, A., Sparr, G., Nielsen, M., Johansen, P. (eds.) ECCV 2002, Part IV. LNCS, vol. 2353, pp. 97–112. Springer, Heidelberg (2002)
5. Guillaumin, M., Mensink, T., Verbeek, J., Schmid, C.: Tagprop: discriminative metric learning in nearest neighbor models for image auto-annotation. In: ICCV (2009)
6. Makadia, A., Pavlovic, V., Kumar, S.: A new baseline for image annotation. In: Forsyth, D., Torr, P., Zisserman, A. (eds.) ECCV 2008, Part III. LNCS, vol. 5304, pp. 316–329. Springer, Heidelberg (2008)
7. Li, X., Snoek, C.G., Worring, M.: Learning social tag relevance by neighbor voting. IEEE Trans. Multimedia **11**, 1310–1322 (2009)
8. Zhu, G., Yan, S., Ma, Y.: Image tag refinement towards low-rank, content-tag prior and error sparsity. In: ACM MM (2010)
9. Goldberg, A., Recht, B., Xu, J., Nowak, R., Zhu, X.: Transduction with matrix completion: three birds with one stone. In: NIPS (2010)

10. Wu, L., Jin, R., Jain, A.K.: Tag completion for image retrieval. IEEE Trans. Pattern Anal. Mach. Intell. **35**, 716–727 (2013)
11. Feng, Z., Feng, S., Jin, R., Jain, A.K.: Image tag completion by noisy matrix recovery. In: Fleet, D., Pajdla, T., Schiele, B., Tuytelaars, T. (eds.) ECCV 2014, Part VII. LNCS, vol. 8695, pp. 424–438. Springer, Heidelberg (2014)
12. Feng, Z., Jin, R., Jain, A.: Large-scale image annotation by efficient and robust kernel metric learning. In: ICCV (2013)
13. Niu, Z., Hua, G., Gao, X., Tian, Q.: Semi-supervised relational topic model for weakly annotated image recognition in social media. In: CVPR (2014)
14. Zhao, R., Grosky, W.I.: Narrowing the semantic gap-improved text-based web document retrieval using visual features. IEEE Trans. Multimedia **4**, 189–200 (2002)
15. Jin, Y., Khan, L., Wang, L., Awad, M.: Image annotations by combining multiple evidence & wordnet. In: ACM MM (2005)
16. Cilibrasi, R.L., Vitanyi, P.: The google similarity distance. IEEE Trans. Knowl. Data Eng. **19**, 370–383 (2007)
17. Wu, L., Hua, X.S., Yu, N., Ma, W.Y., Li, S.: Flickr distance. In: ACM MM (2008)
18. Mikolov, T., Chen, K., Corrado, G., Dean, J.: Efficient estimation of word representations in vector space. arXiv preprint arXiv:1301.3781 (2013)
19. Torralba, A., Fergus, R., Freeman, W.T.: 80 million tiny images: a large data set for nonparametric object and scene recognition. IEEE Trans. Pattern Anal. Mach. Intell. **30**, 1958–1970 (2008)
20. Zhang, H., Berg, A.C., Maire, M., Malik, J.: SVM-KNN: discriminative nearest neighbor classification for visual category recognition. In: CVPR (2006)
21. Huiskes, M.J., Lew, M.S.: The MIR Flickr retrieval evaluation. In: MIR 2008: Proceedings of the 2008 ACM ICMI (2008)
22. Candès, E.J., Li, X., Ma, Y., Wright, J.: Robust principal component analysis? J. ACM **58**, 11 (2011)
23. Chung, F.R.: Spectral Graph Theory. American Mathematical Society, Providence (1997)
24. Gammerman, A., Vovk, V., Vapnik, V.: Learning by transduction. In: UAI (1998)
25. Toh, K.C., Yun, S.: An accelerated proximal gradient algorithm for nuclear norm regularized linear least squares problems. Pac. J. Optimiz. **6**, 615–640 (2010)
26. Cai, J.F., Candès, E.J., Shen, Z.: A singular value thresholding algorithm for matrix completion. SIAM J. Optim. **20**(4), 1956–1982 (2010)
27. Donahue, J., Jia, Y., Vinyals, O., Hoffman, J., Zhang, N., Tzeng, E., Darrell, T.: Decaf: a deep convolutional activation feature for generic visual recognition. arXiv preprint arXiv:1310.1531 (2013)
28. Jeon, J., Lavrenko, V., Manmatha, R.: Automatic image annotation and retrieval using cross-media relevance models. In: ACM SIGIR (2003)
29. Feng, S., Manmatha, R., Lavrenko, V.: Multiple bernoulli relevance models for image and video annotation. In: CVPR (2004)
30. Sigurbjörnsson, B., Van Zwol, R.: Flickr tag recommendation based on collective knowledge. In: ACM WWW (2008)
31. Lee, S., De Neve, W., Plataniotis, K.N., Ro, Y.M.: Map-based image tag recommendation using a visual folksonomy. Pattern Recogn. Lett. **31**, 976–982 (2010)
32. Chen, M., Zheng, A., Weinberger, K.: Fast image tagging. In: ICML (2013)
33. Metzler, D., Manmatha, R.: An inference network approach to image retrieval. In: Enser, P.G.B., Kompatsiaris, Y., O'Connor, N.E., Smeaton, A.F., Smeulders, A.W.M. (eds.) CIVR 2004. LNCS, vol. 3115, pp. 42–50. Springer, Heidelberg (2004)

Extracting Surface Geometry from Particle-Based Fracture Simulations

Chakrit Watcharopas[1,3]($\boxtimes$), Yash Sapra[2], Robert Geist[1],
and Joshua A. Levine[1]

[1] Clemson University, Clemson, USA
cwatcha@clemson.edu
[2] McMaster University, Hamilton, Canada
[3] Kasetsart University, Bangkok, Thailand
chakrit.w@ku.ac.th

Abstract. This paper describes an algorithm for fracture surface extraction from particle-based simulations of brittle fracture. We rely on a tetrahedral mesh of the rest configuration particles and use a simple, table-lookup approach to produce triangulated fracture geometry for each rest configuration tetrahedron based on its configuration of broken edges. Subsequently, these triangle vertices are transformed with a per particle transformation to obtain a fracture surface in world space that has minimal deformation and also preserves temporal coherence. The results show that our approach is effective at producing realistic fractures, and capable of extracting fracture surfaces from the complex simulation.

1 Introduction

To animate solids undergoing fracture, computer graphics researchers frequently use physically-based simulation techniques. These techniques can be grouped into two types of approaches: finite element methods or particle-based methods. While both approaches have advantages and disadvantages, one major distinction between the two is how the domain of interest is discretized for simulation. In finite element methods, a computational mesh is used to represent the solid and naturally provides a description of the fracture surface using a subset of mesh elements. Particle-based methods instead use only a vertex set to represent the solid, and thus a representation of the fracture surface needs to be recovered through a post-processing technique.

This work focuses on a new approach for extracting fracture surfaces from particle-based simulations of brittle fracture. Our approach takes some inspiration from mesh-based approaches, in that we build a tetrahedral mesh of the initial particle set in rest configuration. As the particles move during simulation, we dynamically update information on this mesh. In any given time step, we use a marching tetrahedra approach to locally extract a fracture surface in the rest space configuration. Each triangle on this fracture surface is then mapped into the world space using a best fit linear transformation.

© Springer International Publishing Switzerland 2015
G. Bebis et al. (Eds.): ISVC 2015, Part I, LNCS 9474, pp. 82–91, 2015.
DOI: 10.1007/978-3-319-27857-5_8

The result is a simple, easily parallelizable approach to extracting fracture surface geometry that is practical to implement on GPUs and surprisingly effective. Under the assumption of rigidity, our experiments show this technique can be employed in a variety of complex simulations. Building the input tetrahedral mesh is effectively no more overhead than producing the input point cloud (our approach leverages standard algorithms, such as Delaunay triangulations). The resulting fracture geometry enables new visualizations of particle-based fracture, and it also provides some flexibility in defining where the fracture surface occurs.

2   Related Work

Some of the first approaches for animating fracture involved using deformable models for sheets and cloth [1] and spring-mass networks for 3d solids [2]. O'Brien and colleagues showed that the finite element method could be effective for both brittle [3] and ductile [4] fracture. More recent work has improved the computational efficiency, using quasi-static analysis [5,6] and simplifications of the finite element method for realtime [7]. In these approaches, meshing becomes a concern, especially remeshing near the fracture [8]. Mesh-free methods offer alternatives that avoid remeshing [9–11].

Our work employs the peridynamics-based method of Levine et al. [10] for fracture simulation, but we provide a new solution for fracture surface extraction. Instead of building a predefined piece of geometry for each particle, we extract the surface from a single input mesh. Hirota and colleagues employ a similar approach, but they require that elements must disconnect when a spring breaks (compare Fig. 4 [9] to our Fig. 2). We make no such requirements, which leads to a more complicated case table but further decouples the simulation from the geometry. This also avoids the need to build an implicit-function for the geometry, as is commonly done when skinning particles for fluid simulations [12–14]. Skinning typically builds smooth surfaces that are appropriate for fluids, but struggles to model the sharp features that occur during fracture.

Cutting a single input mesh allows us to use marching methods for fast surface extraction. Our approach is similar in spirit to the well-known Marching Cubes algorithm [15]. Marching cubes focuses only on the representation of manifolds that are level sets, but when fracture occurs it is challenging to model the collection of surfaces as a single level set. Conceptually, our problem is closer to a multi-material representation. Nielson and Franke [16] first proposed a technique for calculating a separating surface in a tetrahedron where its vertices are classified into various classes. Bronson and colleagues [17] extend this idea to lattice cleave multiple material domains. A key difference in our approach is that in any given tetrahedron, an edge can be broken but might still be in the same connected "material" for fracture, necessitating new cases to be developed herein.

3   Algorithm Overview

We use a discretized model of the solid object in the form of tetrahedral mesh where its vertices are the particles we will simulate. To compute a fracture surface, we run a particle simulation, producing time steps $k \in 0 \ldots N$, where each time step tracks the positions of particles and whether any mesh edge is "broken" during simulation. For each time step k, we march through all tetrahedra in the input mesh and perform two operations:

1. look up the case associated with the broken edge pattern (more details in Sect. 3.3) to obtain a set of triangles per tetrahedron, and
2. transform each triangle's vertices (more details in Sect. 3.4)

3.1   Surface Modeling

In peridynamics, like many spring-mass simulations, particles are simulated, move around, and may break *bonds* connecting them. We use bonds to describe pairs of particles that interact during the simulation, which by design are a superset of the mesh edges. When a bond that also exists as an edge in the input tetrahedral mesh is broken (i.e. during fracture), we consider that edge to be broken. Our conceptual surface model is that each tetrahedron goes through states of transitions in the rest space. The transition is categorized by the number of broken edges in the tetrahedron. Since a tetrahedron has 6 edges, there are $2^6 = 64$ possibilities of how edges of a tetrahedron can be broken. Collapsing cases by removing rotational symmetries, we can group these possibilities into 11 cases, described as subcases of 7 transitional states – Case 0 through Case 6 – indicating the number of broken edges.

3.2   Face Topologies

Our 11 cases are best described by first examining the possible configurations for a triangular face of a tetrahedron. The set of face topologies that we use is shown in Fig. 1. Using the number of broken edges on a tetrahedral face, we design the four types of face topologies, F0 through F3. The F0 face topology is used when no broken edge has occurred on a tetrahedral face. This topology may make a transition to the F1 when one edge of the tetrahedral face becomes broken, and so on. Note that in the figure we intentionally make the edge breaks wider in the F1, F2, and F3 topologies so that these broken edges can be seen easily. Nevertheless, in general these edge cut-points are placed exactly in the middle of edges and face cut-points are placed inside the face in the rest position, waiting to be transformed to the world position.

3.3   Case 0 to Case 6

Figure 2 shows all the possible tetrahedral topologies. These can be grouped into a collection of cases based on the number of broken edges.

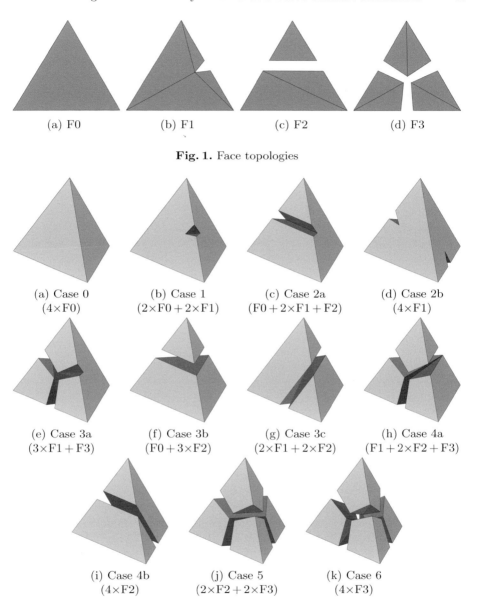

(a) F0   (b) F1   (c) F2   (d) F3

Fig. 1. Face topologies

(a) Case 0
(4×F0)

(b) Case 1
(2×F0 + 2×F1)

(c) Case 2a
(F0 + 2×F1 + F2)

(d) Case 2b
(4×F1)

(e) Case 3a
(3×F1 + F3)

(f) Case 3b
(F0 + 3×F2)

(g) Case 3c
(2×F1 + 2×F2)

(h) Case 4a
(F1 + 2×F2 + F3)

(i) Case 4b
(4×F2)

(j) Case 5
(2×F2 + 2×F3)

(k) Case 6
(4×F3)

Fig. 2. Generic geometry for the 7 transitional cases with their subcases.

Case 0: Its tetrahedral topology is depicted in Fig. 2a; all four faces have the face topology of F0. For this case, we only generate a boundary surface, without an interior surface. For other cases, the interior surfaces are shown.

Case 1: The topology of this case is illustrated in Fig. 2b where only one edge is broken. As seen in the figure, the four faces of the tetrahedral topology become the following: the back and the bottom faces have the F0 face topology, and the left and the right faces both have the F1 face topology.

Case 2: Two edges are broken. There are two cases for where the two broken edges may appear on a tetrahedron: (a) the two broken edges are on the same face, and (b) the two broken edges are *not* on the same face.

An example of Case 2a is shown in Fig. 2c. When occurring on the same face, these two broken edges create a crack path across the face of the tetrahedron. Its topology contains one F0 face, two F1 faces, and one F2 face. For Case 2b, the topology is different. A crack only appears on each broken edge, as exhibited in Fig. 2d; all four faces have the topology of F1.

Case 3: This case may lead to three subcases. The first subcase (3a) arises when all three broken edges are on the same face (as seen in Fig. 2e). One face topology of this case becomes the F3 topology, which has the "Y" crack pattern on the face. Another subcase (3b) is where all three edges emanating from a single vertex are broken. Under this circumstance, the tetrahedron breaks into two pieces, separating that one vertex from the other vertices. Figure 2f illustrates this subcase. A third subcase (3c) exists when two adjacent faces of the tetrahedron have two broken edges on each face, and the other two faces only have one broken edge. Figure 2g illustrates this subcase.

Case 4: There are two subcases for this situation. The first subcase (4a) is similar to Case 3b where three broken edges happen to share the same vertex. This subcase extends to have one more broken edge on the tetrahedron (Fig. 2h). Its topology contains one F1 face, two F2 faces, and one F3 face. The scenario for the second subcase (4b) is comparable to a cutting plane slicing through a tetrahedron as shown in Fig. 2i. The tetrahedron is separated into two pieces.

Case 5: Fig. 2j depicts this scenario. The broken edges cut a tetrahedron into three separate pieces. Since only one edge is not yet broken (appears as the bottom-back edge in the figure), the back and bottom faces of the tetrahedron have the face topology of F2, while the other two faces have the topology of F3.

Case 6: An illustration of this case is shown in Fig. 2k. The tetrahedron breaks into four disjoint pieces, separating four vertices of the tetrahedron completely. The face topologies of this case are all F3.

3.4   Surface Transformation

The goal of our surface transformation is to make sure that the transformation of the fracture surface from the rest space to the world space will not lead to any gaps or discontinuities in the output surface. Our approach guarantees that any two triangles adjacent in the rest space will stay adjacent in the world space. We assign every triangle vertex of the fracture surface to a transformation from only a single particle. Instead of applying the transformation per tetrahedron, which could create cracks in the output surface, we apply the transformation per tetrahedral vertex. By doing this we establish a one-to-one relationship between simulated particle and triangle vertex of the fracture surface. This may cause some minimal stretching and deformation to the fracture surface, but the result is guaranteed to be crack-free.

After the fracture surface is computed in rest space, we use particle positions from dynamic simulation to find appropriate transformation for the fracture surface. Using Procrustes superposition [18, 19], we compute a rigid transformation for each simulated particle. During the dynamic simulation, we use the information on particle bonds to obtain particle connectivity. The direct connectivity from particle i to its neighbors is used to calculate transformation for the particle i. From the connectivity between these particles, we compute the initial particle position in rest space $\mathbf{P}$ and the current particle position in world space $\mathbf{Q}$. Instead of setting transformation from the centroids of $\mathbf{P}$ and $\mathbf{Q}$, we translate with the positions of particle i in rest space and world space, respectively. We then compute the transformation from translated $\mathbf{P}$ to translated $\mathbf{Q}$ using Procrustes superposition. If particle i no longer has any neighbors, we translate its position and use the rotation matrix from the last simulation time step with neighbors.

4   Results

In this section, we present the results produced by our technique. We evaluate the technique by creating four experimental setups. The first experiment is a solid marble sphere flying into a concrete Stanford Armadillo as shown in Fig. 3. The model of the sphere has roughly twice the particle density of the Armadillo. Upon the impact, the sphere propelled the Armadillo backward and caused the body to break apart and the limbs (right arm and tail) to separate from the body.

The second experiment is a projectile shot through a glass plate. This experiment is similar to one described in [10]. Nevertheless, we didn't selectively weaken particle bonds in the glass plate in order to predetermine the fracture pattern. The result in Fig. 4 captures a characteristic spider pattern on the fracture surface. In addition, Fig. 4c shows separately the interior and the boundary triangles of the frame 23. The interior fractures appear at the top of the figure, while the boundary surface is presented at the bottom. Although, from the outside, some regions of the glass surface appear untouched, there were fractures that occurred inside these regions, and our method was able to capture these interior fractures.

The third experiment is a vase dropping onto a table. This experiment shows the effectiveness of per particle transformation over per component transformation. By using the information on how edges are broken inside the tetrahedral mesh, as opposed to which connected component defines the fracture surface, we can compute fractures between pieces of the component that are still attached to each other. Figure 5 shows the result of the experiment, where Fig. 5b displays the simulated particles in world space at frame 70. The particles with the same color belong to the same connected component. As the figure shows, the majority of the vase body at this frame is still in the same component. We compare the per particle transformation shown in Fig. 5c with the per component transformation shown in Fig. 5a. The per component transformation fails to separate

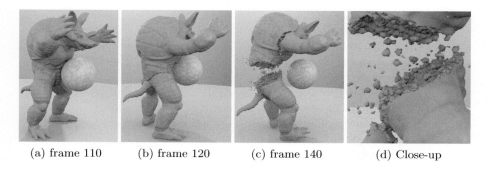

(a) frame 110   (b) frame 120   (c) frame 140   (d) Close-up

Fig. 3. A solid sphere is flying into the Stanford Armadillo.

the vase's body and the right handle, producing an inconsistency with the particle simulation. Using the per particle transformation, we see the fracture pieces of the vase are transformed to practically the same locations of the simulated particles in the world space.

Another advantage of using per particle transformation is that we can avoid the temporal discontinuities seen in the case of per component transformations. This problem occurs when a connected component at one frame breaks into multiple components in the following frame. In such situations, each component acquires a new transformation that may be completely different, leading to sudden changes in the transformation. This situation creates so-called popping artifacts, as described in [10]. We perform this per particle transformation in parallel on the GPU, and the added expense is negligible over per component transformations.

The fourth experiment is a hand chopping a stack of plates, demonstrating a complex simulation of multiple object interactions (Fig. 6). The fracture patterns occurred inside the plates in the rest position (from the top to the bottom plates) are also shown in Fig. 6b. The top plate appeared to have more fractures caused by the direct impact from the hand.

5   Discussion

In Sect. 3.2, we identified a set of face topologies and used them throughout our experiments. The face topology of F3 (three broken edges occurring on the face topology) in this set has some cut-points inside the face, and they can hamper a smooth topology transition from the F2 topology to the F3 topology, since the face cut-points begin to appear in the F3 while no face cut-point exists in the F2. An alternative, second set of the face topologies can help prevent the recreation of geometry when this topology transition occurs. A comparison between the previous and the new F3 topologies is shown in Fig. 7. Although the first set has more cut-points and preserves object mass, the second set has a simpler topology which contains 3 triangles instead of 6, and can also provide the smooth topology transition.

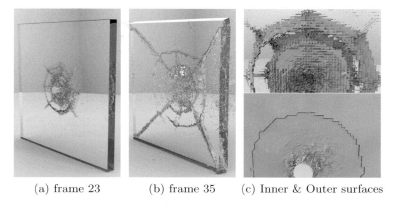

(a) frame 23   (b) frame 35   (c) Inner & Outer surfaces

Fig. 4. A projectile is shot through a glass plate.

(a) Per component   (b) frame 70   (c) Per particle

Fig. 5. A vase is dropped on a table.

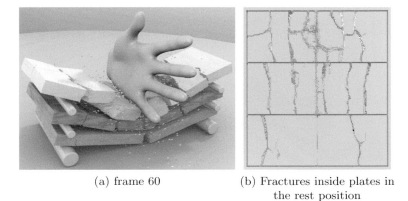

(a) frame 60   (b) Fractures inside plates in
                          the rest position

Fig. 6. A stack of plates is smashed by a hand.

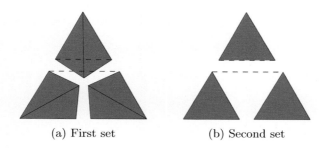

(a) First set   (b) Second set

Fig. 7. The F3 face topology of the first and second sets.

Our fracture geometry extraction approach has some limitation. Since our transformation for the fracture geometry does not perform collision detection, we rely on a best fit linear transformation, which assumes rigidity. As a result, if the collision is not carefully evaluated in the simulation, or if the simulation is not rigid, the geometry can have self-intersection. Nevertheless, in practice, we have never encountered this problem. We believe that our approach may tolerate some ductility in fracture, but the simulated tetrahedral mesh may deform significantly. Therefore, we may need a more sophisticated approach than per particle transformations.

We currently place cut-points in the middle of broken edges, which can limit artistic control of the fracture appearance. For example, a planar cut through the material is difficult to obtain if it does not align with the mesh. We plan to extend our approach to allow placing cut-points more effectively so that manipulation of roughness or smoothness on the fracture cuts becomes feasible.

Similar to many marching algorithms, our algorithm can create a large number of triangles. One of our cases will produce as many as 48 triangles for a single tetrahedron. We are currently investigating adaptive approaches that allow us to coarsen the mesh in regions that do not require detailed geometry.

Acknowledgments. This material is based upon work supported by the National Science Foundation under Grant Nos. IIS-1314757 and CNS-1126344. The School of Computing at Clemson University is an NVIDIA GPU Research Center and an NVIDIA GPU Teaching Center. Thanks to David Leubke, Cliff Woolley, and Chandra Cheij of NVIDIA for their support in providing hardware and technical consultation. We also thank Benafsh Husain for reviewing an early draft of this manuscript. The hand model was created by user monatsend and distributed on www.blendswap.com.

References

1. Terzopoulos, D., Fleischer, K.W.: Modeling inelastic deformation: viscolelasticity, plasticity, fracture. Comput. Graph. **22**(4), 269–278 (1988)
2. Norton, A., Turk, G., Bacon, R., Gerth, J., Sweeney, P.: Animation of fracture by physical modeling. Visual Comput. **7**, 210–219 (1991)

3. O'Brien, J.F., Hodgins, J.K.: Graphical modeling and animation of brittle fracture. In: SIGGRAPH, pp. 137–146 (1999)
4. O'Brien, J.F., Bargteil, A.W., Hodgins, J.K.: Graphical modeling and animation of ductile fracture. ACM Trans. Graph. **21**, 291–294 (2002)
5. Müller, M., McMillan, L., Dorsey, J., Jagnow, R.: Computer animation and simulation 2001. In: Magnenat-Thalmann, N., Thalmann, D. (eds.) Real-time Simulation of Deformation and Fracture of Stiff Materials. Eurographics, pp. 113–124. Springer, Vienna (2001)
6. Bao, Z., Hong, J.M., Teran, J., Fedkiw, R.: Fracturing rigid materials. IEEE Trans. Visual. Comput. Graphics **13**, 370–378 (2007)
7. Parker, E.G., O'Brien, J.F.: Real-time deformation and fracture in a game environment. In: Proceedings of the 2009 ACM SIGGRAPH/Eurographics Symposium on Computer Animation, pp. 165–175. ACM (2009)
8. Koschier, D., Lipponer, S., Bender, J.: Adaptive tetrahedral meshes for brittle fracture simulation. In: Proceedings of the 2014 ACM SIGGRAPH/Eurographics Symposium on Computer Animation, Eurographics Association (2014)
9. Hirota, K., Tanoue, Y., Kaneko, T.: Simulation of three-dimensional cracks. Visual. Comput. **16**, 371–378 (2000)
10. Levine, J., Bargteil, A., Corsi, C., Tessendorf, J., Geist, R.: A peridynamic perspective on spring-mass fracture. In: Proceedings of the ACM SIGGRAPH/Eurographics Symposium on Computer Animation. (2014)
11. Pauly, M., Keiser, R., Adams, B., Dutré, P., Gross, M., Guibas, L.J.: Meshless animation of fracturing solids. ACM Trans. Graph. (TOG) **24**, 957–964 (2005). ACM
12. Akinci, G., Ihmsen, M., Akinci, N., Teschner, M.: Parallel surface reconstruction for particle-based fluids. Comp. Graph. Forum **31**, 1797–1809 (2012)
13. Bhattacharya, H., Gao, Y., Bargteil, A.W.: A level-set method for skinning animated particle data. In: Symposium on Computer Animation, pp. 17–24 (2011)
14. Yu, J., Turk, G.: Reconstructing surfaces of particle-based fluids using anisotropic kernels. ACM Trans. Graph. (TOG) **32**, 5 (2013)
15. Lorensen, W.E., Cline, H.E.: Marching cubes: a high resolution 3d surface construction algorithm. In: ACM siggraph computer graphics, vol. 21, pp. 163–169. ACM (1987)
16. Nielson, G.M., Franke, R.: Computing the separating surface for segmented data. In: IEEE Proceedings on Visualization 1997, pp. 229–233 (1997)
17. Bronson, J., Levine, J., Whitaker, R., et al.: Lattice cleaving: a multimaterial tetrahedral meshing algorithm with guarantees. IEEE Trans. Visual. Comput. Graphics **20**, 223–237 (2014)
18. Kabsch, W.: A discussion of the solution for the best rotation to relate two sets of vectors. Acta Crystallogr. A **34**, 827–828 (1978)
19. Twigg, C.D., Kačić-Alesić, Z.: Point cloud glue: constraining simulations using the procrustes transform. In: Proceedings of the ACM SIGGRAPH/Eurographics Symposium on Computer Animation, pp. 45–54 (2010)

Time-Varying Surface Reconstruction
of an Actor's Performance

Ludovic Blache[1]([✉]) , Mathieu Desbrun[2,3], Céline Loscos[1], and Laurent Lucas[1]

[1] University of Reims Champagne-Ardenne, Reims, France
ludovic.blache@univ-reims.fr
[2] Caltech, Pasadena, USA
[3] INRIA, Sophia-Antipolis, France

Abstract. We propose a fully automatic time-varying surface reconstruction of an actor's performance captured from a production stage through omnidirectional video. The resulting mesh and its texture can then directly be edited in post-production. Our method makes no assumption on the costumes or accessories present in the recording. We take as input a raw sequence of volumetric static poses reconstructed from video sequences acquired in a multi-viewpoint chroma-key studio. The first frame is chosen as the reference mesh. An iterative approach is applied throughout the sequence in order to induce a deformation of the reference mesh for all input frames. At first, a pseudo-rigid transformation adjusts the pose to match the input visual hull as closely as possible. Then, local deformation is added to reconstruct fine details. We provide examples of actors' performance inserted into virtual scenes, including dynamic interaction with the environment.

1 Introduction

Multi-view reconstruction of an actor's performance is an innovative, non-invasive technique for computing a 3D avatar of an actor and placing it as an animated character in a virtual scene. It involves a *virtual cloning* system with a set of multi-viewpoint cameras in an indoor studio that generate an animated 3D model of an actor's performance, without the need for the traditional markers typically used in motion capture. From this 3D data, a temporally-coherent surface mesh needs to be constructed to facilitate post-production editing.

Model-based multi-view reconstruction approaches use a template model representing an actor – typically, an articulated mesh of a generic human body. A high-quality template model is often obtained through reconstruction from an actor's 3D scan [1]. Multi-view reconstruction is then achieved by deforming this template in time according to a set of directives (optical flow or silhouette matching) extracted from the multi-viewpoint video inputs. Vlasic *et al.* [2] and Gall *et al.* [3] use a predefined skeleton to match the template model with a set of poses defined by silhouettes or visual hulls, before applying local deformation of the template to match free-form elements such as clothes or hair.

© Springer International Publishing Switzerland 2015
G. Bebis et al. (Eds.): ISVC 2015, Part I, LNCS 9474, pp. 92–101, 2015.
DOI: 10.1007/978-3-319-27857-5_9

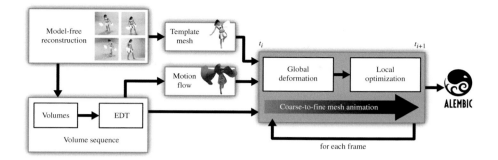

Fig. 1. General pipeline. After a model-free reconstruction, we compute a volume sequence and a template mesh. A motion flow is computed from the volume sequence (Sect. 2.1). The mesh is then animated, according to this motion flow, by a two-step mesh deformation (Sects. 2.2 and 2.3). The animated mesh is finally saved into a file to be used in a post-production software.

However, assuming a specific template model is often too restrictive to capture arbitrary motion sequences: for instance, skeleton-based approaches lead to strong limitations when applied to actors wearing loose costumes (dresses, coats) or accessories (bags, hats) without ad-hoc handling of additional features. This restrictive setup is closer to a markerless motion capture than a reconstruction of the actual scene's content. In addition, TV production stages are often only equipped with video cameras, preventing the use of markers or scans.

In this paper we propose a new approach to reconstruct a time-varying mesh with a fixed connectivity in time from an actor's performance on a typical TV production stage with omnidirectional video capture. Moreover, while many multi-view reconstruction studios use a *model-free* approach that generates a 3D model for each frame of the multi-view video sequence, we facilitate subsequent editing by providing a temporally consistent triangle mesh of the performance through incremental mesh deformation guided by the estimated motion flow.

1.1 Previous Work

Several approaches have been proposed to achieve temporal consistency from this kind of model-free reconstruction, which usually produce a sequence of poses (*i.e.*, a static reconstruction of the scene at each frame of the multi-viewpoint videos). Li *et al.* [4] developed a temporally consistent completion of scanned meshes' sequences, using a deformation graph, to establish pairwise correspondences between consecutive frames. A mesh-tracking method [5–7] can match several meshes according to curvature or texture criteria, from which one can compute the motion flow describing the movements of an actor between two frames [8]. Nobuhara and Matsuyama [9] computes a motion flow from volumetric data, based on a voxel matching algorithm. This voxel-based approach can be made drastically more robust for visual hulls if one considers voxels' orientation

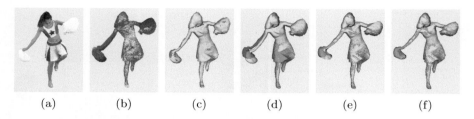

(a)   (b)   (c)   (d)   (e)   (f)

Fig. 2. Our approach at a glance. Starting from two consecutive frames in a sequence of colorized volumes (a), we first compute a motion flow (b); we then sample a set of anchor vertices (c) on the current mesh (in grey). To deform the current mesh towards the next visual hull in yellow (d), we perform a global deformation of the mesh based on the displacements of the anchors (e), before applying a local optimization to finely match the target visual hull (f). This process is repeated throughout the whole sequence to match successive poses (Color figure online).

and texture to help with matching as proposed in Blache *et al.* [10]. Once the motion flow is determined, a template structure needs to be animated. Several types of mesh deformation approaches could be applied in this context. Skeleton-based animation techniques have been shown especially efficient for motion capture. However, it cannot handle complex, non-rigid motions. Recently, 3D surface registration based on energy minimization have been also proven robust for mesh deformation [11]. Our approach is inspired from these different classes of methods, but resolves some of their limitations such as noise sensitivity and lack of genericity.

1.2   Outline

We propose to animate a template mesh derived from the first frame of the sequence of captured volumes to match the subsequent captured frames. Our input data are described in Sect. 2. In order to reconstruct high-frequency movements and to be stable even for less accurate input models, we combine motion flow to ARAP surface modeling, adapted with automatic anchors' selection (Sect. 2.1). The global deformation based on ARAP matching [12] will ensure robustness of the mesh deformation even in the presence of noisy motion flow (Sect. 2.2). Local adjustments will improve quality both of the mesh and of the match between the mesh and the volumetric input data (Sect. 2.3). This deformation process is iterated between pairs of frames until the end of the sequence to consider. Our approach is markerless, fully automatic through the entire pipeline, and is robust to a variety of scenes involving actors wearing costumes and accessories. The different steps are illustrated in Fig. 2.

2   Method Overview

Our input is a sequence of digital volumes obtained by a *visual hull* reconstruction from a multi-view video sequence [13]. Each volume of this sequence represents the actor at time t_i, so that the set of n volumes represents the motion of the actor from t_0 to t_{n-1}. The reconstructed volumes are binary digital volumes with each voxel tagged as 0 for empty or 1 for covering or straddling the actor's spatial occupancy. Straddling (surface) voxels are also assigned an RGB color based on the video inputs.

We begin by computing a Euclidean distance transform (EDT [14]). Each voxel of the volume is thus assigned a positive value which corresponds to the Euclidean distance to the closest boundary of the actor, so that this volumetric description can be seen as a grey-level 3D picture. We construct an initial (template-like) mesh based on the first volume of the sequence by extracting its zero levelset using a *marching cubes* algorithm. Laplacian smoothing and mesh simplification are then applied to ensure that each resulting triangle is non-degenerate, thus avoiding numerical artifacts in our subsequent tracking.

2.1   Motion Flow and Anchors' Selection

In order to compute a motion flow from our set of volumetric images, we use the method described in [10] to compute voxel matching based on both local geometry and color between consecutive poses. The result is a set of motion vectors for each surface voxel throughout the volume sequence. The matching is performed with a distance function computed between two surface voxels from two successive frames. This distance is computed according to several criteria: normal orientation, color and Euclidean distance. We use this matching score between two voxels as a degree of confidence associated to each vector of the motion flow, noted w_a. We select a set of vertices at time t_i to be *anchor* points. As these anchors will drive the global deformation of the character, we select the vectors associated with the largest displacements and highest confidence. Mesh vertices at t_i are thus ordered according to their corresponding $EDT_{i+1} + w$ values. We then select a fixed percentage (10 % of the total number of vertices in our implementation) of the highest ranked vertices. A few anchors in static regions (1 %) are also randomly added to guarantee that the immobile parts of the actor's body will not be deformed. This subsampling of the motion flow allows robustness since a global deformation is derived from two subsequent 3D volumes by removing the high frequency noise that typically impairs proper tracking. The scores w_a will still be used as weights to adjust how strongly we enforce the matching of these anchors in the global deformation step for additional robustness. Note that while this anchors' sampling method is particularly suited to our context, it could easily be adapted to other model-based approaches.

2.2   Pseudo-Rigid Mesh Animation

The mesh at t_i now needs to be advected in the motion flow based on the displacement of the reduced set of sampled anchors. We use a variational approach to our global mesh deformation by searching for a As-Rigid-As-Possible (ARAP [12]) deformed mesh M' with locally rigid transformations, while retaining the final positions of anchor points as much as possible (Fig. 2e).

Formulation. We minimize the following energy:

$$E(M') = E_{ARAP}(M') + E_{ANC}(M'),$$

where E_{ARAP} is the *As-Rigid-As-Possible* energy:

$$E_{ARAP}(M') = \sum_{i=1}^{n} \sum_{j \in N(i)} \gamma_{ij} \left\| (p'_i - p'_j) - R_i(p_i - p_j) \right\|^2, \qquad (1)$$

with $N(i)$ denoting the one-ring neighborhood of i. The terms p_i and p'_i represent the 3D positions of the vertex i, before and after applying the local transformation R_i. Note that if p_i is an anchor, the position p'_i is initialized by applying the associated motion vector to the initial position of the vertex. The weight γ_{ij}, associated with the edge between p_i and p_j, can be computed according to the cotangent weight method, or simply set to 1. Moreover, E_{ANC} is a quadratic energy measuring the error in the displacement of the n_a anchors:

$$E_{ANC}(M') = \sum_{i=1}^{n_a} w_{a_i} \left\| p'_i - p_i \right\|^2, \qquad (2)$$

where the weight w_{a_i} of the anchors represents the degree of confidence given to an anchor point, as described in Sect. 2.1.

Solver. The optimality condition for the minimum of our energy basically mirrors the result of [12], to which terms coming from the quadratic form (Eq. 2) are added. That is, the optimal positions p' must satisfy:

$$\sum_{j \in N(i)} \gamma_{ij}(p'_i - p'_j) + w_{a_i} p'_i = \sum_{j \in N(i)} \frac{\gamma_{ij}}{2}(R_i - R_j)(p_i - p_j) + w_{a_i} p_i \qquad (3)$$

where R_i is a local rotation best matching p_i and its one ring to p'_i. The global deformation is thus computed by iteratively solving a linear system and an optimal set of rotations matrices: we begin by computing the set of $\{p'_i\}_i$ that satisfy the optimal condition for a fixed set of initial rotations $\{R_i\}_i$ by solving a linear system of the form:

$$Lp' = b$$

where L corresponds to the Laplacian operator applied to the mesh M' in which we add the w_{a_i} weights related to each anchor point (Eq. 2) on the diagonal, and b is a column matrix which contains the righthand side of Eq. 3. Optimal rotations R_i are computed through singular value decomposition (SVD) from the positions of p_i and p'_i as derived in [12]. These two steps are repeated until convergence.

2.3   Local Optimization

After the global deformation step, details of the pose due to non-rigid deformation (such as cloth folds or hair) can still be missing. Mesh quality may also degrade over time as large deformation occurs. Local optimization and regularization are thus still necessary for the mesh to better adjust to the new pose's silhouette. We therefore

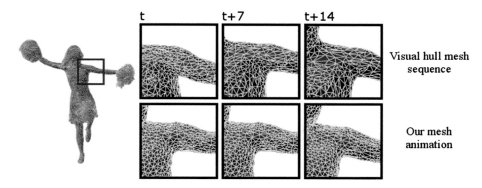

Fig. 3. Comparison between the temporally inconsistent mesh sequence from a model-free reconstruction (top) and our animated mesh (bottom).

compute local vertex displacements based on both fitting accuracy and regularization as follows.

Regularization. We regularize the mesh by applying a spring force per vertex to favor equi-length edges:

$$f_r(p_i) = \alpha \sum_{j \in N(i)} (\|p_i - p_j\| - \bar{r}_i) \frac{p_i - p_j}{\|p_i - p_j\|}$$

where α is a fixed stiffness coefficient and p_j is a vertex from the one-ring neighborhood of p_i, while the rest length $\bar{r}_i$ is set to the current average length of the edges adjacent to p_i. We prevent shrinking of the shape by using only the tangential component of the resulting vector.

Silhouette fitting. Using the EDT zero isovalues (Sect. 2.1), we also locally inflate or deflate the mesh towards the visual hull surface by adding a "balloon" force expressed as:

$$f_s(p_i) = \sum_{j \in N(i)} EDT(p_j) \boldsymbol{n}_{p_j}$$

with $\boldsymbol{n}_{p_i}$ and $EDT(p_i)$ being the normal vector and the EDT value at p_i, respectively. Only the normal component of this force term is used.

Integration. The resulting vectors f_r and f_s are added to obtain a displacement in time for each vertex. This displacement is integrated over a fixed 200 time steps between pose t_k and t_{k+1} by updating position and velocity of each vertex (assumed to be all of unit mass) using a simple Runge-Kutta explicit integrator.

3   Results and Discussion

We tested our method on several datasets obtained through volumetric visual hull reconstruction. The *cheerleader* sequence contains 200 frames with an average $180 \times 270 \times 170$ voxel resolution and a 19234-vertex template. The *astronaut* sequences contains 25 volumes, with an average $150 \times 120 \times 330$ resolution and a 8048-vertex mesh.

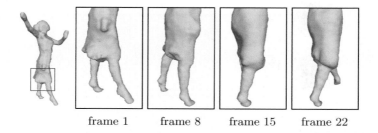

frame 1   frame 8   frame 15   frame 22

Fig. 4. Example of free-moving clothes' tracking in the *dancer* sequence.

These two datasets come from an indoor studio shoot using a 24-camera rig. The *dancer* sequence was generated using the multi-viewpoint images provided by the GrImage platform[1] with an average $150 \times 100 \times 300$ voxel resolution. Due to the low number of cameras (8 viewpoints) and their low resolution, this dataset produces coarse visual hulls. We also tested our method on the *capoeira* sequence, using the multi-view video inputs described in [1][2] with a resolution of $200 \times 275 \times 200$. All timings were achieved on a 64 bit Intel Core i7 CPU 2.20 GHz. Results from these sequences are presented in Fig. 6, demonstrating robustness of our approach in light of the coarseness of the input volumes.

The global deformation (Sect. 2.2) needs at most 200 iterations to converge. The local mesh optimization (Sect. 2.3) is applied with equal weights for the two forces f_r and f_s. A maximum of 200 iterations is necessary for the numerical integration. The total mesh processing is performed in an average of 110 seconds for each frame of the *cheerleader* sequence, and 80 seconds for *astronaut* and *dancer*. Our mesh animation approach described in Sect. 2.2 leads to a locally rigid deformation, which preserves the mesh structure during the whole sequence. It should also be noted that our use of weights based on the reliability of the anchors nicely extends the ARAP modeling technique, rendering it particularly robust to the inherent noise present in the motion flow. This improvement does not require higher computational costs since the added anchor energy we proposed (Eq. 2) only adds diagonal elements in the Laplacian-like matrix involved in the original ARAP method. The final local optimization step (Sect. 2.3) then adapts the mesh to the non-rigid part of the motion, allowing the recovery of detail in clothes and accessories after the global deformation has been properly recovered. The *cheerleader* dateset shows that shape of the pom-poms is correctly adjusted after the global deformation phase (Fig. 2e and f). With the *dancer* sequence, we show that the mesh correctly tracks the shape of the moving dress (Fig. 4). Our mesh animation method leads to an adaptation of the template during the sequence, avoiding some of the model-based inconveniences (as in, e.g., [1]) where the tracked model retains surface details (clothing folds) from the initial pose during the whole sequence. With the *capoeira* sequence, the lower quality of the multi-view images and silhouettes produces a noisy and damaged reconstruction. The visual hulls of this sequence contains many irregularities such as occlusion artifacts and holes in the character's shape. Yet, our template deformation matches the silhouettes better (Fig. 5) than [1], which offers a high quality reconstruction but mostly misses the clothes' deformation. Quantitatively,

---

[1] http://4drepository.inrialpes.fr/.

[2] http://resources.mpi-inf.mpg.de/siggraph08/perfcap/.

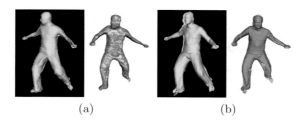

(a)   (b)

Fig. 5. Capoeira sequence. Our method (a), despite the low quality of the visual hull, matches the silhouette better than the model-based method proposed in [1] (b). Left: silhouette overlap. Right: Comparison between visual hull (yellow) and 3D mesh (grey) (Color figure online).

the average Hausdorff distance computed between the visual hull of the target pose and the animated template resulting from the model-based method [1] is 0.011933, while our method obtains an average of 0.003291.

Our temporally coherent mesh is perfect for postproduction editing as it can be easily placed in a virtual environment as a simple animated character, directly exported through, e.g., the Alembic file format. A virtual camera (with an arbitrary trajectory) can then be used without being limited by the characteristics of the shooting studio. The rendering of the virtual scene is noticeably easier with this animated object than with mesh sequences when each pose of the sequence needs to be loaded before the rendering of the corresponding frame. Generating a mesh for which only mesh vertices evolve in time (Fig. 3) has multiple additional advantages. First, it noticeably reduces the flickering effect of visual hull reconstruction. Second, vertices can be used to anchor virtual accessories (e.g. virtual makeup). Third, collision with virtual objects (clothes or other) and environment is easy to detect and handle as one can rely on temporal coherence of the vertices. At last, the animated mesh can keep the same texture during the animation, instead of computing a new one for each frame. For long sequences,

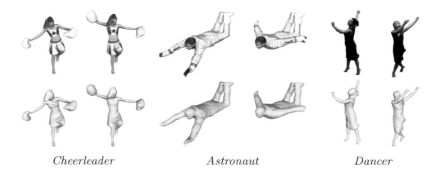

Cheerleader   *Astronaut*   *Dancer*

Fig. 6. Results of our approach for the *Cheerleader*, *Astronaut* and *Dancer* sequences. Comparison between the original visual hull reconstruction (top) and the temporally consistent mesh (bottom).

it is preferable to keep a single UV map for the whole sequence, associated with an animated texture to prevent a visual texture *sliding* effect.

In several extreme cases, our local regularization step degrades the details of very thin features as they are of a size too close to the size of a grid element. A local optimization with subgrid accuracy could solve this issue, most likely at the cost of a significantly increased computational complexity. Our approach also assumes that the initial topology is kept throughout the sequence. However, changes in the topology of the visual hull could occur in the captured sequence, possibly due to occultations if not enough view angles are available. Currently, these events are not supported by our system and requires user interaction to correct. In the future, stereo-matching could be used to improve the accuracy and quality of the volume sequences. Alternatively, one could also handle topology changes through, for instance, the method proposed by Bojsen-Hansen *et al.* [15]: they proposed a surface-tracking based on a non-rigid registration and address the issue of topology changes by partially resampling the mesh.

4   Conclusion

In this paper, we have proposed a new approach for generating a time-evolving triangle mesh representation from a sequence of binary volumetric data representing an arbitrary, possibly complex and unstructured motion of an actor with arbitrary costumes and accessories. We first compute a motion flow of the sequence from a voxel-matching algorithm. Using the visual hull as a prior, we then animate a template mesh, generated by a surface reconstruction of the first volume, via as-rigid-as-possible, detail-preserving transformations guided by the motion flow and based on a sparse set of weighted anchors. A final local optimization adjusts the mesh to better match the mesh shape to the current visual hull, leading to a robust, temporally-consistent mesh reconstruction of the motion.

Acknowledgments. M. Desbrun is partially funded by the National Science Foundation (CCF-1011944 grant), and gratefully acknowledges being hosted by the TITANE team in the context of an INRIA International Chair. We would like to thank our partner XD Productions. This work has been carried out thanks to the support of the RECOVER3D project, funded by the *Investissements d'Avenir* program. Some of the captured performance data were provided courtesy of the Max-Planck-Center for Visual Computing and Communication (MPI Informatik/Stanford) and Morpheo research team of INRIA.

References

1. de Aguiar, E., Stoll, C., Theobalt, C., Ahmed, N., Seidel, H.P., Thrun, S.: Performance capture from sparse multi-view video. ACM Trans. Graph. **27**, 98:1–98:10 (2008)
2. Vlasic, D., Baran, I., Matusik, W., Popović, J.: Articulated mesh animation from multi-view silhouettes. ACM Trans. Graph. **27**, 97:1–97:09 (2008)
3. Gall, J., Stoll, C., de Aguiar, E., Theobalt, C., Rosenhahn, B., Seidel, H.P.: Motion capture using joint skeleton tracking and surface estimation. In: IEEE Conference on Computer Vision and Pattern Recognition (CVPR), pp. 1746–1753 (2009)

4. Li, H., Luo, L., Vlasic, D., Peers, P., Popović, J., Pauly, M., Rusinkiewicz, S.: Temporally coherent completion of dynamic shapes. ACM Trans. Graph. **31**, 2:1–2:11 (2012)
5. Starck, J., Hilton, A.: Correspondence labelling for wide-timeframe free-form surface matching. In: Proceedings of the IEEE International Conference on Computer Vision (ICCV), pp. 1–8 (2007)
6. Varanasi, K., Zaharescu, A., Boyer, E., Horaud, R.: Temporal surface tracking using mesh evolution. In: Forsyth, D., Torr, P., Zisserman, A. (eds.) ECCV 2008, Part II. LNCS, vol. 5303, pp. 30–43. Springer, Heidelberg (2008)
7. Tung, T., Matsuyama, T.: Dynamic surface matching by geodesic mapping for 3D animation transfer. In: Proceedings of the IEEE Conference on Computer Vision and Pattern Recognition (CVPR), pp. 1402–1409 (2010)
8. Petit, B., Letouzey, A., Boyer, E., Franco, J.S.: Surface flow from visual cues. In: International Workshop on Vision, Modeling and Visualization (VMV), pp. 1–8 (2011)
9. Nobuhara, S., Matsuyama, T.: Heterogeneous deformation model for 3D shape and motion recovery from multi-viewpoint images. In: Proceedings of the 2nd International Symposium on 3D Data Processing, Visualization, and Transmission, pp. 566–573 (2004)
10. Blache, L., Loscos, C., Nocent, O., Lucas, L.: 3d volume matching for mesh animation of moving actors. In: EG Workshop on 3D Object Retrieval, pp. 69–76 (2014)
11. Bouaziz, S., Tagliasacchi, A., Pauly, M.: Dynamic 2D/3D registration. In: Holzschuch, N., Myszkowski, K. (eds.) Eurographics Tutorial. The Eurographics Association (2014)
12. Sorkine, O., Alexa, M.: As-rigid-as-possible surface modeling. In: Proceedings of Symposium on Geometry Processing, pp. 109–116 (2007)
13. Laurentini, A.: Visual hull concept for silhouette-based image understanding. IEEE Trans. Pattern Anal. Mach. Intell. **16**, 150–162 (1994)
14. Saito, T., Toriwaki, J.I.: New algorithms for euclidean distance transformation of an n-dimensional digitized picture with applications. Pattern Recogn. **27**, 1551–1565 (1994)
15. Bojsen-Hansen, M., Li, H., Wojtan, C.: Tracking surfaces with evolving topology. ACM Trans. Graph. **31**, 53:1–53:10 (2012)

Interactive Procedural Building Generation Using Kaleidoscopic Iterated Function Systems

Tim McGraw[(✉)]

Purdue University, West Lafayette, USA
`tmcgraw@purdue.edu`

Abstract. We present an approach to designing and generating buildings at interactive rates. The system can run entirely on the GPU in a fragment shader and results can be viewed in real time. High quality raycast or raytraced results can be efficiently visualized because the buildings are represented as distance fields. By exploiting the visual complexity of a class of fractals known as kaleidoscopic iterated function systems (KIFS) we can generate detailed buildings reminiscent of ornate architectural styles, such as Gothic and Baroque, with simpler rules than grammar based methods.

1   Introduction

Many fractal images and surfaces are characterized by complex patterns that repeat over multiple scales. Techniques such as escape-time formulas and iterated function systems (IFS) can be used to create and explore these patterns. Familiar fractals, such as the Sierpinski gasket and Menger sponge, have a structure that can be easily deduced from the geometric rules for their construction. The Menger sponge is constructed by starting with a single cube which is subdivided into 27 equal sized smaller cubes. The cube from the center of the original cube and the cubes from the middle of each face of the original cube are removed. This process is repeated indefinitely for each remaining cube. At all scales the sponge has the characteristic perforated box-like appearance shown in Fig. 1. An IFS construction of the sponge requires recursively applying affine mappings to a set of points. It can also be generated using distance estimation methods which rely on folding and scaling operations. But, as we shall see, a variety of complex shapes can be obtained by subtly changing the generation rules. It is less intuitive how the recursive construction of these modified Menger sponges lead to the resulting array of patterns.

For many 3D fractals it is possible to estimate the distance to the surface from a given point. This permits accelerated ray tracing algorithms and extraction of isosurfaces of the distance estimate. Modeling arbitrary structures with fractals requires the solution of a difficult inverse problem: finding the iterative process which results in a given shape. The Collage Theorem [1] states general conditions for the existence of a solution to this inverse problem and describes a general approach to its construction. In 2D this has led to fractal image compression algorithms, but general 3D solutions have been elusive.

© Springer International Publishing Switzerland 2015
G. Bebis et al. (Eds.): ISVC 2015, Part I, LNCS 9474, pp. 102–111, 2015.
DOI: 10.1007/978-3-319-27857-5_10

Our approach to utilizing fractals in modeling 3D buildings is to warp a selected volume of an IFS fractal into another volume bounded by a rough building shape. In this work a periodic mapping with local symmetry is specified in terms of mod functions. The resulting shape can be visualized using ray tracing, or by rasterizing a triangle mesh generated by isosurface extraction. Our approach permits the user to select building parameters and instantly see the resulting structure.

2 Related Work

Procedural modeling methods permit the generation of graphical content (e.g. meshes or textures) by means of automatic or interactive algorithms. The reader is referred to the survey by Smelik et al. [2] for an overview of the various approaches to generating natural phenomena (such as terrain, plants, bodies of water) and man made objects (buildings, roads, cities). The scale and scope of building generation methods ranges from entire cities [3], to individual building facades [4], and building interior layouts.

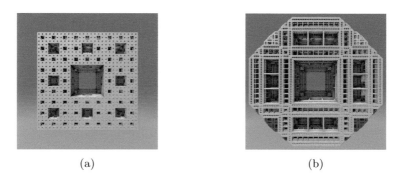

(a)   (b)

Fig. 1. Menger sponge (a) and KIFS (b) with $c = (0.93, 0.93, 0.33)$

Much previous work on building generation has focused on grammar-based approaches [5,6]. A common framework is to extrude a footprint shape into volume mass-model that defines the overall shape of the building. Then a sequence of substitution rules governs how that model is divided into floors and how each floor is divided to produce the building facade. The substitution process terminates at simple primitives such as bricks, windows and doors. By contrast, our method uses a simple arithmetic equation to divide the facade and the visual complexity is achieved by using regions of fractals as our terminal primitives.

Our building modeling method is similar to the shape modeling process described by McGraw and Herring [7], with several important distinctions. We use a specific class of fractal IFS that is well-suited to creating building detail, rather than the Mandelbox and Mandelbulb fractals. The periodic and symmetric structure of most buildings, along with the simple bounding volumes that can be expressed as unions of geometric primitives simplifies the process of mapping the fractal onto the surface. By contrast, McGraw and Herring require the user to interactively define a spline-based warp from the fractal domain onto a mesh by positioning individual vertices.

Iterated function systems [8] are a method of generating points in a fractal set by taking an input point set and repeatedly applying affine transformations to it. Early computer graphics applications of IFSs [9] showed that relatively small sets of transformations could approximate natural objects, such as leaves and ferns as well as reproducing classical fractals such as the Cantor set and Sierpinski gasket.

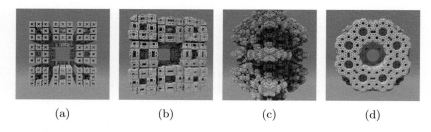

(a)   (b)   (c)   (d)

Fig. 2. KIFS with $s = 3.5$ (a), $R_1 = R_y(\pi/25)$ (b), $R_1 = R_y(\pi/4)$ (c), and $R_2 = R_y(\pi/8)$

The Kaleidoscopic IFS (KIFS) fractal described in Algorithm 1 was developed by Knighty [10] while developing distance estimates for the 3D fractal Sierpinksi tetrahedron and Menger sponge. The operations can be seen as an iterated sequence of folding operations (lines 5–10), rotations (lines 3, 11) and uniform scaling about a center point given by x_c, y_c, z_c (lines 12–14). This algorithm generates the Menger sponge for $x_c = y_c = z_c = 1$, $s = 3$ and $R_1 = R_2 = I$. For other parameter values a rich set of features emerges, both organic and synthetic looking, depending on parameter values. As can be seen in Fig. 2 the surface becomes sparse and disconnected for values of $s > 3$. For R_1 with small rotation angles the surface becomes less regular and resembles an ancient crumbling structure. Matrix R_2 can change the rectilinear structure into one with polygonal and star-like features. As with most fractal systems, a good way to get a sense of the range of shapes is to experiment and explore the parameter space. This is facilitated by a fast GPU implementation and a UI which permits parameter specification.

Hart et al. [11] introduced the idea of determining bounds on the distance to a fractal surface to accelerate ray tracing. Knowing that the distance to a surface is *at least* d we can safely step along a ray by distance d when iteratively searching for the ray-surface intersection. The search is terminated when d falls below some threshold. The process of real time rendering using such a raycasting process is described by Quilez [12] in the context of modern graphics hardware. The distance function representation also allows us to easily perform constructive solid geometry (CSG) operations on shapes.

Algorithm 1. Algorithm for computing the distance estimate to a KIFS fractal from the point (x, y, z). The scale center parameters, x_c, y_c, z_c, and scale factor, s, are scalar values, and R_1, R_2 are rotation matrices. In our experiments we use maximum iterations $M = 6$ and bailout threshold $b = 1.5$.

> **function** $d_{KIFS}(x, y, z, x_c, y_c, z_c, s)$
>   **for** $i = 0$ to M **do**
> $[x\, y\, z]^T = R_1 [x\, y\, z]^T$
> $x = |x|, \; y = |y|, \; z = |z|$
>   **if** $x - y < 0$ **then** swap(x,y)
>   **end if**
>   **if** $x - z < 0$ **then** swap(x,z)
>   **end if**
>   **if** $y - z < 0$ **then** swap(y,z)
>   **end if**
> $[x\, y\, z]^T = R_2 [x\, y\, z]^T$
> $x = s(x - x_c) + x_c, \; y = s(y - y_c) + y_c, \; z = sz$
>   **if** $z < z_c(s - 1)/2$ **then** $z = z - z_c(s - 1)$
>   **end if**
> $r = x^2 + y^2 + z^2$
>   **if** $r > b$ **then** break
>   **end if**
>   **end for**
>   **return** $(r^{1/2} - 2)s^{-i}$
> **end function**

3 Methods

Our building creation system is integrated into a realtime raycasting renderer, and the buildings are volumetrically represented as distance functions. Rays are traced in a glsl fragment shader until they intersect the building, and then lighting is computed. Building mass models are built from simple primitives, such as 3D boxes. A box primitive has distance function

$$\max(|x| - h_x, |y| - h_y, |z| - h_z) \tag{1}$$

where x, y, z are point coordinates and h_x, h_y, h_z are the half-widths of the box along the x,y,z axes.

We give our users the selection of several building mass models created from CSG operations on boxes. The union ($\cup$) and intersection ($\cap$) operations are given by

$$\cup (A, B) = \min(d_A, d_B) \tag{2}$$
$$\cap (A, B) = \max(d_A, d_B), \tag{3}$$

where A, B are shapes and d_A, d_B are distances to A and B.

The mass models used in our system consist of an inner and outer shell. The outer shell is the outermost extent of the building. Since we will later define an

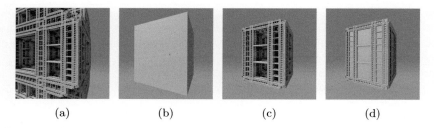

(a)	(b)	(c)	(d)

Fig. 3. KIFS detail (a), mass model outer shell (b), intersection of KIFS and outer shell $\cap(KIFS, Outer)$ (c), union of previous result with inner shell $\cup(Inner, \cap(KIFS, Outer))$

infinite tiling of the KIFS fractal, computing the intersection with this outer shell restricts the building to a finite domain. The inner shell represents windows and external walls. Computing the union of the fractal and the inner shell prevents the user from seeing into the interior of the fractal, as demonstrated in Fig. 3.

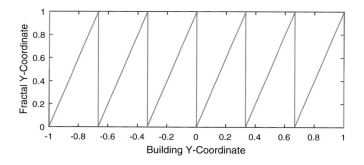

Fig. 4. Y-coordinate mapping function $f_2(y)$

Our building generation technique is based on distance and coordinate transformation of the KIFS system described in Algorithm 1. If the distance to the fractal surface is given by $d_{KIFS}(x, y, z)$ then the distance to the modeled building is $g(d_{KIFS}(f_1(x, z), f_2(y), f_3(x, z)))$ where y is the vertical axis of the structure. The distance transformation $g()$ is a series of CSG operations which define the architectural mass model. Functions $f_1(x, z)$, $f_2(y)$ and $f_3(x, z)$ are coordinate transformations which map regions of the KIFS to the surface of the building.

The vertical coordinate transformation is given by

$$f_2(y) = s_y \mod (y - y_0, h)/h \tag{4}$$

where the mod operation results in a periodic repetition of a part of the fractal which creates the stories or levels of the building. The parameters y_0, h

define a vertical slab of the KIFS, and s_y defines how that slab is mapped onto the building. In Fig. 4 a graph of $f_2(y)$ is shown for a 6 story building with $s_y = 1, y_0 = 0, h = 1/3$.

The building facade within each level is also periodic, but should permit symmetry and repetition of regions. These features are realized by a more complicated function using mod functions which is given in Algorithm 2. The symmetry is evident in the appearance of triangle waves in the plot of fractal coordinates in Fig. 5, as opposed to the nonsymmetric repetition represented by the sawtooth waves in Fig. 4.

Algorithm 2. Algorithm for computing the building transformed coordinates, $x_f = f_1(x, z), z_f = f_3(x, y)$, and facade ids, id_x, id_z. The coordinates being transformed are x, z; w_x, w_z is the repetition period of the pattern on the x- and z-faces of the building, and $\mathbf{r_x} = (r_{x,1}, r_{x,2}, r_{x,3})$ control the widths and ids within the pattern on the building x-faces. $\mathbf{r_z}$ operates similarly.

---

function $f_{xz}(x, z, w_x, w_z, \mathbf{r_x}, \mathbf{r_z})$
  $x_f = |2 \mod (x, w_x)/w_x - 1/2|$
  $z_f = |2 \mod (z, w_z)/w_z - 1/2|$
  $id_x = 0, id_z = 0$
  for $i = 1$ **to** 3 **do**
    $x_f = |x_f - r_{x,i}| - r_{x,i}$
    $z_f = |z_f - r_{z,i}| - r_{z,i}$
    if $x_f > 0$ **then** $id_x = id_x + 1$
    end if
    if $z_f > 0$ **then** $id_z = id_z + 1$
    end if
    $x_f = |x_f|, z_f = |z_f|$
  end for
  return x_f, z_f, id_x, id_z
end function

An optional transformation of x, z coordinates supported by our system is conversion to polar coordinates to create curvilinear building shapes. This transformation, which generates cylindrical and curved buildings, is given by

$$x' = s_\theta(\arctan(z, x) + c_\theta) \tag{5}$$
$$z' = s_r(\sqrt{x^2 + (z - z_0)^2} + c_r), \tag{6}$$

where c_θ is a rotation angle, s_θ determines how much of a circular arc the building subtends, c_r, s_r control the inner and outer radii of the building. Examples of the curvilinear building mass model and facade ids are shown in Fig. 6.

The x_f, z_f coordinates are subject to further scaling and translation based on the building facade id, and then the distance function to the KIFS is evaluated. These id dependent transformations permit each facade region to have a different appearance by selecting from a different region of the KIFS fractal.

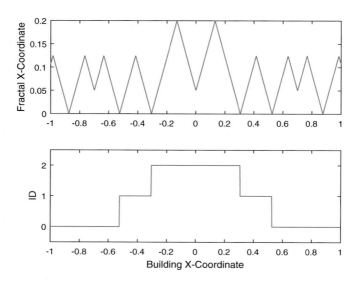

Fig. 5. Fractal coordinates (top) and facade ids (bottom) computed from building coordinates

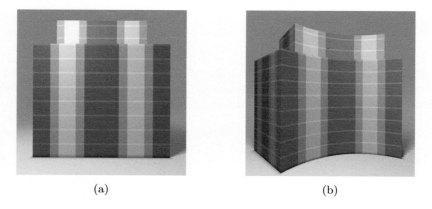

(a)   (b)

Fig. 6. Building facade ids and floors colored on rectilinear mass model (a) and curvilinear mass model (b) for $\mathbf{r}_x = \mathbf{r}_z = (0.25, 0.125, 0.4)$.

Finally, the distance to the fractal is processed using CSG operations. Let the building mass model be represented by inner and outer shells, and let the distance to those shells be given by d_i, d_o respectively. Then $g(d_{KIFS}, d_i, d_o) = \min(d_i, \max(d_o, d_{KIFS}))$. An outline of the entire modeling process is summarized below.

1. Compute inner and outer shell distances d_i, d_o to building mass model.
2. Optionally warp ray coordinates (x, y, z) to polar coordinates.
3. Compute $x_f = f_1(x, z), y_f = f_2(y), z_f = f_3(x, z)$ and facade ids.
4. Perform facade dependent translation and scaling $x_f = s_x x_f + t_x, z_f = s_z z_f + t_z$.
5. Compute KIFS distance, d_{KIFS}, to (x_f, y_f, z_f).
6. Perform CSG operations with building mass model $g(d_{KIFS}, d_i, d_o)$.

4   Results

Our KIFS building generation system was implemented in C++ and OpenGL on a Dell Optiplex workstation with 3.4 GHz Intel Core i7-3770 CPU and 8 GB RAM, Nvidia GeForce GTX 750 Ti with 640 shader cores and 2 GB GDDR5 dedicated video RAM.

The set of parameters which determine the building appearance, and the values used by our system are given below:

– Select mass model from list (3 choices)
– Enable/disable curvilinear coordinates
  • If enabled pick $s_\theta, c_\theta, s_r, c_r$
– Pick $w_x, w_z, \mathbf{r}_x, \mathbf{r}_z$ for facade division
  • We simplify by letting $w_x = w_z$ and $\mathbf{r}_x = \mathbf{r}_z$ which make the x- and z-faces the same.
– Pick (s_x, t_x, s_z, t_z) for each id for facade transformation
  • We simplify by letting $s_x = s_z = a_1 id + a_2, t_x = t_z = b_1 id + b_2$
– Pick KIFS parameters $(x_c, y_c, z_c, s, R_1, R_2)$.
  • We fix $s = 3, R_1 = I, R_2 = R_y(\theta)$

This results in 13 parameters to specify a rectilinear building, and 4 more for a curvilinear building.

KIFS Buildings generated in rectilinear and curved coordinates are shown in Figs. 7, 8 and 9. Results were generated and rendered at between 24 ms and 133 ms per frame (about 8 to 42 fps) in a 640 × 640 viewport.

The technique we presented has several limitations. Roofs and other features are not automatically handled. Physically impossible structures with floating blocks can be created. Our system can, however, be extended to handle more general building styles, such as those with a ground floor which doesn't match the appearance of the other floors, at the expense of requiring more user input.

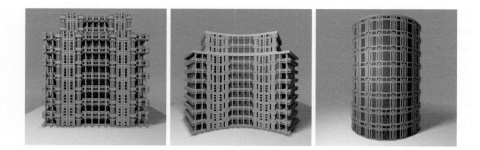

Fig. 7. Rectilinear (left) and curved (middle, right) KIFS buildings

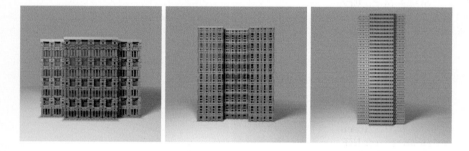

Fig. 8. KIFS buildings with t-shaped (left, right) and h-shaped mass model footprints.

Fig. 9. Additional KIFS building results

5   Conclusion and Future Work

In this paper we have described a method for procedurally generating complex buildings by harnessing the ornate architectural patterns generated by kaleidoscopic iterated function systems (KIFS). We described a domain transformation method which permits periodic patterns and symmetry to be specified and controlled by the designer. Our system can interactively generate and raycast the resulting structures entirely on the GPU. The methods can be implemented in a few hundred lines of fragment shader code. Preliminary results show that this technique can efficiently produce plausible buildings. Future work will involve developing additional tools and interfaces to support building generation; improving support for irregular features, such as entrances and roofs; and user studies to assess the usability of the system.

References

1. Barnsley, M.F., Ervin, V., Hardin, D., Lancaster, J.: Solution of an inverse problem for fractals and other sets. Proc. Natl. Acad. Sci. U.S.A. **83**, 1975–1977 (1986)
2. Smelik, R.M., Tutenel, T., Bidarra, R., Benes, B.: A survey on procedural modelling for virtual worlds. Comput. Graph. Forum **33**, 31–50 (2014)
3. Demir, I., Aliaga, D.G., Benes, B.: Proceduralization of buildings at city scale. In: 2014 2nd International Conference on 3D Vision (3DV), vol. 1, pp. 456–463. IEEE (2014)
4. Müller, P., Zeng, G., Wonka, P., Van Gool, L.: Image-based procedural modeling of facades. In: ACM Transactions on Graphics, vol. 26, p. 85. ACM (2007)
5. Aliaga, D.G., Rosen, P., Bekins, D.R.: Style grammars for interactive visualization of architecture. IEEE Trans. Vis. Comput. Graph. **13**, 786–797 (2007)
6. Müller, P., Wonka, P., Haegler, S., Ulmer, A., Van Gool, L.: Procedural modeling of buildings. ACM Trans. Graph. **25**, 614–623 (2006)
7. McGraw, T., Herring, D.: Shape modeling with fractals. In: Bebis, G., et al. (eds.) ISVC 2014, Part I. LNCS, vol. 8887, pp. 540–549. Springer, Heidelberg (2014)
8. Barnsley, M.F., Demko, S.: Iterated function systems and the global construction of fractals. In: Proceedings of the Royal Society of London A: Mathematical, Physical and Engineering Sciences, vol. 399, pp. 243–275. The Royal Society (1985)
9. Demko, S., Hodges, L., Naylor, B.: Construction of fractal objects with iterated function systems. In: ACM SIGGRAPH Computer Graphics, vol. 19, pp. 271–278. ACM (1985)
10. Knighty: Kaleidoscopic (escape time) IFS (2010). http://www.fractalforums.com/ifs-iterated-function-systems/kaleidoscopic-(escape-time-ifs)
11. Hart, J.C., Sandin, D.J., Kauffman, L.H.: Ray tracing deterministic 3-D fractals. ACM SIGGRAPH Comput. Graph. **23**, 289–296 (1989)
12. Quilez, I.: Modeling with distance functions (2008). http://www.iquilezles.org
13. Togelius, J., Yannakakis, G.N., Stanley, K.O., Browne, C.: Search-based procedural content generation: a taxonomy and survey. IEEE Trans. Comput. Intell. AI Games **3**, 172–186 (2011)

Motion and Tracking

Motion Priors Estimation for Robust Matching Initialization in Automotive Applications

Nolang Fanani[1]([✉]), Marc Barnada[1,2], and Rudolf Mester[1,2]

[1] Visual Sensorics and Information Processing Lab, C.S. Department,
Goethe University, Frankfurt, Germany
`nolang.fanani@vsi.cs.uni-frankfurt.de`
[2] Computer Vision Laboratory, ISY, Linköping University, Linköping, Sweden

Abstract. Tracking keypoints through a video sequence is a crucial first step in the processing chain of many visual SLAM approaches. This paper presents a robust initialization method to provide the initial match for a keypoint tracker, from the $1st$ frame where a keypoint is detected to the $2nd$ frame, that is: when no depth information is available. We deal explicitly with the case of long displacements. The starting position is obtained through an optimization that employs a distribution of motion priors based on pyramidal phase correlation, and epipolar geometry constraints. Experiments on the KITTI dataset demonstrate the significant impact of applying a motion prior to the matching. We provide detailed comparisons to the state-of-the-art methods.

1 Introduction

Vision systems which are able to robustly perform a task under uncontrolled scenarios are complex and consist of a large number of interacting modules which are developed for particular purposes. Our work specifically aims at automotive applications and driving scenes. This paper does not describe a complete system, but regards several components which serve critical functions in a practically useful system.

We focus on the crucial task of matching and tracking of discrete keypoints in a typical automotive application using a forward-looking monocular camera. The final purposes of such matching and tracking modules is obviously monocular egomotion estimation and the estimation of the dynamical 3D structures in the field of view of the moving camera. We intend to show that there are useful and advantageous alternatives to standard approaches, allowing a fresh view on the system design.

The typical situation in many monocular SLAM scenarios is that a large percentage of the image area can be associated with a static scene and the other moving objects are only covering a small percentage of the image area, typically in the order of 5–20 %. This means that there is a dominating epipolar relation, which can be used to impose a constraint on the motion field from 2 degrees of freedom per pixel in an unconstrained (and highly unrealistic) *general motion* scenario to 1 degree of freedom (= depth) in 80–95 % in a typical SLAM scenario.

© Springer International Publishing Switzerland 2015
G. Bebis et al. (Eds.): ISVC 2015, Part I, LNCS 9474, pp. 115–126, 2015.
DOI: 10.1007/978-3-319-27857-5_11

Furthermore, in an automotive scenario (different from the hand-held camera case) the relative ego-motion from image to image varies only very smoothly and can be very well predicted. This means that it makes sense to predict the ego-motion when the next frame comes in and estimate the motion using a *soft* epipolar constraint. These facts have been observed before by Trummer et al. [1], but we exploit them in a different way.

The imposed epipolar constraint gives meaningful information to a tracker on where to search for matches. A differential tracker, such as the Kanade-Lucas-Tomasi (KLT) tracker [2–4], has a limited convergence range, requiring the initial guess of the match being within a certain range around the true correspondence [5, p. 84]. This means that for a successful matching, in particular in the case of fast 2D motion, we need initialization information which we denote as a *motion prior*.

An initialization for the KLT tracker is usually described as a displacement vector which only applies for single pixels and not for patches. For all patches of pixels there will always be a *set* of true motion vectors. As long as the set of motion vectors obey a parametric model (affine motion model, etc.), this can be expressed by a set of parameters, but in case of multiple motions in a patch, this is not sufficient. In our approach, we describe motions, and in particular: motion priors, as *motion distributions* parametrized by a mean vector and a 2×2 covariance matrix.

Extensions of the KLT were proposed in [1,6,7] where epipolar geometry is exploited on the classical KLT optimization to enable tracking not only of corner points but also edge pixels (edgels) [7,8]. We build on Piccini et al. [7] and address an open problem from that paper, namely, how to get the tracking started. As soon as we have matched a point once and the epipolar relation is known, we have an indication of its 3D position and can propagate this information, making the transition from *matching* to *tracking*.

We propose a new method to estimate the first match of newly detected keypoints (as opposed to tracking keypoints from the 2nd frame into subsequent ones). This method generates motion priors by using a novel variant of the phase correlation method, and uses these priors in an epipolar-geometry guided matching scheme. Our scheme explicitly exploits the fact that the camera (ego)motion is very restricted in automotive applications (cf. [9]). Phase correlation, which is a classic method for image alignment and image registration, is employed to estimate the displacement between two image patches from two consecutive frames of the video stream. Here, we apply a recently proposed enhanced variant of phase correlation [10]. The particular advantage of using that method is that it outputs not just a displacement vector, but distribution of vectors, or moments of a distribution. Therefore this method is able to consider multiple motions inside a patch and thus it can handle the important issue of motion discontinuities which often occurs for example in the case of occlusion.

Our work offers several novel contributions:

- The motion of a patch is described by a motion distribution, which can represent the uncertainty introduced by multiple motions.

– Given a set of keypoints, we calculate the motion prior for each keypoint to provide an accurate initialization condition for the tracker.
– We run KLT-style differential matching in integer-grid steps first before going for sub-pixel precision on the last step to prevent expensive computation.
– The epipolar search is restricted on a segment of the epipolar line corresponding to positive depth rather than a full line search.

2 Approach

In automotive scenarios, the tracking of keypoints is often difficult since long and even ultra-long displacements may occur due to the high speed and rotations rates of the vehicle. Typical multi-resolution approaches help only partially, since reducing the resolution mixes different motions and makes small moving elements disappear [11]. Phase correlation can systematically deal with multiple motion, as shown in [10]. We consider this particular implementation of the phase correlation and show how to exploit its advantageous features.

The flowchart of the proposed method is depicted in Fig. 1. We estimate the frame-to-frame motion distribution on a subdivision of each incoming gray scale image into cells using pyramidal phase correlation as described in Sect. 2.1. These per-cell motion distributions are subsequently employed to find the keypoint correspondences between the two frames by imposing an epipolar constraint.

The incremental movement of a keypoint after motion prior estimation and during epipolar KLT matching are shown in Fig. 1(c). From a starting position

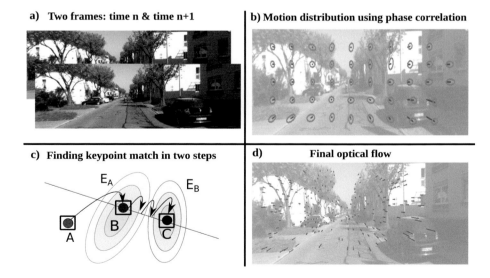

Fig. 1. Overall view: (a) a pair of frames as input, (b) motion distribution using pyramidal phase correlation, (c) keypoints matching in two steps, (d) final optical flow

A, the keypoint moves to point B through a matching initialization method. Point B is the solution of the matching initialization problem using optimization method between the motion prior covariance (ellipse E_A) with the epipolar line constraint. The epipolar KLT matching method subsequently uses matching initialization result and it will iteratively bring the keypoint from point B to point C which is the optimized position of the *mean squared difference* (MSD) ellipse E_B with the epipolar line constraint as discussed in Sect. 2.2.

2.1 Computing the Motion Prior Distribution Using Phase Correlation

We use two-level pyramid for each image by having the original size image, I_{ORI}, and a sub-sampled image (I_{SUB}) which has been down-sampled by a factor of 2.

We apply the phase correlation method to analyze the distribution of spatial shift between two consecutive frames in a sequence of images, namely $I(n)$ and $I(n+1)$. Prior to the phase correlation step, we calculate the average global 2D displacement between $I(n)$ and $I(n+1)$ using the method as proposed in [12] as a preliminary shift information. We use two image patches respectively from $I(n)$ and $I(n+1)$ referred as phase correlation (PhC) cells as the input of the phase correlation method. In case of using the KITTI image sequence (376×1241 pixel resolution), the size of all the cells is 64×128 pixels, since we have mostly horizontal motions in the scene. The cells are distributed partially overlapping each other covering all the image area as can be observed in Figure 2.

The output of phase correlation for I_{SUB} is propagated to the bottom layer of the pyramid (original size, I_{ORI}) obtaining the final shift after adding all the inherited shifts.

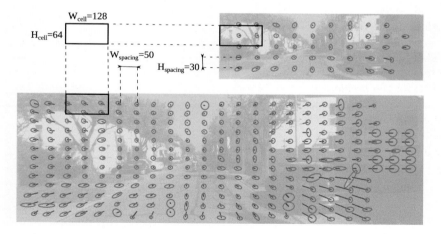

Fig. 2. The size of the PhC cells is the same for both images as well as the spacing between cells. Due to the initial shift propagation, the spacing (or location) of the cells might vary from the top layer to the bottom layer.

Validity Checks. We employ the latest improvement of phase correlation method using multiple checks as proposed in [10]. The output of each phase correlation step is a displacement array containing all the estimated shifts between the two image patches. There are several cases when phase correlation correctly detects and reports that it cannot provide any output, in particular when there is not enough structure inside PhC cells, e.g. for bright sky and plain road surfaces.

The advantages of using PhC methods as proposed in [10] are:

- It provides a distribution of motion (mean and covariance).
- It can detect multiple motions.
- It is insensitive to affine photometric transforms.
- The method as implemented according to [10] has built-in tests that allow us to detect failure situations.

Uncertainty Measurement Using Covariance Ellipse. The phase correlation method gives us the displacement array which could approximately be regarded as a probability mass for each of the displacements. We can compute the mean displacement vector and the spatial uncertainty expressed by the covariance matrix of that distribution as computed in [13].

2.2   Keypoints Matching

The proposed method for determining the correspondence of keypoints is described in Figure 1.

Relative Pose Estimation. The relative pose is composed by a rotation matrix and a translation vector in the 3D space. The scaling of the translation vector (the scale parameter) is typically ambiguous for monocular structure from motion algorithms. We estimate the rotation parameters as proposed in [12] to have a robust pitch-yaw-roll angles estimation which has been correctly verified on KITTI odometry sequences [14,15]. As an estimate of the current translation vector, we use the relative frame-to-frame translation of [16] based on an stereo system for the *previous* frame pair. We assume that the previous translation vector with the estimated relative rotation provides a reasonable *prediction* of the *current* relative frame-to-frame 3D pose. However on a real implementation, the world scaling factor could be provided by a speedometer sensor or by estimating the height of the camera from the ground plane.

Keypoints Matching Initialization. Each selected keypoint will take motion prior distribution from the nearest PhC cell. Each keypoint is associated with a 15×15 patch centred at the keypoint location; these patches are used as the input of our matching method. We estimate the matching initialization for all the selected keypoints on the image by combining the distribution of PhC prior motion for each keypoint and the epipolar constraint from the camera intrinsic and extrinsic (=relative pose) parameters.

Although the case of a completely static vehicle will make the epipolar constraint invalid during that static time, it is actually not a problem for our proposed method since the keypoints will stay at the same place in such situations, hence a correct match is found instantly at the beginning of the search.

Given the covariance ellipse for each PhC cell and given also the epipolar constraint for each keypoint, we solve the constrained optimization problem using Lagrange multipliers. We approximate the given PhC displacement array by a Gaussian distribution which has the same mean vector and the same covariance matrix. This means that the approximated motion prior distribution has the form

$$f(v) = \frac{1}{\sqrt{2\pi \hat{\mathbf{C}}_d}} \exp\left(-\frac{1}{2}(\boldsymbol{v} - \boldsymbol{m})^T \cdot \hat{\mathbf{C}}_d^{-1} \cdot (\boldsymbol{v} - \boldsymbol{m})\right) \tag{1}$$

where $\boldsymbol{m}$ and $\hat{\mathbf{C}}_d$ are the mean vector and the covariance matrix and $\boldsymbol{v}$ is the sought matching initialization displacement vector.

We regularize the obtained covariance matrix $\hat{\mathbf{C}}_d$ by enforcing the variance in each spatial direction to be at least equal to σ_{min}^2:

$$\tilde{\mathbf{C}}_d = \hat{\mathbf{C}}_d + \sigma_{min}^2 \cdot \mathbf{I}_2 \tag{2}$$

The next step is to find the value of the displacement vector $\boldsymbol{v}$ that maximizes the probability density given that $\boldsymbol{v}$ lies on the epipolar line, which is equivalent to finding a point on the epipolar line which has minimum Mahalanobis distance to the center of the covariance ellipse. The solution can be obtained by solving the following optimization problem

$$(\boldsymbol{v} - \boldsymbol{m})^T \cdot \tilde{\mathbf{C}}_d^{-1} \cdot (\boldsymbol{v} - \boldsymbol{m}) \rightarrow \min \tag{3}$$

under the epipolar constraint

$$\boldsymbol{y}_h^T \cdot \mathbf{F} \cdot \boldsymbol{x}_h = 0 \tag{4}$$

where $\mathbf{F}$ is the *fundamental matrix*, $\boldsymbol{x}_h$ and $\boldsymbol{y}_h$ are the homogeneous version of the vector $\boldsymbol{x}$ and $\boldsymbol{y}$ respectively and $\boldsymbol{y} = \boldsymbol{x} + \boldsymbol{v}$. Let us also define $\mathbf{F}'$ be a truncated $\mathbf{F}$ matrix which consists only of its first two rows.

Using a Lagrange multiplier λ_1, we obtain the following equation system:

$$\begin{pmatrix} \tilde{\mathbf{C}}_d^{-1} & \mathbf{F}' \cdot \boldsymbol{x}_h \\ (\mathbf{F}' \cdot \boldsymbol{x}_h)^T & 0 \end{pmatrix} \cdot \begin{pmatrix} \boldsymbol{v} \\ \lambda_1 \end{pmatrix} = \begin{pmatrix} \tilde{\mathbf{C}}_d^{-1} \cdot \boldsymbol{m} \\ -(\boldsymbol{x}_h^T \cdot \mathbf{F} \cdot \boldsymbol{x}_h) \end{pmatrix} \tag{5}$$

Considering the symmetric nature of the 3×3 pre-multiplying matrix, the displacement vector $\boldsymbol{v}$ can be expressed in a closed form solution.

Soft Constraint Optimization. The relative pose between two image frames can only be an estimate in this early phase of processing (recall that we are still in the process of finding the set of keypoint matches). Hence it is certainly not 100 % accurate. We propose a soft epipolar constraint as a further improvement of the keypoint matching initialization. The optimization step will allow the keypoint to be slightly off of the epipolar line, instead of forcing them to lie exactly on the epipolar line.

We recall the loss function in Eq. 3 as $Q_1(v)$ and the imposed epipolar soft constraint $Q_2(v)$ is added to complete the loss function with a well-defined balancing factor λ_2 empirically set at 0.3.

$$Q(v) = Q_1(v) + \lambda_2 Q_2(v) \tag{6}$$

where $Q_2(v)$ is a function of the squared distance to an epipolar line and it is similar to Eq. 4 multiplied by a normalizing factor.

$$Q(v) = (v - m)^T \tilde{C}_d^{-1}(v - m) + \lambda_2 \left(\frac{1}{\|F' \cdot x_h\|} y_h^T \cdot F \cdot x_h \right)^2 \tag{7}$$

In order to minimize $Q(v)$, we calculate the first derivative with respect to v and equal it to zero. This way, the the matching initialization for each keypoint is given by the displacement vector v as follows

$$v = \left[\tilde{C}_d^{-1} + \frac{\lambda_2}{\|z\|^2} z z^T \right]^{-1} \left[\tilde{C}_d^{-1} \cdot m - \frac{\lambda_2}{\|z\|^2} (x_h^T \cdot F \cdot x_h) z \right] \tag{8}$$

where $z = F' \cdot x_h$.

Epipolar KLT Matching. In order to convert the provisional estimate into a final estimate, we employ a matching method based on the KLT optimization with an epipolar constraint. Based on the classical KLT tracker method (see e.g. [17, p. 393]), the matching optimization problem can be written as

$$Q(v) = v^T \cdot A \cdot v + b^T \cdot v + c \tag{9}$$

where $v = (v_x, v_y)^T$ is the sought displacement vector, A is a symmetric 2×2 matrix built from the outer product of the gradient vectors in a patch, b is a 2×1 vector, and c is a scalar.

For our epipolar constraint tracker we add the epipolar constraint to the quadratic optimization function loss function Eq. 9. This problem is solved like in Eq. 5 with a Lagrange multiplier α_1, thus yielding Eq. 10 which leads to a closed form solution of v.

$$\begin{pmatrix} A & F' \cdot x_h \\ (F' \cdot x_h)^T & 0 \end{pmatrix} \cdot \begin{pmatrix} v \\ \alpha_1 \end{pmatrix} = \begin{pmatrix} -b \\ -(x_h^T \cdot F \cdot x_h) \end{pmatrix} \tag{10}$$

We propose a new approach to iteratively optimize the constrained KLT problem. Due to the small range where the differential optimization problem is valid, we moderate the real value displacement vector (v) from the KLT constrained optimization iteration to a single integer pixel step (v_{unit}) vertically and/or horizontally until it converges. Only then, the final real-valued sub-pixel refinement is used. We also take into account the rounded displacement vector (v_{int}) whenever v_{unit} leads to a patch with higher MSD. The complete procedure of our epipolar KLT matching is illustrated in Fig. 3.

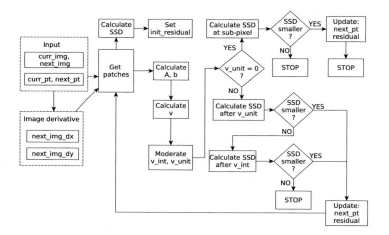

Fig. 3. Block diagram for epipolar KLT matching

Soft Constraint Optimization. Similar to the soft constraint imposed in Eq. 6, we can derive a closed form solution of v with a balancing factor α_2 empirically set at 0.7.

$$v = -0.5 \left[\mathbf{A} + \frac{\alpha_2}{\|z\|^2} z z^T \right]^{-1} \left[b + 2 \frac{\alpha_2}{\|z\|^2} (x_h^T \cdot \mathbf{F} \cdot x_h) z \right] \qquad (11)$$

Mismatch Detection. The final matching position based on the resulting displacement vector will be assessed further by multiple steps of the mismatch detection procedure. The mismatch detection module is used twice: after the motion prior estimation and after the epipolar KLT matching. Its role is to identify and eliminate mismatches. Each mismatch detection module consists of two checks: depth projection check and MSD check.

Depth Projection Check: Given the camera intrinsic matrix ($\mathbf{K}$) and the extrinsic parameters ($^n\mathbf{R}_{n+1}, ^n t_{n+1}$), if the scale of $^n t_{n+1}$ is provided by an odometry sensor or by estimating the ground plane, we can calculate the pixel position in $I(n+1)$ when the pixel position in $I(n)$ and the depth of the 3D point (d) are known. It is calculated as

$$\tilde{x}_{n+1} = \mathbf{K} \cdot ^n \mathbf{R}_{n+1} \cdot \mathbf{K}^{-1} \cdot \tilde{x}_n + \mathbf{K} \cdot ^n t_{n+1}/d \qquad (12)$$

where $\tilde{x}$ is in homogeneous coordinates. Assuming that all the keypoints are further than one meter from the camera, we can determine a segment of the epipolar line on which the match of the keypoint should lie on the $I(n+1)$.

Hence any motion estimate which brings the keypoint to a position outside of the desired epipolar line segment can be regarded as a bad result and thus discarded.

MSD Check: The MSD between two image patches A and B with size $p \times q$ is calculated as

$$MSD = \frac{\sum_{i,j}(A(i,j) - B(i,j))^2}{p \cdot q} \tag{13}$$

The MSD check ensures that the new MSD value after shifting is lower than the old MSD, which indicates that a better match has been found. We only retain the motion vectors which bring the keypoint patch into a new patch with a lower MSD value.

$$MSD_{new} \underset{reject}{\overset{accept}{\lessgtr}} MSD_{old} \tag{14}$$

3   Experiments

In order to evaluate our method, we use KITTI image sequences [14,15] from the *odometry training set*, which are taken from a moving car in an open traffic environment. In addition, we use the estimated translation for the previous frame and its keypoints locations kindly provided by the authors of [16] in order to test our method. The system proposed by Persson et al. [16] computes a high quality estimate of egomotion as well as very precise trajectories of the tracked keypoints. The keypoint correspondences have been extensively filtered by RANSAC and computed from stereo images of the KITTI odometry benchmark. It is to date one of the best vision-only approaches to odometry on that benchmark, even though the initial matches are restricted to the integer pixel grid. We use the keypoints provided by the FAST detector after being filtered internally in [16] and we estimate the initial match for each of these keypoints into the next frame using the method described above.

We tested our method on 11 different sequences of KITTI dataset, comprising of 23,157 image frames and millions of keypoints. The analysis of the matching between our keypoint matches and corresponding matches from [16] is performed regarding two values: Euclidean match distance and MSD of a patch.

3.1   Motion Priors

We analyzed the MSD value of patches after applying the estimated motion prior. Out of all keypoints matches in a frame, we can calculate the average MSD for each frame, as presented in Figs. 4 and 5, where the comparison of the MSD value after each step can be observed. It shows gradual decrease of MSD values: from the starting position to a new position after applying motion priors, and finally to the final position after epipolar matching. It can be easily noticed that most of the time the information provided by the motion priors is so crucial that it leads the patch very close to the optimal matching position, as illustrated by the closeness of the MSD value after the motion prior and the final MSD value after epipolar matching.

3.2   Epipolar Matching

The motion prior estimation is followed by epipolar matching step in order to further find a better match for each of the keypoints patches. Figures 4 and 5 compare the MSD between our results and the results using the (far more complex) method by Persson *et al.* [16] which serves here as a state-of-the-art reference.

Table 1 shows the overall results of our method on the KITTI dataset. Based on the keypoints that passed the mismatch detection, our matching method has a lower MSD on more than 50 % of keypoints than method [16]. Note that [16] strongly exploits multi-frame consistency and the keypoints depth information, unlike our proposed method which does not have access to this information.

The distance values between our keypoint match and the results from [16] for each KITTI sequence are shown in Table 1. The overall distance mean is about 3 pixels, showing a high degree of closeness to the matches of [16].

Our mismatch detection method is capable of identifying strong false match candidates. Table 1 shows that the keypoints rejection rate is between 14.45 %–68.45 %. There are several factors causing the rejection of keypoints. Most rejection cases happen when the phase correlation method cannot provide reliable motion prior information, typically due to insufficient structure in the image, very long jumps of keypoints, and repetitive/periodic structures in the image. In many cases, this can be detected *in the process*. If a bad motion prior is used anyway, the keypoints may move to the wrong direction and they are rejected

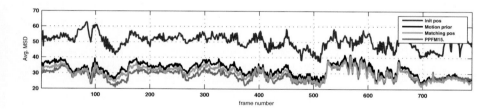

Fig. 4. MSD comparison of keypoint patches after different stages of the proposed method: initial position (blue), after motion priors (black), final matching (green), and MSD using method from [16] (red) on KITTI sequence No. 3 (Color figure online)

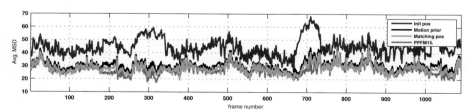

Fig. 5. MSD comparison of keypoint patches after different stages of the proposed method: initial position (blue), after motion priors (black), final matching (green), and MSD using method from [16] (red) on KITTI sequence No. 6 (Color figure online)

Table 1. The statistics of keypoints rejection, number of keypoints with lower MSD on KITTI sequences and keypoints match distance *w.r.t.* [16].

Seq	Frames Total	Keypoints Total	Rejection #points	Percentage	With better MSD #points	Percentage	Distance Mean	Std
0	4,537	1,626,656	535,295	32.91 %	511,549	46.87 %	2.66	1.21
1	1,097	225,059	102,074	45.35 %	72,615	59.04 %	5.01	3.07
2	4,657	1,894,521	1,028,222	54.27 %	449,335	51.87 %	3.22	2.11
3	797	430,102	62,140	14.45 %	198,847	54.04 %	2.37	0.70
4	267	89,669	61,381	68.45 %	16,833	59.51 %	3.56	1.99
5	2,757	1,060,460	359,691	33.92 %	351,031	50.09 %	2.84	1.38
6	1,097	348,949	195,367	55.99 %	81,019	52.75 %	3.12	1.53
7	1,097	474,541	106,592	22.46 %	213,505	58.03 %	2.56	3.75
8	4,067	1,723,007	574,475	33.34 %	600,295	52.27 %	2.75	1.16
9	1,587	590,631	299,963	50.79 %	150,833	51.89 %	2.93	1.55
10	1,197	462,561	126,667	27.38 %	176,362	52.51 %	3.26	1.51

either due to high distance to the epipolar line segment or due to a too high MSD value.

Higher image frame rate or lower vehicle speed increase the quality of our results since the motion between consecutive frames is smaller and the cases of keypoint with a very long jump can be avoided as it is happening in Fig. 4 for frames 550 to 800.

4   Summary and Conclusion

We have demonstrated that a good initialization by estimating *motion priors* significantly helps keypoints matching towards the final correct matches. In particular, we have shown that it is beneficial to employ pyramidal phase correlation to obtain motion probability distributions for each regarded cell of the image frame.

In future extensions of this work, we will extend from two-frame matching into multi-frames matching, thus obtaining keypoints depth and 3D position estimation. The 3D position of the keypoints can be extracted and utilized to get a better matching results, especially on long jumps and repetitive areas such as tiles and fence which cannot be solved by using only differential matching approaches.

References

1. Trummer, M., Munkelt, C., Denzler, J.: Extending GKLT tracking—feature tracking for controlled environments with integrated uncertainty estimation. In: Salberg, A.-B., Hardeberg, J.Y., Jenssen, R. (eds.) SCIA 2009. LNCS, vol. 5575, pp. 460–469. Springer, Heidelberg (2009)
2. Lucas, B.D., Kanade, T.: An iterative image registration technique with an application to stereo vision. IJCAI **81**, 674–679 (1981)
3. Tomasi, C., Kanade, T.: Detection and tracking of point features. School of Computer Science, Carnegie Mellon University, Pittsburgh (1991)
4. Shi, J., Tomasi, C.: Good features to track. In: Proceedings of the CVPR 1994. pp. 593–600. IEEE (1994)
5. Sutton, M.A., Orteu, J.J., Schreier, H.: Image Correlation for Shape, Motion and Deformation Measurements: Basic Concepts, Theory and Applications. Springer, New York (2009)
6. Ochoa, B., Belongie, S.: Covariance propagation for guided matching. In: Proceedings of the Statistical Methods in Multi-image and Video Processing (SMVP) (2006)
7. Piccini, T., Persson, M., Nordberg, K., Felsberg, M., Mester, R.: Good edgels to track: beating the aperture problem with epipolar geometry. In: Agapito, L., Bronstein, M.M., Rother, C. (eds.) ECCV 2014 Workshops. LNCS, vol. 8926, pp. 652–664. Springer, Heidelberg (2015)
8. Birchfield, S.T., Pundlik, S.J.: Joint tracking of features and edges. In: CVPR 2008, pp. 1–6. IEEE (2008)
9. Bradler, H., Wiegand, B., Mester, R.: The statistics of driving sequences - and what we can learn from them. In: ICCV Workshop on Computer Vision for Road Scene Understanding and Autonomous Driving, Santiago de Chile (2015)
10. Ochs, M., Bradler, H., Mester, R.: Enhanced phase correlation for reliable and robust estimation of multiple motion distributions. In: Pacific Rim Symposium on Image and Video Technology, Auckland, New Zealand (2015)
11. Brox, T., Bregler, C., Malik, J.: Large displacement optical flow. In: Proceedings of the CVPR 2009, pp. 41–48. IEEE (2009)
12. Barnada, M., Conrad, C., Bradler, H., Ochs, M., Mester, R.: Estimation of automotive pitch, yaw, and roll using enhanced phase correlation on multiple far-field windows. In: Proceedings of the IEEE Intelligent Vehicles Symposium, Seoul (2015)
13. Mester, R., Hötter, M.: Robust displacement vector estimation including a statistical error analysis. In: 5th Internernational Conference on Image Processing and its Applications, Edinburgh, UK, pp. 168–172 (1995)
14. Geiger, A., Lenz, P., Urtasun, R.: Are we ready for autonomous driving? the KITTI vision benchmark suite. In: Proceedings of the Conference on Computer Vision and Pattern Recognition (CVPR) (2012)
15. Geiger, A., Lenz, P., Stiller, C., Urtasun, R.: Vision meets robotics: the KITTI dataset. Int. J. Robot. Res. (IJRR) **32**, 389–395 (2013)
16. Persson, M., Piccini, T., Felsberg, M., Mester, R.: Robust stereo visual odometry from monocular techniques. In: Proceedings of the Intelligent Vehicles Symposium, Seoul (2015)
17. Szeliski, R.: Computer Vision: Algorithms and Applications. Springer, London (2010)

Multi-target Tracking Using Sample-Based Data Association for Mixed Images

Ting-hao Zhang[1], Hsiao-Tzu Chen[2], and Chih-Wei Tang[1(✉)]

[1] Department of Communication Engineering, National Central University,
Taoyuan City, Taiwan
`cwtang@ce.ncu.edu.tw`
[2] Ambarella, Suite C1, No. 1, Li-Hsin Road 1, Science-Based Industrial Park,
Hsinchu 30078, Taiwan

Abstract. The ubiquitous specular reflections arose from glasses seriously degrade accuracy of previous visual trackers. Although there have been few visual tracking schemes developed for mixed images with reflections, none focuses on the issue of multi-target tracking. Thus, this paper proposes a multi-target tracking scheme on mixed images with reflections. In the framework of particle filter, the proposed scheme combines the sample-based joint probabilistic data association filter (SJPDAF) with a single target based tracker that uses co-inference and maximum likelihood for visual cue integration to improve tracking accuracy of multiple targets. The co-inference predicted states are used for measurement validation of the SJPDAF. Experimental results show that the proposed scheme works well compared with the SJPDAF tracker.

1 Introduction

Visual tracking plays an important role in applications of computer vision. Many tracking schemes have been proposed to tackle problems such as illumination variations, scale variation, occlusion, deformation, motion blur, in-plane rotation, out-of-plane rotation, and background clutters [1]. Due to the change of the target appearance, specular reflections easily raise tracking errors. Currently, there are few schemes developed for single target tracking on mixed images. In [2], layer separation [3] is applied to temporally aligned frames for foreground extraction. Then the Kanade-Lucas-Tomasi tracker tracks the target in the foreground layer. Layer separation [3] is also adopted in [4] to extract the dynamic layer of a mixed image. At the correction stage of particle filter, measurements from both the mixed images and the reflectance image (i.e., the dynamic layer) are fused using maximum likelihood to optimize particle weights. The major problem with [4] is that tracking accuracy will decrease if the camera moves. Thus, the single target based visual tracking scheme in [5] proposes motion compensation layer separation and combines co-inference tracking and maximum likelihood for information fusion.

The main issues of multi-target tracking include data association and state estimation. Data association finds the optimal association between multiple measurements and multiple targets. It can be further divided into sample based algorithms (e.g., [7]) and global optimization algorithms (e.g., [8]). For sample based data association, most

© Springer International Publishing Switzerland 2015
G. Bebis et al. (Eds.): ISVC 2015, Part I, LNCS 9474, pp. 127–137, 2015.
DOI: 10.1007/978-3-319-27857-5_12

schemes can be extended from multiple hypothesis tracking (MHT) or joint probability data association filter (JPDAF) [6, 9]. JPDAF is extended from probability data association filter (PDAF) [6, 9], using the efficient Kalman filter for state estimation. In [10], the tracker uses an independent Kalman filter for each object and switches to the Viterbi algorithm as objects are close to each other. However, non-linear and non-Gaussian models are common in the real world. The Kalman filter is not optimal in these cases while particle filter works well. To handle merging and splitting, the multi-target tracker in [11] uses the JPDAF to obtain the location observations. Then the particle filter fuses the location, color, and background observations. In [7], the sample-based joint probability data association filter (SJPDAF) is proposed. The SJPDAF uses the JPDAF for data association and particle filters for state estimation of multiple targets. In [12], the game theory based data association is combined with the particle filter for multi-target tracking. Each tracker is treated as a player while measurements are modeled as strategies.

Although many works have been proposed for multi-target tracking, none of them is dedicated to mixed images with reflections. Thus, this paper proposes a multi-target tracking scheme for mobile mixed images by extending the work in [5]. In the framework of particle filter, the proposed scheme combines the single target based co-inference tracker with sample based SJPDAF to improve tracking accuracy of multiple objects. The co-inference tracking predicted state is used for measurement validation of SJPDAF. The remainder of this paper is organized as follows. Section 2 reviews the single object based robust tracker using visual cue integration for mixed images with reflections [5]. Section 3 reviews the SJPDAF [7]. Section 4 proposes the multi-target tracking scheme using SJPDAF on mixed images. Section 5 analyses the experimental results. Finally, Sect. 6 concludes this paper.

2   Overview of the Single Target Based Tracker Using Visual Cue Integration for Mixed Images

Since our work is the extension of [5], this section provides a brief review of it. The focus of the single target tracking scheme in [5] is improving tracking accuracy under the condition of reflections. It combines co-inference tracking [13] and maximum likelihood [4] for multiple cue integration. To avoid the particle degeneracy problem, the scheme realizes co-inference tracking using sequential importance resampling (SIR) [14] in the framework of the particle filter. To enhance the robustness of the tracker against camera motion, motion compensated layer separation is proposed to extract objects with active motion.

The scheme in [5] includes system initialization, pre-processing, and tracking. At the system initialization stage, the target is manually detected on the first mixed frame. To build the templates for tracking, the RGB color histogram and the [I、R-G、Y-B] color histograms of the target on the first mixed frame are constructed using the method in [15]. At the pre-processing stage, camera motion is estimated for motion compensated layer separation and compensated motion prediction in each mixed image. Initially, camera motion is estimated as the average of motion vectors of SURF [16] corresponding feature point pairs. The motion vectors that are quite different from the average one are eliminated and then camera motion is estimated again [17]. After motion compensated layer separation, the illumination image (i.e., dynamic layer) is separated from the mixed image. The illumination image is just the dynamic layer of the mixed image. Image binarization constructs the binary motion mask to indicate active objects at the correction stage of tracking.

The interesting phenomenon of co-inference is proposed in 2001 and is refined in [13]. By using the structured variational inference, Wu et al. propose that the variational parameters of one modality can be inferred by the other modalities to maximize the observation likelihood [13]. At the tracking stage of [5] at each time instant, co-inference tracking [15] and maximum likelihood [4] is combined for information fusion. (1) Resampling generates the new set of color cue based color states at time k from the set of motion cue based states at time $k - 1$ with the associated set of motion cue based weights. Then the particle filter predicts the RGB color based state using the motion compensated model [17]. (2) The set of color cue based weights is updated using the RGB color cue at the correction stage of the particle filter. (3) Resampling generates the new set of motion cue based color states at time k from the set of color cue based states at time k with the associated set of color cue based weights. Then the particle filter predicts the motion based state without consideration of camera motion. (4) The set of motion cue based weights is updated using the motion cue at the correction stage of particle filter. (5) Particle weights are optimized by maximum likelihood to select the more reliable cue. (6) The target state is estimated.

3   Overview of the Sample-Based Joint Probabilistic Data Association Filter (SJPDAF)

The SJPDAF [7] combines the merits of the particle filter [14] and JPDAF [6, 9]. Assume N_T targets are tracked with the set of states $X(k) = \{x_1(k), \ldots, x_{N_T}(k)\}$ at time k. Let $Z(k) = \{z_1(k), \ldots, z_{m(k)}(k)\}$ denote the set of measurements at time k, where $m(k)$ is the number of measurements. Let Z^k denote the sequence of all measurements up to time k. A validation matrix $\Omega = [\omega_j^t]$ is constructed using the validated measurements [6, 9], where $j = 1, \ldots, m(k)$ and $t = 0, 1, \ldots N_T \cdot \omega_j^t$ indicates if the jth measurement is inside the validation region of the tth target and $\omega_j^t = \{0, 1\}$. Feasible events $\hat{\Omega}(\theta(k)) = [\hat{\omega}_j^t(\theta(k))]$ are generated from the validation matrix. By summing the probability of all feasible events $\theta(k)$ at time k, the posterior probability that the jth measurement comes from the tth object is computed by [6, 9]

$$\beta_j^t(k) = \sum_\theta p(\theta(k)|Z^k)\hat{\omega}_j^t(\theta(k)). \tag{1}$$

Using Bayes' theory [6],

$$p(\theta(k)|Z^k) = \frac{1}{c_1}p\left[Z(k)|\theta(k), m(k), Z^{k-1}\right] \times p(\theta(k)|m(k)), \tag{2}$$

where c_1 is the normalization constant. $p\left[Z(k)|\theta(k), m(k), Z^{k-1}\right]$ is [6, 9]

$$p(Z(k)|\theta(k), m(k), Z^{k-1}) = \prod_{j=1}^{m(k)} p[z_j(k)|\theta(k), Z^{k-1}], \tag{3}$$

$$p(z_j(k)|\theta(k), Z^{k-1}) = \begin{cases} f_j^t[z_j(k)] & \text{if } \tau_j(\theta(k)) = 1 \\ V^{-1} & \text{if } \tau_j(\theta(k)) = 0 \end{cases}, \tag{4}$$

where $\tau_j(\theta(k))$ is the number of targets that associate with the jth measurement in $\theta(k)$, $f_j^t[z_j(k)]$ is the probability density of the tth target given $z_j(k)$, V is the volume of the validated range. $p(\theta(k)|m(k))$ is the prior probability of $\theta(k)$ [6, 9]

$$p(\theta(k)|m(k)) = \frac{\phi(\theta)! \mu_F(\phi(\theta))}{m(k)! p(m(k))} \times \prod_{t:\delta^t=1} P_D^t \prod_{t:\delta^t=0} (1 - P_D^t), \tag{5}$$

where P_D^t is the probability of detection of the tth target, $\phi(\theta(k))$ denotes the total number of false measurements in $\theta(k)$

$$\phi(\theta) = \sum_{j=1}^{m(k)} \{1 - \min[1, \tau_j(\theta)]\}, \tag{6}$$

δ^t is the target detection indicator

$$\delta^t(\theta) \triangleq \sum_{j=1}^{m(k)} \hat{\omega}_j^t(\theta), t = 1, \ldots, N_T, \tag{7}$$

and $\mu_F(\phi(\theta))$ is the probability mass function of $\phi(\theta)$. For parametric JPDA, $\mu_F(\phi(\theta))$ is assumed Poisson distribution. Thus,

$$p(\theta(k)|Z^k) = \frac{\lambda^{\phi(\theta)}}{c_2} \prod_{j:\tau_j=1} f_j^t[z_j(k)] \prod_{t:\delta^t=1} P_D^t \prod_{t:\delta^t=0} (1 - P_D^t), \tag{8}$$

where c_2 is a normalization constant.

The particle filter approximates the distribution of a target state using the set of particles and associated weights [14]. The SJPDA computes the validation region of each target to keep the validated measurements. As SJPDA computes the weight of nth particle of the tth target $\omega^{t,n}(k)$, it considers the association between the tth target and the jth measurement based on the feasible events [7]

$$\omega^{t,n}(k) = \alpha \sum_{j=0}^{m(k)} \beta_j^t(k) p(z_j(k)|x^{t,n}(k)), \tag{9}$$

where α is a normalization constant, $x^{t,n}(k)$ is the nth particle of the tth target, $p(z_j(k)|x^{t,n}(k))$ is the measurement model, and $\beta_j^t(k)$ is defined as

$$\beta_j^t(k) = \sum_{\theta} [\alpha \gamma^{\phi(\theta)}] \prod_{(j,t)\epsilon\theta} \frac{1}{N_s} \sum_{n=1}^{N_s} p(z_j(k)|x^{t,n}(k))], \tag{10}$$

where N_s is the number of particles.

4   The Proposed Multi-target Tracking for Mixed Images

Our work combines SJPDAF [7] with co-inference tracking [13] and maximum likelihood [4] in the framework of particle filter for multi-target tracking. Additionally, we adopt the HSV color space [18] to measure the color cue. HSV color space interprets the color the same way as human perception and thus it improves tracker robustness against illumination variations. The initialization and pre-processing stages are the same as those of [5] except that multi-target templates need to be built at the initialization stage. The steps of our proposed tracking stage are stated as follows. Let $\mathbf{x}^t(k) = \left(\mathbf{x}_c^t(k), \mathbf{x}_m^t(k)\right)$ denote the state of the tth target at time k, where $\mathbf{x}_c^t(k)$ and $\mathbf{x}_m^t(k)$ are color cue and motion cue based states, respectively. Let $s^{t,n}(k) = (s_c^{t,n}(k), s_m^{t,n}(k))$ denote the state of the nth particle of the tth target at time k with the associated set of weights $(\pi_c^{t,n}(k), \pi_m^{t,n}(k))$, where $s_c^{t,n}(k) = [S_{c,x}^{t,n}(k), S_{c,y}^{t,n}(k), V_{c,x}^{t,n}(k), V_{c,y}^{t,n}(k), H_{c,x}^{t,n}(k), H_{c,y}^{t,n}(k)]^T$, the HSV color cue estimated state at time $k - 1$. $s_m^{t,n}(k) = [S_{m,x}^{t,n}(k), S_{m,y}^{t,n}(k), V_{m,x}^{t,n}(k), V_{m,y}^{t,n}(k), H_{m,x}^{t,n}(k), H_{m,y}^{t,n}(k)]^T$ us the co-inference fused state that refers to both HSV color and motion cues at time k. $(S_{c,x}^{t,n}(k), S_{c,y}^{t,n}(k))$ and $(S_{m,x}^{t,n}(k), S_{m,y}^{t,n}(k))$ are the 2-D locations in the image, $(V_{c,x}^{t,n}(k), V_{c,y}^{t,n}(k))$ and $(V_{m,x}^{t,n}(k), V_{m,y}^{t,n}(k))$ are the velocities, and $(H_{c,x}^{t,n}(k), H_{c,y}^{t,n}(k))$ and $(H_{m,x}^{t,n}(k), H_{m,y}^{t,n}(k))$ are the horizontal and vertical axes of the ellipses. Assume the number of targets does not vary with time. Follow the work in [5], Steps 1 and 2 are co-inference tracking [13] for information fusion. Using the JPDAF [6, 9], Step 3 constructs the validation matrix for computation of joint association event probability of the SJPDAF in Step 4. Step 5 is motion cue based correction and is the last step of co-inference. Step 6 selects the more reliable cue at each time instant by maximum likelihood [4] for state estimation of each target.

Step 1 Color Cue Based Estimation. A new set $\left\{s'^{t,n}(k - 1|k - 1)\right\}_{n=1}^{N_s}$ is generated by resampling from the posterior density $p(\mathbf{x}_m^t(k - 1)|Z^{k-1})$ [14], where $s'^{t,n}(k - 1|k - 1)$ is the state of the nth particle of the tth target at time $k - 1$ with weight $1/N_s$ and N_s is the number of particles to estimate one target. Then the HSV color cue based state of the nth particle of the tth target at time k is predicted using the motion compensated model [17]

$$s_c^{t,n}(k|k - 1) = As'^{t,n}(k - 1|k - 1) + Bu(k) + W_t \tag{11}$$

where A is the state transition model, B is the control input model, $u(k)$ is the 2-D camera motion vector, and W_t is the additive Gaussian noise of the tth target. The normalized color cue based weight is

$$\pi_c^{t,n}(k) = \frac{\pi_c'^{t,n}(k)}{\sum_{n=1}^{N_s} \pi_c'^{t,n}(k)} \tag{12}$$

$$\pi_c'^{t,n}(k) = \frac{1}{\sqrt{2\pi}\sigma_c} e^{-\frac{1 - \rho_c^{t,n}(k)}{2\sigma_c^2}} \tag{13}$$

where σ_c is the standard deviation of $\rho_c^{t,n}(k)$, the Bhattacharyya distance between the HSV color histogram of the nth particle of the tth target and that of the template of the tth target [15, 19]

$$\rho_c^{t,n}(k) = \sum_{u=1}^{b} \sqrt{p^{t,n}(u)q^t(u)} \tag{14}$$

where $p^{t,n}(u)$ is the uth bin of the HSV color histogram of the nth particle of the tth target, $q_t(u)$ is the uth bin of the HSV color histogram of the tth target template, and b is the number of bins in the HSV color histogram.

Step 2 Motion Cue Based Prediction. A new set $\left\{s'^{t,n}(k|k)\right\}_{n=1}^{N_s}$ is generated by resampling from the posterior density $p(x_c^t(k)|Z^k)$ with the discrete representation approximated by Step 1. The weight of each particle is $1/N_s$. The motion cue based state (i.e., the co-inference fused state) of the nth particle of the tth target at time k is predicted by

$$s_m^{t,n}(k|k-1) = s'^{t,n}(k|k) + W_t, \tag{15}$$

where W_t is the additive Gaussian noise of the motion model of the tth target.

Step 3 Measurement Validation. The validation region of the tth target is computed by [9]

$$V(k,\gamma) = \left\{z_j(k) : [z_j(k) - \hat{z}^t(k|k-1)]^T [z_j(k) - \hat{z}^t(k|k-1)] \le \gamma\right\}, \tag{16}$$

where γ is the gate threshold and $\hat{z}^t(k|k-1)$ is the predicted measurement. In our proposed scheme, $\hat{z}^t(k|k-1)$ is defined as

$$\hat{z}^t(k|k-1) = \frac{1}{N_s} \sum_{n=1}^{N_s} Hs_m^{t,n}(k|k-1), \tag{17}$$

where H is the observation model. Only the x and y positions of $\hat{z}^t(k|k-1)$ will be used in our measurement validation. Using the x and y positions of $\hat{z}^t(k|k-1)$, we define $z_j(k)$ as the expected 2-D vector of the 2-D position of the pixels that are indicated by the motion mask [4] and are inside the validation region of the tth target simultaneously. Thus, if the jth measurement at time k, $z_j(k)$, is inside the validation range of the tth target, the corresponding element in the validation matrix will be 1. Otherwise, it will be 0. In this design, the number of measurements is equal to the number of targets, and each target has only one measurement.

Step 4 Computation of Joint Association Event Probability. To compute the joint association probability, feasible events are generated from the validation matrix that is obtained by Step 3. Next, the joint association event probability between the tth target and the jth measurement at time k is computed by summing over all joint association events with (10) of SJPDAF [7]

$$\beta_j^t(k) = \sum_\theta [\alpha\gamma^{\phi(\theta))} \prod_{(j,t)\epsilon\theta} \frac{1}{N_s} \sum_{n=1}^{N_s} p\left(z_j^{t,n}(k)|x_m^t(k) = s_m^{t,n}(k|k-1)\right)], \qquad (18)$$

where $z_j^{t,n}(k)$ is the measured at $s_m^{t,n}(k|k-1)$, and α is a normalization constant

$$\alpha = (\sum \prod_{(j,t)\epsilon\theta} \frac{1}{N_s} \sum_{n=1}^{N_s} p\left(z_j^{t,n}(k)|x_m^t(k) = s_m^{t,n}(k|k-1)\right))^{-1}, \qquad (19)$$

and $p\left(z_j^{t,n}(k)|x_m^t(k) = s_m^{t,n}(k|k-1)\right)$ is assumed Gaussian distribution [20]

$$p\left(z_j^{t,n}(k)|x_m^t(k) = s_m^{t,n}(k|k-1)\right) = \frac{1}{2\pi\sigma} e^{-\frac{1-\rho_{m1}^{t,n}(k)}{2\sigma^2}}. \qquad (20)$$

where $\rho_{m1}^{t,n}(k)$ is the Bhattacharyya distance between the HSV color histogram measured at the nth motion cue based particle of the tth target and that of the template of the tth target. Since each target has only one measurement in our design (Step 3), $\beta_j^t(k)$ will be denoted by $\beta''(k)$ in the following steps.

Step 5 Motion Cue Based Correction. Using the method in [15], the [I、R-G、Y-B] color histogram of the nth particle of the tth target on the mixed image is computed. The histogram is from the intersection of the motion mask indicated region and the rectangular region centered at the nth particle of the tth target. The normalized motion cue based weight is

$$\pi_m^{t,n}(k) = \frac{\pi'^{t,n}_m(k)}{\sum_{n=1}^{N_s} \pi'^{t,n}_m(k)} \qquad (21)$$

$$\pi'^{t,n}_m(k) = \frac{1}{\sqrt{2\pi}\sigma_m} e^{-\frac{1-\rho_m^{t,n}(k)}{2\sigma_m^2}} \qquad (22)$$

where σ_m is the standard deviation of $\rho_m^{t,n}(k)$ $\rho_m^{t,n}(k)$ is the Bhattacharyya distance between the [I、R-G、Y-B] color histogram of the nth particle of the tth target and that of the template of the tth target.

Step 6 State Estimation of Multiple Targets. $\{\pi_c^{t,n}(k)\}_{n=1}^{N_s}$ is re-evaluated with respect to $\{s_m^{t,n}(k)\}_{n=1}^{N_s}$ generated by Step 3. Then the reliable cue is optimally selected using maximum likelihood [4]

$$\tilde{\pi}^{t,n}(k) = \pi_c^{t,n}(k)^{\lambda_n} \pi_m^{t,n}(k)^{1-\lambda_n}, \qquad (23)$$

where $\lambda_n \in [0, 1]$. By applying natural log, (23) is maximized and λ_n is either 1 or 0. Thus,

$$\tilde{\pi}^{t,n}(k) = \begin{cases} \pi_m^{t,n}(k), & \text{if } 1 - \frac{\rho_c^{t,n}(k)}{2\sigma_c^2} > \frac{1 - \rho_m^{t,n}(k)}{2\sigma_m^2} \\ \pi_c^{t,n}(k), & \text{otherwise} \end{cases}. \tag{24}$$

By considering association between the tth target and its measurement,

$$\bar{\pi}^{t,n}(k) = \beta'^t(k)\tilde{\pi}^{t,n}(k). \tag{25}$$

Applying weight normalization to $\bar{\pi}^{t,n}(k)$, the estimated state of the tth target at time k using the normalized weight $\pi^{t,n}(k)$ is

$$s_m^t(k|k) = \sum_{n=1}^{N_s} \pi^{t,n}(k) \times s_m^{t,n}(k|k-1). \tag{26}$$

5 Experimental Results

The set of test data for people tracking includes four self-recorded videos (Table 1) since the public databases of mixed images with reflections are not available. Comparisons between the proposed scheme and the SJPDAF [7] are made to demonstrate the performance of the proposed scheme. Tracking accuracy is measured by the root mean square error (RMSE) between the estimated object position and the ground truth. For each method, five randomized tests are applied to each video. Each color channel of the color histogram is divided into eight bins. The number of particles is 50.

Table 1. Specification of the adopted dataset (spatial resolution: 240×135).

Video No.	Camera	Reflection	The 1st Frame	Number of Frames
Video #1	Static	Weak		55 Frames
Video #2	Static	Strong		60 Frames
Video #3	Moving	None		60 Frames
Video #4	Moving	Strong		55 Frames

Table 2 provides the tracking results of the last frames of videos #1 with weak reflections and video #2 with strong reflections by the proposed scheme and the SJPDAF. The tracker estimated states are marked with rectangles. Each white dot denotes the central position of the ground truth of a target. Our test result shows that tracking accuracy of the proposed scheme is similar to that of the SJPDAF in the first half of video #2. However, the proposed scheme significantly outperforms the SJPDAF

[7] in the second half of video #2 where targets are moving in regions with strong reflections (Table 2). Table 3 provides comparisons of tracking accuracy of two targets between the proposed scheme and SJPDAF. In most cases, the proposed scheme significantly outperforms the SJPDAF. However, the RMSE of the proposed scheme is slightly larger than that of the SJPDAF for target #1 of video #2 due to inaccurate motion masks from the 10th to the 30th frame. Figure 1 provides the RMSE value of each frame of videos #1–#4. The solid and dashed lines represent the proposed scheme and the SJPDAF, respectively. In most cases, the proposed scheme is more accurate than SJPDAF for both mixed (i.e., videos #1, #2, #4) and non-mixed videos (i.e., videos #3). For frames with the larger RMSE tracked by the proposed scheme, the inaccurate motion mask is still the main cause.

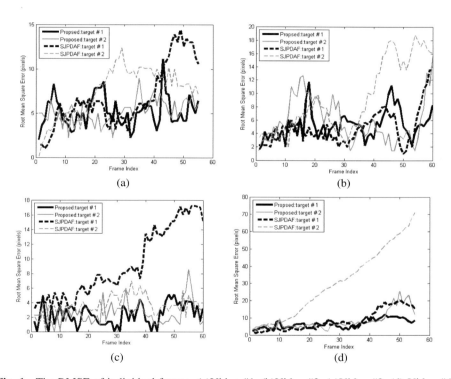

Fig. 1. The RMSE of individual frames. (a)Video #1. (b)Video #2. (c)Video #3. (d) Video #4.

Table 2. Comparisons of the tracking results on the last frames of videos #1 and #2 between the proposed scheme and the SJPDAF [7]. (Rectangle: The tracker estimated state. White dot: the central position of the ground truth of a target.)

Video #1	Proposed	SJPDAF	Video #2	Proposed	SJPDAF
The last Frame			The last Frame		

Table 3. Comparisons of the average RMSE (unit: pixels) of target #1 and target #2 between the proposed scheme and SJPDAF [7].

Video No.	Target #1		Target #2	
	Proposed	SJPDAF	Proposed	SJPDAF
Video #1	5.1711	6.6987	5.1487	7.4109
Video #2	5.2304	4.8945	5.8708	9.1084
Video #3	2.3177	9.1508	2.5331	3.6574
Video #4	5.3746	5.2637	10.4522	33.6263

6   Conclusions

This paper combines the single target based co-inference tracker with SJPDAF for multi-target tracking. The co-inference tracking predicted states are used for measurement validation of SJPDAF. Test results show that the proposed multi-target tracker outperforms the SJPDAF based tracker on mobile mixed images. Accuracy of motion masks will be enhanced in our future work.

Acknowledgement. This work was supported by the Ministry of Science and Technology of Taiwan under the Grants MOST-103-2221-E-008-061 and MOST-104-2221-E-008-059.

References

1. Wu, Y., Lim, J.W., Yang, M.-H.: Online object tracking: a benchmark. In: Proceedings of the IEEE International Conference on Computer Vision and Pattern Recognition, pp. 2411–2418 (2013)
2. Elgharib, M.A., Pitie, F., Kokaram, A., Saligrama, V.: User-assisted reflection detection and feature point tracking. In: Proceedings of the European Conference on Visual Media Production (2013)
3. Weiss, Y.: Deriving intrinsic images from image sequences. In: Proceedings of the IEEE International Conference on Computer Vision, vol. 2, pp. 68–75 (2001)
4. Chen, H.-T., Tang, C.-W.: Visual tracking using blind source separation for mixed images. In: Proceedings of the International Conference on Acoustics, Speech, and Signal Processing, pp. 6548–6552 (2014)
5. Chen, H.-T., Tang, C.-W.: Robust tracking using visual cue integration for mobile mixed images. J. Vis. Commun. Image Represent. **30**, 208–218 (2015)
6. Chang, K.-C., Bar-Shalom, Y.: Joint probabilistic data association for multitarget tracking with possibly unresolved measurements and maneuvers. IEEE Trans. Automat. Control. **29** (7), 585–594 (1984)
7. Schulz, D., Burgard, W., Fox, D., Cremers, A.B.: People tracking with mobile robots using sample-based joint probabilistic data association filters. Int. J. Robot. Res. **2**(2), 99–116 (2003)
8. Berclaz, J., Fleuret, F., Turetken, E., Fua, P.: Multiple object tracking using K-shortest paths optimization. IEEE Trans. Pattern Anal. Mach. Intell. **33**(9), 1806–1819 (2011)
9. Bar-Shalom, Y., Willett, P.K., Tian, X.: Tracking and Data Fusion. A Handbook of Algorithms. Yaakov Bar-Shalom, Storrs (2011)

10. Krausbling, A., Schulz, D.: Tracking extended targets - a switching algorithm versus the SJPDAF. In: Proceedings of the IEEE International Conference on Information Fusion, pp. 1–8 (2006)
11. Pham, N.T., Leman, K., Wong, M., Gao, F.: Combining JPDA and particle filter for visual tracking. In: Proceedings of the IEEE International Conference on Multimedia and Expo, pp. 1044–1049 (2010)
12. Chavali, P., Nehorai, A.: Concurrent particle filtering and data association using game theory for tracking multiple maneuvering targets. IEEE Trans. Sig. Process. **61**(20), 4934–4948 (2013)
13. Wu, Y., Huang, T.S.: Robust visual tracking by integrating multiple cues based on co-inference learning. Int. J. Comput. Vis. **58**(1), 55–71 (2004)
14. Arulampalam, M.S., Maskell, S., Gordon, N., Clapp, T., et al.: A tutorial on particle filters for online nonlinear/non-Gaussian Bayesian tracking. IEEE Trans. Sig. Process. **50**(2), 174–188 (2002)
15. Nummiaro, K., Koller-Meier, E., Gool, L.V.: An adaptive color-based particle filter. Image Vis. Comput. **21**(1), 99–110 (2003)
16. Bay, H., Tuytelaars, T., Van Gool, L.: SURF: Speeded Up Robust Features. In: Leonardis, A., Bischof, H., Pinz, A. (eds.) ECCV 2006, Part I. LNCS, vol. 3951, pp. 404–417. Springer, Heidelberg (2006)
17. Lu, J.-Y., Wei, Y.-C., Tang, C.-W.: Visual tracking using compensated motion model for mobile cameras. In: Proceedings of the IEEE International Conference on Image Processing, pp. 489–492 (2011)
18. Li, S., Gaizhi, G.: The application of improved HSV color space model in image processing. In: Proceedings of International Conference on Future Computer and Communication, vol. 2, pp. 10–13 (2010)
19. Aherne, F., Thacker, N., Rockett, P.: The Bhattacharyya metric as an absolute similarity measure for frequency coded data. Kybernetika **34**(4), 363–368 (1998)
20. Naqvi, S.M., Mihaylovay, L., Chambers, J.A.: Clustering and a joint probabilistic data association filter for dealing with occlusions in multi-target tracking. In: Proceedings of IEEE International Conference on Information Fusion, pp. 1730–1735 (2013)

A Hierarchical Frame-by-Frame Association Method Based on Graph Matching for Multi-object Tracking

Sourav Garg[1], Ehtesham Hassan[1], Swagat Kumar[1(✉)], and Prithwijit Guha[2]

[1] Tata Consultancy Services, New Delhi, India
{sourav.garg,ehtesham.hassan,swagat.kumar}@tcs.com
[2] Indian Institute of Technology Guwahati, Guwahati, Assam, India
pguha@iitg.ernet.in

Abstract. Multiple object tracking is a challenging problem because of issues like background clutter, camera motion, partial or full occlusions, change in object pose and appearance etc. Most of the existing algorithms use local and/or global association based optimization between the detections and trackers to find correct object IDs. We propose a hierarchical frame-by-frame association method that exploits a spatial layout consistency and inter-object relationship to resolve object identities across frames. The spatial layout consistency based association is used as the first hierarchical step to identify easy targets. This is done by finding a MRF-MAP solution for a probabilistic graphical model using a minimum spanning tree over the object locations and finding an exact inference in polynomial time using belief propagation. For difficult targets, which can not be resolved in the first step, a relative motion model is used to predict the state of occlusion for each target. This along with the information about immediate neighbors of the target in the group is used to resolve the identities of the objects which are occluded either by other objects or by the background. The unassociated difficult targets are finally resolved according to the state of the object along with template matching based on SURF correspondences. Experimentations on benchmark datasets have shown the superiority of our proposal compared to a greedy approach and is found to be competitive compared to state-of-the-art methods. The proposed concept of association is generic in nature and can be easily employed with other multi-object tracking algorithms.

1 Introduction

Multiple object tracking is often challenged by the phenomena of cluttered background, unpredictable and non-smooth trajectories, occlusions and variations in appearance on account of illumination, deformations and articulations. The last two decades have seen the exploration of contour, point and region based methods [1] for handling such challenges. However, the recent years have witnessed the tracking-learning-detection [2] based algorithms given the advancement made

© Springer International Publishing Switzerland 2015
G. Bebis et al. (Eds.): ISVC 2015, Part I, LNCS 9474, pp. 138–150, 2015.
DOI: 10.1007/978-3-319-27857-5_13

in the arena of object detection and on-line learning algorithms [3]. In these approaches, a detector is used to locate objects in each frame and then associate them across frames to generate long trajectories. These methods can be broadly classified into batch processing and on-line methods.

Batch processing methods are primarily off-line methods that require detections over all the frames. These detection responses are linked together to form short trajectories called, tracklets which may be fragmented due to occlusions. These tracklets are then globally associated to generate longer trajectories [4–7]. In some cases, the long term trajectories are built directly from individual detection responses as in [8–10]. In either case, the global association is crucial in these methods and a number of approaches [4,11,12] have been proposed in the literature to achieve this. The computational requirement for these methods are huge as they require iterative associations for generating globally optimized trajectories. It is therefore difficult to apply these methods to real-time applications and leads to latency when used with a sliding time window [5].

On the other hand, on-line methods [13–16] can be used for real-time applications as the trajectories are build sequentially by resolving associations on a frame-by-frame basis using only current and past information. While these methods are comparatively simpler, they suffer from frequent ID switches and drifting under occlusions.

In this work, our main focus is to improve tracking performance under occlusions. A complete categorization of such occlusions in to 14 different static and/or dynamic forms can be found in [17]. We aim to improve the performance of on-line methods by using a hierarchical framework that uses spatial layout consistency and an inter-object relationship model to resolve associations among the detected objects between two consecutive frames. As a first step, easy to resolve targets are associated using spatial layout consistency. This is done by finding a MRF-MAP solution for a probabilistic graphical model using a minimum spanning tree over the object locations and finding an exact inference using belief propagation. The targets which remain unassociated in the first stage are processed through the second stage where a Kalman Filter based relative motion model is used to predict the occlusion state for each target. This along with the information of immediate neighbors is used to resolve the identities of objects under static or dynamic occlusion.

The main contribution made in this paper are as follows. First, a new hierarchical framework is proposed for resolving frame-by-frame associations among the detection responses that is non-greedy, fast and non-iterative. Second, in contrast to existing graphical methods [18,19] which use dense representations of objects, we use a sparse representation where the target locations are connected through a minimum spanning tree and the association is resolved through MRF-MAP belief propagation. Third, Unlike methods [20] that use relative motion model for tracking objects, we use predict the occlusion state using the same. This along with the information about immediate neighbors is used to detect IDs of objects under static or dynamic occlusion.

The rest of paper is organized as follows. A brief review of related work is presented in Sect. 2. The proposed approach is described in Sect. 3. The experimental results are presented in Sect. 4. Finally, we conclude the work in Sect. 5 and sketch the future extensions.

2   Related Work

In the related works, significant amount of research contributions exist in the field of multi-object tracking which have attempted different issues and challenges. Nevertheless, in this section we focus on some of the recent and significant works which are related to the problem of association resolution in multiple object tracking which is the major focus of this paper.

In the recent works, use of probabilistic graphical models for tracking has shown some interesting results. In [4,19] novel formulations of CRF has been applied for tracking in multi-object scenario. In [4], authors proposed online learning of CRF models for solving the multi-target tracking which uses tracklet pairs as node and correlation among pairs for edge potential computation. Subsequently in [5], the authors have extended the concept in multiple instance learning framework for learning a non-linear motion map. In [19], authors proposed single CRF model for joint tracking and segmentation of multiple objects. They solve a multi-label problem where node potentials are computed with detection responses and likelihood of super-pixels belonging to a target, and edge potentials are computed as overlap between super-pixels in spatial and temporal domain. The approach is similar to Poiesi et al. [18] which uses a particle filter based tracker and applies a MRF for resolving the IDs between multiple objects where each particle carrying an ID acts as a node in MRF model.

All of these methods use dense object representations of an object such as multiple particles/super-pixels, while we use sparse detections as the nodes with edges defined over their spatial connectivity in our MRF based graphical model.

The association based approaches defined using a probability distribution as an optimization problem are solved using energy minimization as in [4–6,19] or a MAP solution as in [16,18]. Depending on the type of underlying graph, kinds of constraints involved in the formulation and the number of variables, these optimization routines differ from each other. We make use of Belief Propagation as a MAP inferencing technique to find an optimal assignment for our probability distribution. The underlying graph in our case being a Minimum Spanning Tree allows for an exact inference in two iterative steps.

The use of motion model for predicting the trajectories due to lack of a proper assignment is done in different ways. [4,5] use CRF model for learning the motion model, [18] uses particle filter and means shift algorithm, [6,19] use motion cues in the energy minimization formulation. We use Kalman Filter for individual motion and relative motion between tracker pairs. The relative motion model also present in [5] is based on long term trajectories by defining head-close and tail-close tracklet pairs for discriminating them among the other possible pairs, whereas, we define it based on Kalman Filter relative motion learning for predicting occlusions and determining occluder-occludee pairs.

We define <u>states</u> for the tracked object for it being completely visible or partially or fully occluded by either other objects or background. A similar definition for occlusion scenarios is also mentioned in [19] where occlusion is defined using visibility and depth of the objects, cues from which become penalties in the energy minimization problem. On the other hand, we define these <u>states</u> based on the relative motion model as a result of which we explicitly handle the occlusion recovery from the <u>occluder-occludee</u> pairs.

3   Proposed Approach

In order to explain our approach, we will make use of the following notations. A given video sequence is represented by the symbol I_k, $k = 1, 2, \ldots, N$ indicating that the video has a total of N frames. Each detected object is represented by a bounding box (BB) and is labeled with a global ID. Each detected object for a given frame I_k is represented by the symbol D_i^k, $i = 1, 2, \ldots, n$, where n is the number of objects detected in the frame. Each detector D_i^k is represented by a rectangle $\{c_i, w_i, h_i\}$, where c_i is the two-dimensional image coordinate of the center of the rectangle. w_i and h_i are the width and the height of the rectangle respectively. The corresponding tracker windows available in the frame is represented by the symbol T_j^k, $j = 1, 2, \ldots, m$. The trackers are the bounding boxes obtained from those in the previous frame using a local tracker based on SURF matching, color histograms or simply, motion predictors.

The objective of our work is to grow trajectories of detected objects by resolving frame-by-frame associations in an on-line fashion. The various components of the proposed scheme is shown in Fig. 1. It has three modules: Detection, Association and Tracking. The detection module uses object detection algorithm to find bounding boxes for all objects available in each query frame. These detections are associated with the detections from previous frame in the association module. The association module makes use of a two stage hierarchical approach to resolve these associations. The first stage makes use of spatial layout consistency to resolve identities of easier targets. The second stage makes use of a Kalman filter-based relative motion model to deal with difficult cases of occlusion. This is the main contribution of our approach. Finally, there is the histogram and SURF matching based tracking module to track some objects locally which cannot be identified otherwise especially the ones which reappear after occlusion. These modules are explained next in this section.

The Detection Module – An object detector is run on every frame to find out the detection responses. These responses act as an observation for the graphical model to find out the associations between trackers and detections. New trackers are initialized for the unassociated responses that occur consecutively for more than three times.

The Association Module – This module resolves the association among the detections between two consecutive frames. This is done by using a two step process. In the first step, easier targets are identified by using spatial layout

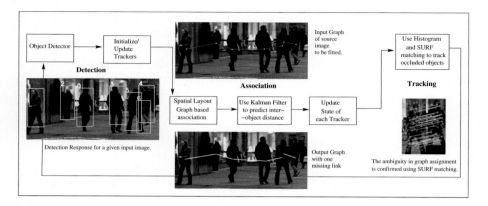

Fig. 1. Block diagram depicting various components of our approach. It can be broadly categorized into Detection, Association and Tracking modules.

consistency. The difficult cases like occlusion and missed detections are dealt in the second stage by using a relative motion model and is described next.

Spatial Layout Consistency – can be used to identify wrong or missed detections by exploiting the fact that the arrangement of objects may not change immediately between consecutive frames. The arrangement of objects is captured in the form of a graph which is formed by fitting a minimum spanning tree (MST) over the centers of tracker windows.

Considering the nodes as the random variable X_i over the bounding box locations c_i, the edges of the tree is represented by constructing an adjacency matrix A_{ij}. We define a joint probability distribution over the graph (MST) G with each of its node represented by X_i and edge by A_{ij}

$$P(G) = 1/Z \prod_{i,j} \psi(\mathbf{X}_i, \mathbf{X}_j) A_{ij} \tag{1}$$

where Z is the normalizing factor. The function $\psi(X_i, X_j)$ is the potential function for an edge between nodes X_i and X_j and is defined as

$$\psi(X_i, X_j) = \exp\left(-w_1 \Delta\theta_{ij} - w_2 \Delta\mathbf{r}_{ij}\right)\phi_1\phi_2\Delta s_1\Delta s_2 \tag{2}$$

where $\Delta\theta_{ij}$ is the change of angle between the edges and $\Delta\mathbf{r}_{ij}$ is the change in edge length between the source and destination graph. ϕ_1 and ϕ_2 are the measures of appearance matching coefficient (Bhattacharya Coefficient in this case) for respective nodes as compared to the source nodes. Δs_1 and Δs_2 are the measures of change in scale of the nodes compared to the source nodes.

Source Graph – The underlying source graph (minimum spanning tree) as shown in Fig. 2(a) is the spatial layout with random variables defined for each node that represents a tracked object. The edges are specified by the adjacency matrix. The use of tree is intuitively justified as we do not really want to model the edges between the pairs of targets which are far off from each other. Also in real life scenarios it is highly likely that moving objects like humans or vehicles interact and change their motion according to other similar moving objects in their immediate neighborhood. This in turn leads to simplification of the optimization problem that we solve using Belief Propagation in two iterative steps as a MAP inferencing technique for the probabilistic graphical model.

Observations – The objects detected by the detector in the next frame I_{k+1} are the observations which will be used for resolving associations through graph matching. These observations are shown as green rectangles in Fig. 2(b). The yellow rectangles are the dummy observations which are obtained from the tracker locations in the previous frame. The dummy observations help in reducing the false assignments and facilitate one-to-one matching. These dummy variables are deliberately given weights which are 4 times less than the actual observations, so that there is high chance of assigning a true observation than dummy than a false one to the random variable in a respective order. Hence, there is no scope for many-to-one associations. Also, one-to-many associations do not occur because the optimization formulation searches for only the maximum value assignment. Thus, the two-sided mutex constraints are implicitly taken care of.

Resultant Graph – Now a MRF-MAP solution is sought for the graph matching taking the source graph as the prior and the current observations as the observation likelihood. The resulting graph is shown in Fig. 2(c). This helps in getting rid of the rightmost detection window which wrongly detected a tree as a human. Secondly, several missed detections were recovered such as objects with IDs 25 and 32. These missed detections are shown using ellipses in Fig. 2(c) as the tracker location is obtained only from prediction. The right most detected window is assigned a new object ID 35.

Relative Motion Model and Occlusion State of the Target – As stated earlier, the more difficult cases of static and/or dynamic occlusions are resolved by using relative motion model and neighborhood information of each object in a group. We use relative motion model to predict if an object is going to be occluded in the immediate future. This is unlike [20] where the relative motion model is used for tracking objects. We define a Kalman Filter K_{ij} for each tracker pair (T_i, T_j) using the inter-object distance d_{ij} as the state variable. The relative inter-object motion is approximated using a constant velocity state model given by the following equation

$$\begin{bmatrix} d_{ij}^{k+1} \\ \dot{d}_{ij}^{k+1} \end{bmatrix} = \begin{bmatrix} 1 & \Delta t \\ 0 & 1 \end{bmatrix} \begin{bmatrix} d_{ij}^{k} \\ \dot{d}_{ij}^{k} \end{bmatrix} + \mathbf{w}_k \tag{3}$$

where, $\dot{d}_{ij}^{k}$ is the corresponding velocity and $\mathbf{w}_k$ is the process noise with covariance matrix given by

$$\mathbf{Q}_k = \begin{bmatrix} \Delta t^2 & \Delta t \\ \Delta t & 1 \end{bmatrix} \tag{4}$$

(a) Spatial Layout of source image (I_k). The graph is a Minimum Spanning Tree for the tracked pedestrians.

(b) The detection responses in frame I_{k+1}. The actual detections are shown in green color. The dummy observations that are implanted based on Kalman Filter motion predictor are shown in yellow. These are the candidates for assignment in the probabilistic graphical model.

(c) Resultant Spatial Layout is the output of MRF-MAP based selected edges. The left-most edge is not selected as detection response is not available for that.

(d) The confirmed pedestrians are shown as a rectangle and the predicted ones are shown as an ellipse.

Fig. 2. Resolving IDs for detection responses through graph matching (Color figure onliine).

And, the measurement model is given by

$$\left[d_{ij}^{k\ m}\right] = \begin{bmatrix} 1 & 0 \end{bmatrix} \begin{bmatrix} d_{ij}^k \\ \dot{d}_{ij}^k \end{bmatrix} + \mathbf{v}_k \tag{5}$$

where, $\mathbf{v}_k$ is the measurement noise with covariance matrix

$$\mathbf{R}_k = \left[\sigma_{\Delta d_{ij}}^2\right] \tag{6}$$

The predicted state d_{ij}^{k+l}, $l = 1, 2, 3, \ldots$ is used to determine the state of the tracking object as explained next. A target can be any of the following states:

$$S_i = \begin{cases} 1, \text{if target is available and is confirmed} \\ 2, \text{if target is occluded by another target} \\ 3, \text{if target is occluded by the background} \\ 4, \text{if target cannot be confirmed by any means} \end{cases} \tag{7}$$

Any target whose ID is resolved and confirmed by spatial layout consistency (the first step) is said to be in state $S_i = 1$. If the target matches with the dummy observation then it can have any of the remaining three states. It belongs to inter-object occlusion if the following condition is satisfied.

$$S_i = 2 \iff \exists\ l \in \{1, 2, 3\} : d_{ij}^{k+l} < \epsilon \tag{8}$$

where, d_{ij}^{k+l} represents the predicted values of the inter-object distance d_{ij}^k. The target belongs to the state of object-background occlusion if the following condition is satisfied.

$$S_i = 3 \iff \phi_i < \delta \tag{9}$$

where ϕ_i is the appearance matching value computed through Bhattachharya Coefficient for color distributions. The target belongs to the fourth state if it is not confirmed whether it is present or not, that is, there is no sufficient evidence to prove its belongingness to any of the other three states.

Merge and Split – The dynamic occlusions of objects are usually dealt by considering the concepts of merge and split [21]. The merges are implicitly dealt with in our approach due to the spatial layout consistency as explained in Fig. 3. The two objects that are about to merge are constrained to be apart such that it is better for the missing target to match with its dummy observation than with its occluder. Hence, ID switches can be easily avoided during a merge.

On the other hand, splits are handled with the knowledge of state of the target. Any occluded target, when reappears, is searched in the neighborhood of the occluder. If the appearance of the target is similar to the new detection response in the neighborhood of occluder then it is confirmed to be the occludee. This way we are able to avoid unnecessary fragments for targets which are occluded even for longer durations. This is explained in the Fig. 4. In this figure, the trajectories of objects 1 and 3 merge together and get occluded as shown in Figs. 4(a)–(c). The pedestrian 1 is correctly identified by using neighborhood relationship information as shown in Fig. 4(e).

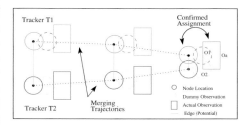

Fig. 3. The availability of a dummy observation facilitates correct assignment of detection response during merging of trajectories.

Tracking – Association of targets is sometimes limited by the availability of detection responses and the detector's accuracy and precision. This makes it necessary to localize some of the targets by means of appearance based local region tracking. We use SURF correspondences for localizing targets that are in state $S_i = 4$.

4   Experiments and Results

The performance of the proposed algorithm is tested on three datasets, namely, ETH [22], TUD [12] and PETS 2009 [23] datasets. The ETH dataset comprises of two sequences, that is, Bahnhoff and Sunny Day, both having a dynamic camera movement, but a similar view angle. Videos of both TUD and PETS dataset are recorded using a static camera. The TUD dataset has a view from a low-lying

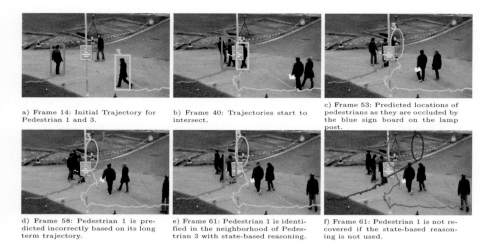

a) Frame 14: Initial Trajectory for Pedestrian 1 and 3.

b) Frame 40: Trajectories start to intersect.

c) Frame 53: Predicted locations of pedestrians as they are occluded by the blue sign board on the lamp post.

d) Frame 58: Pedestrian 1 is predicted incorrectly based on its long term trajectory.

e) Frame 61: Pedestrian 1 is identified in the neighborhood of Pedestrian 3 with state-based reasoning.

f) Frame 61: Pedestrian 1 is not recovered if the state-based reasoning is not used.

Fig. 4. Occlusion recovery based on knowledge of correct states of target. Pedestrian 1 is identified again once it recovers from occlusion by making use of the information it was in the neighborhood of Pedestrian 3 before occlusion.

camera with humans of relatively larger size whereas, the PETS dataset has a camera placed far from the crowd with a slanted top view. Both ETH and TUD datasets have frequent dynamic occlusions while PETS dataset has inconsistent trajectories and small human sizes. These datasets are publicly available. The performance of the proposed algorithm is compared with several state-of-the-art algorithms. The proposed algorithm was implemented on an Intel Core $i7$ CPU with 2.90 GHz clock speed and 8 GB RAM. The average computation time for processing each frame depends on the number of objects in each frame. For instance, the per frame computation time for ETH dataset is around 100 ms as there are high number of objects per frame. For TUD dataset, the per frame computation time is 70 ms. The PETS dataset takes very less time, that is, 38 ms, because of less number of trajectories and smaller number of objects being tracked. The computational complexity of our algorithm is polynomial in number of trajectories in a given frame. Apart from that, few peaks occur for pedestrians of larger size due to SURF matching. There is no off-line processing involved and hence, no latency is noted. The computation time for the detection of humans is not taken into account in any of these cases.

Evaluation Metrics – In order to compare the performance of our algorithm with others, we have used the evaluation metrics as defined in [24–26]. We make use of 12 evaluation parameters for comparing our method with other algorithm as shown in Table 1. The parameters have the usual meaning and are not described here due to space constraints. The average rank of each tracker is computed as described in [26] to indicate the overall performance.

The performance of our algorithm is compared with four other state-of-the-art methods on three datasets, namely, ETH, TUD and PETS 2009 as mentioned above. Out of these four, two methods [5,6] use global optimization to resolve associations. In [5], authors have used temporal window of 8 frames to make the process on-line but, with a latency of 2 s. It is also mentioned that there is a trade-off between accuracy and latency in their approach. The work reported in [7] is one of the most recent approach in the category of frame-by-frame association. In this work, the authors have reported results with four different settings with each one of them having different evaluation parameters and performing well in different scenarios. We have chosen to compare our results with the one condition that does not use global optimization. This is closest to our approach, though our results are competitive even for other settings. Finally, we have compared our method to a greedy approach [27] where the associations between detectors and trackers are resolved by using the Hungarian algorithm. The affinity matrix is constructed by using histogram matching and overlap between the object windows.

The performance comparison of the above methods with the proposed method is provided in Table 1. In terms of average ranking, our algorithm gives best performance for PETS dataset, second best performance for TUD dataset and worst performance for ETH dataset. The poor performance on ETH dataset is mostly due to lack of a better appearance model than due to camera motion as proven by the high number of trajectory fragments. This means that the objects

Table 1. Comparison of results with state of the art methods on standard datasets. The average rank shows that our approach has the best rank for PETS, second best for TUD and last for ETH.

Dataset	Method	MOTP	MOTA	Recall	Precision	FAF	GT	MT	PT	ML	Frag	IDS	FPS	Avg Rank
PETS	Nevatia et al. 2014 [5]	-	88.38	93.0	95.3	0.268	19	89.5	10.5	0.0	13	0	14	2.8
	Bae et al. 2014 [7]	69.59	83.04	-	-	0.204	23	100.0	0.0	0.0	4	4	5	2.6
	Anton et.al. 2014 [6]	80.2	90.6	-	-	0.074	23	91.3	4.35	4.35	6	11	1	2.9
	Garg et al. 2015 [27]	86.5	90.4	97.46	93.65	0.380	19	94.73	5.27	0.00	27	22	20	2.7
	Proposed approach	84.30	91.24	95.3	95.6	0.234	19	89.5	10.5	0.0	12	2	26	2.0
TUD	Nevatia et al. 2014 [5]	-	84.0	87.0	96.7	0.184	10	70.0	30.0	0.0	1	0	10	1.6
	Anton et.al. 2014 [6]	65.5	71.1	-	-	0.513	9	77.8	22.2	0.0	3	4	1	2.5
	Garg et al. 2015 [27]	80.0	73.4	88.5	85.8	0.90	10	80.00	20.0	0.0	9	5	10	2.4
	Proposed approach	81.7	79.1	80.5	98.7	0.067	10	60.0	40.0	0.0	7	3	14	2.0
ETH	Nevatia et al. 2014 [5]	-	70.6	79.0	90.4	0.637	125	68.0	24.8	7.2	19	11	10	1.8
	Bae et al. 2014 [7]	64.0	72.0	-	-	0.035	126	73.8	23.8	2.4	38	18	2	2.0
	Garg et al. 2015 [27]	72.5	62.0	86.0	78.4	7.65	124	75.0	20.2	4.8	70	31	10	2.4
	Proposed approach	70.1	58.7	77.4	80.6	1.422	124	56.5	35.4	8.1	62	14	10	2.9

which reappear after occlusion is not alloted the same ID as the appearance matching fails to identify the target.

We also calculated results for a greedy association based approach [27] that solves frame to frame associations between detection and tracking results. This is done by using an affinity matrix calculated by histogram matching and overlap coefficients. The matrix is solved for one to one associations using Hungarian algorithm. This can be considered as a bare minimum framework required for resolving frame-to-frame associations. The table shows that the proposed algorithm helps in reducing the ID switches and trajectory fragmentation to a greater extent as compared to the greedy approach. As the concepts are generic in nature, this can be applied in conjunction with other existing methods to improve their performance.

5   Conclusion and Future Work

In this paper, we have made an attempt to improve the performance of frame-by-frame association methods in multiple object tracking by using concepts like spatial layout consistency and neighborhood relationship information of objects in a group. The spatial layout consistency is imposed through MRF-based graph matching scheme which is used to resolve the association of easily detected objects in a frame. The difficult cases of occlusions and missed detections are dealt with by using a relative motion model which is used to predict if the objects are going to be occluded in the near future. The association for such cases are resolved using information of immediate neighbors of each object tracked in a group. Through experiments on standard datasets, it is shown that the proposed approach works better when compared to the greedy method of association. The results are also competitive as compared to the other state of the art methods. Moreover, our approach doesn't have any latency or off-line processing and uses higher level information to take the right decision for an object's state instead of every time relying on the trajectories. In our future work, we would combine our current approach with a better a appearance model which could be used for tracking individual targets.

References

1. Li, X., Hu, W., Shen, C., Zhang, Z., Dick, A., Hengel, A.V.D.: A survey of appearance models in visual object tracking. ACM Trans. Intell. Syst. Technol. **4**, 58 (2013)
2. Smeulders, A., Chu, D., Cucchiara, R., Calderara, S., Dehghan, A., Shah, M.: Visual tracking: an experimental survey. IEEE Trans. Pattern Anal. Mach. Intell. **36**, 1442–1468 (2014)
3. Zhang, X., Yang, Y.H., Han, Z.: Object class detection: a survey. ACM Comput. Surv. **46**, 10 (2013)
4. Yang, B., Nevatia, R.: An online learned CRF model for multi-target tracking. In: 2012 IEEE Conference on Computer Vision and Pattern Recognition (CVPR), pp. 2034–2041. IEEE (2012)

5. Yang, B., Nevatia, R.: Multi-target tracking by online learning a CRF model of appearance and motion patterns. Int. J. Comput. Vis. **107**, 203–217 (2014)
6. Milan, A., Roth, S., Schindler, K.: Continuous energy minimization for multitarget tracking. IEEE Trans. Pattern Anal. Mach. Intell. **36**, 58–72 (2014)
7. Bae, S.H., Yoon, K.J.: Robust online multi-object tracking based on tracklet confidence and online discriminative appearance learning. In: 2014 IEEE Conference on Computer Vision and Pattern Recognition (CVPR), pp. 1218–1225. IEEE (2014)
8. Berclaz, J., Fleuret, F., Turetken, E., Fua, P.: Multiple object tracking using k-shortest paths optimization. IEEE Trans. Pattern Analy. Mach. Intell. **33**, 1806–1819 (2011)
9. Ben Shitrit, H., Berclaz, J., Fleuret, F., Fua, P.: Tracking multiple people under global appearance constraints. In: 2011 IEEE International Conference on Computer Vision (ICCV), pp. 137–144. IEEE (2011)
10. Pirsiavash, H., Ramanan, D., Fowlkes, C.C.: Globally-optimal greedy algorithms for tracking a variable number of objects. In: 2011 IEEE Conference on Computer Vision and Pattern Recognition (CVPR), pp. 1201–1208. IEEE (2011)
11. Brendel, W., Amer, M., Todorovic, S.: Multiobject tracking as maximum weight independent set. In: 2011 IEEE Conference on Computer Vision and Pattern Recognition (CVPR), pp. 1273–1280. IEEE (2011)
12. Andriyenko, A., Schindler, K.: Multi-target tracking by continuous energy minimization. In: 2011 IEEE Conference on Computer Vision and Pattern Recognition (CVPR), pp. 1265–1272. IEEE (2011)
13. Breitenstein, M.D., Reichlin, F., Leibe, B., Koller-Meier, E., Van Gool, L.: Online multiperson tracking-by-detection from a single, uncalibrated camera. IEEE Trans. Pattern Anal. Mach. Intell. **33**, 1820–1833 (2011)
14. Shu, G., Dehghan, A., Oreifej, O., Hand, E., Shah, M.: Part-based multiple-person tracking with partial occlusion handling. In: 2012 IEEE Conference on Computer Vision and Pattern Recognition (CVPR), pp. 1815–1821. IEEE (2012)
15. Song, X., Cui, J., Zha, H., Zhao, H.: Vision-based multiple interacting targets tracking via on-line supervised learning. In: Forsyth, D., Torr, P., Zisserman, A. (eds.) ECCV 2008, Part III. LNCS, vol. 5304, pp. 642–655. Springer, Heidelberg (2008)
16. Wu, B., Nevatia, R.: Detection and tracking of multiple, partially occluded humans by bayesian combination of edgelet based part detectors. Int. J. Comput. Vis. **75**, 247–266 (2007)
17. Guha, P., Mukerjee, A., Subramanian, V.K.: Ocs-14: You can get occluded in fourteen ways. In: 22nd International Joint Conference on Artificial Intelligence (IJCAI 2011), pp. 1665–1670 (2011)
18. Poiesi, F., Mazzon, R., Cavallaro, A.: Multi-target tracking on confidence maps: an application to people tracking. Comput. Vis. Image Underst. **117**, 1257–1272 (2013)
19. Milan, A., Leal-Taixé, L., Schindler, K., Reid, I.: Joint tracking and segmentation of multiple targets. In: Proceedings of the IEEE Conference on Computer Vision and Pattern Recognition, pp. 5397–5406 (2015)
20. Yoon, J.H., Yang, M.H., Lim, J., Yoon, K.J.: Bayesian multi-object tracking using motion context from multiple objects. In: 2015 IEEE Winter Conference on Applications of Computer Vision (WACV), pp. 33–40 (2015)
21. Guha, P., Mukerjee, A., Subramanian, V.K.: Formulation, detection and application of occlusion states (oc-7) in the context of multiple object tracking. In: 8th IEEE International Conference on Advanced Video and Signal-Based Surveillance (AVSS 2011), pp. 191–196. IEEE (2011)

22. Ess, A., Leibe, B., Schindler, K., Van Gool, L.: Robust multiperson tracking from a mobile platform. IEEE Trans. Pattern Anal. Mach. Intell. **31**, 1831–1846 (2009)
23. Pets 2009 (2009). http://www.cvg.reading.ac.uk/PETS2009/
24. Li, Y., Huang, C., Nevatia, R.: Learning to associate: hybridboosted multi-target tracker for crowded scene. In: IEEE Conference on Computer Vision and Pattern Recognition, CVPR 2009, pp. 2953–2960. IEEE (2009)
25. Bernardin, K., Stiefelhagen, R.: Evaluating multiple object tracking performance: the clear mot metrics. J. Image Video Process. **2008**, 1 (2008)
26. Leal-Taixé, L., Milan, A., Reid, I., Roth, S., Schindler, K.: MOTChallenge 2015: Towards a Benchmark for Multi-Target Tracking. ArXiv e-prints (2015)
27. Garg, S., Rajesh, R., Kumar, S., Guha, P.: An occlusion reasoning scheme for monocular pedestrian tracking in dynamic scenes. In: 12th IEEE International Conference on Advanced Video and Signal based Surveillance (AVSS 2015) (2015)

Experimental Evaluation of Rigid Registration Using Phase Correlation Under Illumination Changes

Alfonso Alba[(✉)] and Edgar Arce-Santana

Facultad de Ciencias, Universidad Autónoma de San Luis Potosí,
Av. Salvador Nava Mtz. S/N, Zona Universitaria, 78290
San Luis Potosí, SLP, Mexico
fac@fc.uaslp.mx, arce@fciencias.uaslp.mx

Abstract. The phase correlation method is a computationally-efficient technique for image alignment. Presently, the method is capable of performing rigid image registration with sub-pixel accuracy, and is fairly robust to noise and long translations. However, there are also cases when the images to be aligned were taken at different times or come from different sensors, and may present differences in intensity values or illumination. Many algorithms exist to deal with these issues; however, most of them are computationally expensive. In this article, we explore the robustness of the phase correlation method to illumination and/or intensity changes by means of a quantitative evaluation using artificially-generated rigid transformations. Our results suggest that rigid registration using phase correlation may be fairly robust to gamma correction, quantization and multi-spectral acquisition, but more sensitive to differences in illumination and lighting conditions between the input images.

1   Introduction

Rigid image registration is the problem of aligning two images of the same scene using a rigid transformation, which only consists of rotation and translation, although mild scale changes are often present and some algorithms can correct them as well. This is often an essential pre-processing step in many applications where a rigid registration model is plausible, for example: image stitching, aerial imaging, super-resolution imaging, camera stabilization, shape and pattern matching, biometrics, and brain image analysis. [2,7,10,11]

One of the most popular techniques for rigid image registration is based on the phase correlation method, which can be used to estimate the translational shift between one image and a translated version of it [5]. In order to estimate the rotation and scaling, Reddy and Chatterji proposed to apply a log-polar transformation to the log-spectra of the input images, and then apply the phase-correlation method. Image spectra are insensitive to translations, but they preserve rotation and scale; when these spectra are transformed to a log-polar space, any rotation and change of scale between the images in the original

© Springer International Publishing Switzerland 2015
G. Bebis et al. (Eds.): ISVC 2015, Part I, LNCS 9474, pp. 151–161, 2015.
DOI: 10.1007/978-3-319-27857-5_14

space will correspond to a translation in the log-polar space, so that they can be easily recovered using the classic phase-correlation method. Once the rotation angle and scale factor are known, one can use them to rectify the images in the original space, and apply the phase correlation method to the rectified images in order to recover the translation [9]. According to Google Scholar, this method has been cited over 1500 times since its publication in 1996.

The main advantages of the phase correlation method are its low computational complexity, easy implementation and tuning, and its robustness to difficult transformations and moderate noise levels. Furthermore, various enhancements have been proposed to improve the accuracy of the estimations since the original phase correlation method can only estimate integer displacements. See [1] for a review of some of these methods.

In most applications, it is often assumed that the gray levels or colors of corresponding pixels are more or less preserved between the two images to be registered, and that noise is the only factor that may cause perturbations in the pixel intensities. However, there are many cases where this assumption is wrong; for example, when illumination conditions vary from one image to the other, or when the two images were acquired using different sensors, such as when registering multi-spectral or multi-modal images (e.g., a magnetic resonance image with a computed tomography image). While various methods have been devised to deal with these problems [2,3,6,13], most of them are computationally expensive and may be more difficult to implement and/or calibrate. For this reason, it is interesting to assess the robustness and applicability of classical registration methods, such as phase correlation, to different illumination conditions and intensity changes.

In this work, a series of experiments are performed in order to determine the extent to which the phase correlation method is robust to illumination and intensity changes, both natural and synthetic. Quantitative evaluation is achieved by estimating the registration error and success rate over several artificial rigid transformations with known ground truth. In this way, this work is intended to complement previous studies where the robustness of phase correlation to noise and data loss has been already evaluated [1,4,12].

The rest of this paper is organized as follows: Sect. 2 summarizes the rigid registration framework based on Reddy and Chatterji's method, but with a refinement stage to achieve sub-pixel accuracy. Section 3 presents the different types of illumination and intensity transformations under study, as well as the error measures used for quantitative evaluation. Results from the evaluation will be presented and discussed in Sect. 4. Finally, Sect. 5 will summarize our conclusions.

2   Rigid Registration Framework

We follow the approach used in [1] for rigid image registration with sub-pixel accuracy. Consider two images $f(r)$ and $g(r)$ where $r = (x, y)$ denote the coordinates of a pixel. The normalized cross-spectrum C of F and G is given by

$$C(k) = \frac{F(k)G^*(k)}{|F(k)G^*(k)|}, \tag{1}$$

where $F(k)$ and $G(k)$ are the Fourier transforms of f and g, respectively, $k = (k_x, k_y)$ is the frequency index in Fourier space and $G^*(k)$ is the complex conjugate of $G(k)$. The phase-only correlation (POC) $c(r)$ between f and g is obtained as the inverse Fourier transform of $C(k)$. It can be easily shown that if G is a translated version of F, that is $G(r) = F(r - d)$, then $c(r) = \delta(r + d)$, where δ is the impulse function. Therefore, the translation vector $\hat{d}$ required to re-align the images can be easily recovered as $\hat{d} = \arg\max c(r)$. In the case of digital signals or images, the translation between g and f is assumed to be circular and integer-valued, and the Fourier transform and its inverse are computing using FFTs. To achieve sub-pixel accuracy, one can estimate the maximum of the POC function by finding the zero of its derivative around the integer-valued solution $\hat{d}$ [1]. Since the POC function is the inverse DFT of the normalized cross spectrum, it can be written as

$$c(x, y) = \sum_{k_x=0}^{N_x-1} \sum_{k_y=0}^{N_y-1} C(k_x, k_y) \exp\{j2\pi(k_x x/N_x + k_y y/N_y)\}, \qquad (2)$$

for a digital image of size $N_x \times N_y$. Note that the right hand side of (2) is actually a continuous, band-limited version of the phase correlation function, where x and y can take non-integer values. Let us denote this function by $\tilde{c}(x, y)$, where the tilde indicates that x and y may vary continuously. The gradient $\nabla \tilde{c}$ of this function is given by

$$\nabla \tilde{c}(x, y) = \left[\frac{\partial \tilde{c}}{\partial x}(x, y), \ \frac{\partial \tilde{c}}{\partial y}(x, y) \right], \qquad (3)$$

where

$$\frac{\partial \tilde{c}}{\partial x}(x, y) = -\frac{2\pi}{N_x} \sum_{k_x=0}^{N_x-1} \sum_{k_y=0}^{N_y-1} k_x \Im\{C(k_x, k_y) \exp\{j2\pi(k_x x/N_x + k_y y/N_y)\}\}, \quad (4)$$

and

$$\frac{\partial \tilde{c}}{\partial y}(x, y) = -\frac{2\pi}{N_y} \sum_{k_x=0}^{N_x-1} \sum_{k_y=0}^{N_y-1} k_y \Im\{C(k_x, k_y) \exp\{j2\pi(k_x x/N_x + k_y y/N_y)\}\}. \quad (5)$$

In order to find the zeros of $\nabla \tilde{c}$, one can minimize the squared norm $\tilde{h}(x, y) = ||\nabla \tilde{c}(x, y)||^2$ around the integer valued solution d, for instance, using the well-known Nelder-Mead optimization heuristic [8]. In this case, the Nelder-Mead method requires three starting points: one of them is the integer-valued solution $b = (b_x, b_y)$, and the other two are chosen as $(b_x \pm 0.5, b_y)$ and $(b_x, b_y \pm 0.5)$, where the signs are chosen depending on which points yield the lowest values for $\tilde{h}(x, y)$. This method usually converges in less than 100 iterations.

The approach discussed above is capable of estimating translations with sub-pixel accuracy. To estimate rotations and scales, we apply the method proposed in [9], but with sub-pixel accuracy as proposed in [1]. Consider two input images f_1 and f_2 of size $N_x \times N_y$. The log-magnitude of their spectra is computed as $M_i(k) = W(k) \log |F_i(k)|$, where F_i is the Fourier transform of f_i, $k = (k_x, k_y)$ is the frequency index in Fourier space, and $W(k)$ is the frequency response of a high-pass filter given in [9]. The filtered log-spectra are then transformed to log-polar coordinates (ρ, θ) given by $\rho = K \log \sqrt{(k^T k)}$ and $\theta = \text{atan2}(k_y, k_x)$, where K is a suitable magnitude scaling factor which controls the resolution of the log-radius (ρ) axis; following [1], we use $K = (N_x + N_y)/8$. The phase correlation between M_1 and M_2 is computed and the location $(\hat{\rho}, \hat{\theta})$ of its maximum is estimated with sub-pixel accuracy using the method described above. From these, the rotation angle $\hat{\phi}$ and scaling factor $\hat{s}$ between the original images can be recovered as $\hat{\phi} = 2\pi\hat{\theta}/N_y$ and $\hat{s} = \exp(\hat{\rho}/K)$. Then one can rectify the input images using the linear transformation

$$T_{\hat{s},\hat{\phi}} = \begin{bmatrix} \hat{s}\cos\hat{\phi} & -\hat{s}\sin\hat{\phi} \\ \hat{s}\sin\hat{\phi} & \hat{s}\cos\hat{\phi} \end{bmatrix} \tag{6}$$

to align f_1 to f_2. Specifically, one can compute the rectified f_1 as $\hat{f}_1(r) = f1(T_{\hat{s},\hat{\phi}}r)$ using bi-cubic interpolation. Finally, the translation parameters are estimated by computing the phase correlation between $\hat{f}_1$ and f_2 and finding the location $(\hat{d}_x, \hat{d}_y)$ of its maximum with sub-pixel accuracy.

3   Evaluation Methodology

The rigid registration method described in the previous section has been applied to several grayscale image pairs where either artificial or real illumination or intensity transformations have been applied to one or both images. Artificial changes are applied by means of non-linear tone transfer functions, whereas real changes are the result of different illumination conditions, or different acquisition conditions. In all cases, the image pairs are originally aligned and a random rigid transformation, composed of rotation and translation, is applied to one of the images in order to obtain a test case with a known ground truth.

3.1   Artificial Intensity Transformations

Assuming that the pixel intensities are normalized between 0 and 1, a tone transfer function is any function $t : [0,1] \rightarrow [0,1]$ that is simply applied in a pixel-by-pixel basis; i.e., given an image $f(r)$, the intensity-transformed image is $g(r) = t(f(r))$. For these tests, a grayscale 512×512 version of the well-known Lena image has been used (see Fig. 1a). The artificial tone transfer functions under study are the gamma correction function and quantization, which are defined as follows:

Fig. 1. Results of applying different tone transfer functions to the Lena image: (a) original image, (b) gamma correction with $\gamma = 1/2$, (c) gamma correction with $\gamma = 2$, (d) quantization at 8 levels, (e) quantization at 2 levels.

Gamma Correction. Gamma correction is often used to compensate for the different sensitivities of image acquiring devices, or to compress or expand the dynamic range of an image so it is suitable for a given output device. It is defined by the tone transfer function $t(x) = x^\gamma$ for $\gamma > 0$, where $\gamma = 1$ represents a linear transfer. Here we perform tests with $\gamma = 2^k$ for $k = -3, \ldots, 3$. Figure 1b and c show the result of applying gamma correction to the Lena image with $\gamma = 1/2$ and $\gamma = 2$, respectively.

Quantization. Quantization reduces the number of gray levels, and therefore the intensity resolution. It is implemented by the transfer function $t(x) = \lfloor qx \rfloor / q$, where q is the number of quantization levels. Here we have tested for $q = 256$ (no quantization), 128, 64, 32, 16, 8, 4 and 2. Note that, unlike gamma correction, this transfer function is not invertible. Figure 1d and e show the result of applying quantization to the Lena image with 8 and 2 levels, respectively.

3.2   Real Illumination Changes

Further tests have been performed with the image sets described below with different illumination or acquisition conditions. The original images are already aligned and a known random rigid transformation, composed of rotation and translation, is applied to one of the images in order to generate a test case. All of these are shown in Fig. 2. Again, 100 random test cases were generated for each image pair.

Outdoor Images. This is an outdoor scene of a backyard obtained at three different illumination conditions: daylight, dusk and with a flash. The flash picture was taken during dusk and it introduces shadows and specular reflections not present in the other images. All images have been cropped and re-sampled to 512 × 512 pixels. Grayscale versions of the images were used for registration.

Indoor Images. An indoor scene of a living/dining room obtained at night under six different illumination conditions: back lights, front lights, no lights, exterior lights, all lights and flash. Different lights produce shadows and very inhomogeneous illumination in different regions; also note that the window in the back produces different reflections depending on the lighting conditions. All images were cropped and re-sampled to 512 × 512 pixels. Grayscale versions of the images were used for registration.

Multispectral Images. Two sets of 512 × 512 multispectral images were used in this test. Each set consists of 31 images of the same scene displaying the reflectance at different wavelengths from 400 nm to 700 nm at 10 nm steps. The sets are labeled as "Balloon" and "Jelly beans", respectively (for details, please see [14]). For this test, we have performed the registration of the first image of each set (400 nm) with all the 31 images from the same set in order to evaluate the performance of the registration with respect to the difference in wavelengths between both images to be aligned. Figure 2 shows an sRGB composition of the images, plus the 400 nm and 700 nm components.

3.3   Evaluation Measures

Quantitative evaluation of the rigid registration algorithm is achieved by computing the True Mean Relative Error (TRME) [2], which is defined as follows:

$$\text{TRME} = \frac{1}{4} \left[\left| \frac{(d_x - \hat{d}_x)}{d_x} \right| + \left| \frac{(d_y - \hat{d}_y)}{d_y} \right| + \left| \frac{(\phi - \hat{\phi})}{\phi} \right| + \left| \frac{(s - \hat{s})}{s} \right| \right] \times 100\,\%, \quad (7)$$

where (d_x, d_y, ϕ, s) are the parameters of the true transformation (ground truth) and $(\hat{d}_x, \hat{d}_y, \hat{\phi}, \hat{s})$ are the parameters of the estimated transformation. A registration was considered successful if the corresponding TRME was below 10 %.

One hundred random rigid transformations were generated using the following parameters: translation parameters d_x and d_y between -64 and 64 pixels and rotation angle ϕ between $-30°$ and $30°$. The true scaling factor is always set to $s = 1$, since rigid transformations do not include a change of scale; however, the registration algorithm described above does estimate a scaling factor between the images, which may be different than 1, and may contribute negatively to the TRME. In order to avoid numerical instability in the TRME computation, the d_x, d_y and ϕ parameters were forced to be greater or equal than 1.0 (with ϕ given in degrees). For each image pair, the 100 known transformations were applied to one of the images to obtain 100 test cases to be registered. After

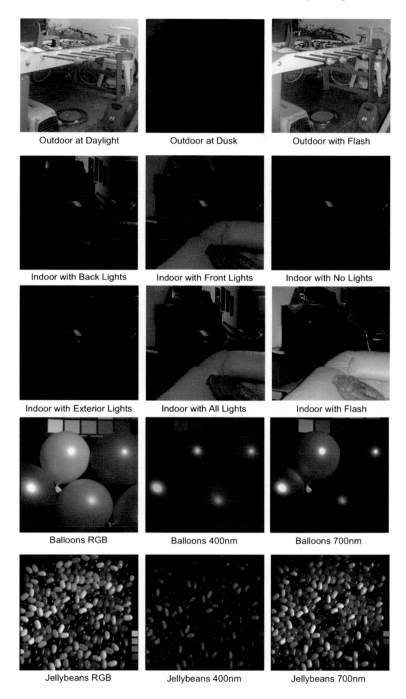

Fig. 2. Real image sets with scenes under different illumination or acquisition conditions.

registering each case, the TRME was estimated for each case, and the median TRME (across the 100 cases) was computed; the median was chosen because unsuccessful registrations may have an arbitrarily large TRME which could bias the average. We also obtained the *success rate* for each image pair, which is defined as the percentage of successfully registered cases; that is, the percentage of cases for which the TRME was below 10 %.

4   Results and Discussion

Gamma Correction. Registration results for gamma correction are shown in Fig. 3a for gamma values between 2^{-3} and 2^3 (the horizontal axis is in base-2 logarithmic scale). Both the Success Rate and the Median TRME are shown together. Peak performance is expected for $\gamma = 1$, where no correction is actually applied. Acceptable performance (success rate >80 %) is achieved for all tested gamma values except for $\gamma = 8$, which yields a success rate of 79 %. Also, median TRME is below 1 % for all gamma values except again for $\gamma = 8$, which has a TRME of 1.29 %. Therefore, the phase correlation method seems to be fairly robust to gamma correction.

Quantization. Fig. 3b shows the results for quantization varying the number of bits per pixel (which corresponds to 2^b intensity levels, where b is the bits per pixel). Peak performance is expected for $b = 8$ (256 levels), where no quantization takes place. Acceptable performance is obtained with as few as 2 bits per pixel (4 levels), although a competitive 78 % success rate is obtained even with 2 quantization levels. In all cases, median TRME is below 1 %. Again, this suggests that the phase correlation method is fairly robust to quantization.

Outdoor Images. Table 1 shows the results (Success rate and median TRME) for image pairs from the Outdoor set. Peak performance is expected when the two images to be aligned have the same illumination conditions, corresponding to the diagonal of Table 1. The registration between Daylight and Dusk images is highly successful (success rate between 89 % and 92 %), but registering any of those two images against the Flash image turned out to be more difficult (success rate between 36 % and 66 %). Note that the changes in illumination between the three images are very severe but the phase correlation method still manages to align them in most cases.

Indoor Images. Table 2 shows the results for the Indoor images. The performance is similar to the Outdoor set, with very good results when both images to be aligned have the same illumination conditions (diagonal of Table 2). When the images have different illumination, results are highly variable with success rates that range from 0 % (No Lights vs. Flash) to 91 % (All Lights vs. Back Lights). Once more, the worst results are obtained when registering the Flash image with any of the others.

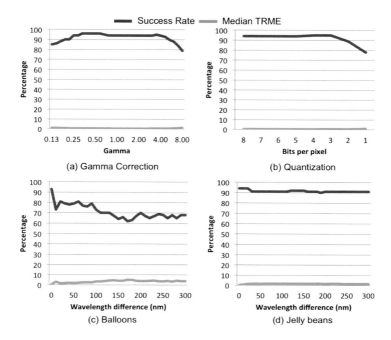

Fig. 3. Rigid registration results for image pairs with different intensity transformations: (a) results with gamma correction for gamma values between 0.125 and 8.0, (b) results with quantization from 8 to 1 bits per pixel, (c) results for the Balloon multispectral images, (d) results for the Jellybean multispectral images.

Table 1. Results from the rigid registration of image pairs from the Outdoor set.

	Daylight		Dusk		Flash	
	SR	TRME	SR	TRME	SR	TRME
Daylight	100	0.38	92	0.91	66	3.12
Dusk	89	0.73	100	0.35	47	14.69
Flash	59	3.73	36	83.26	99	0.38

Multispectral Images. Results for multispectral images are shown in Fig. 3c and d. The reference image corresponds to a wavelength of 400 nm, while the transformed image had wavelengths between 400 nm and 700 nm. The horizontal axis represents the difference in wavelength between the images to be aligned; therefore, peak performance is expected when this difference is zero. The Balloons images turned out to be more difficult to register than the Jellybeans set, with a success rate between 62 % and 93 %. The Jellybeans images, on the other hand, produced very good results, with a success rate above 90 % regardless of the wavelength difference. Note that there does not seem to be a steady decrease of the registration performance with respect to the wavelength difference, nor

Table 2. Results from the rigid registration of image pairs from the Indoor set.

	Back lights		Front lights		No lights		Ext. lights		All lights		Flash	
	SR	TRME	SR	TRME	SR	TRME	SR	TRME	SR	TRME	SR	TRME
Back lights	95	0.48	33	49.06	35	42.6	32	37.61	85	1.04	8	112.77
Front lights	20	56.9	92	0.59	45	17.07	28	60.29	76	1.59	52	7.66
No lights	22	74.3	45	21.27	95	0.43	77	2.67	15	148.38	0	250.51
Ext. lights	29	42.25	17	80.03	78	2.23	99	0.35	55	8.19	1	325.04
All lights	91	0.85	75	1.79	27	73.92	63	3.21	100	0.44	29	61.93
Flash	8	85.25	40	35.01	1	206.13	1	316.48	20	87.33	97	0.53

the performance falls abruptly at some point, as in the case of gamma correction and quantization. This suggests that, for multispectral images, the performance of the phase correlation method may depend more on the scene itself than the fact that the images were acquired at different spectral bands.

5   Conclusions

The phase correlation method has been extensively used for rigid image registration. In this work, a series of tests using artificially-generated transformations were performed to assess the applicability and robustness of phase correlation to image pairs with different intensity and/or illumination conditions. From these tests, we conclude that the phase correlation method is robust to some artificial tone-transfer functions such as gamma correction and quantization, and may also be applied in the registration of multi-spectral images. However, it seems to be more sensitive to changes in illumination and lighting conditions, particularly in the case of images acquired with a flash, although it can still produce acceptable results in many cases.

There are other cases where two images to be registered may show differences in intensities and/or illumination; for instance, multi-modal medical images acquired from different types of sensors, or other non-linear tone transfer functions such as saturation or encryption. Unfortunately, due to length restrictions it was not possible to include them all in this article. An extended version of this work is currently being developed, which will include other types of images, transformations with scalings, and a comparison against other state-of-the-art algorithms.

Acknowledgements. This work was supported by CONACyT grant 154623.

References

1. Alba, A., Vigueras-Gomez, J.F., Arce-Santana, E.R., Aguilar-Ponce, R.M.: Phase correlation with sub-pixel accuracy: a comparative study in 1D and 2D. Computer Vision and Image Understanding (2015)

2. Arce-Santana, E., Alba, A.: Image registration using markov random coefficient and geometric transformation fields. Pattern Recogn. **42**(8), 1660–1671 (2009)
3. Arce-Santana, E.R., Campos-Delgado, D.U., Alba, A.: Affine image registration guided by particle filter. IET Image Process. **6**(5), 455–462 (2012)
4. Argyriou, V., Vlachos, T.: A study of sub-pixel motion estimation using phase correlation. In: BMVC, pp. 387–396. Citeseer (2006)
5. Kuglin, C.D., Hines, D.C.: The Phase correlation image alignment method. In: Proceedings of the IEEE International Conference on Cybernetics and Society, pp. 163–165 (1975)
6. Maes, F., Collignon, A., Vadermeulen, D., Marchal, G., Suetens, P.: Multimodality image registration by maximization of mutual information. IEEE Trans. Med. Image **16**(2), 187–198 (1997)
7. Nakajima, H., Kobayashi, K., Higuchi, T.: A fingerprint matching algorithm using phase-only correlation. IEICE Trans. Fundam. Electron. Commun. Comput. Sci. **87**(3), 682–691 (2004)
8. Nelder, J.A., Mead, R.: A simplex method for function minimization. Comput. J. **7**, 308–313 (1965)
9. Reddy, B.S., Chatterji, B.N.: An FFT-based technique for translation, rotation, and scale-invariant image registration. IEEE Trans. Image Process. **5**(8), 1266–1271 (1996)
10. Szeliski, R.: Image alignment and stitching: a tutorial. Found. Trends Comput. Graph. Vis. **2**(1), 1–104 (2006)
11. Tong, X., Ye, Z., Xu, Y., Liu, S., Li, L., Xie, H., Li, T.: A novel subpixel phase correlation method using singular value decomposition and unified random sample consensus. IEEE Trans. Geosci. Remote Sensing, **53**(8), 4143–4156 (2015)
12. Vera, E., Torres, S.: Subpixel accuracy analysis of phase correlation registration methods applied to aliased imagery. In: Proceedings of the European Signal Processing Conference (EUSIPC0 2008), Lausanne, Switzerland (2008)
13. Wells III, W.M., Viola, P.A., Atsumi, H., Nakajima, S., Kikinis, R.: Multi-modal volume registration by maximization of mutual information. Med. Image Anal. **1**, 35–51 (1996)
14. Yasuma, F., Mitsunaga, T., Iso, D., Nayar, S.: Generalized assorted pixel camera: post-capture control of resolution, dynamic range and spectrum. Technical report, November 2008

Multi-modal Computer Vision for the Detection of Multi-scale Crowd Physical Motions and Behavior in Confined Spaces

Zoheir Sabeur[1(✉)], Nikolaos Doulamis[2,3], Lee Middleton[1], Banafshe Arbab-Zavar[1], Gianluca Correndo[1], and Aggelos Amditis[2]

[1] Department of Electronics and Computer Science,
University of Southampton IT Innovation Centre, Southampton, UK
zas@it-innovation.soton.ac.uk
[2] Institute of Communication and Computer Systems, Athens, Greece
[3] National Technical University of Athens, Athens, Greece

Abstract. Crowd physical motion and behaviour detection during evacuation from confined spaces using computer vision is the main focus of research in the eVACUATE project. Its early foundations and development perspectives are discussed in this paper. Specifically, the main target in our development is to achieve good rates of correct detection and classification of crowd motion and behaviour in confined spaces respectively. However, the performance of the computer vision algorithms, which are put in place for the detection of crowd motion and behaviour, greatly depends on the quality, including causality, of the multi-modal observation data with ground truth. Furthermore, it is of paramount importance to take into account contextual information about the confined spaces concerned in order to confirm the type of detected behaviours. The pilot venues for crowd evacuation experimentations include: (1) Athens International Airport, Greece; (2) An underground train station in Bilbao, Spain; (3) A stadium in San Sebastian, Spain; and (4) A large cruise ship in St. Nazaire, France.

1 Introduction

The eVACUATE project investigates on dense crowd behaviour detection in various confined environments [1]. It is supported with the development of a number of specific experimental scenarios with targeted crowd behaviour of interest. These experiments are illustrated in Fig 1 below. They specifically concern a subway station, Football stadium, Cruise liner, and an International Airport. In each of the cases the crowds are dense and potentially very large. The camera angles in these environments are non-ideal since they are challenged by lighting issues. Occlusion is severe within crowds and is not uncommon, due to the camera placements, for significant portions of the crowd that are occluded by one or more individuals.

© Springer International Publishing Switzerland 2015
G. Bebis et al. (Eds.): ISVC 2015, Part I, LNCS 9474, pp. 162–173, 2015.
DOI: 10.1007/978-3-319-27857-5_15

Fig. 1. eVACUATE scenario environments: (a) Subway station (b) Football stadium (c) Cruise liner (d) International Airport

1.1   Meaning of "Crowd"

The design and subsequent validation of crowd motions and behaviour analyses require deeper insights into the most important features of pedestrian (human) crowds who are viewed as a living complex system. In general terms, a crowd can be considered to be a collection of loosely coordinated individuals, who share a common and temporarily bound interest. This could cover spectators and moving people.

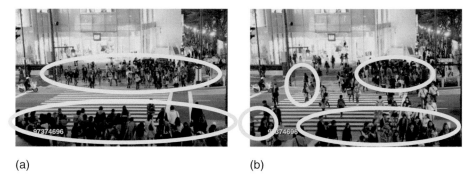

(a)   (b)

Fig. 2.   Pedestrian crossing with dynamic motions of crowds (a) t = 0 s (b) t = 24 s

From a practical sense there are nuances in the simple definition presented above. Figure 2 shows two images of a pedestrian crossing at two different times. Initially, there are two distinct crowds each of which desire to cross to the other side of their original positions. This is shown by the two large red ellipses under Fig. 2(a). After a length of time, the situation of these crowds evolves interestingly. There are number of various possibilities of crowd groupings as shown in the red ellipses under Fig. 2(b). Figure 2a and b show that the notion of crowd is a dynamic concept that may change in time through the split and merger of groups of people. The overall intention of each individual in the crowd when crossing the road may have not changed in the two images, but it became much clearer though the course of time. Their distribution in space evolves interestingly in time, and as a result, a multi-scale approach of crowd detection was adopted in our investigation in order to advance our understanding of crowd descriptions, dynamic motions and behaviour using computer vision. The multi-scale approach of crowd descriptions is listed below: (*a*) *Micro-scale description*: Pedestrians are identified individually. The state of each of such individual is delivered by position and velocity; (*b*) *Meso-scale description*: The micro-scale description of pedestrians is still identified by position and velocity but it is represented statistically through a distribution function. Such views of the crowd at meso-scale really focusses on the grouping of individuals within the crowd; and (*c*) *Macro-scale description*: The crowd is considered as a continuum body. Furthermore, it is described with average quantities (observables) such as spatial density, momentum, kinetic energy and collectiveness. These quantities are space and time dependent and conceptually local averages of the micro-scale properties of the crowd. The crowd descriptions at multiple scales are important to put in place in our investigation of crowd motions with multiple levels of spatial granularity and deeper crowd motions detection using computer vision. This will be the best way forward to understand crowd behaviour, when attempting to couple it with context modelling, fusion and reasoning. [2].

2   Multi-modal Crowd Motion and Behaviour Detection

As we have previously stated, computer vision is used together with context modelling and reasoning for achieving deep understanding of crowd motions and behaviour. The purpose of this is to really advance the safety of existing methods concerning *active crowd evacuation routes* from confined spaces in crisis situations. Crowd contextual behavior distinguishes three main components (Fig. 3). They include: *Behaviours; Effects; and Features.* These operate at the so-called *micro-scale or macro-scale* domains [2].

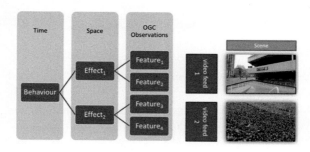

Fig. 3.   Crowd behavior model components.

Features: These are a set of constraints over observable sensory properties in a scene within a venue of interest and manifesting a particular effect. Sensory data are captured from visible, thermal and hyper-spectral imaging devices. Optical Cameras are triggered on visible spectra and thus they can be used to identify visually detectable objects (humans) of interest. Thermal imaging allows efficient determination of people from the background even under highly dynamic scenes, illumination, occlusions or content alterations since it exploits humans' temperature compared with the environment. Finally, hyper-spectral imaging triggers different spectral bands with respect to the material properties of an object, yielding an improvement of the detection accuracy. *Effects*: are the tangible (or intangible) results of the act of performing some behavior. In this model's interpretation effects are the direct causes of assuming a given behavior is performed by the agents monitored around the venue. *Behaviors*: are described as templates of human interaction and are described in the model as system assumptions to check against the available data. Each behavior causes a set of expected effects on different part of a venue at different times.

Our early analysis faces two different scales: T*he micro and the macro-scale. Micro-scale Behavior Recognition*: The analysis consists of understanding individuals or group interactions within the crowd; group formation, split or merger of individuals; and event detection for groups. The usual approach to solving these processes consists of object-detection, object-tracking, object-classification and behavior-recognition [3] mainly detecting motion characteristics. That may represent the overall state of crowd behavior in case of relatively less dense conditions [4–6]. *Macro-scale Crowd Behavior Recognition:* Crowds from a distance exhibit a variety of high level fluid flow motions, which include both steady and turbulent flows. These flows are motions between large scale groups acting together within a crowd [7] and are usually discriminated into two classes [8]. The first class is to use techniques to isolate individuals, track them and estimate the crowd density and typical motions, while the second class models the crowd motion using fluid flow theory [9].

2.1   Information Fusion from Multi-sensors

In this paper, features are extracted by processing optical, thermal and hyper-spectral imaging data. The visible sensory data are analyzed to extract the foreground from the background and then to estimate crowd properties for each detected seed. These algorithms, however, cannot be directly applied in case of thermal imaging analysis since thermal data present (a) low signal to noise ratio (noisy data) and (b) contain almost continuous pixel values (objects' temperature).

On the other hand, the recent advantages in sensory technology has emerged the development of frame hyper-spectral sensors with the capability of acquiring hypercubes at video rates. Hyper-spectral sensors offer enhanced discrimination capabilities for the characterization of subtle spectral features and important object reflectance properties. To this end, a generic framework is designed, developed and validated for multiple object tracking in hyper-spectral video sequences. The background estimation is efficiently addressed through advanced scale-space filtering and dimensionality reduction. The object recognition task exploits certain spectral and geometric features which were associated with rule-based classification.

All the information is fused together to increase the reliability of micro-scale analysis. Then, unsupervised clustering is supported for organizing the low-level information at higher scales of hierarchy, providing input paradigms for object behavior recognition.

2.2   Crowd Features Measurement

Spatial crowd density is among the parameters that can be used to measure and represent crowd seeds. However, the question of "how to measure a crowd" can have many different answers. It may depend on the type of information required and the level of granularity which is of interest. As a result, the inner workings and state of a crowd of humans has been investigated. A set of features are defined for the crowd. These are purposely chosen with an aim to characterize the state and the type of crowds. Further, the crowd is assumed homogeneous. In this, while the micro-scale motions within the crowd are observed, the defined properties and features describe the crowd as a whole (macro-scale).

Different physical analogies and modelling approaches have been used in crowd and traffic modelling. A physical model imposes a hypothetical structure that is controlled by a set of parameters. Thus, in order to fit each model, one sets out to find these parameters. Some of the more popular modelling analogies in this domain include: *cellular automata* [10, 11], *social force model* [12, 13] *and fluid flow mechanics* [14, 15]. However, the modelling analogies from statistical mechanics for gases thermodynamic theories have been considered in this work.

Thermodynamics is concerned with heat and temperature and their relation to energy and work. Furthermore, the thermodynamics which defines the macroscopic properties of gases can be elegantly derived from statistical mechanics fundamental theories, which postulate the gas molecular behavior.

It is this concept which directly links the mechanics of the constituents with the overall macroscopic properties of the fluid which are of great interest in this work. In this sense, statistical mechanics based theories that derive the macroscopic features are adopted for measuring crowds, simply from their microscopic constituents. These are represented by individual people within the crowd.

Holistic/macro features in a way that would enable the description and differentiations between different kinds of crowds are defined. These also shall differentiate between different states of crowds.

As will be further discussed in this section, three parameters are defined. These are: *structure; energy; and translation*. A three dimensional crowd space will map every crowd to a point in the structure-energy-translation space. The aim is to achieve a good separation between different types of crowd in this space.

In our model, a force which keeps the individual members of the crowd together is assumed. The strength of connections between the members of the crowd will be referred to as '*structure*'. Irrespective of the strength of connections, the crowd may be in an excited state (high energy) or a calm state (low energy); this feature of the crowd will be called '*energy*'. It is also possible to imagine that the whole crowd may travel in space; this is referred to as '*translation*'. The notion of structure described here bears notable similarities with the concept of entropy of different states of matter, while the energy resembles the internal energy described in statistical mechanics. In thermody-namics, entropy, S, is a measure of disorder:

$$S = -K \sum_i p_i \, lnp_i$$

Where, for a classic system with a discrete set of microstates, p_i is the probability of occurrence for microstate i and K is the Boltzmann constant. One hypothesize that a pattern is formed in the crowd if each individual is bounded by the same pattern inde-pendently. Although the independence assumption may be too strong, one proposes that when analyzing the crowd formation through a few correlated frames, the available information is not sufficient to infer complex dynamics between the members of the crowd. Thus a simpler model which can exhibit similar outcomes is more plausible.

Since the locations of individuals are independent of one another, the joint entropy of the crowd, $S\left(X_1, \ldots, X_{N_p}\right)$ simplifies to:

$$S\left(X_1, \ldots, X_{N_p}\right) = \sum_{k=1}^{N_p} S\left(X_k\right)$$

Where, N_p is the average population of the crowd in N_f frames. Let $n_{i,j}$ be the number of times that individual j has been observed in bin l_i in N_f frames. The probability of selecting this bin, l_i, by individual j is:

$$P\left(x_j = l_i\right) = \frac{n_{i,j}}{N_f}$$

Given that the location of individuals is considered as independent, an estimate of the probability of selecting bin l_i, $P\left(x = l_i\right)$, can be given by:

$$P\left(x = l_i\right) = \frac{\sum_{k=1}^{N_p} P\left(x_k = l_i\right)}{N_p} = \frac{n_i}{N_f N_p} = p_i$$

n_i, is the sum of all density counts at bin l_i in N_f frames. In order to calculate entropy, the internal position of each pedestrian within the crowd x_i is required. If the crowd is stationary, then the observed position, x_o, is equal to the internal position ($x_i = x_o$ if $v_f = 0$). However, if the crowd is moving with a flow velocity v_f, then the change in internal position in a time step dt can be calculated as:

$$dx_i = dx_o - v_f dt$$

Entropy is then normalized with respect to the crowd population and the extent of the crowd spatial area. The Internal energy is computed as sum of the microscopic kinetic energy of internal motion of the individuals in the crowd.

3   The Multi-modal Crowd Motion Detection Algorithms

3.1   Visual Imaging Algorithms

A useful metric by which crowds can be analyzed is the spatial density. This can provide regions where there are too many people (above some threshold) which may lead to the compromise of their safety in case of an emergency and evacuation. Furthermore, it can be used to inform safety studies of regions and lines, together with emerging trends at various spatial scales in the crowd [16]. The approach outlined in this work first finds an estimate of the numbers of people in the confined environment and then via projective geometrical techniques, a measure of density is computed. The individual steps are shown in Fig. 4 below. The individual steps will now be briefly elaborated.

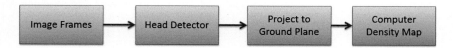

Fig. 4.   Flow of processing steps for density map generation.

The work described here is underpinned by the use of a calibrated camera. Calibration is the process of estimating internal camera parameters and the cameras position in the world. The internal parameters are computed via a process where a known metric geometry (such as a chessboard) is moved within the cameras field of view and planar homographies can be computed. This approach is generally attributed to Zhang [17]. If it is not possible to use this approach (for example the footage is provided not captured), then architectural features in the environment may be used in place of the chessboard. In our case, features within the image such as banisters and flooring tiles which conform to very specific sizes were used. Checking the calibration can be performed by back-projection after the fact. The external parameters of the camera are computed by picking an origin and computing the rotation and translation of the "Camera Centre".

A "Person Detector" in crowds is a difficult problem to overcome, due to the significant amount of occlusion bother between people and the environment. For this reason, a typical person detector will not work. As a result, a custom head and shoulder detector was purposely built in this study. We followed the approach of Dalal and Triggs [18] and built a model based on Histogram of Gradients. In our case, images from synthetic and real imagery were to achieve it. The use of synthetic imagery allows the generation of a larger dataset than one would achieve in the field [19].

Once the heads are found, they are projected to a ground 2D plane via a previously computed homography. This homography uses the calibrated camera to compute points of a known height within the image. By computing 8 such points: 4 on the ground plane and 4 at a known height, a homography can be processed. This is used to map points in the image from a fixed height to the ground plane under the assumption that all people are of similar heights. In reality, this assumption is fine as most people in a population are normally distributed about a population mean. Furthermore any associated errors can be removed further up the chain by subsequent processing.

The final step in the process is to compute a density map based upon the projected location for people in the ground plane. Specifically, kernel density estimation based approach is employed in the computation. For every projected head, a Gaussian is projected into the image which has a variance that is based on an estimate of the expected individual personal space. In reality, this value is slightly greater than the average size of a person and the person's size with a small neighborhood can be employed in such case. Because the density map is created in a coordinate space relative to the world origin it is straightforward to further project over the image of interest. Figure 5 shows an example of the processing chain in action on a synthetic sequence. Figure 5(a) shows the original frame with crowd. Figure 5(b) shows the heads of the people labelled and the associated density map.

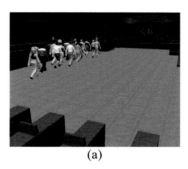 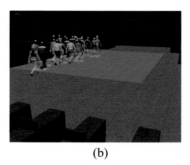

(a)   (b)

Fig. 5. An example of the processing chain in action on a synthetic sequence. (a) Original data, (b) head of people labelled and the respective associate map.

3.2   Thermal Imaging Algorithms

Thermal imagery presents some properties that are not present in RGB data. For instance, thermal detection can be used for both day and night scenarios. Therefore, they are a prime candidate for a persistent (24-7) video system for surveillance and monitoring and thus it is a very important toolkit for an active evacuation routing. In thermal sensors, a person has a different radiation compared to the background (higher or lower). Therefore, foreground extraction is more reliable in thermal imagery rather than in RGB cameras.

Micro-Scale Crowd Analysis: Foreground extraction is performed through background modelling by taking into consideration the particular properties of the thermal imagery. In particular, our method exploits a highly adjustable mixture model where the structure and parameters can be directly estimated from the data distribution, allowing dynamic model adaptation to uncontrolled and changing environments. A variational inference framework is adopted as a learning paradigm for the model parameters. Variational inference associates the functional structure of the model with real data distributions as obtained from the infrared images. The performance of background modeling for the thermal data of Athens airport is illustrated in Fig. 6. Adaptive algorithms can be also incorporated similar to the work of [20].

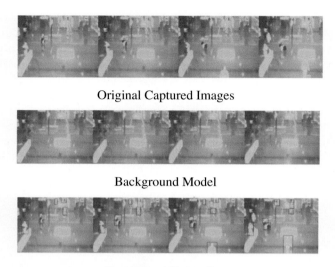

Original Captured Images

Background Model

Foreground Extraction

Fig. 6. The performance of detecting individuals from thermal imagery using the adjustable mixture modelling.

Crowd Measurements: In the framework of this paper, we use a dense optical flow technique, namely Farneback's algorithm. Farneback algorithm uses Polynomial Expansion to approximate the neighbors of a pixel. The Expansion could be seen as a quadratic equation with matrices and vectors as variable and coefficients. This dense optical flow analysis produces a displacement field from two successive video frames. Each displacement vector in the field is estimated by solving a minimization problem subject to constraints derived from the polynomial expansion. The error during minimization is the weighted sum of differences in the pixel neighborhood between the images. Image Pyramids is used to detect large displacements and Gaussian filter to smooth out the neighboring displacements. Other measurements addressing 3D dimension can be incorporated like the ones presented in [21].

Figure 7 visually presents the performance of Farneback's method when applied on thermal data captured from the Athens airport. The first column presents the original

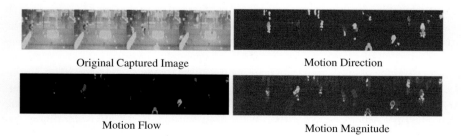

Original Captured Image   Motion Direction

Motion Flow   Motion Magnitude

Fig. 7. Performance of Farneback's method.

captured frames and motion flows which show the presence of dense optical flow in the scene, this is the fusion of motion intensity and motion direction. The second column presents the direction of motion, where different colors represent different directions, and finally the last set of images represent the intensity of the motion. The blue color corresponds to no motion, while the red color represents motion of high intensity.

4   Results and Discussions

The datasets used to evaluate the multi-modal crowd motion detection algorithms have been obtained within the early research experimentations in eVACUATE [1]. These concern the design of an active evaluation routing in the premises of Athens International Airport. The experiment has been conducted at the satellite terminal of the airport at which we have installed optical, thermal and hyper-spectral imaging sensors for capturing different types of crowd evacuation activities. To map the detected objects and their respective properties from camera view field onto 2D terminal plan, ortho-photos of the area are exploited following geometric transformations. Figure 8 presents the detected seeds (humans) as being seen by the thermal sensor using the methodology of the algorithm described in Sect. 3.2. Figure 8 also shows the ortho-mapping of the detected seeds onto the terminal plan. Velocities and directions of the crowd have also been extracted through the Farneback's method (see Sect. 3.2) and projected onto the terminal map.

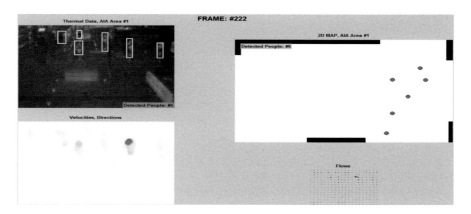

Fig. 8. Ortho-mapping of the detected individuals and their crowd properties from thermal sensors onto Athens international airport terminal plan.

Figure 9(a) groups crowd properties to extract spatial effects, which is the pre-requisite step towards crowd behavior recognition. We depict results at different number of clusters to indicate the sensitivity of the algorithm as the number of clusters increases. Figure 9(b) presents the clustering outcomes assuming spatial coherency among the crowd properties clusters. In both cases, grouping is performed by applying the spectral clustering algorithm.

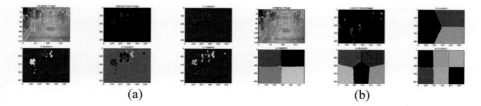

(a)   (b)

Fig. 9. Crowd properties grouping for spatial effects extraction

5   Future Development

In this early research work, the multi-scale conceptualization of crowds' description in confined spaces and the multi-modal computer vision algorithms for the detection of their motions has been discussed. These constitute the foundations for the detection of crowd physical motion and behavior at multiple scales. This will lead to further analyses of crowd measurements, their further interpretations with reasoning in context of the confined spaces in which crowds are evacuated. The contextual reasoning on crowd behaviour is currently ongoing and will be finalized and validated across all pilots using further experimentations in a subsequent paper.

References

1. A holistic scenario –independent, situation-awareness and guidance System for sustaining the Active Evacuation Route for large Crowds, European Integrated Project, eVACUATE. http://www.evacuate.eu/. Accessed 22 October 2015
2. Correndo, G., Arbab-Zavar, B., Zlatev, Z., Sabeur, Z.A.: Context ontology modelling for improving situation awareness and crowd evacuation from confined spaces. In: Denzer, R., Argent, R.M., Schimak, G., Hřebíček, J. (eds.) ISESS 2015. IFIP AICT, vol. 448, pp. 407–416. Springer, Heidelberg (2015)
3. Brostow, G.J., Cipolla, R.: Unsupervised Bayesian detection of independent motion in crowds. In: 2006 IEEE Computer Society Conference on Computer Vision and Pattern Recognition, vol. 1, pp. 594–601 (2006)
4. Chang, M.-C., Krahnstoever, N., Lim, S., Yu, T.: Group level activity recognition in crowded environments across multiple cameras. In: 2010 Seventh IEEE International Conference on Advanced Video and Signal Based Surveillance (AVSS), pp. 56–63 (2010)
5. Ge, W., Collins, R.T., Ruback, R.B.: Vision-based analysis of small groups in pedestrian crowds. IEEE Trans. Pattern Anal. Mach. Intell. **34**(5), 1003–1016 (2012)
6. Lalos, C., Voulodimos, A., Doulamis, A., Varvarigou, T.: Efficient tracking using a robust motion estimation technique. Multimed. Tools Appl. **69**(2), 277–292 (2012)
7. Coscia, V., Canavesio, C.: First-order macroscopic modelling of human crowd dynamics. Math. Models Methods Appl. Sci. **18**(supp01), 1217–1247 (2008)
8. Marana, A.N., Velastin, S.A., Costa, L.F., Lotufo, R.A.: Automatic estimation of crowd density using texture. Saf. Sci. **28**(3), 165–175 (1998)

9. Pellegrini, S., Ess, A., Schindler, K., Van Gool, L.: You'll never walk alone: modeling social behavior for multi-target tracking. In: 2009 IEEE 12th International Conference on Computer Vision, pp. 261–268 (2009)
10. Burstedde, C., Klauck, K., Schadschneider, A., Zittartz, J.: Simulation of pedestrian dynamics using a two-dimensional cellular automaton. Phys. Stat. Mech. Its Appl. **295**(3–4), 507–525 (2001)
11. Kirchner, A., Nishinari, K., Schadschneider, A.: Friction effects and clogging in a cellular automaton model for pedestrian dynamics. Phys. Rev. E **67**(5), 056122 (2003)
12. Helbing, D., Molnár, P.: Social force model for pedestrian dynamics. Phys. Rev. E **51**(5), 4282–4286 (1995)
13. Mehran, R., Oyama, A., Shah, M.: Abnormal crowd behavior detection using social force model. In: IEEE Conference on Computer Vision and Pattern Recognition, CVPR 2009, pp. 935–942 (2009)
14. Helbing, D.: A fluid-dynamic model for the movement of pedestrians. Complex Syst. **6**, 391–415 (1992)
15. Moore, B.E., Ali, S., Mehran, R., Shah, M.: Visual crowd surveillance through a hydrodynamics lens. Commun. ACM **54**(12), 64–73 (2011)
16. Doulamis, A.: Event-driven video adaptation: a powerful tool for industrial video supervision. Multimed. Tools Appl. **69**(2), 339–358 (2012)
17. Zhang, Z.: A flexible new technique for camera calibration. IEEE Trans. Pattern Anal. Mach. Intell. **22**(11), 1330–1334 (2000)
18. Dalal, N., Triggs, B.: Histograms of oriented gradients for human detection. In: IEEE Computer Society Conference on Computer Vision and Pattern Recognition, CVPR 2005, vol. 1, pp. 886–893 (2005)
19. Shotton, J., Sharp, T., Kipman, A., Fitzgibbon, A., Finocchio, M., Blake, A., Cook, M., Moore, R.: Real-time human pose recognition in parts from single depth images. Commun. ACM **56**(1), 116–124 (2013)
20. Doulamis, A.: Dynamic tracking re-adjustment: a method for automatic tracking recovery in complex visual environments. Multimed. Tools Appl. **50**(1), 49–73 (2009)
21. Makantasis, K., Doulamis, A.: 3D measures computed in monocular camera system and SVM-based classifier for humans fall detection. TMC Acad. J. **7**(2), 1–14 (2012)

HMM Based Evaluation of Physical Therapy Movements Using Kinect Tracking

Carlos Palma[1]([✉]), Augusto Salazar[1,2], and Francisco Vargas[1]

[1] Grupo SISTEMIC, Facultad de Ingenierías, Universidad de Antioquia UdeA,
Calle 70 No. 52 - 21, Medellín, Colombia
{carlos.palma,augusto.salazar,jesus.vargas}@udea.edu.co
[2] Grupo de Investigación AEyCC, Facultad de Ingenierías,
Instituto Tecnológico Metropolitano ITM,
Carrera 21 No. 54 - 10, Medellín, Colombia

Abstract. Recognition of human activities in videos has experienced considerable changes with the introduction of cost-effective technology that allows for the tracking of individual body parts. This has led to the development of numerous tele-health applications that aim to help patients in their recovery process. Most of these systems are based on techniques to measure the degree of similarity of time series, together with thresholds to evaluate whether the movement satisfies the specification. This means that sequences similar enough to a template, but containing deviations from the correct form, may be considered correct, and thus the quality of movement incorrectly assessed. In this paper we propose the use of Hidden Markov Models as novelty detectors to evaluate the quality of movement in human beings. The results show the potential of this approach in detecting the sequences that deviate from normality for a wide range of activities common in physical therapy and rehabilitation.

1   Introduction

Assessing the quality of movement in humans has multiple applications, especially in the fields of sports-training and healthcare. Recently, thanks to the development of cost-effective technology capable of giving information about the location of relevant body parts, several tele-health applications have been developed that use these devices as tools to improve rehabilitation [1,2]. Most of these applications make use of the coordinate information as input to a system based on finite state machines or as input to a recognition module that makes use of the dynamic time warping algorithm. Finite State Machines-based systems depend heavily on the nature of the conditions used to transition between states, and these conditions are usually expressed as thresholds [3,4]. This means that outlier values associated with instants in which the tracking of the selected body parts is uncertain lead the system to believe that an error has occurred.

Hidden Markov Models (HMM) have proven to be valuable tools when modeling sequential processes, and have been widely used on the field of human

© Springer International Publishing Switzerland 2015
G. Bebis et al. (Eds.): ISVC 2015, Part I, LNCS 9474, pp. 174–183, 2015.
DOI: 10.1007/978-3-319-27857-5_16

action recognition [5,6], due to the sequential nature of the data obtained when observing a specific movement, which can be thought of as going through a series of states. Most applications focus on the use of such models to perform multi-class classification of gestures, based on the probabilities calculated when an instance of a movement is observed. Some other statistical tools have been used to detect abnormal movements, by means of generating a probability and comparing it to a threshold.

This paper presents a method to detect deviations from normal repetitions in activities common in physical therapy by using Hidden Markov Models. Each repetition of a movement is divided into two phases, because most movements for therapy consist of a gesture and a return to the initial position, and sequences of incorrect repetitions are analyzed by using the probability that the models assign to them to check the ability of the model to determine if there was an error.

The rest of the paper is organized as follows: Sect. 2 reviews works that have proposed algorithms to detect when a human being is executing a movement according to a certain specification. Section 3 presents the calculations performed to train several HMM models for detecting a correct movement. Section 4 presents the experiments performed to test the approach, Sect. 5 presents and discusses the results and Sect. 6 presents the conclusions.

2   Related Work

Most proposals in the field of tele-health based on the tracking offered by the Kinect device make use of Finite State Machines or Dynamic Time Warping (DTW) [7] to determine if the performed movement is close to the specification by the expert. An example of the first approach is Velloso et al. [8], who developed a system to help evaluate movements in a physical therapy context, by recording ten repetitions of a movement and then obtaining a model for the activity. Then a direct comparison is used between such models and the observed values to determine whether the activity has been correctly executed. The second approach has been used by Su et al. [9] who proposed recording a single instance of an activity under supervision and using DTW to compare sequences of coordinates of the hands. The degree of similarity is then determined by using fuzzy membership functions, one for the DTW result and one for the speed of the movement. Cuellar et al. [10] developed a system that aligns a repetition of an activity with a template recorded by a therapist, obtaining a score that is the average of scores for each frame of the sequence after aligning them with DTW, the result becomes the input of an evaluation function which is a Gaussian Function or a Gaussian Bell Function, used to assign a score to the movement.

The use of classification methodologies has also been studied as a means to detect mistakes, this is the case of Staab [11], who uses Support Vector Machines with different kernels to recognize common mistakes for the case of three physical activities. The main downside of this approach is the fact that it becomes unfeasible to train classes for all the different kinds of errors that can appear while performing a certain activity, as stated in [12].

Alternatives based on the use of statistical models to determine deviations from normality have also been proposed, Paiement et al. [13] use such a model to detect abnormalities in a sequence of movements of a person moving on stairs, and determine empirically the threshold to consider an instance of observations as normal.

Outlier recognition based on HMM is applied to fault detection in antennae by Smyth et al. [14]. Yan et al. [15] propose a methodology based on the Wavelet transform and HMM to detect outliers and test it on data coming from depth measurements. Zhu et al. [16] use a modified HMM model to detect faults in industrial processes. The advantages of a system based on HMM to detect abnormal movement sequences are that it can be trained only with repetitions of movements considered to be normal, that it can model sequential data and that depending on the training sequences it can be adjusted to be more or less tolerant to deviations from normality.

3   HMM for Error Detection in Human Movements

This section shows the calculations performed to estimate the features of interest of the studied movements, which are the angles formed with the motion planes, and quantize them, in order to train three independent HMM models (one for each angle) for the limb of interest, and how the model can be customized to be more or less strict in enforcing the limits for the movements. Experimentally it was found that using more than three states for each model yielded no significant increase in the accuracy of the system to detect such deviations.

3.1   Angles of Interest

The standard in kinesiology and physical therapy is to define human movements in terms of the planes of motion, these are three planes perpendicular to each other, as shown in Fig. 1. The sensor information can be used to estimate vectors perpendicular to these planes, the first one is perpendicular to the floor and is given by the SDK, the vector perpendicular to the frontal planes is obtained taking the cross product of the vector connecting the left shoulder to the shoulder center and the left and right shoulders, respectively. Finally, the vector perpendicular to the sagittal plane is calculated by taking the cross product of the vectors perpendicular to the floor and the frontal plane.

3.2   Quantization and Training

A Hidden Markov Model is trained for specific movements of the limbs, these movements correspond to the stages of common activities during physical therapy. In order for a HMM to be trained it is necessary to generate a discrete sequence from the data provided by the sensor; this is done by estimating the angles formed with respect to the three planes of motion and then quantizing them according to Table 1. The quantization is done so that it covers the whole

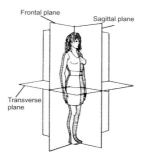

Fig. 1. Planes of motion

range of values of the angles between a vector and a plane. In the case of movements characterized by smaller angles the quantization is modified to consider intervals of five degrees.

Table 1. Quantization of angles formed with the planes of motion

Range	Symbol	Range	Symbol
$[-90\ -80]$	0	$[0\ 10]$	9
$[-80\ -70]$	1	$[10\ 20]$	10
$[-70\ -60]$	2	$[20\ 30]$	11
$[-60\ -50]$	3	$[30\ 40]$	12
$[-50\ -40]$	4	$[40\ 50]$	13
$[-40\ -30]$	5	$[50\ 60]$	14
$[-30\ -20]$	6	$[60\ 70]$	15
$[-20\ -10]$	7	$[70\ 80]$	16
$[-10\ 0]$	8	$[80\ 90]$	17

The HMM models are described in terms of a set of matrices, a matrix A to model the probability of transition between states, a matrix B to model the probability of emission of a symbol in each state and a matrix π containing the initial probabilities of being in each state. The reason why the model is suitable to be used to recognize incorrect repetitions lies in the structure of the matrix B, which after training with example sequences has a structure like the following (this is the emission matrix for the ascent of the leg during a hip flexion):

$$B = \begin{bmatrix} 0\ 0\ 0\ 0\ 0\ 0\ 0\ 0\ 0.9714\ 0.0286\ 0\ 0\ 0\ 0\ 0\ 0\ 0\ 0 \\ 0\ 0\ 0\ 0\ 0\ 0\ 0\ 0\ 0.0012\ 0.9988\ 0\ 0\ 0\ 0\ 0\ 0\ 0\ 0 \\ 0\ 0\ 0\ 0\ 0\ 0\ 0\ 0\ 0.0224\ 0.9776\ 0\ 0\ 0\ 0\ 0\ 0\ 0\ 0 \end{bmatrix}. \tag{1}$$

Each row of the matrix B describes the probability that one of the symbols of Table 1 is emitted in each of the states considered. b_{ij} is the probability that

178   C. Palma et al.

in state i the symbol j is emitted. In the case shown it is clear that most symbols are not emitted for the specific phase of the activity. Three states were selected to model the movement. As an example, the symbol 12 has a zero probability of being emitted in any of the three states, because all the elements of the twelfth column of matrix B are zero,and this heavily penalizes the probability of observing a sequence containing such a symbol. Given the sequential nature of the data used, the model also penalizes an anomalous order of symbols, this is because the model trained is a left-right HMM, as can be seen from the structure of the State Transition Matrix:

$$A = \begin{bmatrix} 0.9050 & 0.0950 & 0.0000 \\ 0.0031 & 0.9919 & 0.0050 \\ 0.0000 & 0.0706 & 0.9294 \end{bmatrix} . \qquad (2)$$

Where a_{ij} is the probability of transition from state i to state j. The structure shows that it is much more likely to remain in the present state or transition to the next state than to go back to previous states.

The model can be modified manually so that the presence of a single outlier does not affect detecting the movement as occurring according to the specification. This is achieved by modifying the structure of matrix B and recalculating the probabilities for the sequences. Figure 2 shows a synthetic error sequence containing a single outlier for the sagittal angle during a Hip Flexion movement.

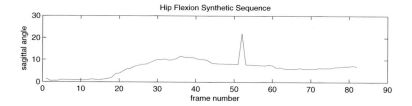

Fig. 2. Modified sequence containing an outlier

Calculation of the probability of observing this sequence yields a numerical error result on Matlab, owing to the fact that it is an extremely small number, but upon modifying matrix B to make it probable to emit 12 as a symbol like this

$$B = \begin{bmatrix} 0 & 0 & 0 & 0 & 0 & 0 & 0 & 0 & 0.9714 & 0.0286 & 0 & 0 & 0 & 0 & 0 & 0 & 0 & 0 \\ 0 & 0 & 0 & 0 & 0 & 0 & 0 & 0 & 0.0012 & 0.9988 & 0 & 0 & 0 & 0 & 0 & 0 & 0 & 0 \\ 0 & 0 & 0 & 0 & 0 & 0 & 0 & 0 & 0.0224 & 0.8 & 0.1776 & 0 & 0 & 0 & 0 & 0 & 0 & 0 \end{bmatrix} . \qquad (3)$$

It is found that the logarithm of the probability of the synthetic sequence is −27.8073. By taking all the sequences considered to be correct and calculating their probabilities according to the trained HMM, an interval can be constructed of two standard deviations around the mean, for this particular case the interval found is $[-28.9661\ 8.7130]$. This means that the manually modified model is robust to the presence of a few outliers, in a scenario where a threshold detection would have failed. Introducing two outliers, however, yields a numerical error, meaning that the number is extremely small.

The identification of mistakes can be done by analyzing the symbols emitted in the erroneous repetitions and comparing them with the symbols emitted in the training repetitions. Symbols that are greater or smaller than the ones present in the training repetitions indicate angles that are either too big or too small. This allows for feedback expressed in terms of the angular quantities, at the end of the recognition of each phase of the activities.

3.3   Implementation

The evaluation system was implemented on C# using the Accord.net library to evaluate the quantized sequences [17]. Once the final stage has been reached the sequence of angles is processed using the forward algorithm to determine the probability that it was generated by the model. A probability interval is determined for each activity, so that any probability value less than the inferior limit is considered anomalous and the repetition of the activity mistaken.

4   Experimental Setup

Experiments were conducted in order to validate the ability of the proposed approach to detect deviations from normality on sequences of movements. For each of the ten activities considered, two errors were taken into account, which consist mainly in deviations from one of the movement planes. Ten people were asked to perform ten repetitions of each of the two different kind of errors for each of the activities, resulting in 100 erroneous sequences for each type of error.

For abduction movements the errors considered were deviations from the frontal plane. *Error* 1 consists in extending the limb of interest away from the frontal plane towards the front of the body. *Error* 2 consists in deviating from the frontal plane towards the back of the body.

For extension and flexion movements deviations from the sagittal plane were considered. *Error* 1 consists in the limb deviating from the sagittal plane getting away from the body, while *Error* 2 consists in the limb deviating from the sagittal plane moving towards the body.

For external and internal rotations deviations from the transversal plane were considered. *Error* 1 consists in the limb moving under the transversal plane, while *Error* 2 consists in the limb moving in a trajectory above such plane.

In order to train the models, a database of 14 people performing the 10 physical activities was gathered, they were instructed to perform the movement according to the specifications of an expert in physical therapy, three times. This means that a total of 42 repetitions of the movements of interest are taken into account for training. Some repetitions had to be excluded in the cases they were

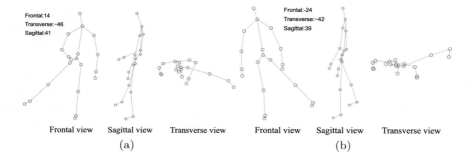

Fig. 3. Examples of incorrect movement sequences. (a) Erroneous performance consisting in moving the limb towards the front of the body. (b) Erroneous performance consisting in moving the limb towards the back of the body

very noisy. All subjects were standing at 1.80 m in front of the sensor. The sensor was standing on a table 70 cm above the floor (Fig. 3).

In order to test the validity of the proposed approach, the data of the erroneous repetitions was segmented and separated into two phases for each activity, and evaluated using the Baum-Welch Algorithm [18] on Matlab, with a model trained using the correct repetitions.

5   Results

The results are summarized in Table 2, and show the ability of the proposed algorithm to detect deviations for each of the activities. A sequence is considered erroneous when its probability of appearance does not lie in an interval constructed by taking two standard deviations from the mean of the probability of the correct repetitions as evaluated by the model.

Table 2 shows the percentage of incorrect repetitions detected by the trained models for each of the characteristics of interest. Each row shows the percentage for the frontal, transversal and sagittal models, and for each of the errors considered, grouped according to the phase of the movements that was considered. The first phase of the movement is usually a movement increasing the distance between the limb of interest and the rest of the body, while the second phase is a return to the initial position.

It was observed that the proposed method works well in the case of shoulder abduction for both kinds of deviations. The movements for which deviations from normality where detected less accurately were the extension ones, specially shoulder and elbow extension for the initial phase of the activity. In general, Error 2 was more difficult to identify for extension and flexion exercises, possibly owing to the magnitude of the deviation from normality in the sagittal plane.

Figure 4 shows two examples of erroneous movements for which the performance of the method is significantly different, in the first movement subjects were asked to deviate the movement of the limb of interest in the frontal plane.

Table 2. Successful error recognition percentage for the activities of interest in the phases of the movements.

Activity	Ascent						Descent					
	Error 1 [%]			Error 2 [%]			Error 1 [%]			Error 2 [%]		
Shoulder Abd	90.0	92.9	21.4	34.4	96.7	26.7	77.1	75.7	82.9	52.2	78.9	92.2
Hip Abd	32.5	62.5	41.3	36.7	30.4	8.9	22.5	60.0	41.2	10.1	21.5	7.6
Shoulder Ext	14.8	38.63	64.8	45.0	3.0	7.0	25.0	36.4	76.1	51.0	1.0	9.0
Hip Ext	78.7	40.4	71.9	72.0	30.0	25.0	21.3	28.9	95.5	37.0	24.0	57.0
Elbow Ext	17.9	35.8	54.7	26.0	57.3	21.9	63.1	28.4	87.4	79.2	32.3	98.9
Shoulder Flex	17.0	54.0	68.0	10.3	63.9	77.3	33.0	72.0	25.0	30.9	78.3	47.4
Hip Flex	12.0	7.0	99.0	27.3	16.2	29.3	0	7.0	97.0	22.2	11.1	28.3
Elbow Flex	89.0	89.0	99.0	90.0	81.0	25.0	19.0	46.0	78.0	18.0	78.0	55.0
Internal Rot	68.2	77.3	32.9	43.3	75.3	39.2	48.9	73.9	27.3	39.2	77.3	24.7
External Rot	16.0	88.0	25.0	17.0	68.0	38.0	30.0	91.0	43.0	38.0	38.0	59.0

In the second case the deviation was present with respect to the sagittal plane. The first deviation was always detected while the second one was not. This is due to the fact that the angles calculated with respect to the sagittal plane were found to be much more similar to the ones observed for the incorrect repetitions in the database.

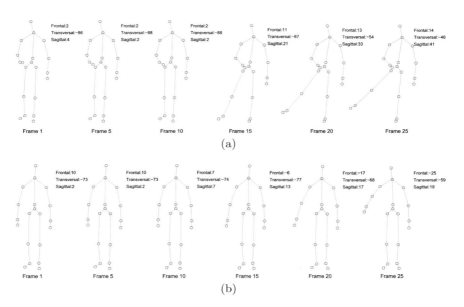

Fig. 4. Examples of incorrect movement sequences. (a) Erroneous sequence correctly identified by the algorithm. (b) Erroneous sequence not identified by the algorithm

6   Conclusion

This paper presents a method for the recognition of deviations from normal performance in physical therapy scenarios by means of the use of Hidden Markov Models. A wide variety of movements common in exercise routines and common deviations from these movements were taken into account to test the effectiveness of the method.

Experiments show that it is possible to detect deviations from normality in selected physical therapy movements by using Hidden Markov Models trained with correct repetitions of the activities.

Due to the sequential nature of the coordinate data produced by the sensor, it is possible to train a model for each phase of the movements considered, allowing for detection of deviations both at the beginning and the end of the activity.

HMM based models for the movements presented have an advantage over threshold detection given the possibility of modifying the structure of the model to make it more robust to single outliers in the sequences of interest.

It remains interesting to improve the recognition of mistakes in the cases where there was reduced performance, either by selecting different characteristics or by using a more elaborate statistical model, like a Conditional Random Field.

Acknowledgment. This work was funded by Ruta N (Regalías de la Nación), número del convenio: 512C-2013. Código SUI (Viceinvestigaciones UdeA): 20139080.

References

1. Ravi, A.: Automatic gesture recognition and tracking system for physiotherapy. In: Electrical Engineering and Computer Sciences University of California at Berkeley (2013)
2. Kayama, H., et al.: Efficacy of an exercise game based on kinect in improving physical performances of fall risk factors in community-dwelling older adults. Games Health: Res. Dev. Clin. Appl. **2**(4), 247–252 (2013)
3. Zhao, W., et al.: Rule based realtime motion assessment for rehabilitation exercises. In: IEEE Symposium on Computational Intelligence in Healthcare and e-health (CICARE 2014), pp. 133–140. IEEE (2014)
4. Zhao, W., et al.: A feasibility study of using a single kinect sensor for rehabilitation exercises monitoring: a rule based approach. In: IEEE Symposium on Computational Intelligence in Healthcare and e-health (CICARE 2014), pp. 1–8. IEEE (2014)
5. Song, Y., et al.: A kinect based gesture recognition algorithm using GMM and HMM. In: 6th International Conference on Biomedical Engineering and Informatics (BMEI 2013), pp. 750–754. IEEE (2013)
6. Godoy, V., et al.: An HMM-based gesture recognition method trained on few samples. In: IEEE 26th International Conference on Tools with Artificial Intelligence (ICTAI 2014), pp. 640–646. IEEE (2014)

7. Sakoe, H., Chiba, S.: Dynamic programming algorithm optimization for spoken word recognition. IEEE Trans. Acoust. Speech Signal Process. **26**(1), 43–49 (1978)
8. Velloso, E., Bulling, A., Gellersen, H.: MotionMA: motion modelling and analysis by demonstration. In: Proceedings of the SIGCHI Conference on Human Factors in Computing Systems, pp. 1309–1318. ACM (2013)
9. Su, C., Huang, J., Huang, S., et al.: Ensuring home-based rehabilitation exercise by using kinect and fuzzified dynamic time warping algorithm. In: Proceedings of the Asia Pacific Industrial Engineering and Management Systems Conference, pp. 884–895 (2012)
10. Cuellar, M.P., Ros, M., Martin-Bautista, M.J., Le Borgne, Y., Bontempi, G.: An approach for the evaluation of human activities in physical therapy scenarios. In: Agüero, R., Zinner, T., Goleva, R., Timm-Giel, A., Tran-Gia, P. (eds.) MONAMI 2014. LNICST, vol. 141, pp. 401–414. Springer, Heidelberg (2015)
11. Staab, R.: Recognizing specific errors in human physical exercise performance with Microsoft Kinect (2014)
12. Velloso, E., et al.: Qualitative activity recognition of weight lifting exercises. In: Proceedings of the 4th Augmented Human International Conference, pp. 116–123. ACM (2013)
13. Paiement, A., et al.: Online quality assessment of human movement from skeleton data. Computing **27**(1), 153–166 (2009)
14. Smyth, P.: Markov monitoring with unknown states. IEEE J. Sel. Areas Commun. **12**(9), 1600–1612 (1994)
15. Yan, Z., Chi, D., Deng, C.: An outlier detection method with wavelet HMM for UUV prediction following (2013)
16. Zhu, J., Ge, Z., Song, Z.: HMM-driven robust probabilistic principal component analyzer for dynamic process fault classification. IEEE Trans. Industr. Electron. **62**(6), 3814–3821 (2015)
17. Souza, C.R.: The Accord.NET Framework, Decemeber 2014. http:// accord-framework.net
18. Rabiner, L.R.: A tutorial on hidden Markov models and selected applications in speech recognition. In: Proceedings of the IEEE 77.2, pp. 257–286 (1989)

Segmentation

Segmentation of Partially Overlapping Nanoparticles Using Concave Points

Sahar Zafari[1]([⊠]), Tuomas Eerola[1], Jouni Sampo[2],
Heikki Kälviäinen[1,3], and Heikki Haario[2]

[1] Machine Vision and Pattern Recognition Laboratory,
School of Engineering Science, Lappeenranta University of Technology,
Lappeenranta, Finland
{tuomas.eerola,sahar.zafari}@lut.fi
[2] Mathematics Laboratory, School of Engineering Science,
Lappeenranta University of Technology, Lappeenranta, Finland
{Jouni.Sampo,Heikki.Haario}@lut.fi
[3] School of Information Technology, Monash University Malaysia, Selangor, Malaysia
heikki.kalviainen@monash.edu

Abstract. This paper presents a novel method for the segmentation
of partially overlapping nanoparticles with a convex shape in silhouette
images. The proposed method involves two main steps: contour evidence
extraction and contour estimation. Contour evidence extraction starts
with contour segmentation where contour segments are recovered from
a binarized image by detecting concave points. After this, contour segments which belong to the same object are grouped by utilizing properties of fitted ellipses. Finally, the contour estimation is implemented
through a non-linear ellipse fitting problem in which partially observed
objects are modeled in the form of ellipse-shape objects. The experiments
on a dataset consisting of nanoparticles demonstrate that the proposed
method outperforms two current state-of-art approaches in overlapping
nanoparticles segmentation. The method relies only on edge information
and can be applied to any segmentation problems where the objects are
partially overlapping and have an approximately elliptical shape, such
as cell segmentation.

1 Introduction

Segmentation of overlapping objects aims to address the issue of representation
of multiple objects with partial views. Overlapping or occluded objects occur in
various applications, such as morphology analysis of molecular or cellular objects
in biomedical and industrial imagery where quantitative analysis of individual
objects by their size and shape is desired [1–3]. In many such applications, the
objects can often be assumed to have approximately elliptical shape. For example, the most commonly measured properties of nanoparticles are their length
and width, which can correspond to the major and minor axis of an ellipse fitted
over the particle contour [4].

© Springer International Publishing Switzerland 2015
G. Bebis et al. (Eds.): ISVC 2015, Part I, LNCS 9474, pp. 187–197, 2015.
DOI: 10.1007/978-3-319-27857-5_17

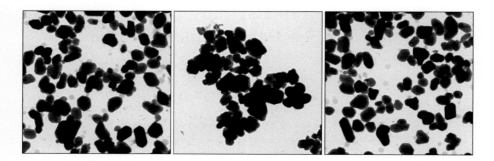

Fig. 1. Overlapping nanoparticles.

Even with rather strong shape priors, segmentation of overlapping objects remains a challenging task. Deficient information from the objects with occluded or overlapping parts introduces considerable complexity into the segmentation process. For example, in the context of contour estimation, the contours of objects intersecting with each other do not usually contain enough visible geometrical evidence, which can make contour estimation problematic and challenging. Frequently, the segmentation method has to rely purely on edges between the background and foreground, which makes the processed image essentially a silhouette image (see Fig. 1). Furthermore, the task involves simultaneous segmentation of multiple objects. A large number of objects in the image causes a large number of variations in pose, size and shape of the objects, and leads to a more complex segmentation problem.

Several approaches have been proposed for the segmentation of overlapping objects in various applications. The watershed transform is one of the commonly used approaches in overlapping cell segmentation [5–7]. However, methods based on the watershed transform suffer from a poor or inadequate initialization and may experience difficulties with segmentation of highly overlapped objects in which a strong gradient is not present.

Several approaches have resolved the segmentation of overlapping objects within the variational framework through the use of active contours [8,9]. The efficiency of the active contour based methods depends highly on the accuracy of the model initialization. Active contour based methods are also computationally heavy when the amount of objects is large.

Also morphological operations have been used for overlapping object segmentation. In [1], an automated morphology analysis coupled with a statistical model for contour inference to segment partially overlapping nanoparticles was proposed. The method is prone to under-segmentation with highly overlapped objects.

In [2], the problem of overlapping objects segmentation was approached using concave points extraction through the polygonal approximation and ellipse fitting. Although this approach is efficient for regular shaped objects, such as bubbles, objects with a shape that deviates from elliptical shape, such as nanoparticles, cause problems for the method. In [10], the problem of incomplete cells with non-elliptical shapes was addressed by proposing a method combining certain heuristics with concave point extraction. In [11], a modified generalized hough transform for recognizing partially occluded objects was proposed. The method is, however, computationally expensive if the number of objects is large.

In this paper, a novel and efficient method is proposed for the segmentation of partially overlapping nanoparticles with a convex shape. The nanoparticles are assumed to be clearly distinguishable from the background of the image and their contours form approximately elliptical shapes. The proposed method relies on two sequential steps of contour evidence extraction and contour estimation. The contour evidence extraction step is further divided into two sub-steps: contour segmentation and segment grouping. In the contour segmentation step, object contours are divided into separate contour segments. In the segment grouping step, contour evidences are built by joining the contour segments that belong to the same object. Once the contour evidence is obtained, contour estimation is performed using numerically stable direct ellipse fitting. The proposed method relies only on edge information and can be applied also to other segmentation problems where the objects are partially overlapping and have approximate elliptical shapes, such as cell segmentation.

The main contribution of this work is a novel combined model for contour evidence extraction that relies on detection of concave points through curvature analysis, the concavity test and an efficient search procedure. The proposed method is shown to outperform earlier methods in the task of overlapping nanoparticles segmentation based on detection rate and segmentation accuracy.

2   Overlapping Object Segmentation

The proposed method consist of two consecutive main step: contour evidence extraction and contour estimation. Figure 2 summarizes the method. Given a gray-scale image as input, the segmentation process starts with pre-processing to build an image silhouette and the corresponding edge map. The binarization of the image is obtained by background suppression based on the Otsu's method [12] along with morphological opening to smooth the object boundaries. The edge map is constructed using the Canny edge detector [13]. In the contour evidence extraction steps, edge points that belonged to each object are grouped using concave points and properties of fitted ellipses. Once the contour evidence has been obtained, contour estimation is carried out to infer the missing parts of the overlapping objects.

2.1   Contour Evidence Extraction

The first step of the proposed method is to extract the contour evidence containing the visible parts of the objects boundaries that can be used to inference the occluded parts of overlapped objects. The contour evidence extraction involves two separate tasks: contour segmentation and segment grouping.

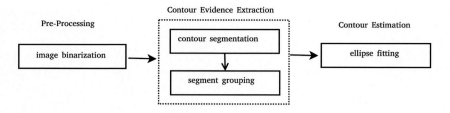

Fig. 2. Proposed method.

Contour Segmentation. A partial overlap between two or more elliptic-shape objects leads to a concave shape with concave edge points that correspond to the intersections of the object boundaries. It is a common practice to utilize these concave points to segment the contour of overlapping objects. Different methods such as polygonal approximation [2,10], curvature [14], and angle [10,15] have been applied to determine the location of concave points in the image.

In this work, after extracting the image edge by canny edge detector [13], the concave points are obtained through the detection of corner points followed by the concavity test. The corner points are detected using the modified curvature scale space (CSS) method based on curvature analysis [16]. The output of the corner detector includes the points with the maximum curvature lying on both concave and convex regions of object contours. Since being only interested in the concave points joining the contours of overlapping objects, the detected corner points are examined if they lie on concave regions.

Let us denote a detected corner point by p_i, and its two kth adjacent contour points by p_{i-k} and p_{i+k}. The corner point p_i is qualified as concave if the line connecting p_{i-k} to p_{i+k} does not reside inside the object. The obtained concave points are used to split the contours into contour segments.

Figure 3 shows an example of concave point extraction and contour segmentation.

Segment Grouping. Due to the overlap between the objects and the irregularities in the object shapes, a single object may produce multiple contour segments. Segment grouping is needed to merge all the contour segments belonging to the same object. The basic idea behind the proposed method for segment grouping is to find a group of contour segments that together form an object with elliptical shape. Segment grouping in its naive form, iterates over each pair of contour segment, examining if they can be combined. In this work, to optimize the grouping process, a limited search space is applied and the contour segment under grouping process is only examined with the neighbouring segments. Two segments are neighbour if the Euclidean distance between their center points is less than the predefine threshold value.

The contour segment grouping is carried out through the process of ellipse fitting. Given a pair of contour segments, s_i and s_j, and a function measuring the goodness of ellipse fitting, the segment s_i is grouped to s_j if the goodness of

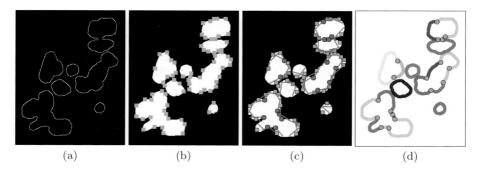

(a)	(b)	(c)	(d)

Fig. 3. Contour segmentation: (a) Edge map; (b) Corner detection by [16]; (c) Concavity test to extract concave corners (green circle) and removed convex corners (pink square); (d) Contour segmentation by concave points (the colors are used only for illustrative purpose to visualize the segmented contour by concave points) (Color figure online).

ellipse fitted to the joint segments is higher compared to the goodness of ellipses fitted to each individual contour segments separately.

The goodness of fit is described as average distance deviation (ADD) [2] which measures the discrepancy between the fitted curve and the candidate contour points. The lower value of ADD indicates higher goodness of fit and, therefore the joint rule to perform segment grouping in terms of ADD is defined as

$$
\begin{aligned}
\text{ADD}_{s_i \cup s_j} &\leq \text{ADD}_{s_i}, \\
\text{ADD}_{s_i \cup s_j} &\leq \text{ADD}_{s_j},
\end{aligned}
\tag{1}
$$

where the definition of ADD is as follows:

Given the contour segment s_i consisting of n points, $s_i = \{p_k(x_k, y_k)\}_{k=1}^{n}$, and the corresponding fitted ellipse points, $s_{f,i} = \{p_{f,k}(x_{f,k}, y_{f,k})\}_{k=1}^{n}$, ADD_{s_i} is defined as

$$
\text{ADD}_{s_i} = \frac{1}{n} \sum_{k=1}^{n} \sqrt{(x_k - x_{f,k})^2 + (y_k - y_{f,k})^2}.
\tag{2}
$$

Within the transformed coordinate system

$$
\begin{bmatrix} x'_k \\ y'_k \end{bmatrix} = \begin{bmatrix} \cos\theta & -\sin\theta \\ \sin\theta & \cos\theta \end{bmatrix} \begin{bmatrix} x_k - x_{eo} \\ y_k - y_{eo} \end{bmatrix},
\tag{3}
$$

Equation (2) can be simplified to

$$
\text{ADD}_{s_i} = \frac{1}{n} \Big[\sum_{k=1}^{n} \sqrt{x'^2_k + y'^2_k} \big(1 - \frac{1}{|D_k|}\big) \Big],
\tag{4}
$$

where D_k is given by

$$
D_k^2 = \frac{x'^2_k}{a^2} + \frac{y'^2_k}{b^2},
\tag{5}
$$

and a, b, (x_{eo}, y_{eo}) and θ are the ellipse parameters, the semi-major axis length, the semi-minor axis length, the ellipse center point, and the ellipse orientation angle with respect to x axis, respectively.

The plain ADD criterion for segment grouping often leads to undesired results if the contour points do not strictly fit to the ellipse model. In order to address this issue, additional rules are needed.

Given a pair of contour segments to be processed for grouping, the segment with longer length usually provides a more reliable clue to the object than the shorter one. Based on this assumption, a weighing scheme using the length of contour segments is added to the grouping process where the ADD of contour segment with longer length is down-weighted by the ratio of its length with respect to the total length of contour segments to be grouped. Assuming the contour segment s_i is longer than contour segment s_j, Eq. (1) is replaced by

$$\text{ADD}_{s_i \cup s_j} \leq w_i \text{ADD}_{s_i},$$
$$\text{ADD}_{s_i \cup s_j} \leq \text{ADD}_{s_j}. \tag{6}$$

where

$$w_i = \frac{l_i}{l_i + l_j},$$

and l_i and l_j are the lengths of contour segments s_i and s_j, respectively.

The contour segments in far proximity are less likely to represent a single object and should not be merged. As the result, the two contour segments whose ellipse models are at very far distance from each other should not grouped. Either, ellipse fitted to the combined contour segments should not be at far distance from the ellipses fitted to each individual contour segments. Following these conventions and being interested in grouping of close contour segments, two additional rules are applied similarly to [10].

Let us denote the centroids of the fitted ellipse for the contour segments s_i, s_j and $s_{i \cup j}$ by e_i, e_j and $e_{i \cup j}$, respectively. The contour segments s_i and s_j should not be grouped as a single segment provided that, first, the distance from the ellipse centroid of the combined contour segments $e_{i \cup j}$ to the center of its members, e_i and e_j, is larger than the preset threshold t_1:

$$d(e_i, e_{ij}) > t_1$$
$$d(e_j, e_{ij}) > t_1, \tag{7}$$

and second, the distance between their corresponding ellipse centroids is larger than the predefined threshold t_2

$$d(e_i, e_j) > t_2, \tag{8}$$

where $d(p_1, p_2)$ is the Euclidean distance between points p_1 and p_2.

The value of t_1 can be determined using the object properties [10] and is usually close to the length of the minor axis of fitted ellipses to the smallest

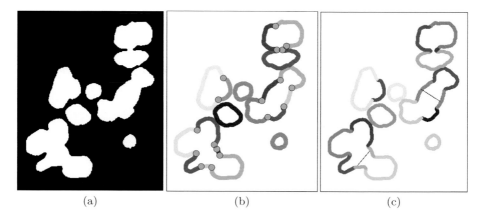

(a)   (b)   (c)

Fig. 4. Segment grouping: (a) Original binary image; (b) Contour segmentation; (c) Segment grouping (the thin gray lines are added to illustrate the grouping of non-adjacent segments).

object in the image. The value of t_2 should be set in such way that prevents the grouping of the contour segment belong to different objects or as [10] proposed 2.5 to 4 times higher than the threshold t_1. Figure 4 shows an example of segment grouping.

2.2   Contour Estimation

The last step of proposed method is the contour estimation, where, by means of the visual information produced from the previous step, the missing parts of the object contours are estimated. Ellipse fitting is a very common approach in overlapping object segmentation, especially in the medical and industrial applications.

The most efficient recent ellipse fitting methods based on shape boundary points are generally addressed through the classic least square fitting problem. In this work, the contour estimation is addressed through a stable direct least square fitting method [17] where the partially observed objects are modeled in the form of ellipse-shape objects. Figure 5 shows an example of contour estimation applied to contour evidences.

3   Experiments

3.1   Data

The experiments were carried out using a dataset consisting of nanoparticles images captured by transmission electron microscopy. In total, the dataset contains 11 images of 4008×2672 pixels. Around 200 particles were marked manually in each image by an expert. The annotations consist of manually drawn contours

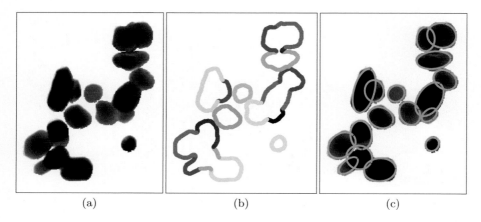

Fig. 5. Contour estimation: (a) Original image; (b) Contour evidence extraction; (c) Contour estimation.

of the objects. Since not all the objects are marked, a pre-processing step was applied to eliminate the unmarked objects from the images. It should be noted that the images consist of dark objects on a white background and, therefore, pixels outside the marked objects could be colored white without making the images considerably easier to analyze.

3.2   Results

To evaluate the method performance and to compare the methods, two specific performance measures, True Positive Rate (TPR) and Positive Predictive Value (PPV), were used:

$$TPR = \frac{TP}{TP + FN} \tag{9}$$

$$PPV = \frac{TP}{TP + FP} \tag{10}$$

where True Positive (TP) is the number of correctly segmented objects, False Positive (FP) is the number of incorrectly segmentation results and False Negative (FN) is the number of missed objects.

To decide whether the segmentation result was correct or incorrect, Jaccard Similarity coefficient (JSC) [18] was used. Given a binary map of the segmented object O_s and the ground truth particle O_g, JSC is computed as

$$JSC = \frac{O_s \cap Og}{O_s \cup Og}. \tag{11}$$

The threshold value for the ratio of overlap in the JSC was set to 0.5. The average JSC (AJSC) value was used as the third measure to evaluate the segmentation performance.

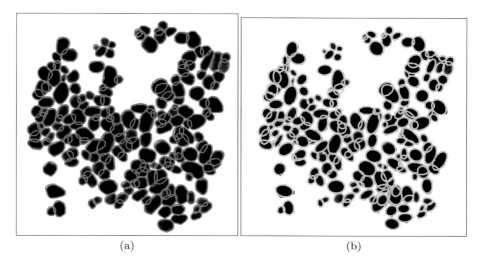

(a)   (b)

Fig. 6. An example of the proposed method segmentation result on nanoparticles dataset: (a) Ground truth; (b) Proposed method.

The method parameters, k, t_1 and t_2 were set experimentally to 10, 23 and 45, respectively. Figure 6 shows an examples of proposed method segmentation result in the nanoparticles dataset. The performance of the proposed segmentation method was compared to two existing state-of-the-art methods, Nanoparticles Segmentation (NPA) [1] and Concave-point Extraction and Contour Segmentation (CECS) [2]. The NPA and CECS methods are particularly chosen as previously applied for segmentation of overlapping convex and elliptical shape objects, respectively. The implementation made by the corresponding authors was used for NPA [1]. CECS was implemented by ourselves based on [2]. Examples of typical segmentation results are presented in Fig. 7. NPA suffers from under-segmentation while CECS tends to over-segment the objects. The proposed method neither under- or over segments the objects.

The corresponding performance statistics of the competing methods applied to the dataset are shown in Table 1. As it can be seen, the proposed method outperforms the other two with respect to the TPR and JSC and achieves a comparable performance with NPA in terms of PPV. The high JSC value of the proposed method indicates its superiority with respect to the resolved overlap ratio.

3.3   Computation Time

The proposed method was implemented in MATLAB, using a PC with a 3.20 GHz CPU and 8 GB of RAM. With the selected combination of parameters the computational time was 21 s per image, while NPA demanded 200 s and CECS 77 s. The computational time breakdown was as follows: contour segmentation 3 %, segment grouping 95 %, and ellipse fitting 2 %. However, it should

Table 1. Comparison of the performance of the proposed method for the nanoparticles dataset.

Methods	TPR[%]	PPV[%]	AJSC[%]
Proposed	**85**	84	**73**
NPA	62	**90**	58
CECS	66	73	53

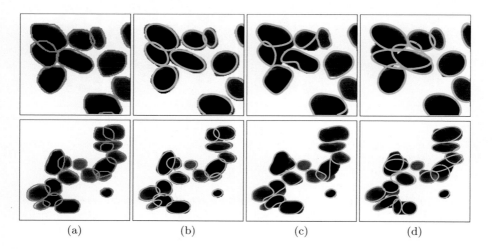

| (a) | (b) | (c) | (d) |

Fig. 7. Example segmentation results on slice of nanoparticles dataset: (a) Ground truth; (b) Proposed method; (c) NPA; (d) CECS.

be noted that the method performance was not optimized and the computation time could be improved.

4   Conclusions

This paper presented a method to segment multiple partially overlapping convex shaped nanoparticles in silhouette images using concave points and ellipse properties. The proposed method consisted of two main steps: contour evidence extraction to detected the visible part of each object and contour estimation to estimate the final objects contours. The experiments showed that the proposed method achieved high detection and segmentation accuracies and outperformed two competing methods on a dataset of nanoparticles images. The proposed method relies only on edge information and can be applied also to other segmentation problems where the objects are partially overlapping and have an approximately elliptical shape, such as cell segmentation.

References

1. Park, C., Huang, J.Z., Ji, J.X., Ding, Y.: Segmentation, inference and classification of partially overlapping nanoparticles. IEEE Trans. Pattern Anal. Mach. Intell. **35**, 669–681 (2013)
2. Zhang, W.H., Jiang, X., Liu, Y.M.: A method for recognizing overlapping elliptical bubbles in bubble image. Pattern Recogn. Lett. **33**, 1543–1548 (2012)
3. Kothari, S., Chaudry, Q., Wang, M.: Automated cell counting and cluster segmentation using concavity detection and ellipse fitting techniques. In: IEEE International Symposium on Biomedical Imaging, pp. 795–798 (2009)
4. Fisker, R., Carstensen, J., Hansen, M., Bødker, F., Mørup, S.: Estimation of nanoparticle size distributions by image analysis. J. Nanopart. Res. **2**, 267–277 (2000)
5. Shu, J., Fu, H., Qiu, G., Kaye, P., Ilyas, M.: Segmenting overlapping cell nuclei in digital histopathology images. In: 35th International Conference on Medicine and Biology Society (EMBC), pp. 5445–5448 (2013)
6. Cheng, J., Rajapakse, J.: Segmentation of clustered nuclei with shape markers and marking function. IEEE Trans. Biomed. Eng. **56**, 741–748 (2009)
7. Jung, C., Kim, C.: Segmenting clustered nuclei using h-minima transform-based marker extraction and contour parameterization. IEEE Trans. Biomed. Eng. **57**, 2600–2604 (2010)
8. Zhang, Q., Pless, R.: Segmenting multiple familiar objects under mutual occlusion. In: IEEE International Conference on Image Processing (ICIP), pp. 197–200 (2006)
9. Ali, S., Madabhushi, A.: An integrated region-, boundary-, shape-based active contour for multiple object overlap resolution in histological imagery. IEEE Trans. Med. Imaging **31**, 1448–1460 (2012)
10. Bai, X., Sun, C., Zhou, F.: Splitting touching cells based on concave points and ellipse fitting. Pattern Recogn. **42**, 2434–2446 (2009)
11. Tsai, D.M.: An improved generalized hough transform for the recognition of overlapping objects. Image Vis. Comput. **15**, 877–888 (1997)
12. Otsu, N.: A threshold selection method from gray-level histograms. Automatica **11**, 23–27 (1975)
13. Canny, J.: A computational approach to edge detection. IEEE Trans. Pattern Anal. Mach. Intell. **8**, 679–698 (1986)
14. Wu, X., Kemeny, J.: A segmentation method for multi-connected particle delineation. Proc. IEEE Workshop Appl. Computer Vis. **1992**, 240–247 (1992)
15. Wang, W.: Binary image segmentation of aggregates based on polygonal approximation and classification of concavities. Pattern Recogn. **31**, 1503–1524 (1998)
16. He, X., Yung, N.: Curvature scale space corner detector with adaptive threshold and dynamic region of support. In: Proceedings of the 17th International Conference on Pattern Recognition, ICPR 2004, vol. 2, pp. 791–794 (2004)
17. Fitzgibbon, A., Pilu, M., Fisher, R.B.: Direct least square fitting of ellipses. IEEE Trans. Pattern Anal. Mach. Intell. **21**, 476–480 (1999)
18. Choi, S.S., Cha, S.H., Tappert, C.C.: A survey of binary similarity and distance measures. J. Systemics Cybern. Informatics **8**, 43–48 (2010)

Temporally Object-Based Video Co-segmentation

Michael Ying Yang[1], Matthias Reso[2]([✉]), Jun Tang[2],
Wentong Liao[2], and Bodo Rosenhahn[2]

[1] Computer Vision Laboratory, TU Dresden, Dresden, Germany
[2] Institut für Informationsverarbeitung,
Leibniz Universität Hannover, Hannover, Germany
`reso@tnt.uni-hannover.de`

Abstract. In this paper, we propose an unsupervised video object
co-segmentation framework based on the primary object proposals to
extract the common foreground object(s) from a given video set. In
addition to the objectness attributes and motion coherence our framework exploits the temporal consistency of the object-like regions between
adjacent frames to enrich the original set of object proposals. We call
the enriched proposal sets temporal proposal streams, as they are composed of the most similar proposals from each frame augmented with
predicted proposals using temporally consistent superpixel information.
The temporal proposal streams represent all the possible region tubes
of the objects. Therefore, we formulate a graphical model to select a
proposal stream for each object in which the pairwise potentials consist
of the appearance dissimilarity between different streams in the same
video and also the similarity between the streams in different videos.
This model is suitable for single (multiple) foreground objects in two
(more) videos, which can be solved by any existing energy minimization method. We evaluate our proposed framework by comparing it to
other video co-segmentation algorithms. Our method achieves improved
performance on state-of-the-art benchmark datasets.

1 Introduction

Video object segmentation aims to group the pixels over frames of a video into
spatial-temporal coherent regions, i.e., to find those pixels belonging to the same
foreground object(s) in each frame. Most of the algorithms for video object
segmentation focus on single object scenarios in one video sequence, e.g. [1–3].

However, in practical scenarios, the video contents are much more complicated and diverse. For instance, most videos contain more than one object; some
foreground objects arise in indistinguishable motion or are surrounded by the
background which is similar in appearance to the objects. In such circumstances,
using the joint information from other videos containing the same objects, can
help us to discover the foreground objects much more precisely. This method is

---

M.Y. Yang and M. Reso—The first two authors contribute equally to this paper.

G. Bebis et al. (Eds.): ISVC 2015, Part I, LNCS 9474, pp. 198–209, 2015.
DOI: 10.1007/978-3-319-27857-5_18

known as *video co-segmentation*, firstly introduced by Rubio et al. in [4], which segments the common regions appearing over all the frames of two or more given video sequences containing the same objects.

While the results look promising, the task to distinguish and extract the foreground objects by using only the joint appearance among the videos still remains unsolved. In this paper, we propose a general framework for video co-segmentation based on the object proposals, which is a graphical model for single or multiple foreground objects in two or more videos. Our algorithm differs significantly from others mainly in two parts:

- we refine and expand the original set of object proposals from [5] by predicting them onto adjacent frames using temporal coherence information to create temporal proposal streams as well as
- formulate the selecting problem of the object proposal streams as a portable conditional random field (CRF) model.

Our first contribution exploits that the appearance and shape of objects are assumed to vary slowly over frames. We use the information of temporally consistent superpixels to create temporal proposal streams which represent a temporally consistent object-like region in the video. By the second contribution multiple foreground objects can be dealt with more easily. Our model can be solved by any existing energy minimization method. We validate our framework using two public benchmark datasets, MOViCS [6] and ObMiC [7], and compare our results with the state-of-the-art methods. An overview of our framework is illustrated in Fig. 1.

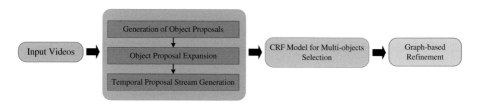

Fig. 1. The overview of our proposed framework. Firstly, given a video set, a group of primary object proposals for each video frame is generated. Next, we expand the original proposal set with the predicted ones based on temporal information from adjacent frames, and then combine the similar proposals from each frame in a video as the temporal proposal streams. Then the CRF model selects the best proposal stream for each object in each video. Finally, the segments of each stream is refined by a spatial-temporal graph model.

2   Related Work

Image Co-segmentation. The concept of co-segmentation was firstly applied to image pairs by Rother et al. in [8]. They proposed a method to discover the common object by considering the joint information from a pair of images. Vicente et al. [9] extended the method to image based object co-segmentation by using the object proposal from [5] to segment similar objects from image pairs. Moreover, the methods proposed by [10,11] dealt with multiple object classes using discriminative clustering.

Video Object Segmentation. A lot of research has been done on video object segmentation, i.e. to separate the objects from the background in videos. In contrast to the methods based on low-level features, several approaches adopted the 'objectness' measure to seek for the object-like proposals for primary video object segmentation in a single video [1,2,12]. In addition, Grundmann et al. [13] clustered a video into spatio-temporal consistent supervoxels, and Jain and Grauman [14] used the consistency of these supervoxels as a higher order potential for semi-supervised foreground segmentation.

Video Object Co-segmentation. Recently, an increasing number of methods focus on video object co-segmentation. The method proposed in [4] grouped the pixels into two levels: the higher level consists of space-time tubes, and the lower one is composed of region segments within the frames. Based on initial foreground and background estimation and dense feature extraction of the regions and tubes, they constructed a probabilistic model of the foreground and background, and iteratively refined the results and updated the model. The supervoxel based method proposed in [15] employed dense optical flow to derive the intra-video relative motion and Gaussian mixture models to characterize the common object appearance. Both methods can only deal with the videos containing a single common object. In [6], Chiu and Fritz proposed a multi-class video object co-segmentation method using a non-parametric Bayesian model to learn a global appearance model, which connected all the segments of the same object. However, it is based on low-level descriptors for grouping the foreground pixels into classes. Fu et al. [7] built a standard multi-state selection graph model (MSG) in view of the intra- and inter-video coherence. Guo et al. [16] also considered the persistence of different parts of the foreground during the video and also proposed automatic model selection while binding them together. In all of these methods, [7] achieves the best segmentation results. However, the MSG certainly assigns each node (frame) an optimal label (object proposal), which means they can not find the objects when it does not appear in the first frames. Besides, if the object is totally covered in some frames, there would be no selectable proposal to represent the object, even so the MSG still chooses one for these frames, which fulfils the lowest MRF energy. Although they used a graph-based segmentation for refinement, it is still unrecoverable in case of the wrong proposals characterized by low-level features which are similar to the object of interest. Furthermore, their graph model is constructed with fully-connected states of each node between the multiple videos, which costs lots of time in comparison between different states.

Our proposed method based on temporal proposal streams overcomes aforementioned challenges. All the streams are generated by the detected similar proposals from each frame without the limitation of starting or ending point. Similar streams are merged into a single stream via spectral clustering, even if they are not completely consecutive. Moreover, our framework is more efficient to obtain the final results, due to the fewer comparisons between the states (temporal proposal streams).

3   Proposed Method

Given a set of N videos as $\{V^1, \ldots, V^N\}$ we primarily achieve a group of object-based proposals using [5] in each frame f_t^n, $n \in N$, $t \in F_n$. These proposals p_t^n are generated by performing graph cuts based on a seed region and a learned affinity function. They are also scored from best to worst based on a ranking system. These candidates are used as input of our proposed video co-segmentation method.

3.1   Object Proposal Expansion

In order to find the object-like candidates among them, we define a score as proposed by [1] based on appearance cues and salient motion patterns relative to their surroundings:

$$A(p_{t_i}^n) = O(p_{t_i}^n) + M(p_{t_i}^n), \tag{1}$$

where the score $A(p_{t_i}^n)$ of ith proposal in frame f_t^n is constituted by the static intra-frame objectness score $O(p_{t_i}^n)$ and the dynamic inter-frame motion score $M(p_{t_i}^n)$. The objectness score $O(p_{t_i}^n)$ is the original score in the proposal-generating process from [5]. It reflects how likely the proposal $p_{t_i}^n$ is a whole object. The motion score $M(p_{t_i}^n)$, as defined in (2), measures the confidence that the proposal $p_{t_i}^n$ corresponds to a coherently moving object in the video.

$$M(p_{t_i}^n) = 1 - exp(-\frac{1}{M}\chi_{flow}^2(p_{t_i}^n, \overline{p_{t_i}^n})), \tag{2}$$

where $\overline{p_{t_i}^n}$ denotes the pixels around the proposal $p_{t_i}^n$ within a loosely fit bounding box, and $\chi_{flow}^2(p_{t_i}^n, \overline{p_{t_i}^n})$ is the χ^2-distance between L_1-normalized optical flow histograms with $\overline{M}$ denoting the mean of the χ^2-distance.

In [5] only the local information of each individual frame is considered, thereby neglecting the temporal information. Taking this into consideration, we adopt the idea of [17] to create temporally consistent superpixels (TCS) to map all the proposals onto the adjacent frames. In consequence, the TCS labels of each proposal may guide us to predict an additional successive proposal in these frame.

As illustrated in Fig. 2, each proposal of frame f_t^n is warped by selecting the superpixels with the same TCS labels on frame f_{t+1}^n. Therefore, the new predicted proposal contains the TCS labels from $p_{t_i}^n$. We refine it using graph-based image segmentation [18]. With this predicted proposal, we seek for a proposal

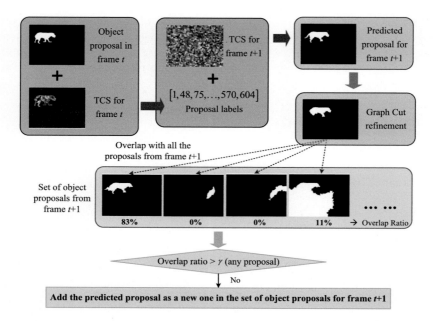

Fig. 2. Object proposal expansion procedure. For each object proposal in frame t, we predict its temporally consistent one for frame $t+1$ based on the TCS labels and overlap it with all the original existing proposals. If none of them has an overlap ratio higher than the threshold, we add the predicted proposal as an additional one in the proposal set of frame $t+1$.

in frame f_{t+1}^n which is similar to the newly predicted one. The intersection-over-union overlap ratio is defined as the judgment criteria as follows:

$$o = \frac{|Warp_{t \to t+1}(p_{t_i}^n) \cap p_{t+1_j}^n|}{|Warp_{t \to t+1}(p_{t_i}^n) \cup p_{t+1_j}^n|}. \tag{3}$$

If any proposal $p_{t+1_j}^n$ in frame f_{t+1}^n does not have an overlap ratio larger than the threshold γ (which is set to 0.7 in this paper), we use the predicted $Warp_{t \to t+1}(p_{t_i}^n)$ as an additional proposal and add it to the proposal set of frame f_{t+1}^n. In practice, this procedure is carried out in a consecutive fashion in both forward and backward direction. This ensures that any missing proposal is properly propagated onto every frame.

3.2 Temporal Proposal Streams

Based on the expanded proposals, we discover the groups of temporal proposal streams which may represent a consistent foreground object-like region in the video.

Primarily, we start to generate the streams for video V^n from its first frame f_1^n. The x most highly ranked proposals in the first frame are assigned as

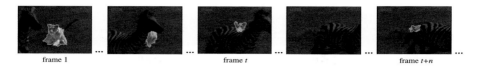

frame 1   ...   frame t ...   frame $t+n$

Fig. 3. Object occlusion occurring in a video of 'LionsAll' video set in MOViCS dataset [11]. We retain the proposal before the occlusion and use its appearance to compare with the proposals when the lion shows up again.

the beginning of the x initial proposal streams. Then we seek a similar proposal in the next frame for each of the stream with the overlap of the TCS labels, as mentioned in Sect. 3.1. The one with the highest overlap ratio will be regarded as a new member in the corresponding proposal stream. If a proper proposal can not be found in this frame, this stream ends up here; otherwise, the process moves on to next frame. Meanwhile, we also consider the x most highly ranked proposals in the following frames f_k^n. Some of them may be already connected with the existing streams, and the rest are used to start new streams. So, in practice, the set of the streams grows over frames. But with the limitation of x, it will not grow too much, because most of the highly ranked proposals of each frame should just continue the already started streams. On the other hand, this growing process helps us to find new objects which maybe does not show up in the first frame.

In some cases, the object is totally occluded in some frames and then shows up again, as shown in Fig. 3. Our aforementioned method treats it as two different streams which are supposed to represent the same object. To solve this problem, we need to bond some of the generated streams. Before the combination, the streams which span all the frames are retained unchanged, while the ones containing only one frame are abandoned. For the rest of them, we adopt the spectral clustering based on their colour appearance to group them in y clusters.

3.3   CRF Model for Multi-object Selection

Since the graphical model provides a standard framework for capturing complex dependencies among random variables, it helps us to select the most object-like temporal proposal stream for each video as the segmented object. In this paper, the problem is formulated as a graphical model in the form of a conditional random field (CRF), as illustrated in Fig. 4.

Each node represents an object in a video and the possible states comprise the corresponding temporal proposal streams. We seek a proper stream to represent the object for each node. The energy function of the graphical model is defined as:

$$E = \sum_{n=1}^{N} \sum_{k=1}^{C_n} E_{unary}(s_k^n) + \alpha_1 \cdot \sum_{n=1}^{N} \sum_{\substack{k,h=1 \\ k \neq h}}^{C_n} E_{intra}(s_k^n, s_h^n) + \alpha_2 \cdot \sum_{\substack{n,m=1 \\ n \neq m}}^{N} \sum_{k=1}^{C_n} \sum_{l=1}^{C_m} E_{inter}(s_k^n, s_l^m),$$

$$(4)$$

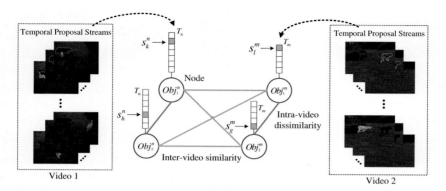

Fig. 4. Our multi-object selection graphical model for the 'Dog and Deer' video set from ObMiC dataset [7], which contains two video sequences with two objects.

where α_1 and α_2 are weighting coefficients.

The unary energy uses the aforementioned score $A(s_k^n)$ from Sect. 3.1 and the saliency score $S(s_k^n)$ of all the regions in each stream to represent its likelihood belonging to the foreground:

$$E_{unary}(s_k^n) = -log[max(\bar{A}(s_k^n), S(s_k^n))], \tag{5}$$

where $\bar{A}(s_k^n)$ is the mean score of all the proposals in stream s_k^n. Due to the irregular movements of the foreground objects, we also consider their saliency as a supplementing static cue. For the whole video, we compute the co-saliency map based on all the frames using [19] and get the saliency score from the overlap between the region and the corresponding map.

The pairwise term includes two parts, the intra- and the inter-video energies. $E_{intra}(s_k^n, s_h^n)$ is the intra-video energy between the streams s_k^n and s_h^n in video V^n, which represents the stream similarity penalty. Each object has its own stream in a video, which means a stream should not be assigned to different objects. Thus, the stream similarity penalty is described by the dissimilarity between the streams:

$$E_{intra}(s_k^n, s_h^n) = -log(D_f(s_k^n, s_h^n)), \tag{6}$$

where D_f is the low-level feature similarity between them, which is defined as:

$$D_f(s_k^n, s_h^n) = \frac{1}{M_m}\chi_f^2(s_k^n, s_h^n). \tag{7}$$

In practice, we compare all the regions of both streams and average the scores. $\chi_f^2(s_k^n, s_h^n)$ is the weighted combination of the χ^2-distances between the normalized colour histograms and the shape histograms of s_k^n and s_h^n. M_m denotes the mean value of the χ^2-distance. In our work, shape is represented by the HOG descriptor [20] within a minimum bounding box enclosing the region.

The other pairwise term $E_{inter}(s_k^n, s_l^m)$ measures the object consistency among different videos. In this graph, each stream from one video is connected to those in the other videos. We define the inter-video energy as:

$$E_{inter}(s_k^n, s_l^m) = D_f(s_k^n, s_l^m), \qquad (8)$$

in which D_f is the low-level feature similarity computed by (7).

For inference, we employ TRW-S [21] to find the approximated labelling that minimizes the energy function. Since the original object proposals generated by [5] are only roughly segmented, we refine the final results as [1] with a pixel-level spatio-temporal graph-based segmentation to achieve a better segmentation.

4   Experiments

We implement our proposed method in MATLAB and compare it against four state-of-the-art methods related to video co-segmentation: Multi-class video co-segmentation (MVC) [6], Object-based multiple foreground video co-segmentation (ObMiC) [7], Extracting primary objects by video co-segmentation (EPOVC) [22] and the latest Consistent foreground co-segmentation (CFC) [16]. For the comparison we use two state-of-the-art datasets: Multi-Object Video Co-segmentation (MOViCS) dataset [6] for single object video co-segmentation and Object-based Multiple Foreground Video Co-segmentation (ObMiC) dataset [7] for the multiple objects case. Same as in [6], the *intersection-over-union metric* (IOU), defined as $\frac{R \cap GT}{R \cup GT}$, is used as evaluation metric in this paper.

Implementation Details. For both datasets, the number of TCS in each frame of each video sequence is around 1500, which makes sure that each TCS represents a region with a proper size containing consistent appearance. The threshold γ in the propagation procedure of object proposals is defined as 0.7, which judges whether a new additional proposal should be added to the proposal set of the next frame. When we discover the temporal proposal streams for each video, we use the $x = 40$ most highly ranked proposals in the first frame to initialize the streams. In addition, the 10 most highly ranked proposals in the following frames are considered as the candidates to start new streams. After generating the streams, all the incomplete streams are grouped into 20 or 5 clusters, depending on the amount of incomplete streams. All the low-level features leveraged in the framework consist of colour information and shape information. The colour feature is computed in CIELab colour space and RGB colour space with 117 bins and the shape information from HOG descriptor is presented in 81 bins. To combine these two features, we set the weighting coefficient for the colour as 2 to increase the weight. As for the graphical model, the weighting coefficients α_1 and α_2 to balance the two pairwise potential terms is set empirically.

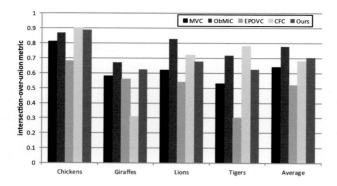

Fig. 5. The IOU metric on MOViCS dataset [6] (Color figure online).

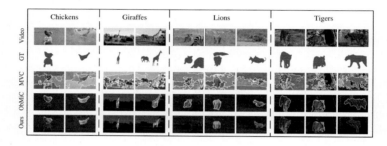

Fig. 6. Single object co-segmentation results on MOViCS dataset. First row is the sample frames of given videos; second row represents the ground truth; from the third to fifth row are the segmentation results from MVC [6], ObMiC [7] and our framework, respectively.

Evaluation on MOViCS Dataset. We test our framework on the MOViCS dataset [6], which includes four different video sets and 11 videos in total. Each video set contains one or two objects, and for five frames of each video a ground truth labelling is provided. All objects appear in the videos of this dataset in an irregular way. Although some videos comprise more than one foreground object, we only consider the object appearing in each video of the video set.

As shown in Fig. 5, our proposed framework outperforms the multi-class video co-segmentation method of [6] significantly. Using the temporal coherence between the adjacent frames improves the segmentation results. Comparing with the ObMiC from [7], we have better results in one video set. The reason for the difference in the other video sets is that they employed all object proposals in each frame as candidates for their graphical model, which chooses the proper proposal for each frame separately. This low-level method keeps more details for each proposal, but loses some temporal relevance between the proposals. Besides, the computational overhead for the fully connected graphical model is much higher as more similarities have to be evaluated. In comparison to EPOVC [22] which has a similar structure as ObMiC our method produces a higher

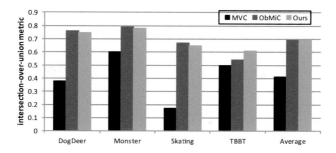

Fig. 7. The IOU metric on ObMiC dataset [7] (Color figure online).

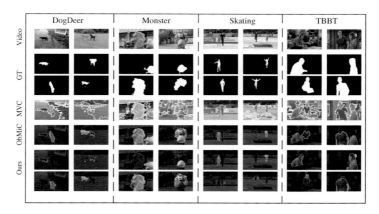

Fig. 8. Multiple objects co-segmentation results on ObMiC dataset. First row is the sample frames of given videos; second row represents the ground truth; from the third to fifth row are the segmentation results from MVC [6], ObMiC [7] and our framework, respectively.

average accuracy. They applied only the low-level feature of each proposal to build the graphical model, which is restricted by its initial configuration. Although the recently published CFC method of [16] automatically chooses a suitable model for each video set and performs well on the 'Tigers' sequence we achieve on par or better accuracy on the other three video sets. Figure 6 shows some qualitative segmentation examples.

Evaluation on ObMiC Dataset. The ObMiC dataset [7] comprises four video pairs, each containing two common foreground objects. The scenarios of these video sets are completely different and a ground truth labelling is provided for all frames.

In the first three video sets our accuracy is better than MVC but lower than ObMiC as shown in Fig. 7. But in the last video set, our accuracy is superior to theirs. The segmentation results of the **DogDeer** and **Monster** sequences in the fist and second column of Fig. 8 show that ObMiC segments the object boundaries slightly better than our method. More complicated environments can be seen in the reality scenes with human beings. In the third column, our segmentation results of the **Skating** sequence are better than MVC and comparable to ObMiC. MVC segments the bodies in many pieces and ObMiC can only find partial region of the objects, in which the colour appearance is consistent. In the last **TBBT** video set, our framework outperforms the two other methods in the intersection-over-union metric. In contrast to the ObMiC method our resulting segments comprise the clothes and the heads of the characters in this video set. From this aspect, our results are better for the woman, but yet to be satisfied for the man.

5   Conclusion

We propose a video co-segmentation framework to extract the common foreground object(s) from given video sets. The procedure consists of two key steps: based on the basic object proposals, we firstly use the temporal information between the frames to combine the consistent proposals together as temporal proposal streams; secondly, a stream for each object is selected in each video by the CRF model depending on their appearance. Our framework is not restricted in the number of objects or videos, and it outperforms most of the state-of-the-art methods in terms of accuracy with a lower computational burden on both state-of-the-art benchmark datasets for the video object co-segmentation task.

Acknowledgements. The work is partially funded by DFG (German Research Foundation) YA 351/2-1. The authors gratefully acknowledge the support.

References

1. Li, Y.J., Kim, J., Grauman, K.: Key-segments for video object segmentation. In: ICCV, pp. 1995–2002 (2011)
2. Zhang, D., Javed, O., Shah, M.: Video object segmentation through spatially accurate and temporally dense extraction of primary object regions. In: CVPR, pp. 628–635 (2013)
3. Yang, M., Rosenhahn, B.: Video segmentation with joint object and trajectory labeling. In: WACV, pp. 831–838 (2014)
4. Rubio, J.C., Serrat, J., López, A.: Video co-segmentation. In: Lee, K.M., Matsushita, Y., Rehg, J.M., Hu, Z. (eds.) ACCV 2012, Part II. LNCS, vol. 7725, pp. 13–24. Springer, Heidelberg (2013)
5. Endres, I., Hoiem, D.: Category independent object proposals. In: Daniilidis, K., Maragos, P., Paragios, N. (eds.) ECCV 2010, Part V. LNCS, vol. 6315, pp. 575–588. Springer, Heidelberg (2010)
6. Chiu, W.C., Fritz, M.: Multi-class video co-segmentation with a generative multi-video model. In: CVPR, pp. 321–328 (2013)
7. Fu, H., Xu, D., Zhang, B., Lin, S.: Object-based multiple foreground video co-segmentation. In: CVPR, pp. 3166–3173 (2014)

8. Rother, C., Minka, T., Blake, A., Kolmogorov, V.: Cosegmentation of image pairs by histogram matching-incorporating a global constraint into mrfs. In: CVPR, pp. 993–1000 (2006)
9. Vicente, S., Rother, C., Kolmogorov, V.: Object cosegmentation. In: CVPR, pp. 2217–2224 (2011)
10. Joulin, A., Bach, F., Ponce, J.: Discriminative clustering for image co-segmentation. In: CVPR, pp. 1943–1950 (2010)
11. Joulin, A., Bach, F., Ponce, J.: Multi-class cosegmentation. In: CVPR, pp. 542–549 (2012)
12. Ma, T., Latecki, L.J.: Maximum weight cliques with mutex constraints for video object segmentation. In: CVPR, pp. 670–677 (2012)
13. Grundmann, M., Kwatra, V., Han, M., Essa, I.: Efficient hierarchical graph-based video segmentation. In: CVPR, pp. 2141–2148 (2010)
14. Jain, S.D., Grauman, K.: Supervoxel-consistent foreground propagation in video. In: Fleet, D., Pajdla, T., Schiele, B., Tuytelaars, T. (eds.) ECCV 2014, Part IV. LNCS, vol. 8692, pp. 656–671. Springer, Heidelberg (2014)
15. Chen, D., Chen, H.T., Chang, L.: Video object cosegmentation. In: ACM International Conference on Multimedia, pp. 805–808 (2012)
16. Guo, J., Cheong, L.-F., Tan, R.T., Zhou, S.Z.: Consistent foreground co-segmentation. In: Cremers, D., Reid, I., Saito, H., Yang, M.-H. (eds.) ACCV 2014. LNCS, vol. 9006, pp. 241–257. Springer, Heidelberg (2015)
17. Reso, M., Jachalsky, J., Rosenhahn, B., Ostermann, J.: Temporally consistent superpixels. In: ICCV, pp. 385–392 (2013)
18. Boykov, Y., Kolmogorov, V.: An experimental comparison of min-cut/max-flow algorithms for energy minimization in vision. PAMI **26**, 1124–1137 (2004)
19. Fu, H., Cao, X., Tu, Z.: Cluster-based co-saliency detection. IEEE Trans. Image Process. **22**, 3766–3778 (2013)
20. Dalal, N., Triggs, B.: Histograms of oriented gradients for human detection. In: CVPR, pp. 886–893 (2005)
21. Kolmogorov, V.: Convergent tree-reweighted message passing for energy minimization. PAMI **28**, 1568–1583 (2006)
22. Lou, Z., Gevers, T.: Extracting primary objects by video co-segmentation. IEEE Trans. Multimedia **16**, 2110–2117 (2014)

An Efficient Non-parametric Background Modeling Technique with CUDA Heterogeneous Parallel Architecture

Brandon Wilson[(✉)] and Alireza Tavakkoli

Computation and Advanced Visualization Engineering Lab,
University of Houston–Victoria, Victoria, TX, USA
{WilsonBJ1,TavakkoliA}@uhv.edu

Abstract. Foreground detection plays an important role in many content based video processing applications. To detect moving objects in a scene, the changes inherent to the background need to be modelled. In this work we propose a non-parametric statistical background modeling technique. Moreover, the proposed modeling framework is designed to utilize Nvidia's CUDA architecture to accelerate the overall foreground detection process. We present three main contributions: (1) a novelty detection mechanism capable of building accurate statistical models for background pixels; (2) an adaptive mechanism for classifying pixels based on their respective statistical background model; and (3) the complete implementation of the proposed approach based on the Nvidia's CUDA architecture. Comparisons and both qualitative and quantitative experimental results show that the proposed work achieves considerable accuracy in detecting foreground objects, while reaching orders of magnitude speed-up compared to traditional approaches.

1 Introduction

With the increase in the availability and computational power of digital imaging devices, it is natural to think about integrating artificial intelligence models with processing of digital images and videos in order to improve their performance. However, for such applications to be efficient, there is a need for addressing a number of significant challenges. Accounting for the presence of regions that do not belong to objects of interest, global ambient illumination variations in the environment over long periods of time, and the real-time constraints inherent to live video processing applications are among such obstacles.

The detection of foreground objects in videos is an important step to facilitate fundamental computations inherent to content-based video processing applications. As such, this stage may be considered as a pre-processing component employed to detect regions of interest or to model the background within the video. The majority of the state-of-the-art foreground detection techniques utilize background subtraction; where a statistical (or analytical) model for the background of a sequence is trained. Thereafter, each pixel is tested on how well

© Springer International Publishing Switzerland 2015
G. Bebis et al. (Eds.): ISVC 2015, Part I, LNCS 9474, pp. 210–220, 2015.
DOI: 10.1007/978-3-319-27857-5_19

it fits this model to generate foreground regions of interest to be used by further processes in the pipeline.

The simplest way to model the background is to store the most common background value pixel(s) and measure their differences from the current pixel. Benezeth *et al.*[1], and Wang and Dudek [2] have used this approach as a fast mechanism to detect background objects in videos. However, this method fails to fully consider periodic, slow changes, or dynamic backgrounds.

Gaussian Mixture Model(GMM) approaches use parametric models to account for inherent changes in the background [1, 3–6]. These techniques make an assumption on the probability model of background pixel to be composed of a mixture of normal (Gaussian) distributions and aim at training the parameters of such Gaussian Mixture Models.

A background model can also be achieved with neural networks. Maddalena and Petrosino [7] introduced a self-organizing background subtraction algorithm that relies on weighted neural networks to model the background. Moreover, spatial coherence was added to the proposed scheme to remove false positive results from the foreground [8]. An approach using weightless neural networks was introduced by Gregorio and Giordano [9], capable of detecting foreground while ignoring sudden changes such as global lighting changes.

Co-occurrence Probability-based Pixel Pairs can also be used for background subtraction by using a single Gaussian to model the changes between a pixel and its group of supporting pixels. Liang and Kaneko [10] proposed improvements to this method by adding a multichannel model and limiting the rate of change of intensity values with an on-line approach. This method allows for global illumination changes and dynamic backgrounds.

Elgammal *et al.*[11] proposed a non-parametric approach to background subtraction by modeling pixel intensity through a mixture of three to five Gaussian distributions. A novel physically-based approach to background subtraction was introduced by Sedky *et al.*[12]. This method employs a dichromatic color reflectance model to compute the spectral reflectance of visible surfaces to measure the similarities between pixels and segment out the foreground. Tavakkoli *et al.*[13] proposed a recursive learning approach for training non-parametric background models adaptively.

The highest performing method to date is an approach proposed by Wang *et al.*[14]. This technique uses a mixture of flux tensor motion detection to detect foreground pixels and split Gaussian models to detect background pixels. A rule-based system is used to combine these results, and then object level classifications are made to improve these results.

As discussed above, modeling video backgrounds has been subject to intensive research over the past decades. A large number of approaches have been proposed in the literature with varying degrees of accuracy. However, one major issue still remains to be addressed in implementing an efficient background modeling framework with utility in a wide range of video processing application. This issue relates to the processing speed required for a successful and efficient background modeling architecture.

With the advent of General Purpose Graphics Processing Units (GPGPU's), new computing models are available by which massive amounts of acceleration may be achieved for traditional applications. In this paper, we propose an architecture for the decoupling of step complexity from work complexity in a non-parametric background modeling framework. This enables us to implement different aspects of the proposed background modeling computation on suitable hardware to efficiently carry out the computation to accelerate the overall process of foreground detection in videos.

Whereas the traditional background modeling and subtraction methods aim for complex multi-class learning techniques to model the background/ foreground object(s) in videos, we approach a novelty detection mechanism as the core process in our proposed framework. Moreover, the proposed novelty detection approach is designed in such a way as to facilitate the parallelization required for the acceleration in support of the overall computation. We target the heterogeneous computing model afforded by the Nvidia CUDA's Kepler architecture.

2   Methodology and Approach

This paper proposes a novel non-parametric approach for foreground detection, capable of exploiting parallelism available through Nvidia CUDA. Our proposed mechanism is comprised of three main stages and two optional post-process modules. Figure 1 shows the workflow utilized in the proposed framework. The update, training, and classification phases of the proposed framework in Fig. 1 are represented as the lateral sections. The horizontal sections represent stages of the proposed architecture as preprocess, post-process or the detection stage.

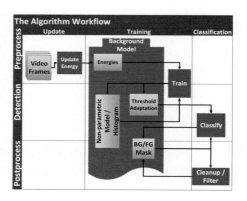

Fig. 1. Proposed framework workflow.

The first stage is an update phase, executed on every frame to update the background information based on the newly available data. This data is then analyzed in the training phase to produce a model for each background pixel in the frame. The training phase is the slowest, but only needs to be executed on a fraction of the available blocks every frame. Finally, a classification phase is executed on every frame to determine if each pixel belongs to the background or foreground using strict and loose classification criteria. Two post-process modules are also employed in real-time to refine the foreground detection results and remove undesirable artifacts and noise.

2.1   Modeling the Background

The background model begins with the update phase, which stores a sum of every time a pixel value is encountered since the previous training phase and utilized as a measure of the energy contained within the pixel's history. In the training phase, these sums are normalized to build a statistical non-parametric model for the pixel's distribution in the form of its histogram based solely on the newly available data. This histogram is linearly interpolated with the previous histogram based on a given interpolation rate. This interpolation rate will act as a learning rate to adapt the model to gradual as well as sudden changes. Therefore, it determines how fast the model learns new changes to the background.

The training phase uses per pixel histograms to calculate all significant components and the corresponding strict and loose classification ranges. These background model components are employed by the classification phase to quickly determine if a pixel is strictly, loosely, or not at all in the foreground. Two postprocess

---

Algorithm 1. The Train Kernel

Data: pixel energy
Result: strict / loose decision maps
find relevant pixel column and row;
if *this is a valid pixel* **then**
  foreach *classification criteria* **do**
    normalize the pixel energies;
    $\sum(t) = \alpha(t)\beta(t) + (1 - \alpha(t))\beta(t-1)$;
    reset the pixel energies;
    get the global min and max;
    lerp(Min, Max);
    classify each pixel value and store the results in the strict map / loose map;

---

modules are further applied on the final foreground mask to cleanup the foreground regions and remove erroneously detected regions due to noise. These postprocess modules include a large median filter to remove small misclassified regions due to salt-and-pepper noise, and a morphological closing filter used to remove any holes in detected objects.

The proposed approach is geared towards optimizing speed. Training is the slowest phase of the framework, but it is not required to be performed on all pixels, every frame. Our proposed training algorithm will be performed on the required number of pixels in such a way to ensure real-time speed while maintaining the background model accuracy. This allows the algorithm to be used in virtually any situation. In real-time video applications that require foreground detection as a pre-process, the training module will be performed only on critical pixels whose models require training. This will allow the overall framework to meet the frame rate requirements for live, real-time applications. In video compression or other non-real-time applications, all the data can be trained at first, requiring only a single training phase for the entire video.

2.2   The GPU Acceleration

There are five kernels that are executed on the GPU. All of these kernels execute once per frame for each pixel, with the exception of the training kernel,

---

Algorithm 2. Classify Kernel

Data: strict and loose decision maps
Result: strict and loose foreground mask
get the relevant pixel column and row;
if *this is a valid pixel* **then**
    foreach *classification criteria* **do**
        if *this pixel value's strict decision map is foreground* **then**
            classify this pixel as strictly foreground;

    if *this pixel was not classified as strictly foreground* **then**
        classify this pixel as strictly background;

    foreach *classification criteria* **do**
        if *this pixel value's loose decision map is foreground* **then**
            classify this pixel as loosely foreground;

    if *this pixel was not classified as loosely foreground* **then**
        classify this pixel as loosely background;

---

and the update and training kernels must execute an additional time for each classification criteria. For this implementation we used luminance, chroma blue, and chroma red as classification criteria.

For every frame, the first kernel to execute is the update kernel. This kernel keeps track of the temporal measure for each pixel. This represents the pixel energy, which is used by the training phase to update each pixel's histogram. Since this kernel is executed every frame, it is optimized for speed. In addition, the data is subsampled from 256 bins down to 64 bins to reduce memory requirements. The only data this kernel needs to execute is the raw frame, as shown in Algorithm 3.

The training phase of our proposed algorithm is shown in Algorithm 1. In order to allow for varying speeds of execution, this kernel is only executed on a predetermined number of pixels each frame. Due to the limited amount of times this kernel is executed each frame, most of the costly operations execute on this kernel.

The training kernel first normalizes the pixel energy stored since the previous trained kernel. This normalized data forms a histogram of the data not included in the current histogram. These two histograms are linearly interpolated by a learning rate, and then normalized to form the current frame's histogram. The pixel energy is reset, and the global maximum and minimum of the resulting histogram are stored. The strict and loose adaptive thresholds are calculated by linearly interpolating the global maximum and minimum by a predetermined strict and loose ratio, respectively. These adaptive thresholds are tested against the largest value across a range of neighboring bins, and the outcomes across the range of all possible pixel values, represented by the number of sub-sampled bins. These values are then stored as the strict and loose decision map.

The classify kernel handles the segmentation of the background from the foreground. This process, shown in Algorithm 2 requires the raw frame and its corresponding strict and loose decision maps. The current frame's pixel value for each criteria is classified by looking up the predetermined decision stored in this pixel's strict and loose decision maps. If any of the criteria is considered foreground, then the entire pixel is considered foreground, which results in the strict and loose foreground masks. Similar to the update kernel, this kernel is optimized for speed at the cost of GPU memory due to its necessity each frame.

Algorithm 3. The Update Kernel

Data: raw frame
Result: pixel energy
get the relevant pixel column and row;
if *this is a valid pixel* **then**
  foreach *classification criteria* **do**
    update the current pixel's
    energy value;

Algorithm 4. Cleanup Kernel

Data: rough foreground mask
Result: improved foreground
        mask
get the relevant pixel column and
row;
if *this is a valid foreground pixel*
then
  forall the *pixels in the stencil*
  do
    if *this pixel is foreground*
    then
      update the foreground
      count;
    else
      update the
      background count;

  if *there are more background*
  pixels than foreground pixels
  then
    change this pixel to
    background;

Algorithm 5. Fill Kernel

Data: strict and loose mask
Result: final foreground mask
get the relevant pixel column and
row;
if *this is a valid strictly*
background pixel that is loosely
foreground **then**
  forall the *pixels in the stencil*
  do
    if *this pixel is strictly*
    foreground **then**
      update the foreground
      count;
    else
      update the
      background count;

  if *there are more strict*
  foreground pixels than
  background pixels **then**
    change this pixel to
    foreground;

The cleanup kernel, shown in Algorithm 4, is the first post-process and executes after the classify kernel on the strict foreground mask. For each pixel in the strict foreground mask, this post-process sums up all neighboring foreground and background pixels across a square filter. If the majority of these neighboring pixels are background, then this pixel is also considered background.

The second and final post-process, shown in Algorithm 5, is a fill operation. For each pixel that is loosely foreground but not strictly foreground, this operation sums all of the neighboring pixels across a square filter that are considered

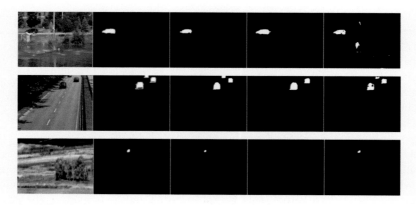

Fig. 2. Sample qualitative results comparisons from three sequences. From left: original frame, ground truth, detected foreground by our approach, the FTSG [1,14].

Table 1. Comparison of quantitative results with all changedetection.net methods

Method	Avg ranking	Re	Sp	FPR	FNR	PWC	Precision	F-Meas
FTSG [14]	1.364	0.7826	0.9930	0.0070	0.2174	1.2470	0.7750	0.7427
CwisarDH [9]	2.364	0.6816	0.9950	0.0050	0.3184	1.4159	0.7832	0.7010
Spectral-360 [12]	3.818	0.7497	0.9844	0.0156	0.2503	2.3318	0.7191	0.6932
The proposed method	6.000	0.6475	0.9900	0.0010	0.3525	2.2831	0.7356	0.6286
Bin Wang [2]	6.455	0.7103	0.9799	0.0201	0.2897	2.8450	0.7165	0.6606
KNN [3]	6.545	0.6695	0.9785	0.0215	0.3305	3.4389	0.6770	0.5985
SC SOBS [8]	8.000	0.7617	0.9551	0.0449	0.2383	5.0853	0.6083	0.6000
RMoG [6]	8.182	0.6035	0.9849	0.0151	0.3965	3.0785	0.6987	0.5799
KDE [11]	9.455	0.7391	0.9523	0.0477	0.2609	5.5292	0.5738	0.5689
CP3-online [10]	9.909	0.7286	0.9728	0.0272	0.2714	3.1668	0.5708	0.5966
Mahalanobis distance [1]	10.091	0.1678	0.9921	0.0079	0.8322	3.4981	0.7523	0.2329
GMM - Stauffer &Grimson [4]	10.182	0.6881	0.9733	0.0267	0.3119	3.8848	0.6053	0.5764
GMM - Zivkovic [3]	11.182	0.6633	0.9695	0.0305	0.3367	4.2342	0.6104	0.5674
MSST BG model [16]	11.909	0.6676	0.9544	0.0456	0.3324	5.4765	0.5509	0.5183
Euclidean distance [1]	13.909	0.6828	0.9443	0.0557	0.3172	6.5279	0.5397	0.5397

strictly foreground, and if there are mostly foreground pixels, then this pixel is switched to foreground. This is more beneficial than the morphological close operation because it avoids filling in holes that are unlikely to be foreground.

3   Experimental Results

Qualitative and Quantitative comparisons have been made with other state-of-the-art foreground detection methods shown on the www.changedetection.net website [15]. These comparisons range from 11 categories; each containing four to six videos. Each video has been scored based on seven measurements against the ground truths provided by the changedetection.net website.

Table 2. Result of the proposed approach on all sequences organized into 11 categories

Category	Re	Sp	FPR	FNR	PWC	Precision	F-Meas
Night videos	0.3268	0.9777	0.0223	0.4524	3.1623	0.3268	0.3268
Baseline	0.9115	0.9978	0.0022	0.2287	1.3726	0.9115	0.9115
Dynamic BG	0.7700	0.9981	0.0019	0.2166	0.4240	0.7700	0.7700
Bad weather	0.9509	0.9996	0.0004	0.3058	0.5123	0.9509	0.9509
PTZ	0.2529	0.9676	0.0324	0.4630	3.6992	0.2529	0.2529
Thermal	0.9352	0.9981	0.0019	0.4341	3.9745	0.9352	0.9352
Camera jitter	0.7148	0.9763	0.0237	0.2428	3.1738	0.7148	0.7148
Shadow	0.7355	0.9877	0.0123	0.2048	2.1410	0.7355	0.7355
Intermittent motion	0.8339	0.9958	0.0042	0.5698	4.6680	0.8339	0.8339
Turbulence	0.7522	0.9933	0.0067	0.2756	0.8389	0.7522	0.7522
Low Frame rate	0.9086	0.9983	0.0017	0.4835	1.1476	0.9086	0.9086

3.1 Qualitative Comparisons

We compared our results to the original frame, ground truth, and two other state-of-the-art foreground detection methods. Figure 2(a), (b) and (c) shows some of these results. Each row shows from right to left; The sequence frame, the ground truth, our proposed method's results, the FTSG method's results [14], and the Euclidean Distance method's results [1].

3.2 Quantitative Comparisons with Change Detection Results

Our proposed method has been ranked with all of the published state-of-the-art methods of background segmentation on the `changedetection.net` website using all available data, and the results have been posted in Table 1. These results show that our approach is comparable in quality to the traditional methods.

Most of traditional methods report a processing speed at below 24 fps on a 320×240 video [15], where our algorithm has achieved a maximum processing speed of 1146 fps on the same 320×240 sequences using an Nvidia Tesla K80 cluster. The processing speeds of the top four methods based on average ranking is shown in Fig. 3. The proposed method used less than 3.5 % of the device on average when processing all of the

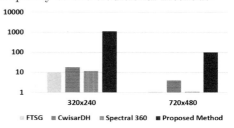

Fig. 3. Speed comparisons between the top 4 average ranking methods – [9,12,14] and the proposed technique.

`changedetection` videos, and we have achieved similar results on a commercial Nvidia graphics card. In comparison, the sequential version of our

Table 3. Showcase of changedetection.net sequences across several categories

Sequence	Re	Sp	FPR	FNR	PWC	Precision	F-Meas
Office	0.9576	0.9985	0.0015	0.5303	3.8048	0.9576	0.9576
Pedestrians	0.9595	0.9996	0.0004	0.0425	0.0816	0.9595	0.9595
Fountain02	0.9751	1.0000	0.0000	0.1192	0.0305	0.9751	0.9751
Skating	0.9922	0.9996	0.0004	0.1401	0.7269	0.9922	0.9922
Lakeside	0.9861	0.9999	0.0001	0.6411	1.2381	0.9861	0.9861
Badminton	0.9673	0.9991	0.0009	0.2165	0.8332	0.9673	0.9673
Bus station	0.9197	0.9981	0.0019	0.4187	1.7325	0.9197	0.9197
Street light	0.9963	1.0000	0.0000	0.7969	3.8698	0.9963	0.9963
Parking	0.9939	0.9997	0.0003	0.4022	3.1390	0.9939	0.9939
Turbulence2	0.9993	1.0000	0.0000	0.1022	0.0038	0.9993	0.9993
Turnpike	0.9377	0.9956	0.0044	0.1781	1.7314	0.9377	0.9377

algorithm, using an enterprise Intel Xeon E5-2620 v3 CPU and the same parameters, processed 38 times slower on a small resolution (320×240) sequence, and 82 times slower on a large resolution (720×480) sequence.

In addition, we have calculated the average metrics across all 53 videos organized into eleven categories based on seven measurements. These results are shown in Table 2, and some additional results from a handful of individual videos are shown in Table 3. The proposed technique provides superior performance for the sequences in which a static camera is employed, such as the baseline, bad weather, and thermal categories. In addition, despite this method being designed for static cameras, it still receives competitive results on videos with panning, tilting, or zooming (PTZ) and environments with global illumination changes, such as the videos in the camera jitter and shadow categories. This demonstrates that the proposed method can handle videos with PTZ and global illumination changes, given that it has enough time to learn these changes. Moreover, the proposed technique is designed to provide the least amount of false positives and the highest specificity ratings. As such, even for the videos in which a dynamic camera is employed, such as the videos in the camera jitter and turbulence categories, the proposed method still provides superior specificity and false positive rates. For such scenarios, a post-process module may be employed to compensate for the possible loss of foreground regions on an object level.

4 Conclusions and Future Work

This paper presents a novel parallelized non-parametric background modeling technique, designed for efficiently detecting foreground objects in videos with quasi-stationary or dynamic backgrounds. The proposed technique utilizes an adaptive background model based on pixel intensity/color histograms and employs three stages of training, classification, and update, designed for the heterogeneous parallel architecture of Nvidia CUDA.

Our results show that massive improvements to the processing speed of background segmentation algorithms can be achieved through massive parallelization while maintaining highly competitive results. The parallel version of our algorithm processed at incredibly fast speeds when compared to the sequential version while yielding the same results. In the future, we seek to exploit the parallelizable nature of more accurate change detection algorithms to achieve orders of magnitude faster processing speeds than their sequential versions.

Acknowledgements. This material is based upon work supported in part by the U. S. Army Research Laboratory and the U. S. Department of Defense under grant number W911NF-15-1-0024 and W911NF-15-1-0455, and by the NASA MUREP ASTAR program under grant number NNX15AU31H. This support does not necessarily imply endorsement by the DoD or NASA.

References

1. Benezeth, Y., Jodoin, P.M., Emile, B., Laurent, H., Rosenberger, C.: Comparative study of background subtraction algorithms. J. Electron. Imaging **19**, 033003–033003 (2010)
2. Wang, B., Dudek, P.: A fast self-tuning background subtraction algorithm. In: 2014 IEEE Conference on Computer Vision and Pattern Recognition Workshops (CVPRW), pp. 401–404. IEEE (2014)
3. Zivkovic, Z., van der Heijden, F.: Efficient adaptive density estimation per image pixel for the task of background subtraction. Pattern Recogn. Lett. **27**, 773–780 (2006)
4. Stauffer, C., Grimson, W.E.L.: Adaptive background mixture models for real-timetracking. In: IEEE ComputerSociety Conference on Computer Vision and Pattern Recognition, 1999, vol. 2. IEEE (1999)
5. Zivkovic, Z.: Improved adaptive gaussian mixture model for background subtraction. In: Proceedings of the 17th International Conference on Pattern Recognition, 2004. ICPR 2004, vol. 2, pp. 28–31. IEEE (2004)
6. Varadarajan, S., Miller, P., Zhou, H.: Spatial mixture of gaussians for dynamic background modelling. In: 2013 10th IEEE International Conference on Advanced Video and Signal Based Surveillance (AVSS), pp. 63–68. IEEE (2013)
7. Maddalena, L., Petrosino, A.: A fuzzy spatial coherence-based approach to background/foreground separation for moving object detection. Neural Comput. Appl. **19**, 179–186 (2010)
8. Maddalena, L., Petrosino, A.: The sobs algorithm: what are the limits? In: 2012 IEEE Computer Society Conference on Computer Vision and Pattern Recognition Workshops (CVPRW), pp. 21–26. IEEE (2012)
9. Gregorio, M.D., Giordano, M.: Change detection with weightless neural networks. In: 2014 IEEE Conference on Computer Vision and Pattern Recognition Workshops (CVPRW), pp. 409–413. IEEE (2014)
10. Liang, D., Kaneko, S.: Improvements and experiments of a compact statistical background model. arXiv preprint arXiv:1405.6275 (2014)

11. Elgammal, A., Harwood, D., Davis, L.: Non-parametric model for background subtraction. In: Vernon, D. (ed.) ECCV 2000. LNCS, vol. 1843, pp. 751–767. Springer, Heidelberg (2000)
12. Sedky, M., Moniri, M., Chibelushi, C.C.: Spectral-360: a physics-based technique for change detection. In: 2014 IEEE Conference on Computer Vision and Pattern Recognition Workshops (CVPRW), pp. 405–408 (2014)
13. Tavakkoli, A., Nicolescu, M., Bebis, G.: Non-parametric statistical background modeling for efficient foreground region detection. Int. J. Mach. Vis. Appl. **20**, 395–409 (2008)
14. Wang, R., Bunyak, F., Seetharaman, G., Palaniappan, K.: Static and moving object detection using flux tensor with split gaussian models. In: 2014 IEEE Conference on Computer Vision and Pattern Recognition Workshops (CVPRW), pp. 420–424. (2014)
15. Goyette, N., Jodoin, P.M., Porikli, F., Konrad, J., Ishwar, P.: changedetection.net: a new change detection benchmark dataset (2012)
16. Lu, X.: A multiscale spatio-temporal background model for motion detection. In: 2014 IEEE International Conference on Image Processing (ICIP), pp. 3268–3271. IEEE (2014)

Finding the N-cuts of Watershed Partitions for Image Segmentation

Chao Zhang and Sokratis Makrogiannis[✉]

Mathematical Imaging and Visual Computing Group, Department of Mathematics,
Delaware State University, Dover, DE, USA
czhang11@students.desu.edu, smakrogiannis@desu.edu

Abstract. The normalized cut (N-cut) algorithm uses an algebraic graph optimization technique for image segmentation. Although N-cut produces good results for a variety of images, it has some weaknesses, such as high computational cost and sub-optimal partitions. In this paper we adopt the watershed transform to address these problems. Watershed can improve slow computing speed and produce closed object boundaries. However, watershed itself has the drawback of over-segmentation. Therefore, we propose to first apply watershed, then build a graph from the watershed regions, and find the N-cuts of the watershed region graph to improve segmentation accuracy. The objective of this paper is two-fold; the first goal is to reduce the complexity of this problem by optimizing region-based graph structures. The second goal is to validate the performance of the existing and proposed methods, and to test the hypothesis that region-based analysis reduces the complexity of optimization problem and improves segmentation accuracy.

1 Introduction

Image segmentation is a key image analysis procedure that divides an image into a set of meaningful regions or objects that are not overlapping. In general, the accuracy of segmentation depends on whether the objects in the image scene are accurately segmented or not. For example, in automated inspection of electronic assemblies, interest lies in analyzing images of standardized products with the objective of determining the presence or absence of specific anomalies, such as missing components or broken connection paths [1]. Segmentation has been widely applied in aerospace technology, transportation control systems, human face recognition, fingerprint identification, machine vision, and medical imaging, such as B ultra-sound, CT, and X-ray. In medical image analysis applications, segmentation can help locate tumors and other pathologies, assist in measuring tissue volume, perform computational neuroanatomy studies, and operating computer-guided surgery [2]. Successful segmentation is crucial for subsequent high level image analysis tasks. Therefore, image segmentation is a very important step for image processing.

Since the introduction of segmentation research field, an abundance of methods have been proposed for image segmentation. One of the early methods is

© Springer International Publishing Switzerland 2015
G. Bebis et al. (Eds.): ISVC 2015, Part I, LNCS 9474, pp. 221–230, 2015.
DOI: 10.1007/978-3-319-27857-5_20

the classic minimum spanning tree proposed by Zahn in 1971 [3]. This algorithm maps the image onto a graph, deletes the edges with the smallest weight, and partitions the image into different sub-images for segmentation. In the 1970s, Beucher and Lantuéjoul proposed the watershed method, which was successfully applied to grayscale images. This method has attracted a lot of attention [4–8]. In 2000, Shi and Malik proposed the normalized cut method using graph theory [9]. To accomplish segmentation, this approach computes the dissimilarity between different partitioned areas and the similarity in the same area. This solves the flaw of the principle of MST (minimum spanning tree) [3]. A group of related approaches propose to form superpixels before performing higher level computer vision tasks. In 2003, Xiaofeng Ren et al. introduced the concept of superpixel [10], which is an image region consisting of pixels with similar texture, color, brightness, and spatial proximity. It categorizes the pixels by the degree of similarity, which can help to reduce redundant visual information and the complexity in the following image analysis operations to a great extent [11].

This paper has the following objectives: the first is to introduce joint region- and graph-theoretic segmentation techniques to reduce the computational complexity and improve segmentation accuracy. The second goal is to validate the performance of the existing and the proposed methods and to test the hypothesis of reduced complexity and increased accuracy. We propose two methods for watershed map graph partitioning to improve the performance of conventional N-cut partitioning. In the Results section we analyze the performance of the proposed and some existing segmentation methods. In the final section we summarize the major keypoints in this paper and present some goals for future work.

2   Methods

Here we introduce two region- and graph-based cooperative approaches to image segmentation that utilize the watershed transform and N-cut optimizations. We denote the N-cuts of watershed ridges that define the region boundaries by the term water-cuts.

2.1   Water-Cuts Clustering

We consider segmentation to be a graph partitioning problem. We first give definitions of a graph and its bi-partitioning. A graph G is a finite set that consists of a set of vertices V and a set of edges E which connect the vertices. A graph $G = (V, E)$ can be partitioned into two disjoint sets. i.e.: $A, B, A \bigcup B = V, A \bigcap B = \emptyset$. We partition the graph by removing edges connecting the two parts. The degree of similarity between these two parts can be computed as the total weight of edges that have been removed. In graph theoretic language it is called the cut:

$$cut(A, B) = \sum_{u \in A, v \in B} w(u, v) \qquad (1)$$

where $w(u, v)$ is the weight of the edges connecting vertices u and v. It represents the similarity degree between two points.

Shi and Malik's normalized cuts (N-cuts) algorithm [9] is a spectral graph clustering approach that seeks the optimal graph cut that bi-partitions the graph G by solving the generalized eigenvalue problem of the Laplacian matrix defined from the input image. Iterative bi-partitionings are applied to reach the desired number of regions. Shi and Malik proposed to partition V into A and B by minimizing the normalized cut:

$$Ncut(A, B) = \frac{cut(A, B)}{assoc(A, V)} + \frac{cut(A, B)}{assoc(B, V)} \tag{2}$$

where $assoc(A, V) = \sum_{u \in A, t \in V} w(u, t)$ is the total connection from nodes in A to all nodes in the graph, and $assoc(B, V)$ is similarly defined for set B. Let W be an $N \times N$ symmetrical matrix with $W(i, j) = w_{ij}$ and $d(i) = \sum_j W(i, j)$ be the total connection from node i to all other nodes. Let $D = diag(d_1, d_2, ...d_N)$ be an $N \times N$ diagonal matrix with $d(i)$ on its diagonal. If y is relaxed to take on real values, we can minimize (2) by solving the generalized eigenvalue system $(D - W)y = \lambda Dy$. This can be rewritten as:

$$D^{-\frac{1}{2}}(D - W)D^{-\frac{1}{2}}z = \lambda z, \tag{3}$$

where λ is an eigenvalue and $z = D^{-\frac{1}{2}}y$, $z_0 = D^{-\frac{1}{2}}I$ is an eigenvector of (3) with eigenvalue of 0. $(D - W)$ is Laplacian matrix, it is also semi-positive definite therefore $D^{-\frac{1}{2}}(D - W)D^{-\frac{1}{2}}$ is symmetric semi-positive definite. We use the eigenvector with second smallest eigenvalue (because the first smallest eigenvalue always equals 0) to bi-partition the graph by finding the splitting point such that N-cuts is maximized.

The watershed algorithm evolved from the field of mathematical morphology and has found applications in many segmentation problems. The main concept of watershed is to consider the image matrix as a topographic surface that is a function of the spatial coordinates. The watershed algorithm is completed in two steps: sorting of gradient magnitude minima and flooding [12]. Fast implementations of this algorithm have been developed and used extensively for image analysis. Another advantage is that watershed regions possess closed boundaries with single-pixel width. However, due to the fact that the watershed method is very sensitive to noise in the image, it frequently produces segmentation redundancy known as over-segmentation [13–15]. Overall, Watershed has the following advantages: (1) it is computationally efficient; (2) watershed regions have closed boundaries; (3) it produces accurate delineation.

Building upon the concept presented in [16] for a region-based graph partitioning method, in this paper, we will combine the watershed technique with the N-cuts spectral clustering method. In this approach we use the watershed method first to segment the target image, and construct a piece-wise constant region-based representation of the image. We then use the watershed regions as vertices for building a weighted graph. We define graph edge weight as the product of feature similarity term and a spatial similarity term as in [9]. Next,

Fig. 1. Main stages of Water-cuts Clustering.

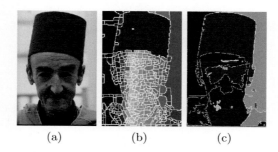

(a)　　　　　　(b)　　　　　　(c)

Fig. 2. (a) Original image; (b) watershed regions; (c) segmentation produced by Water-cuts Clustering.

we compute the normalized cuts of the watershed-region graph. In this way, we will not only address over-segmentation produced by the watershed method, but also reduce the computational load of N-cuts clustering. For example, for an image of N pixels we will have an $N \times N$ system in the calculation. It is obvious that if the image size is large, then the computational and memory requirements may become intractable. Following the proposed approach, we can reduce the cardinality of node set and vertex set and decrease computational complexity.

The key stages are shown in Fig. 1. N-cuts Clustering ignores spatial connectivity, so we included a connected component labeling step that identifies and labels the spatially connected regions. A segmentation example is displayed in Fig. 2. We observe that Water-cuts Clustering image segmentation can produce a good segmentation result as it reduces the complexity of the optimization problem. The reduced cardinality of graph vertex set also implies reduction of the computational time.

2.2　Water-cuts RAG

The RAG (Region adjacency graph) [17] is a data structure suitable for storing graphs with adjacency relations. Suppose we have a collection of objects, some of which are found to be adjacent to others. Define a graph by taking a vertex per object and joining two vertices with an edge if the objects are adjacent. This procedure produces an adjacency graph. In an RAG, vertices correspond to regions and neighboring regions are connected by an edge. The segmented image consists of regions with similar properties (intensity, texture, color,etc) that correspond to some entities in the scene, and the neighborhood relation is fulfilled, if the regions have some common boundary. We build an RAG using the watershed flooding result. Graph topology is based on the watershed partition.

Fig. 3. The Water-cuts RAG algorithm.

Each graph vertex is mapped to a site on the image plane; in our work this is the mass center of a watershed basin. The site value is set to the mean color of the corresponding basin. Likewise, each graph edge is mapped to a site, here a pair of basin mass centers. In mathematical methods we use an adjacency matrix A to represent which vertices (or nodes) of a graph are adjacent to which other vertices. If the target image has n vertices, then the adjacency matrix must be a $n \times n$ matrix. We marked the nodes as $V_1, V_2, \ldots, V_n$. If $(V_i, V_j) \in E(G)$, $A_{ij} = w_{ij}$, otherwise $A_{ij} = 0$.

The Water-cuts RAG method also seeks the optimal N-cuts of a watershed region graph with strict spatial constraints. Use of RAG further simplifies the graph by keeping the edges between adjacent regions only. The key stages are shown in Fig. 3. In several applications the image that we need to segment has a large matrix size. Then the dimensions of the Laplacian matrix computed over image pixel entities by traditional N-cuts would be very large.

In the example of Fig. 4, after applying watershed segmentation the image in Fig. 4(a) is divided into 493 regions (Fig. 4(b)). We can treat the 493 regions as 493 graph vertices. This implies significant simplification of topological complexity, i.e. fewer vertices in the graph, and reduced computational burden. Final segmentation is displayed in Fig. 4(c). This method also reduces the complexity of this problem by optimization of region-based graph structures, decreases calculation time, and yields more accurate segmentation results.

3 Results

3.1 Computational Time Comparison

In our first set of experiments we compared the computational costs of watershed, normalized cuts, Water-cuts Clustering and Water-cuts RAG. We first measured

(a)   (b)   (c)

Fig. 4. (a) Original image; (b) watershed regions; (c) segmentation produced by Water-cuts RAG method.

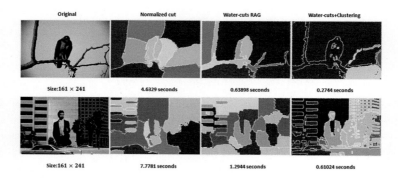

Fig. 5. Computational times for N-cuts, Water-cuts RAG, Water-cuts Clustering.

computational times using repeated experiments over the training set of BSD (Berkeley Segmentation Dataset) with 100 images [18]. Next, we discuss the computational complexity of each method.

From the results in Fig. 5 and Table 1, it is clear that our two proposed methods are executed in significantly shorter time than the traditional N-cuts technique. For example, in Fig. 5, top row, N-cuts takes 4.63 s, while Water-cuts RAG takes merely 0.64 s. In Table 1 we have set in N-cuts and Water-cuts RAG the same number of final regions so we can compare the computational time over the images of the BSD dataset. Solving a standard eigenvalue problem for all eigenvectors takes $O(n^3)$ time, where n is the number of pixels in the image. However, the watershed method just takes $O(n^2)$. Therefore the use of watershed for pre-segmentation significantly reduces the computational complexity. Meanwhile, we observe that the two proposed methods produce more accurate delineation of the visual scene than traditional N-cuts.

Table 1. Computational time comparison for 10 regions.

	Watershed	Normalized cut	Water-cuts Clustering	Water-cuts RAG
mean(s)	0.053	12.796	2.145	4.165
std.dev(s)	0.005	1.709	0.722	1.090

In the example of Fig. 6, which displays a screenshot of our segmentation application and the computed graph structures, watershed produces an over-segmented region map. Water-cuts Clustering and Water-cuts RAG partitions have greatly reduced over-segmentation, without affecting the accuracy. Thus, this method has lowered calculation complexity, and obtained a relatively satisfactory segmentation result.

Fig. 6. GUI (Graphical User Interface) for segmentation evaluation.

3.2 Segmentation Accuracy Comparison

Here we use a segmentation accuracy evaluation metric to compare the performance of watershed, N-cuts, and region-graph cooperative techniques. The Segmentation Cost Function (SCF) was proposed by Yang and Liu to evaluate segmentation accuracy [19]. SCF quantifies the error between the piecewise constant image model after segmentation and the original image. Smaller SCF values indicate more accurate segmentation with a smaller number of regions and larger region areas.

The SCF measure is expressed by the relation:

$$SCF = \sqrt{\frac{N_R}{h \cdot w \cdot c}} \cdot \sum_{i=1}^{N_R} \frac{\sigma_i^2}{\sqrt{card_i}} \tag{4}$$

In this equation, h, w and c are the height, width, and number of the image channels, respectively. N_R is the number of final regions, σ_i^2 is the color error over region i, and $card_i$ is the number of pixels in region i.

In Fig. 7 and Table 2, we have set the same final number of regions in N-cuts and Water-cuts RAG. Table 2 compares mean and standard deviations of SCF values after segmenting the original image into 10, 20, 30, 40 and 50 regions over the training set of BSD (100 images). In Table 2, SCF_W, SCF_{WNC}, SCF_N, and SCF_{WC} denote SCF values for watershed, Water-cuts Clustering, N-cuts, and Water-cuts RAG techniques respectively. Figure 7 displays the corresponding bar graphs of SCF values produced by N-cuts and Water-cuts RAG only, because we can not precisely control the final number regions for the other methods.

Table 2. SCF values versus the number of final regions.

Regions		SCF_W	SCF_{WNC}	SCF_N	SCF_{WC}
10	mean	2917.52	183.68	2.65	2.64
	std.dev	1890.98	198.66	1.65	1.33
20	mean	2917.52	182.83	8.10	7.96
	std.dev	1890.98	202.75	4.21	3.80
30	mean	2917.52	185.40	16.30	15.72
	std.dev	1890.98	200.48	8.08	7.71
40	mean	2917.52	180.97	26.30	24.38
	std.dev	1890.98	201.74	12.47	11.30
50	mean	2917.52	181.46	38.96	34.95
	std.dev	1890.98	201.51	17.74	16.17

Thus it is meaningless to compare SCF value under the condition of different regions. In Table 2, when the segmentation has $N = 10$ regions, the SCF mean value and standard deviation calculated by N-cuts are 2.65 and 1.65, while the SCF mean value and standard deviation calculated by Water-cuts RAG are 2.64 and 1.33; both are smaller than the former two numbers. We note that this trend increases with increasing number of final regions. Furthermore, Table 3 lists the p-values produced by two-sample paired t-tests between SCF_N and SCF_{WC} for different numbers of final regions. These p-values indicate whether there is a systematic difference between the two compared methods. We observe that p-values for N=40, and N=50 show statistically significant differences in segmentation performance (p-value ≤ 0.05). These results imply that our Water-cuts RAG method produced a systematic improvement in SCF scores compared to traditional N-cuts.

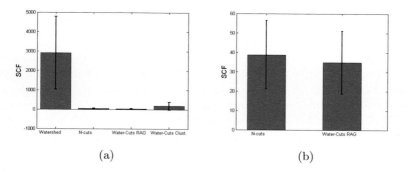

(a)   (b)

Fig. 7. Plots of SCF values for 50 regions comparing (a) watershed, N-cuts, Water-cuts RAG, and Water-cuts Clustering, and (b) N-cuts, and Water-cuts RAG only.

Table 3. Paired t-test for segmentation comparison.

Regions	p-value (SCF_N, SCF_{WC})
10	0.47
20	0.35
30	0.17
40	0.025
50	1.4×10^{-3}

4   Conclusion

Image segmentation has remained a popular research topic over the past few decades because it deals with a challenging and significant computer vision problem. Successful segmentation is a key process for identifying and interpreting the content of static or dynamic visual scenes. In this paper we proposed a combined approach of using watershed and N-cuts together, without or with spatial constraints, named as Water-cuts clustering and Water-cuts RAG techniques respectively. The watershed approach is used to produce the region elements for building a weighted graph that we seek to partition using N-cut optimization.

The extensive experiments that we conducted to evaluate performance of the proposed methods in comparison to traditional watershed and N-cut techniques have demonstrated that joint segmentation reduces the computational and memory requirements. Moreover, Water-cuts RAG produced encouraging segmentation accuracy results compared to traditional N-cuts, implying an improved graph partitioning solution. Due to factors such as computational and space complexity, there is still much room for further studies. The main goals are to further improve the segmentation accuracy and to extend these methods to biomedical data of higher dimensionality including volumetric images and image sequences (3-D, 2D+t, 3D+t).

Acknowledgments. This research was supported by the National Institute of General Medical Sciences of the National Institutes of Health (NIH) under Award Number SC3-GM113754 and by the Intramural Research Program of National Institute on Aging, NIH. We acknowledge the support of the Center for Research and Education in Optical Sciences and Applications (CREOSA) of Delaware State University funded by NSF CREST-8763. We also acknowledge the US Department of Defense through the grant "Center for Advanced Algorithms" (W911NF-11-2-0046) for their support.

References

1. Gonzalez, R., Woods, R.: Digital Image Processing, 3rd edn. Prentice Hall, Upper Saddle River (2007)
2. Pham, D., Xu, C., Prince, J.: Current methods in medical image segmentation. Annu. Rev. Biomed. Eng. **2**, 315–337 (2000)

3. Zahn, C.: Graph-theoretical methods for detecting and describing gestalt clusters. IEEE Trans. Comput. **C–20**(1), 68–86 (1971)
4. Digabel, H., Lantuejoul, C.: Iterative algorithms. In: Chermant, J.-L. (ed.) 2nd European Symposium on Quantitative Analysis of Microstructures in Material Science, Biology and Medicine, pp. 85–99 (1977)
5. Haris, K., Efstratiadis, S., Maglaveras, N., Katsaggelos, A.: Hybrid image segmentation using watersheds and fast region merging. IEEE Trans. Image Process. **7**(12), 1684–1699 (1998)
6. Salembier, P., Pardas, M.: Hierarchical morphological segmentation for image sequence coding. IEEE Trans. Image Process. **3**(5), 639–651 (1994)
7. Paul, R., Canagarajah, C., David, R.: Image segmentation using a texture gradient based watershed transform. IEEE Trans. Image Process. **12**(12), 1618–1633 (2003)
8. O'Callaghan, R., Bull, D.: Combined morphological spectral unsupervised image segmentation. IEEE Trans. Image Process. **14**(1), 49–62 (2005)
9. Shi, J., Malik, J.: Normalized cuts and image segmentation. IEEE Trans. Pattern Anal. Mach. Intell. **22**(8), 888–905 (2000)
10. Ren, X., Malik, J.: Learning a classification model for segmentation. In: Proceedings of the 9th IEEE International Conference on Computer Vision, pp. 10–17 (2003)
11. Achanta, R., Shaji, A., Smith, K., Lucchi, A., Fua, P.: Sabine: slic superpixels compared to state-of-the-art superpixel methods. IEEE Trans. Pattern Anal. Mach. Intell. **34**(11), 2274–2281 (2012)
12. Luc, V., Pierre, S.: Watersheds in digital spaces: an efficient algorithm based on immersion simulations. IEEE Trans. Pattern Anal. Mach. Intell. **13**(6), 583–598 (1991)
13. Yaakov, T., Amir, A.: Automatic segmentation of moving objects in video sequence: a region labeling approach. IEEE Trans. Circuits Syst. Video Technol. **12**(7), 597–612 (2002)
14. Haifeng, X., Akmal, A., Mansur, R.: Automatic moving object extraction for content based applications. IEEE Trans. Circuits Syst. Video Technol. **14**(6), 796–812 (2004)
15. Demin, W.: Unsupervised video segmentation based on watersheds and temporal tracking. IEEE Trans. Circuits Syst. Video Technol. **8**(5), 539–546 (1998)
16. Makrogiannis, S., Economou, G., Fotopoulos, S., Bourbakis, N.: Segmentation of color images using multiscale clustering and graph theoretic region synthesis. IEEE Trans. Syst. Man Cybern. Part A Syst. Hum. **35**(2), 224–238 (2005)
17. Sonka, M., Hlavac, V., Boyle, R.: Image Processing, Analysis, and Machine Vision. Thomson Engineering, Toronto (2008)
18. http://www.eecs.berkeley.edu/Research/Projects/CS/vision/bsds/
19. Yang, Y.H., Liu, J.: Multiresolution image segmentation. IEEE Trans. Patt. Anal. Mach. Intell. **16**(7), 689–700 (1994)

A Novel Word Segmentation Method Based on Object Detection and Deep Learning

Tomas Wilkinson$^{(\boxtimes)}$ and Anders Brun

Department of Information Technology, Uppsala University, Uppsala, Sweden
{tomas.wilkinson,anders.brun}@it.uu.se

Abstract. The segmentation of individual words is a crucial step in several data mining methods for historical handwritten documents. Examples of applications include visual searching for query words (word spotting) and character-by-character text recognition. In this paper, we present a novel method for word segmentation that is adapted from recent advances in computer vision, deep learning and generic object detection. Our method has unique capabilities and it has found practical use in our current research project. It can easily be trained for different kinds of historical documents, uses full gray scale information, does not require binarization as pre-processing or prior segmentation of individual text lines. We evaluate its performance using established error metrics, previously used in competitions for word segmentation, and demonstrate its usefulness for a 15th century handwritten document.

1 Introduction

Segmentation of individual words in documents, in particular historical handwritten documents, is a challenging task that is often crucial for further processing and data mining. Applications include for instance word spotting (image retrieval), clustering, and full text recognition where the document images are converted to a string of characters. Unconstrained handwriting can vary significantly, over time and between different writers. Challenges include words touching each other, both vertically and horizontally; words with no apparent space between them; Non-uniform letter sizes; words with excessive amounts of skewing or shearing. This is particularly challenging in historical documents, where damages and unconventional writing habits add to the variability. For these reasons, we have based our approach in this paper on learning. Given a relatively small amount of training data, we can adopt and optimize our method to the material at hand.

1.1 Prior Work

The majority of the approaches for word segmentation operate on the text line level [1–4]. That is, the text line segmentation is a precursor to word segmentation. Most methods also require binary images [1–3], though there are exceptions [4]. The proposed method in this paper has neither of these constraints

© Springer International Publishing Switzerland 2015
G. Bebis et al. (Eds.): ISVC 2015, Part I, LNCS 9474, pp. 231–240, 2015.
DOI: 10.1007/978-3-319-27857-5_21

and instead works on images of entire document pages in gray scale. In fact, our approach is more similar to "text spotting in the wild" approaches like [5] that use a two-step paradigm to detect words based on proposal sampling and subsequent classification. However, text spotting in the wild focus on different kinds of data, detecting and recognizing text in natural images, where the difficulty lies in finding text in complex scenes rather than separating several lines of text into distinct words.

2   Background

To understand the method that we propose, we here briefly describe the concepts of convolutional neural networks and the bounding box overlap metric.

2.1   Convolutional Neural Networks

Because the method we present consists of two steps, where the second step is a classification of image region proposals (word candidates), we have a need to classify images. Recent advances in computer vision and object detection has shown that Convolutional Neural Networks (CNN) is a powerful tool for this. In particular, methods using CNNs have won several competitions lately and is widely considered to be current state of the art [6–8].

A (CNN) is a special kind of network specifically designed for image data. A typical CNN is a multi-layered model consisting of an input layer, into which data is fed, several hidden layers, and an output layer. What characterizes a CNN is that one or more of the hidden layers are so-called convolutional layers. A convolutional layer consists of a filter bank, where each filter is applied to the input using convolution, generating a set of output feature maps. A non-linear function is then applied to produce the output. When training a CNN, the weights being learned are the filter coefficients. A convolutional layer is typically followed by a max-pooling layer. This layer subsamples an image using the max operation, that is by keeping the max value over a small region (e.g., 2×2 pixels) for each region position in an image. The hidden layers of a CNN usually has several convolutional layers, each of which followed by a max-pooling layer, then one to two fully connected layers. CNNs are easier to train than regular neural networks due to convolutional layers having significantly fewer parameters, which is a partial explanation behind their recent success in a variety of different tasks. For more, see for example [9].

2.2   Bounding Box Overlap

When we evaluate the result of our method, there is a need to measure how close the resulting segmentation resembles ground truth, i.e. bounding boxes of words in a document that have been annotated by an expert. For this we need a metric to compare bounding boxes. Intersection over union (IoU) [5,10] is a common way to calculate how much two rectangular boxes overlap as a percentage. It is defined as the area of intersection between two boxes divided by the area of their union. It is commonly used in the detection of objects or text in images to determine how precise the algorithm in question can localize the objects you are interested in. This is typically calculated using only the box representations, e.g., a point and a width and height. Though if you have binary data, it is possible to calculate the IoU using only the foreground information. That is, you count how many foreground pixels are present in the intersection of the two boxes divided by how many foreground pixels there are in the union of the two boxes.

3   Method

The method we present in this paper is adapted from the R-CNN framework [10], Regions with Convolutional Neural Network Features. It is a method used in computer vision for generic object detection. The R-CNN framework has proven to work well in object detection, where it achieved state-of-the-art results when it was introduced and it has continually been improved upon since then. For example, removing a lot of unnecessary computation by making use of the Spatial Pyramid Pooling layer is introduced in [11]. Word segmentation can be seen as a special case of object detection that will work better if we adapt the framework for some specific properties of regions containing handwritten words in historical documents.

The method consists of two parts or modules. The first module produces bounding box region proposals. It is completely separate from the second part, which classifies the proposals into two classes, whether or not a bounding box is centered on a word. The algorithm used in [7] failed to generate word-like region proposals on the document images from the datasets used in this paper. Consequently, we have re-designed a region proposal algorithm specifically for word segmentation in historical document images.

3.1   Generating Region Proposals for Handwritten Words

The algorithm to generate tentative proposals for handwritten words takes its point of departure in the uncertainty of gray levels and the exact scale and shape of the morphology operators necessary to separate words.

1. Given a gray scale image, generate a unique set of l binary images, with the foreground being white, by globally thresholding the image at different multiples of the image's mean intensity. Call this set of thresholds L.
2. Generate a set of k rectangular structuring elements by forming all combinations of heights and widths (in pixels) from two sets H and W. Typically H and W are ranges of natural numbers, sampled at even intervals.
3. For all the l binary images resulting from step 1, apply each of the k structuring elements from step 2 using a morphological closing operation. This results in a new set of $k \cdot l$ binary images.
4. For each of the $k \cdot l$ binary images, label its foreground objects and extract a bounding box for each labeled object.
5. Remove all duplicate boxes.

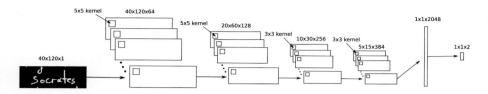

Fig. 1. The CNN architecture used for classifying whether or not a bounding box is centered on a word.

The choice of L, H and W is discussed in the experiments section and given for each dataset that we have tested our algorithm on.

3.2   Classification

The second module consists of a deep convolutional neural network (CNN) trained to classify whether bounding boxes are centered on words or not. The network architecture is based on the one used in [5] to recognize text, though it has been simplified slightly. The network consists of four convolutional layers followed by two fully connected layers, where the last layer is an SVM output layer [12]. See Fig. 1 for an image of the architecture used. Zero-padding is used to keep the sizes intact when performing convolutions. After the first three convolutional layers, subsampling by max pooling is done in a 2×2 region, which corresponds to downsampling by a factor of 2 in each dimension. A dropout layer [8] was used after the first fully connected layer with dropout ratio of 0.5. The network is trained using the L1-hinge loss. Rectified linear units were used for all activation functions.

The region proposals come in a large variety of sizes and since the CNN requires a predefined fixed size input, the region proposals are resized to 40×120 before being inputted into the network. Before resizing, we include a surrounding context of 16 pixels vertically and 48 pixels horizontally for each bounding box.

We create training data for the network by labeling each of the region proposals extracted from the images belonging to the training set. For a given region proposal, we calculate the IoU overlap between the it and all ground truth boxes. If the one of the overlaps exceeds a predefined threshold t_u, it is assigned a positive label, otherwise it is assigned a negative label. The training details are described in the experiments section.

For testing, all of the region proposals are scored by the CNN, where the score is the probability of the box being centered on a word. Then non-max suppression is applied as a post-processing step in which boxes that have an overlap with a higher scoring box over a threshold value is removed. Finally, boxes with a score lower than threshold are discarded (typically noise from the binarization and small parts of characters).

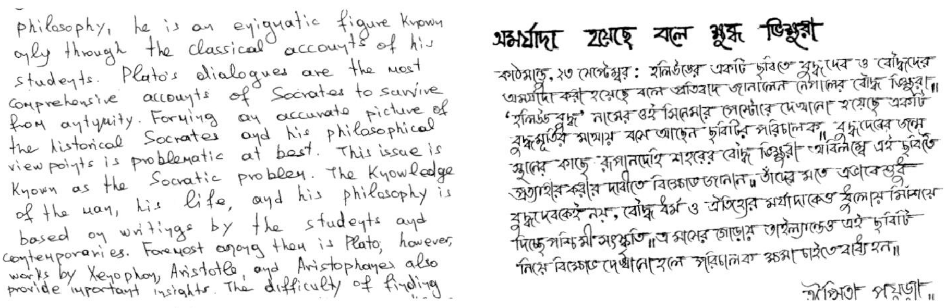

(a) English   (b) Bangla

Fig. 2. Two sample images from the ICDAR2013 dataset in two languages.

4 Data

Here we describe the datasets we have used to evaluate the proposed method.

4.1 ICDAR2013

The first dataset is the ICDAR2013 handwriting segmentation contest[1] [13], the latest installment in a well known series of handwriting segmentation contests. Along with the contest, a dataset (hereafter referred to as ICDAR2013) was released for both text line and word segmentation. It consists of 200 training pages and 150 testing pages and contains several writers, writing in English, Greek and Bangla languages. We use 10%, or 20 pages, of the training data as a validation set. The images of the dataset are binary and does not contain any non text elements. Two sample images can be seen in Fig. 2

4.2 C61

The second dataset consist of five spreads or ten pages from the C61 collection, written in the 15th century. It is written in Swedish and although the entire C61 collection is written by either four or five writers, the subset we use is written by a single writer. The text is tightly written (as is often the case with historical manuscripts, as parchment was expensive) with little space between text lines and words, making word segmentation for this data quite challenging. A sample of an image from C61 can be seen in Fig. 3. Contrary to the ICDAR2013 dataset, C61 is available in colour. The annotations for this dataset were manually done for the five images. We split the data into a training set of three images, a validation set of one image and a testing image to show the results.

5 Evaluation

In this section we describe how we evaluate the proposed approach.

---

[1] http://users.iit.demokritos.gr/~nstam/ICDAR2013HandSegmCont/.

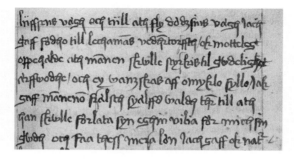

Fig. 3. A part of a page from the C61 dataset.

5.1 The Choice of Overlap Threshold

The choice of overlap threshold is critical when evaluating a word segmentation algorithm. In this paper, the data can be divided into categories; Overlap calculated using all the pixels in the bounding boxes and overlap calculated using only the foreground pixels. The two categories need different thresholds to achieve a similar overlap requirement in practice. This is evident in Fig. 4, where the two boxes have an overlap of 96 % calculated using only foreground pixels and 66 % using all pixels. The problem is particularly evident with words that contain ascenders and descenders (The parts of letters that go below and above the baseline, e.g., g and h).

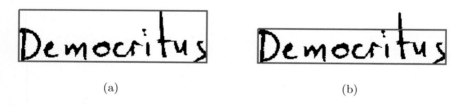

(a)   (b)

Fig. 4. IoU overlap using all the pixels between 4a and 4b is 66 % while using only foreground pixels is 96 %.

5.2 Competition Metrics

To be comparable to the previous results we adopt the evaluation metrics used in the ICDAR2013 handwriting segmentation competition [13]. According to the competition rules, a word is correctly segmented if the IoU of the foreground pixels of the ground truth and a proposal is higher than 90 %. Let N be the number of ground truth elements, M be the number of resulting proposals from a segmentation algorithm and $o2o$ be the number of one-to-one matches between the results and the ground truth. Then the detection rate (dr) and recognition accuracy (ra) are defined as

$$dr = \frac{o2o}{N}, ra = \frac{o2o}{M} \tag{1}$$

and the final metric is the harmonic mean between dr and ra.

$$fm = \frac{2 \cdot dr \cdot ra}{dr + ra} \tag{2}$$

The three metrics vary between 0–100%, where 100 % is a perfect score.

6   Experiments

In this section, we experimentally evaluate our word segmentation approach using two datasets. The experiments were performed using the Cuda/C++ library Caffe [14] and Python. The region proposer is implemented in pure Python and processing an image takes about 15 to 30 min, depending on the sizes of L, H, W and the size of the image

6.1   ICDAR2013

Since the images in this dataset are already binary, the first step in 3.1 is simplified. Let the triple *(start, stop, step)* represent a range of numbers starting at *start*, ending at *stop* with step length *step*. We found that $H = (1, 28, 3)$ and $W = (2, 42, 4)$, giving $k = 110$ gives a good trade-off between running time and recall. Using these settings we get an average test set recall of 93.28 %. The upper IoU threshold, t_u, of 90 % is already defined by the competition rules. A non-max suppression overlap of 0.08 and score threshold of 0.3 is determined by doing a coarse grid search on a validation set.

The weights of the network are initialized from a zero mean Gaussian distribution with a standard deviation of 0.01. Training was done using an initial learning rate of 0.001 and multiplying it after 22000 iterations with 0.1. Training is stopped at 42000 iterations. A momentum of 0.9 was used as well as an $l2$ weight decay of 0.0005. We use three times as many negative examples as positive in the training set.

As can be seen in Table 1, the proposed method achieves the second to highest score on the ICDAR2013 dataset, coming in below the state-of-the-art by 0.7 %.

6.2   C61

Since this dataset is in color, we first convert it to gray scale by averaging the R, G and B color channels. Then five binary images were generated from the gray-scale image with $L = 0.6, 0.7, ..., 1.0$. We found that $H = (3, 28, 5)$ and $W = (3, 28, 5)$, giving $k = 36$ structuring element sizes worked sufficiently well. These parameters give a recall of 89.6 % on the test set. Since we work in gray

Table 1. The results of our method compared to other methods on the ICDAR2013 dataset

Method	dr (%)	ra (%)	fm (%)
GOLESTAN-a from [13]	89.66	90.44	90.04
Method from [1]	90.50	91.55	91.03
Proposed method	87.09	93.82	90.33

scale with this dataset, IoU overlap is counted using all pixels, requiring a lower u_t. We found $u_t = 70\%$ reasonable. As above, a non-max suppression overlap of 10% and a score threshold of 0.07 is determined by doing a coarse grid search on the validation set.

We initialize weights of the network using the weights from the network trained on the ICDAR2013 dataset, this is called pre-training and is a common practice in deep learning. We use an initial learning rate of 0.001 and multiply it every 3000 iterations with 0.1 until training is stopped at 9000 iterations. Note that though the images are binarized when extracting region proposals, the image classification network is trained using the gray scale images.

Once again using the metrics from above, we get dr = 77.85%, ra = 82.41%, and fm = 80.07%. The results are visualized in Fig. 5.

7   Discussion

The choice of structuring elements to use is a trade-off between recall and running time. For a document image with a width or height between words of at most 100, setting both H and W to (1, 50, 1) would yield highest recall. This would give $k = 50 \cdot 50 = 2500$, which is probably unnecessarily high. In practice, you can significantly reduce k with a marginal loss in recall by increasing *step* and

(a) Result   (b) Ground Truth

Fig. 5. Result from the proposed methods for the test set with its corresponding ground truth from the C61 dataset.

reducing *stop*. We have found that we consistently achieve a recall of over 90 % using a k in the range of 30–100. Another important choice is the selection of L. We opted for a simple approach, inspired by [15]. Exchanging this method to state-of-the-art binarization approaches would probably increase performance, but at the cost of making the algorithm more complex.

8 Conclusions

In this paper, we have introduced a new method adopted from R-CNN, a generic object detection framework, to the new task of handwritten word segmentation in document images. The proposed method achieves results competitive to the state-of-the-art methods, as evaluated on the ICDAR2013 dataset, while at the same time removing the requirement of pre-segmented text lines and binary images. A possible downside of the method is a reliance on training data and, for the current implementation, a GPU for fast training. However, we also show that the amount of training data needed can be quite low, thanks to pre-training on other datasets. Only three spreads (six pages), were needed as training data to get a reasonable result for a challenging image from this particular 15th century document, consisting of over 1100 pages in total.

Acknowledgment. This project is a part of q2b, From quill to bytes, a framework program sponsored by the Swedish Research Council (Dnr 2012-5743) and Uppsala university. The work is done in part as a collaboration with the Swedish Museum of Natural History (Naturhistoriska riksmuseet).

References

1. Ryu, J., Koo, H.I., Cho, N.I.: Word segmentation method for handwritten documents based on structured learning. IEEE Signal Process. Lett. **22**, 1161–1165 (2015)
2. Stafylakis, T., Papavassiliou, V., Katsouros, V., Carayannis, G.: Robust text-line and word segmentation for handwritten documents images. In: IEEE International Conference on Acoustics, Speech and Signal Processing, ICASSP 2008, pp. 3393–3396 (2008)
3. Varga, T., Bunke, H.: Tree structure for word extraction from handwritten text lines. In: Proceedings Eighth International Conference on Document Analysis and Recognition, vol.1, pp. 352–356 (2005)
4. Manmatha, R., Rothfeder, J.L.: A scale space approach for automatically segmenting words from historical handwritten documents. IEEE Trans. Pattern Anal. Mach. Intell. **27**, 1212–1225 (2005)
5. Jaderberg, M., Simonyan, K., Vedaldi, A., Zisserman, A.: Reading text in the wild with convolutional neural networks. arXiv preprint arXiv:1412.1842 (2014)
6. Gidaris, S., Komodakis, N.: Object detection via a multi-region & semantic segmentation-aware cnn model. arXiv preprint arXiv:1505.01749 (2015)
7. Havaei, M., Davy, A., Warde-Farley, D., Biard, A., Courville, A., Bengio, Y., Pal, C., Jodoin, P.M., Larochelle, H.: Brain tumor segmentation with deep neural networks. arXiv preprint arXiv:1505.03540 (2015)

8. Krizhevsky, A., Sutskever, I., Hinton, G.E.: Imagenet classification with deep convolutional neural networks. In: Advances in neural information processing systems, pp. 1097–1105 (2012)

9. Bengio, Y., Goodfellow, I.J., Courville, A.: Deep Learning. Book in preparation for MIT Press (2015)

10. Girshick, R., Donahue, J., Darrell, T., Malik, J.: Rich feature hierarchies for accurate object detection and semantic segmentation. In: 2014 IEEE Conference on Computer Vision and Pattern Recognition (CVPR), pp. 580–587. IEEE (2014)

11. He, K., Zhang, X., Ren, S., Sun, J.: Spatial pyramid pooling in deep convolutional networks for visual recognition. In: Fleet, D., Pajdla, T., Schiele, B., Tuytelaars, T. (eds.) ECCV 2014, Part III. LNCS, vol. 8691, pp. 346–361. Springer, Heidelberg (2014)

12. Tang, Y.: Deep learning using linear support vector machines. arXiv preprint arXiv:1306.0239 (2013)

13. Stamatopoulos, N., Gatos, B., Louloudis, G., Pal, U., Alaei, A.: ICDAR 2013 handwriting segmentation contest. In: 2013 12th International Conference on Document Analysis and Recognition (ICDAR), pp. 1402–1406. IEEE (2013)

14. Jia, Y., Shelhamer, E., Donahue, J., Karayev, S., Long, J., Girshick, R., Guadarrama, S., Darrell, T.: Caffe: convolutional architecture for fast feature embedding. In: Proceedings of the ACM International Conference on Multimedia, pp. 675–678. ACM (2014)

15. Kovalchuk, A., Wolf, L., Dershowitz, N.: A simple and fast word spotting method. In: 2014 14th International Conference on Frontiers in Handwriting Recognition (ICFHR), pp. 3–8 (2014)

Recognition

Estimating the Dominant Orientation of an Object Using Image Segmentation and Principal Component Analysis

Sravan Bhagavatula$^{(\boxtimes)}$ and Nashlie Sephus

Partpic, Inc., 1040 W Marietta St. NW, Atlanta, GA 30318, USA
{sravan,nashlie}@partpic.com

Abstract. An object's orientation can often be a hurdle in computer vision applications. Assuming the object has a major axis, i.e., is longer in one of its dimensions than in others, the object's dominant orientation can be found. Knowing and compensating for an object's orientation may simplify processes such as recognition, segmentation, template matching, etc. However, solving this problem with no prior knowledge of the object's properties is not trivial. A solution is proposed which uses an image segmentation process that requires minimal prior information of the object, followed by feature extraction, and finally principal component analysis. Once the object's orientation is computed, one can easily rotate the image as needed.

Keywords: Object orientation · Image segmentation · Principal component analysis · Binary thresholding

1 Introduction

An object's orientation can often be a hurdle in computer vision applications. Assuming the object has a major axis, i.e., is longer in one of its dimensions than in others, the object's dominant orientation can be found. Knowing and compensating for an object's orientation may simplify processes such as recognition, segmentation, template matching, etc. However, solving this problem with no prior knowledge of the object's properties is not trivial. A solution is proposed which uses an image segmentation process that requires minimal prior information of the object, followed by feature extraction, and finally principal component analysis (PCA). Once the object's orientation is computed, one can easily rotate the image as needed.

When using matching algorithms, an object's orientation may often affect the technique's efficiency. For feature matching techniques such as Scale Invariant Feature Transform [1] or SURF: Speeded Up Robust Features [2], often times a portion of the process is assigned to counter this issue. However, if the image is aligned such that the object is correctly oriented, i.e., similar to how the object is aligned in the comparison image, the algorithm can focus on the issue of matching without having to delegate any additional memory or processing time.

© Springer International Publishing Switzerland 2015
G. Bebis et al. (Eds.): ISVC 2015, Part I, LNCS 9474, pp. 243–252, 2015.
DOI: 10.1007/978-3-319-27857-5_22

The same orientation-correcting technique may be applied to the comparison image to ensure that both are similarly aligned such that matching may happen faster.

Performing this task with as little prior knowledge of the object as possible can be challenging, since the techniques for segmentation and feature extraction often use several parameters and heuristics that are usually determined from empirical metrics. These metrics are often unique to a user's requirements and scenarios. A method that works on as few of these heuristics as possible may be applied to more scenarios without the need for re-adjusting.

Common methods used for segmentation such as binary thresholding and graph cuts [3] are techniques that use several parameters. However, when using such techniques, one may use information from each image to ensure that these parameters are not set constant numbers, but are automatically generated based on the images' conditions. For this purpose, a few assumptions need to be set to ensure that these parameters may be generated or used to lesser extent.

The five main assumptions made in the proposed solution are as follows -

1. The object of interest is the largest one in the scene.
2. Uniform intensity or color along the object.
3. A uniform background that does not merge with the object.
4. The object of interest has a dominant orientation, i.e., longer in one of its dimensions than the others.
5. The image is of a good size, i.e., an HD image.

Some of the main techniques used in this paper are Otsu's thresholding [4] and PCA [5], which are both used for alternative applications that are mostly different from how they are more commonly utilized in computer vision. The proposed solution uses Otsu'e method as an object segmentation technique. Additionally, PCA is well-known as a facial recognition tool in computer vision [6] but is used here for identifying the object's dominant orientation given a dataset of extracted features.

Given the five aforementioned assumptions, the contributions of the proposed solution are summarized as below -

- We use a segmentation algorithm that performs a thresholding operation to isolate the object. Prior knowledge of the object is not required.
- We then use this information to find the features of the object. If needed, an outlier removal technique may be applied to the dataset of feature points. PCA is then used to find the dominant orientation of the object.

The paper is organized in seven sections. The details regarding segmenting the object is presented in Sect. 3. Feature Detection techniques and recommendations are described in Sect. 4. The various outlier removal methods used based on each feature detection technique are presented in Sect. 5. The significance and use of PCA is detailed in Sect. 6. In Sect. 7, we discuss the parameters used and present example images and the corresponding angles found. Finally, the conclusions and future steps are provided in Sect. 8.

2   Related Work

In the past, facial recognition algorithms attempt to account for rotations of faces [7]. Since the development of these techniques, other computer vision applications have emerged. Determining angle rotations with PCA led to a solution for tilt correction for recognizing vehicle license plates [8]. Another application involves estimating rotation angles of textures for automatically detecting problems with textile machinery, making them more faster and accurate than using standard sensors and hardware [9]. One other application involve similar techniques for detecting forgery of signatures [10]. Since such methods have been developed to work for specific applications, we aim to propose a pipeline of techniques that may be utilized to detect the dominant orientation of an object for several applications. Thus, our method involves a minimal amount of parameters and heuristics.

3   Object Segmentation

Segmenting the object from its background is a crucial step in the process, especially if there is noise or clutter present in the background, as noise can add unnecessary details to the data before PCA is applied. Removing any additional clutter that remain after this step may be done using an outlier removal method, which will be covered in more detail in Sect. 5. We know that the object is fairly uniform in intensity. As a result, using as few parameters or other heuristics as necessary is important when segmenting it from the background.

The primary objective when segmenting the object is to obtain a mask around the object. Since there is a possibility that the features in this region are not exclusive to the object, we can then apply an outlier removal technique if needed to ensure that the features used for PCA do not include any noise.

In order to obtain a mask, we may use Otsu's threshold since the main constraint for this technique is that the background and the object are different enough from one another. However, using Otsu's thresholding for a simple gray image is not sufficient as elements such as shadows, reflections, and other lighting-induced artifacts could be included in the object's segmentation. In order to remove the effect of these elements, some pre-processing must be applied. The pre-processing technique used for our method is based on creating an image that is largely illuminant invariant. The method proposed by Tan and Triggs [11] for creating an image that is illuminant invariant works well for this purpose.

The pre-processing technique involves a few steps, the first and most important of which is a difference-of-Gaussian (DoG) image that highlights the edges of any object present in the image. The next steps involve normalizing the image and increasing its contrast as the DoG image tends to be largely similar in intensity. Once we have a clearer image that removes almost all lighting-based noise and primarily contains all the necessary details, i.e., edges of the object, the next step may proceed.

The illuminant invariant image now needs a blur operation in order to find a threshold. Blurring must also be implemented to suppress any noise created from the Tan-Triggs pre-processing. A good kernel size for this purpose has been estimated based on experience, and is mentioned in more detail in Sect. 7.

There are two possible blurring techniques that one may use for this purpose, the first of which is a simple Gaussian blur, which has the advantage of speed as it does not take too much processing time. The second technique would be to use a bilateral filter, which has the advantage of primarily blurring background noise without largely influencing the object's edges. A Gaussian blur affects both the background noise as well as an object's edges, as seen in Fig. 1. However, a bilateral filter usually takes a longer time to process compared to a Gaussian blur. Once this blurring is applied, we can then apply Otsu's thresholding.

Fig. 1. A visual comparison of the blurring methods. Gaussian blur is shown on the left, and bilateral filter on the right

Since Otsu's thresholding may segment the object completely, the intermediate image requires a connected component analysis to find the image consists largely of regions of the object of interest that are not connected to each other, as illustrated in Fig. 2(a). Since these portions are all from the same object and the aim of the segmentation is to obtain a good mask, these are grouped together to form a single uniform segment as shown in Fig. 2(b).

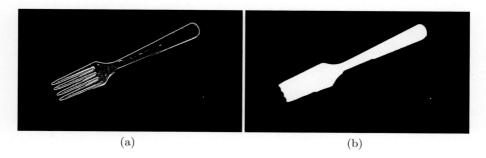

(a)   (b)

Fig. 2. The intermediate contours generated are shown in 2(a), whereas the results of the grouping are in 2(b)

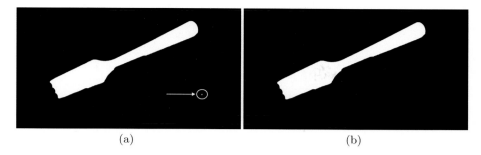

(a)   (b)

Fig. 3. Mask with and without noisy regions. The noise present marked in the bottom right corner of 3(a) is removed in 3(b)

The grouping is performed using morphological closing [12]. A suitable size may be used for a morphological closing element to ensure that the remaining connected components are grouped together. However, a large enough size must be chosen such that the connected components are grouped together but at the same time, it should not group in noisy regions of the images into the object of interest's mask.

Once this is done, we must also ensure that the mask does not contain noisy regions. In order to remove these regions, we can simply choose the largest region that remains in the image. The assumption here is that the largest object in the image is the object of interest. The largest region serves as a good mask for the object, as shown in Fig. 3.

4   Feature Extraction

The pre-processed image created from Tan-Triggs' preprocessing is not just useful for the segmentation step, but also for the feature extraction portion. Finding feature points is far better when performed on the preprocessed image rather than the raw grayscale image as shown in Fig. 4.

Choosing a feature type is an important step in this process, as different techniques yield different types of features. For example, among a couple of the freely available feature extraction algorithms are ORB [13] and Good Features To Track [14] (GFTT). For our purposes, using ORB features compared to GFTT gives us varying features that must be treated differently. In GFTT's case, the technique may create noise where not needed. Removing this noise would need a different outlier removal technique as outlined in Sect. 5. ORB, on the other hand, tends to find most of its features on the object also requiring far fewer parameters compared to GFTT. For these reasons, ORB is used as the primary technique for extracting features.

Once we use ORB features and the preprocessed image, the mask obtained from the segmentation step is applied towards the feature extracted image. The masking helps to create a subset of the features found that are mostly present in

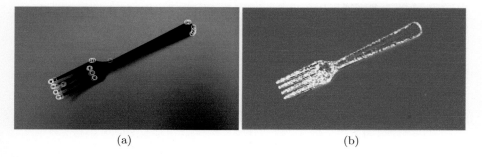

(a)   (b)

Fig. 4. ORB features for a standard grayscale image (4(a)) and the ones found for a pre-processed image (4(b))

the object. In rare cases, some outliers may remain that could potentially affect the angle measured by PCA. Based on the type of background present in the image, one may use an outlier removal technique as outlined in Sect. 5.

5   Outlier Removal

An outlier removal process may be required to remove features that do not belong to the object. The method used may vary based on the techniques used in the above sections as well as the nature of the background. Since the combination of using the object segmentation process listed above, the number of outliers will be minimal. However, if there are any that remain, this section explains techniques which may be used to remove them.

Detecting and removing outlier from a dataset is non-trivial as the algorithm needed can change from one dataset to the next. In most cases, however, a distance metric is chosen that helps differentiate between outliers and relevant data. Once the distribution of all data points is known, a simple way one may classify a point as an outlier is to measure how many standard deviations the distance of this point is from the mean.

5.1   Using GFTT

When one uses GFTT for feature extraction, the data points that count as outliers are often isolated and further apart whereas the points on the object are closer together. Given this criteria, using the adjacent distances in both dimensions between two points are used as the two metrics over which a distribution is found. An outlier may be defined as a point that is a multiple of the standard deviation away from the mean.

5.2   Using ORB

When ORB is used as the feature detector, one may use a Euclidean distance based metric for the data points. Since ORB's features are primarily on the

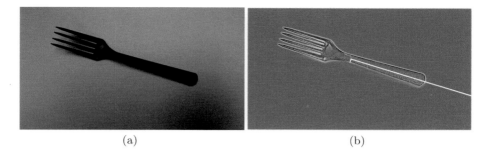

(a)   (b)

Fig. 5. PCA performed on features found

object, any features found after masking will be on the object. In addition, since the background's features are mostly suppressed, the few remaining outliers are the ones that are in the background within the mask. As such, using ORB ensures that the features found will most likely not contain enough outliers to disrupt the process.

5.3   Density Based Clustering

Another possible approach for finding the object of interest is using clustering. If an assumption can be made that the features of an object are denser than the features of any outliers, then a density based clustering approach may be used. A well-known method is the DBSCAN technique [15]. The downside to using this method is that there are several parameters involved, and these tend be very specific to particular types and sizes of parts. Clustering also takes more processing time than the other techniques.

6   PCA Application

Once all outliers are removed, the process may continue with the features found in the object. After the previous sections' steps are completed, the points that remain in our dataset are the features of the object. This set of points may now be used as a dataset for PCA. Since the goal is to find the object's orientation, one may simply use PCA to find the dominant eigenvector. The eigenvector represents the direction in which the points are oriented in, as shown in Fig. 5.

7   Experimental Results

For the first step of object segmentation, the image is first resized to approximately 2 Megapixels. The bilateral filter used for smoothing has a kernel size of 15 pixels along both dimensions. A morphological closing element was used to group together the connected components and find the object of interest which

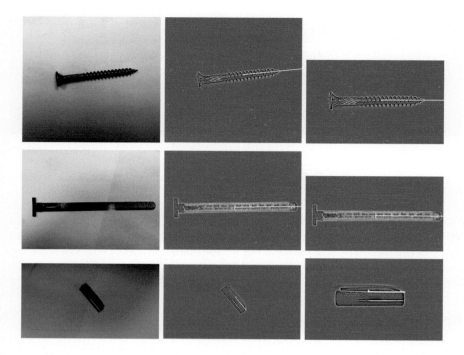

Fig. 6. Example objects, their respective angles, and the rotated image

is 31 pixels along both dimensions. When performing feature extraction, the ORB feature extractor is used. The extractor requires the maximum number of features to be found as a parameter, which was set to 2500.

Once the ORB features are found and outlier removal is still required, the proposed outlier removal methods are used to build the statistics based on the feature detector used. In ORB's case, not enough outliers are detected for this step. As mentioned in Sect. 5, the measurement used for GFTT is the inter-point distance for adjacent points. Outliers are identified as being 1.5 times the standard deviation away from the mean. Once we have a data set of features that contained no noisy information, and only features that belonged to the object, we may perform PCA on the points to find the dominant orientation.

As the image is only resized to ensure that the remaining parameters work well with the resized image, and ORB is the primary feature extractor used, we use three parameters, namely -

1. Bilateral filter kernel size (15).
2. Morphological closing element size (31).
3. Maximum number of ORB features (2500).

A few example objects' images are provided in Fig. 6 along with the corresponding angle and the corresponding rotated image. Please note that the rotated image is cropped for clarity purposes.

8   Conclusions and Future Steps

As demonstrated in the results, the proposed method to find an object's orientation serves well for several cases. Using PCA for finding the dominant orientation for the object works well for many types of objects, and from a computer vision perspective, is a unique application of the technology. Applying PCA on the set of features that are exclusive to the object of interest ensures that the points used are precisely the ones needed to find the orientation. This task is non-trivial and the steps performed in the proposed solution each play a role to ensure that the data used does not contain any noise. The minimal use of hard parameters also ensures that the solution may be applied to most scenarios with as little prior information. In the future, methods that can simplify the task of removing extraneous feature points, i.e., data for the PCA step, would serve as a stepping stone towards improving the efficiency of the solution.

References

1. Lowe, D.G.: Distinctive image features from scale-invariant keypoints. Int. J. Comput. Vis. **60**, 91–110 (2004)
2. Bay, H., Ess, A., Tuytelaars, T., Van Gool, L.: Speeded-Up robust features (SURF). Comput. Vis. Image Underst. **110**, 346–359 (2008)
3. Boykov, Y., Veksler, O., Zabih, R.: Fast approximate energy minimization via graph cuts. IEEE Trans. Patt. Anal. Mach. Intell. **23**, 1222–1239 (2001)
4. Otsu, N.: A threshold selection method from gray-level histograms. IEEE Trans. Syst. Man. Cybern. **9**, 62–66 (1979)
5. Jolliffe, I.: Principal Component Analysis. Springer Series in Statistics. Springer-Verlag, New York (2002)
6. Turk, M., Pentland, A.: Eigenfaces for recognition. J. Cogn. Neurosci. **3**, 71–86 (1991)
7. Beymer, D., Poggio, T.: Face recognition from one example view. In: Fifth International Conference on Computer Vision, 1995, Proceedings, pp. 500–507 (1995)
8. Pan, M., Yan, J., Xiao, Z.: An approach to tilt correction of vehicle license plate. In: International Conference on Mechatronics and Automation, 2007, ICMA 2007, pp. 271–275 (2007)
9. Ulas, C., Demir, S., Toker, O., Fidanboylu, K.: Rotation angle estimation algorithms for textures and their real-time implementation on the fu-smartcam. In: 5th International Symposium on Image and Signal Processing and Analysis, 2007, ISPA 2007, pp. 469–475 (2007)
10. Wei, W., Wang, S., Zhang, X., Tang, Z.: Estimation of image rotation angle using interpolation-related spectral signatures with application to blind detection of image forgery. IEEE Trans. Inf. Foren. Secur. **5**, 507–517 (2010)
11. Tan, X., Triggs, B.: Enhanced local texture feature sets for face recognition under difficult lighting conditions. IEEE Trans.Image Process. **19**, 1635–1650 (2010). A Publication of the IEEE Signal Processing Society
12. Maragos, P.: Handbook of Image and Video Processing. Elsevier, Burlington (2005)
13. Rublee, E., Rabaud, V., Konolige, K., Bradski, G.: ORB: An efficient alternative to SIFT or SURF. In: 2011 International Conference on Computer Vision, pp. 2564–2571. IEEE (2011)

14. Shi, J., Tomasi, C.: Good features to track. In: Proceedings of IEEE Conference on Computer Vision and Pattern Recognition CVPR-94, pp. 593–600. IEEE Computer Society Press (1994)
15. Ester, M., Kriegel, H.-P., Sander, J., Xu, X.: A density-based algorithm for discovering clusters in large spatial databases with noise, pp. 226–231. AAAI Press (1996)

Label Propagation for Large Scale
3D Indoor Scenes

Keke Tang$^{(\boxtimes)}$, Zhe Zhao, and Xiaoping Chen

University of Science and Technology of China, Hefei, China
{kktang,zhaozhe}@mail.ustc.edu.cn, xpchen@ustc.edu.cn

Abstract. RGB-D mapping or semantic mapping is becoming more and
more important for computer vision and robotics. However, manually
segmenting and generating semantic labels for RGB-D image sequence
or global point cloud will cost a lot of human labors. That is why there
still lacks a satisfactory indoor dataset for testing semantic mapping
system. While automatic label propagation can help, almost all existing
methods were designed for 2D videos which ignore the 3D characteristics
of RGB-D images. In this paper, we build a global map for RGB-D
image sequence firstly, and then propagate labels on the global map.
In this way, we can enforce label consistency over the global scene and
require fewer frames to be manually labeled. Also we model the overlap
information between images and use a greedy algorithm to automatically
choose frames for manual labeling. Experiments demonstrate that our
method can reduce manual efforts greatly. For a scene which contains
1831 images, only 22 labeled images can achieve 93 % accuracy for label
propagation.

1 Introduction

With affordable RGB-D sensors, many researchers start to explore 3D techniques of computer vision, such as 3D reconstruction ([1–4]), RGB-D semantic
segmentation ([5–10]), 3D semantic mapping ([11–15]). Such point cloud data
has shown great potential for solving significant problems in computer vision and
robotics. Recently, building a semantic map for robots attracts a lot of attention.
One common approach to get such semantic representation is by building 3D
reconstruction of the environment and augment the reconstruction with semantic labels. Such representation is required by robots to do high-level reasoning,
plan and execute complex tasks.

While important, labeling reconstructed model or global map is not an easy
task at all. It often involves two steps: (1) Build the 3D reconstruction from
RGB-D sequences. (2) Label the 3D reconstructed model. In the first step, we
exhibit sufficient overlapping views to compute the 3D transformation between
neighboring frames and register frames according to the transformation. However, such overlapping views between frames cause two main problems for the

---

Keke Tang and Zhe Zhao — These two authors contributed equally to this work.

© Springer International Publishing Switzerland 2015
G. Bebis et al. (Eds.): ISVC 2015, Part I, LNCS 9474, pp. 253–264, 2015.
DOI: 10.1007/978-3-319-27857-5_23

second step. (1) Typically, semantic labels need to be computed for each of the image. This means that many parts of the scene get labeled multiple times. It not only wastes a lot of computing time, but also causes inconsistent labels for the same parts of the scene. Recently, some researchers have noticed this problem and proposed methods to solve this inconsistent problems ([16,17]). (2) In order to get the ground truth labels for the reconstructed scene, we have to manually label all of the RGB-D images. It requires intensive human labors. However, a lot of works are to label the same regions because of overlapping views exist in the frames. This problem is often ignored by most researchers. In this work, we try to generate ground truth for the global map and every image with minimal manual effort. That means we will manually label some images and propagate labels for the remaining point cloud and images. This method will save a lot of human efforts and the propagated results can be used as training data or evaluation data for some related systems.

Some works try to build a RGB-D indoor dataset, such as NYU indoor dataset [6], Cornel indoor dataset [14] and SUN 3D dataset [18]. However, NYU indoor dataset only contains ground truth for some sample images which are taken from different indoor scenes. It has no ground truth for the global map and every frame in an indoor scene. Cornel dataset only has ground truth for point cloud which does not include images and many indoor scenes are very small. SUN 3D dataset gives a LabelMe style toolbox for labeling all of the images used for reconstruction. However, it still needs human to label a lot overlapping images and decide which image should be labeled or not. In fact, we do not need to label all of the images to get the semantic labels for the reconstructed model. A lot of images have the overlap regions with some other images.

In our work, we try to generate labels for each image and a global map with minimal manual effort. Towards this goal, we select some active frames to be labeled by human and propagate labels to the remaining point cloud and images. Active frames are chosen according to the overlap relations between frames. We want selected active frames to maximally occupy the global scene. For label propagation, most approaches use optical flow information and propagate labels frame by frame [19,20]. In this way, each unlabeled image can only get information from neighboring frames. Different from the traditional approaches

Fig. 1. An example of our method. From left to right: 3D reconstructed map, sample views of RGB-D keyframes, labels of the global map which are generated only from 7 active frames, labels of the global map which are created through label propagation.

which propagate labels frame by frame, we build the 3D global map firstly and infer labels on the global map immediately. In this way, each point can get information from many other points and have a consistent label over the global scene. After we infer labels for the global map, each pixel of RGB-D image can get their soft labels according to their 3D locations. We infer their final labels by CRF model. As we use the global map to infer labels, we do not need so much active frames to be handled. An example of our method is shown in Fig. 1. This indoor scene contains 2227 frames, and there exists a lot of view overlaps among keyframes. By our method, 7 active frames are selected to be labeled by humans. Through label propagation, final labels of global map are shown in this figure. Even though only 7 active frames are labeled, we obtain a 91 % accuracy for label propagation.

In summary, our work consists of two main contributions:

– As many frames have overlapping views, it is not necessary to label all of images. According to the view of overlapping information, we propose a method to choose active frames. The goal is to select some images to maximally occupy the global scene.
– We propose a higher-order Dense CRF method to propagate labels on the global map. Appearance feature is used as unary potential. Tracking feature is used to generate higher-order cliques. A new inference method is proposed. After the labels of global map are obtained, final labels of each RGB-D image can be computed by pairwise CRF model.

In the rest of the paper, we introduce the details of the active frame selection method in Sect. 2. Section 3 discusses label propagation approach. Experimental results and statistics are described in Sect. 4 and we conclude in Sect. 5.

2   Active Frame Selection

In order to select active frames, we firstly reconstruct global models from the image sequence. Through the reconstruction step, we will get the 3D relationship and overlap information between keyframes. Using the overlap information, active frames are selected to occupy the global scene maximally. This problem can be formulated as an optimization problem. In the following, we will introduce the reconstruction approach to get keyframes and the optimization method to get active frames.

2.1   3D Reconstruction from RGB-D Images

We compute 3D transformations of RGB-D images and merge the point cloud of each image into a global scene. Instead of processing all of the frames, we detect keyframes to reconstruct the global scene. Shi-Tomasi [21] corner points and Lucas-Kanade [22] optical flow method are adopted to track points between frames. Then RANSAC algorithm is used to find a 3D transformation according to corner pairs. As long as average 2D distance between corner pixels and 3D transformation distance are above some thresholds, we do not add this frame as a new keyframe. For all finding keyframes, we use both SURF and corner points to track feature pairs and estimate the final 3D transformation. The last 3D global map is built by merging point cloud of keyframes according to 3D transformation and voxelized into voxels by Octree.

2.2   Active Frame Selection

Similarity Matrix. In order to measure overlap information, we define similarity matrix M for keyframes. M_{ij} means how many pixels in the image j can be propagated from image i. From the reconstruction approach, we know that each pixel in the keyframe is associated with a voxel in the global map. If we label the image i, some voxels in the global map will get labeled. The pixels in the image j which are located in these labeled voxels are the overlapping pixels with image i. We use M_{ij} to measure the overlap information between image i and image j. It is computed as: $M_{ij} = N_{ij}/N_j$. N_{ij} is the number of pixels in the image j which can be propagated from image i (Note that $N_{ij} \neq N_{ji}$). N_j is the total number of pixels which have depth information of image j (Note that some pixels in the RGB-D image do not have depth information). So M_{ij} means that if we label image i, what percentage of pixels in the image j get labels ($M_{ii} = 1$).

Active Frame Selection. If we use $x_i \in \{0, 1\}$ to denote whether we select image i. The largest proportion of labeled pixels in the image j will be $\max_i M_{ij} x_i$. We want to select the least amount of active frames to ensure every image to have at least α percent pixels getting labeled. So we define an optimization problem like below to select active frames:

$$\min_{\mathbf{x}} \quad \sum_i x_i$$
$$\text{s.t.} \quad x_i \in \{0, 1\}$$
$$\forall j, \max_i M_{ij} x_i >= \alpha \tag{1}$$

Now we define a set $S_i = \{k | M_{ik} >= \alpha\}$, it represents that if we select image i, these images in the set S_i satisfy the constraints (1). So the optimization problem in the above is equal to a classic Set Covering Problem (SCP): we need to select the least amount of sets to ensure that $\bigcup_i S_i x_i = U$. $U = \{1, 2, ..., N\}$, N is the number of keyframes.

$$\min_{\mathbf{x}} \quad \sum_i x_i$$
$$\text{s.t.} \quad x_i \in \{0, 1\}$$
$$\bigcup_i S_i x_i = U$$

The classic Set Covering Problem is NP hard problem which has greedy algorithms to solve it. α represents that what percentage of the global scene need to be labeled by active frames. Different values of α will generate different numbers of active frames. When $\alpha = 1$, we need to select all keyframes to be active frames. In experiments, different values of α are tested.

3   Label Propagation

After active frames are chosen and labeled, some voxels get labels directly. We need a method to propagate labels to the remaining voxels and images. In this section, we use a Dense CRF [23] model to propagate labels. The energy of the Dense CRF is defined as the sum of the unary, pairwise and higher-order potentials:

$$E(x) = \sum_{i \in V} \psi_i(x_i) + \sum_{i,j \in V} \psi_{ij}(x_i, x_j) + \sum_{c \in C} \psi_c(x_c) \tag{2}$$

3.1   Unary Potential

The manual labels of the active frames yield C possible object labels and some labeled voxels. C Gaussian mixture models are learnt based on RGB values of labeled voxels. To compute the likelihood that which object category each voxel belongs, we define the following equation:

$$\psi_i(x_i) = -log P(F(x_i)|N(\mu_c, \Sigma_c)) \tag{3}$$

where $F(x_i)$ is the RGB color value for the voxel, $N(\mu_c, \Sigma_c)$ is the c-th Gaussian mixture model. In this way, unary potential encodes appearance feature for each voxel.

3.2   Pairwise Potential

We compute pairwise potentials by a linear combination of Gaussian kernels like [23]:

$$\psi_{ij}(x_i, x_j) = \mu(x_i, x_j) \sum_{m=1}^{K} w^{(m)} k^{(m)}(f_i, f_j) \tag{4}$$

where $\mu(x_i, x_j)$ is a label compatibility function, the vector f_i, f_j are feature vectors and $k^{(m)}$ are Gaussian kernel $k^{(m)}(f_i, f_j) = exp(-\frac{1}{2}(f_i - f_j)^T \Lambda^{(m)}(f_i - f_j))$. Each kernel $k^{(m)}$ is characterized by a symmetric, positive-definite precision matrix $\Lambda^{(m)}$, which defines its shape.

Two Gaussian kernels are defined: appearance kernel and smoothness kernel. The appearance kernel is computed as follow:

$$w^{(1)} \exp(-\frac{|p_i - p_j|^2}{2\theta_\alpha^2} - \frac{|I_i - I_j|^2}{2\theta_\beta^2}) \tag{5}$$

Fig. 2. Two examples of some corresponding object parts among images. We use these object parts to generate higher-order cliques which are drawn by different colors.

where p is the 3D location of the point and I is the color vector. This kernel is used to model the interactions between points with a similar appearance. The smooth kernel is defined as follow:

$$w^{(2)} \exp(-\frac{|p_i - p_j|^2}{2\theta_\gamma^2})$$ (6)

This kernel is used to removes small isolated regions.

3.3   Higher-Order Potential

In addition to using appearance feature, we will utilize the tracking feature to propagate labels, too. Through the 3D reconstruction process, we know the 3D transformation among frames. With the 3D transformation, we can track image parts (superpixels) frame by frame and get the correspondence of object parts. Using the tracking parts, we can not only propagate labels, but also enforce the label consistency among object parts. Figure 2 shows two examples of some corresponding object parts found by our method.

Firstly, we segment images into superpixels using a fast graph-based segment algorithm [24]. This algorithm produces superpixels by building a graph connecting the pixels in a local neighborhood and segmenting the graph according to local decisions. Secondly, we discover the correspondence of superpixels among images. For two images I_t, I_{t+1}, we back project image I_{t+1} into image I_t by 3D transformation and compute overlap regions for all superpixels. If a superpixel S_i in the image I_{t+1} has the biggest overlap region with a superpixel S_j in the image I_t, we treat them as corresponding superpixels.

We find all the corresponding superpixels for all neighboring frames like Fig. 2 shows. In fact, these corresponding superpixels are kind of different views of the same object and need to have the same label. So, higher-order cliques are generated by voxels which belong to corresponding superpixels. And the higher-order potential is defined as below:

$$\psi_c(x_c) = \begin{cases} y_l & \text{if } \forall i \in c, x_i = l \\ y_{max} & \text{otherwise} \end{cases}$$ (7)

This adds a special penalty y_l, if all the voxels in a clique take class label l, or y_{max}, if one or more voxels have a different label. We use higher-order potentials to ensure corresponding superpixels to have the same label.

3.4   Inference of the Higher-Order Dense CRF

Instead of computing the exact distribution $P(x)$, the mean field approximation method is used to compute a distribution $Q(x)$ that minimizes the KL-divergence $D(Q||P)$ among all distributions Q that can be expressed as a product of independent marginal, $Q(x) = \prod_i Q_i(x_i)$. In order to handle the inference of higher-order Dense CRF, we extend the mean-field method of pairwise Dense CRF [23] and the mean-field updates can be expressed as:

$$Q_i(x_i) \leftarrow \exp\{-\psi_i(x_i) - \hat{Q}_i(x_i) - higher_i(x_i)\} \tag{8}$$

where $\hat{Q}_i(x_i)$ computes the pairwise update, $higher_i(x_i)$ computes the higher-order potential update. The pairwise update is computed as follow:

$$\hat{Q}_i(x_i) = \sum_{\hat{l} \in L} \sum_{j \neq i} Q_j(x_j = \hat{l})\psi_{ij}(x_i, x_j) \tag{9}$$

where L is the label set. The higher-order potential update can be computed as:

$$higher_i(x_i = l) = \sum_{c \in C, i \in c} \{(\prod_{j \in c, j \neq i} Q_j(x_j = l))y_l \\ +(1 - \prod_{j \in c, j \neq i} Q_j(x_j = l))y_{max}\} \tag{10}$$

However, this equation does not work as expected. Given the distribution Q_i, the different probabilities for the classes will often have rather small values. With larger higher-order clique, consisting for example of 2000 voxels, the computation of $\prod_{i \in c} Q_i(x_i = l)$ will typically underflow a double variable and thus result into a max penalty in almost all cases. So, for the sake of simplicity, we replace the product operation with average operation and the final computation are as follows:

$$higher_i(x_i = l) = \sum_{c \in C, i \in c} \{(\frac{1}{N_c} \sum_{j \in c, j \neq i} Q_j(x_j = l))y_l \\ +(1 - \frac{1}{N_c} \sum_{j \in c, j \neq i} Q_j(x_j = l))y_{max}\} \tag{11}$$

where N_c is the number of the voxels of each higher-order clique.

3.5   Label Propagation for Images

When global map gets labels through higher-order Dense CRF, each pixel of RGB-D keyframe can obtain its label from its corresponding voxel. However, due to Kinect sensor noises, some pixels do not have depth information and corresponding voxel. We need to infer labels for these pixels. We use CRF to infer label assignments again. Different from higher-order Dense CRF, we smooth labels of each RGB-D image by pairwise CRF. For those pixels which do not have corresponding voxels, their unary potentials can be obtained by Eq. 3. For other pixels, their unary potentials are represented by $Q(x_i)$ which is the label probability of higher-order Dense CRF for the corresponding voxel. Notice that if the image is not a keyframe, we can obtain its location by computing its 3D transformation with its neighboring keyframe.

4   Experiments

4.1   Dataset

We test our system on the NYU v2 RGB-D dataset[1]. This dataset is recorded using Microsoft Kinect cameras at 640*480 RGB and depth image resolutions. It contains 464 RGB-D scenes taken from 3 cities and totally 407024 unlabeled frames. Labeling all of the frames will cost a lot of human efforts. In our experiments, we build a global map for each RGB-D scene, select active frames according to the overlap information, propagate labels from active frames by CRF model. From experiments, we discover that for a scene which contains 1831 frames, only 22 active frames can achieve about 93 % accuracy for label propagation.

4.2   Active Frame Selection

We choose 5 indoor scenes from NYU dataset and build global maps for each of them. Active frames are chosen from keyframes using the method in Sect. 2. Different values of α will generate different number of active frames according to Sect. 2. The number of raw images, keyframes, active frames and label accuracy of propagation on the global map are shown in Table 1. For example, the living room contains 2272 RGB-D images and 89 keyframes are extracted to generate our global map. When $\alpha = 0.2$, we obtain 7 active frames and 91 % accuracy for label propagation on the global map. These results illustrate that a lot of overlapping views exist among RGB-D image sequence. It is not necessary to manually label all of images. For this living room, only 17 human labeled frames (active frames) can achieve 95 % accuracy for label propagation on the global map.

4.3   Label Propagation for Global Maps

When active frames are chosen, some voxels are labeled through active frames and others need to be propagated through higher-order Dense CRF. Our Dense CRF encodes appearance feature by Gaussian mixture models and tracking feature through higher-order term. The correspondence of object parts are discovered by tracking through back-projected which is illustrated in Sect. 3.3. Figure 3 shows an example of superpixels among images. Corresponding superpixels are drawn by the same color in this figure. These corresponding superpixels can

---

[1] http://cs.nyu.edu/~silberman/datasets/nyu_depth_v2.html.

Table 1. Results about active frame selection and label propagation for the global map.

Scene	Frames	Keyframes	$\alpha = 0.2$		$\alpha = 0.4$		$\alpha = 0.6$	
Office	1108	18	3	0.89	4	0.91	8	0.93
Kitchen	1261	16	2	0.67	4	0.97	5	0.98
Bedroom	1114	38	5	0.83	9	0.90	23	0.98
Dining room	1831	73	8	0.80	22	0.93	57	0.96
Living room	2272	89	7	0.91	17	0.95	44	0.97

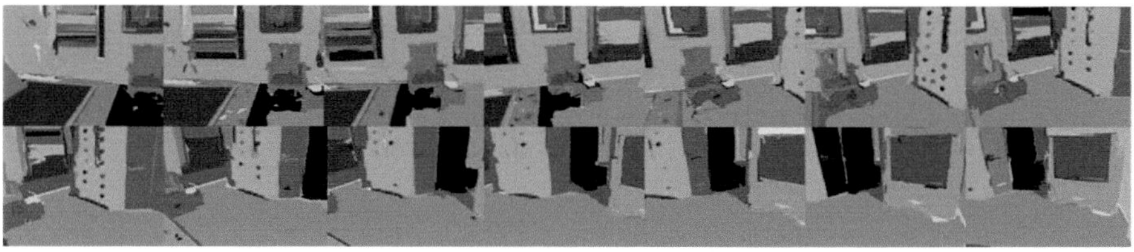

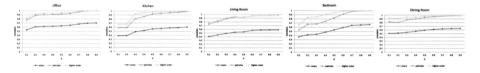

Fig. 3. Examples of superpixels among images. Corresponding superpixels are drawn by the same color.

Fig. 4. Comparisons of three different label propagation methods.

be seen as different views of the same object and should have the same label. We treat these parts as higher-order cliques. These higher-order potentials can ensure label consistency for object parts, as appearance model may induce some wrong labels for some points in object parts. We test label propagation by three methods: (1) Unary potential. It only uses unary potential to infer labels. (2) Pairwise Dense CRF. It encodes appearance feature and pairwise potential. (3) Higher-order Dense CRF. It both encodes appearance and tracking features. Figure 4 gives the label accuracy for all three methods according to different values of α. Higher-order CRF is always better than others. Figure 5 shows some visualizations of the global maps and labels of the global maps through propagation.

4.4 Label Propagation for Images

After the global map is labeled, we use pairwise CRF to infer labels of each RGB-image as Sect. 3.5. For comparison, we use an existing algorithm of [19], which uses a pixel-flow for propagation. This is the only video segmentation

Fig. 5. Visualizations of global maps and label propagation for global maps.

Table 2. Results about label propagation for images.

Scene	Active frames	Forward [19]	Near [19]	Pixelflow [19]	Our
Office	3	55.67	64.74	60.31	**92.27**
	4	66.18	74.36	70.25	**94.31**
Dining room	8	66.46	62.85	70.70	**87.95**
	22	83.20	85.32	84.95	**96.14**
Kitchen	2	71.36	85.48	88.43	**97.55**
	4	92.11	87.49	92.54	**98.53**
Bedroom	5	69.37	69.38	71.97	**81.33**
	9	76.05	70.45	78.18	**89.89**
Living room	7	61.64	60.56	62.47	**89.71**
	17	61.60	71.15	64.92	**91.82**

propagation algorithm which has the same purpose with us[2]. For each scene, we adopt $\alpha = 0.2$ and 0.4 to generate our active frames. [19] will use same active frames with our method, but to propagate labels by pixel-flow based method. Detailed results of two methods are displayed in Table 2. The number of active frames, average label accuracy of RGB-D images are shown in the table. From the table, we can see that our method is better than pixel-flow based method.

---

[2] http://vision.cs.utexas.edu/projects/activeframeselection/.

The main reason for this situation is that: [19] propagates labels frame by frame like many other optical flow based method. However, when the number of active frames are small, we need to propagate labels for a long distance. In this way, errors may accumulate over time and "drift" effect will become more dramatic. Instead, we propagate labels on the global map firstly, and then infer labels for each RGB-D image based on their corresponding voxels. In this way, we enforce label consistency on the global scene and no errors cumulate.

5   Conclusion

In this paper, we propose a method to automatically generate ground truth for 3D point cloud and RGB-D images. Our approach models overlap information between keyframes, and adopts a greedy algorithm to choose active frames. Then higher-order Dense CRF which encodes appearance and tracking features is adopted to propagate labels to the remaining point cloud and images. Results show that our methods can reduce human efforts for labeling effectively. This method has great potential to improve the work of building RGB-D indoor datasets and other learning based RGB-D applications. The limitation is that it only works for static indoor scenes. In the future, we will plan to extend it into dynamic scenes.

References

1. Henry, P., Krainin, M., Herbst, E., Ren, X., Fox, D.: Rgb-d mapping: Using kinect-style depth cameras for dense 3d modeling of indoor environments. I. J. Robotic Res. **31**, 647–663 (2012)
2. Newcombe, R.A., Izadi, S., Hilliges, O., Molyneaux, D., Kim, D., Davison, A.J., Kohli, P., Shotton, J., Hodges, S., Fitzgibbon, A.W.: Kinectfusion: Real-time dense surface mapping and tracking. In: ISMAR, pp. 127–136 (2011)
3. Whelan, T., Kaess, M., Fallon, M., Johannsson, H., Leonard, J., McDonald, J.: Kintinuous: spatially extended KinectFusion. In: RSS Workshop on RGB-D: Advanced Reasoning with Depth Cameras, Sydney, Australia (2012)
4. Engelhard, N., Endres, F., Hess, J., Sturm, J., Burgard, W.: Real-time 3d visual slam with a hand-held rgb-d camera. In: Proceedings of the RGB-D Workshop on 3D Perception in Robotics at the European Robotics Forum, Vasteras, Sweden, vol. 2011 (2011)
5. Ren, X., Bo, L., Fox, D.: Rgb-(d) scene labeling: features and algorithms. In: CVPR, pp. 2759–2766 (2012)
6. Silberman, N., Hoiem, D., Kohli, P., Fergus, R.: Indoor segmentation and support inference from RGBD images. In: Fitzgibbon, A., Lazebnik, S., Perona, P., Sato, Y., Schmid, C. (eds.) ECCV 2012, Part V. LNCS, vol. 7576, pp. 746–760. Springer, Heidelberg (2012)
7. Silberman, N., Fergus, R.: Indoor scene segmentation using a structured light sensor. In: ICCV Workshops, pp. 601–608 (2011)
8. Banica, D., Sminchisescu, C.: CPMC-3D-O2P: Semantic segmentation of rgb-d images using cpmc and second order pooling, CoRR abs/1312.7715 (2013)

9. Gupta, S., Arbelaez, P., Malik, J.: Perceptual organization and recognition of indoor scenes from rgb-d images. In: 2013 IEEE Conference on Computer Vision and Pattern Recognition (CVPR), pp. 564–571. IEEE (2013)

10. Couprie, C., Farabet, C., Najman, L., LeCun, Y.: Indoor semantic segmentation using depth information, (2013). arXiv preprint arXiv:1301.3572

11. Nüchter, A., Hertzberg, J.: Towards semantic maps for mobile robots. Robotics Auton. Syst. **56**, 915–926 (2008)

12. Stuckler, J., Biresev, N., Behnke, S.: Semantic mapping using object-class segmentation of rgb-d images. In: 2012 IEEE/RSJ International Conference on Intelligent Robots and Systems (IROS), pp. 3005–3010. IEEE (2012)

13. Hermans, A., Floros, G., Leibe, B.: Dense 3d semantic mapping of indoor scenes from rgb-d images. In: ICRA (2014)

14. Koppula, H.S., Anand, A., Joachims, T., Saxena, A.: Semantic labeling of 3d point clouds for indoor scenes. In: NIPS, pp. 244–252 (2011)

15. Valentin, J.P., Sengupta, S., Warrell, J., Shahrokni, A., Torr, P.H.: Mesh based semantic modelling for indoor and outdoor scenes. In: 2013 IEEE Conference on Computer Vision and Pattern Recognition (CVPR), pp. 2067–2074. IEEE (2013)

16. Floros, G., Leibe, B.: Joint 2d–3d temporally consistent semantic segmentation of street scenes. In: 2012 IEEE Conference on Computer Vision and Pattern Recognition (CVPR), pp. 2823–2830. IEEE (2012)

17. Miksik, O., Munoz, D., Bagnell, J.A., Hebert, M.: Efficient temporal consistency for streaming video scene analysis. In: 2013 IEEE International Conference on Robotics and Automation (ICRA), pp. 133–139. IEEE (2013)

18. Xiao, J., Owens, A., Torralba, A.: Sun3d: A database of big spaces reconstructed using sfm and object labels. In: 2013 IEEE International Conference on Computer Vision (ICCV), pp. 1625–1632. IEEE (2013)

19. Vijayanarasimhan, S., Grauman, K.: Active frame selection for label propagation in videos. In: Fitzgibbon, A., Lazebnik, S., Perona, P., Sato, Y., Schmid, C. (eds.) ECCV 2012, Part V. LNCS, vol. 7576, pp. 496–509. Springer, Heidelberg (2012)

20. Fauqueur, J., Brostow, G.J., Cipolla, R.: Assisted video object labeling by joint tracking of regions and keypoints. In: ICCV, pp. 1–7 (2007)

21. Shi, J., Tomasi, C.: Good features to track. In: 1994 IEEE Computer Society Conference on Computer Vision and Pattern Recognition, 1994, pp. 593–600. IEEE (1994)

22. Lucas, B.D., Kanade, T., et al.: An iterative image registration technique with an application to stereo vision. IJCAI **81**, 674–679 (1981)

23. Krähenbühl, P., Koltun, V.: Efficient inference in fully connected crfs with gaussian edge potentials, CoRR abs/1210.5644 (2012)

24. Felzenszwalb, P.F., Huttenlocher, D.P.: Efficient graph-based image segmentation. Int. J. Comput. Vis. **59**, 167–181 (2004)

Symmetry Similarity of Human Perception to Computer Vision Operators

Peter M. Forrest[(⊠)] and Mark S. Nixon

VLC Group, School of Electronics and Computer Science,
University of Southampton, Southampton, UK
{pmf1g09,msn}@ecs.soton.ac.uk

Abstract. Symmetry occurs everywhere around us and is key to human visual perception. Human perception can help guide the improvement of computer vision operators and this is the first paper aiming to quantify that guidance. We define the Degree of Symmetry (DoS) as the measure of 'how symmetrical' a region is as human perception sees symmetry in a continuous manner. A new dataset of symmetry axes, the Degree of Symmetry Axis Set, is compiled for ordering by DoS. A human perception rank order is found by crowd-sourced pairwise comparisons. The correlation of two ranked orders is defined as the Symmetry Similarity which we use to evaluate symmetry operators against human perception. No existing symmetry operator gives a value for DoS of a reflection axis. We extend three operators to give a value for the DoS of an axis: the Generalised Symmetry Transform, Loy's interest point operator, and Griffin's Derivative-of-Gaussian operator. The highest Symmetry Similarity of a symmetry operator to human perception revealed they are a poor approximation of human perception of symmetry.

1 Introduction

Symmetry is a property of the structure of an object or shape which is believed to play a significant role in the human vision processes of image segmentation and object recognition [1], motivating computer vision approaches aiming to replicate its properties [2–4]. As yet, no approaches appear to have been explicitly guided by the way human perception works, and only recently has perception of symmetry been studied in the context of computer vision [5]. Furthermore, performance evaluations and comparisons of symmetry operators for identifying symmetry axes (hereafter referred to as axes) have only recently been introduced [6,7].

Symmetry exists as a continuous property not just in space but also in strength [5,8–10]: comparing any two regions of symmetry one can be evaluated as more, or less (or the same) symmetric than the other. We term this concept of 'how symmetrical' a region is as its Degree of Symmetry (DoS). We define two measures of DoS: absolute Degree of Symmetry and relative Degree of Symmetry. Absolute DoS is a measure on the scale 0 to 1 from lacking any symmetry to perfectly symmetric. Relative DoS can take any value and is only comparable

© Springer International Publishing Switzerland 2015
G. Bebis et al. (Eds.): ISVC 2015, Part I, LNCS 9474, pp. 265–274, 2015.
DOI: 10.1007/978-3-319-27857-5_24

within a set. It is worth noting that absolute DoS is a scale normalised form of relative DoS.

We investigate how automated operators compare in performance with human perception on the same images. To enable this we calculate the Symmetry Similarity for three extended symmetry operators. The Symmetry Similarity measures the Spearman's rank correlation between two sets of axes, rank-ordered by relative DoS [5]. We use crowd-sourcing to investigate human perception for the same dataset of images so we can then determine to what extent the automated DoS correlates with the human perceived version.

2 Dataset Formation

A suitable image dataset was selected to accurately measure the Symmetry Similarity to human perception of symmetry operators. This was formed from real world photographs containing a range of symmetry axes. As the dataset is being used to measure the performance of DoS the dataset will be a collection of pre-labelled symmetry axes. The symmetry axes cover as wide a range of DoS as possible. The axes also vary in scale within the image from covering near to the whole image to small parts of the image. They cover symmetry of a whole object, parts of an object and between multiple objects. These axes should also not be based entirely on human vision as this would bias the dataset. There was no existing dataset that covered all of these specification requirements so a new dataset was formed, combining images already used for symmetry analysis with additional randomly selected low symmetry images.

In order to achieve the goals of the dataset, multiple sources of symmetry axes needed to be used. To form the dataset 74 axes on 52 images were taken from the training reflection set provided by the symmetry competition dataset [6]. Each image in the dataset was originally picked because it contained strong symmetry. Each axis is the result of labelling by 20 trained students. This means that the axes positions are biased towards human-perceived symmetry. In order to minimise bias a second set of 55 images were randomly selected from Flickr[1] with a suitable Creative Commons licence[2]. As there is no way of calculating DoS random images are likely to provide the widest range of DoS. 75 axes were then generated from the images using the highest confidence results from Loy's SIFT symmetry algorithm for each image [11]. The randomly selected images may contain a low DoS giving a wide range of DoS across the whole dataset. However, as there are many strong symmetry axes from the competition dataset there will also be finer accuracy in the most important part of the scale. The test reflection set from the symmetry competition was not included due to the high complexity of gathering human responses to form an accurate ranked list, in the order N^2 where N is the number of symmetry axes within the dataset.

---

[1] http://www.flickr.com.
[2] http://creativecommons.org/licenses/.

The complete dataset, Degree of Symmetry Axis Set (DoSAS), of 107 images as well as their labelled 149 symmetry axes is available online[3]. The ranked-order from the following pairwise comparisons of human-perceived symmetry is also provided with the dataset.

3   Ranking by Human Perception

Following an earlier approach a ranked-order of the dataset is formed from pairwise comparisons of images [5]. To form an accurate rank of the entire dataset each axis needs to be compared with each other axis at least once; for the dataset of 149 axes this is 11026 pairwise comparisons. In order to achieve this number of responses crowd-sourcing was used. Candidates were presented the question "Which axis initially appears to have stronger mirror symmetry?" along with two images side by side each with a symmetry axis superimposed. Four response options were given: 'Left image', 'Right image', 'Same' and 'Don't know'.

As a monetary gain is involved in crowd-sourcing a number of checks ensured the highest quality data was kept and could be validated. Firstly, 30 images were selected at random from the synthetic dataset devised previously for Symmetry Similarity analysis were also added to the compared images. By including images which already had an accurate rank, labelled by non-crowd-sourced means, the results could be correlated providing validation that the comparisons obtained were accurate. The crowd-sourcing platform used was CrowdFlower[4] which itself has a number of tools which were utilised to maximise the accuracy of the responses. Each candidate responding to questions must achieve a 70 % or greater accuracy on test questions with gold standard answers; if their accuracy falls below this threshold all of their responses are discounted. Questions which should have a binary response were chosen to be test questions; where one axis clearly had a higher value of DoS than the other. The user was still not forced to respond to symmetry a certain way as a threshold less than 100 % allowed for people to disagree with some of the gold standard answers. Further quality control included limiting the number of responses from a single candidate and limiting the percentage of responses which could be 'Don't Know' to 20

In total 20,441 accepted responses were collected from 203 participants. These responses were then ranked using the Bradley-Terry model extended to accommodate the occurrence of ties (Fig. 1) [12,13]. The correlation of the resulting ranked-order for the synthetic images with the ranked-order obtained previously without crowd-sourcing [5] is 0.964. This high correlation validates the reliability of the responses obtained from the crowd-sourcing platform. To show the results are also consistent with themselves the pairwise comparisons were randomly split into two halves and each ranked. The correlation between the two ranked-orders is 0.966 showing high consistency.

---

[3] http://users.ecs.soton.ac.uk/pmf1g09/symsimilarity.
[4] http://www.crowdflower.com.

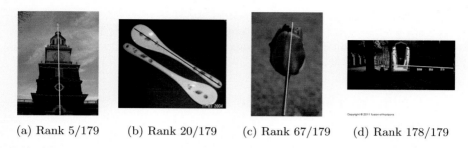

(a) Rank 5/179   (b) Rank 20/179   (c) Rank 67/179   (d) Rank 178/179

Fig. 1. Example axes from the new dataset, DoSAS, and their human perceived DoS rank

4   Symmetry Similarity of Computer Vision Operators

This new ranked dataset of real world images can be used to measure the Symmetry Similarity to human perception of existing symmetry operators. However, other than the Continuous Symmetry Distance operator [3] no symmetry operator has been specifically designed to give a DoS value. Although the Continuous Symmetry Distance operator already provides a continuous value for symmetry it was not evaluated because it requires the input to be in the form of vertex sets: our dataset does not provide sets of vertices for each axis and it is not feasible to calculate the vertices of a symmetric object from the image as the object or part of the image in question for each axis is unknown.

Some symmetry operators can be adapted to give a value for DoS. Of these three of the most promising were chosen to be studied in more detail; the Discrete Symmetry Transform [2], Loy's SIFT interest point operator [11], and Griffin's new Derivative-of-Gaussian Orbifold operator [14]. All values for the Symmetry Similarity were calculated using the human perceived DoS rank-ordered dataset. The Symmetry Similarity can range from –1 to 1; inversely correlated to correlated.

4.1   Generalised Symmetry Transform

One of the earliest symmetry operators is the Generalised Symmetry Transform (GST) [2]. The GST was designed in order to replicate attention mechanisms of human vision although is not guided by it; beyond the aim however, human perception is never referred to and no comparison to it is made. The GST calculates a DoS map from the phase and magnitude of the intensity gradient by a voting process. The intensity gradient of the image focuses the GST on the shape of objects and so could be expected to perform comparably with perception.

Using the relative DoS map for the entire image, I_S, a single relative DoS value for a symmetry axis, S, can be calculated by averaging the value of DoS over the whole axis. This is a simple calculation if the symmetry axis is parallel to one of the image axes however becomes more complex if the symmetry axis is diagonal, as is most likely. In order to determine the most accurate average DoS for a diagonal axis a line is used to represent the axis location. The anti-aliased line drawing algorithm proposed by Wu [15] was adapted to be used as a mask, M, to calculate the DoS for an axis, S. The DoS is averaged over the entire axis rather than summed in order to normalise the DoS for different length axes (Fig. 2).

$$S = \frac{\sum_{i,j \in I_S} I_S(i,j)M(i,j)}{\sum_{i,j \in I_S} M(i,j)} \tag{1}$$

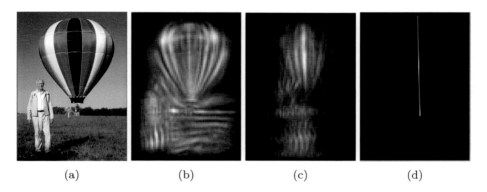

(a)   (b)   (c)   (d)

Fig. 2. (a) Input image (b) GST, $\sigma = 5$ (c) GST angular accumulator, $\sigma = 20$ (d) Dataset axis mask

Symmetry Similarity of the GST was calculated after all images in the dataset were scaled so the longest dimension of the image was 100px for necessary speed. As the GST calculates the relative DoS in an accumulator space based upon edges the DoS is also dependent upon the density of edges. A symmetric object with two pairs of edges will score twice as highly as a symmetric object with only a single pair of edges. It is not possible to normalise the DoS by the number of contributing points using the standard GST as many points contribute to an accumulator bin which do not form part of the axis. By extending the GST to include an extra accumulator dimension for the angle normal to the vertex between each pair of points, the number of points contributing to an axis can be isolated. This allows for normalising the DoS for each axis. We used 18 angular bins each representing 10°. For greater accuracy due to the low resolution the accumulator was interpolated.

The Symmetry Similarity results of both the un-normalised GST and the normalised GST are shown in Fig. 3. The un-normalised GST has a negative Symmetry Similarity for low σ and tends towards to a value of 0.1 as σ reaches the size of the image. This is almost the inverse of the normalised GST which initially has a positive Symmetry Similarity with low σ values and drops to a Symmetry Similarity of -0.176 at a σ of 22.2 before finally tending towards -0.07. Although the negative value shows an inverse correlation it would be easy to reverse the values to form a positive correlation.

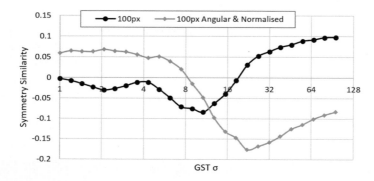

Fig. 3. Symmetry Similarity of GST operator through changing scale

4.2 SIFT Interest Points

Loy's symmetry operator matches pairs of symmetric SIFT (Scale-Invariant Feature Transform) interest points to find symmetry axes [11]. Symmetric pairs of interest points vote towards possible symmetry axes in a Hough style accumulator space. The weighting of the votes is very similar to the Generalised Symmetry Transform: the weighting based uses the features phase coherence, scale similarity, and separating distance. Although Loy's interest point operator does not directly provide a value for DoS it has been included as it has been considered as the current best operator for reliably finding a symmetry axis [6].

The accumulator value is considered as a confidence value that an axis is symmetric and for our purposes is considered as the relative DoS of the symmetry axis. In order to match symmetry axis found by Loy's interest point operator to the axes within the dataset the same algorithm was used as proposed by Liu et al. for axis matching [6]. When a positive match is found the dataset axis is assigned a DoS equal to that of the confidence value of the matched operator output axis and is normalised by dividing by the number of pairs of matched pairs of interest points. This normalises for differences in the density of interest points. In the case where multiple operator output axes are matched to a single dataset axis the confidence value all matched axes are averaged. If no matches are found for an axis it is excluded from the dataset.

Small adjustments to any of the input parameters had large affects on the Symmetry Similarity of the Loy's interest point operator. Whilst calculating the Symmetry Similarity all images were resized so their maximum dimension was constant. Since the resolution affects the interest points found the resolution was one of the varied input parameters. A simple hill-climbing algorithm designed to find the best input parameters struggled due to the large number of local maxima. A frequency distribution of many sets of input parameters, Fig. 4a, shows that the Symmetry Similarity results begin to form a gamma distribution reflecting randomness. There is also a wide range in the number of axes which are successfully matched. The hill-climbing algorithm aimed to optimise the product of the Symmetry Similarity with the number of matched axes. Using both variables ensured that as many images from the dataset were used as possible. The resulting distribution of the number of matched axes is shown in Fig. 4b. The percentage of matched axes from the dataset never exceeded 70 %, for some axes a match was never found. Many of the axes not matched have a low DoS which Loy's algorithm is not designed to find so cannot be assigned a DoS value.

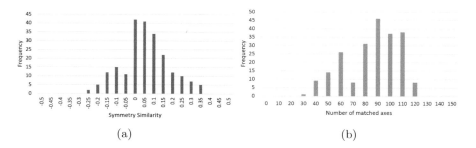

(a)   (b)

Fig. 4. Frequency distribution of Symmetry Similarity and number of matched axes with range of parameters to Loy's interest point operator

Loy's operator depends largely on the number of interest points found within an image, specifically the mirrored area. The more interest points that are paired the more accurate relative DoS can be assigned to the axis. Small changes in parameters have a large affect upon the operator changing which interest points are matched and so the DoS assigned. As symmetry axis with a low DoS are ignored early on for computational optimisation many axes are not assigned a DoS. Although these are ignored when calculating the Symmetry Similarity they still have a large affect. Without the low DoS axes the range being ranked is much smaller, with finer differences, making them harder to order correctly.

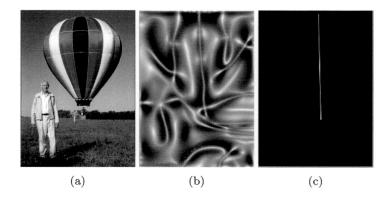

(a)   (b)   (c)

Fig. 5. (a) Input image (b) DtG continuous symmetry, $\sigma = 8$ (c) Dataset axis mask

4.3   Derivative-of-Gaussian Filters

Griffin's Derivative-of-Gaussian (DtG) filters can be used to calculate information about local image structure including symmetries [14,16]. Griffin's DtG operator poses the advantage to us that it can most easily provide a normalised DoS value and the only one which can calculate absolute DoS. This operator is naturally much faster to compute as the filters can be applied in the frequency domain. By combining the six DtG filters up to the second order extrinsic properties of the image can be removed leaving only the intrinsic properties, such as symmetry, in an orbifold space (l, b, a). Each area corresponding to a different symmetry properties are defined by Griffin et al. [14]. Absolute DoS can be calculated for each point within the orbifold using a distance function from the areas which correspond to the property we are interested in. The exterior surface of the orbifold (Eq. 2) represents points which have a finite number of mirrors, intersecting at that point.

$$|l| \leq \frac{\pi}{2} \wedge b \in \left[0, \frac{\pi}{2}\right] \wedge a \in \left\{0, \frac{\pi}{2}\right\} \tag{2}$$

In our new distance function the surface of the orbifold has a DoS of 1 and the furthest point from the surface, the centre of the orbifold, has a DoS of 0. In order for the whole surface to represent a perfect symmetry value (1) and the furthest point from any surface, the origin, to represent no symmetry (0) the orbifold surface can be mapped into a normalised unit-radius sphere. This can be done based upon the metric tensor and Euclidean embedding of the orbifold space. Equation 3 transforms the orbifold space (l, b, a) to a unit circle which can then be easily used to calculate absolute DoS, S; as shown in Eq. 4.

$$\begin{pmatrix} x \\ y \\ z \end{pmatrix} = \begin{pmatrix} \frac{4}{\pi} \left(b - \frac{\pi}{4}\right) \cos l \\ \frac{4}{\pi} \left(a - \frac{\pi}{4}\right) \cos l \sin 2b \\ -\frac{2l}{\pi} \end{pmatrix} \tag{3}$$

$$S = \sqrt{x^2 + y^2 + z^2} \tag{4}$$

Using the absolute DoS for each pixel the absolute DoS for a symmetry axis can be calculated by averaging over an anti-aliased line in the same way as defined in Sect. 4.1. It was found that a non-linear mapping $S' = S^2$ gives for a marginally higher Symmetry Similarity and is faster to calculate (Fig. 5).

Each image was resized so that the largest dimension was a constant value. Griffin's DtG operator is a scale space operator, changing the resolution shifts the σ by which the same response is obtained. Figure 6 shows this, at each tested resolution (200px, 400px, and 600px) the σ required to gain comparable Symmetry Similarity shifts. Of the parameters tested a resolution of 600px with a σ of 26.62 achieved the highest Symmetry Similarity of 0.161. Griffin's DtG operator is a local symmetry operator so at high values of σ, close to the image resolution, the DtG performs poorly due to boundary conditions. Many symmetry axes within the dataset are the focus of the image so cover the majority of the image. For these axes the DtG struggles to optimally calculate the DoS for the whole object as a human would.

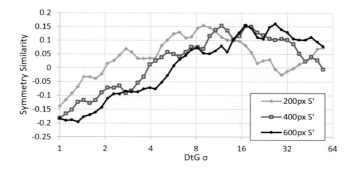

Fig. 6. Griffin's DtG operator Symmetry Similarity with varied resolution and σ

Griffin's DtG operator should perform similarly to the mathematical model MI (direct comparison of pixel intensities) as both operators calculate symmetry based on pixel intensity [5]. MI compares corresponding pixel intensities either side of an axis calculating absolute DoS. MI was shown to have a Symmetry Similarity of 0.170 for a synthetic dataset so it can be expected Griffin's DtG operator achieves a similar maximum Symmetry Similarity.

The correlation between the highest Symmetry Similarity points for the unnormalised GST (σ of 95.4) and Griffin's DtG operator (σ of 26.6) is -0.07. This distinct lack of correlation shows that although both algorithms do not conform to human perceived DoS they do not follow the same alternate model.

5 Discussion

We generated a consistent ranked order of symmetry axes to represent human perceived relative DoS using crowd-sourcing. The collected responses when validated against previously collected data have a very high correlation showing the quality of data remains high even when collected by crowd-sourcing. The newly formed dataset contains a wide range of DoS, allowing for performance evaluation of operators.

We show that calculating the DoS by existing symmetry operators does not have high correlation against the dataset ranked by human perception. Three symmetry operators were evaluated for Symmetry Similarity; the Generalised Symmetry Transform, Loy's interest point operator, and Derivative of Gaussian operator. All three evaluated operators have at best a low Symmetry Similarity. Both the GST and Loy's interest point operator do not calculate a normalised DoS meaning raw values should not be compared directly to one another. The chance of consistency by Loy's operator appears to be low; small changes on the inputs can lead to large changes in the resulting Symmetry Similarity.

There are a wide range of axis lengths and mirrored areas within the dataset meaning the scale space operators, GST and DtG, do not use the optimal scale for each individual axis rather only the optimal overall scale. It may be possible to combine the results of multiple scales when calculating a DoS of an axis to find the best scale for each axis. It is likely however that this will only provide small improvements to the Symmetry Similarity.

Of these operators Griffin's DtG operator appears the most promising for providing a normalised absolute DoS value it however performs poorly on a global scale. Each operator has inherent issues with the way they calculate DoS: in order to closely replicate human perception a new approach is required.

References

1. Treder, M.S.: Behind the looking-glass: a review on human symmetry perception. Symmetry **2**, 1510–1543 (2010)
2. Reisfeld, D., Wolfson, H., Yeshurun, Y.: Context-free attentional operators: the generalized symmetry transform. IJCV **14**, 119–130 (1995)
3. Zabrodsky, H., Peleg, S., Avnir, D.: Symmetry as a continuous feature. IEEE TPAMI **17**, 1154–1166 (1995)
4. Zavidovique, B., Di Gesù, V.: The s-kernel: a measure of symmetry of objects. Pattern Recogn. **40**, 839–852 (2007)
5. Forrest, P.M.: Symmetry stability: Learning from human perception, ECS, University of Southampton, VLC (2015)
6. Liu, J., Slota, G., Zheng, G., Wu, Z., Park, M., Lee, S., Rauschert, I., Liu, Y.: Symmetry detection from real world images competition 2013: summary and results. In: IEEE CVPR Workshops, pp. 200–205 (2013)
7. Rauschert, I., Brocklehurst, K., Kashyap, S., Liu, J., Liu, Y.: First symmetry detection competition: summary and results. CSE Dept., The Penn. State University, Technical report (2011)
8. Csathó, Á., van der Vloed, G., van der Helm, P.A.: The force of symmetry revisited: symmetry-to-noise ratios regulate (a) symmetry effects. Acta psychol. **117**, 233–250 (2004)
9. Masuda, T., Yamamoto, K., Yamada, H.: Detection of partial symmetry using correlation with rotated-reflected images. Pattern Recogn. **26**, 1245–1253 (1993)
10. Zabrodsky, H., Algom, D.: Continuous symmetry: a model for human figural perception. Spat. Vis. **8**, 455–467 (1994)
11. Loy, G., Eklundh, J.-O.: Detecting symmetry and symmetric constellations of features. In: Leonardis, A., Bischof, H., Pinz, A. (eds.) ECCV 2006. LNCS, vol. 3952, pp. 508–521. Springer, Heidelberg (2006)
12. Bradley, R.A., Terry, M.E.: Rank analysis of incomplete block designs: I. the method of paired comparisons. Biometrika **39**, 324–345 (1952)
13. Caron, F., Doucet, A.: Efficient Bayesian inference for generalized Bradley-Terry models. J. Comput. Graph. Stat. **21**, 174–196 (2012)
14. Griffin, L.D., Lillholm, M.: Symmetry sensitivities of derivative-of-gaussian filters. IEEE TPAMI **32**, 1072–1083 (2010)
15. Wu, X.: An efficient antialiasing technique. In: ACM SIGGRAPH Computer Graphics, vol. 25, pp. 143–152. ACM (1991)
16. Griffin, L.D.: The second order local-image-structure solid. IEEE TPAMI **29**, 1355–1366 (2007)

UT-MARO: Unscented Transformation and Matrix Rank Optimization for Moving Objects Detection in Aerial Imagery

Agwad ElTantawy[(✉)] and Mohamed S. Shehata

Electrical and Computer Engineering, Faculty of Engineering and Applied Science,
Memorial University of Newfoundland, St. John's, NL, Canada
{agwad.eltantawy,mshehata}@mun.ca

Abstract. Aerial imagery is widely used in many civilian and military applications, as it provides a comprehensive view and real-time surveillance. Automated analysis is an essential task of aerial imagery to detect moving objects, however, the shakiness of these images and the small size of the moving objects are major challenges facing such task. This paper proposes UT-MARO, a novel moving object detection technique. UT-MARO achieves high accurate detection of small-size moving objects in shaky aerial images with low computation complexity and is composed of two phases: (1) UT-alignment and (2) MARO-extraction. UT-alignment utilizes unscented transformation to first align shaky images, then in the second phase, MARO-extraction detects small moving objects by extracting the background using low rank matrix optimization. The robustness of the proposed technique is tested on DARPA and UCF aerial images datasets; and the obtained results prove that UT-MARO has the best performance with lowest complexity compared to relevant current state of the art techniques.

1 Introduction

Airborne vehicles (e.g., drones, balloons or aircrafts) are typical platforms that can have a camera mount to them to capture aerial imagery. Such imagery provides bird-eye view that allows more comprehensive vision of a scene than ground-based images while still providing real-time performance. Therefore, many civilian and military applications heavily depend on such imagery to accomplish different missions such as traffic monitoring [1], oil spill detection [2] search and rescue [3], and reconnaissance missions [4].

One of the most essential and challenging tasks of automated analysis in aerial imagery is moving object detection. The objects in aerial imagery appear very small because of the high altitude of the airborne vehicles. Moreover, the continuous motion of such vehicles causes captured images to be shaky and consequently, distinguishing between real moving objects and static background elements (which will appear as moving) is very challenging.

Generally, aerial images consist of background and moving objects. The background is considered the main structure of the image while moving objects can be considered as elements corrupting this structure. Given a sequence of images arranged in a matrix, where each image is a column vector in the matrix, the original structure of the matrix

© Springer International Publishing Switzerland 2015
G. Bebis et al. (Eds.): ISVC 2015, Part I, LNCS 9474, pp. 275–284, 2015.
DOI: 10.1007/978-3-319-27857-5_25

(i.e. the background) is extracted by calculating the optimal low rank matrix (i.e., the minimum number of linearly independent columns of the matrix). Therefore, the moving object detection task can be formulated as the following optimization problem:

$$\min_{B} \quad \text{Rank}(B), \quad \text{s.t.} \quad F = B + O \tag{1}$$

where *"B"* is the background, *"F"* is the matrix of images, and *"O"* is the matrix of moving objects.

The background cannot be calculated by optimizing the former equation directly due to the shaky nature of aerial images that corrupts the background, i.e. the background elements are not linearly correlated. Therefore, image alignment is required to overcome the shakiness effect and ensure that the background elements of the images are linearly correlated.

This paper proposes a novel technique based on Unscented Transformation and MAtrix Rank Optimization (UT-MARO) to detect moving objects in aerial imagery. UT-MARO achieves high detection rate with low computations compared to current state of the art techniques and it consists of two phases: (1) UT-alignment and (2) MARO-extraction. UT-alignment is a novel technique for image alignment based on unscented transformation [5] to overcome existing image alignment techniques short-comings, such as low accuracy and high complexity. In the proposed UT-alignment, SURF feature points [6] are initially detected and matched between images. Then, the matched points are propagated through unscented transformation, which will reduce the number of needed matched points, to get an accurate affine transformation matrix used in image alignment. In the second phase, MARO-extraction detects moving objects, based on phase one aligned images, by obtaining the optimal low rank matrix that represents the background. The proposed MARO-extraction is novel as it utilizes inexact augmented langrage multiplier to achieve accurate and fast extraction of the low rank matrix that represents the background.

The remainder of this paper is organized as follows: related work is provided in Sect. 2. Then Sect. 3 illustrates the proposed moving objects detection technique. The experiments and results are presented in Sect. 4. Finally, Sect. 5 concludes the paper as well as provides the future directions.

2   Related Work

Calculating optimal low rank matrix of aerial imagery can be used for detecting small size moving objects from moving camera platforms instead of tradition techniques such as background modeling and motion-based techniques [7, 8]. Oreifej et al. [9] demonstrated the use of low rank matrix optimization for moving object detection from static ground camera platforms. A video is divided into background, noise and moving objects based on their intrinsic properties. The decomposition is achieved using augmented Lagrange multiplier. Another algorithm, called DEtecting Contiguous Outliers in the Low-rank Representation (DECOLOR), was proposed in [10] to detect moving objects from moving camera platforms. It decomposes a video into moving objects and background using Lagrange multiplier. To overcome the shakiness of images captured by moving camera platform, it uses iterative first order Taylor series [11] for image alignment within the video decomposition. More accurate and faster detection is proposed by ElTantawy et al. [12] who used inexact augmented Lagrange multiplier (IALM), instead of the traditional Lagrange multiplier used in DECOLOR, to decompose a video into moving object and background. However, the execution time in [12] is still high due to the iterative first order Taylor series used for image alignment which is a substep in the video decomposition. Iterative first order Taylor series starts by considers an initial guess for the transformation matrix then uses Taylor series iteratively to correct the matrix values using matched feature points between the input image and the reference image. This technique has many drawbacks because it heavily depends on guessing proper initial values for the transformation matrix; also it has high execution time due to the fact that it iteratively corrects the transformation matrix values. In this paper, the MARO-extraction phase successfully lowers the video decomposition complexity compared to the method in [12], DECOLOR [14], and 3-term decomposition [13] by removing the alignment process from the iterative calculations and performing it once at the beginning and by also using a simpler technique for alignment based on unscented transformation.

3   UT-MARO

3.1   UT-Alignment

In the first phase, the alignment of images is achieved using five steps, shown in Fig. 1, to calculate the transformation matrices that map all images to a reference image (initially it is the first image in the sequence) using unscented transformation.

Step 1-SURF feature detection. SURF feature points are detected in the current image.

Step 2-SURF feature matching. The detected SURF feature points in the current image are matched to the SURF features of the reference image (SURF feature points are assumed to be perfectly matched using RANSAC [13]) generating two lists of matched SURF feature points: "Mtch_crnt" which is a list that contains matched SURF feature points in the current image, and "Mtch_ref" which is a list that contains matched SURF feature points in the reference image.

Step 3-Unscented transformation. The two lists, Mtch_crnt and Mtch_ref, are propagated through unscented transformation to encode the statistical properties of both lists including the mean and the covariance matrix for each list. The calculated means and covariance matrices are then used to generate two sets of five sigma points: "Core_sigma_crnt" and "Core_segma_ref" which are considered as reduced representation of Mtch_crnt and Mtch_ref lists and have the same statistical properties.

The calculation of "Core_sigma_crnt" and "Core_segma_ref" is done using the following Eqs. 2–4.

$$\sigma_0 = \mu \tag{2}$$

$$\sigma_i = \sigma_0 + (\sqrt{(d)} \sum), \quad i = 2,4 \tag{3}$$

$$\sigma_i = \sigma_0 - (\sqrt{(d)} \sum), \quad i = 1,3 \tag{4}$$

where σ, μ, $\sum$ and d are the sigma point, mean, covariance matrix and matched feature points dimensions, respectively.

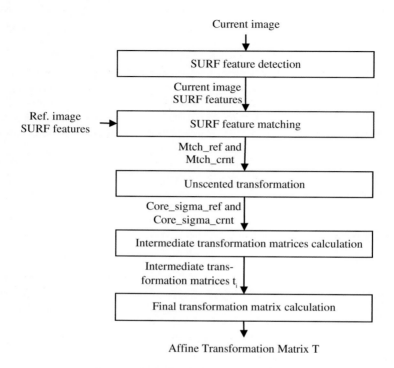

Fig. 1. Image alignment technique pipeline

Step 4-Intermediate transformation matrices calculation. The corresponding pairs of Core_sigma_crnt and Core_sigma_ref are used to calculate five intermediate similarity transformation matrices "t_i" between the current frame and the reference frame. Since the minimum number of points required for calculating an affine transformation matrix is four points, extra four sigma points are calculated around each core sigma point in Core_sigma_crnt and Core_sigma_ref. To keep the statistical properties of Mtch_crnt and Mtch_ref unchanged, the same covariance matrix is used to calculate the extra sigma points. Figure 2 provides graphical illustration of step 4.

Step 5-Final transformation matrix calculation. The final affine transformation matrix "T" which will be used for image alignment is calculated as a weighted average of the five intermediate transformation matrices, as shown in the following equation:

$$T = \frac{1}{5} \sum_{i=0}^{5} w_i t_i. \tag{5}$$

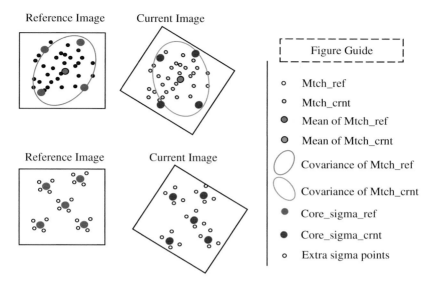

Reference Image   Current Image

Figure Guide

- ○ Mtch_ref
- ○ Mtch_crnt
- ◉ Mean of Mtch_ref
- ◉ Mean of Mtch_crnt
- ⬭ Covariance of Mtch_ref
- ⬭ Covariance of Mtch_crnt
- ● Core_sigma_ref
- ● Core_sigma_crnt
- ○ Extra sigma points

Reference Image   Current Image

Fig. 2. Propagating matched SURF feature points through unscented transformation; calculating core sigma points and extra sigma points

where t_i is intermediate affine transformation matrix. The weights w_i are:

$$w_0 = \zeta / (d + \zeta) \tag{6}$$

$$w_i = \zeta / 2(d + \zeta), \quad i = 1, \dots, 4 \tag{7}$$

where ζ is a scaling parameter and d is matched feature points dimensions. To adapt with high shakiness in the images, a new reference image is taken each time it's the transformation matrix is found to be larger than a threshold.

3.2   MARO-Extraction

Lin *et al.* describes in [14] the solution to the low rank matrix optimization problem using different methods such as iterative threshold method, accelerated proximal gradient method, dual approach, or augmented Lagrange multiplier methods. The best method for solving such problem is obtained using augmented Lagrange multiplier (ALM) methods which calculate the exact optimal solution, instead of an approximation of the optimal solution (e.g. iterative threshold approaches and accelerated proximal gradient approaches) and also converge to the optimal solution Q-linearly [15]. The inexact augmented Lagrange multiplier (IALM) [14] has computation complexity of five times lower than other augmented Lagrange multiplier methods and hence, UT-MARO proposed in this paper utilizes IALM to calculate the low rank matrix.

To solve the optimization problem using IALM, UT-MARO reformulates Eq. 1 in Lagrange function form as follows:

$$\ell(B,O,Y) = \|B\|_* + \lambda \|O\|_1 + \langle Y, F - B - O \rangle + \frac{\mu}{2} \|F - B - O\|_f^2 \tag{8}$$

where Y is the Lagrange multiplier matrix while μ and λ are weighting parameters. $\|x\|_*, \|x\|_1, \|x\|_f^2$ and $<x,y>$ denote nuclear norm, l1-norm and Frobenious norm, and inner product, respectively.

The optimal low rank matrix cannot be solved directly from Eq. 8 due to its high complexity. To solve this issue, the Lagrange function is optimized over its parts, i.e. background *"B"*, moving objects *"O"* and Lagrange multiplier matrix *"Y"* individually as follows:

$$B = \arg\min_{B} \|B\|_* + \langle Y, - B \rangle + \frac{\mu}{2} \|F - B - O\|_f^2 \tag{9}$$

$$O = \arg\min_{O} \|O\|_1 + \langle Y, - O \rangle + \frac{\mu}{2} \|F - B - O\|_f^2 \tag{10}$$

$$Y = \arg\min_{Y} Y + \frac{\mu}{2} \|F - B - O\|_f^2 \tag{11}$$

The background *"B"* and the moving objects *"O"* are obtained by iteratively minimizing Eqs. 9, 10 and 11. The minimization of Eq. 3 is calculated using singular value decomposition (SVD), as shown in Eqs. 12 and 13.

$$(U,S,V) = \text{svd}(F + \mu^{-1} Y - O) \tag{12}$$

$$B = US_{\mu^{-1}}[S]V^T \tag{13}$$

Equation 14 is minimized using convex optimization theory as follows:

$$O = S_{\lambda\mu^{-1}}[F + Y - B] \tag{14}$$

The moving objects detection technique pseudo code is as follows:

```
Function Moving_Object_Detection(F)
   Y[0]  = sgn(F)/J(F);
   µ[0]  = 1.25/norm(F);
   λ     = 1000;
   //detection moving Objects
   While not converge do
      (U,S,V) = SVD(F[i]+Y[i]/µ[i]-O[i]);
      B[i+1]  = US_{1/µ[i]}[S]V^T
      O[i+1]  = S_{λ/µ[i]}(F[i]+Y[i]/ µ[i]-B[i+1]);
      Y[i+1]  = Y[i]+µ[i](F[i]-B[i+1]-O[i+1]);
      µ[i+1]  = ρµ[i];
   End while
   Output O;
End Function
```

4   Experiments and Results

Extensive testing based on three experiments has been conducted to evaluate UT-MARO:

Experiment 1. The detection accuracy of UT-MARO is evaluated using challenging datasets of aerial imagery, namely: DARPA dataset [16] and UCF aerial action dataset [17]. DARPA dataset contains multiple sequences captured in different rural area and highways outdoor environments. The sizes of moving objects vary from 15×15 pixels to 50×50 pixels. UCF aerial action dataset contains smaller size moving, i.e. about 20×40 pixels, objects compared to DARPA as well as the images are more shaky.

Experiment 2. The relevant current state of the art techniques, namely: DECOLOR [10], the method in [12], and 3-tem decomposition [9] are evaluated using same datasets (DARPA and UCF). Then the results of these techniques are compare to the proposed algorithm, UT-MARO.

Experiment 3. The execution time is calculated and compared for the proposed UT-MARO, DECOLOR [10], the method in [12], and 3-tem decomposition [9]. The used computer in this experiment has the following specifications: processor Core (TM) i5-4200U CPU@ 1.60 GHz 2.30 GHz, x64-based processor, RAM 8 GB. The execution time is recorded at five points based on the number of the size of "F" the matrix of images, i.e. 50, 100, 200, 300, and 400 frames, respectively. All methods are implemented using Matlab code.

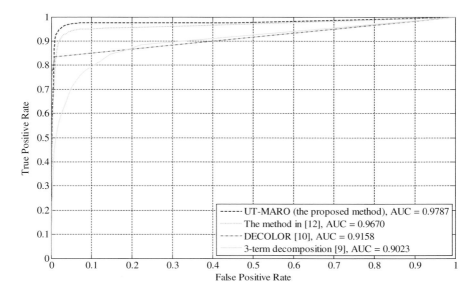

Fig. 3. Receiver operating characteristic (ROC)

The performances of the experiments are depicted using receiver operating characteristic (ROC), with threshold step size 0.02, which shows the relationship between true positive rate (TPR) and false positive rate (FPR). As shown in Fig. 3, TPR of UT-MARO is very high, while it FPR is low. The TPR of the method in [12] is decreased. A noticeable reduction of the TPR occurred in DECOLOR and 3-term decomposition, however, DECOLOR has a very low FPR compared with 3-term decomposition. In summary, the best performance is achieved by UT-MARO with an area under the curve (AUC) in the ROC figure of 0.9787. The method in [12] has lower AUC in the ROC figure with 0.9670. Finally, the AUC for DECOLOR and 3-tem decomposition are 0.9158 and 0.9023 respectively.

Figure 4 shows the execution time of each method. In DECOLOR, larger frame matrix "F" have higher execution time, due to the iterative calculations of the transformation matrices between consecutive frames. However, DECOLOR execution time is better than the 3-term decomposition. The method in [12] has a better execution time with lower effect of changes in "F" matrix size (5.8 s on average). The best execution time is 1.7 s on average and is achieved by UT-MARO and also is only slightly affected by "F" matrix size.

Figure 5 shows sample results of UT-MARO, the method in [12], DECOLOR, 3-term decomposition. UT-MARO has high detection rates, although, the method in [12] has low false detection, there are larger miss detections compared with UT-MARO. DECOLOR has low detection rate with low false detections, in contrary, 3-term decomposition has a bit high detection rate with high false detections.

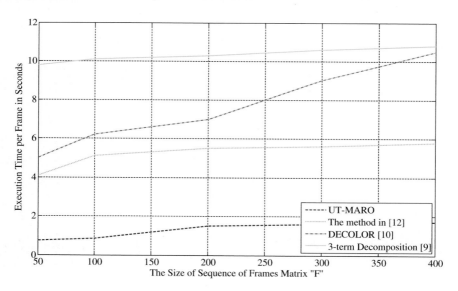

Fig. 4. The execution time of low rank based methods

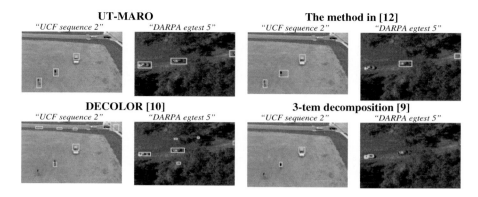

Fig. 5. Sample results of evaluated detection methods

5   Conclusion and Future Directions

In this paper, a novel technique is proposed for accurate and fast moving objects detection from aerial images. It is composed of two phases; UT-alignment that utilizes SURF feature point with unscented transformation to align shaky aerial images, and MARO-extraction that detects small moving objects by obtaining low rank matrix optimization using IALM. The experimental results show that UT-MARO has best accuracy and lowest execution time compared to the current state of the art techniques. The future directions include testing on more challenging datasets to evaluate UT-MARO in different environments.

References

1. Puri, A., Valavanis, K.P., Kontitsis, M.: Statistical profile generation for traffic monitoring using real-time UAV based video data. In: Mediterranean Conference on Control and Automation, MED 2007, pp. 1–6 (2007)
2. Kristiansen, R., Oland, E., Narayanachar, D.: Operational concepts in UAV formation monitoring of industrial emissions. In: IEEE 3rd International Conference on Cognitive Infocommunications (CogInfoCom), 2012, pp. 339–344 (2012)
3. Ryan, A., Hedrick, J.K.: A mode-switching path planner for UAV-assisted search and rescue. In: 44th IEEE Conference on Decision and Control, 2005 and 2005 European Control Conference. CDC-ECC 2005, pp. 1471–1476 (2005)
4. Wei-Long, Y., Luo, L., Jing-Sheng, D.: Optimization and improvement for multi-UAV cooperative reconnaissance mission planning problem. In: 2014 11th International Computer Conference on Wavelet Active Media Technology and Information Processing (ICCWAMTIP), pp. 10–15 (2014)
5. Julier, S.J., Uhlmann, J.K.: Unscented filtering and nonlinear estimation. Proc. IEEE **92**, 401–422 (2004)

6. Bay, H., Tuytelaars, T., Van Gool, L.: SURF: speeded up robust features. In: Leonardis, A., Bischof, H., Pinz, A. (eds.) ECCV 2006, Part I. LNCS, vol. 3951, pp. 404–417. Springer, Heidelberg (2006)
7. Ali, S., Shah, M.: COCOA - tracking in aerial imagery. In: Computer Vision (2005)
8. Sheikh, Y., Javed, O., Kanade, T.: Background subtraction for freely moving cameras. In: IEEE 12th International Conference on Computer Vision, pp. 1219–1225. IEEE (2009)
9. Oreifej, O., Xin, L., Shah, M.: Simultaneous video stabilization and moving object detection in turbulence. IEEE Trans. Pattern Anal. Mach. Intell. **35**, 450–462 (2013)
10. Xiaowei, Z., Can, Y., Weichuan, Y.: Moving object detection by detecting contiguous outliers in the low-rank representation. IEEE Trans. Pattern Anal. Mach. Intell. **35**, 597–610 (2013)
11. Foy, W.H.: Position-location solutions by Taylor-series estimation. IEEE Trans. Aerosp. Electron. Syst. 187–194 (1976)
12. ElTantawy, A., Shehata, M.: Moving object detection from moving platforms using lagrange multiplier. In: International Conference on Image Processing (ICIP 2015). IEEE (2015)
13. Fischler, M.A., Bolles, R.C.: Random sample consensus: a paradigm for model fitting with applications to image analysis and automated cartography. Commun. ACM **24**, 381–395 (1981)
14. Lin, Z., Chen, M., Wu, L., Ma, Y.: The augmented lagrange multiplier method for exact recovery of corrupted low-rank matrice. In: UIUC (2009)
15. Bertsekas, D.: Enlarging the region of convergence of Newton's method for constrained optimization. J. Optim. Theory Appl. **36**, 221–252 (1982)
16. http://vision.cse.psu.edu/data/vividEval/datasets/datasets.html
17. http://crcv.ucf.edu/data/UCF_Aerial_Action.php. Accessed 11 June 2015

Architectural Style Classification of Building Facade Towers

Gayane Shalunts$^{(\boxtimes)}$

SAIL LABS Technology GmbH, Vienna, Austria
Gayane.Shalunts@sail-labs.com

Abstract. Architectural styles are phases of development that classify architecture in the sense of historic periods, regions and cultural influences. In the scope of an image-based architectural style classification system of building facades the current paper presents the first approach, addressing the problem of architectural style classification of facade towers. Towers are architectural structural elements symbolizing power, characteristic for ecclesiastical and secular monumental buildings, such as churches and city halls. The architectural styles classified are Romanesque, Gothic and Baroque, each spanning a few centuries and geographically widely spread in Europe. The approach is based on clustering and learning of local features. Experiments, conducted on an image database of automatically segmented towers of the observed architectural styles, achieve high classification rate.

Keywords: Building facade towers · Classification · Clustering and learning of local features

1 Introduction

Geographical, geological, climate, religion, social, political and historical factors influenced the formation of architectural styles [1]. Building facade images from online image databases usually do not have any labels related to architectural styles. To know the style of an observed building, one can search for the building name. Whereas this is a solution for famous buildings with well-known names, it is not applicable to buildings lacking names. In this case facade visual information holds the only clue to its style. Here an alternative to an architectural style visual classification tool could be the visual search engines, like Google visual search[1] or landmark recognition engines, like [2]. Nevertheless observations show that whereas those visual engines work well for world famous landmarks, succeeding in mining those buildings and all information about them, they fail for ordinary buildings, not even delivering facades belonging to the same architectural style. An automatic tool for image-based classification of architectural styles will solve this problem, not depending on building popularity or its importance as a cultural heritage object.

---

[1] Google Visual Search Engine http://images.google.com.

© Springer International Publishing Switzerland 2015
G. Bebis et al. (Eds.): ISVC 2015, Part I, LNCS 9474, pp. 285–294, 2015.
DOI: 10.1007/978-3-319-27857-5_26

The research related to architectural style classification of facades emerged in computer vision community recently. The approach introduced in [3], is a four-stage method, incorporating steps of scene classification, image rectification, facade splitting and style classification [3]. The authors in [4] adopt Deformable Part-based Models to capture the morphological characteristics of architectural components and propose Multinomial Latent Regression that introduces the probabilistic analysis and tackles the multi-class problem in latent variable models. [5] performs classification using mining of word pairs and semantic patterns. The authors in [6] propose a hierarchical sparse coding algorithm to model block-lets, representing basic architectural components.

Building facade architectural style classification permits to cluster building image databases by historic periods. Such a semantic categorization limits the search of building image databases to certain category portions and may be useful in building recognition [2,7], Content Based Image Retrieval (CBIR) [8], 3D reconstruction and 3D city-modeling [9]. Inverse procedural modeling aims to reconstruct a detailed procedural model of a building from a set of images or even from a single image [3]. Considering the vast diversity of buildings and their appearances in images, the underlying optimization problem easily becomes intractable if the search space of all possible building styles had to be explored [3]. Thus, the inverse procedural modeling algorithms narrow down their search by implicitly assuming an architectural style [3]. Architectural style classification system may also find an interesting application in virtual tourism [10], as well as in real tourism provided with smart phones.

The current work, being the first approach addressing the classification of building facade towers, is a module in a novel computer vision system, called STYLE, the objective of which is to classify the architectural style of a building facade given its image. Architectural elements, such as windows, towers, domes, columns, are the component parts of buildings. Tower is a building or part of a building that is exceptionally high in proportion to its width and length [11]. The STYLE project proposed the first general algorithm [12] for solving the problem of image-based architectural style classification. The algorithm is 'smart', modeling the human cognitive process of architectural style classification, that is the search of style typical architectural elements on a facade.

The algorithm of the STYLE architectural style classification system comprises three major steps:

(1) Semantic segmentation of facade architectural elements is performed.
(2) Each segmented element is passed to an appropriate module for architectural style classification.
(3) Architectural style voting of the classified elements determines the architectural style of the whole facade.

Among numerous architectural styles there are such that have influenced large regions and dominated a few centuries. The presented article classifies building facade towers of such influential pan-European architectural styles:

Romanesque (8th - 12th centuries)
Gothic (12th - 16th centuries)
Baroque (17th - mid 18th centuries)

Samples of Romanesque, Gothic and Baroque facades featuring towers are shown respectively in Fig. 1a (St. Anton church in Vienna), Fig. 1b (Vienna City Hall) and Fig. 1c (Mariahilf church in Vienna). The towers, segmented automatically by another unit of the STYLE project, are displayed under each image correspondingly in Fig. 1d, e and f. Certain gradient directions are dominating in each tower class. The approach of the tower classification is based on clustering and learning of local features to find out the image dominant gradient directions and thus categorize the classes of diverse architectural styles. The approach yields acceptable classification rate (Sect. 3) on a wide variety of facade towers. The experimental image dataset of towers is an additional contribution of this paper (Appendix A).

In the scope of architectural style classification problem it should be noted about architectural revivalism, which is a phenomenon of imitation of past architectural styles. The singularity of 19th century revivalism, as compared with earlier revivals, was that it revived several kinds of architecture at the same time [13]. These revived styles are also referred to as neo-styles, e.g. Gothic Revival is also referred to as neo-Gothic. The STYLE project does not intend to differ between original and revival architectural styles, as only visual information is not enough for such a discrimination. If there be a need to differ original styles from the revived ones, building date should be provided as an additional feature together with the image. Thus while saying the project addresses the classification of Romanesque, Gothic and Baroque styles, the revivals of those should be also understood by default.

The publications within the scope of the STYLE project tackled the problems of classification of windows [12], classification of architectural elements called tracery, pediment and balustrade [14], as well as the classification [15] and segmentation [16] of domes.

The article is organized as follows. Section 2 explains the methodology of the architectural style classification of towers. The experiments and results of are detailed in Sect. 3. The conclusions are drawn in Sect. 4. The experimental image database with the listing of the buildings is presented in Appendix A.

2 The Classification of Towers

The visual classification of facade towers by architectural styles is a complex problem because of the following factors:

– the big intra-class variety of visually intricate historic architectural styles,
– the unlimited quantity of possible 2D projections (images) of 3D facades.

The tower classification approach is illustrated on the examples of Fig. 1d, e and f, which display the segmented towers of the corresponding Romanesque,

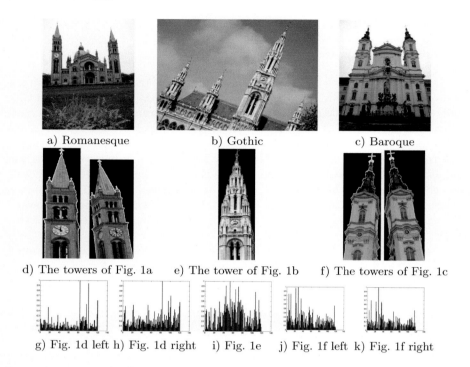

a) Romanesque   b) Gothic   c) Baroque

d) The towers of Fig. 1a   e) The tower of Fig. 1b   f) The towers of Fig. 1c

g) Fig. 1d left h) Fig. 1d right   i) Fig. 1e   j) Fig. 1f left k) Fig. 1f right

Fig. 1. Facades of Romanesque, Gothic, Baroque styles, the segmented towers and the histograms of visual words.

Gothic and Baroque facades, located above (Fig. 1a, b and c). The classification is performed by the same algorithm, which was successfully employed for classification of architectural elements window [12] and tracery, pediment, balustrade [14]. The bag of visual words (BoW) approach [17] (Fig. 2) is chosen for classification. The Scale Invariant Feature Transform (SIFT) [18] is used in the learning phase to extract the information of gradient directions. After performing the difference of Gaussians on different octaves and finding minimas/maximas, i.e. finding interest points, rejection of interest points with low contrast is performed by setting a low threshold. All interest points that lie on tower edges are kept. Note that the original work in [18] is not followed in this step, as the response of the filter along the edges is not suppressed. The last phase is finding the interest points and local image descriptors (SIFT image descriptors) and performing their normalization. Since the number of local features is usually large, learning of visual vocabulary (codebook) from the training set is done by clustering. An unsupervised clustering method (k-means) is used to find the visual cluster centers and to create the codebook of separate classes (Fig. 2).

The classification of a query image begins likewise with the extraction and normalization of SIFT descriptors. The next step is to cluster the SIFT descriptors using the codebook of visual cluster centers, generated during the

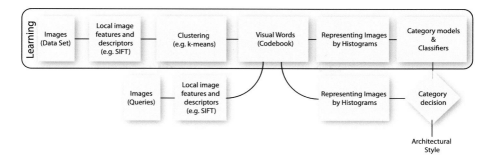

Fig. 2. Learning visual words and classification scheme [12].

training stage. The simple, yet efficient Nearest Neighbor (NN) classifier, which is a special case of the KNN classifier, when $K^2=1$, is applied for clustering. The Eucleadian distances of the query SIFT descriptor to all visual cluster centers are calculated. The SIFT descriptor is assigned to the nearest cluster center. Afterwards all image descriptors are clustered, the image histogram is made, showing the distribution of descriptors among visual cluster centers. The histogram representations of the segmented towers in Fig. 1d are shown in Fig. 1g and h. Similarly Fig. 1i exhibits the histogram of the tower in Fig. 1e, j and k – of the towers in Fig. 1f. The histograms are built using a codebook of 40 cluster centers for each class and are normalized in the interval [0, 1]. The codebook represents the Baroque, Gothic and Romanesque classes by the first, second and third 40 visual cluster centers correspondingly. Note that though the images also share similar visual features, for the Baroque towers the histogram high responses are located on the bins from 1 to 40 (Fig. 1j, k), for the Gothic tower - from 41 to 80 (Fig. 1i) and for the Romanesque towers - from 81 to 120 (Fig. 1g, h). The class, to which the majority of image descriptors belong, is chosen as the image class, i.e. the image class is decided by finding the maxima of integrated responses of the 3 classes. E.g. for the histogram presented in Fig. 1i, the sum of all responses of Baroque class is 8.51, Gothic class – 19.58 and Romanesque class – 8.86. Thus the image is correctly classified as Gothic.

Besides using the KNN classifier with K=1, likewise in [12,14], the author conducts an additional experiment to see, if the classification accuracy may increase on the training dataset by applying K>1 values. But simply using the KNN classifier with equal weights to all K nearest neighbors in a 3-class classification scheme may lead to classification ambiguity, when the classes have equal number of nearest neighbors. To exclude the ambiguous cases and to weigh the closer neighbors more heavily than the farther ones, the distance-weighted KNN rule (WKNN), proposed by [19], is chosen. According to the rule the votes of the different members among K nearest neighbors are weighed by a function of

---

[2] The KNN classifier parameter is noted K (upper case), not to confuse with k parameter of k-means clustering during codebook generation.

their distances to the query object (Eq. 1).

$$w_i = \begin{cases} (d_K^{NN} - d_i^{NN})/(d_K^{NN} - d_1^{NN}) & ; d_K^{NN} \neq d_1^{NN} \\ 1 & ; d_K^{NN} = d_1^{NN} \end{cases} \qquad (1)$$

where d_i^{NN} is the distance of the i-th nearest neighbor to the query object, d_1^{NN} is the distance of the nearest neighbor, d_K^{NN} is the distance of the K-farthest neighbor. In Eq. 1 the nearest neighbor's weight is equal to 1, the farthest neighbor's – to 0, the rest of the neighbors get their weights in the interval (0,1), so that a neighbor with smaller distance scores a heavier weight than one with greater distance.

Nevertheless the experiment (Sect. 3) shows that with the increase of K, the classification precision on the training set decreases, thus the classification of the testing dataset will be led by NN classifer, i.e. K=1 value.

The category model based on the maxima of the integrated class responses proves to be effective, as it makes the vote for the right class strong by integration of the high responses and suppresses the false class peaks.

3   The Evaluation and Results of Tower Classification

An extensive study was conducted to test and evaluate the tower classification approach. One of the challenges for testing the approach is the lack of building databases labelled by architectural styles. Thus the author compiled an image dataset featuring towers from own and Flickr[3] image databases. The tower classification approach receives as input tower images, automatically segmented by another unit of the STYLE project (e.g. Fig. 1d, e, f). The image database, used to evaluate the performance of tower segmentation and classification modules, includes 325 images of 35 buildings, among those famous world landmarks, like Notre Dame Cathedral in Paris, city halls of Vienna and Brussels. In the dataset are represented original and revived Romanesque, Gothic and Baroque architectural styles. The list of the buildings is presented in the Appendix A. Running the tower segmentation tool on 325 input images delivered 544 segmented towers. As a training set for codebook generation are used 102 of the segmented towers and the testing set is formed from the remaining 442 images. Both training and testing image datasets are labelled by architectural styles, the training dataset – for building the visual codebook, the testing set – as ground truth for the evaluation of the classification rate.

An experiment, exercising the values of k for k-means clustering algorithm and SIFT peak threshold (p), is carried out to choose the codebook best fitting the tower image database. Table 1 and Fig. 3a display the results of the experiment, pointing how NN classifier classification rate on the training set depends on k and p. On one hand the value of k should big enough, so that the number of clusters is sufficient to discriminate the visual words of the 3 classes. On the other hand raising the value of k too much tends to make image histograms sparse,

---

[3] http://www.flickr.com.

Table 1. Tower classification accuracy on the training set with different codebook sizes.

Peak Threshold (p)	k = 25	k = 30	k = 35	k = 40	k = 45
0,01	88,24	92,16	85,29	82,35	88,24
0,02	86,27	84,31	88,24	91,18	88,24
0,03	88,24	82,35	88,24	92,16	94,12

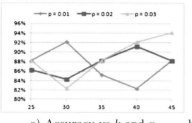 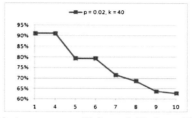

a) Accuracy vs k and p b) Accuracy vs K ($p = 0.02$, $k = 40$)

Fig. 3. Classification accuracy on the training set. Finding the best codebook size (k), SIFT peak threshold (p) and KNN classifier K.

i.e. non-representative visual words take place. The experiment shows that the optimal value of k lies in the range 25–45 (Table 1 and Fig. 3a). Now coming to the analysis of the p parameter: the bigger p value, the less SIFT descriptors are extracted. So while choosing the optimal value of p it should be kept in mind, that a value too big will not extract enough discriminative descriptors, whereas a value too small will overdo by extracting descriptors located on tower construction material textures. The sought value of p is in the range 0.01–0.03 (Table 1 and Fig. 3a). The chosen values to generate the final codebook are $k=40$ and $p=0.02$. Here the preference not in the favor of the pair $k=45$ and $p=0.03$, yielding the highest classification rate, is justified by th fact, that the codebook best performing on the training set may be not general enough to succeed the classification on the testing set.

The goal of the second experiment is to find out, whether classification precision on the training set may increase with application of K>1 values on KNN classifier. K=2 and K=3 values will apparently repeat the result of K=1, since according to Eq. 1 $w_1=1$ and $w_K=0$. Figure 3b shows that whereas K=4 value provides the same classification rate as K=1, with further raise of K the classification rate descends. Thus the classification on the testing database will be performed by NN classifier (K=1), like in [12]. Running the classification with the mentioned codebook on the testing dataset of 442 images results in 86 false classified images, which yields an average classification rate of **80.54 %**. A confusion matrix, with true positives, is given in the Table 2. As the approach uses SIFT features for classification, it is rotation and scale invariant [18].

Table 2. Tower classificaion. Confusion matrix and the accuracy rate.

	Baroque	Gothic	Romanesque	Sum
Baroque	84 (73.0 %)	24	7	115
Gothic	8	171 (87.69 %)	16	195
Romanesque	0	31	101 (76.52 %)	132
Sum	92	226	124	442

4   Conclusion

The article presented the first method in computer vision literature for classification of Romanesque, Gothic and Baroque building facade towers. The approach is based on clustering and learning of local features. The work was carried out in the scope of a computer vision system, targeting the architectural style classification problem of building facades and proposing a general solution as successive phases of segmentation, classification and voting of architectural elements. The performance evaluation on a self-collected image database, including a vast variety of buildings, reported acceptably high classification rate.

A   Appendix: The List of the Database Buildings

The image database, gathered from the author's own and Flickr image datasets to evaluate the performance of the tower segmentation approach, includes images of 35 cathedrals, churches, basilicas and city halls spread in Austria, Germany, Sweden, Czech Republic, Hungary, France, Spain, Luxemburg, England, Belgium, Switzerland and China. The author's own share of the images is freely available to the scientific community here:

https://www.flickr.com/photos/lady_photographer/sets/7215763614955 0844/ The building names are listed below:

1. Vienna City Hall, Austria
2. Votiv church, Vienna, Austria
3. Maria Treu church, Vienna, Austria
4. Jesuit Church, Vienna, Austria
5. Mariahilf church, Vienna, Austria
6. Maria of Siege church, Vienna, Austria
7. Anton church, Vienna, Austria
8. Breitenfelder church, Vienna, Austria
9. Franz of Assisi church, Vienna, Austria
10. Altlerchenfelder church, Vienna, Austria
11. Salzburg Cathedral, Austria
12. Cologne Cathedral, Germany
13. St. Pantaleon Church, Cologne, Germany
14. Bremen Cathedral, Germany

15. Fulda Cathedral, Germany
16. Basilica St. Castor, Koblenz, Germany
17. St. Gall Abbey, Switzerland
18. Notre Dame Cathedral, Paris, France
19. Reims Cathedral, France
20. Abbaye Aux Hommes Caen, France
21. York Minster, England
22. Westminster Abbey, London, England
23. Barcelona Cathedral, Spain
24. Burgos Cathedral, Spain
25. St. Michael and St. Gudula Cathedral, Brussels, Belgium
26. Brussels City Hall, Belgium
27. Church of Our Lady of Laeken, Brussels, Belgium
28. St. Petrus And St. Paulus Church, Ostend, Belgium
29. Loreta church, Prague, Czech Republic
30. St.Peter and St.Paul Church Vysehrad, Prague, Czech Republic
31. St. Mary Magdalene church, Karlovy Vary, Czech Republic
32. St. Stefan's Cathedral, Budapest, Hungary
33. Church Saints Cosmas and Damian, Luxemburg
34. Lund Cathedral, Sweden
35. St. Michael's Cathedral, Qingdao, China

References

1. Fletcher, B.: A History of Architecture on the Comparative Method, 5th edn. Scribner's, Batsford (1920)
2. Zheng, Y.T., Zhao, M., Song, Y., Adam, H., Buddemeier, U., Bissacco, A., Brucher, F., Chua, T.S., Neven, H.: Tour the world: building a web-scale landmark recognition engine. In: Proceedings of the 20th International Conference on Computer Vision and Pattern Recognition (CVPR), Miami, Florida, USA, pp. 1085–1092. IEEE (2009)
3. Mathias, M., Martinovic, A., Weissenberg, J., Haegler, S., Gool, L.V.: Automatic architectural style recognition. In: Proceedings of the 4th International Workshop on 3D Virtual Reconstruction and Visualization of Complex Architectures. International Society for Photogrammetry and Remote Sensing, Trento, Italy, pp. 280–289 (2011)
4. Xu, Z., Tao, D., Zhang, Y., Wu, J., Tsoi, A.C.: Architectural style classification using multinomial latent logistic regression. In: Fleet, D., Pajdla, T., Schiele, B., Tuytelaars, T. (eds.) ECCV 2014, Part I. LNCS, vol. 8689, pp. 600–615. Springer, Heidelberg (2014)
5. Goel, A., Juneja, M., Jawahar, C.V.: Are buildings only instances?: exploration in architectural style categories. In: Proceedings of the ICVGIP 2012, Mumbai, India, pp. 1–8 (2012)
6. Zhang, L., Song, M., Liu, X., Sun, L., Chen, C., Bu, J.: Recognizing architecture styles by hierarchical sparse coding of blocklets. Inf. Sci. **254**, 141–154 (2014)
7. Zhang, W., Kosecka, J.: Hierarchical building recognition. Image Vis. Comput. **25**(5), 704–716 (2004)

8. Li, Y., Crandall, D., Huttenlocher, D.: Landmark classification in large-scale image collections. In: Proceedings of 12th International Conference on Computer Vision (ICCV), Kyoto, Japan, pp. 1957–1964. IEEE (2009)
9. Cornelis, N., Leibe, B., Cornelis, K., Gool, L.V.: 3d urban scene modeling integrating recognition and reconstruction. IJCV **78**, 121–141 (2008)
10. Snavely, N., Seitz, S.M., Szeliski, R.: Photo tourism: exploring photo collections in 3d. ACM Trans. Graph. **25**, 835–846 (2006)
11. Illustrated architecture dictionary: Tower (2015). http://www.buffaloah.com/a/DCTNRY/t/tower.html. Accessed 4 August 2015
12. Shalunts, G., Haxhimusa, Y., Sablatnig, R.: Architectural style classification of building facade windows. In: Bebis, G., Boyle, R., Parvin, B., Koracin, D., Wang, S., Kyungnam, K., Benes, B., Moreland, K., Borst, C., DiVerdi, S., Yi-Jen, C., Ming, J. (eds.) ISVC 2011, Part II. LNCS, vol. 6939, pp. 280–289. Springer, Heidelberg (2011)
13. Collins, P.: Changing Ideals in Modern Architecture, pp. 1750–1950. McGill-Queen's University Press, Montreal (1998)
14. Shalunts, G., Haxhimusa, Y., Sablatnig, R.: Classification of gothic and baroque architectural elements. In: Proceedings of the 19th IWSSIP. LNCS, Vienna, Austria, pp. 330–333. IEEE (2012)
15. Shalunts, G., Haxhimusa, Y., Sablatnig, R.: Architectural style classification of domes. In: Bebis, G., et al. (eds.) ISVC 2012, Part II. LNCS, vol. 7432, pp. 420–429. Springer, Heidelberg (2012)
16. Shalunts, G., Haxhimusa, Y., Sablatnig, R.: Segmentation of building facade domes. In: Alvarez, L., Mejail, M., Gomez, L., Jacobo, J. (eds.) CIARP 2012. LNCS, vol. 7441, pp. 324–331. Springer, Heidelberg (2012)
17. Csurka, G., Dance, C.R., Fan, L., Willamowski, J., Bray, C.: Visual categorization with bags of keypoints. In: Proceedings of International Workshop on Statistical Learning in Computer Vision, ECCV, Prague, Czech Republic, pp. 1–22 (2004)
18. Lowe, D.G.: Distinctive image features from scale-invariant keypoints. IJCV **60**(2), 91–110 (2004)
19. Dudani, S.A.: The distance-weighted k-nearest neighbor rule. IEEE Trans. Syst. Man Cybern. **SMC–6**, 325–327 (1976)

Visualization

Visualizing Document Image Collections Using Image-Based Word Clouds

Tomas Wilkinson$^{(\boxtimes)}$ and Anders Brun

Department of Information Technology, Uppsala University, Uppsala, Sweden
{tomas.wilkinson,anders.brun}@it.uu.se

Abstract. In this paper, we introduce image-based word clouds as a novel tool for a quick and aesthetic overviews of common words in collections of digitized text manuscripts. While OCR can be used to enable summaries and search functionality to printed modern text, historical and handwritten documents remains a challenge. By segmenting and counting word images, without applying manual transcription or OCR, we have developed a method that can produce word or tag clouds from document collections. Our new tool is not limited to any specific kind of text. We make further contributions in ways of stop-word removal, class based feature weighting and visualization. An evaluation of the proposed tool includes comparisons with ground truth word clouds on handwritten marriage licenses from the 17th century and the George Washington database of handwritten letters, from the 18th century. Our experiments show that image-based word clouds capture the same information, albeit approximately, as the regular word clouds based on text data.

1 Introduction

In the last decade, word clouds have evolved to become a common tool on blogs and websites. They have the power to compile a large amount of text into a single image that grants the viewer an overview of the text's contents from which properties of the text may be inferred, such as genres or important keywords. In libraries around the world today, there are many collections of old handwritten manuscripts that few people alive today have read or are able to read. The number of experts able to read these manuscripts are few and far between and hence heavily outnumbered by the amount of texts. Because of this, selecting which manuscripts to study can be an overwhelming process. Access to a qualified guess as to what a book contains would help alleviate this problem. Image-based word clouds would allow an expert to get a feel for what a manuscript contains prior to actually working with the material.

Text-based word clouds can be described as doing basic histogram processing (i.e., counting word occurrences) and can be seen as one of the most basic examples of data mining. Word clouds scales very well with text size, in principle it allows you to reduce all the text on the entire Internet to a single image. The typical word cloud processing is as follows. Given some text as input, the word

© Springer International Publishing Switzerland 2015
G. Bebis et al. (Eds.): ISVC 2015, Part I, LNCS 9474, pp. 297–306, 2015.
DOI: 10.1007/978-3-319-27857-5_27

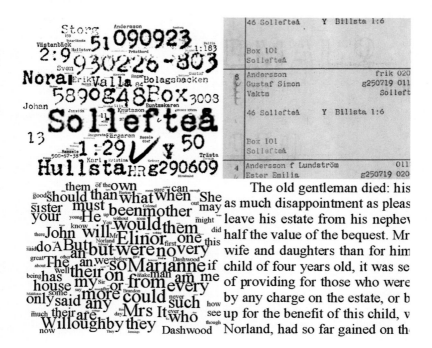

Fig. 1. Two image-based word clouds and their respective source material. The top left image depicts an image-based word cloud created from Swedish tax registries from the mid 20th century, shown to the top right. The bottom left image shows an image-based word cloud created from the book "Sense and Sensibility" and a sample image is shown to the bottom right.

occurrences are counted. With the help of a dictionary, common and uninformative words (called stop words) are removed. The final step is then to place the informative words on a canvas and rescale the size of the words so that the word size is proportionate to the word count. The placement can be done in many different ways, spanning from something simple like a circle or spiral to complex shapes like elephants or guitars. Two different examples of image-based word clouds can be seen in Fig. 1.

2   Image-Based Word Clouds

The purpose of image-based word clouds is to try and capture the same information contained in regular word clouds but using only image data, and without doing any recognition. Some of the processing steps have an exact analogue to the text version, e.g., clustering corresponds to calculating the histogram, whereas other steps are new, such as word segmentation and feature extraction. The layout step is similar to the text-based word cloud except for the fact that the images that are laid out are bitmaps, not vectorial. In some cases, this can

Fig. 2. The processing pipeline for image-based word cloud.

cause severe pixelation in the images due to excessive rescaling of small words. The processing pipeline consists of five stages, visualized in Fig. 2.

2.1   Word Segmentation

In our current pipeline for image based word clouds, word segmentation is a necessary step. Using a word segmentation method adapted from [1], we have repeatedly created word clouds for both good and poor quality typewritten text, see Fig. 1. For handwritten material in general, we have so far only been able to find stable word segmentations by manual tuning. For this reason, this paper does not present a solution to the word segmentation problem for handwritten sources in general and it remains a topic for future research. Instead, in our quantitative experiments, we have used challenging handwritten datasets where the word segmentation has been known beforehand.

2.2   Feature Extraction

For each word image, a set of features is extracted using the approach described in [2]. First, a simple pre-processing step is applied wherein the images are padded with 8 pixels and resized to a fixed size (56×160 in this case). The image is then split into cells of size 8×8 pixels and for each cell, HOG [3] (31 dimensional) and Local Binary Pattern (LBP) [4] (58 dimensional) descriptors are extracted, using the VlFeat library [5]. The HOG and LBP features are then normalized so they have unit l^2 norm and then concatenated. The resulting feature vector, $\mathbf{v} \in \Re^k$, where $k = (31 + 58) \cdot 7 \cdot 20 = 12460$.

After extracting HOG and LBP features for each image in the dataset, choose a random subset of your data set of size $n = 3750$ and call it $\mathbf{M} \in \Re^{n \times k}$. Define a partition P as the interval $[1, ..., n]$ split into groups of size $p = 15$ (the values for n and p are adopted from [2]). Then, project all of your feature vectors onto M, i.e., $\mathbf{u} = \mathbf{Mv}$. Finally, split each $\mathbf{u}$ using the partition P and do max pooling. That is, for each group of size 15 keep the max value, giving you the final vector $\mathbf{x} \in \Re^{250}$.

Given the extracted features, their optimal weighting is unknown. Yet the weighting will be of great importance for the following unsupervised clustering step. Some feature dimensions are correlated, leading to an emphasis of certain word characteristics, while others contain noise. One recent approach in query-by-string word spotting is to use Canonical Correlation Analysis (CCA) [6], which aligns and de-correlates the feature dimensions in a query string feature

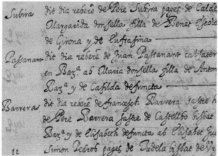 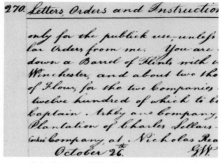

(a) Marriage License dataset   (b) George Washington dataset

Fig. 3. Sample images of the three datasets used for the experiments, going from easiest to most difficult.

space and a word image feature space. In our work here, we use a related but different method to learn an optimal representation suitable for unsupervised clustering, which has previously been described in [7]. In this method, we perform CCA between image features and their class labels, encoded using the one-hot encoding. The gain from this reweighting scheme is two-fold; we reduce the feature dimension to 100 from 250, and to put our performance gain into perspective, get about 10 % higher Mean Average Precision on the Washington dataset compared to [2], using only 20 % of the available data as training data. In particular, this method does not require training on fully transcribed words. In the training step of this method, each word image needs a label, but there is no need to submit any string data representing a transcription of the word.

2.3   Clustering

The clustering step in the image-based word cloud method corresponds to counting the occurrences of each word in a given text. In the ideal case, each cluster would correspond to a unique word and only contain instances of that word, i.e., be completely homogeneous. However, in the general case the number of unique words in a text is unknown and therefore also the number of clusters to use. There are two ways to get around this problem. One approach is to use algorithms where the number of clusters is not a parameter [8]. However, they often have other sensitivity parameters that more or less correspond to confining the number of clusters to an interval. Another approach would be to try and estimate the number of clusters somehow. This has been done before in the clustering of word images using Heaps' Law [9]. For a given natural language, Heaps' Law provides an estimate of the number of unique words in a document given its length. However, in an environment where the language is constrained, e.g., scanned forms, Heaps' Law might not perform as well.

Using Heaps' Law to estimate the number of clusters unlocks the use of many powerful clustering algorithms. The algorithms that have shown previous success [10] and also performs the best in our experiments are Hierarchical Agglomerative Clustering (HAC) [11] techniques with different linkage-criteria. In our case, we found that average linking together with cosine distance measure used in [2] gave the best result.

2.4   Stop Word Removal

This step has an exact analogue with the regular word cloud, though in the text case, it is very simply solved by using a dictionary of words to remove for each language. The image case is more complex. The goal is to remove images of words that are frequently occurring and bear no information. For English, these include common words that occur often in text such as "and", "the", and "of". For the image case, words that occur often should correspond to largest clusters from the clustering. Related to this are some early results in information retrieval that suggest that one can split the types of words in a text into three categories.

1. Frequently recurring stop words, corresponding to the largest clusters. These words hold little to no information.
2. Sporadically occurring words, noncrucial to describe the content of the text. These correspond to the smallest clusters.
3. Moderately recurring words that correspond to the middle sized clusters are the words that are informative and make up the content of a text.

This property of text can be described by Zipf's Law [12]. It states that for a given text, the plot of term frequencies exhibits a distribution where the kth most frequent term has a frequency of k/f_0, where f_0 is the most frequent term. Figure 4 shows a plot of the term frequencies for the Washington dataset along with Zipf's Law. The sporadically occurring words typically cause no trouble for word clouds since they are usually so small you can barely see them, if they are even in the image. The stop words on the other hand, cause problems. A simple, approximate way to remove them is to discard the largest clusters before proceeding to the next stage. This is not a perfect solution and therefore the efficacy of this may vary from quite high to relatively low depending on the quality of the clustering and the text that you are working with. To choose an appropriate number of clusters to discard, we look at what words the largest clusters contain and manually choose the cutoff. However, this method will not eliminate all stop words, as stop word clusters smaller than the largest informative word cluster will not be discarded.

2.5   Layout

The layout procedure that is used in this paper can be described as a greedy search over a set of coordinates generated in a specific shape, where an image is placed in the first place that it fits. More concretely, for each cluster, the image closest to its cluster's centroid is selected as a representative for that cluster.

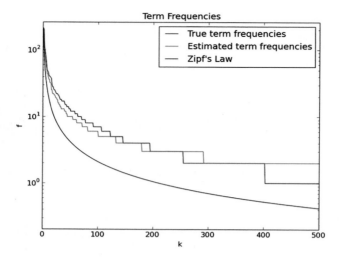

Fig. 4. Term frequencies using ground truth data, estimated term frequencies (cluster sizes), and Zipf's Law in the Washington dataset. The frequencies for words 500 to 1204 are one.

They are then sorted with respect to the sizes of the clusters in descending order. First, an empty canvas is created, then starting with the image representing the largest cluster, the image is (non-linearly) scaled according to the cluster size, then we iterate over the set of coordinates and place the image at the first position that is available. There are many possible scaling functions. We have used $c^\alpha + 1$, where c is the cluster size for that word and α is a tunable parameter. Typically $\alpha = 0.3$ works well. This procedure is repeated for all images until there is no more room on the canvas. To increase the density of the canvas, in addition to scaling the images according to their cluster size, they are also scaled down when placing an image rescaled by cluster size fails to find an empty spot in the set of coordinates, due to overlap with another image. While this does create a false sense of proportion, we feel that it improves the final look of the word cloud. However, more sophisticated layouts may eliminate the need for this step. The final step is some basic contrast enhancement to remove small variations in the background. This is done using gray level transforms with the parameters manually chosen for each dataset.

3   Experiments

To evaluate the image-based word clouds, we perform experiments on two handwritten datasets manuscripts from the 17th and 18th century for which transcriptions are available, allowing us to create ground truth word clouds using correct counts as a comparison. The first dataset is the Barcelona Historical Handwritten Marriages database (BH2M) [13]. It is comprised of a book of marriage licenses, 174 pages long, written between 1617 and 1619 by a single author

(a)   (b)

Fig. 5. Word clouds of the BH2M dataset. Left: the estimated word cloud. Right: the true word cloud, where the ground truth clustering has been used. For both word clouds, Zipfs law has been applied to remove stop words.

in old Catalan. However, in interest of keeping computation time and memory requirements low, we only make use of the first 79 pages, or 26087 words. Since the dataset is comprised of a large amount of marriage licenses (Fig. 3a), and each license follows a kind of grammar (essentially they are historical forms), the language is quite repetitive. Therefore, Heaps' Law does not necessarily hold. As it turns out, this is not such a big problem. Using Heaps Law works quite well, with parameters estimated on running English text.

The second dataset is the George Washington letters [14]. This is a well known and widely used dataset for word spotting [2]. It consists of 20 pages, written during the mid 18th century by George Washington and his secretary, which correspond to 4860 words (Fig. 3b).

To quantitatively evaluate the estimated word clouds, we measure the cosine similarity between them and their ground truth counterparts. The information present in a word cloud is essentially a histogram, or bag-of-words model, and is therefore represented as a high dimensional vector, with as many dimensions as there are unique words in the word clouds. The cosine similarity measure calculates the angle between these two vectors and produces a number $s \in [0,1]$, where 1 means that the word clouds contain the same information and 0 means that they contain completely different information. The cosine measure can be defined as

$$s = \frac{\mathbf{u^T} \cdot \mathbf{v}}{\sqrt{\mathbf{u^T} \cdot \mathbf{u}}\sqrt{\mathbf{v^T} \cdot \mathbf{v}}} \tag{1}$$

where u and v are two bag-of-words vector representations of word clouds. Another measure that is simpler but slightly more intuitive is the ratio between the number of words present in the two clouds and the max of the number of words in each cloud, i.e. the overlap percentage defined as

$$o = \frac{M}{max(O,P)} \tag{2}$$

where M is the number of words present in both clouds, and O and P is the number of words rendered in their respective clouds.

3.1 Marriage Licenses

The purpose of this experiment is to show the that even though the assumptions of Heaps' Law and Zipf's Law do not necessarily hold for this type of data, results are still interpretable and satisfactory. As mentioned above, Heaps' Law tuned for English seems to work well for old Catalan. We use 10 % of the data, or 2608 words, as training data for the CCA, and we use the remaining data to generate the word clouds. Heaps' Law estimates that there are 3356 clusters, which is around 57 % too many. However, the estimated and the ground truth cloud look quite similar, see Fig. 5. A lot of the words correspond to common names like "Juan" and "Antoni".

3.2 George Washington Letters

For the second experiment, we create a word cloud of a more difficult dataset, even though Heaps' Law should work much better. The main difficulty stems from the fact that it is a lot smaller than the BH2M dataset, resulting in less data for the CCA feature weighting. Using 20 %, or 972 words, as training data for CCA works sufficiently well. We estimate the number of clusters to be 1190, which is about 14 % too many. The generated clouds can be seen in Fig. 6. By inspecting Fig. 6a, one can guess a genre or topic for the text. Words like "Captain" and "Orders" indicates some sort of military text.

(a)   (b)

Fig. 6. Word clouds for the Washington dataset. Left: the estimated word cloud. Right: the true word cloud. For both word clouds, Zipfs law has been applied to remove stop words.

The results from the two experiments, as well as the Sense and Sensibility dataset can be seen in Table 1, the cosine similarity and overlap is presented for the three datasets. The quantitative results reinforce the notion that similar information is captured by the two word clouds.

Table 1. Results from word cloud evaluation.

Dataset	Cosine similarity	Overlap
Sense and Sensibility	0.98	0.93 (94 of 101)
BH2M	0.88	0.76 (56 of 74)
Washington	0.76	0.64 (38 of 59)

4   Conclusion

In this paper, we have presented a novel framework for generating image-based word clouds. We have quantitatively and qualitatively shown that image-based word clouds generate word clouds that look similar to ordinary word clouds. Meaningful visualizations of large document collections can be generated using this framework, despite the lack of a proper stop word list and without advanced recognition algorithms that decode the text letter by letter. From a theoretical point of view, this algorithm is conceptually close to earlier work in word spotting. However, from an application point of view, the introduction of image based word clouds could potentially open up completely new ways for users to experience large document collections.

As of right now, the word segmentation algorithm is the limiting factor in this framework. Better word segmentation for handwritten documents and segmentation free approaches to image-based word clouds are interesting directions for future research. Interactive tools that allow users to quickly correct, merge clusters and remove obvious stop words manually, are also possible extensions. The current evaluation focuses on comparing the quality of the clustering compared to the true word counts. Another interesting comparison for the future would be to compare the word clouds to fully text-based word clouds, i.e., using a dictionary for stop word removal.

Acknowledgments. This project is a part of q2b, From quill to bytes, a framework program sponsored by the Swedish Research Council (Dnr 2012-5743) and Uppsala university. The work is done in part as a collaboration with the Swedish Museum of Natural History (Naturhistoriska riksmuseet). We would also like to thank Alicia Fornés and the Document Analysis group of the Computer Vision Center at Universitat Autnoma de Barcelona for access to the BH2M dataset.

References

1. Zagoris, K., Pratikakis, I., Antonacopoulos, A., Gatos, B., Papamarkos, N.: Handwritten and machine printed text separation in document images using the bag of visual words paradigm. In: ICFHR, pp. 103–108 (2012)

2. Kovalchuk, A., Wolf, L., Dershowitz, N.: A simple and fast word spotting method. In: 2014 14th International Conference on Frontiers in Handwriting Recognition (ICFHR), pp. 3–8 (2014)

3. Dalal, N., Triggs, B.: Histograms of oriented gradients for human detection. In: IEEE Computer Society Conference on Computer Vision and Pattern Recognition, CVPR 2005, vol. 1, pp. 886–893. IEEE (2005)

4. Ojala, T., Pietikainen, M., Maenpaa, T.: Multiresolution gray-scale and rotation invariant texture classification with local binary patterns. IEEE Trans. Pattern Anal. Mach. Intell. **24**, 971–987 (2002)

5. Vedaldi, A., Fulkerson, B.: Vlfeat: an open and portable library of computer vision algorithms. In: Proceedings of the international conference on Multimedia, pp. 1469–1472. ACM (2010)

6. Almazán, J., Gordo, A., Fornés, A., Valveny, E.: Handwritten word spotting with corrected attributes. In: 2013 IEEE International Conference on Computer Vision (ICCV), pp. 1017–1024. IEEE (2013)

7. Johansson, B.: On classification: simultaneously reducing dimensionality and finding automatic representation using canonical correlation (2001)

8. Frey, B.J., Dueck, D.: Clustering by passing messages between data points. Science **315**, 972–976 (2007)

9. Heaps, H.S.: Information Retrieval: Computational and Theoretical Aspects. Academic Press Inc, Orlando (1978)

10. Rath, T.M., Manmatha, R.: Word spotting for historical documents. IJDAR **9**, 139–152 (2007)

11. Hastie, T., Tibshirani, R., Friedman, J., Hastie, T., Friedman, J., Tibshirani, R.: The Elements of Statistical Learning, vol. 2. Springer, New York (2009)

12. Zipf, G.K.: Human Behavior and the Principle of Least Effort. Addison-Wesley Press, Cambridge (1949)

13. Fernández-Mota, D., Almazán, J., Cirera, N., Fornés, A., Lladós, J.: Bh2m: the barcelona historical, handwritten marriages database. In: 2014 22nd International Conference on Pattern Recognition (ICPR), pp. 256–261. IEEE (2014)

14. Lavrenko, V., Rath, T.M., Manmatha, R.: Holistic word recognition for handwritten historical documents. In: Proceedings of the First International Workshop on Document Image Analysis for Libraries, pp. 278–287. IEEE (2004)

Guided Structure-Aligned Segmentation of Volumetric Data

Michelle Holloway[1]([⊠]), Anahita Sanandaji[2], Deniece Yates[2], Amali Krigger[5],
Ross Sowell[4], Ruth West[3], and Cindy Grimm[2]

[1] Washington University in St. Louis, St. Louis, USA
mavaugha@wustl.edu
[2] Oregon State University, Corvallis, USA
[3] University of North Texas, Denton, USA
[4] Cornell College, Vernon, USA
[5] University of the Virgin Islands, Charlotte Amalie, USA

Abstract. Segmentation of volumetric images is considered a time and resource intensive bottleneck in scientific endeavors. Automatic methods are becoming more reliable, but many data sets still require manual intervention. Key difficulties include navigating the 3D image, determining where to place marks, and maintaining consistency between marks and segmentations. Clinical practice often requires segmenting many different instances of a specific structure. In this research we leverage the similarity of a repeated segmentation task to address these difficulties and reduce the cognitive load for segmenting on non-traditional planes. We propose the idea of guided contouring protocols that provide guidance in the form of an automatic navigation path to arbitrary cross sections, example marks from similar data sets, and text instructions. We present a user study that shows the usability of this system with non-expert users in terms of segmentation accuracy, consistency, and efficiency.

1 Introduction

Revealing useful knowledge from 3D/4D biological imaging usually starts with a segmentation of the structure(s) of interest. Segmentation is critical in a wide range of applications such as rendering visualizations, quantitative analysis, treatment planning, and virtual simulations. Thus, effective segmentations need to be accurate, consistent, and created efficiently. When segmentation becomes a bottleneck in the data-to-knowledge pipeline, it hinders research and could even result in health risks in clinical practice [4].

A variety of research in automatic segmentation techniques is ongoing. While these methods are becoming more reliable [16], many segmentations are still created manually to ensure the necessary quality and accuracy. Manual segmentation entails marking the boundary of the structure throughout the volume commonly done by drawing contours on multiple 2D slices. In standard practice, the slicing planes are either parallel along, or orthogonal to, the scanning direction of the image data. Delineating enough contours to create an accurate

© Springer International Publishing Switzerland 2015
G. Bebis et al. (Eds.): ISVC 2015, Part I, LNCS 9474, pp. 307–317, 2015.
DOI: 10.1007/978-3-319-27857-5_28

segmentation can be both difficult and time intensive, depending on the shape of the structure and the size and quality of the image data.

In many settings the same type of structure is repeatedly segmented (e.g. patient livers, mitochondria in cells, etc.). In this work we leverage the fact that the same general shape is segmented in each instance in order to define a *protocol* to guide segmentation of a specific class of structures. This can help to reduce navigation time as well as help users maintain consistency both in how and where they place marks from one instance to another. A protocol is useful in the traditional parallel or orthogonal contouring approach, but more importantly it supports using arbitrarily-oriented contouring planes that can be placed to follow the global structure of the shape — for instance, tracking a tubular structure as it curves in space. The primary finding of this paper is that we can support these structure-specific contouring planes (which reduces both time and error in the segmentation process) without unduly increasing the cognitive load over more traditional contouring methods.

Specifically, we present the first stage of a new conceptual approach for working with volume data – guided 3D segmentation through structure-aligned contouring protocols – and design a novice friendly interface for evaluating this approach.

Contributions of this paper include:

1. A guided, structure-aligned approach to segmentation that reduces the cognitive burden of using non-traditional contouring planes.
2. A real-time visualization of location in a 3D volume in parallel with drawn contours and segmentation surface.
3. Verification that experts can utilize structure-aligned planes to create segmentations comparable to their traditional parallel segmentations.
4. A user study to show that this guided approach allows novices to produce segmentations of comparable quality to experts.
5. Identification of factors that influence segmentation time and quality, enabling comparison of different contouring protocols.

2   Related Work

2.1   Manual 3D Image Segmentation

Manually creating accurate segmentations in an efficient manner is a non-trivial task. Biological tissues are often not clearly separated, in addition to image noise and lack of sufficient resolution and contrast. Within volumetric imaging, this is further compounded by difficulties in maintaining orientation and structural awareness in the data. Therefore, in practice usually only domain experts with years of training in inspecting their specific imaging modality for their specific structures perform segmentation.

Traditionally, experts mark boundaries on parallel slices throughout the volume. Many systems also allow marking contours on orthogonal intersecting slices, as generally less marking is needed and they can better capture surface curvature extrema. Orthogonal contouring has spawned various interactive systems that take user contour input and reconstruct a surface using splines or implicit functions [1,6]. The work in [5] makes use of the live-wire technique for drawing contours on orthogonal planes and for interpolating between the contours. These methods, however, are all dependent on the user's ability to choose planes to contour on and do not provide guidance in this regard. Additionally, methods that rely on detecting boundaries may lead to undesirable results or require more input when the boundaries are unclear.

2.2   Segmentation Guidance

The main difficulty in any interactive segmentation system is navigating and choosing where to mark boundaries. Thus new research has explored using guidance in segmentation systems. The work in [9] uses the random walks algorithm in a seed-based semi-automatic segmentation system where users place marks inside the structure instead of drawing boundary contours. The random walk probability is used to direct users to areas where more seeds could be placed.

Contour-based guidance approaches aim to automatically choose planes to mark contours on. In [13] the authors extend the work in [5] to arbitrary planes, and use the live-wire cost of the segmentation result to suggest planes that need marking. Similarly in [14] the authors have users provide drawn contours to seed the random walks algorithm, and in turn actively suggest new arbitrary planes with high uncertainty in the segmentation. Both of these methods require some initial contour(s) to find an initial segmentation to use as training data. There is little contouring consistency between similar data sets as the algorithms' choices depend on the training input and uncertainty probabilities.

These automatic plane selection methods are quite useful and our work fits nicely in tandem with them. We don't require initial input since we instead define structure specific protocols that encode a preset automatic navigation path to contouring planes. This path encourages consistency, so our protocols could be used to provide consistent training data to these other methods. Additionally, experts could use these methods to aid in protocol creation.

3   Methods

3.1   Motivation

Two major challenges posed by manual volume segmentation are the cognitive strain of free-form data navigation and knowing where to draw contours. It is for these reasons that many segmentation systems limit the selection of planes to only parallel or orthogonal placement. However, arbitrarily placed planes can offer more useful views of the data that are aligned to the structure's shape and capture known features more efficiently [15]. Our approach leverages prior knowledge of the structure's general shape in order to use arbitrary planes without increasing the segmenter's cognitive load to complete the task.

To address 3D free-form navigation we propose to remove the need for large-scale manual navigation. Navigation in 3D data is difficult for domain experts let alone novices unfamiliar with biological structures. Our approach provides a navigation path to a relative set of structure-aligned planes to draw contours on for a given structure. Here "relative" means the planes shift according to the initial position of the first plane placed in the volume. Plane locations incorporate structure shape knowledge, and they describe a set path that increases predictability from data set to data set. In this work we side step the larger problem of choosing the best set of planes for a given structure and have an expert manually choose a reasonable structure-aligned sequence of planes.

In [10] it was shown that references images can substantially improve contour drawing consistency for novice users for a single, arbitrary plane. Thus, to address difficulties knowing where to draw on a given plane, we provide visual cues in the form of images of example expert-drawn contours from similar data sets along with text instructions. Using reference visuals transforms a complex perceptual and cognitive task requiring domain expertise to, essentially, a pattern matching task that can be performed by novices. This allows us to use knowledge of the structure while still taking advantage of people's abilities to adapt to the data.

3.2   Contouring Protocol Definition

The *guided contouring protocol* incorporates a preset navigation path of arbitrary planes within the volume, images of example contours from similar data sets for each plane in the navigation path, and text instructions. It is essentially step-by-step instructions for drawing a set of contours that result in a segmentation surface for a specific type of structure.

The navigation path is defined as a set of successive 3D transformations between the chosen planes in a temporal order. For each step, the system automatically navigates to a plane by means of a smooth animated transition. Text instructions and a small set of expert-drawn example contours from a similar data set provide guidance for drawing the new contour. Additionally, users can make local plane navigation adjustments to better match the example images before drawing. A surface is constructed and progressively refined after the addition of each contour to provide immediate visual feedback. We utilize [8] for surface reconstruction for its speed and reliability in producing a surface with partial or distorted input.

3.3   Interface

Our contouring protocols could easily be implemented into any standard segmentation interface. For novice testing we built an interface, inspired by [10]'s design, that focused on (1) supporting clear orientation and enhancing structural awareness, and (2) exposing all functionality through in-screen elements to minimize required clicks. Figure 1 depicts the four areas of the interface.

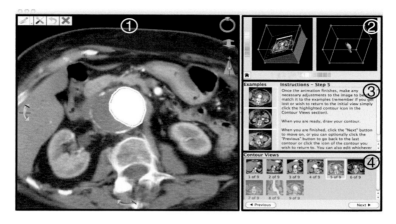

Fig. 1. Interface layout. Area 1 is the main window, 2 is the 3D volume localization, 3 is the protocol instructions, and 4 is the navigation path visualization.

(1) Main Window: The primary focus for interaction that shows the current slice, drawn contours, and if desired, the surface. Includes standard navigation and free hand drawing tools.

(2) Localization: The pair of linked views visualizes the current slice's position in the volume (left) and the location of all drawn contours and the segmentation surface (right). They are presented as YAH (You-Are-Here) cubes, inspired by [7, 12]. Turntable controls rotate the cubes to find a useful view of the volume without losing global orientation.

(3) Instructions: Provides both the image-based and text-based information. Each of the example image icons can be clicked to view it in the main window.

(4) Navigation Path: The navigation path is visualized as a sequence of thumbnails. The user navigates using the next and previous buttons or by clicking on a thumbnail in the enforced order. The thumbnail for a completed step shows their drawn contour; thumbnails for uncompleted steps are grayed out and filled with an example image. The navigation path provides a simple method for "scanning" through the volume and shows the progression through the protocol.

4   User Study

To demonstrate that our system is suitable we tested it with expert users from radiation oncology. To demonstrate that novices are able to complete a valid segmentation using the guided contouring protocol, we conducted a study with novice users and compared the results to expert segmentations. The four hypotheses we tested are:

H1 Given sufficient guidance in the form of example images and predetermined contouring planes, novices can reliably produce valid segmentations for relatively complex structures.

H2 Using a set of structure-aligned planes increases segmentation accuracy and consistency compared to a set of parallel or orthogonal planes with the same amount of work.

H3 Total contouring time is dependent on contour length and curvature and the number of contours.

H4 Experts can produce comparable segmentations and in less time using non-parallel contouring to those using the traditional parallel approach.

For H1, we compare novices both pairwise and to ground truth expert segmentations. For H2 we compare the accuracy, consistency, and completion time for three types of protocols used on a single dataset where each protocol requires approximately the same amount of drawing. For H3 we look at novice completion time and contour drawing times in contrast to their length and curvature. For H4 we examine expert completion time and the inter-expert consistency of segmentations produced using parallel contouring and our guided approach. For mesh comparisons we use Dice's coefficient (DC) [3] and the mean percent distance (MPD), which is the mean distance [2] normalized by the structure size.

4.1   Study Design

Participants: We recruited 20 volunteers from the Washington University in St. Louis community to segment three different data sets using structure-aligned protocols. Only 20 % were moderately familiar with medical images and only 10 % claimed to be moderately or highly experienced with image segmentation. Additionally we recruited 8 volunteers from the Oregon State University community to segment one of the data sets using parallel and orthogonal protocols. Only 1 user claimed to be highly experience with image segmentation.

Image Data: We used CT scans of a liver, an aorta, and a ferret brain. The selected data sets enable us to evaluate our approach sufficiently to demonstrate validity. They range from simple (aorta) to complex (ferret) with differing image/contrast quality representative of data encountered in real segmentation tasks. Figure 2 shows example images from each data set and an expert segmentation. The contours show the planes used in the structure-aligned protocols. Since the aorta is a tube we had users segment slightly more than the expert and then clipped the results.

Training: Users received a verbal introduction and explored the interface using a data set not included in the study. The training averaged 15 to 20 min.

Procedure: The first 20 users used structure-aligned protocols to segment the three data sets in a random order determined pseudo-randomly such that the six orderings were evenly represented. The number of protocol planes were eight for the liver, nine for the aorta, and ten for the ferret brain. The 8 additional users segmented only the liver using both a parallel and orthogonal protocol in a random order consisting of seven and eight planes respectively. The planes for the parallel and orthogonal protocols were chosen such that they required about the same total amount of drawing as the structure-aligned protocol. We only allowed free-hand drawing for comparison purposes. No user failed to complete

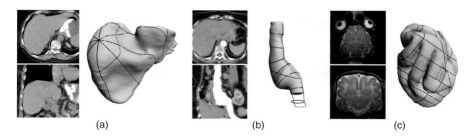

(a)   (b)   (c)

Fig. 2. Example images and segmentations from the three data sets: (a) liver (b) aorta (c) ferret brain.

any tasks. A questionnaire was administered at the completion of the tasks to gauge user background and opinions on the task.

4.2   Expert Evaluation

We invited 7 experts to complete segmentations of only the liver using their traditional parallel contouring and using our structure-aligned guided contouring protocol. The training and comparison metrics were the same as the novice study. The procedure was similar, except they did parallel contouring first followed by the guided approach (using the same interface), and did not complete a questionnaire. Inter-expert consistency was higher for the guided approach versus the traditional one (DC: 0.92 ± 0.016 vs 0.87 ± 0.138, and MPD: 1.62 ± 0.45 versus 1.86 ± 1.71), and overall completion time was faster (469 ± 186 s vs 804 ± 113 s) confirming H4. For more details see thesis [11].

4.3   Novice Results and Discussion

Accuracy: We evaluate the accuracy of novice segmentations by comparing to a ground truth expert segmentation for each structure (Fig. 3 Top). The liver ground truth was the average of the expert segmentations from Sect. 4.2, while the aorta and ferret brain ground truths were provided to us with the data.

Users were able to construct reasonably accurate segmentations for the three structure-aligned data sets. The ferret brain was least accurate due to its complexity. Much of the expert boundaries don't fall on strong gradients, shown with the solid line in Fig. 4, and many users tended to follow the stronger boundaries, shown with the dotted lines. The variance in the aorta was because some planes had a second visible boundary that some chose to follow, as shown in Fig. 4. For reference, we show the liver results from Sect. 4.2 as dashed lines in Fig. 3 and can see that the novice accuracy is close to the variance seen among expert liver segmentations. These results indicate that using our guided contouring protocol novice users are capable of producing segmentations of comparable quality to expert segmentations, confirming H1.

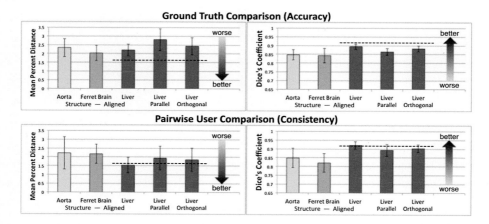

Fig. 3. Top: The average MPD and DC between novice segmentations and the ground truth. Bottom: The average MPD and DC between all pairwise novice segmentations. The dashed line indicates the results from Sect. 4.2.

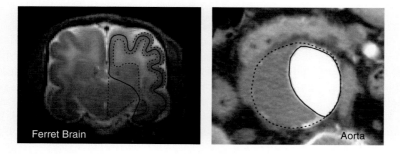

Fig. 4. Examples of ambiguous boundaries. The solid outline is the expert segmentation, and the dotted lines show boundaries that some novices followed instead.

For the liver cases, the structure-aligned protocol produced a significantly (T-Tests MPD: p=0.00007 $\leq$ 0.05, DC: p=0.00016 $\leq$ 0.05) more accurate segmentation compared to the parallel protocol confirming H2. And while the structure-aligned results were slightly more accurate than the orthogonal, they were not significantly so (T-Tests MPD: p=0.1015 $\geq$ 0.05 ,DC: p=0.0836 $\geq$ 0.05).

Consistency: We evaluate the consistency of the novice segmentations by performing pairwise comparisons among all users for each protocol (Fig. 3 Bottom).

Users were reasonably consistent for the three structure-aligned cases. The ferret brain and aorta had more variation for the same reasons discussed in the accuracy section. Looking at the dashed line, the novice liver segmentations were as consistent as the experts, which shows that novice users can reliably produce similar segmentations using our protocols, confirming H1. The protocol approach encourages consistency because it enforces a defined contouring scheme so that segmentations are performed in a relatively uniform manner.

For the liver cases, the structure-aligned protocol produced significantly more consistent segmentations than both the parallel (T-Tests MPD: p=0.00001 $\leq$ 0.05, DC: p=$1.19144E-8 \leq 0.05$) and orthogonal (T-Tests MPD: p=0.00024 $\leq$ 0.05 ,DC: p=0.00005 $\leq$ 0.05) protocols, confirming H2.

Efficiency: We evaluate the efficiency of our guided contouring protocol by computing the average completion time for each protocol. We also examine contouring time in terms of contour length and curvature.

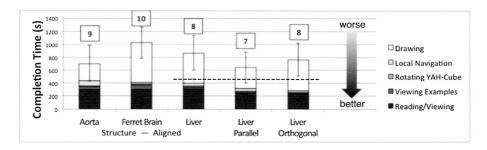

Fig. 5. Average completion time broken down by user actions. The number of protocol planes is shown above each bar. The dashed line on the indicates the results from Sect. 4.2.

The average completion time for each protocol was between 10–15 minutes (Fig. 5) with the exception of the ferret brain due to its complexity. For the liver, novice users took less than double the expert time for the liver on average. We feel that these results are compelling considering the users had little to no prior experience. Using our protocols the segmentation time is firstly dominated by time spent drawing contours and secondly by time spent idling (i.e. reading/observing), confirming H3; little time is used for manual navigation.

For the liver cases, the completion time for the three protocols were very similar since they were designed to have approximately the same amount of drawing. The parallel case was slightly significantly faster than the structure-aligned (T-Test: p=0.04914 $\leq$ 0.05) but it was also less accurate, meaning that more parallel contours are needed which would increase the time.

Because segmentation time is now mostly dependent on the time taken to draw contours, we consider two contour length and curvature, and their influence on drawing time. In Fig. 6 that the general trend shows that as the length or curvature increases so does the drawing time, confirming H3. The aorta contours all have similar length and curvature, so we see them clustered together in both plots. The liver contours have similar curvature but vary in length. The complex ferret brain contours varied fairly equally on both factors.

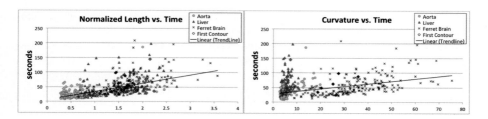

Fig. 6. Scatter plots of contour length and curvature versus drawing time. The contour length was normalized to a unit square drawing window. The curvature for each contour was measured as the sum of the absolute curvature across all contour points.

From these results we can represent the approximate time, T, needed to draw n contours as a function of the contours' length and curvature, plus an uncertainty factor, as shown in Eq. 1 below.

$$T = \sum_{i=1}^{n} \alpha L_i + \beta C_i + U_i \tag{1}$$

For each contour, i, L_i is its length, C_i is its curvature, and U_i is an estimate of the time needed for its uncertainty. We normalize curve length by mapping the visible window to a unit square, and the curvature is the sum of the absolute curvature across the contour. We fit a linear polynomial surface to our novice 3D data points of contour length vs. curvature vs. time (with outliers removed) to find α as 23.85 and β as 0.3135 for our examples.

5   Conclusions

Segmentation quality is defined by what the domain expert deems as useful. This notion drives the usefulness and flexibility of expert designed protocols that is more powerful than an arbitrary "gold standard". Future work will develop tools for protocol creation with automatic plane selection to easily choose the best structure-aligned planes. To further improve accuracy we will make use of our structure protocol knowledge in conjunction with surface reconstruction.

Acknowledgment. The authors would like to thank Dr. Daniel Low, formerly of the Washington University School of Medicine in St. Louis for the liver data sets, Dr. Sandra Rugonyi of Oregon Health & Science University for the aorta data sets, and Dr. Philip Bayly of Washington University in St. Louis for the ferret brain data sets.

This work was supported by the National Science Foundation under grants DEB-1053554 and IIS-1302142.

References

1. de Bruin, P., Dercksen, V., Post, F., Vossepoel, A., Streekstra, G.: Interactive 3D segmentation using connected orthogonal contours. Comput. Biol. Med. **35**, 329–346 (2005)

2. Cignoni, P., Rocchini, C., Scopigno, R.: Metro: measuring error on simplified surfaces. Comput. Graph. Forum **17**, 167–174 (1998)
3. Dice, L.R.: Measures of the amount of ecologic association between species. Ecology **26**(3), 297–302 (1945)
4. Foppiano, F., Fiorino, C., Frezza, G., Greco, C., Valdagni, R.: The impact of contouring uncertainty on rectal 3D dose-volume data: results of a dummy run in a multicenter trial. Int. J. Radiat. Oncol. Biol. Phys. **57**(2), 573–579 (2003)
5. Hamarneh, G., Yang, J., McIntosh, C., Langille, M.: 3d live-wire-based semi-automatic segmentation of medical images. SPIE Med. Imaging **5747**, 1597–1603 (2005)
6. Heckel, F., Konrad, O., Hahn, H.K., Peitgen, H.O.: Interactive 3D medical image segmentation with energy-minimizing implicit functions. Comput. Graph. Spec. Issue Vis. Comput. Biol. Med. **35**(2), 275–287 (2011)
7. Khan, A., Mordatch, I., Fitzmaurice, G., Matejka, J., Kurtenbach, G.: Viewcube: a 3d orientation indicator and controller. In: Proceedings of the 2008 Symposium on Interactive 3D Graphics and Games, I3D 2008, pp. 17–25. ACM (2008)
8. Macedo, I., Gois, J., Velho, L.: Hermite interpolation of implicit surfaces with radial basis functions. In: 2009 XXII Brazilian Symposium on Computer Graphics and Image Processing (SIBGRAPI), pp. 1–8, October 2009
9. Prassni, J.S., Ropinski, T., Hinrichs, K.: Uncertainty-aware guided volume segmentation. IEEE Trans. Vis. Comput. Graph. **16**(6), 1358–1365 (2010)
10. Sowell, R., Liu, L., Ju, T., Grimm, C., Abraham, C., Gokhroo, G., Low, D.: VolumeViewer: an interactive tool for fitting surfaces to volume data. In: SBIM 2009: Proceedings of the 6th Eurographics Symposium on Sketch-Based Interfaces and Modeling, pp. 141–148. ACM (2009)
11. Sowell, R.T.: Modeling Surfaces from Volume Data Using Nonparallel Contours. Ph.D. thesis, Washington Univ. in St. Louis (2012)
12. Stoakley, R., Conway, M.J., Pausch, R.: Virtual reality on a wim: interactive worlds in miniature. In: Proceedings of the SIGCHI Conference on Human Factors in Computing Systems, CHI 1995, pp. 265–272. ACM (1995)
13. Top, A., Hamarneh, G., Abugharbieh, R.: Active learning for interactive 3d image segmentation. In: Proceedings of the 14th International Conference on Medical Image Computing and Computer-assisted Intervention - Part III, pp. 603–610 (2011)
14. Top, A., Hamarneh, G., Abugharbieh, R.: Spotlight: automated confidence-based user guidance for increasing efficiency in interactive 3D image segmentation. In: Menze, B., Langs, G., Tu, Z., Criminisi, A. (eds.) MICCAI 2010. LNCS, vol. 6533, pp. 204–213. Springer, Heidelberg (2011)
15. Weiss, E., Richter, S., Krauss, T., Metzelthin, S.I., Hille, A., Pradier, O., Siekmeyer, B., Vorwerk, H., Hess, C.F.: Conformal radiotherapy planning of cervix carcinoma: differences in the delineation of the clinical target volume. A comparison between gynaecologic and radiation oncologists. Radiother Oncol. **67**(1), 87–95 (2003)
16. Wirjadi, O.: Survey of 3D image segmentation methods. Technical report 123, Fraunhofer ITWM (2007)

Examining Classic Color Harmony Versus Translucency Color Guidelines for Layered Surface Visualization

Sussan Einakian[(✉)] and Timothy S. Newman

Department of Computer Science, University of Alabama in Huntsville,
Huntsville, USA
{seinakia,tnewman}@cs.uah.edu

Abstract. This paper considers certain aspects of the role of color in isosurface-based volume visualization. A primary concern is examining guidelines for colors in layered translucent surface visualization introduced by House et al. [1] and Ebert et al. [2,3]. These guidelines are considered against an existing classic rule (guideline) about color choices in visual presentations to determine their applicability for isosurface volume renderings. The classic guideline is based on the rules of color harmony introduced by Itten [4]. The examinations are based on evaluation by human observers.

Keywords: Volume visualization · Isosurfacing · Color theory · Color harmony

1 Introduction

Volume data visualization can provide key insights into science and engineering data. One common way to visualize scalar values arranged on a 3D rectilinear lattice (henceforth, "scalar volumetric dataset") is isosurfacing. In isosurfacing, the surface corresponding to the set of locations in the volume whose values equal an isovalue (denoted herein as α) is determined. Of special interest to us are situations in which multiple isosurfaces must be visualized simultaneously (i.e., in a single rendering).

One common solution for such visualizations is to use a different color for each isosurface. Color choices can be a challenge, however. First, some sets of colors may be too similar, producing difficulty in determining relative locations of structures in space. Second, some combinations of colors may clash to the point of inhibiting gaining of insights. Yet, a variety of rule sets for selecting color combinations for artistic or presentation applications have been suggested by color theorists. In addition, some prior visualization research has yielded some alternatives to the artistic rule sets. Here, we consider certain aspects of value for three sets of color selection rules, focusing on the multiple isosurface visualization application domain. To our knowledge, the work here is the first comparative evaluation of these three rule sets for this application domain.

G. Bebis et al. (Eds.): ISVC 2015, Part I, LNCS 9474, pp. 318–327, 2015.
DOI: 10.1007/978-3-319-27857-5_29

One rule set (Itten color combination guidelines [4,5]) has been examined previously for selecting colors for multiple isosurface visualization [6]. In this paper, we evaluate color selection for such visualization *based on certain visualization color guidelines identified by House et al. [1] and Ebert et al. [2,3]* and compare these versus the Itten guidelines. Our evaluations are via user studies.

1.1 Organization

The paper is structured as follows. Section 2 describes background. Sections 3 and 4 discuss our studies and results. Results are analyzed in Sect. 5. Section 6 concludes the paper and describes possible future works.

2 Background and Related Works

Color combinations have been used often as a visual cue to visualize dimensions (variables) of a dataset. Sometimes, the combinations used are simply well-known ones. For example, using a rainbow color map is popular in visualization [7]; one survey found that over half of the visualizations in the IEEE Visualization Conference proceedings from 2001 to 2005 used a rainbow color map [7]. However, that color map may confuse viewers due to lack of perceptual ordering [7].

A number of rules and guidelines about color use in art exist. Some of these may be useful in visualization. One type of artistic rule set relates to harmonious color combinations. Itten [4] is one color theorist who has developed such rules (originally for pigments, based on the relative positions of colors on his hue wheel and the relationships among them [4,8]). His rules specify that sets of two, three, four, or six colors whose hues form an equal-spaced layout on his hue wheel are harmonious. As examples, any pair of complementary colors (i.e., separated by 180° on his hue wheel) are *two-color harmonies* and any three colors separated by 120° on his hue wheel are *three-color harmonies.*

Color harmony rules based on hue templates [5,9,10] also exist. They have an advantage of being defined independently from underlying hue wheels [11].

Color harmony may be advantageous; the use of aesthetically-pleasing color palettes can increase user interest in examining multivariate visualizations [12]. Contrariwise, jarring or unpleasing color palettes may reduce user interest, limiting understanding [12].

One use of color in isosurface visualization has been to display uncertainty. For example, Rhodes et al. [13] have explored isosurface renderings that present uncertainty in the data using either hue, saturation, or brightness (depending on user choice). They found hue to be the most reasonable one to use, especially considering the visual system's sensitivity to hue changes.

Some volume visualization techniques have previously made use of transparency, especially for displaying surfaces simultaneously. For example, colorization using transparency has been applied in computed tomography (CT) data for medical visualization [14], in particular by tying tissue density to opacity. Maximizing luminance contrast has also been used with opacity to generate colors that produce accurate and realistic visualizations [14].

As House and Ware [15] have pointed out, displaying surfaces simultaneously is a difficult perceptual problem in certain visualization applications. In medical imaging, for example, viewing the tissue and skeleton at the same time is important. However, structures can occlude one another in such applications, which sometimes makes it hard to see critical parts of certain structures. There are methods that can reduce this occlusion, such as using transparency. Other research on transparency for layered surface-based visualization by Interrante et al. [16] suggests that using transparency inhibits depth perception. Therefore, they added opaque texture elements (such as strokes) to improve perception of transparent surface shape and depth. They also used texture element color to indicate distance between the outer and inner surface at each point.

House et al. [1] have developed a framework, based on perceptual concerns, for visualizations of two layered surfaces. Based on tests using the framework, they developed guidelines to reduce surface occlusion. Some parts of these guidelines are explored here: (1) saturation should be higher for the bottom surface than for the top one, (2) color value for the bottom surface should be about either 50 % or 80 % of full-scale brightness, and (3) the opacity for the top surface should be between 20 % and 60 % of the opacity of the bottom surface.

Ebert et al. [2,3] have proposed a new volume visualization scheme that uses direct volume rendering and non-photorealistic rendering. This scheme aims at improving perception of structure, shape, orientation, and depth relationships. In particular, it uses warm colors (i.e., in color temperature) for surfaces facing a light source and cool colors for other surfaces. Warm colors are also used to increase the sense of depth that something is close by, while cool colors are used to promote sensation that something is far away.

3   Work Description

Our work here examines applicability of three sets of color selection guidelines to layered surface visualization. Via user evaluations, we focus on (1) House et al.'s [1] saturation, value, and opacity rules and (2) Ebert et al.'s [2,3] color temperature rules, as they intersect with Itten's guidelines [4,5]. We also examine these two versus (3) Itten-based guidelines not following all House et al. and Ebert et al. rules. One longer-term aim is to seek evidence of factors that undergird Itten's rules' success.

Specifically, we consider renderings based on subsets of the Itten, House et al., and Ebert et al. guidelines mentioned in Sect. 2 (also shown in Table 1).

The Itten harmonies considered here are his two- and three-color harmonies (described in Sect. 2). For disharmonies, we selected color combinations based on prior work [6] that found colors separated by 150° (or 210°) on the Itten hue wheel were the most disharmonious ones. The opponent color combinations considered here are chromatic opponents (i.e., red-green and blue-yellow).

Next, we report the color rule set evaluation results.

Table 1. Guideline subsets considered

Guidelines	Explanation
Harmonious-House	Itten's harmonies that also satisfy House et al. guidelines
Disharmonious-House	Disharmonies that also satisfy House et al. guidelines
Opponent-House	Opponency that also satisfy House et al. guidelines
Harmonious-Ebert	Itten's harmonies that also satisfy Ebert et al. guidelines
Disharmonious-Ebert	Disharmonies that satisfy Ebert et al. guidelines
Opponent-Ebert	Opponency that also satisfy Ebert et al. guidelines
Harmonious-Itten	Itten's harmonies
Disharmonious-Itten	Disharmonies
Opponent-Itten	Opponency

3.1 Study's Presentations

Our user evaluations involved considering renderings done on three volumetric datasets from the Volume Library [17]. Each rendering presented multiple structures, each associated with a certain isovalue, α (i.e., α denotes isovalues). The datasets are the $(256 \times 256 \times 256)$ (CT) Engine (Block), the DTI (a $128 \times 128 \times 58$ Diffusion Tensor MRI brain scan), and the H_2O (a $128 \times 128 \times 128$ simulation of the electron distribution in a water molecule) datasets. Renderings were generated using OpenDX. For Engine and DTI, two isosurfaces were rendered (associated with $\alpha \in \{180, 60\}$ for Engine and $\alpha \in \{1200, 200\}$ for DTI). For H_2O, three isosurfaces were rendered ($\alpha \in \{100, 75, 50\}$). Figure 1 presents example isosurface renderings of these datasets (using opponent colors).

We selected the color combinations using ColorWheel Expert [18], which allows determining color combinations according to the RGB or RYB hue wheel.

For each dataset, three sets of renderings were utilized. One set contained multiple renderings using harmonious color combinations. Another set contained multiple renderings using opponent color combinations. The third set contained multiple renderings using disharmonious color combinations. Figure 2(a)-(c) shows example renderings of the Engine dataset using the three harmonious color schemes tested here.

3.2 Evaluators and Viewing

Our evaluators were 40 students from a variety of disciplines (20 females and 20 males). Some were graduate students, others were undergraduates. About 15 % of them had taken graphics or visualization courses. All evaluators had normal (or corrected-to-normal) vision, with no known visual deficiency. Images were shown to them on a 20" LCD display with resolution $1,680 \times 1,050$. The monitor was color-calibrated at least weekly or after every fifth evaluator, whichever came first. Color calibration was done via Windows 7 monitor color calibration and the PowerStrip calibration software. Ambient room lighting was

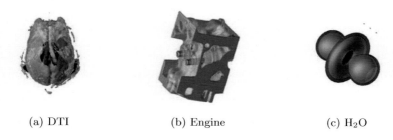

(a) DTI   (b) Engine   (c) H_2O

Fig. 1. Renderings of nested isosurfaces for three datasets

(a) Itten's version   (b) Harmonious-House   (c) Harmonious-Ebert

Fig. 2. Renderings of Engine nested isosurfaces using harmonies

kept identical (experiments were performed in the same environment (i.e., same room, lighting, monitor, location)). Each evaluator also adjusted seating (chair up, down, and distance) and monitor (rotating to the right or left) for best visibility.

Evaluators viewed a set of renderings for each dataset based on each scheme. Views involved multiple steps, a set of images at a time, and the actual tasks performed by evaluators were to look at the renderings and report which one, among each scheme (in each compared set), was (1) most distinct and (2) most preferred. In the first step, the renderings of each dataset were considered together for each scheme (first, four H_2O images, then 8 DTI images, and finally 8 Engine images). Then, all renderings of the next scheme were considered similarly, and so forth. Finally, evaluators considered pairwise top outcomes of earlier steps versus one another in a pairwise way (e.g., each evaluator viewed their most distinct harmonious color rendering versus their most distinct disharmonious color rendering). For each pairwise presentation, the evaluators reported (1) which of the isosurface renderings of the pair they thought was most distinct, based on its color combinations, and (2) which of the isosurface renderings of the pair they preferred, based on its color combinations. The datasets and color combinations were presented in the same order for all participants (to incur common after-image effects, if there were any).

The number of evaluators finding a particular type of rendering to be distinct was the base measurement for distinctness. The number who found a particular type of rendering to be preferred was the base measurement for preference.

4   Results

Study results are reported next. We report summaries of responses and a statistical analysis of significance (via the statistical sign test) for each case. As per standard practice, if an evaluator could not decide between renderings in a set, such outcomes were not considered in the analysis of significance.

4.1   Disharmonies Versus Opponent Colors

Figure 3 summarizes, by dataset, the evaluation of distinctness between the disharmonious and opponent color combinations for the three guidelines. The Disharmonious-Itten colors are (significantly) more distinct than the Opponent-Itten ones for all datasets; the Disharmonious-House ones are more distinct than the Opponent-House ones for DTI and H_2O; and the Opponent-Ebert ones are more distinct than the Disharmonious-Ebert ones for all renderings.

Figure 4 summarizes, by dataset, the pairwise comparisons evaluating preference between the disharmonious and opponent color combinations for the guidelines. Disharmonious are preferred over opponent color combinations for all situations, except the Engine dataset using Disharmonious-House.

4.2   Harmonious vs. Other Schemes

Figure 5 summarizes, by dataset, the evaluation of distinctness between the disharmonious and harmonious color combinations for the three guidelines. The Harmonious-Itten and Harmonious-House schemes were usually found to be less distinct than the disharmonious versions of the same. The opposite is usually the case for the Ebert et al. guidelines.

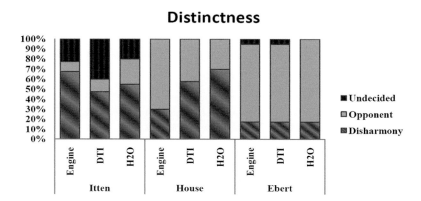

Fig. 3. Summaries of distinctness (disharmonious vs. opponent combinations) (Color figure online)

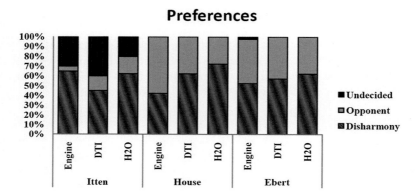

Fig. 4. Summaries of preference (disharmonious vs. opponent combinations) (Color figure online)

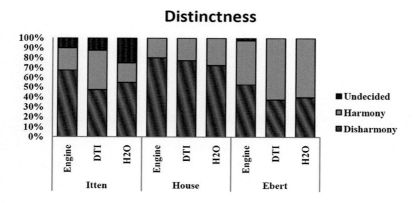

Fig. 5. Summaries of distinctness (disharmonious vs. harmonious combinations) (Color figure online)

4.3   Summary Results

Next, we report in tabular format distinctness and preference counts for all classes of datasets in each guideline. The tables present counts of those finding a class to be most distinct or preferred, number of undecideds (unless 0), and the sign test's Z scores and confidence levels. Included here are comparison results for House-based and Ebert-based schemes versus the Itten-based scheme. (The colors in Itten-based renderings used here violated House- and Ebert-based rules.)

Table 2 summarizes the evaluations of distinctness and preference for disharmonious versus opponent color combination renderings for the schemes. Users preferred (with significance) the disharmonies over the opponencies for all three guidelines. However, the distinctness findings were more mixed.

Table 2. Disharmonious vs. Opponent color combinations

	Distinctness			Preference		
	Itten	House	Ebert	Itten	House	Ebert
Disharmony	68	63	21	69	71	69
Opponent	19	57	95	15	49	50
Undecided	33	0	4	36	0	1
Z Score	5.146	0.456	6.964	5.783	1.917	1.650
Conf. Level	99 %	70 %	99 %	99 %	97 %	95 %

Table 3. Disharmonious vs. Harmonious color combinations

	Distinctness			Preference		
	Itten	House	Ebert	Itten	House	Ebert
Disharmony	68	92	52	67	64	63
Harmony	33	28	67	36	56	55
Undecided	19	0	1	17	0	2
Z Score	3.383	5.751	1.467	2.956	0.639	0.644
Conf. Level	99 %	99 %	92 %	99 %	75 %	75 %

Table 3 summarizes the evaluations of distinctness and preference of disharmonious versus harmonious color combination renderings for the schemes. Here, the disharmonies were distinct (with significance) over the harmonies for Itten- and House-based schemes. However, Disharmonious-House and Disharmonious-Ebert were not preferred (with significance) to Harmonious-House and Harmonious-Ebert, contradicting the Itten outcomes. Yet, more participants did prefer disharmonious to harmonious color combinations.

Table 4 summarizes results for harmonious versus opponent color combinations. Results were mixed for distinctness. Users preferred (with significance) harmonious combinations over opponent color combinations, though. We also found there is a statistically significant difference in distinctness between harmonious and opponent color combinations for Itten- and House-based schemes, but such is not the case for the Ebert-based scheme.

5   Discussion

While prior work [19] has found harmonies (i.e. of Itten) in map-based visualization preferable to disharmonies and opponencies, it seems such is not uniformly the case for layered surface visualization. Instead, for layered surface visualization, disharmonies are often preferred over opponencies and harmonies, especially for the House- and Itten-based schemes, (Itten harmonies may not be successful for layered surfaces.) Also, House- and Itten-based disharmonies are often the most distinct ones. Thus, use of disharmonious color combinations,

Table 4. Harmonious vs. Opponent color combinations

	Distinctness			Preference		
	Itten	House	Ebert	Itten	House	Ebert
Harmony	33	45	53	34	77	73
Opponent	19	75	60	13	43	44
Undecided	68	0	7	73	0	3
Z Score	1.803	2.830	0.753	2.917	3.012	2.59
Conf. Level	95 %	99 %	78 %	99 %	99 %	99 %

especially via Disharmonious-House and Disharmonious-Itten guidelines, may be valuable for visualizing multiple isosurfaces at once in a single rendering, perhaps due to enhancing a surface separation.

Given the human visual system's greater sensitivity to hue changes than to lightness and saturation changes [13], the Itten color guidelines (that vary hue for harmonious, disharmonious, and opponent color combinations while keeping saturation and lightness constant) appear to yield results aligned with human properties. In addition, since the House-based and Ebert-based schemes use differing saturation and lightness in all three color combinations, it seems reasonable that their comparisons produced divergent outcomes (viz., Tables 2, 3 and 4).

6   Conclusion

This paper has compared user perceptions of isosurface renderings utilizing certain coloration guidelines. The results of the studies suggest that House-based coloration schemes appear promising for layered surface visualization, with results about on par with Itten colors. Renderings using disharmonies are more distinct than harmonious colors for both Itten- and House-based schemes for surface-based volume visualization. However, the House-based disharmonies are not as uniformly valuable as Itten-based ones. Also, renderings using harmonies are preferred over opponent color combinations for all three schemes, and disharmonies are preferred (with significance) over the harmonies for Itten-based schemes, but still more participants preferred disharmonies over harmonies for House- and Ebert-based schemes in layered surface visualization.

There are many avenues for future work, such as investigating the use of more datasets, the value of these three guidelines in other visualizations, especially the House et al. guidelines intersection with Itten's guidelines. We also would like to see if experimental results are affected by changing presentation order to be random. Another direction would be to perform an adjunct study of task completion times using the varied visualizations.

References

1. House, D., Bair, A., Ware, C.: An approach to the perceptual optimization of complex visualizations. IEEE Trans. Vis. Comp. Graph. **12**, 509–521 (2006)
2. Ebert, D., Rheingans, P.: Volume illustrative: non-photorealistic rendering of volume models. In: Proceedings of IEEE Visualization 2000, pp. 195–202. Salt Lake City (2000)
3. Ebert, D.: Illustrative and non-photorealistic rendering. IEEE Trans. Vis. Comp. Graph. **11**, 20–27 (2005)
4. Itten, J.: The Art of Color. Van Nostrand Reinhold, New York (1961). Translated by Ernst Van Hagen from Kunst der Farbe
5. Cohen-Or, D., Sorkine, O., Gal, R., Leyvand, T., Xu, Y.Q.: Color harmonization. In: Proceedings of SIGGRAPH 2006, pp. 624–630. Boston (2006)
6. Einakian, S., Newman, T.: Effective color combinations in isosurface visualization. In: Proceedings of Visualization and Data Analysis 2012, vol. 8654, pp. P1–P8. SPIE, San Francisco (2012)
7. Borland, D., Taylor, R.: Rainbow color map (still) considered harmful. IEEE Comp. Graph. Appl. **27**, 509–552 (2007)
8. Itten, J.: The Elements of Color: A Treatise on the Color System of Johannes Itten Based on his Book the Art of Color. Van Nostrand Reinhold, New York (1970). Translated by E. Van Hagen
9. Matsuda, Y.: Color Design. Askura Shoten, Japan (1995)
10. Tokumaru, M., Muranaks, N., Imanishi, S.: Color design support system considering color harmony. In: Proceedings of IEEE International Conference on Fuzzy Systems, pp. 378–383. Honolulu (2002)
11. O'Donovan, P., Agarwala, A., Hertzmann, A.: Color compatibility from large datasets. ACM Trans. Graph. (TOG) **30**, 1–12 (2011)
12. Saunders, P., Interrante, V.: An Investigation into Color Preference for Color Palette Selection in Multivariate Visualization. University of Minnesota, Minneapolis, Minnesota (2005)
13. Rhodes, P., Laramee, R., Bergeron, D., Sparr, T.M.: Uncertainty visualization methods in isosurface volume rendering. In: Eurographics 2003 Short Papers, pp. 83–88, Granada, Spain (2003)
14. Silverstein, J., Parsad, N., Tsirline, V.: Automatic perceptual color map generation for realistic volume visualization. J. Biomed Inform. **41**, 927–935 (2008)
15. House, D., Ware, C.: A method for the perceptual optimization of complex visualizations. In: Proceedings of Advanced Visual Interfaces (AVI 2002), pp. 148–155 (2002)
16. Interrante, V., Fuchs, H., Pizer, S.: Conveying the 3D shape of smoothly curving transparent surfaces via texture. IEEE Trans. Vis. Comp. Graph. **3**, 98–117 (1997)
17. Volume-Library: The volume library. http://www9.informatik.uni-erlangen.de/External/vollib/. Accessed 2014
18. Abitom: Abitom software. http://www.abitom.com/. Accessed 2014
19. Einakian, S., Newman, T.: Experiments on effective color combination in map-based information visualization. In: Proceedings of Visualization and Data Analysis 2010, vol. 7530, pp. 5–12. SPIE, San Jose (2010)

Guidance on the Selection of Central Difference Method Accuracy in Volume Rendering

Kazuhiro Nagai[1] and Paul Rosen[2]([⊠])

[1] University of Utah, Salt Lake City, USA
[2] University of South Florida, Tampa, USA
prosen@usf.edu

Abstract. In many applications, such as medical diagnosis, correctness of volume rendered images is very important. The most commonly used method for gradient calculation in these volume renderings is the Central Difference Method (CDM), due to its ease of implementation and fast computation. In this paper, artifacts from using CDM for gradient calculation in volume rendering are studied. Gradients are, in general, calculated by CDM with second-order accuracy, $\mathcal{O}(\Delta x^2)$. We first introduce a simple technique to find the equations for any desired order of CDM. We then compare the $\mathcal{O}(\Delta x^2)$, $\mathcal{O}(\Delta x^4)$, and $\mathcal{O}(\Delta x^6)$ accuracy versions, using the $\mathcal{O}(\Delta x^6)$ version as "ground truth". Our results show that, unsurprisingly, $\mathcal{O}(\Delta x^2)$ has a greater number of errors than $\mathcal{O}(\Delta x^4)$, with some of those errors leading to changes in the appearance of images. In addition, we found that, in our implementation, $\mathcal{O}(\Delta x^2)$ and $\mathcal{O}(\Delta x^4)$ had virtually identical computation time. Finally, we discuss conditions where the higher-order versions may in fact produce less accurate images than the standard $\mathcal{O}(\Delta x^2)$. From these results, we provide guidance to software developers on choosing the appropriate CDM, based upon their use case.

1 Introduction

Direct volume rendering is one of the most powerful and widely used visualization techniques available for understanding 3D data. Users of these visualizations trust the images they see are true to the underlying phenomena being studied. However, artifacts from every stage of the volume rendering pipeline can potentially create inaccurate images. This inaccuracy may have severe consequences, such as failures to diagnose disease [1]. One important volume rendering pipeline stage is the calculation of gradients. These values are critical because the gradient directions are used to light the volume, and the magnitudes of these gradients are used for classification in multidimensional transfer functions.

When studying gradients, there are many potential types of artifacts. The majority of studies have evaluated artifacts from interpolation of gradients and developed new techniques to minimize these errors. Examples include: calculating gradients from analytical derivatives of filtering equations in order to produce fewer artifacts [2]; using interpolations that produce fewer artifacts than the analytical method [3]; and using global illumination in cases where the magnitude of

© Springer International Publishing Switzerland 2015
G. Bebis et al. (Eds.): ISVC 2015, Part I, LNCS 9474, pp. 328–338, 2015.
DOI: 10.1007/978-3-319-27857-5_30

gradient is too small to accurately calculate a direction, such as in homogeneous materials cases [4].

Although a number of advanced methods are available for calculating the gradient field, far and away, the most popular approach is using Central Difference Method (CDM) with second-order accuracy, $\mathcal{O}(\Delta x^2)$. To our knowledge, no existing work has considered artifacts coming from gradient values calculated using CDM[1].

The CDM is based upon Taylor Series expansion that enables approximating the first derivative of a scalar field with some bounded error, in this case $\mathcal{O}(\Delta x^2)$. Although the error within the CDM is well known, little has been done to measure its influence on the quality and correctness of volume rendered images. To address this shortcoming in the literature, we investigate how this albeit small error may impact the output image. We compare the standard CDM approach with more accurate versions of the gradient calculated using the Taylor Series, those with errors of $\mathcal{O}(\Delta x^4)$ and $\mathcal{O}(\Delta x^6)$. In particular, we look at variations in gradient direction and magnitude in a variety of publicly available datasets. Finally, we provide guidance for volume renderer designers, such that they can know when it is appropriate to select an alternative approach.

2   Prior Work

Volume rendering consists of a wide variety of operations, including sampling, filtering, classification, shading, and integration [6]. Each operation introduces artifacts, which may decrease the correctness and/or quality of images. There are various studies of artifacts in each stage of volume rendering.

In sampling, wood-grain artifacts are found with low sampling rate. According to Nyquist-Shannon sampling theorem, the original continuous data can be reconstructed from digital data if sampling rates are twice the highest frequency [7]. However, implementation of Nyquist-Shannon sampling theorem in volume rendering reduces the speed of interaction (i.e. lower frame rate). Adaptive Sampling [8] or stochastic jittering [9] are used in practice.

The filtering stage interpolates between multiple volume functions, usually through bilinear or trilinear interpolation. Using interpolation functions with C^1-continuity produces better images. Examples include B-spline interpolation [10], Catmull-Rom splines [11], and texture-based convolution.

The classification stage calculates the proper RGBA values by applying a transfer function. The main error created in this stage is determined by whether the transfer function is applied before or after interpolation. Generally speaking, post-interpolation is better than pre-interpolation [12].

In the shading stage, the gradient is the main factor producing artifacts. The main source of error comes from the interpolation of gradients. A common source of problems is the case of low magnitude gradients, where interpolation produces

---

[1] Usman et al. [5] use CDM with $\mathcal{O}(\Delta x^4)$ as a standard to evaluate their calculations of gradients, but they did not establish a justification for the use of $\mathcal{O}(\Delta x^4)$ instead of $\mathcal{O}(\Delta x^2)$.

noisy output. For precomputed normalized gradients, interpolation may produce non-normalized gradients. These can be easily renormalized on-the-fly. In addition, the precomputed method may introduce quantization errors by storing the gradients in 8-bits texture channels. Individually, these quantization errors are small. However, when accumulated, these errors may result in wood-grain artifacts.

There are a variety of methods to calculate gradients, beyond just CDM, such as kernel filter or other techniques mentioned previously. For a variety of reasons, $\mathcal{O}(\Delta x^2)$ CDM is still a common choice to calculate gradients in volume renderers. As far as we know, no one has published research to evaluate artifacts coming from gradients calculated using $\mathcal{O}(\Delta x^2)$ CDM.

One important premise on the study of artifacts in volume rendering is that no single definition of quality exists. Stochastic jittering is a good example. This technique increases the uncertainty by using random noise to erase wood-grain artifacts. The images have high visual quality, but also they may no longer be true to the original data. The result is that many techniques have been published on the premise of "better image quality" without clearly stating whether the basis is aesthetic, correctness, or both.

3 Central Difference Method (CDM)

The finite different stencil associated with CDM is calculated using the Taylor Series of $f(x + \Delta x)$, $f(x)$, and $f(x - \Delta x)$.

$$f(x + \Delta x) = f(x) + f'(x)\Delta x + \frac{f''(x)\Delta x^2}{2!} + \mathcal{O}(\Delta x^3) \tag{1}$$

$$f(x) = f(x) \tag{2}$$

$$f(x - \Delta x) = f(x) - f'(x)\Delta x + \frac{f''(x)\Delta x^2}{2!} + \mathcal{O}(\Delta x^3) \tag{3}$$

In these equations, x represents the current voxel, Δx represents the distance between voxels, $f(x)$ represents the function value, and $f'(x)$ and $f''(x)$ represent the first and second derivatives, respectively. By solving these equations for $f'(x)$, we have found the $\mathcal{O}(\Delta x^2)$ CDM.

$$f'(x) = \frac{f(x + \Delta x) - f(x - \Delta x)}{2} + \mathcal{O}(\Delta x^2) \tag{4}$$

We can also place Eqs. (1), (2), and (3) into a matrix.

$$\begin{bmatrix} f(x + \Delta x) \\ f(x) \\ f(x - \Delta x) \end{bmatrix} = \begin{bmatrix} 1 & \Delta x & \frac{\Delta x^2}{2!} \\ 1 & 0 & 0 \\ 1 & -\Delta x & \frac{\Delta x^2}{2!} \end{bmatrix} \begin{bmatrix} f(x) \\ f'(x) \\ f''(x) \end{bmatrix} \tag{5}$$

To get the CDM equations, the matrix equation can be solved ($A^{-1}B = A^{-1}AX$) and $f'(x)$ selected.

This approach can be generalized to work for both boundary (non-symmetric) cases and higher-order cases by selecting additional input function values (e.g. $f(x+2\Delta x)$, $f(x-2\Delta x)$, etc.) and expanding the Taylor series further, such that the matrix is square.

4 Experiments

For this study, we used a variety of volume types from The Volume Library website [13]. All experiments were run on a Windows-based laptop with an Intel Core i7 processor and NVIDIA GTX 880 m GPU. The experiments were performed using a homegrown volume renderer with the configuration shown in Table 1.

Table 1. Configuration of volume renderer

Sampling rate:	Greater than twice per voxel
Filtering:	Trilinear
Classification:	Pre-interpolation
Shading:	Pre-computed gradients renormalized in fragment shader
Integration:	8-bits used to store each fragment value.

These conditions are relatively common ones. We do place some considerations towards minimizing artifacts from other pipeline stages. For example, the sampling rate is so high that trilinear interpolations will not show wood-gain artifacts. Additionally, using pre-interpolation for classification avoids generating high frequency artifacts. Finally, although gradients are precomputed, interpolated gradients are normalized in the fragment shader.

For our results, we make an assumption that the $\mathcal{O}(\Delta x^6)$ gradient is our ground truth (i.e. no errors). We conducted several experiments to determine the influence of various levels of accuracy for gradients, focusing on magnitude and direction separately. This is because they tend to be used in different stages of the pipeline (i.e. magnitude in classification, direction in lighting), rarely together.

Throughout, the absolute error (d_m) and relative error (d_m^r) in magnitude is calculated by:

$$d_m = |A| - |B|, \tag{6}$$

$$d_m^r = \frac{|A| - |B|}{|A|}, \tag{7}$$

where A is the gradient of the $\mathcal{O}(\Delta x^6)$ voxel and B is the gradient of the same voxel with either $\mathcal{O}(\Delta x^2)$ or $\mathcal{O}(\Delta x^4)$.

The absolute error in the direction of gradients (d_d) is calculated by:

$$d_d = 1 - \frac{A \cdot B}{|A||B|}. \tag{8}$$

Since gradient direction is a normalized vector, relative error for gradient direction cannot be calculate.

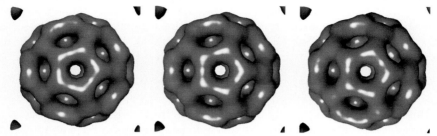

(a) Shown with $\mathscr{O}(\Delta x^2)$ (left), $\mathscr{O}(\Delta x^4)$ (middle), and $\mathscr{O}(\Delta x^6)$ (right), the first difference is in the shape of the specularity. Secondarily, the $\mathscr{O}(\Delta x^4)$ and $\mathscr{O}(\Delta x^6)$ have darker illumination in the holes of the bucky ball.

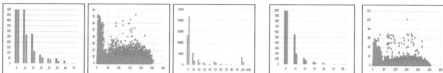
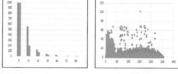

(b) Plots of the error in gradient magnitude for each voxel with $\mathscr{O}(\Delta x^2)$ (blue) and $\mathscr{O}(\Delta x^4)$ (orange). The histogram (left) shows the number of voxels with at different absolute error levels. The center chart plots the volume function value (horizontally) against the absolute error (vertically). The right histogram plots relative error (%) horizontally. Errors are calculated using Equation (6) and (7).

(c) Plots of the error in gradient direction for each voxel with $\mathscr{O}(\Delta x^2)$ (blue) and $\mathscr{O}(\Delta x^4)$ (orange). The histogram (left) shows the number of voxels with at different error levels. The right chart plots the volume function value (horizontally) against the error (vertically). Errors are calculated using Equation (8).

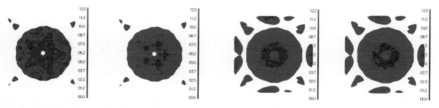

(d) Images showing error in the gradient magnitudes with $\mathscr{O}(\Delta x^2)$ (left) and $\mathscr{O}(\Delta x^4)$ (right).

(e) Images showing error in the gradient directions with $\mathscr{O}(\Delta x^2)$ (left) and $\mathscr{O}(\Delta x^4)$ (right).

Fig. 1. Experimental results using Bucky.pvm (Color figure online).

5 Results

We tested all of pvm files we could obtain from The Volume Library website [13], with the exception of 16-bit XMasTree.pvm and 8-bit Porsche.pvm, which we were unable to load. In total, we tested over 30 datasets. Due to space limitations, we discuss only a few interesting cases here.

5.1   Errors in Gradient Magnitude and Direction

Our results show that the gradients of each voxel calculated from $\mathscr{O}(\Delta x^2)$ have greater error than those calculated using $\mathscr{O}(\Delta x^4)$. The comparison of the magnitude and direction can be seen in Fig. 1. In particularly, differences can be seen in the results of lighting in Fig. 1a.

Gradient Magnitude. Figure 1b shows the magnitude error for each voxel using $\mathscr{O}(\Delta x^2)$ and $\mathscr{O}(\Delta x^4)$. The plots shows $\mathscr{O}(\Delta x^2)$ has greater error than $\mathscr{O}(\Delta x^4)$. For example, 6 % of voxels have 0.78 % error in $\mathscr{O}(\Delta x^2)$. On the other hand, using $\mathscr{O}(\Delta x^4)$, only 1 % of voxels have the same error (see Fig. 1b, center).

Figure 1d shows the distributions of error on the image. Of particular note is the red-colored pentagon shape on the surface of the bucky ball in $\mathscr{O}(\Delta x^2)$. In $\mathscr{O}(\Delta x^4)$, small red spots around the center hole of the bucky ball are observed.

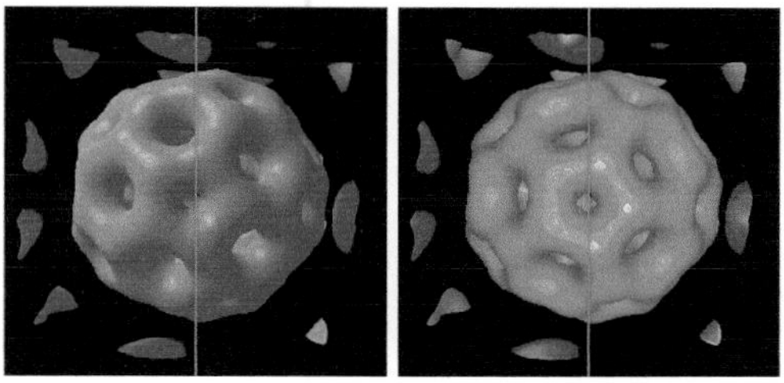

Fig. 2. The left half of each image is calculated by $\mathscr{O}(\Delta x^2)$, while the right uses $\mathscr{O}(\Delta x^4)$. The lighting in the left image is similar, but in the right image, the right half shows specular spots, which are not seen in the left half.

Gradient Direction. Figure 1c shows the error in gradient direction $\mathscr{O}(\Delta x^2)$ and $\mathscr{O}(\Delta x^4)$, both showing greater error in $\mathscr{O}(\Delta x^2)$ than $\mathscr{O}(\Delta x^4)$.

Figure 1e displays the distribution of errors in the gradient direction with $\mathscr{O}(\Delta x^2)$ (left) the $\mathscr{O}(\Delta x^4)$ and $\mathscr{O}(\Delta x^4)$ (right). While there is some difference between $\mathscr{O}(\Delta x^2)$ and $\mathscr{O}(\Delta x^4)$, the difference is not as obvious as it was for the gradient magnitude.

Figure 2 shows rendered images of Bucky.pvm. The left half of each use $\mathscr{O}(\Delta x^2)$ and the right uses $\mathscr{O}(\Delta x^4)$. In the right image, the white specular highlights are observed. These are only vis-

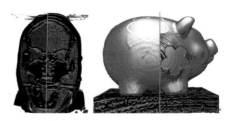

Fig. 3. Images of VisMale.pvm (left) and Pig.pvm (right) where the left half of each is calculated using the $\mathscr{O}(\Delta x^2)$ CDM and the right half uses $\mathscr{O}(\Delta x^4)$ CDM. VisMale.pvm demonstrates what happens with homogenous materials, while Pig.pvm demonstrates what can happen with high frequency edges.

ible under certain lighting directions and not visible under $\mathscr{O}(\Delta x^2)$. This represents a small, but nonetheless important, difference in images produced with less accurate gradients.

(a) Images using $\mathscr{O}(\Delta x^2)$, $\mathscr{O}(\Delta x^4)$, and $\mathscr{O}(\Delta x^6)$, respectively. The images show greater ringing artifacts for CDMs that should produce more accurate results (i.e. $\mathscr{O}(\Delta x^6)$).

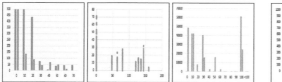

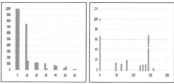

(b) Plots of the error in gradient magnitude for each voxel with $\mathscr{O}(\Delta x^2)$ (blue) and $\mathscr{O}(\Delta x^4)$ (orange). The histogram (left) shows the number of voxels with at different absolute error levels. The center chart plots the volume function value (horizontally) against the absolute error (vertically). The right histogram plots relative error (%) horizontally. Errors are calculated using Equation (6) and (7).

(c) Plots of the error in gradient direction for each voxel with $\mathscr{O}(\Delta x^2)$ (blue) and $\mathscr{O}(\Delta x^4)$ (orange). The histogram (left) shows the number of voxels with at different error levels. The right chart plots the volume function value (horizontally) against the error (vertically). Errors are calculated using Equation (8).

(d) Measurement of error in gradient magnitude for $\mathscr{O}(\Delta x^2)$ (left) and $\mathscr{O}(\Delta x^4)$ (right).

(e) Measurement of error in gradient direction for $\mathscr{O}(\Delta x^2)$ (left) and $\mathscr{O}(\Delta x^4)$ (right).

Fig. 4. Experimental results using Cross.pvm (Color figure online)

5.2   Homogeneous Materials

Materials that are more or less homogeneous (i.e. they have consistent values) are known to be problematic for CDM. Essentially, if all of the voxels surrounding me have approximately the same value, my gradient magnitude is zero and direction is underconstrained. In the $\mathscr{O}(\Delta x^2)$ CDM, the stencil size is quite small, limited to 6 surrounding voxels. By using the $\mathscr{O}(\Delta x^4)$ or higher CDM, the stencil footprint increases, and the voxel gradients becomes more stable.

Figure 3 (left) is an example of this. The left half of the image shows no texture in the brain with $\mathscr{O}(\Delta x^2)$. The right half, with $\mathscr{O}(\Delta x^4)$ begins to show some of the texture detail. Its important to recognize, however, that although

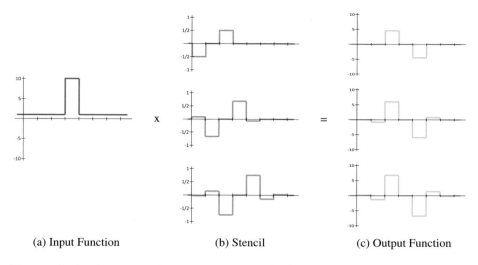

|(a) Input Function|(b) Stencil|(c) Output Function|

Fig. 5. A simple rectangular input function (left), representing a high-frequency feature, has $\mathcal{O}(\Delta x^2)$ (top), $\mathcal{O}(\Delta x^4)$ (middle), and $\mathcal{O}(\Delta x^6)$ (bottom) stencils applied (center). The resulting output (right) shows larger spread for the higher accuracy stencils. Since these output represent the gradient in one direction, the larger footprints lead to the larger ringing for those stencils.

the gradient is better constrained, that does not mean the gradient is true to the underlying data. The real problem here is that the data are either truly homogeneous or undersampled.

5.3 High Frequency Materials and Ringing Artifacts

Although theoretically more accurate, higher-order CDM does not always lead to an accurate. High frequency surfaces, such as those seen in Fig. 3 (right) or Fig. 4a, are just such failure cases. Here, the large sampling area of the $\mathcal{O}(\Delta x^4)$ and $\mathcal{O}(\Delta x^6)$ CDMs have led to a ringing artifacts. Figure 5 demonstrates how this occurs. The larger footprint of the higher order stencils results in capturing the feature in the gradient calculation at further distance.

Figure 4b shows the error in the gradient magnitude for $\mathcal{O}(\Delta x^2)$ and $\mathcal{O}(\Delta x^4)$ CDM versions. Both plots show the error in the gradient magnitude of $\mathcal{O}(\Delta x^2)$ is larger than that of $\mathcal{O}(\Delta x^4)$—the same result as Bucky.pvm. However, the assumption that the $\mathcal{O}(\Delta x^6)$ CDM is the ground truth lead us to this false conclusion. Figure 4d shows the distribution of errors in image space. As we can see, the errors are occurring in regions where the gradients of the high frequency surface voxels are not parallel. This is obvious in the round dent, but it also occurs in the seam where the perpendicular surfaces come together.

Figure 4c shows the error for gradient directions with $\mathcal{O}(\Delta x^2)$ and $\mathcal{O}(\Delta x^4)$ CDM on each voxel. Both plots show $\mathcal{O}(\Delta x^2)$ has greater error than $\mathcal{O}(\Delta x^4)$. Figure 4e displays the distributions of errors in the direction of gradients with

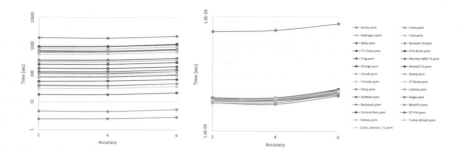

Fig. 6. The time vs. $\mathcal{O}(\Delta x^2)$, $\mathcal{O}(\Delta x^4)$ and $\mathcal{O}(\Delta x^6)$. The left shows the time to calculate 100 iterations. The right is the same result per voxel, normalized by dividing the time by the number of voxels (Color figure online).

$\mathcal{O}(\Delta x^2)$ (left) and $\mathcal{O}(\Delta x^4)$ (right). The $\mathcal{O}(\Delta x^2)$ shows darker than $\mathcal{O}(\Delta x^4)$. However, this $\mathcal{O}(\Delta x^2)$ does not have the red lines, which we could see in the error for the gradient magnitude.

The bottom of the right images of Fig. 4a shows the bottom of the right side of Cross.pvm image. $\mathcal{O}(\Delta x^4)$ and $\mathcal{O}(\Delta x^6)$ show extra lines that do not appear in $\mathcal{O}(\Delta x^2)$. This ringing artifact is well studied [14]. In this case, the higher order version contains more artifacts than its lower order counterpart.

5.4   Times to Compute CDM with $\mathcal{O}(\Delta x^2)$, $\mathcal{O}(\Delta x^4)$, and $\mathcal{O}(\Delta x^6)$

The time to compute gradients for all datasets are presented in Fig. 6. In most cases, the time to calculate $\mathcal{O}(\Delta x^4)$ is about the same as the $\mathcal{O}(\Delta x^2)$. In fact, some of $\mathcal{O}(\Delta x^4)$ cases are less than $\mathcal{O}(\Delta x^2)$. The time to calculate $\mathcal{O}(\Delta x^6)$ is only about 20 % more than $\mathcal{O}(\Delta x^2)$. We speculate that this small variation has something to do with memory locality, since stencil operations tend to be memory bound. This is due to the relatively small number of arithmetic operations performed relative to the large number of memory operations in inner product calculations. For these types of memory bound operations, cache performance is king. Larger stencils result in greater numbers of conflict misses.

6   Discussion and Conclusions

With these results, we give some brief guidance on selecting what type of CDM to use in the real-world.

My CDM choice does matter. Although many images were similar under multiple orders of CDM, there were small, possibly important differences, depending upon your usage scenario. When scientific data is used for decision making, it is critical to keep the images true to the data.

Higher-accuracy is (usually) better because it theoretically produces images true to their underlying data (i.e. with lower error). However, one has to be cautious when blindly applying either a low or high order CDM as the practice does not always live up to theory. The choice may in fact depend upon the data.

Higher-accuracy costs more time, but it does not cost nearly as much as one might expect. If the algorithm is well designed to take advantage of memory locality, a higher order calculation may only suffer a small onetime overhead, in the case of static datasets. If looking to improve performance, focus on aspects that impact the performance of rendering per frame, such as sampling rate.

$\mathscr{O}(\Delta x^4)$ *represents a good compromise* between fast calculation, accuracy, and small ringing artifacts. While renderers should really be adapting to their data, we have settled on the $\mathscr{O}(\Delta x^4)$ CDM as a better default than the commonly selected $\mathscr{O}(\Delta x^2)$ version. This finding supports the decision of Usman et al. [5] to use CDM with $\mathscr{O}(\Delta x^4)$ to evaluate their calculations of gradients.

In conclusion, when it comes to putting together a quality volume renderer, many design decisions must be made. It is clear from our results that, while not the most critical element, the choice of gradient calculation mechanism can have an important impact on the final rendered image. As such we recommend more thought be put into the choice, and, at the very least, volume renderer designers should chose $\mathscr{O}(\Delta x^4)$ as their default method.

References

1. Lundstrm, C., Ljung, P., Persson, A., Ynnerman, A.: Uncertainty visualization in medical volume rendering using probabilistic animation. IEEE Trans. Vis. Comput. Graph. **13**, 1648–1655 (2007)
2. Bentum, M.J., Lichtenbelt, B.B.A., Malzbender, T.: Frequency analysis of gradient estimators in volume rendering. IEEE Trans. Vis. Comput. Graph. **2**, 242–254 (1996)
3. Hossain, Z., Alim, U.R., Möller, T.: Toward high-quality gradient estimation on regular lattices. IEEE Trans. Vis. Comput. Graph. **17**, 426–439 (2011)
4. Kniss, J., Premoze, S., Hansen, C., Shirley, P., McPherson, A.: A model for volume lighting and modeling. IEEE Trans. Vis. Comput. Graph. **9**, 150–162 (2003)
5. Alim, U., Moller, T., Condat, L.: Gradient estimation revitalized. IEEE Trans. Vis. Comput. Graph. **16**, 1495–1504 (2010)
6. Engel, K., Hadwiger, M., Kniss, J.M., Lefohn, A.E., Salama, C.R., Weiskopf, D.: Real-time volume graphics. In: ACM SIGGRAPH 2004 Course Notes. ACM, New York (2004)
7. Nyquist, H.: Certain topics in telegraph transmission theory. Trans. AIEE **47**, 617–644 (1928)
8. Kraus, M., Strengert, M., Klein, T., Ertl, T.: Adaptive sampling in three dimensions for volume rendering on gpus. In: Hong, S.H., Ma, K.L. (eds.) APVIS, pp. 113–120. IEEE (2007)
9. Cook, R.L.: Stochastic sampling in computer graphics. ACM Trans. Graph. **5**, 51–72 (1986)

10. Mihajlovi, Z., Goluban, A., Zagar, M.: Frequency domain analysis of B-spline interpolation. In: Proceedings of the IEEE International Symposium on Industrial Electronics (ISIE 1999), vol. 1, pp. 193–198. IEEE (1999)
11. Möller, T., Machiraju, R., Mueller, K., Yagel, R.: Classification and local error estimation of interpolation and derivative filters for volume rendering. In: Volume Visualization Symposium, pp. 71–78. IEEE (1996)
12. Engel, K., Kraus, M., Ertl, T.: High-quality pre-integrated volume rendering using hardware-accelerated pixel shading. In: Proceedings of the ACM SIG-GRAPH/EUROGRAPHICS workshop on Graphics hardware, pp 9–16. ACM (2001)
13. Roettger, S.: The Volume Library (2015). http://lgdv.cs.fau.de/External/vollib/. Accessed 19 March 2015
14. Marschner, S.R., Lobb, R.J.: An evaluation of reconstruction filters for volume rendering. In: Bergeron, R.D., Kaufman, A.E. (eds.) Proceedings of the Conference on Visualization, pp. 100–107. IEEE Computer Society Press, Los Alamitos (1994)

Deep Learning of Neuromuscular Control for Biomechanical Human Animation

Masaki Nakada[(✉)] and Demetri Terzopoulos

Computer Science Department, University of California, Los Angeles, USA
nakada@cs.ucla.edu

Abstract. Increasingly complex physics-based models enhance the realism of character animation in computer graphics, but they pose difficult motor control challenges. This is especially the case when controlling a biomechanically simulated virtual human with an anatomically realistic structure that is actuated in a natural manner by a multitude of contractile muscles. Graphics researchers have pursued machine learning approaches to neuromuscular control, but traditional neural network learning methods suffer limitations when applied to complex biomechanical models and their associated high-dimensional training datasets. We demonstrate that "deep learning" is a useful approach to training neuromuscular controllers for biomechanical character animation. In particular, we propose a deep neural network architecture that can effectively and efficiently control (online) a dynamic musculoskeletal model of the human neck-head-face complex after having learned (offline) a high-dimensional map relating head orientation changes to neck muscle activations. To our knowledge, this is the first application of deep learning to biomechanical human animation with a muscle-driven model.

1 Introduction

The modeling of graphical characters based on human anatomy is becoming increasingly important in the field of computer animation. Progressive fidelity in biomechanical modeling should, in principle, result in more realistic human animation. Given realistic biomechanical models, however, we must confront a variety of difficult motor control problems due to the complexity of human anatomy.

In conjunction with the modeling of skeletal muscle [1,2], existing work in biomechanical human modeling has addressed the hand [3–5], torso [6–8], face [9–11], neck [12], etc. These prior efforts demonstrated different control schemes for individual parts of the human body. Further research is needed to achieve a fully robust control system for muscle-actuated biomechanical human animation.

The challenge in controlling biomechanical human models stems from anatomical intricacy, which complicates the kinematics, dynamics, and actuation of the character. For example, the neck-head biomechanical model developed in [12] has 10 bones and 72 muscle actuators (Fig. 2), while the full-body biomechanical model presented in [13] includes 103 bones and 823 muscle actuators,

© Springer International Publishing Switzerland 2015
G. Bebis et al. (Eds.): ISVC 2015, Part I, LNCS 9474, pp. 339–348, 2015.
DOI: 10.1007/978-3-319-27857-5_31

Fig. 1. Snapshots from an animation demonstrating robust online neuromuscular control of head orientation in a muscle-actuated biomechanical neck-head-face system. Despite the continuously varying orientation of its shoulders that are fixed to the unsteady wagon base, the character can visually track a moving target (doll) while naturally supporting the mass of its head in gravity atop its anatomically accurate flexible neck. Facial expressions are automatically produced in response to the observed movement of the doll—the character expresses awe when the doll is raised above the head, anger when the doll is shaken, and pleasure when the doll is held calmly at eye level.

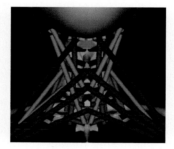

Fig. 2. Close-up views of the musculoskeletal structure of the biomechanical neck-head model, comprising 7 cervical vertebra and 72 actuators: 48 deep muscles (red), 12 intermediate muscles (green), and 12 superficial muscles (blue) (Color figure online).

an order of magnitude greater complexity. Controlling such models requires the specification of muscle innervations (i.e., activation signals) at every time step of the biomechanical simulation.

1.1 Neuromuscular Control

As pioneered by Lee and Terzopoulos [12] in the context of neck-head control, neuromuscular learning approaches are the most biomimetic and promising in tackling high-dimensional, muscle-actuated biomechanical motor control problems. To control the neck-head system in gravity, rather than resorting to traditional inverse dynamics control, which is both unnatural and computationally expensive, these authors employed a conventional, artificial neural network with two hidden layers of sigmoidal units, which they trained offline through backpropagation learning. The training data comprised tens of thousands of random target head orientations as network inputs and the muscle activation levels

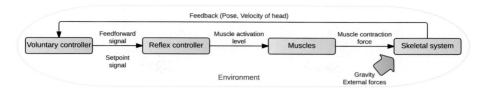

Fig. 3. Control flow for the neck-head model.

required to achieve these target orientations as associated network outputs. Their trained neuromuscular controller was capable of efficiently controlling the biomechanical model online, thereby achieving a real-time neuromuscular neck-head control system. The neck-head biomechanical model is actuated by 72 muscles, so their trained neural network controller maps 3 inputs (a target head orientation) to 72 outputs (the muscle activations required to achieve this orientation) with adequate generalization. Their *offline* training procedure took approximately 10 h to complete the backpropagation learning process.

Conventional neural networks with only a few hidden layers have not proven effective at coping with the higher dimensionalities and necessarily much greater quantities of training data required to learn to control biomechanical models of an order of magnitude greater complexity, such as the one presented in [13]. For this reason, Lee and Terzopoulos [12] made no attempt to generalize their neck-head controller to deal with other than horizontally-oriented shoulders and, unsurprisingly, their trained neuromuscular controller fails to maintain satisfactory control unless the shoulders remain nearly horizontal.

1.2   Deep Learning of Neuromuscular Control

Deep neural networks with many hidden layers could not at first be trained due to technical difficulties with the backpropagation learning algorithm such as local minima and the vanishing gradient problem. Motivated by recent breakthroughs in "deep learning" that have mitigated these problems, we have applied deep neural network architectures to neuromuscular control. The long-range goal of our research is to overcome the aforementioned challenges in the neuromuscular control of complex biomechanical models.

In this paper, we adapt deep learning techniques to the neuromuscular control of the human neck-head biomechanical model described in [12]. Our contributions include the introduction of a new neuromuscular controller, whose deep architecture differs markedly from the one in [12], with dramatically enhanced learning ability and efficiency even when employing much higher dimensional training data. This enables us to push significantly beyond the prior benchmark, training our neck-head neuromuscular controller to work well not merely for horizontal shoulders, but for a considerable range of shoulder orientations.

1.3 Overview

The remainder of this paper is organized as follows: In Sect. 2, we briefly explain the neck-head model and its neuromuscular control system. We then develop our novel neuromuscular learning approach in Sect. 3, and demonstrate the effectiveness of our approach through experiments reported in Sect. 4. Section 5 presents our conclusions and discusses future work.

2   Biomechanical Model and Control System

In our work, we use the anatomically accurate musculoskeletal model of the human neck-head complex implemented by [12], which is illustrated in Fig. 2. The model comprises a skull, whose mass is commensurate with that of an average adult male head, atop a cervical skeleton comprising 7 vertebrae connected by 3 degree-of-freedom rotational joints, and a clavicle base. This skeletal system is actuated by 72 contractile cervical muscles, modeled as Hill-type actuators, arranged in three layers (Fig. 2). Appendix 1 presents additional details about the biomechanical model.[1] The associated neuromuscular control architecture (Fig. 3) involves a high-level voluntary controller that determines the head pose and a lower-level reflex controller that produces the necessary muscle activation inputs. Appendix 2 presents additional details about the hierarchical neuromuscular control system, as well as the gaze control mechanism.

The specific technical challenge that we address in our work is how best to implement and train the voluntary controller. The considerable kinematic and muscular redundancy of the biomechanical model makes it infeasible to control the cervical muscles online in real time using an inverse dynamics approach. Instead, as in [12], we pursue a biomimetic neuromuscular control approach. A neural network controller is trained offline on a large dataset comprising several tens of thousands of random head orientations (inputs) and the associated muscle activations (outputs). To generate training data, each 72-dimensional muscle activation output, which must achieve the associated target head pose and sustain it against the pull of gravity, is synthesized off-line through (computationally expensive) inverse kinematics/dynamics computations as in [12].

Lee and Terzopoulos [12] used a conventional, shallow neural network for their voluntary controller. It typically took on the order of 10 h to train this network, which imposed practical limits on the dimensionality and size of the usable training dataset. To surmount this obstacle in neuromuscular control, we devise a deep neuromuscular control architecture that proves to be more efficient to train and is capable of learning more proficiently from larger and much higher-dimensional datasets. In principle, this makes it amenable to biomechanical models of much greater complexity—potentially even to a full-body biomechanical model with an order of magnitude more articular degrees of freedom and muscle actuators [8,13].

---

[1] Due to space limitations, the appendices are found in a supporting document (see, e.g., https://db.tt/wn37j8rg).

At each time-step of the biomechanical simulation, muscle activations are computed by our trained deep neural network in the voluntary controller, and they are input to the reflex controller. The latter adjusts the muscle activations in accordance with the difference between the current state and the desired state of each muscle. The biomechanical simulator computes the muscle forces induced by the muscle activations, applies the muscle forces to the dynamic skeletal system, and numerically updates the accelerations, velocities, and positions of the articulated bones, thereby advancing the simulation in time.

2.1   Egocentric Head-Eye Control

Humans obtain visual information about objects from their foveated retinal images. Sensing a target of interest in the periphery of the visual field triggers a rapid eye rotation to foveate the target in conjunction with a comparatively sluggish head rotation (due to the greater mass of the head) toward the target. Lee and Terzopoulos [12] synthesized desired head rotation trajectories through the spherical linear interpolation (slerp) of head Euler angles, which is unnatural.

By contrast, in our neuromuscular control system, we define the angular error vector as the difference of the current head orientation and the target head orientation. In each time step, our neuromuscular controller acts to reduce this 2-D angular error vector (with horizontal angle θ and vertical angle ϕ). Our neck-head control system does not require knowledge of the position of the visual target nor of the orientation of the head relative to the world frame. Rather, the relevant variables are represented relative to the egocentric frame whose origin is at the head and which rotates with the head.

3   The Deep Neuromuscular Controller

Our neuromuscular controller differs substantially from the shallow (3-input, 72-output, 2-hidden-layer) network described in [12]. We map to the 72 muscle activation outputs not only the target head orientation inputs, but also the 72 current muscle activations, so that the current muscle activation state affects the computation of the output muscle activations. Incorporating the current muscle activations as inputs is necessary in order to reduce the angular error vector in egocentric coordinates while achieving robust control regardless of shoulder orientation. The higher input dimensionality of our controller along with the much greater quantity of data necessary to train it at a variety of shoulder orientations exceeds the capabilities of the shallow network used in [12], and this problem is unlikely to be overcome without the use of a deeper network.

Our proposed deep network architecture includes 74 inputs, comprising the 2 angles, θ and ϕ, of the head orientation error vector plus the 72 current muscle activations, along with 72 muscle activation outputs. The intermediate, hidden layers of the network implement an n-stage stacked denoising autoencoder (SdA) [14], which will be described in the next section. The bottom layer outputs the modified 72 cervical muscle activations (illustrated in Appendix 3).

A more powerful neuromuscular controller learning method is needed to train our deep network. Our learning method has two phases. The first is a pre-training stage, where we apply unsupervised learning to the SdA. The second is a fine-tuning stage where we apply supervised learning to the multilayer perceptron. The first phase contributes to finding better initial parameters and the second phase tunes those parameters by comparing the output from the network and the training example values.

The overall learning process proceeds as follows: We generate a training dataset comprising many input/output examples by numerically simulating the biomechanical model. Then, we run the SdA pre-training phase on this dataset. After pre-training, we initiate fine-tuning with the multilayer perceptron. Subsequently, we can employ the trained deep network online to control the biomechanical model.

Additional technical details about our deep neuromuscular controller learning process are presented in the following sections.

3.1   Stacked Denoising Autoencoder (SdA)

An autoencoder is a multilayer neural network that minimizes the difference between its input and output after encoding/decoding through its hidden layers. The encoding is as follows:

$$\mathbf{y}_e = \boldsymbol{\sigma}(\mathbf{W}\mathbf{x} + \mathbf{b}), \tag{1}$$

where $\mathbf{x}$ is the input, $\mathbf{y}_e$ is the encoded output, $\boldsymbol{\sigma}$ is a nonlinearity with sigmoidal $\sigma(z) = 1/(1+e^{-z})$ components, $\mathbf{W}$ is the weight matrix, and $\mathbf{b}$ is the bias vector. Likewise for the decoder,

$$\mathbf{y}_d = \boldsymbol{\sigma}(\mathbf{W}'\mathbf{y}_e + \mathbf{b}'), \tag{2}$$

where $\mathbf{y}_d$ is the decoded value, and $\mathbf{W}'$ and $\mathbf{b}'$ are the weight matrix and bias vector. After encoding/decoding, the inputs are compared to the outputs, yielding the error function

$$E = \frac{1}{N}(\sum_{i=1}^{N} ||\mathbf{x}_i - \mathbf{y}_{d_i}||^2), \tag{3}$$

where i indexes over the N training examples. The error gradient ∇E is used to update the weights and biases through gradient descent (in our work, we set the learning rate to 0.1).

In mini-batch stochastic gradient descent, ∇E is estimated using a limited number of training examples. We include 32 training examples in each mini-batch. Processing all the mini-batches in this way constitutes a training epoch. We use a tied-weight scheme, $\mathbf{W}' = \mathbf{W}^T$, which has a beneficial regularization effect and requires less memory. Hence, the parameters to be updated are $\mathbf{W}$, $\mathbf{b}$, and $\mathbf{b}'$.

A Denoising Autoencoder (dA) forces its hidden layers to discover more robust features and prevent learning the identity mapping, which the simple autoencoder may uselessly determine as being the zero-error optimal solution.

To this end, one adds noise to the training data, and the dA tries to reduce the noise and recover the original data from the corrupted data. Associating a probability (0 implies certainty) with each unit in each layer, we set the probabilities to 0.1 for the hidden layers and randomly mask some inputs by setting them to zero.

We stack the dAs to form a Stacked Denoising Autoencoder (SdA), where the outputs of each dA are propagated as inputs to the next dA. The advantage of the SdA is that each dA is trained independently and sequentially, and the dAs may be stacked as deeply as necessary to produce a good output.

3.2   Pre-training Phase

During the pre-training phase, the training starts at the input layer of the SdA and advances to the output layer, completing the training one layer at a time. We run 300 epochs for each layer sequentially such that each dA can learn a good representation and the input data are mapped to the output data with minimal information loss.

The initial weights are uniformly sampled from the interval $[-v, v]$, where $v = \sqrt{6/(n_i + n_o)}$, with n_i being the number of inputs and n_o the number of outputs in a layer. It is known that uniform sampling from this interval works well with sigmoidal units [15]. This prevents the learning process from becoming trapped in a local minimum early in the pre-training phase and it yields better optimization and faster convergence.

3.3   Fine-Tuning Phase

During the fine-tuning phase, we start with the weights and biases obtained in the pre-training phase, and then we update them using the multilayer backpropagation algorithm. This phase is supervised, as we compare the outputs to the labeled training data and reduce the error by updating the weights and biases using mini-batch stochastic gradient descent with (mean sum-of-squared) error backpropagation.

The training process is stopped when it ceases to improve sufficiently on a validation dataset (we set the threshold in each training epoch to 0.005 %). This "early stopping" prevents the training process from overfitting.

4   Experiments and Results

4.1   Fixed, Horizontal Shoulder Orientation

Our first set of experiments evaluates the performance of our deep neuromuscular controller relative to the shallow one reported in [12] with fixed horizontal shoulder orientation. Figure 4 plots the learning performance of different network structures. The plots indicate the superiority of our deep neuromuscular controller learning method with pre-training. Appendix 4.1 presents the details of this experiment.

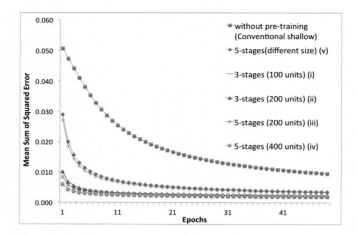

Fig. 4. Training error with different network structures when fine-tuning on the validation dataset.

4.2 Robust Control with Varying Shoulder Orientation

We verified that the shallow neuromuscular controller developed in [12] fails to maintain control of the head when the shoulders are not in a horizontal orientation. To facilitate comparison, the motion of the doll visual target is the same as the one in the head-eye gaze behavior demonstration illustrated in Fig. 10 of reference [12]. The more the orientation of the shoulders deviates from the horizontal, the more difficulty the shallow neuromuscular controller experiences in tracking the visual target. More strikingly, as the shoulders tilt back moderately, the neck-head skeleton collapses backward in gravity, resulting in catastrophic failure. We will next explain how we train our deep neuromuscular controller to accomplish this more challenging control task.

Synthesis of Training Data: First, with horizontal initial shoulder orientation, we synthesize training data by incrementing the head by $1°$ angles around the θ and ϕ axes in the range $-50° < \theta < 50°$ and $-50° < \phi < 50°$. Then, we repeat the same process with different shoulder orientations, varying the orientation of the shoulders by $5°$ from $-25°$ to $25°$ around both the frontal x axis along the shoulders and the sagittal z axis. This yields 1,000,000 example target angular error vectors with associated muscle activations computed through inverse kinematics/dynamics. The training data synthesis took 6.5 days to complete. Finally, we partition the data into training, validation, and testing datasets in the ratio $70\,\%{:}15\,\%{:}15\,\%$, respectively.

We trained our deep neuromuscular controller with the aforementioned dataset and verified that, unlike the shallow neuromuscular controller, our controller enables the neck-head system to smoothly track the visual target while the orientation of the shoulders varies continuously in the range of $-25°$ to $25°$ around both the x and y axes. We used the 5-stage SdA network with 400 units in each stage—network (iv) in Fig. 4—since it proves to learn best as is shown in the figure.

Figure 1 shows snapshots from a demonstration video of our neck control system, demonstrating that it copes very well with continuously varying shoulder orientations. Although our training dataset was 33 times larger (and higher-dimensional) than the one used in [12], our off-line training process completed in only 23.0 h, with the pre-training phase taking 19.3 h and the fine-tuning phase taking 2.7 h.

The pre-training phase worked well with our training data, and our deep neural network learned the appropriate initial weights and biases after 300 epochs of training for each autoencoder layer. Subsequently, the fine-tuning phase started with a relatively small 0.021 error (the error would have been much higher had we simply set random initial weights and biases). This reduced the time for the fine-tuning phase, which completed with 0.003 error after satisfying the early stopping condition in 82 epochs.

For a fair comparison, we replicated the shallow neural network presented in [12] and tested it with our larger dataset. Appendix 4.2 includes a plot of the training progress. It took 2.5 days to complete 500 training epochs, and the error fluctuated between 0.0525 and 0.0545 without decreasing. None of the different parameter settings and shallow network structures that we tried resulted in successful convergence.

5 Conclusion and Future Work

We have introduced a new approach to the design and training of neuromuscular controllers of complex biomechanical human models for use in computer graphics animation. To our knowledge, this is the first application of deep neural networks to the problem. Our approach proves to be superior to previous methods both in terms of the efficiency of the learning process and, potentially more importantly, in the ability to learn effectively from much larger quantities of higher-dimensional training data. These benefits enabled us to train a deep neuromuscular neck-head controller to work well over a continuous range of shoulder orientations.

Our experiments demonstrated that our deep network, which incorporates significantly more units in its hidden layers, can learn effectively from our largest training datasets (in our case, one million input/output examples), reinforcing the notion that deep networks are generally more suitable than shallow networks for the regression, with good generalization, of massive quantities of training data.

We expect our deep learning approach to architecting and training neuromuscular controllers to work for other, similarly complex biomechanical models, such as those for arms and hands. Furthermore, our approach seems promising for the neuromuscular control of full-body human models with a dramatically higher dimensional muscle activation space, on the order of 1,000 muscle actuators.

References

1. Ng-Thow-Hing, V.: Anatomically-based models for physical and geometric reconstruction of humans and other animals. Ph.D. thesis, University of Toronto, Computer Science Department (2001)
2. Irving, G., Teran, J., Fedkiw, R.: Invertible finite elements for robust simulation of large deformation. In: Proceedings of the ACM SIGGRAPH/EG Symposium on Computer Animation, p. 131. ACM Press, New York (2004)
3. van Nierop, O.A., van der Helm, A., Overbeeke, K.J., Djajadiningrat, T.J.: A natural human hand model. Vis. Comput. **24**, 31–44 (2007)
4. Tsang, W., Singh, K., Fiume, E.: Helping hand: an anatomically accurate inverse dynamics solution for unconstrained hand motion. In: Proceedings of the ACM SIGGRAPH/EG Symposium on Computer Animation, pp. 319–328 (2005)
5. Sueda, S., Kaufman, A., Pai, D.K.: Musculotendon simulation for hand animation. ACM Trans. Graph. **27**, 83 (2008)
6. Sifakis, E., Neverov, I., Fedkiw, R.: Automatic determination of facial muscle activations from sparse motion capture marker data. ACM Trans. Graph. **24**, 417–425 (2005)
7. DiLorenzo, P., Zordan, V., Sanders, B.: Laughing out loud: control for modeling anatomically inspired laughter using audio. ACM Trans. Graph. **27**, Article no. 125 (2008). ACM SIGGRAPH Asia 2008 Proceedings
8. Lee, S.H., Sifakis, E., Terzopoulos, D.: Comprehensive biomechanical modeling and simulation of the upper body. ACM Trans. Graph. **28**(99), 1–17 (2009)
9. Lee, Y., Terzopoulos, D., Waters, K.: Realistic modeling for facial animation. In: Proceedings of ACM SIGGRAPH 1995, pp. 55–62 (1995)
10. Kähler, K., Haber, J., Yamauchi, H., Seidel, H.: Head shop: generating animated head models with anatomical structure. In: Proceedings of the ACM SIGGRAPH/EG Symposium on Computer Animation, pp. 55–63 (2002)
11. Sifakis, E., Neverov, I., Fedkiw, R.: Automatic determination of facial muscle activations from sparse motion capture marker data. ACM Trans. Graph. **1**, 417–425 (2005)
12. Lee, S.H., Terzopoulos, D.: Heads up! Biomechanical modeling and neuromuscular control of the neck. ACM Trans. Graph. **23**, 1188–1198 (2006)
13. Si, W., Lee, S.H., Sifakis, E., Terzopoulos, D.: Realistic biomechanical simulation and control of human swimming. ACM Trans. Graph. **34**(10), 1–15 (2014)
14. Bengio, Y., Lamblin, P., Popovici, D., Larochelle, H.: Greedy layer-wise training of deep networks. In: Advances in Neural Information Processing Systems 19 (NIPS 2007), pp. 153–160 (2007)
15. Glorot, X., Bengio, Y.: Understanding the difficulty of training deep feedforward neural networks. In: International Conference on Artificial Intelligence and Statistics, vol. 9, pp. 249–256 (2010)

NEURONAV: A Tool for Image-Guided Surgery - Application to Parkinson's Disease

José Bestier Padilla[1], Ramiro Arango[1], Hernán F. García[2],
Hernán Darío Vargas Cardona[2]([✉]), Álvaro A. Orozco[2],
Mauricio A. Álvarez[2], and Enrique Guijarro[3]

[1] Universidad del Quindío, Armenia, Colombia
{jbpadilla,ramy}@uniquindio.edu.co
[2] Faculty of Electrical Engineering,
Universidad Tecnológica de Pereira, Pereira, Colombia
{hernan.garcia,hernan.vargas,aaog,malvarez}@utp.edu.co
[3] Universitat Politecnica de Valencia, Valencia, Spain
eguijarro@eln.upv.es

Abstract. Deep Brain Stimulation (DBS) is a surgical procedure used to treat symptoms associated with Parkinson's Disease (PD). The success of DBS depends on the correct location of a microelectrode over the subthalamic area. This paper presents the software NEURONAV for image-guided surgery. This software serves as a support for the medical specialist in DBS. NEURONAV has two principal modules. Firstly, it allows to plan the DBS. The application contains options such as 3D image viewer, image registration, slice viewers, landmark medical planning and visualization of 3D brain structures. Secondly, NEURONAV includes an interactive application to perform the tracking of the microelectrode during the DBS. NEURONAV has been tested in procedures performed at a specialized medical center in Colombia. Results show an adequate tracking of the microelectrode implanting process and it is possible to achieve better therapeutic outcomes, and more robustness to the side effects generated by DBS in PD patients.

1 Introduction

Deep Brain Stimulation (DBS) is a surgical option for patients with Parkinson's disease that do not respond efficiently to a drug therapy. During DBS, a microelectrode is implanted over the subthalamic region. This device stimulates electrically a target brain structure, specially the Subthalamic nucleus (STN). Controlled stimulation with the DBS electrode achieves a reduction in all symptoms of PD and improves the quality of life of patients [1].

Image-guided surgery (IGS) systems are often used in complex surgical procedures. IGS systems have the ability to track surgical devices in spatial coordinates during intraoperative procedures. Since the IGS tools have been established as clinical support for specialists during surgical procedures demanding high precision, the scientific community has increased the interest in the development of these systems [2,3].

© Springer International Publishing Switzerland 2015
G. Bebis et al. (Eds.): ISVC 2015, Part I, LNCS 9474, pp. 349–358, 2015.
DOI: 10.1007/978-3-319-27857-5_32

Currently, research teams are developing IGS software for functional localization of brain structures in DBS. For example, [4] proposed an automatic pattern recognition system based on Spikes analysis and frequential transform methods to identify microelectrode recordings (MER) signals from specific subcortical nuclei (i.e. Subthalamic Nucleus-STN) [5]. Similarly in [6], the authors presented a query software to estimate the electrical potential generated by the DBS electrode in the nervous system, and they conducted an analysis of the relationship with clinical outcomes in several patients. These systems proved the profits of automated softwares as a decision support for the medical team during DBS.

Accurate preoperative systems applied to surgical procedures are necessary. A priority of IGS must be the protection of patients against any potential mistake [7]. For this reason, the medical image processing and their applications to the neurosciences field must be explored [8,9]. Image-guided surgery gives new advantages in critical surgical procedures like DBS. For example, Integration of computed tomography (CT) and magnetic resonance imaging (MRI) allow an accurate mapping of the brain in 3D models. Also, it is possible to acquire spatial coordinates for tracking of microelectrode devices during DBS [10].

In this paper, we present a tool for image guide surgery called NEURONAV. NEURONAV is an open-source application that allows interactive neurosurgical planning, volume rendering, rigid and nonrigid registration, 3D image viewer, flexible layouts and slice viewers, atlas registration, and on-line tracking of microelectrode devices during DBS. The goal is to provide a support tool for the DBS surgery in which the neurosurgeons can accurately identify the real location of the microelectrode device over the subthalamic area in a 3D environment. Moreover the main purpose of this system is to reduce the adverse side effects that may occur because of inadequate identification of the target brain areas. Currently, NEURONAV has been tested in DBS's performed in the institute of epilepsy and Parkinson of the Eje Cafetero, located in Pereira, Colombia. Results obtained are promising and satisfactory.

The rest of the paper is arranged as follows. Section 2 provides a detailed discussion of materials and methods. Section 3 presents the experimental results and discussion respectively. The paper concludes in Sect. 4, with a summary and some ideas for future research.

2 Materials and Methods

2.1 Database

The database of Universidad Tecnológica de Pereira (DB-UTP) contains recordings of MRI studies from four patients with Parkinson's disease and it was labeled (brain structures) by neurosurgery specialist from the Institute of Parkinson and Epilepsy of the Eje Cafetero, Pereira, Colombia. This database contains $T1$ and $T2$ sequences with $1\,\mathrm{mm} \times 1\,\mathrm{mm} \times 1\,\mathrm{mm}$ voxel size and slices of 512×512 pixels.

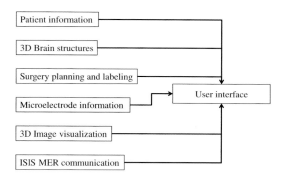

Fig. 1. NEURONAV pipeline.

2.2   Microelectrode Device

The neurosurgical equipment ISIS MER[1] (Inomed Medical GmbH), is used to collect data for derivation of potentials from the trajectory of microelectrodes in deep brain stimulation surgery applied to patients with Parkinson disease. The sampling frequency of the ISIS MER system is 25 KHz and 8 bit-resolution.

2.3   Software Development

NEURONAV is composed of a main interface and two independent modules, which are loaded dynamically during the program initialization. These modules are related by functions and controls defined in the main program. Software modules include:

1. Patient information module
2. 3D Brain structures module
3. Surgery planning and labeling module
4. Microelectrode information module
5. 3D Image visualization module
6. ISIS MER communication module

Figure 1 shows the outline of the NEURONAV system.

Patient Information Module. The patient information module contains the name, age, sex, and neurological and surgery planning information of the medical diagnosis. This information is relevant for purposes of medical history of patients.

3D Brain Structures Module. In this module the user can edit and refine the brain structures segmentation using a geometric transformation panel with translation, rotation and scale options. In addition, this tool allows the selection of a particular brain structure (3D mesh), color change, opacity and visibility.

---

[1] http://www.inomed.com/products/functional-neurosurgery/isis-mer-system/.

Deformable Registration. We use deformable medical image registration based on B-spline transform to perform the atlas reconstruction [11]. To perform the registration task, we took as system parameters: (1) *Moving Image:* we denote this parameter as Atlas Intensity and it contains the MRI $T1$ volume data from SPL-PNL atlas.[2] (2) *Fixed Image:* Here we used the MRI $T1$ data volume from the DB-UTP, because the main goal of our approach is deform the MRI volume that contains the Atlas labels (SPL-PNL) to match them with the MRI volume from a given patient. (3) *Metric:* To assess the quality of the registration process, we have calculated the mean and variance of the mean squared sum of intensity differences (MSE) for similarity measure (error) as well as normalized mutual information (4) *Interpolator:* we proved the performance of the registration process using nearest neighbor. (5) *Optimizer:* for data processing, we used regular step gradient descent and versor rigid 3D transform optimizer to measure the computational cost of the procedure.

Brain Atlas Reconstruction. After performing the registration process, we obtain the registered moving image that matches with the MRI volume data, the next step is to perform a transformation over the labels that describe the brain structures. The 3D brain structure reconstruction process is accomplished by transforming the labels maps of a given atlas using the B-spline method. Finally we used the marching cubes algorithm to build the 3D models of the brain structures. For the marching cubes algorithm we use the available implementation in the visualization toolkit (VTK) [8]. Figure 2 shows the brain atlas reconstruction framework.

Surgery Planning and Labeling Module. This module allows the planning of deep brain stimulation surgery. Here, it is possible to make interactive marks on each of the MRI slices (axial, coronal and sagittal) such as entry location and target location of the microelectrode device (stereo-tactic coordinates). In addition, this module allows the visualization of the AC-PC coordinates. Finally, the user can add another fiducial points[3] in order to perform an atlas registration for a new study of neurosurgery.

Microelectrode Information Module. The microelectrode information module, indicates the real position of the DBS device. But, the most important function of this module is the tracking of the neurostimulator inside the brain during a surgery. This position is indicated in millimeters and it is taken from the depth sensor of ISIS MER.

---

[2] Brain Atlas developed by the Surgical Planning Laboratory in collaboration with the Harvard Neuroscience Laboratory at Brockton VA Medical Center. http://www.spl.harvard.edu/publications/item/view/1265.

[3] A fiducial point is a key mark over the MRI volume used to plan the entry and the target point in the DBS surgery.

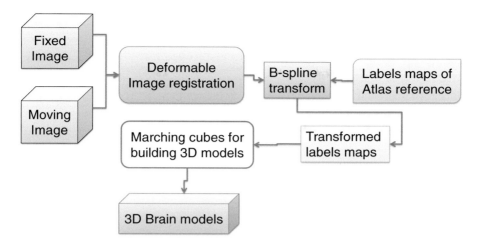

Fig. 2. 3D brain atlas reconstruction scheme

3D Image Visualization Module. This module has an interactive way to analyze medical images by scrolling the MRI and CT slices in three axis (axial, coronal and sagittal). Also, this module has a 3D view for mesh and volume visualization. Some of the tools of this module are zoom, contrast, illumination and slices scrolling.

ISIS MER Communication Module. Within this module, NEURONAV establishes a communication with ISIS MER via the IP port. This module receives the real location of the microelectrode device using IP communication, whereby the neurosurgical device sends the current state of the depth information (microelectrode implant process) at a 25 KHz frequency.

2.4   Toolboxes

To build this system we use powerful open-source toolboxes. First, the Visualization Toolkit (VTK[4]) was used for image processing and visualization. Second, Insight Segmentation and Registration Toolkit (ITK[5]) was used to atlas register and landmark manipulation. ITK is an open-source cross-platform system that provides developers with an extensive suite of software tools for image analysis. Thirdly, the Image-Guided Surgery Toolkit (IGSTK[6]) was used to perform the microelectrode tracking. IGSTK is a high-level, component-based framework which provides a common functionality for image-guided surgery applications. Finally, FLTK[7] toolkit was used to build the graphic user interface. FLTK is a

---

[4] http://www.vtk.org/.
[5] http://www.itk.org/.
[6] http://www.igstk.org/.
[7] http://www.fltk.org/.

cross-platform C++ GUI toolkit multi-platform. FLTK provides modern GUI functionality without the bloat and supports 3D graphics via OpenGL and its built-in GLUT emulation.

3   Experimental Results and Discussion

3.1   Deformable Registration Results

Table 1 summarizes the results of the registration quality of the MRI DB-UTP database in terms of sum of the squared differences between intensity values (MS) and normalized correlation coefficient (NC) for the different metrics, optimizers and interpolator configurations. The results clearly show that the registrations which are based on nearest neighbour interpolator improve the correlation between the images before and after the registration process. The results also show that the computational cost is higher when the mean square metric is used. This is due to the search for similarities values on intensity values.

Table 1. Deformable registration accuracy results. **N**: Nearest Neighbor Interpolator, **NC**: Normalized Correlation, **MS**: Mean Squares, **V**: Versor Rigid Optimizer, **R**: Regular Step Gradient Descent

Interpolator	Metric	Optimizer	Time (s)	Error
N	NC	R	218.852 ± 3.2	0.72 ± 0.023
N	NC	V	229.024 ± 4.1	0.83 ± 0.015
N	MS	R	627.68 ± 8.3	279.2 ± 0.21
N	MS	V	663.67 ± 5.4	279.6 ± 0.16

3.2   NEURONAV System

This is an application in which a specialist can perform the planning and execution of a deep brain stimulation surgery. NEURONAV allows an interactive management and analysis of MRI data volumes, compute an atlas registration and perform real-time tracking of microelectrode devices. Figure 3 shows the interface of the application and the 3D transformation module on brain structures. The system includes functions for translation, rotation, scale, opacity, and color visualization.

3.3   Results Obtained in the Operating Room

Currently, NEURONAV has been tested in Deep Brain Stimulation Surgeries performed at the Institute of Epilepsy and Parkinson of the Eje Cafetero, in Pereira-Colombia. The results show a useful and interactive tool for the medical

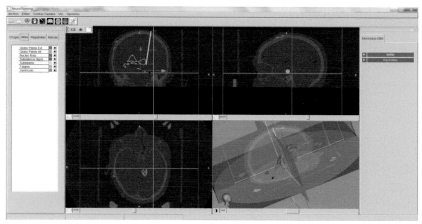

(a) Transformation module

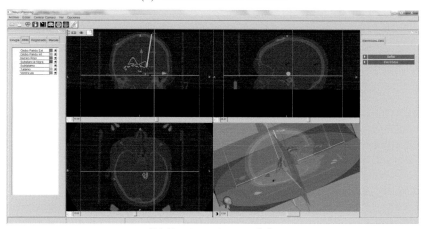

(b) Registration module

Fig. 3. NEURONAV 3D mesh transformation and registration modules. We implement rigid and nonrigid registration methods that deform accurately a given atlas into a new MRI volume [9].

specialist, which allows to identify the real location of the microelectrode device during DBS surgery. The results show that NEURONAV fulfill the medical protocols for planing an image-guided surgery. Figure 4 shows the operating room interface. The results shows the 3D environment in which the neurosurgeon can control the depth of the microelectrode device to perform the stimulation.

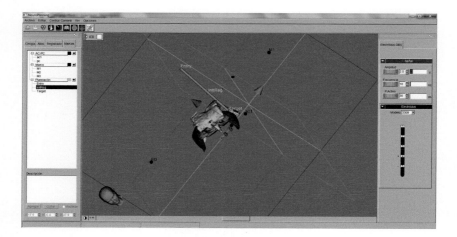

Fig. 4. Operating room interface and microelectrode information modules.

3.4 Discussion

We give some comments about results:

- NEURONAV is an application developed with open-source platforms for neuronavigation applied to Parkinson's disease surgery. The user can perform planning of deep brain stimulation, where it is defined the boarding path and the target point (Subthalamic Nucleus). Moreover, during surgery the software is linked to the medical equipment ISIS MER for tracking the location of the electrode during the insertion process. For this purpose, we use the IP port and the depth sensor information is transmitted online to NEURONAV. Moreover, the user has the possibility of analyzing MRI data volumes and performing Atlas registration for automatic segmentation of key brain structures in DBS.
- Another attribute that this software has is the transformation and registration models of 3D brain structures. For this purpose we developed functions of translation, rotation, scaling, opacity and color display as well as rigid and nonrigid image registration methods [9]. These tools provide a significant contribution to medical specialists due to the ability of tuning certain 3D structures according to their convenience.
- Figure 4 shows the operating room interface and the microelectrode tracking device. The figure shows the fiducial points related to entry and target marks and manual landmarks. The 3D environment shows an illustrative way to visualize a brain structure while the surgical procedure is accomplished. Right GUI panel, shows the real location of the depth information of the microelectrode. The results show high robustness in the IP communication between the ISIS MER device and NEURONAV tool.

4   Conclusions and Future Work

In this paper, a tool for image-guided surgery called NEURONAV was presented. NEURONAV is a open-source application that allows interactive neurosurgical planning, volume rendering, rigid and nonrigid registration, 3D image viewer, flexible layouts and slice viewers, atlas registration, and real-time tracking of surgical instruments (microelectrode device).

The results show a medical support tool for the DBS surgery in which the neurosurgeons can accurately identify the correct location of the microelectrode device in the thalamus area whereby the success of surgical procedures increases. The developed application has great potential for planning and execution in surgeries resulting from neurosurgical procedures.

Three main tasks are let as future work: Firstly, the proposed scheme can be used to reconstruct 3D brain structures from 2D MRI slices. Secondly, NEURONAV tool pretends to be available for free access to the academic and medical community. Thirdly, NEURONAV should be continued using in DBS surgeries in order to collect some data tests which can improve the performance of the developed application. Furthermore from this tool, we are developing a Volume Tissue Activation module to control the stimulation parameters in Parkinson's treatment.

Acknowledgments. H.F García and H.D. Vargas Cardona are funded by Colciencias under the program: *formación de alto nivel para la ciencia, la tecnología y la innovación - Convocatoria 617 de 2013*. This research has been developed under the project: *Estimación de los parámetros de neuro modulación con terapia de estimulación cerebral profunda en pacientes con enfermedad de Parkinson a partir del volumen de tejido activo planeado*, financed by Colciencias with code 1110-657-40687.

References

1. Benabid, A.: Deep brain stimulation for Parkinson's disease. Curr. Opin. Neurobiol. **13**, 696–706 (2003)
2. Metson, R., Cosenza, J., Cunningham, M.: Physician experience with an optical image guidance system for sinus surgery. Laryngoscope **110**, 972–976 (2000)
3. Eliashar, R., Sichel, J.Y., Gross, M.: Image guided navigation system: a new technology for complex endoscopic endonasal surgery. Postgrad. Med. J. **79**, 686–690 (2003)
4. Vargas Cardona, H., Padilla, J., Arango, R., Carmona, H., Alvarez, M., Guijarro Estelles, E., Orozco, A.: Neurozone: on-line recognition of brain structures in stereotactic surgery - application to parkinson's disease. In: 2012 Annual International Conference of the IEEE Engineering in Medicine and Biology Society (EMBC), pp. 2219–2222 (2012)
5. García, H.F., Álvarez, M.A., Orozco, A.: Bayesian shape models with shape priors for MRI brain segmentation. In: Bebis, G., et al. (eds.) ISVC 2014, Part II. LNCS, vol. 8888, pp. 851–860. Springer, Heidelberg (2014)

6. Miocinovic, S., Noecker, A., Butson, C., McIntyre, C.: Cicerone: stereotactic neu- rophysiological recording and deep brain stimulation electrode placement software system. In: Sakas, D.E., Simpson, B.A. (eds.) Operative Neuromodulation: Neural Networks Surgery, vol. 97/2, pp. 561–567. Springer, Heidelberg (2007)

7. Wise, S., DelGaudio, J.: Computer-aided surgery of the paranasal sinuses and skull base. Expert Rev. Med. Devices **2**, 395–408 (2005)

8. Cosa, A., Canals, S., Valles-Lluch, A., Moratal, D.: Unsupervised segmentation of brain regions with similar microstructural properties: application to alcoholism. In: 35th Annual International Conference of the IEEE Engineering in Medicine and Biology Society (EMBC 2013), pp. 1053–1056 (2013)

9. Cabezas, M., Olive, A., Llado, X., Freixenet, J., Cuadra, M.: A review of atlas- based segmentation for magnetic resonance brain images. Comput. Methods Pro- grams Biomed. **104**, e158–e177 (2011)

10. Schmelzeisen, R., Nils-Claudius, G., Schramm, A., Schon, R., Otten, J.: Navigation-guided resection of temporomandibular joint ankylosis promotes safety in skull base surgery. Int. J. Oral Maxillofac. Surg. **60**, 1275–1283 (2002)

11. Rueckert, D., Sonoda, L.I., Hayes, C., Hill, D.L.G., Leach, M.O., Hawkes, D.J.: Nonrigid registration using free-form deformations: application to breast MR images. IEEE Trans. Med. Imaging **18**, 712–721 (1999)

ST: 3D Mapping, Modeling and Surface Reconstruction

Generation of 3D/4D Photorealistic Building Models. The Testbed Area for *4D Cultural Heritage World* Project: The Historical Center of Calw (Germany)

José Balsa-Barreiro[✉] and Dieter Fritsch

Institut für Photogrammetrie (IFP), University of Stuttgart,
Geschwister-Scholl-Str. 24D, 70174 Stuttgart, Germany
jose.balsa-barreiro@ifp.uni-stuttgart.de

Abstract. 3D/4D photorealistic models present great advantages in different applications like preservation of the cultural heritage, urban planning, public management, etc. For getting precise and reliable models, the geometries should be accurate and the textures wrapped with optimal quality imagery. The systems of *terrestrial laser scanner* (TLS) are one of the most reliable techniques for capturing accurate data. However, these systems have some drawbacks like the missing data in certain areas or the lack of information about textures. Photogrammetric images contain the required textures and can be used to generate dense point clouds by dense image matching. In this way, the combined use of photogrammetric images and laser systems is symbiotic, being thus able to exploit the advantages of both technical methods. If the textures are wrapped using historical photos, the temporal dimension is incorporated to the model (4D), making it possible to reconstruct the past. However, historical photos have some additional drawbacks like the lack of metadata and the fact that they have been captured from only a few positions.

1 Introduction

Photogrammetry and laser scanning present complementary aspects for 3D model reconstruction. Nowadays, the systems of *terrestrial laser scanning* (TLS) are one of the most important and reliable technical methods to measure the position and shape of objects in the three dimensional space [1]. TLS systems can quickly acquire very dense point clouds fitted to the surface of the object swept, offering a high accuracy. Thus, these systems have a number of key advantages such as the massive acquisition of data, the speed of data collection, the reduction of staff and time required in surveying, among others. These features make these systems one of the most valued to perform 3D reconstructions by *high definition surveying* (HDS).

However, not all are advantages. TLS systems have some drawbacks such as the high costs of the instrument, the increasing complexity and the extended times in the phase of the data processing, which requires further investments. Data collection with TLS has several considerations such as the difficulty for extracting information on certain materials and/or surfaces due to irregular circumstances, the occlusions in certain sectors and the absence of

© Springer International Publishing Switzerland 2015
G. Bebis et al. (Eds.): ISVC 2015, Part I, LNCS 9474, pp. 361–372, 2015.
DOI: 10.1007/978-3-319-27857-5_33

data about the object texture. In addition, there are certain limitations to the capture of data because of the equipment size and the restricted access in certain environments.

The limitations of TLS systems need to overcome data losses and gaps using additional techniques. One possibility is the use of photogrammetry, which can be used and combined with TLS [2]. Photogrammetry can get 3D models from point clouds generated from 2D images of an object captured from a variety of different positions and orientations. The *Institut für Photogrammetrie* (IFP) of the *Universität Stuttgart* (Germany) has broad experience in the development of methodologies for generating 3D models from photogrammetric images. Two steps are taken in this approach. Firstly, *Structure from Motion* (SfM) reconstruction is used to derive orientations and to get sparse data from the surface. Secondly, with the resulting orientations, the *SURE software*, which has been developed by the IFP, is applied to obtain dense point clouds [3].

Photogrammetric point cloud generation can be performed using *stereo pairs* (*multiview stereo*) from close-range and aerial imagery [4]. Thus, the 3D coordinates of the surface of an object can be determined performing measurements in two or more images captured from different positions. Image matching needs to be applied to find the common points on each image. The *collinearity equations in photogrammetry* draw a line of sight from the camera location to the object point, where rays determine the 3D point coordinates of the object.

Accurate 3D building models can be generated from the point clouds obtained by photogrammetry and laser scanner. Before 3D modeling, the different point clouds should be brought into a common main reference system by means of a *Helmert (7-parameter) transformation* and the application of the *Iterative Closest Point* (ICP) algorithm, which enable to get a better performance. From the combined point cloud, *Leica Cyclone* software is used to produce geometric (CAD) models, which provide a simplified vision of the urban landscape.

2 Generation of 3D/4D Models in Urban Environments

The complex environment of the urban areas explains the inexistence of optimal procedures for the generation of 3D models. Thus, despite the advantages of the TLS for capturing data, its use in these environments has several drawbacks related to the missing data in certain areas such as roof buildings. The combined use of both aerial and close range photogrammetry, besides TLS, allows to get a more complete and exhaustive process of data acquisition in urban environments.

Techniques derived from close-range photogrammetry permit to add textures from different photos to the geometric model. Adding textures obtained from historical photos enable to generate 4D models, where the temporal perspective is considered. The objective of these models is to recreate and reconstruct environments of the past that are visually realistic.

Fully textured 3D models have great importance in digital preservation as reference scale models. Thus, more effective policies or regulations for the maintenance and conservation of historical buildings can be applied. Other possible applications of these models are related to city planning, building maintenance, urban and infrastructure analysis, among others.

3   The Project *4D-CH-World*

The data used for this conference paper belong to the project *Four Dimensional Cultural Heritage World* (abbreviated as *4D-CH-World*), framed within the 7th *EU Framework Programme PEOPLE-2012-IAPP*. Thus, the results and considerations introduced here are dependent on this project.

The testbed area selected for the *4D-CH-World* project is the historical center of the city of *Calw*, a frequently visited area located 35 km southwest of *Stuttgart* (Germany), close to the northern part of the *Black Forest* (Fig. 1). The old city counts with a broad history and rich cultural heritage. Among the 15th and 17th centuries, *Calw* was a commercial center of wood dissemination and processing, cloth production and salt trade, becoming an urban pole which was more important than *Stuttgart*, the real capital of the Federal State *Baden-Württemberg*, which is the third most extended and populated region (of sixteen) of Germany. At the end of 17th century the medieval city of *Calw* was burned and destroyed by the French soldiers. In the 18th and 19th centuries, the city lived a renaissance phase motivated by cloth processing.

Fig. 1. (Left) location of the historical center of Calw relative to Stuttgart. (Right) Aerial perspective of this area [5].

Besides this, *Calw* is the birth town of one of the most internationally renowned German writers, *Hermann Hesse* (1877-1962), awarded with the *Literature Nobel Prize* in 1946. The writer lived here in his early years, until the year 1890. Concrete locations of the city are described by *Hermann Hesse* in some of his books.

The rich cultural heritage of Calw explains the existence of huge amounts of images, sketches, plans, drawings and maps. The *Calw City Archive* has several collections with high historical value, like the sketches, plans and drawings dating back to the 10th and 11th centuries, and some photos captured in the middle of the 19th century. In total, more than 3,500 photos between the years 1860 and 2014 have been offered to the project *4D-CH-World*, in addition to many plans of the old town, sketches and paintings dating back to the early medieval ages (from 1,100 onwards). Furthermore, the city is still continuously visited and photographed by tourists, allowing to meet thousands of images in *Open Access Image Repositories* (OAIR) accessible online such as on *Flickr*, *Picasa* or *Panoramio*.

3.1 Objective

The aim of the research project is the generation of 3D/4D virtual models of the historical center of Calw (Germany). These models will be integrated on platforms like *Europeana* or *Google Earth* at successive steps of the project. *Europeana* is an European institution to collect and preserve 3D digital reconstructions of cultural heritage [6]. The diffusion of the generated models allows the user to reach a better knowledge of the cultural heritage of this city, based on intuitive representation methods. Furthermore, this project will enable to carry out a more exhaustive catalogue of the cultural heritage of the historical center of Calw. Thus, more efficient strategies and policies for the conservation and preservation of the cultural heritage of this area can be implemented.

The interest of this research project presents a global perspective, involving both the local authorities of Calw and all its visitors, besides all the technical staff who develop this project, from both industry and academia partnerships. In the case of the IFP, this project allows to continue with the development of methodologies and procedures which are more efficient for the generation of 3D/4D photorealistic building models.

3.2 Staff Used

The TLS model used in this research project was the *Leica Scan Station P20*, owned by the IFP. This model scans 360 degrees in horizontal and 270 in vertical direction around the fixed scan station. The maximum measuring range is of 120 meters, with a frequency up to 1 million points per second. The *Leica Scan Station P20* can register large datasets, including maps of intensity values and RGB images. The most representative technical parameters of this TLS model can be found on its product specifications [7].

Considering that the studied area is quite visited, one relevant aspect of this TLS model is that the emitted laser beam is eye-safe. The levels of resolution and accuracy chosen for the scanning were medium-high, obtaining thus an efficient relation between scanning time and point densities captured.

Table 1. Performance parameters related with the lens Carl Zeiss Biogon 2,8/21 ZM [8].

Focal length	21 mm
Aperture range	2.8 – 22 (1/3 steps)
Focusing range	0.5 m – ∞
Coverage at close range	47 × 71 cm
Angular field, diag./horiz./vert.	90°/80°/58°

For image capturing, the photo camera used was the *Ricoh GXR*, a digital compact camera of the Japanese *Ricoh Company*. This model has interchangeable units, each housing a lens, sensor and image processing engine. Each unit has features optimized for different purposes and environments. The lens used was a *Carl Zeiss Lens Biogon 2.8/21 ZM*, whose main performance parameters are exposed in the Table 1 [8]. The photos were captured with infinite focal length and using the function mode dial A (*aperture priority*), which leads to adjust the shutter speed for getting an optimal time exposure of the images.

Furthermore, several aerial images of the historical center of Calw supplied by the *Landesamt für Geoinformation und Landentwicklung* of Baden-Württemberg (LGL-BW) [9] were used to reconstruct the roof landscapes.

3.3 Current Project Status

The IFP and the program of the *M.Sc. in Geomatics Engineering GEOENGINE*, both from the *Universität Stuttgart (Germany), are* entrusting several students for the development of this project in diverse areas of the historical center of Calw (Fig. 2).

Fig. 2. Studied areas within the historical center of Calw.

Two research projects have been performed in the last years. The first one [10] was centered in the *Market Square Calw* and surrounding buildings, including the *Town Hall* and the building where Herman Hesse was born. The second one [11] was centered in the *Badstrasse* area, including the historical bridge of *Nikolaus* over the *Nagold river*. Nowadays four additional research projects are being performed: number 3 (area surrounded by the streets *Lederstrasse, Torgasse, Nonnengasse* and *Biergasse*), number 4 (area surrounded by the streets *Biergasse, Lederstrasse, Kronengasse* and the *Markplatz*), number 5 (area surrounded by the streets *Altburger Strasse, Postgasse* and the *Markplatz*, including the *Herman Hesse Museum*) and number 6 (area surrounded by the streets *Schulgasse, Im Zwinger strasse* and the *Kirchplatz*, including the *Evangelical Church* of Calw). Each one of the studied area comprises a set of historical buildings, around 20, and an area between 5,500 and 7,700 m^2, which is represented in the Fig. 2.

4 Resume of Procedure

The procedures performed in this project can be classified into two phases: acquisition and processing of data. The main aspects related to each one of these phases are exposed below.

4.1 Data Capturing

Two techniques are used for the capturing of data: *terrestrial laser scanning* (TLS) and photogrammetric images. The most relevant aspects from each of these techniques are presented in the following sections:

a. Terrestrial laser scanning (TLS)

The TLS scanner emits a narrow laser beam that sweeps across a target object, gathering millions of points with 3D coordinates (XYZ) closely spaced from each other (Fig. 3). The points measured are collected and clustered to represent the position and geometry of the objects in the three dimensional space.

Fig. 3. (Left) Photography and (right) TLS point cloud obtained in two building facades of the historical center of Calw.

The different studied areas are swept from different scan (fixed) stations. However, each one of the different scan positions has a limited view, which depends on the characteristics of the area surveyed and the surrounding environment. For this reason, the number of scan stations varies for each studied area. Before TLS surveying, laser campaign should be planned. Thus, the technical staff examines the most optimal location for each scan station, as well as the position of the common target points between the successive scan stations. This strategy does not entirely avoid the appearance of missing data areas, although it may reduce them in both number and size.

Scan points are recorded and saved in a (local) coordinate system, which is related to the scanner in each station. For this reason, during the phase of data processing the spatial relationship between the different scan stations has to be determined, also denoted *scan registration*. Thus, the position and distance from each scan station regard to set of control (target) points should be measured during the acquisition of the data.

Normally, three target points manually placed and visible from successive scan stations are required. However, sometimes it is difficult to measure the position of certain target points because of the poor visibility, difficult meteorological conditions, non-optimal geometry or errors in the network settings, among others. In case that the minimum number of target points between successive scan stations could not be recorded during the surveying, the alternative is to locate natural points that are common between successive scan stations during the phase of data processing.

b. Photogrammetric images

By using photogrammetric techniques, the missing area of the building facades and roof landscapes in the TLS point cloud can be made up. These techniques enable to generate 3D point clouds from images, both close range and aerial, previously positioned, oriented and rectified.

The 3D point cloud generation can be performed by using *stereo image pairs*, also called *multi-view stereo*. Thus, the 3D coordinates of the points of an object can be determined by measurements made in two or more images captured from different positions. Image matching is applied to find the common points on each image. By using the *collinearity equations in photogrammetry* (1) a line of sight from the camera location to the object point can be determined. The intersection of these rays determines the position of each point of the image in the 3D real space.

$$x = x_0 - f \frac{r_{11}\left(X - X_0\right) + r_{21}\left(Y - Y_0\right) + r_{31}(Z - Z_0)}{r_{13}\left(X - X_0\right) + r_{23}\left(Y - Y_0\right) + r_{33}(Z - Z_0)}$$

$$\text{(1)}$$

$$y = y_0 - f \frac{r_{12}\left(X - X_0\right) + r_{22}\left(Y - Y_0\right) + r_{32}(Z - Z_0)}{r_{13}\left(X - X_0\right) + r_{23}\left(Y - Y_0\right) + r_{33}(Z - Z_0)}$$

where

x and y the image coordinates
X and Y and Z the object space coordinates
f the focal distance of the photo camera
$r_{11} \dots r_{33}$ the elements of the rotation matrix $(R)_{3x3}$ between both the object and the image reference systems

In this way, a sparse 3D point cloud of the object can be obtained. However, the point density of this initial point cloud is too low for performing 3D modeling. For this reason, in a second step, dense image matching is accomplished to provide for each pixel of the object the disparity in image space. The following forward intersection delivers a very dense point cloud of the object.

b.1   Close-range photogrammetry

This technique is carried out by using images captured at closer range, often less than 300 meters. Close-range photogrammetry is used for detailed three dimensional renderings and plotting of small-scale features and objects. Capturing photos in different positions and orientations, the forward intersection can be done and one point cloud can be generated. Li [11] used techniques from close-range photogrammetry for filling the missing area in the laser point cloud because the scan stations could not be fixed on the opposite side of the *Nagold* river.

For each building, different images from diverse positions are captured. The acquisition of digital images should cover the entire facade of each building. Thus, successive photos partially overlapped should be captured from different positions and orientations to get stereo images. The configuration of the images will affect the result of the 3D reconstruction. For this reason, the places where the images are captured need to be well previously arranged. In order to find common features in the imagery is frequent to use the *Scale-invariant feature transform* (SIFT) operator and the software *VisualSfM*, which is able to get the orientations and positions from where the photos were captured. Afterwards, a *bundle block adjustment* of the diverse photos can be carried out, which enable to compute a sparse surface of the object. The RANSAC robustness estimation can be used for filtering mismatches [12].

However, the sparse point cloud of the surface obtained is not yet sufficient to 3D modeling which requires a much denser one. Here, the *SURE* software, developed by the IFP, is applied to obtain very dense point clouds. *SURE* is a software solution for multi-view stereo, which enables to derive dense point clouds from a given set of images and its orientations. Within a first step, images are undistorted and pair-wise rectified. In a second step, SURE selects suitable image pairs and performs the dense stereo matching similar to the *Semi-Global Matching* (SGM) algorithm. Afterwards, a forward intersection of each stereo pixel can be done using the disparity image. SURE software enables to extract a point cloud (in format LAS) with a resolution of one 3D point per pixel, which can offer a very accurate representation of minimal details.

b.2   Aerial photogrammetry

Aerial photogrammetry can generate point clouds in areas where TLS showed data gaps. Thus, the point clouds obtained from aerial photogrammetry may be used for making up the missing area of the roof landscape, and also for transforming the different point clouds to the same reference system.

Several aerial images captured from a high altitude are used to get data about the missing area of the roof landscapes. The aerial images used for this project were supplied by the LGL-BW. These images were flown with a *Ground Sampling Distance* (GSD) of 10 cm and have georeferenced coordinates for the projection centers as well as orientation angles.

The points cloud generated from the aerial images are georeferenced in the local *Gauss-Kruger* coordinate system. The process in *SURE* software comprises two steps. In the first one, the introduced images are undistorted and pair-wise rectified. In the second one, suitable image pairs are selected and matched using a dense stereo method similar to the *Semi-Global Matching* (SGM) algorithm. Finally, dense point clouds spaced at 10 cm (one 3D point per pixel) are generated.

4.2   Data Processing

Once the phase of data acquisition was carried out, the point clouds obtained are processed in two steps: registration and modeling. The main aspects of each one of these steps are outlined in the following sections:

a. Point clouds registration

Leica Cyclone can perform the scan registration of the laser data captured with the TLS system. For this, the individual point clouds generated by TLS are merged using a minimum of common points between stations. In this project, the scan registration of the most part of the area surveyed was performed locating manually natural points, which were common in successive scan stations. This procedure can origin certain errors, which should have low range. For example, the errors in the scan registration obtained by Li [11] were lower than 15 cm. Once the scan registration is performed, a complete TLS point cloud of the block and the neighboring area can be obtained. However, the inner part of this block is missing due to the fact that the roofs could not be scanned by TLS. Furthermore, missing data can occur in certain parts of the building facades because of occlusions derived from the presence of trees or other elements (cars, pedestrians).

The point cloud generated by aerial photogrammetry, which is georeferenced in *Gauss-Krueger* coordinates, is used as georeference system for the other point clouds through a process called *transformation*. The combination of point clouds obtained with the different techniques (TLS and photogrammetric images) lead to generate a complete (and unique) point cloud, which is used for 3D modeling (Fig. 4). Previous to this combination, the different point clouds should be matched in position, orientation and scale.

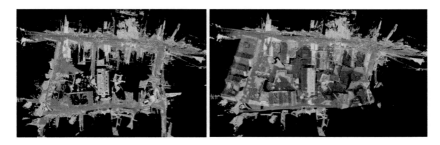

Fig. 4. Scan registration in the study area 3: (left) before and (right) after adding the point clouds generated by photogrammetry.

For this purpose, a *3D Helmert transformation* should be applied. A total of seven parameters are calculated including three translations (T), three rotation angles (R) and a scale factor (λ). In order to calculate these parameters, more than seven equations need to be set, which require at least three common points between the different point clouds.

Generally, the location of these common points in the respective point clouds is performed manually. However, the manual selection of these points can lead to suboptimal results. For this reason, the *Iterative Closest Point* (ICP) algorithm, which does not require the manual selection of common points, is applied to get a better result. It is important to note that this procedure is uniquely valid if the point clouds present similar geometries and are partially alignment [12].

b.   3D Modeling

The 3D reconstruction of the complete point cloud is performed by using *Leica Cyclone*. This software has some modeling tools, which can fit different geometries and surfaces to the original shape of the registered point clouds. Thus, polygonal geometries are used to build the *model primitives* for each single building represented in the point cloud. For getting high accuracies, the primitives drawn should be closely fitted to the point cloud of the object. An optimal performance can generate highly accurate 3D models which are based on very precise data. The resulting 3D model of the buildings is exported as a CAD model (in format DXF).

The 3D modeling in *Leica Cyclone* can perform a simplified reconstruction of each building. Nonetheless, certain details are not represented in this model. For this reason, a more exhaustive 3D reconstruction of the model requires the use of other software like *Trimble's SketchUp* and *Autodesk 3ds Max*. These software incorporate additional menus and optimal tools which enable to perform the modeling of more details and to get more reliable 3D models.

For wrapping the textures onto the model, the images rectified and with less distortion are used. *Adobe Photoshop* can adjust photo settings, and select the suitable area of each building. After having wrapped all the textures onto the model, a photorealistic 3D model can be generated for visualization (Fig. 5).

Fig. 5. (Left) 3D model generated in the Badstrasse area, including the historical bridge of Nikolaus over the Nagold river [11]. (Right) Detailed 3D model obtained for one building.

If the wrapped textures are historical photos, the models generated are *four dimensional* (4D). In this case, the process of addition of textures is similar as explained above, though should be considered that the number, the position and the orientation of the photos is not always the optimal. Thus, photogrammetric procedures of rectification of images should be applied to get the best results. Finally, the photorealistic 4D models generated enable to perform a reconstruction of the past of the historical center of Calw [13].

5   Discussion and Conclusions

The generation of 3D urban models has some complexities due to the large heterogeneity of the materials and elements, further the constant irregularities of buildings shapes. Moreover, in urban areas the presence of noise and disturbing elements is frequent (pedestrian, cars).

Despite their clear advantages, TLS systems present some deficiencies for performing this project. It can be classified in two groups: related (a) with the proper systems and (b) with its technical performance. With regard to the first group, the main drawbacks are the high costs, the physical magnitude of these systems, and the difficulty to access certain areas. Further, these systems require specialist staff for acquiring and processing the data. Considering the second group, the main drawbacks are the missing data because of the occlusions and the absence of the texture information. These constraints limit the volume and quality of data obtained in certain sectors of the respective studied areas.

The combined use of photogrammetric images and laser systems is symbiotic, being thus able to exploit the advantages of both procedures. The photogrammetric images provide the required texture and can be used to fill missing areas. In this research project, texture information of the building facades was extracted by close-range photogrammetry, whereas roof landscapes were rebuilt with aerial photogrammetric images.

The high precision of the data captured enable to obtain 4D/3D photorealistic models, which are geometrically accurate and visually aesthetic. Thus, in this project the massive capture of data and high precision of data obtained by traditional photogrammetry are checked with the visual representation of objects of computer vision.

The objective of the phase of data collection is to obtain a uniform point cloud of the entire object. High levels of resolution and quality scan increase exponentially the scan time and the number of points. However, the existence of redundant information maximizes the processing time and can reduce, paradoxically, the quality of results [14]. Therefore, strategies for reducing the point density in certain areas should be implemented. Furthermore, prior to survey, the level of scan resolution should be previously estimated depending of the research objectives [15].

The texture-wrapped 4D/3D models from historical photos are uploaded and integrated in virtual platforms like *Google Earth*. The internet user can virtually visit the old city of Calw using the *pegman* icon in *Street View*, where the buildings reconstructed will automatically pop-up. An optimal visualization requires files with a reduced size.

The methodology proposed for wrapping the photo-textures from different historical periods implies that there are not geometric variations in the CAD model of the buildings. Thus, this methodology is valid for old urban environments highly preserved as the one introduced here. Changes in the buildings shape require the rectification of both the CAD and the texture-wrapped model.

The existence of large photo repositories increases the potentiality of the proposed methodology. However, the processing and rectification of historical photos can be highly complex because of the fact that these photos were captured from a few positions, further the lack of metadata (position, date, camera, user, etc.), information available in the recent images.

In research terms, the future challenges are the automation of the process, the reduction of the times required and the minimization of the manual workload [12].

The results obtained in this research project raise future achievements as the generation of 4D photorealistic models with a great number of possible applications in archeology, architecture, geography, civil engineering, etc. Furthermore, the proposed methodology can be extrapolated and applied to other areas.

Acknowledgements. The results published in this work are supported by the European project 4D-CH-World framed within the 7th EU Framework Programme PEOPLE-2012-IAPP.

References

1. Heritage, G., Large, A.: Laser Scanning for the Environmental Sciences. Wiley Blackwell Publishing Ltd, Hoboken (2009)
2. Moussa, W., Abdel-Wahab, M., Fritsch, D.: An automatic procedure for combining digital images and laser scanner data. Int. Arch. Photogram. Remote Sens. Spat. Inf. Sci. **XXXIX**(B5), 229–234 (2012)
3. Rothermel, M., Wenzel, K., Fritsch, D., Haala, N.: SURE – surface reconstruction using images. Pres. Paper, LC3D Workshop, Berlin (2012)
4. Luhmann, T., Robson, S., Kyle, S., Harley, I.: Close Range Photogrammetry: Principles, Techniques and Applications. Whittles Publishing, Caithness (2006)
5. Bundesamt für Kartographie und Geodäsie webpage. http://www.bkg.bund.de. Accessed 19 August 2015
6. Europeana webpage. http://www.europeana.eu. Accessed 19 August 2015
7. LEICA company: Product Specifications of the Leica ScanStation P20. http://www.leica-geosystems.com/downloads123/hds/hds/ScanStation_P20/brochures-datasheet/Leica_ScanStation_P20_DAT_en.pdf. Accessed 19 August 2015
8. ZEISS company: Technical Specifications of the Lens Biogon T* 2,8/21 ZM. http://www.zeiss.com/camera-lenses/en_de/camera_lenses/zeiss-ikon/biogont2821zm.html. Accessed 19 August 2015
9. Landesamt für Geoinformation und Landentwicklung of Baden-Württemberg webpage. https://www.lgl-bw.de. Accessed 19 August 2015
10. Bustamante, A., Fritsch, D.: Digital preservation of the Calw market by means of automated HDS and photogrammetric texture mapping. Master thesis in Geomatics Engineering, University of Stuttgart (2013)
11. Li, Y.. Fritsch, D., Khosravani, A.: High definition modeling of Calw, badstrasse and its Google earth' integration. Master thesis in Geomatics Engineering, University of Stuttgart (2014)
12. Han, L., Chong, Y., Li, Y., Fritsch, D.: 3D reconstruction by combining terrestrial laser scanner data and photogrammetric images. In: Proceedings of the Asia Assoc. of Remote Sensing, Nay Pyi Taw, Myanmar (2014)
13. Fritsch, D., Klein, M.: Augmented reality 3D reconstruction of buildings – reconstructing the past. Int. J. Multim. Tools Appl. (2015). (under review)
14. Balsa-Barreiro, J., Pere, J., Lerma, J.L.: Airborne light detection and ranging (LiDAR) point density analysis. Sci. Res. Essays **33**, 3010–3019 (2012)
15. Balsa-Barreiro, J., Lerma, J.L.: A new methodology to estimate the discrete-return point density on airborne lidar surveys. Int. J. Remote Sensing **35**, 1496–1510 (2014)

Visual Autonomy via 2D Matching in Rendered 3D Models

D. Tenorio, V. Rivera, J. Medina, A. Leondar, M. Gaumer, and Z. Dodds[✉]

Harvey Mudd College CS Department, Claremont, CA 91711, USA
zdodds@gmail.com

Abstract. As they decrease in price and increase in fidelity, visually-textured 3D models offer a foundation for robotic spatial reasoning that can support a huge variety of platforms and tasks. This work investigates the capabilities, strengths, and drawbacks of a new sensor, the *Matterport 3D camera*, in the context of several robot applications. By using hierarchical 2D matching into a database of images rendered from a visually-textured 3D model, this work demonstrates that – when similar cameras are used – 2D matching into visually-textured 3D maps yields excellent performance on both global-localization and local-servoing tasks. When the 2D-matching spans very different camera transforms, however, we show that performance drops significantly. To handle this situation, we propose and prototype a map-alignment phase, in which several visual representations of the same spatial environment overlap: one to support the image-matching needed for visual localization, and the other carrying a global coordinate system needed for task accomplishment, e.g., point-to-point positioning.

1 Motivation and Context

This project investigated the strengths and drawbacks of a sensor that is relatively new for robot systems: a *Matterport camera* [1]. As shown in Fig. 1, the camera offers a new level of ease and fidelity to the creation of visually-textured 3D maps. Although made more ubiquitous by the Kinect [2], the use of such rich, visual 3D representations is still relatively new as a foundation for robot spatial reasoning. Considerable work has shown the power of 3D-to-3D matching, e.g., [3–6] and many others, but the underlying 3D visual maps are far less well studied for robot platforms with our era's most accessible and least expensive sensor source: a run-of-the-mill RGB camera. Bridging this 2D-3D gap will allow mobile robots to move autonomously within rich environmental models using inexpensive sensors.

This work developed an image-matching framework to allow robots of several types – with only 2D image input – to recognize their position with respect to a 3D (Matterport) model. Ideally, any robot with a camera would be able to utilize this work's algorithms to determine its position within a visually-textured 3D model; in practice, we found that different imaging transforms can cause significant degradation of matching accuracy. To compensate, we prototyped a system in which local 2D visual maps from different camera sources are aligned to the 3D model, providing both accurate 2D visual matching and corresponding 3D coordinates within the model's global coordinate system.

G. Bebis et al. (Eds.): ISVC 2015, Part I, LNCS 9474, pp. 373–385, 2015.
DOI: 10.1007/978-3-319-27857-5_34

Thus, this work demonstrates both the challenges and the promise emerging from Matterport models and other rich, visually-textured 3D maps. Specifically, we contribute

- the use of Matterport's sensor to create 3D maps suitable for a variety of robot tasks
- a three-tier approach for matching a robot's live-acquired 2D images with a database of known-location renderings from a 3D visual Matterport model
- a quantitative assessment of both the accuracy and speed of that matching process
- task-applications implemented on several different robot platforms, providing insights into the matching-process's strengths (accuracy and speed sufficient to support real-time control) and drawbacks (less robustness across distinct cameras/ image sources) – along with an alignment step that mitigates the drawbacks

Overall, there are compelling reasons for considering Matterport models as a basis for robot control: they match our intuition of our shared environs as a "human visual world" (even though we subconsciously construct it, moment-to-moment), they are the space in which humans plan and execute tasks, and thus they are the space in which humans most naturally specify tasks for *robot* planning and execution. This work helps introduce and inform both roboticists and vision researchers of what Matterport's models can (and can't) contribute to their efforts.

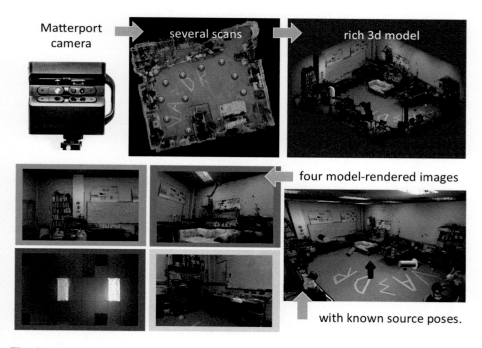

Fig. 1. Matterport camera and a lab-environment model, along with several rendered images from that model and the *known* locations from which those images were rendered.

2   Processing the Matterport Models

Computationally, the fundamental problem we tackle is that of comparing a (live) 2D image with a 3D model created both at an earlier time and from perspectives different than the current image. Our approach to this problem is split into two parts: a preliminary processing of the 3D environmental model(s) and robust 2D image matching.

We made a comprehensive model of several environments beforehand. Figure 1 shows screenshots from one that served as our primary robot workspace; others from our lab and elsewhere are available from [1]. To create these models, we first took multiple scans of the room with the Matterport camera, at about 45 s per scan. Following the company's standard process, the data was sent to Matterport, where it was compiled into a photorealistic 3D model of the environment. The default interface, however, allows navigation only from and to the locations at which scans were taken in the real world, noted in the top row's middle panel. This constraint prevents us from being able to create renderings – with accompanying coordinates – except from those environmental viewpoints.

This problem was solved by downloading the full environmental mesh from and processing it with Unity 3D [7]. Its rendering engine allows arbitrary navigation of the space, rendering a 2D view of the model from any location inside it or out, and even from poses far from source-scan locations, e.g., Figure 1's top-right rendering.

2.1   From 3D Model to 2D Renderings

We leveraged this geometric freedom to create a database of 2D images of the environment. Inside the three-dimensional model, we programmed a virtual camera to move to each lattice point within a large grid, taking one screenshot facing each wall at every position, creating a set of 200 2D images that covered the space: many of those renderings had significant field-of-view overlap. We built several such databases for testing; in the final version, we rotated the camera 90° between each rendering in order to minimize the number of pictures in the database. Certainly, Unity 3D makes it straightforward to create sparser or denser sets of renderings if the additional resolution is needed for a specific task. Crucially, the Unity-based system recorded and remembered the pose at which each rendering was made: thus, each image "knew" where it had been taken. Figure 1 (bottom row) shows four of these rendered images, along with an overview showing the location and direction of the virtual camera's location for each.

3   Matching into the Database of Rendered Images

To facilitate finding the best match between a robot's current image and the images in the database we combined several image comparison algorithms using OpenCV 3.0's Python bindings. The resulting system used three steps in order to match a novel (live) image with the database of renderings described in Sect. 2:

1. First, color-histogram matching is performed in order to narrow the search.
2. Geometric matching with ORB features further narrows the set of possible matches

3. Finally, several geometric constraints are considered: the database image that best satisfies those constraints is used as the best match for the task's current timestep

Where the database's resolution suffices, robot tasks use the location of the best-matched image as an approximation of the current robot location: Sect. 4 shows that for some aerial navigation tasks, this suffices. Further pose refinement is possible, however, from the image information in the corresponding ORB features; Sect. 4 also highlights other tasks, e.g., robot-homing that leverage this additional precision. This section details the three-step matching process and presents its results.

	Bad (%)	Par. Good (%)	Good (%)	Number of Images Used
Color Algs. - Score	13.33	13.33	73.33	15
ORB Algs. - Score	93.33	7.67	0	15
ORB Algs. - Vote	7.67	7.67	86.67	15
All Algs. - Vote	4.5	37	58.5	200
Homography Check	0	0	100	200

Fig. 2. (top) Examples of the scores produced by the four color-histogram-matching algorithms used in the image-matcher's first pass; (bottom) the overall match results both for individual tiers of the image-matching algorithm, top four rows, and the last-row final result. With layers of color-based filtering, keypoint-based matching, and geometric consistency checks, *all* novel rendered images matched very closely to the 200-image database.

3.1   Pixel-Based and Color-Histogram Matching

The first group of algorithms used in the image matching process consisted of seven scoring thods using color-histogram and pixel-by-pixel comparisons. The three pixel-by-pixel comparisons included are mean-squared error (MSE), root MSE, and the structural similarity index [8]. We also use four color-histogram comparison algorithms adapted from [9]. For each image, we discretized its data into N = 16 colors (histogram bins). Four different histogram-comparison methods provided by OpenCV [10], named *Correlation*, *Chi-Square*, *Intersection*, and *Hellinger*, were then used in order to measure the similarity of a novel image with each database image. Figure 2 (top) shows the resulting scores of a one-vs-all comparison using a small seven-image database obtained both from [9] and our environment's images. Here, the test image (a lake scene) is included in the comparison in order to show how a perfect match score compares to the imperfect match scores. In all subsequent tests, however, including those summarized in the results in Fig. 2 (bottom), the test image was *not* included in the comparison set.

As Fig. 2's lower table shows, even a small test database of 15 images from our lab environment showed only mediocre accuracy, when based on color-based matching alone (top row). This prompted the addition of a second tier of algorithms, which added geometric information to the matching process via small-patch image features.

3.2   Feature-Based Matching

The second group of algorithms relied on an image's geometric information. A menagerie of feature types exist with SIFT [11] and SURF [12] two of the most commonly used. Following [13], ORB keypoints, i.e., Oriented FAST and Rotated BRIEF keypoints, detect and compute distinctive image locations at a speed more than ten times greater than that of SURF (and SIFT). That speed guided our choice of ORB. Using OpenCV's ORB implementation, our image-matcher considers five geometric scores: (a) the sum of all ORB-matched visual distances, (b) the median of those distances, (c) their mean, (d) the sum of the visual distances of the top 50 % of matches by visual similarity, and (e) the number of inlier matches based on geometric (homography) consistency, as described in Sect. 3.3.

Preliminary tests involving the accuracy of the image matching system (Fig. 3, at right) showed that, like the color matching, the geometric matching alone did not provide accuracy suitable for supporting robot navigation. Having obtained a database of 200 images from the 3D model of the lab environment, a subset of fifteen images was run against the other images in the set and found the best match as determined by the system. The result was then assessed by eye, noting each as *good*, *partially good*, or *bad*. The (human assessment) metric the team used for these three categories, along with examples of each, appear in Fig. 3, below. Again, no image was included within the database set against which it was matched.

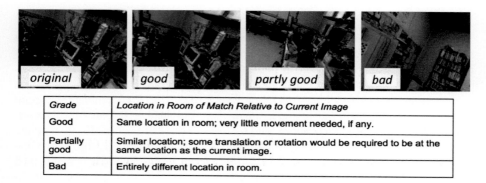

Grade	Location in Room of Match Relative to Current Image
Good	Same location in room; very little movement needed, if any.
Partially good	Similar location; some translation or rotation would be required to be at the same location as the current image.
Bad	Entirely different location in room.

Fig. 3. Examples and criteria for judging the categories *good*, partly or *partially good*, and *bad* image matches: Good image matches are one of the nearest neighbors in a sampled direction; partly good matches overlap less closely; bad matches do not overlap.

In addition to comparing color-only and feature-only matching, we compared two different methods for determining the "best" match: a "voting" system and a "scoring" system. A total of 12 match-scores are computed (7 pixel- and color-based and 5 ORB-based, as noted above): those 12 are our feature vector for how well two images match. The "voting" system gives a single point, or vote, to the best-matching image for each of these 12 algorithms, with the winner having the largest number of votes. By contrast, the normalized "scoring" approach first normalizes each algorithm on a commensurate scale of 0.0 to 1.0 and then sums the 12 individual scores. The highest overall score is named the best match.

For each tier image-matching layer by itself, the "voting" system yielded more accurate results – this is due to the fact that a single bad score can sink an otherwise excellent match due to the color or feature differences in the small portion of the two images that do *not* overlap. However, even the voting system produced a completely incorrect match for 20 % or more of the images examined. To improve this result, we post-processed the best matches to determine the amount of visual overlap they contained.

3.3 Improving Match Quality via Geometric Constraints

Upon examining the results carefully, the source of all of the poor and many middling matches was an accidentally well-matched set of image keypoints across two frames with similar color compositions. Figure 4 shows a typical example of a poorly-aligned and a well-aligned set of keypoints from example matches. As long as there are four matched keypoints – and every possible match included more than four – the transformation of those keypoints' locations from the test image to the database image yields a 2D-2D homography that transforms the first image plane into the second. When the keypoints are inconsistently matched, the result is a dramatic – and obviously incorrect – warping of the original scene, as Fig. 4's top example illustrates.

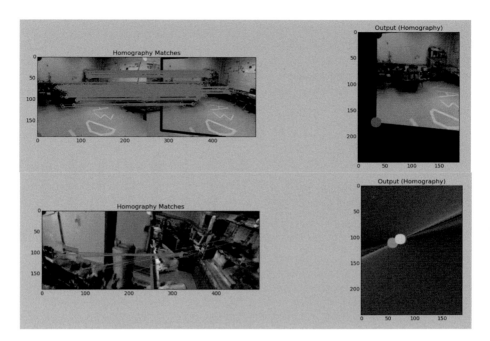

Fig. 4. (top) A poorly-aligned group of keypoints, yielding to a homography-transformed original that has "collapsed" into a very small image, a sign that the match is incorrect; (bottom) a well-aligned group of keypoints, with significant overlap: a strong match.

As such, we could determine when a bad match had been made by examining the Euclidean distance of the four corners post-homography. If the transformed corners pull too close to (or too far from) each other, the transformed image has experienced a "collapse" (or its reverse), and the system eliminates it from contention, continuing on in its examination of other candidate matches from the database.

With this homography-based geometric-consistency check, the results were encouraging. With the same database of 200 images as in the preliminary tests, now in a leave-one-out analysis, the test of every one of the 200 images produced an accurate, or *good* match. Thus, this homography check eliminated the bad matches without affecting the good ones, which ware those with significant image overlap, e.g., Figure 4's second example.

Though his paper's approach does benefit from not requiring camera calibration, these geometric constraints are *not* the only ones that might benefit this application. For instance, the P3P algorithm [17], among other 2D/3D relationships such as the epipolar constraint [16, 18], leverage the 3D model more deeply in order to support image-feature matching.

3.4 Results: From Accuracy to Speed

Encouraged by the accuracy of these results, we also considered the time required to match a novel image with a set of K database images. For use in robotic applications, speed translates directly into task-capability and -responsiveness. In addition, time-per-database-image determines how large, either in spatial extent or pose resolution, the environment's rendered database can be, while still supporting the ultimate goal of the project: autonomous, visual navigation. Thus, we examined how different sets of algorithms performed in terms of time per image, with the results shown in Fig. 5:

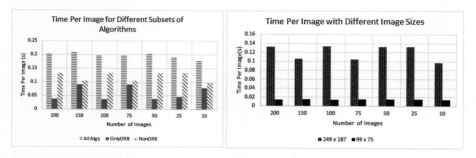

Fig. 5. (left) The time used in second to execute the full matching system on full-size (249 × 187) images, as well as its two component layers: ORB indicates the geometric components; NonORB, the color-based components of the algorithm; (right) the speedup obtained when decimated (99 × 75) images are used for the color-based matching.

Noticing the smaller time per image for ORB algorithms when compared with non-ORB algorithms for the default image size of 249 × 187 (Fig. 5, left), we explored whether smaller images could significantly improve the efficiency of the color-based matching. Although the ORB algorithms did not perform well at a smaller image size, the color algorithms did. The change to 99 × 75 pixel images – only for the color-based matching – improved the run-time by approximately an entire order of magnitude (Fig. 5, right). The significantly smaller time-per-image solidified the design decision to use color-based matching as a first pass over the database. The final image-matching system, as well as both of these timing tables, employ precomputed histograms and precomputed ORB features for the database images. For the test (novel) image, however, the time to compute those features is included.

4   Validation via Visual Navigation Tasks

With accurate and fast results, we deployed and evaluated the image-matching system in service of several autonomous, visual-navigation robot tasks:

1. Image-based localization using a pan/tilt rig and a webcam: two live-image cameras *distinct* from the one used to image the database
2. Image-based homing on a differential-drive wheeled vehicle, comparing odometric homing accuracy with the matching system's image-based feedback
3. Point-to-point navigation of an aerial vehicle whose only sensor is a forward-facing camera

Each of these tasks – and its results – is detailed in the following subsections. For these tasks, the system used a 3D visually-textured model, in .obj format, via the widely used Unity3d rendering engine. Though obtained by a Matterport camera, the approaches detailed here apply to any such model in a compatible format.

4.1 Localization Using Different Source Cameras

We sought to test the functionality of the 2D image-matching under imaging conditions very different than the source, the Matterport sensor. To so this, input images were drawn from a pan/tilt rig with an inexpensive webcamera (as a side note, this platform is also a Nerf-based missile launcher) attached and the higher-quality camera built-in to a desktop iMac. Using images from either the launcher's built-in camera or the computer's webcam, the system compares those images to the database of 2D images from the model using our image matching program, and indicates the location at which the image was taken in the 3D model. Building from ROS's infrastructure [14], a client/server system connected the novel images (taken and compared on the server side) with the rendering of the resulting estimated camera position (by an arrow, as done in Fig. 1).

As Fig. 6 attests, the results were far less accurate than those obtained using the same live- and database-source cameras. Here again, three examples are provided: one each of *bad*, *partially good*, and *good*, this time equivalence classes for the **pose** of the result. In each case the best-matched database image is on the left, and the novel image is on the right (these, taken from the iMac's built-in camera):

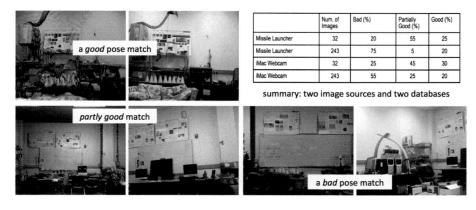

	Num. of Images	Bad (%)	Partially Good (%)	Good (%)
Missile Launcher	32	20	55	25
Missile Launcher	243	75	5	20
iMac Webcam	32	25	45	30
iMac Webcam	243	55	25	20

summary: two image sources and two databases

Fig. 6. Example pose matches and their results when using two different source cameras, each distinct from the original Matterport model's. Both a small (32-image) and a larger (243-image) database of environmental 2D renderings were compared. The strong accuracy results, above, are **not** replicated when the novel-image sources differ from the model's.

Combined with the excellent accuracy of the final matching system presented in Sect. 3 and Fig. 3, these results suggest several things. First, with different source cameras, the system is less accurate for larger databases: this results from the significantly different color response of each camera's sensor. The larger underlying database seems to provide more opportunities for distractors, causing a lower accuracy there than for the smaller set of source images. Most surprising to us was that, when the iMac's camera and even the low-quality launcher's webcam were used as *both* the source of both novel *and* stored images, the system performed at 100 % (or near-100 %) accuracy: it was not the absolute quality of the imaging sensor that mattered but its *consistency*.

4.2   Visual Homing Versus Odometric Homing on a Wheeled Robot

To leverage that fact for the two subsequent tasks (wheeled and aerial navigation), we overlaid the 3d environment's rendered images in a task-dependent way with images from the wheeled robot's camera, which was another launcher camera riding atop an iRobot Create, and with images from the drone's camera – the built-in sensor in a Parrot AR.Drone quadcopter. The wheeled vehicle was then run through eight *homing* tasks: from a given initial position – and initial image – the robot was driven randomly for several seconds in order to move far from that starting location.

The goal from there is to return home, i.e., to return to the initial position through a combination of odometry and image matching. First, the wheel-rotation sensors, which are quite noisy on the differential-drive Create robot [15], bring the robot back to the location believed to be the odometric starting point –not entirely accurate due to wheel

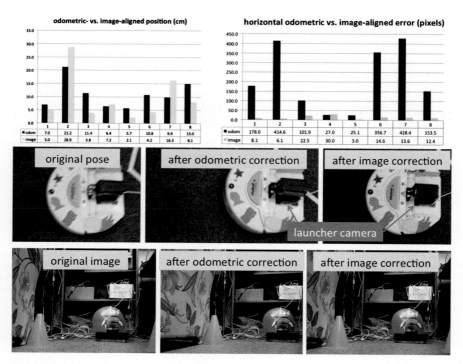

odometric- vs. image-aligned position (cm)	1	2	3	4	5	6	7	8
odom	7.0	21.2	11.4	6.4	5.7	10.8	9.9	15.0
image	5.0	28.9	3.8	7.2	2.1	4.2	16.3	8.1

horizontal odometric vs. image-aligned error (pixels)	1	2	3	4	5	6	7	8
odom	178.0	414.6	101.9	27.0	25.1	356.7	428.4	153.5
image	8.1	6.1	22.5	30.0	3.0	14.6	13.6	12.4

Fig. 7. Overall pose-accuracy results and an example of a wandered and corrected trajectory (top row), along with an example of a bird's-eye view of the odometry-vs-image-matching alignment (middle row) and the alignment of the images themselves (bottom row)

slippage. From there, the image-matching program is used to further refine the robot's position, by minimizing two components of the image-error: the average scale-change, Δs, among the ORB features and their average horizontal image-translation, Δx, to align the current image as close as possible to the initial image's view. Here, the goal is investigating the use of the images, not the control strategy: the robot alternates making small Δs and Δx corrections until it is no longer improving. Figure 7 summarizes all 8 and illustrates one homing run.

4.3 A Point-to-Point Task on a Vision-Only Quadcopter

A third spatial-reasoning task used to test the image-matching system combined the insight from the first localization trials (the significantly improved performance of having similar source and live cameras) with the coordinate system provided by the original image-matching task (using the 3D model of the lab environment). To do this,

(1) The quadcopter lifts off and takes an image;

(2) it is matched – and aligned to the lab's map.

(3) The drone then can rotate to face its goal, the sofa,
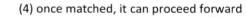
(4) once matched, it can proceed forward

(5) and land once it's traveled the distance to its goal.

Fig. 8. Shows the use of aligned image-matching to accomplish a point-to-point aerial task. (**first image**, at upper left) Here, a quadrotor helicopter ascends and hovers in an unknown location within its environment. (**second image**, upper right) The image from the quadcopter (or "drone") appears at left and is matched with the image rendered from the lab map at right. That matching process (detailed in Sect. 3) results in a known, well-localized pose for the drone. From there (**third image**, middle left), the drone rotates in place until it is facing its desired destination (the lab couch). The process repeats (**fourth image**, lower right) with the drone-taken image matching a map-rendered image to trigger a state-change to proceed forward. Finally (**fifth image**, lower left), the drone lands at its desired pose.

an overlay of images from the forward-camera of a Parrot AR.Drone quadcopter was placed atop the coordinate system of the 3D model. Although the quadcopter's source images did not carry any information about their location in the lab environment, the 3D model to which those images were aligned did. In addition, because the images were regularly spaced, only one image needed to be hand-aligned in each drone location; the position of the rest could be inferred from the first. Although one could imagine a fully automatic alignment from the drone camera to the environment-model's camera, that process was not done for this set of tests. Figure 8 summarizes the results of a point-to-point task in which the drone sought out the lab environment's sofa, used image-matching to determine its orientation in the room, rotated to face the goal, again matched to confirm its rotational pose and determine its distance, and subsequently made the final motion required to land on the couch.

5   Verdict

Inspired by the capability and potential of the Matterport camera's 3D visual models as a basis for robot spatial-reasoning tasks, this work sought to (1) develop a 2D image-matching system to allow (robots with) ordinary cameras to localize within a Matterport model (2) assess the performance of that image-matching, and (3) validate the system through three different robot tasks. The three-tier image-matching system developed here demonstrated and validated both the accuracy and speed required to support auton-omous vision-based robot tasks, but *only when the model and live source cameras were the same.* Here, we accommodated that limitation by aligning images from the robot itself to those portions of the 3D model necessary for task success, e.g., to a single location in the case of robot homing and with several rotational scans of the lab for the point-to-point navigation task with the aerial robot.

Although heartened by these successes, we believe this work highlights an impor-tant, perhaps underappreciated, challenge in using 3D models such as Matterport's for robot applications. As such models become more common, more opportunities will arise to query them with distinct sources of image information. Automatic *photometric* calibration between a Matterport camera and a (different) robot camera thus seems a tantalizing – and worthwhile – challenge. When combined with a robust 2D image-matching system, such as investigated in this work, that combina-tion would tap even more of the robotics and spatial-reasoning potential of those richly-textured 3D maps.

Acknowledgments. We acknowledge and thank the generous support of the NSF, though CISE REU Site project #1359170, as well as from Harvey Mudd College and its Computer Science department.

References

1. Matterport. matterport.com. Accessed 20 October 2015
2. Oliver, A., Kang, S., Wunsche, B.C., MacDonald, B.: Using the Kinect as a navigation sensor for mobile robotics. In: Proceedings of IVCNZ 2012, pp 509–514, Dunedin, New Zealand. ACM, NY
3. Olesk, A., Wang, J.: Geometric and error analysis for 3D map-matching. In: Proceedings of IGNSS Symposium, 14 p, Queensland, Australia (2009)
4. Pinto, M., Moreira, A.P., Matos, A., Sobreira, H., Santos, F.: Fast 3D map matching localisation algorithm. J. Autom. Control Eng. 1(2), 110–114 (2013)
5. Endres, F., Hess, J., Engelhard, N., Sturm, J.: In: Proceedings of ICRA 2012, pp. 1691–1696, 14–18 May 2012
6. Wu, C., Clipp, B., Li, X., Frahm, J.-M.: 3D model matching with viewpoint-invariant patches (VIP). In: Proceedings of CVPR 2008, pp. 1–8, 23–28 June 2008
7. Unity. unity3d.com. Accessed 20 August 2015
8. Wang, Z., Bovik, A.C., Sheikh, H.R., Simoncelli, E.P.: Image quality assessment: from error visibility to structural similarity. IEEE Trans. Image Process. 13(4), 600–612 (2004)
9. Rosebrock, A.L Pyimagesearch. www.pyimagesearch.com/
10. The OpenCV Library [http://opencv.org/]; Bradski, G. The OpenCV Library Dr. Dobbs Journal, November 2000
11. Lowe, D.G.: Distinctive image features from scale-invariant keypoints. Int. J. Comput. Vis. 60(2), 91–110 (2004)
12. Bay, H., Ess, A., Tuytelaars, T., Van Gool, L.: SURF: speeded up robust features. Comput. Vis. Image Underst. (CVIU) 110(3), 346–359 (2008)
13. Rublee, E., et al.: ORB: an efficient alternative to SIFT or SURF. In: 2011 IEEE International Conference on Computer Vision (ICCV). IEEE (2011)
14. Quigley, M., Gerkey, B., Conley, K., Faust, J., Foote, T., Leibs, J., Berger, E., Wheeler, R., Ng, A.Y.: ROS: an open-source robot operating system. In: Proceedings of Open-Source Software Workshop of the International Conference on Robotics and Automation (ICRA) (2009)
15. Watson, O., Touretzky, D.S.: Navigating with the Tekkotsu Pilot. In: Proceedings of FLAIRS-24, pp. 591–596, Palm Beach, FL, USA, 18–20 May 2011
16. Hartley, R., Zisserman, A.: Multiple View Geometry. Cambridge University Press, Cambridge (2004)
17. Haralick, R., Lee, C., Ottenberg, K., Nolle, M.: Review and analysis of solutions of the three point perspective pose estimation problem. Int. J. Comput. Vis. 13(3), 331–356 (1994)
18. Lynen, S., Sattler, T., Bosse, M., Hesch, J., Pollefeys, M., Siegwart, R.: Get Out of My Lab: Large-scale, Real-Time Visual-Inertial Localization. Science and Systems, Rovotics (2011)

Reconstruction of Face Texture Based on the Fusion of Texture Patches

Jérôme Manceau$^{(\boxtimes)}$, Renaud Séguier, and Catherine Soladié

CentraleSupélec/IETR UMR6164, Team FAST (Facial Analysis, Synthesis and Tracking), Avenue de la Boulaie, 35576 Cesson-Sévigné, France
jerome.manceau@centralesupelec.fr

Abstract. 3D face clones can be used as pretreatments in many applications, such as emotion analysis. However, such clones should model facial shape accurately, while keeping the attributes of individuals; and they should be semantic. A clone is semantics when we know the position of the different parts of the face (eyes, nose...). The main problem of texture reconstruction methods is the seam appearance on fusion texture data. In our technique, we use a low cost RGB-D sensor to get an accurate and detailed facial unfolded texture. We use shape and texture patches to preserve the person's characteristics. They are detected using an error distance and the direction of the normal vectors computed from the depth frames. The tests we perform show the robustness and the accuracy of our method.

1 Introduction

Facial clones are an important area in Computer Vision and Graphics. They are frequently used in Human Computer Interaction. For example, clones of neutral faces are used as a pretreatment to find the pose of the head in applications such as gaze detection [1] or people identification [2]. Moreover, in such application the texture must be well-defined, without blurring, and it must contain the attributes of the people.

We want to reconstruct the texture of a face using the 3D data of a low resolution camera. The system must be cheap and shall be available for end-users at their home. It must also be fully automatic. Moreover, the texture must be mapped on a semantic clone to be used in applications such as gaze detection. A semantic clone is a mesh of which we know the position of different parts of the face (eyes, nose...).

The main contribution of this paper is that we automatically reconstruct an accurate and detailed unfolded texture, that is seamless, automatic and is obtained with a low-cost RGB-D sensor. The main problem of texture reconstruction methods is the appearance of seams. The originality of this article is the use of texture patches to preserve the person's characteristics (beauty spot, beard ...) and to resolve the seams problem. We use several texture frames to retrieve all the facial texture information. Each texture frame from the RGB-D sensor contains details of the face but also noise and biased information. We do

© Springer International Publishing Switzerland 2015
G. Bebis et al. (Eds.): ISVC 2015, Part I, LNCS 9474, pp. 386–395, 2015.
DOI: 10.1007/978-3-319-27857-5_35

not want to merge these bad texture information, which can reduce the quality of the final texture. Therefore, we detect the parts of frames (patches) that are accurate and appropriate using depth frames. We know the correspondence between the shape and the texture of each frame. Depth frames are used to detect the color data that are not correctly captured by the sensor. Indeed, the precise texture areas are located where the normal vectors are parallel to the optical axis of the camera. To detect these patches, we calculate the normal vectors of each 3D point and the error between each depth frame and the semantic clone. The error distance eliminates the rigid alignment errors and detects the places where the sensor does not give the correct information (hole ...). The normal vectors are used to detect the accurate texture areas. Finally, we merge the texture patches detected. The second problem of texture reconstruction methods is the alignment of the texture data from the different frames. In our method, we use both depth and texture data to perform the alignment. We use the ICP algorithm to achieve the data alignment.

This article is organized as follows. In the next section, we review some previous work in the area of facial texture reconstruction. Section 3 presents the various components of our patches detection and fusion method. Section 4 demonstrates the accuracy of our results. The last section concludes the paper.

2   Related Works

A first set of texture reconstruction approaches in the literature use Morphable Face Models to reconstruct 3D facial texture from RGB images [3,4]. The texture is reconstructed using an optimization of the model parameters. These methods do not allow to retrieve the attributes of a person (mustache, beard ..). They are limited by the model database. A beauty spot can not be found. Indeed, they can be placed anywhere on the face and they never have the same appearance.

A second set of techniques in the literature uses RGB data provided by a sensor to map 2D texture on 3D shape [5–7]. These methods consist of three steps. First, the 3D mesh is unfolded to create a texture map and texture coordinates. Then the texture map is completed with different RGB images provided by the sensor. Finally, several treatments are applied to improve the quality of the texture. More precisely, in first step, He et al. [5] use the software FaceGen to create a 3D face clone, the texture map and the associate texture coordinates. To create the texture map, Zhang and Luo [6] and Hwang et al. [7] project the 3D mesh on a cylindrical 2D plane. We also use this technique because it gives good results to unfold a mesh.

For the second step, He et al. [5] rebuild the texture map from a single RGB input image. They detect the characteristic points on the RGB image with an AAM algorithm [8] and they calculate the pose of the head using the characteristic points of the input image and of the 3D mesh. All the points of the 3D mesh are then projected onto the RGB input image. But the detected characteristic points need to be very accurate. It is difficult to obtain good results automatically. Hernandez et al. [9] and Sun et al. [10] project only a single facial image on the 3D model. But they can not recover the entire face information with a single view (sides of the nose,...). Lee and Magnenat-Thalmann [11] propose a method for mapping a texture on a 3D face from a face image and an image of each face of the profile. They use a manual method to align the images by matching feature points and facial lines. Then the 3 images are merged from a pyramidal decomposition method [12]. Finally, the texture is projected onto the 3D mesh. But this technique is not completely automatic. Zhang and Luo [6] also use a face image and a facial profile image to improve the texture map. They generate the texture using a multi-resolution technique and merge a facial image and a facial profile using a reference line (hairline, eyebrows, around the eyes, mouth and chin) defined on the images. Then, they determine the correspondence between the texture and the unfolded depth map. For this, they use the 2D warping algorithm according to line characteristics pairs [13]. This technique provides satisfying results but it reveals seams. Xu et al. [14] describe a method for generating a texture from multiple calibrated images. They use a calibration matrix to find the relationship between the images and the 3D mesh. Then, they project the 3D mesh on the different RGB images and they use the method of Lempitsky and Ivanov [15] to find the most suitable texture for each face of the mesh. For this technique to be effective, they need calibrated images. The main drawback of these methods is the seams produce by the fusion of different frames, which require the use of a third step.

At the last step, treatments are performed on the completed texture map to increase the quality of the clone and remove the seams. There are differences in color and brightness on the texture obtained (seam) due to change of position of the head during acquisition. He et al. [5] use a smoothing filter, Zhang and Luo [6] use the image decomposition method, Xu et al. [14] use a gradient mixing method to filter the differences in intensity and improve the quality of the texture, Hwang et al. [7] use the gradient of the texture and use a morphological operation to complete the holes. Ge et al. [16] and Dessein et al. [17] present a texture fusion method from Poisson equations. It limits the differences in color and lighting that can appear on a reconstructed texture. These techniques are effective in removing the seams, but they can delete texture details. Indeed, they smooth the texture and therefore the details of the face (beauty spot...).

The methods that use a Morphable Model provide high resolution facial textures. But these methods are limited. They depend on the quality of the Morphable Model unfortunately. The database must be diversified in order to find some attributes of a face. This is why, the physical characteristics (beauty spot ...) cannot be found by the model. The second set of techniques has the advantage of reconstructing the attributes of a person. The main challenge is to eliminate the seams. The third step eliminates the seams but adds blur to the texture. In our method, we use of texture patches to preserve the person's characteristics (beauty spot, beard ...) and to resolve the seams problem. The second challenge is to match the texture map with different RGB images automatically. Yet it is difficult to align multiple RGB images so that the results are correct. In our method, we use depth data to align the texture.

3   Method

Our method is composed of 3 iterative steps: the alignment of each frame and the clone will be described in Subsect. 3.1, the texture patches detection (Subsect. 3.2) and texture patches fusion on unfolded mesh (Subsect. 3.3). We optimize a Morphable Face Model to obtain the semantic clone. Figure 1 describes our overall process.

3.1 D_p and C_s Rigid Alignment

We use the Iterative Closest Point algorithm to achieve the alignment between each depth frame D_p and the clone C_s. This algorithm is slow and requires many iterations to converge. To reduce the number of iterations of our method, we initialize the angle of rotation with the pose computed by Intraface process [18]. The method of ICP [19] consists of 3 iterated stages. First, each point in the 3D clone C_s is matched with the closest point of the depth frame D_p. Douadi et al. [20] compare different distances (euclidean, color...) for matching points. They show that the use of color in the ICP increases the algorithm's performance. Here, we reject pairs of points with a color distance above a certain threshold. That is to say, DS_p and CS_s are subsets of D_p and C_S. To limit the impact of brightness, we work in the YIQ color space. The color distance is calculated from the two chrominance components. Note that at the first iteration of the algorithm, we do not have any texture information.

Then, we estimate the transformation matrices (rotation and translation) minimizing the error distance E_{icp} between the pair of points. We use the Point Plan ICP from [21].

$$E_{icp}(\hat{M}) = argmin \sum_{p=1}^{N} W_p.||(\hat{M}*DS_p - CS_s).n||^2, DS_p \subset D_p, CS_s \subset C_s. \quad (1)$$

E_{icp} is the sum of the squared distance between each source point (DS_p) and the tangent plane at its corresponding destination point (CS_s). In this equation, $\hat{M}$ is the estimated transformation matrix and n is the normal vector at CS_s. DS_p and CS_s are paired points according to their distance. The weights W_p are computed from the distance between CS_s and DS_p. Pairs of points with a short distance are the most important ones, so that W_p are inversely proportional to the distance between the matched points.

The algorithm is more efficient when there is a large overlap area between the two clouds to align. Indeed, it is easier to align a frame front view with a clone than a facial profile frame. That is why, the first processed frame I_1 in our method is a front view frame. Moreover, the pupil moves during the acquisition. That's why, to avoid blur at the level of pupil, we use data of the first frame to complete the areas of eyes. We also use data of the first frame to complete the areas of the mouth and eyebrows because we observed that the results are better. The information of these three zones is contained accurately in the front view frame. For they are perpendicular to the axis of the camera.

3.2   Patches Detection

We want to detect on each texture frame I_p, the parts of texture that are adequate and accurate. For example (Fig. 2), for a texture frame in right profile (b_1) we want to keep the pixels that match the right profile. We call "patch" all isolated pixels of each texture we want to keep. We use two criteria for the patch detection: the error distance between the clone C_s and the depth frame D_p and the direction of normal vectors.

Distance Error. The use of a distance error eliminates sensor noise and alignment errors. This criterion is based on the point-to-point error distance between each depth frame D_p and the semantic clone C_s. We keep the texture patches P_t corresponding to the detected points of the depth frames. This distance error have to be smaller than a threshold (1 mm). On the front view of Fig. 2 (a_1), there is a hole in the nose because the sensor does not give the information. The data of texture T_p for this part of the face do not reflect the identity of the subject. The distance error removes these information (a_2).

Normal Vector. The RGB-D camera captures more precisely the zones where the optical axis is perpendicular to the surface object. Therefore, we only keep the points of the depth frames D_p that have a normal vector practically parallel to the optical axis of the camera (threshold: $\pm 10°$). In Fig. 2, we observe that this selection criterion eliminates the points that the camera does not capture well $(a_3$ and $b_3)$. For example, for a depth frame in the front view (a_1), we do not obtain all the information of the nose. Indeed, the points on the sides of the nose are not well captured by the camera.

In our process, we use the two criteria detailed above. For a texture patch to be preserved, the corresponding points must have a normal vector parallel to the optical axis of the camera and the distance between the semantic clone C_s and the depth frame D_p has to be smaller than a threshold. We tolerate an angle of $\pm 20°$ for the normal vectors. These two criteria are used to detect areas that are not captured by the camera in a single view and we also eliminate sensor noise and fitting error. Figure 2 shows that only the appropriate texture patch is preserved $(a_4$ and $b_4)$.

3.3   Texture Patches Fusion on Unfolded Mesh

The third section of our 3D texture reconstruction method is texture patches fusion on unfolded mesh. This step consists of 3 parts. First, we create the texture map T_c. Then we warp the detected texture patches P_t to complete this map. Finally, we merge the detected texture patches of Sect. 3.2 (see overall process in Fig. 1).

To create the texture map T_c, we unfold the 3D clone C_s by projecting it onto a cylinder. We get the texture map T_c of the 3D clone (1024 * 1024 pixels). The resolution of the texture frames I_p (960 * 1280) is greater than the resolution of the depth frames D_p (480 * 640). We use an unfolded texture with a higher resolution (1024 * 1024) to obtain a better definition of the texture clone. Each point of the clone C_s, called the anchor points, has a match in the texture map T_c. We complete the map with the texture of different frames obtained with the RGB-D camera.

In Sect. 3.1, we found the correspondence between the anchor points of texture map T_c and texture frames I_p. To fill the pixels between anchor points, we warp each texture patches P_t onto the clone texture map T_c. For each pixel of the texture map T_c, there may be several overlapping patches. When a patch pixel overlaps with several patches pixels, we fuse these different pixels value. We tested four types of fusion: average, median, weighted average and robust average. For the weighted average, we use the distances calculated in the detection patches step (Sect. 3.2). For the robust average we do not take into account the aberrant pixels. Before performing the average, we eliminate the pixels away from the median using a threshold (2 mm).

4   Results

In this section, we first explain our acquisition protocol. Then, we compare our 4 types of fusion. Finally, we compare our method with [22] and [9]. The Figs. 3, 4 and 5 show that our method is more efficient.

Experimental Protocol. We use a Kinect camera version 1, which is equipped with a color sensor (960 * 1280) and a depth sensor (480 * 640). The subject performs a rotational movement of the head in front of the camera at 0.5 m. He must do a neutral expression during the acquisition of data. Our database of test consists of 15 subjects (see Fig. 5).

Type of Fusion. We compare four types of fusion: average, median, weighted average and robust average. We obtain a blurred image of the texture when we use the average and the weighted average. The robust average slightly improves the results. Fusion with a median eliminates blur part of the texture and gives the best results. The Fig. 5 shows the reconstruction of the 3D texture obtained with median fusion.

Lighting Condition. The Fig. 3 shows that our method is more efficient with bad lighting than [22]. We deliberately used a side light, so that the light is not uniform. We compare our method with KinectFusion [22] because it also uses data provided by a Kinect camera. The results obtained with [22] are noisy (left side of the face). Indeed, they retain texture information that are noisy. Moreover, Kinect Fusion [22] does not provide semantic 3D clones and therefore cannot be directly used as a pretreatment in applications requiring knowledge of the correspondence of the mesh points with the face. Our method avoids this type of noise because we only keep the areas perpendicular to the camera. Our patch system fuses that adequate information. But we see the appearance of seams at eye level, eyebrows and mouth because we use the information of the first frame for the eyes, the mouth and eyebrows. Indeed, the poor lighting is the cause of the seams. These seams can be eliminated by using Poisson equation as in [16]. Indeed, we know the position of the seams because we use semantic

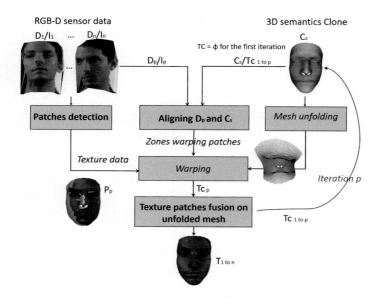

Fig. 1. Overall process

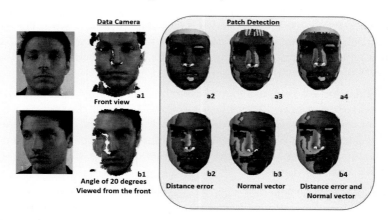

Fig. 2. Examples of patches detection on the 3D shape: a_1: a depth frame in the front view. b_1: a depth frame in right profile. We select the parts of the mesh (patch) where the error is small (a_2 and b_2) and normals vectors are relevant (a_3 and b_3). In our process, we use both criteria (a_4 and b_4).

clone. We performed this test on 15 different faces (on the website). The results provided by [22] are always noisy. Our method is more robust and provides better results.

Attributes of the Individuals. The Fig. 4 shows that our method performs better than [9] to represent the attributes (beauty spot...). We have taken an

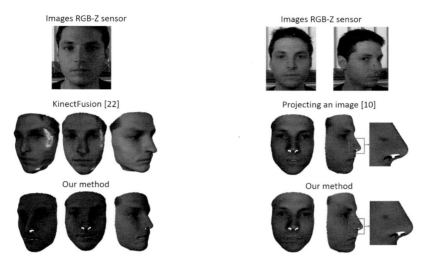

Fig. 3. Comparison of our method with [22] in poor lighting conditions (a side light).

Fig. 4. Representation of attributes of people. We compare our method with [9]

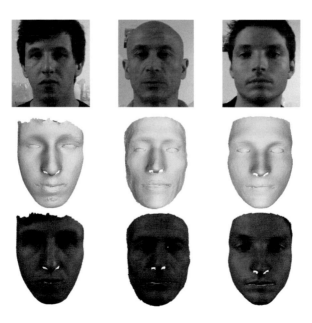

Fig. 5. Results of our method on 3 people. Results of 15 people on the website: http://www.rennes.supelec.fr/ren/perso/manceau_jer/recherche.php.

example of a person with a beauty spot on the nose. In [9], they project a single front view image on the 3D clone. We note that the beauty spot is deformed on their results because the front view frame does not contain the correct information of the sides of the nose. Indeed, the curvature of the nose is very important. Therefore, on the result obtained with [9], the beauty spot is very extensive in the direction of the nose. In our method, we use the depth frames to detect texture patches. The direction of the normal vectors can eliminate the facial areas with a large curvature. Therefore, we achieves better results for we eliminate the sides of the nose for a front view frame. In our results, the beauty spot is not extended and is placed at the right place (see Fig. 2).

5   Conclusion

We propose a system, that reconstructs a face texture using a low-cost depth sensor. Our method keeps the local morphological attributes of individuals. The use of patches makes it easier to find the attributes of an individual's face and eliminate the seams. In our future work, we want to use super resolution methods to increase the resolution of the Kinect data and improve the quality of our results.

Acknowledgment. This research has been conducted with the support of Miles (FUI project) and Brittany Region (ARED).

References

1. Funes Mora, K.A., Odobez, J.M.: Gaze estimation from multimodal kinect data. In: IEEE Conference in Computer Vision and Pattern Recognition, Workshop on Gesture Recognition (2012)
2. Paysan, P., Knothe, R., Amberg, B., Romdhani, S., Vetter, T.: A 3d face model for pose and illumination invariant face recognition. In: Tubaro, S., Dugelay, J.L. (eds.) AVSS, pp. 296–301. IEEE Computer Society (2009)
3. Zollhofer, M., Thies, J., Colaianni, M., Stamminger, M., Greiner, G.: Interactive model-based reconstruction of the human head using an RGB-D sensor. Comput. Animat. Virtual Worlds **25**, 213–222 (2014)
4. Blanz, V., Vetter, T.: Face recognition based on fitting a 3d morphable model. IEEE Trans. Pattern Anal. Mach. Intell. **25**, 1063–1074 (2003)
5. He, X., Yuk, S., Chow, K., Wong, K., Chung, R.: Super-resolution of faces using texture mapping on a generic 3d model. In: Fifth International Conference on Image and Graphics, ICIG 2009, pp. 361–365 (2009)
6. Zhang, J., Luo, S.: Image-based texture mapping method in 3d face modeling. In: IEEE/ICME International Conference on Complex Medical Engineering, CME 2007, pp. 147–150 (2007)
7. Hwang, J., Yu, S., Kim, J., Lee, S.: 3D face modeling using the multi-deformable method. Sens. **12**, 12870–12889 (2012)
8. Cootes, T.F., Edwards, G.J., Taylor, C.J.: Active appearance models. In: Burkhardt, H., Neumann, B. (eds.) ECCV 1998. LNCS, vol. 1407, pp. 484–498. Springer, Heidelberg (1998)

9. Hernandez, M., Choi, J., Medioni, G.: Laser scan quality 3-d face modeling using a low-cost depth camera. In: Proceedings of the 20th European Signal Processing Conference (EUSIPCO 2012), pp. 1995–1999 (2012)

10. Sun, Q., Tang, Y., Hu, P., Peng, J.: Kinect-based automatic 3d high-resolution face modeling. In: The 4th International Conference on Image Analysis and Signal Processing (IASP) (2012)

11. Lee, W.S., Magnenat-Thalmann, N.: Fast head modeling for animation. Image Vis. Comput. **18**, 355–364 (2000)

12. Burt, P.J., Adelson, E.H.: A multiresolution spline with application to image mosaics. ACM Trans. Graph. **2**, 217–236 (1983)

13. Beier, T., Neely, S.: Feature-based image metamorphosis. SIGGRAPH Comput. Graph. **26**, 35–42 (1992)

14. Xu, L., Li, E., Li, J., Chen, Y., Zhang, Y.: A general texture mapping framework for image-based 3d modeling. In: 17th IEEE International Conference on Image Processing (ICIP 2010), pp. 2713–2716 (2010)

15. Lempitsky, V., Ivanov, D.: Seamless mosaicing of image-based texture maps. In: IEEE Conference on Computer Vision and Pattern Recognition, CVPR 2007, pp. 1–6 (2007)

16. Ge, Y., Yin, B., Sun, Y., Tang, H.: 3d face texture stitching based on Poisson equation. In: IEEE International Conference on Intelligent Computing and Intelligent Systems (ICIS 2010), vol. 2, pp. 809–813 (2010)

17. Desssein, A., Smith, W.A.P., Wilson, R.C., Hancock, E.R.: Seamless texture stitching on a 3D mesh by Poisson blending in patches. In: 2014 IEEE International Conference on Image Processing (ICIP), pp. 2031–2035, 27–30 October 2014

18. Xiong, X., De la Torre, F.: Supervised descent method and its applications to face alignment. In: IEEE Conference on Computer Vision and Pattern Recognition (CVPR) (2013)

19. Besl, P.J., McKay, N.D.: A method for registration of 3-d shapes. IEEE Trans. Pattern Anal. Mach. Intell. **14**, 239–256 (1992)

20. Douadi, L., Aldon, M., Crosnier, A.: Pair-wise registration of 3d/color data sets with ICP. In: 2006 IEEE/RSJ International Conference on Intelligent Robots and Systems, IROS 2006, 9–15 October 2006, Beijing, China, pp. 663–668 (2006)

21. Chen, Y., Medioni, G.: Object modelling by registration of multiple range images. Image Vis. Comput. **10**, 145–155 (1992)

22. Newcombe, R.A., Izadi, S., Hilliges, O., Molyneaux, D., Kim, D., Davison, A.J., Kohli, P., Shotton, J., Hodges, S., Fitzgibbon, A.: KinectFusion: real-time dense surface mapping and tracking. In: Proceedings of the 2011 10th IEEE International Symposium on Mixed and Augmented Reality, ISMAR 2011, pp. 127–136. IEEE Computer Society, Washington, D.C. (2011)

Human Body Volume Recovery from Single Depth Image

Jaeho Yi[1], Seungkyu Lee[1](✉), Sujung Bae[2], and Moonsik Jeong[2]

[1] Department of Computer Engineering, Kyung Hee University, Seoul, South Korea
seungkyu@khu.ac.kr
[2] Samsung Electronics, DMC R&D Center, Suwon, South Korea

Abstract. We propose on-line human body volume recovery framework using only single depth image. Depth image contains partial 3d geometry information of human body surface seen from the sensor viewpoint. Previous volume reconstruction methods require multiple images from different viewpoints to reconstruct complete closed body volume. They have limitation in real-time application with dynamic objects or require multiple sensors. In this paper, we propose a generic model based human body volume recovery. First, we register the pose of the generic model to partial body surface observation from single depth image. And then remaining 3d points on unseen surface is optimized by propagating confidence of the partial observation. Experimental result shows our method captures reasonable human body volume from single depth image on-line.

1 Introduction

Recently, consumer depth cameras have been widely applied in computer vision, graphics and interaction. Instant capturing of 3D geometry benefits traditional applications such as 3d scene and object reconstruction, augmented reality, interactive games and robot navigation. However, depth cameras capture only partial 3D surface of a target object that can be seen from the camera viewpoint. In order to reconstruct the complete surface of an object including the occluded region such as rear surface, we have to put additional camera or rotate either single camera or target object to collect complete surface information from multiple viewpoints. These approaches have limitations in practical applications. Multiple calibrated cameras surrounding a target object requires spacious place, but has limitation in capturing an object bigger than its size [8]. Single moving camera capturing multiple images from changing viewpoints [5] reconstructs complete surface of a static scene. This brings a practical limitation with dynamic or deforming objects in on-line volume reconstruction. In order to apply this approach for 3d human volume reconstruction, target person turn around taking multiple depth images. However, it is difficult to keep their pose while turning around. It requires additional prior knowledge or model to alleviate the deformation distortion [7].

Existing human volume reconstruction methods reconstruct relatively accurate and clean volume using multiple viewpoint images of target body. However,

© Springer International Publishing Switzerland 2015
G. Bebis et al. (Eds.): ISVC 2015, Part I, LNCS 9474, pp. 396–405, 2015.
DOI: 10.1007/978-3-319-27857-5_36

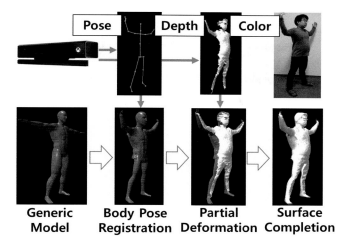

Fig. 1. Proposed body volume recovery using single depth image

they require complicated registration scheme or multiple cameras that has difficulty in on-line capturing of a dynamic object. Newcombe et al. [5] capture a series of depth images from changing viewpoints to build the 3d model of a static object. Similarly, Barmpoutis [3] method reconstructs human body on-line while the human subject moves arbitrarily in front of the camera. Therefore, the algorithm requires multi-viewpoint depth images to build a full volume of human body that cannot be obtained in real time. Zeng et al. [1] capture 3d shapes from multiple depth cameras and build a complete human body volume. In order to address the misalignment problem of deforming body, a deformable object registration method is proposed that requires around 15 min to generate single model. Malleson et al. [4] propose a method for geometry reconstructing of indoor scenes containing dynamic human body using a single depth sensor. This method is online in the sense that each frame is processed sequentially but it also requires multiple views to build a volume and their reported optimal processing time is 2 frames per second. Chen et al. [2] build a tensor-based human body model. Based on the depth observation, they estimate both shape and pose parameters of a human body and try to deform their model. However, they cannot reflect every details of the observed shape from depth camera and cannot be run in real time (average 1.78 s to fit one point cloud). Recently, Zhang et al. [7] have proposed a dynamic human body capturing method that requires a prior model that is built based on multiple depth observations. Zollhfer et al. [8] propose a dynamic human body modeling method. But is requires eight depth cameras for moving body. Although, recovering human body volume using single depth camera is an ill-posed problem, human body volume reconstructed at reasonable accuracy benefits practical applications such as volumetric human interaction, volumetric haptic feedback, body information analysis like body physique, fatness, abdominal obesity, etc., mixed reality and game employing human body volume and virtual cloth fitting.

In this paper, we propose an on-line human body volume reconstruction using single depth image. In order to infer unseen surface of rear part of human body, we propose to adopt a generic human volume model as an initial guidance of general body shape. We reflect the observed shape from single depth image onto the corresponding part of our generic body model. For the completion of remaining unseen surface, we infer the optimal location of each surface point based on the global shape of generic body model and local shape confidence propagation from the surface points having depth observation. Proposed method is summarized in Fig. 1.

2   Proposed Method

Our method has three major steps. Single depth image of human body and its pose are obtained using consumer depth camera. Our volume recovery starts from a given generic body model. We assume that observed target human body is parallel to the coronal plane and doesn't observe any occlusion. By registering each joint in our generic body model to the corresponding joint observation from depth camera, transformation for each rigid body part (rotation and translation in 3d coordinates) is calculated. Based on the transformations we track the pose and reflect it on the generic body model. This step reflects the real human body size and ratio. In the second step, vertex points on our generic body model having depth observations are relocated reflecting observed body shape. Finally, remaining unseen body surface such as side and back regions are completed by our confidence propagation method in the last step.

2.1   Body Pose Registration

We assume that human body consists of connected rigid body parts. First, we capture the pose of target body from single depth image and reflect it onto the generic body model. As illustrated in Fig. 2, each body part is specified by at least two connected joints. Based on the correspondences between the joint pairs, we specify the transformation of i_{th} body part as follows.

$$Tr(i) = R_i(S_i p + T_i) \tag{1}$$

where p is the point on the i_{th} body part, $T_i \in \mathbb{R}^3$ is three dimensional translation registering proximal joint of model to the observed joint. $R_i \in \mathbb{R}^1$ is one dimensional rotation in the plane defined by corresponding body parts registered in their proximal joints. S_i is the scaling factor based on the joint distance ratio calculated at the generic body model and observed human body respectively. Note that each vertex in our body model has its index indicating which body part it belongs to. In this step, model size will adjusted reflecting the size of target human body. Figure 1 (second image on the bottom row) shows before and after the registration of every body parts reflecting the pose and size observed in depth image.

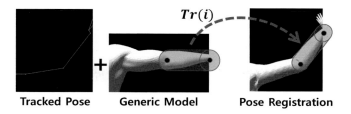

Fig. 2. Rigid parts transformation for pose registration

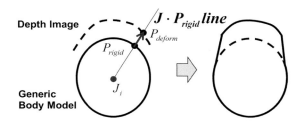

Fig. 3. Partial model deformation toward depth observation

2.2   Partial Model Deformation Based on Depth Image

Once our generic model is registered to pose observation, we perform partial model deformation reflecting observed shape of target human on the generic model. As shown in Fig. 3, however, not all model points can be projected onto the depth observation because significant part of the model has occluded and has no corresponding observation in single depth image. Our projection scheme is as follows. First we draw a line between closest joint J_i and a point P_{rigid} on the rigid registered model. And then move the point P_{rigid} to the intersecting point between the J_i-P_{rigid} line and the depth surface of human body. If there is no such intersection, we conclude that the point P_{rigid} has no corresponding observation in the depth image. Usually, our human body model consists of dense 3d points and registered 3d model have higher 3d resolution compared to a sparse depth image. In real implementation, however, finding a tangent plane and calculating the intersecting point for every 3d points on our body model are computationally expensive to apply for any on-line application. In order to avoid any brute-force search for intersecting point, we propose an approximation method. Point P_{rigid} is an initial guess of this approximation and iteratively moved along the J_i-P_{rigid} line. At each iteration current 3d point $p_n(x_n, y_n, z_n)$ jumps to next approximation $p_{n+1}(x_{n+1}, y_{n+1}, z_{n+1})$ using Eq. (2). Let 'hat' represent the coordinate in the image plane compared to the world coordinate. Then, $(\hat{x_n}, \hat{y_n})$ denotes the image plane coordinate of (x_n, y_n) of current model point P_n and $\hat{D}_n$ is corresponding depth value in the image plane.

$$y_{n+1} = y_n + \alpha(D_n - y_n)$$
$$x_{n+1} = x_n + \alpha(D_n - x_n)$$

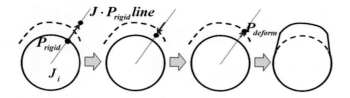

Fig. 4. Iterative approximation of the model point

$$z_{n+1} = D_n \tag{2}$$

where D_n is z-coordinate value converted from $\hat{D}_n$ and α is the slope of J_i-P_{rigid} line. This approximates next z_{n+1} by the depth values at $(\hat{x_n}, \hat{y_n})$ in image coordinate. Stopping criterion of the approximation is $|z_{n-1} - z_n| < \beta$ and the final approximation of the model point becomes P_{deform}. This iteration is illustrated in Fig. 4. Figure 5 shows examples of converging and diverging cases of the approximation. Diverging happens when the absolute slope of body surface in depth image is bigger than the J_i-P_{rigid} line (Fig. 5(b)). In most cases, the absolute slope of J_i-P_{rigid} line is bigger than the body surface (Fig. 5(a)). In order to eliminate diverging or any other noisy approximation, we perform three steps of post-processing. First, if the image coordinate $(\hat{x_n}, \hat{y_n})$ of an approximation falls out of body segment during our approximation process, we conclude the point is diverging. We discard those points in our final reconstruction results. Secondly, we perform a median filtering within each two neighbor triangles (four vertexes) after the approximation step. Finally, fast cumulative temporal filtering [6] is applied on the model vertexes of multiple time frames (Eq. (3)).

$$p_t = F(\frac{1}{N}p_t, \frac{N-1}{N}p_{t-1}) \tag{3}$$

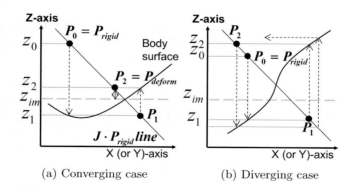

(a) Converging case   (b) Diverging case

Fig. 5. Converging and diverging cases of the depth observation approximation

2.3   Surface Completion by Confidence Propagation

After the observed depth is reflected on the frontal surface of the generic model, we perform the completion step for remaining unseen surface (side and back). We propose to propagate frontal observation to backward iteratively inferring new location of vertex points on the surface. Initial location of unseen surface is the point on the generic model that provides reasonable initial shape before the propagation. In our framework, these initial points on the generic model prevent our propagation being diverge forcing to keep original shape of human body. On the other side, depth observation of frontal surface pulls their neighbors to reflect real depth iteratively propagating backward. Overall, starting from such initial guidance points, we iteratively propagate confident frontal surface shape from the boundary to backward surface by minimizing the energy at each surface point p as defined in the following equation.

$$E(p) = E_s(p) + \alpha E_c(p) \tag{4}$$

where $E_s(p)$ is directional smoothness energy and $E_c(p)$ is shape confidence energy. Let $W_c(p)$ $(0 \le W_c(p) \le 1)$ be a confidence at point p and set to 1 for all surface points having depth observation. All remaining unseen points get 0 confidence. Directional Smoothness is defined as follows.

$$E_s(p) = \sum_{i \in neighbor} \beta_i \cdot \frac{d^2 \Phi_i}{dp_i^2} \tag{5}$$

where β_i is an weight along the confidence of its neighbor and Φ_i is the signed angle (concave gets minus sign) between the edges connecting current pixel p to neighbor p_i and the neighbor p_i to its next neighbor p_{ij} along the p_i direction. Shape confidence term is defined as follows.

$$E_c(p) = exp(-\frac{\sum_{i \in neighbor} exp(-\sqrt{(p - p_i)^2}) \cdot W_{c_i}}{\sum_{i \in neighbor} exp(-\sqrt{(p - p_i)^2})}) \tag{6}$$

where W_{c_i} is confidence of i_{th} neighbor and $\sqrt{(p - p_i)^2}$ is Euclidean distance to the neighbour. We iteratively minimize the energy for each pixel and relocate it. After each iteration, confidence of each vertex is updated as follows.

$$\hat{W}_c = \frac{\sum_{i \in neighbor} exp(-\sqrt{(p - p_i)^2}) \cdot W_c}{\sum_{i \in neighbor} exp(-\sqrt{(p - p_i)^2})} \tag{7}$$

3   Experiments

In our experiment, we use Kinect depth camera V2. Depth image resolution is 512×424. We take depth image of human body located 2 m away from the camera. Average number of vertex points observed on the human body is around

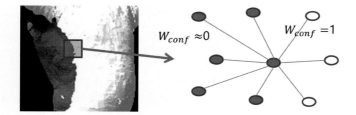

Fig. 6. Vertex model and confidence: brighter region of frontal surface has higher confidence than darker region of rear or side surfaces.

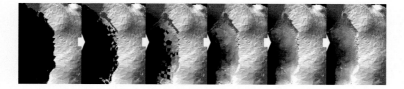

Fig. 7. Sample surface completion by confidence propagation: brighter region has higher confidence

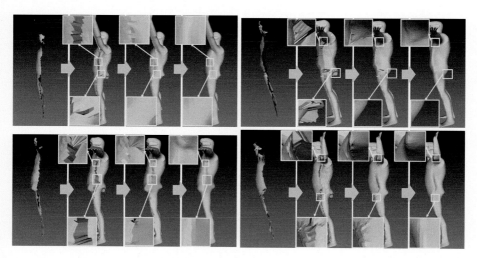

Fig. 8. Body volume recovery experimental results

20,000 points. Our generic body model has 120,000 3d points and 23 joints. Any other more precise, dense or different body models can be used. Qualitative and quantitative evaluations are performed to verify the proposed recovery method. Previous methods for the completion of missing surface of human body [9,10] are not aimed to recover such global and significant missing data and fail to recover the whole rear surface. Figure 8 shows several on-line volume recovery results of

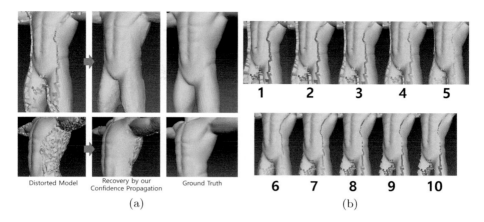

Fig. 9. (a) Body volume recovery experimental results (b) Quantitative evaluation model set having gradually changing dialation/erosion levels in rear surfaces

our proposed method. First column shows observed partial surface. Second column is 3d generic model registered to the pose of each target depth observation. Observed depth is reflected on the frontal surface of respective generic model. Around the boundary of frontal surface, serious discontinuity is observed due to the surface discrepancy between depth observation and the generic model. Third column is the generic model after the confidence propagation step. Surface on the generic model have been deformed based on the confidence influence from foreground surface optimizing unseen surface location. Last column shows final recovery result after further noise elimination and remeshing (Figs. 6 and 7).

We also perform a quantitative evaluation using graphics model of human body. First we separate human body model into two regions; front side surface and rear side surface. We consider that the front surface has high confidence. Now we deform rear surface with several variations such as Gaussian noise, salt and paper noise and erosion/dialation. After that we apply our proposed method to recover the distorted rear surface based on the confident shape of front surface. Finally, average Euclidean distance from each recovered surface point to the original surface is calculated to evaluate how well our proposed method infer the original surface shape of the rear part. Figure 9(a) shows examples of the graphic models before and after our recovery showing that the randomly distorted rear surface is recovered to the original surface. Ground truth model on the third column is compared. In order to perform a quantitative evaluation, we have created a set of evaluation graphic body models having gradually changing dialation/erosion levels in rear surfaces (Fig. 9(b)). Average recovery error and gain is calculated as summarized in Figs. 10 and 11. When there is a significant error in rear surface (such as the cases #1 - #4 and #8 - #10), proposed recovery method infers improved surface that is close to the ground truth shape.

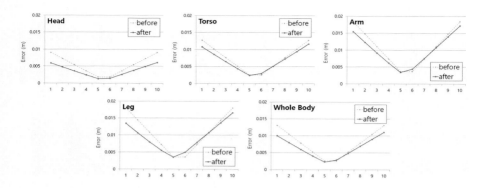

Fig. 10. Average error for each body part before and after the proposed recovery

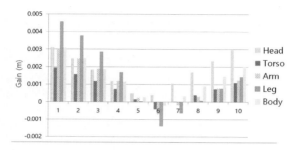

Fig. 11. Average gain after the proposed recovery for all cases

3.1   Body Volume Analysis

Based on our proposed method, we measure the height and waist size of the reconstructed body volume and quantitatively evaluated the accuracy. Height is inferred from the length of joint-joint distances from head to foot of both left and right sides. Based on a training measurement, a polynomial regression model is obtained between the joint-joint distances and height. In a similar manner, waist size is also inferred from the reconstructed body volume around torso joint. Based on 30 test people, average height measurement error is around 1.4 cm (std: 0.96) and average waist size error is around 1.2 inches. (std: 0.92) showing that our proposed method infers reasonable complete human body volume using single depth image.

4   Conclusion

In this paper, we propose a human body volume recovery method using only single depth image. In order to complete unseen surface of rear part of human body, we have a generic human volume model as an initial guidance of body shape. We infer the local shape of unseen surface point based on the global shape of generic body model and local shape confidence propagation. This framework allows us fast and stable recovery of human body volume using only single depth image.

References

1. Zeng, M., Zheng, J., Cheng, X., Liu, X.: Templateless quasi-rigid shape modeling with implicit loop-closure. In: IEEE Conference on Computer Vision and Pattern Recognition, pp. 145–152 (2013)
2. Chen, Y., Liu, Z., Zhang, Z.: Tensor-based human body modeling. In: IEEE Conference on Computer Vision and Pattern Recognition, pp. 105–112 (2013)
3. Barmpoutis, A.: Tensor body: real-time reconstruction of the human body and avatar synthesis from RGB-D. IEEE Trans. Cybern. **43**(5), 1347–1356 (2013)
4. Malleson, C., Klaudiny, M., Hilton A., Guillemaut, J.-Y., Single-view RGBD-based reconstruction of dynamic human geometry. In: IEEE International Conference on Computer Vision Workshops, pp. 307–314 (2013)
5. Newcombe, R.A., Izadi, S., Hilliges, O., Molyneaux, D., Kim, D., Davison, A.J., Kohli, P., Shotton, J., Hodges, S., Fitzgibbon, A.: KinectFusion: real-time dense surface mapping and tracking. In: IEEE ISMAR (2011)
6. Lee, S.: ToF depth camera accuracy enhancement. SPIE Opt. Eng. **51**, 083203 (2012)
7. Zhang, Q., Fu, B., Ye, M., Yang, R.: Quality dynamic human body modeling using a single low-cost depth camera. In: IEEE Conference on Computer Vision and Pattern Recognition (CVPR), pp. 676–683 (2014)
8. Zollhfer, M., Niessner, M., Izadi, S., Rehmann, C., Zach, C., Fisher, M., Wu, C., Fitzgibbon, A., Loop, C., Theobalt, C., Stamminger, M.: Real-time non-rigid reconstruction using an RGB-D camera. ACM Trans. Graph. **33**(4), 156 (2014)
9. Kwok, T.H., Yeung, K.Y., Wang, C.C.L.: Volumetric template fitting for human body reconstruction from incomplete data, special issue on depth cameras techniques and applications on design, manufacturing and services. J. Manufact. Syst. **33**(4), 678–689 (2014)
10. Kraevoy, V., Sheffer, A.: Template based mesh completion. In: Proceedings of Symposium on Geometry Processing, pp. 13–22 (2005)

Dense Correspondence and Optical Flow Estimation Using Gabor, Schmid and Steerable Descriptors

Ahmadreza Baghaie$^{(\boxtimes)}$, Roshan M. D'Souza, and Zeyun Yu

College of Engineering and Applied Science, UW-Milwaukee,
Milwaukee, WI, USA
A.Baghaie@uwm.edu

Abstract. In this paper, the use of three dense descriptors, namely Schmid, Gabor and steerable descriptors, is introduced and investigated for optical flow estimation and dense correspondence of different scenes and compared with the well-known dense SIFT/SIFTFlow. Several examples of optical flow estimation and dense correspondence across scenes with high variations in the intensity levels, difference in the presence of features and different misalignment models (rigid, deformable, homography etc.) are studied and the results are quantitatively/qualitatively compared with dense SIFT/SIFTFlow. The proposed dense descriptors provide comparable or better results than dense SIFT/SIFTFlow which shows the high potential in this area for more thorough investigations.

Keywords: Dense correspondence · Optical flow · SIFT · Schmid filters · Gabor filters · Steerable filters

1 Introduction

The problem of image alignment/registration/correspondence and the range of its applications can go beyond the boundaries of different disciplines and examples can be found in medical image processing [1–4], remote sensing [5,6] etc. Generally speaking, image registration problems can be categorized into four major categories: *multi-view analysis*, *multi-temporal analysis*, *multi-modal analysis* and *scene to model registration* [7,8]. However, when thinking about *correspondence* or *alignment*, even this general categorization is not *general enough*. An instance of this argument can happen in object recognition where the goal is to align different samples of the same object category, like buildings, windows etc. [29]. In this case, one should try to find a way to represent the images beyond their pixel values and extract more meaning for better alignment. This is where feature descriptors are proven to be useful. One big category of image alignment methods using descriptors involves defining descriptors for sparse features detected from the scenes. In this category, at first features (corners, blobs, T-junctions etc.) are extracted from the input images. This is followed by assigning

© Springer International Publishing Switzerland 2015
G. Bebis et al. (Eds.): ISVC 2015, Part I, LNCS 9474, pp. 406–415, 2015.
DOI: 10.1007/978-3-319-27857-5_37

descriptor vectors to the neighborhood of each feature point and finally matching these descriptors between different scenes [9]. These local feature descriptors are very useful in applications like wide baseline matching, object and texture recognition, image retrieval, robot localization, video data mining, image mosaicing and recognition of object categories [10]. Examples of use of such techniques in other areas like 3D surface reconstruction can be found in [11–13]. Spin images [14], Gabor filters [20], steerable filters [24], speeded up robust features (SURF) [15] and scale invariant feature transform (SIFT) [27] are few examples from the many in this category. Going towards more abstract problems (form the viewpoint of a computer, of course) like action recognition, pose or expression detection, object recognition/categorization, correspondence across different scenes, image retrieval and many more, local features/descriptors may not be the proper choice. In these cases, the need for dense descriptors are more suitable. Examples of this category can be found in [16–19,29].

In this paper, the use of three new dense descriptors, namely Schmid descriptors, Gabor descriptors and steerable descriptors, is introduced and investigated for optical flow estimation and dense correspondence and compared with the well-known dense SIFT/SIFTFlow [29]. Examples shown here for the proposed descriptors revolve around optical flow estimation, image matching and computing the needed transformations for image alignment and the results show promising potential for further explorations and investigations. The paper is organized as follows. Section 2 contains comprehensive details about the three dense descriptors (Gabor, Schmid, steerable) proposed for dense correspondence as well as an overview of SIFT, which is followed by describing the mathematical framework used here for dense descriptor matching. Results of the use of these dense descriptors for optical flow estimation and dense correspondence are presented in Sect. 3. Finally, Sect. 4 concludes the paper with some remarks regarding future directions of the work.

2   Methods

2.1   Feature Descriptors

Gabor Filter Bank (G). The general equation for the complex Gabor filter can be defined as $G(x, y) = s(x, y)w_r(x, y)$, where $s(x, y)$ represents a complex sinusoid defined as $s(x, y) = e^{j(2\pi(u_0 x + v_0 y) + P)}$ in which (u_0, v_0) is the spatial frequency and P is the phase of the sinusoid [20,21]. As it is obvious this function has two components, one real and one imaginary. Therefore the general Gabor filter also is consisted of real and imaginary parts. The second term in the right hand side, $w_r(x, y)$, represents a Gaussian envelope which can be defined as $w_r(x, y) = Ke^{-\pi(a^2(x-x_0)_r^2 + b^2(y-y_0)_r^2)}$, where (x_0, y_0) represents the location of the peak while a and b are scaling parameters and the r subscript is for a rotation operation which is represented by

$$\begin{aligned}
(x - x_0)_r &= (x - x_0)\cos\theta + (y - y_0)\sin\theta \\
(y - y_0)_r &= -(x - x_0)\sin\theta + (y - y_0)\cos\theta
\end{aligned} \tag{1}$$

For this work, the implementation of the Gabor filter in the frequency domain, elaborated in [22] is used.

Schmid Filters (S). For computing Schmid descriptors, the response of each pixel in the image to a set of isotropic *Gabor-like* filters is computed [23]. These descriptors are rotationaly invariant and combine frequency and scale information. The general form of such filters is:

$$S(x, y, \tau, \sigma) = S_0(\tau, \sigma) + cos(\frac{\sqrt{x^2 + y^2}\pi\tau}{\sigma})e^{-\frac{x^2+y^2}{2\sigma^2}} \tag{2}$$

in which τ is the number of cycles of the harmonic function within the Gaussian envelope of the filter, which is the same as the one used in Gabor filters. The term $S_0(\tau, \sigma)$ is added for obtaining the zero DC component. Following the same approach as [23], 13 filters with scales σ between 2 and 10 and τ between 1 and 4 are used for creating dense descriptors.

Steerable Filters (St). The idea behind the design of steerable filters comes form this question: What are the conditions for any function $f(x, y)$ to be written as a linear sum of rotated versions of itself [24]? This can be represented as follows:

$$f(x, y) = \sum_{j=1}^{M} k_j(\theta)f^{\theta_j}(x, y) \tag{3}$$

where M is the number of required terms and $k_j(\theta)$ is the set of interpolation functions. In polar coordinates $r = \sqrt{x^2 + y^2}$ and $\phi = arg(x, y)$ and considering f as a function that can be expanded in a Fourier series in polar angle ϕ we have $f(r, \phi) = \sum_{n=-N}^{N} a_n(r)e^{in\phi}$. It has been proven that this holds if and only if the interpolation functions are solutions of $[e^{in\theta}]_{N\times1} = [e^{in\theta_m}]_{N\times M}[k_m(\theta)]_{M\times1}$ for $n = 0, 1, 2..., N$ and $m = 1, 2, ..., M$. As stated and proved in [24], all bandlimited functions in angular frequency are steerable, given enough basis filters. Steerable filters, 2D or 3D, are proven to be very useful in different fields of computer vision over the years [25, 26]. Here, steerable filters of up to order five are used for creating dense descriptors. Odd orders are edge detectors while even orders detect ridges.

SIFT. Four stages of SIFT method are [27]: (1) scale-space extrema detection, (2) keypoint localization, (3) orientation assignment and (4) keypoint descriptors. For the first step, a Gaussian function is considered as the scale-space kernel. By finding the scale-space extrema in the response of the image to difference-of-Gaussian (DoG) masks, not only a good approximation for the scale-normalized Laplacian of Gaussian function is provided, but also as pointed out in [28] the detected features are more stable. These local extrema are found in a $3 \times 3 \times 3$ neighborhood of the interest point. For accurate localization of the keypoints in the set of candidate keypoints a 3D quadratic function is fitted to the local sample points. After eliminating poor keypoints that are usually located in low contrast regions and near edges, orientations can be assigned by creating a 36-bin histogram for orientations in the keypoint's neighborhood. Each neighbor contributes to the histogram by a weight computed based on its gradient magnitude and also by a Gaussian weighed circular window around the keypoint. Using the location, scale and orientation determined for each keypoint up until now the local descriptor is created for each keypoint in a manner which makes it invariant to differences in illumination and viewpoint. This is done by combining the gradients at keypoint locations weighted by a Gaussian function over each 4×4 sub-region in a 16×16 neighborhood around the keypoint into 8-bin histograms. This results in a $4 \times 4 \times 8 = 128$ element vector for each keypoint. Normalizing the feature vectors to unit length will reduce the effect of linear illumination changes. This is usually followed by thresholding the normalized vector an re-normalizing it again to reduce the effects of large gradient magnitudes. Here, the descriptors are computed for all of the image pixels to create a dense descriptor for the image.

2.2   Problem Statement

The problem of matching/registering between images is modeled as a dual-layer factor graph, with de-coupled components for horizontal/vertical flow to account for sliding motion. This model is based on the work of Liu et al. [29] which takes advantage of L_1 truncated norm for achieving higher speeds in matching. Assuming F_1 and F_2 as two multi-dimensional feature images and $\mathbf{p} = (x, y)$ as the grid coordinates of the images, the objective function to be minimized can be written as:

$$E(\mathbf{w}) = \sum_{\mathbf{p}} min(||F_1(\mathbf{p}) - F_2(\mathbf{p} + \mathbf{w}(\mathbf{p}))||, t)$$

$$+ \sum_{\mathbf{p}} \eta(|u(\mathbf{p})| + |v(\mathbf{p})|) \tag{4}$$

$$+ \sum_{(\mathbf{p},\mathbf{q}) \in \epsilon} min(\alpha|u(\mathbf{p}) - u(\mathbf{q})|, d) + min(\alpha|v(\mathbf{p}) - v(\mathbf{q})|, d)$$

in which $\mathbf{w}(\mathbf{p}) = (u(\mathbf{p}), v(\mathbf{p}))$ is the flow vector at point $\mathbf{p}$.

The three summations in this equation are *data, small displacement* and *smoothness* terms respectively. The data term is for minimizing the difference between the feature descriptors along the flow vector, while the small displacement term keeps the displacements as small as possible when no information is available. Finally the smoothness term guaranties that the flow vectors for neighbor pixels are similar. As is obvious in this formulation, the horizontal and vertical components are de-coupled. This is mainly for reducing the computational complexity as mentioned in [29]. But this gives additional benefit of being able to account for sliding motions during the procedure of image matching. The objective is considered as a factor graph, with $\mathbf{p}$ and $\mathbf{q}$ as the variable nodes while the factor nodes represent the data, small displacement and smoothness terms. The flow is then computed by using a dual-layer loopy belief propagation algorithm [30].

3   Results and Discussion

For assessment of the performance of the three dense descriptors, two different experiments are designed: (1) optical flow estimation and (2) dense correspondence between scenes with different intensity levels and various transformations (rigid, deformable and homography). For the first test, four image pairs (*Grove2*, *RubberWhale, Urban3, Venus*) as well as their true optical flow fields acquired from the Middlebury database [31] are used. Table 1 reports the performance of the four dense descriptors using Interpolation Error (IE), Normalized Interpolation Error (NIE) and Average Angular Error (AAE) as metrics. Overall, dense steerable descriptors show better results in comparison to the rest of the descriptors while dense Schmidt descriptors perform poorly. Of course the difference between the best and the worst is not very big.

Moving towards a more challenging test, three image pairs as depicted in Fig. 1 are used for dense correspondence. The first pair consists of satellite images of *Badafshan*[1] acquired from Google Maps (Google) and Bing Maps (Microsoft). As it is obvious, these are taken at different times and the intensity levels as well as the terrain features contained in the images are different. The misalignment between them is relatively small and can be represented by a rigid translation. The checkerboard view depicts the misalignment better. Figure 2 shows the results of using the four dense descriptors for representation and alignment of this image pair. The first row is the computed flow field for dense correspondence while the second row shows the warped template after applying the flow filed. The last row shows the checkerboard view. The results are relatively identical for all of the dense descriptors, with the exception of having small perturbations. This is due to presence of small features in the input images that causes the dense descriptors to not be able to have a uniform representation across different scales and orientations.

The second pair consists of two images captured from the surface of *Mars* within 6 years [32]. Even though the changes between these two images are mostly changes in the intensity levels, a fracture possibly created by an artifact in the stitching procedure was detected during dense matching in the work of [29]. Therefore, other than the overall deformable transformation, detection of this fracture can be considered as a metric for assessing the performance of the used dense descriptors. The checkerboard view in Fig. 1 depicts the misalignment more obviously, even though the fracture is not obvious in this view. Figure 3 shows the results of the four descriptors for this pair. Even though, from the checkerboard views, the misalignment is corrected in the results of all of the descriptors, the fracture is not detected for dense Gabor and the flow field is relatively smooth while in the other three the fracture is detected.

---

[1] The first author's village in central part of Iran.

Table 1. Optical flow comparisons

Image pairs	Metric\Method	dense-G	dense-S	dense-St	dense-SIFT
Grove2	IE	9.36	9.64	**9.28**	9.75
	NIE	0.95	0.99	**0.94**	1.00
	AAE	**8.00**	8.52	8.03	8.31
RubberWhale	IE	3.31	3.30	**3.12**	3.27
	NIE	0.47	0.49	**0.46**	0.47
	AAE	11.68	13.94	12.67	**11.66**
Urban3	IE	5.49	5.97	5.67	**5.26**
	NIE	0.81	0.91	**0.79**	0.83
	AAE	15.33	13.59	10.47	**10.17**
Venus	IE	6.24	6.35	**6.15**	6.37
	NIE	**0.68**	0.72	0.70	0.70
	AAE	8.15	8.19	**5.70**	6.04

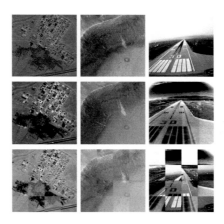

Fig. 1. Some example pictures for dense correspondence with different intensity levels, misalignment and imaging modalities: *Badafshan* pair and the checkerboard view (left column), *Mars* pair and the checkerboard view (center column) and *NASA* pair and the checkerboard view (right column)

The third pair consists of images that have large misalignment due to change in the homography as well as large differences in the intensity levels [33]. The checkerboard view in Fig. 1 demonstrates the misalignment as well as the large difference in the intensity levels more obviously. Figure 4 shows the results for this pair. Despite having different points of view as well as large differences in the intensity levels, all of the descriptors compensate the misalignment to some extent. Among them, dense steerable descriptors' result is the best while dense

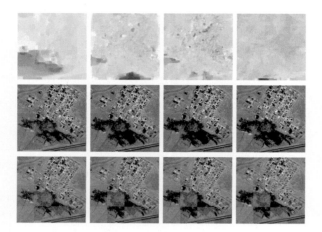

Fig. 2. Dense correspondence results for the *Badafshan* pair using dense-G (column 1), dense-S (column 2), dense-St (column 3) and dense-SIFT (column 4) descriptors; rows represent the computed flow (first), the warped template (second) and the checkerboard view after alignment (third)

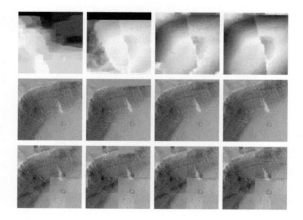

Fig. 3. Dense correspondence results for the *Mars* pair using dense-G (column 1), dense-S (column 2), dense-St (column 3) and dense-SIFT (column 4) descriptors; rows represent the computed flow (first), the warped template (second) and the checkerboard view after alignment (third)

Gabor descriptors' result is the second best. The performances of dense SIFT and Schmid descriptors are relatively similar.

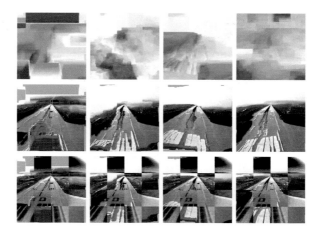

Fig. 4. Dense correspondence results for the *NASA* pair using dense-G (column 1), dense-S (column 2), dense-St (column 3) and dense-SIFT (column 4) descriptors; rows represent the computed flow (first), the warped template (second) and the checkerboard view after alignment (third)

4   Conclusion

The problem of dense correspondence can be defined as the process of finding matching points between images. This problem appears in different scenarios, from image matching between images of the same scene for image stitching or stereo matching to more complicated cases like object recognition which involves matching instances of objects from the same category (buildings, sky, windows etc.). In this paper, the use of three dense descriptors, namely Schmid descriptors, Gabor descriptors and steerable descriptors, is introduced and investigated for dense correspondence of different scenes as well as optical flow estimation. Starting from the image domain, dense descriptors are computed and then used for dense correspondence and optical flow estimation. This transformation, going from the image domain to descriptor domain, makes the problem of matching less affected by changes in the intensity levels and more sensitive to the features contained in the images. Several examples of optical flow estimation and dense correspondence across scenes with high variations in the intensity levels, difference in the presence of features and different misalignment models are studied and the results are quantitatively/qualitatively compared with dense SIFT/SIFTFlow. The proposed dense descriptors showed comparable or better results which shows the high potential in this area for more thorough investigations.

References

1. Baghaie, A., Yu, Z.: Curvature-based registration for slice interpolation of medical images. In: Zhang, Y.J., Tavares, J.M.R.S. (eds.) CompIMAGE 2014. LNCS, vol. 8641, pp. 69–80. Springer, Heidelberg (2014)
2. Baghaie, A., D'souza, R.M., Yu, Z.: Sparse And Low Rank Decomposition Based Batch Image Alignment for Speckle Reduction of retinal OCT Images (2014). arXiv preprint arXiv:1411.4033
3. Baghaie, A., Yu, Z., D'souza, R.M.: State-of-the-art in retinal optical coherence tomography image analysis. Quant. Imaging Med. Surg. 5(4), 603–617 (2015). doi:10.3978/j.issn.2223-4292.2015.04.08
4. Baghaie, A., D'souza, R.M., Yu, Z.: Application of Independent Component Analysis Techniques in Speckle Noise Reduction of Single-Shot Retinal OCT Images (2015). arXiv preprint arXiv:1502.05742
5. Lillesand, T., Kiefer, R.W.: Remote Sensing and Image Interpretation. Wiley, New York (2014)
6. Li, Q., Wang, G., Liu, J., Chen, S.: Robust scale-invariant feature matching for remote sensing image registration. IEEE Geosci. Remote Sens. Lett. 6(2), 287–291 (2009)
7. Zitova, B., Flusser, J.: Image registration methods: a survey. Image Vis. Comput. 21(11), 977–1000 (2003)
8. Baghaie, A., Yu, Z., D'souza, R.M.: Fast mesh-based medical image registration. In: Bebis, G., et al. (eds.) ISVC 2014, Part II. LNCS, vol. 8888, pp. 1–10. Springer, Heidelberg (2014)
9. Tafti, A.P., Hassannia, H., Yu, Z.: siftservice.com-Turning a Computer Vision algorithm into a World Wide Web Service (2015). arXiv preprint arXiv:1504.02840
10. Mikolajczyk, K., Schmid, C.: A performance evaluation of local descriptors. IEEE Trans. Pattern Anal. Mach. Intell. 27(10), 1615–1630 (2005)
11. Tafti, A.P., Kirkpatrick, A.B., Alavi, Z., Owen, H.A., Zeyun, Y.: Recent advances in 3D SEM surface reconstruction. Micron 78, 54–66 (2015)
12. Tafti, A.P., Kirkpatrick, A.B., Owen, H.A., Yu, Z.: 3D microscopy vision using multiple view geometry and differential evolutionary approaches. In: Bebis, G., et al. (eds.) ISVC 2014, Part II. LNCS, vol. 8888, pp. 141–152. Springer, Heidelberg (2014)
13. Tafti, A.P., Baghaie, A., Kirkpatrick, A.B., Owen, H.A., D'Souza, R.M., Yu, Z.: A Comparative study on the application of SIFT, SURF, BRIEF and ORB for 3D surface reconstruction of electron microscopy images. In: Computer Methods in Biomechanics and Biomedical Engineering: Imaging & Visualization (2016)
14. Johnson, A.E., Hebert, M.: Recognizing objects by matching oriented points. In: 1997 IEEE Computer Society Conference on Computer Vision and Pattern Recognition, Proceedings. IEEE (1997)
15. Bay, H., Tuytelaars, T., Van Gool, L.: SURF: speeded up robust features. In: Leonardis, A., Bischof, H., Pinz, A. (eds.) ECCV 2006, Part I. LNCS, vol. 3951, pp. 404–417. Springer, Heidelberg (2006)
16. Wang, H., Klser, A., Schmid, C., Liu, C.-L.: Action recognition by dense trajectories. In: IEEE Conference on Computer Vision and Pattern Recognition (CVPR). IEEE (2011)
17. Tola, E., Lepetit, V., Fua, P.: Daisy: an efficient dense descriptor applied to wide-baseline stereo. IEEE Trans. Pattern Anal. Mach. Intell. 32(5), 815–830 (2010)

18. Sangineto, E.: Pose and expression independent facial landmark localization using dense-SURF and the Hausdorff distance. IEEE Trans. Pattern Anal. Mach. Intell. **35**(3), 624–638 (2013)
19. Dalal, N.,Triggs, B.: Histograms of oriented gradients for human detection. In: IEEE Computer Society Conference on Computer Vision and Pattern Recognition, CVPR 2005, vol. 1. IEEE (2005)
20. Gabor, D.: Theory of communication. part 1: the analysis of information. J. Inst. Electr. Eng. III. Radio Commun. Eng. **93**(26), 429–441 (1946)
21. Movellan, J.R.: Tutorial on Gabor filters. Open Source Document (2002)
22. Ilonen, J., Kmrinen, J.-K., Kllvinen, H.: Efficient computation of Gabor features. Lappeenranta University of Technology, Lappeenranta (2005)
23. Schmid, C.: Constructing models for content-based image retrieval. In: Proceedings of the 2001 IEEE Computer Society Conference on Computer Vision and Pattern Recognition, CVPR 2001, vol. 2. IEEE (2001)
24. Freeman, W.T., Adelson, E.H.: The design and use of steerable filters. IEEE Trans. Pattern Anal. Mach. Intell. **9**, 891–906 (1991)
25. Jacob, M., Unser, M.: Design of steerable filters for feature detection using canny-like criteria. IEEE Trans. Pattern Anal. Mach. Intell. **26**(8), 1007–1019 (2004)
26. Aguet, F., Jacob, M., Unser, M.: Three-dimensional feature detection using optimal steerable filters. In: IEEE International Conference on Image Processing, ICIP 2005, vol. 2. IEEE (2005)
27. Lowe, D.G.: Distinctive image features from scale-invariant keypoints. Intern. J. Comput. Vis. **60**(2), 91–110 (2004)
28. Mikolajczyk, K., Schmid, C.: An affine invariant interest point detector. In: Heyden, A., Sparr, G., Nielsen, M., Johansen, P. (eds.) ECCV 2002, Part I. LNCS, vol. 2350, pp. 128–142. Springer, Heidelberg (2002)
29. Liu, C., Yuen, J., Torralba, A.: Sift flow: dense correspondence across scenes and its applications. IEEE Trans. Pattern Anal. Mach. Intell. **33**(5), 978–994 (2011)
30. Pearl, J.: Probabilistic Reasoning in Intelligent Systems: Networks of Plausible Inference. Morgan Kaufmann, San Mateo (2014)
31. Baker, S., Scharstein, D., Lewis, J.P., Roth, S., Black, M.J., Szeliski, R.: A database and evaluation methodology for optical flow. Intern. J. Comput. Vis. **92**(1), 1–31 (2011)
32. Malin, M.C., Edgett, K.S., Carr, M.H., Danielson, G.E., Davies, M.E., Hartmann, W.K., Ingersoll, A.P., James, P.B., Masursky, H., McEwen, A.S., Soderblom, L.A., Thomas, P., Veverka, J., Caplinger, M.A., Ravine, M.A., Soulanille, T.A., Warren, J.L.: New Gully Deposit in a Crater in Terra Sirenum: Evidence That Water Flowed on Mars in This Decade? In: NASA's Planetary Photojournal, MOC2-1618, 6 December 2006. http://photojournal.jpl.nasa.gov/
33. Yang, G., Stewart, C.V., Sofka, M., Tsai, C.-L.: Registration of challenging image pairs: initialization, estimation, and decision. IEEE Trans. Pattern Anal. Mach. Intell. **29**(11), 1973–1989 (2007)

ST: Advancing Autonomy
for Aerial Robotics

Efficient Algorithms for Indoor MAV Flight Using Vision and Sonar Sensors

Kyungnam Kim[⊠], David J. Huber, Jiejun Xu, and Deepak Khosla

HRL Laboratories, LLC, 3011 Malibu Canyon Rd, Malibu, CA 90265, USA
kkim@hrl.com

Abstract. This work describes an efficient perception-control coupled system and its underlying algorithms that enable autonomous exploration of indoor environments by a Micro Aerial Vehicle (MAV) equipped with a monocular camera and sonar sensors. The perception subsystem uses inputs from the camera to detect the vanishing point and doors in corridors. It detects the vanishing point by grid-based line-intersection voting (GLV) and Mixture-of-Gaussians (MoG)-based classification, while doors are detected by using simple but effective geometric scene properties (GSP) with template matching and temporal filtering. It also detects distance to obstacles, for example walls, using inputs from one forward-looking and two side-looking sonar sensors. These algorithms are accurate, computationally efficient, and suitable for real-time operation on offboard and onboard power-constrained computing platforms. The control subsystem employs a priority-based planner that combines outputs from the perception subsystem to compute high-level direction and velocity commands for the MAV. We evaluate our perception-control system on a commercially available AR.Drone 2.0 MAV with offboard processing and successfully demonstrate collision-free autonomous exploration and flight in building corridors and rooms at approximately 2 m/s speed.

1 Introduction

Micro Aerial Vehicles (MAVs) require real-time accurate perception and control capabilities to autonomously navigate in many applications, such as search-and-rescue, mapping, and unmanned surveillance. Perception and control technologies for MAVs are an active research area especially due to their widespread availability, flexibility, and low cost. However, the current state of the art approaches are generally computationally complex and require *a priori* mapping and knowledge of the environment to navigate. In indoor environments, detection of vanishing point and doors in corridors and rooms is a necessary capability to enable the MAV to autonomously explore (e.g., enter a room through a door) and plan its flight in real-time. This requires efficient perception and control algorithms that can be mapped to power-constrained computing platforms.

This work describes a simple yet efficient coupled perception and control approach primarily designed to work for MAV in indoor environments. The perception subsystem uses inputs from the camera to detect the vanishing point and doors in corridors and rooms, while the control subsystem uses this information to plan the flight path and automatically navigate the MAV.

© Springer International Publishing Switzerland 2015
G. Bebis et al. (Eds.): ISVC 2015, Part I, LNCS 9474, pp. 419–431, 2015.
DOI: 10.1007/978-3-319-27857-5_38

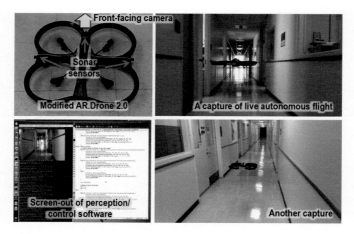

Fig. 1. Our modified AR.Drone 2.0 used in the experiments. From the laptop display, we can monitor the perception outputs and control commands being executed. Two sample screen snapshots of live MAV flight are shown here.

Our approach to detecting a vanishing point begins with fast edge detection and linking followed by grid-based line-intersection voting and Mixture-of-Gaussians (MoG)-based classification. This is accurate and robust in a variety of illumination and conditions, and yet more efficient than previous approaches that use computationally intensive algorithms, such as the Hough transform and Markov modeling [1]. Based on the detected vanishing point, our door detection algorithm chooses door candidates using the geometric scene properties, such as door width and relative door orientation, and intensity profile information, such as the mean and variance of intensities in the door and its adjacent areas. If no vanishing point is detected, a template matching-based approach is used to detect door lines that look similar to the last detected door boundaries. The final door candidates are selected by temporal filtering to eliminate inconsistent detections.

The topic of door detection using 2D and other sensors has been previously investigated [2–5]. The work in [2] detects door corners first and then groups the four corners to form door candidates. The grouped door-corner candidates are then verified by matching and combining edges and corners. Other work [3, 4] rely on both a camera and a 2D laser range finder. The method in [3] relies on detection of doorknobs and frames as additional information. However, these features are not always observable and limit the applicability of this method. The approach in [5] depends on not only shape information, but also appearance information, and uses a Bayesian model of shape and appearance likelihood to verify door candidates formed from detected corners. The methods, such as in [2, 5], usually need to deal with validation of several door candidates since there are hundreds of corner groups. Therefore, they are computationally intensive and not suited for real-time applications.

Our door detection approach is different from prior work in several aspects. First, it is based on linked edge information obtained from a single monocular camera and does not depend on the appearance of a door or detection of specific door parts, such as doorknobs and doorframes. Second, it leverages the scene geometry information about the orientation of door lines and intensity profile of doors compared to its neighboring areas. This makes our approach work under different illumination conditions and/or viewing angles in the corridors. Third, it is efficient and can run onboard resource-limited computing platforms. Finally, it does not require any training.

The perception algorithms described above have been implemented and tested using a laptop and a quadrotor MAV (AR.Drone 2.0 [6]) for autonomous flight in a corridor (see Fig. 1). In order to accommodate indoor navigation challenges, such as unexpected air drifts, self-turbulence, and obstacles during flight, our control strategy combines a visual sensor for primary navigation with front and side-mounted sonar sensors to detect obstacles and avoid collision. A sonar sensor measures distances to obstacles. The control subsystem assigns different priorities to the sensor modalities and employs a planner that combines outputs from the perception subsystem to compute high-level direction and velocity commands for the MAV in a closed-loop manner. Our approach does not require optical flow calculation, complicated learning strategies, or additional external sensors.

The rest of the paper is organized as follows. Section 2 describes our proposed perception subsystem and its underlying algorithms. Section 3 described the details of our control subsystem. Section 4 presents experimental results of the proposed algorithm on MAV flight in real-worlds. Finally, Sect. 5 discusses the results and conclusion.

2 Perception Subsystem

Our perception subsystem consists of several processes to detect vanishing point (VP) and doors. Figure 2 illustrates the perception subsystem block diagram. Its input is video from a single front-facing monocular camera mounted on the MAV. An edge detector is first applied to each frame of the incoming video. The detected edge points are then connected to longer edges through edge linking. Horizontally-oriented edges are used to calculate the vanishing point, while vertically-oriented edges provide the initial candidates for door lines. If the

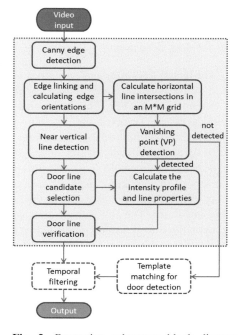

Fig. 2. Perception subsystem block diagram for vanishing point and doors detection using a single monocular camera.

vanishing point is available, the door candidates are evaluated using their geometric properties (e.g., length, location, distance) and intensity profiles. The detected vanishing point and doors are used by the MAV to autonomously explore the indoor environment. Details of the vanishing point and door detection algorithms are described below.

Vanishing Point Detection. The vanishing point (VP) is a useful destination for a MAV in a corridor, but can also be used for guiding the perception subsystem in finding doors. Due to its importance in our approach, the VP detection must perform accurately in real-time with high probability of detection and low false positives. To satisfy these requirements, we have developed an efficient VP detection algorithm that starts with the edge detection and linking described above and follows with grid-based line-intersection voting (GLV) and MoG-based classification.

A. Edge Detection

Given each frame in the incoming video, the Canny edge detector [11] is used to detect possible edge pixels whose intensity gradients fall within a preselected range. In addition, non-maximum suppression is used within the Canny detector to retain only thin lines (i.e., edge-thinning). The gradient directions at all edge pixels are calculated and discretized into N bins. Edge pixels with gradient orientations belonging to the same bin are linked using connected component analysis. In this way, shorter edges are connected to form longer lines. Each line's orientation is calculated and stored as its geometric feature along with its two end points and length. These line features are later used for door detection.

B. Grid-based line-intersection voting (GLV) and MoG classification

Based on the long edges detected in the image frame, we now estimate the position of the vanishing point (VP). This is the point where corridor horizontal lines intersect in the image frame. To detect the vanishing point in each frame, the previously detected edge lines are first filtered to remove lines that are too close to the horizontal direction (left to right). Specifically, if the angle between an edge line and the image horizontal line is below a threshold (e.g., 5°), this line is ignored for the VP calculation. The intersections of remaining lines are then calculated. To find the VP, the entire image area is split into an $M \times M$ grid and the number of line intersections within each grid cell is calculated. The cell with the maximum number of line intersections is assumed to contain the true vanishing point. This is based on the observation that there are many lines parallel to the floor of the corridor floor and they mostly intersect at the end of the corridor.

We define the centroid of intersections within the chosen cell as the vanishing point. To further reduce false alarms, a mixture of Gaussian (MoG) model is trained to classify a centroid of the intersections within a grid cell as the VP. The MoG model is represented as

$$p(v|\lambda) = \sum_{i=1}^{k} w_i g\left(v \middle| \mu_i, \sum_i\right)$$

where v is a 2-by-1 vector consisting of the number of intersections in the selected VP grid and the variance of their 2D locations, K is the number of Gaussian components (e.g., 3), w_i is the Gaussian component's weight and $g(v|\mu_i, \sum_i)$ is the Gaussian component with a mean μ_i and covariance matrix $\sum_i$. Given a set of training data with labeled VP grids, we learn the parameters w_i, μ_i and $\sum_i$. Once a new grid cell is detected as a candidate VP grid in the image frame, we use the above MoG model to calculate its probability of containing a true VP.

Our VP detection algorithm is unique in (1) using the grid-based voting to find the cluster of near-horizontal edge intersections, and (2) using MoG classification for estimating the confidence of detecting a VP. In addition, our algorithm is computationally efficient. The first step of Canny edge detection is extremely fast. For a typical 720×480 image, it only requires about 2 ms to detect edges. The edge linking step depends on the complexity of the scene. Its computation is on the order of the number of edges in the image and typically takes about 20 ms. The other steps in VP detection require only a few additional milliseconds. Typically our VP detection for a 720×480 image takes about 30 ms per frame and therefore achieves real-time 30 fps speed. The closest prior work to our VP detection approach is described in [12]. It uses the Hough Transform (HT) to detect lines (HT actually has Canny operators inside) and additionally requires estimation of parameters. We use the MoG model for estimating the probability of a VP. In contrast, [12] uses a Markov Model to estimate the probability of the location of the vanishing point and reduce false alarms, and its inference was done using the Viterbi algorithm, which increases its computational complexity/time. We implemented and evaluated the Markov Model according to [12] as a baseline experiment. Our MoG results are presented in Sect. 5.

Door Detection. We propose a computationally-efficient door detection algorithm that is based only on edge and geometric shape information. Our algorithm assumes doors are mostly imaged upright in the video which is a reasonable assumption for a stable MAV flight. It uses the near-vertical lines obtained from the previous VP detection as the initial door candidates, and then filters them based on their geometric scene properties (GSP) to detect doors.

A. Edge Based Door Detection

When a MAV is flying down a corridor, doors are usually aligned along the vertical direction in each frame. Using this assumption, we first select the edges that are close to the vertical direction, allowing some small angular deviation of α degree (for example $\pm 10°$). In addition, the door line candidates are filtered according to their lengths and locations of their lower and upper end points. In typical indoor situations, the image location that divides the left and right walls of a corridor is at the vanishing point and is usually close to the image center. For simplicity, we detect doors separately on the left side and right side of this location. For detecting doors on the left side, only edges in the leftmost portion of the image are considered. In our approach, we used the leftmost fraction $\lambda\%$ of the image width (e.g., 35 % of the image width). A similar approach is used for detecting doors on the right side.

The initially filtered vertical edges are further merged by connecting short edges to form longer lines. Only edges with similar distances to the middle point of the left-most image border and similar edge orientations are merged, which reduces the number of candidate door lines and saves computation. This process generates longer vertical lines that may better represent the candidates of door lines. These candidate lines are further validated by a series of filtering (validation) processes described below to detect the true door that is closest to the MAV on each side of the corridor:

(1) If the candidate line intersects the top or bottom of the image, but the intersection is outside of the image, it is not a valid candidate and removed.

(2) The distance from the vanishing point to the candidate door line is calculated. If the door line is too close to the vanishing point (e.g., 10 % of image width), we assume that it does not belong to a door and is removed.

(3) Each candidate line is paired with another candidate line to form a candidate door. If the distance between a pair of lines (i.e., the candidate door's width) falls outside of a typical door width (e.g., 15 % to 35 % of image width), this candidate door is invalid.

(4) We use the intensity profile of the candidate door and its adjacent sides to further filter door hypotheses even further. Specifically, the image area around the pair of door lines is split into three parts as seen in Fig. 3: the portion to the left of the door (A), the door (B), and the portion to the right of the door (C). An intensity profile is calculated for these parts. As illustrated by Fig. 3, a horizontal line is crossed over the door lines and the pixels along the horizontal line are used to calculate the intensity profile. One can see that the horizontal line is split into three segments by the pair of door lines. The mean and variance of intensities within each segment are calculated. The candidate pair of door lines is filtered according to their intensity profile and the prior knowledge of the corridor. For example, if the door is open, its mean intensity is often lower than the other two parts for a white wall. In addition, the intensity variance within an open door area is often larger than the other two parts because of the cluttered objects in an office. This filtering process needs to be adjusted for different scenarios because the prior knowledge may change.

Fig. 3. Calculation of the intensity profile at the adjacent areas of a candidate door.

(5) The above step 4 is processed repeatedly at multiple vertical locations for more reliable decision, as shown with the red lines in Fig. 3, along the candidate door lines. The number of iterations that pass the intensity profile filtering is accumulated. If the total number is above a predefined threshold (e.g., 50 %), the pair of door lines is declared as valid and a door is reported to be detected.

In the current implementation, we only find a single door at each side of the corridor, which usually corresponds to the door closest to the MAV. However, the above process can be applied iteratively to find more doors.

B. Template Matching for Door Detection

The edge-based door detection method described above requires the vanishing point to validate door hypotheses. Therefore, it can only be applied when a vanishing point is successfully detected. However, there might be cases where neither a vanishing point is detected nor are door hypotheses generated due to low contrast caused by illumination changes and motion. In such cases, we need other methods to detect doors. One such method is to apply template matching based on previously detected doors. When a vanishing point and door are successfully detected in a frame, we extract the narrow image patch around the door's boundary as a template and store such templates in a buffer. This buffer is updated online and it automatically clears old templates after a few frames (e.g., 5 frames). When a new door is detected, its template is added to the buffer.

When no vanishing point is detected or no door hypotheses are generated in a new frame, the latest template in the buffer is used to compare against the new image frame to find a matching image patch. This is achieved using a standard template matching approach based on normalized correlation [13]. If the matching score is above a predefined threshold (e.g., 0.95), a door line that looks similar to the previously detected door line is detected. The previously detected door line is then updated with the new location.

The above template matching method can generally find the new location of a recently detected door line. However, the template buffer size must be carefully controlled because template matching may fail when the actual door is already too far from the previously detected location. In such cases, the door hypothesis generated by template matching should be rejected. When no door is detected for a long period, the template buffer will become empty and no template matching can be applied.

C. Temporal Filtering of Detected Doors

The door detection generated by all processes discussed above may still be noisy, considering the large variations and clutter in typical indoor settings. To further reduce errors, we use a temporal filtering to ensure that the detections are temporally consistent. The filtering process checks if a door is continuously detected in each frame within a predetermined temporal window. If the door is detected in most of the frames within this window (e.g., 3 out of 5 frames), it is considered a valid detection. Otherwise it is rejected as a valid door. This temporal filtering is an optional process for refining door detection.

Our GSP-based door detection approach depends only on the geometric scene properties of door edges (e.g., intensity profile, door width, relative location of the top/bottom points of door lines) instead of complex structures like door corners [2, 5] or door knobs [3], which are sometimes not observable in the image. Since our approach only requires the commonly and reliably occurring features of a door (i.e., the two long edges) to generate door hypotheses, it is readily applicable to real-world settings. The validation of door hypotheses is based on these general geometric properties. For corridor scenarios, our algorithm also relies on the generally reliable vanishing point position to filter door candidates. In addition, we have added a template matching process to deal with situations where the vanishing point is not available or no door hypotheses are found. Finally, our algorithm is efficient because it re-uses the edges detected by the VP detection step. The only additional computation is the selection of near-vertical line pairs and validation of the selected pairs. Since we limit the width of the door with respect to its height, the complexity of the door candidate selection and validation is on the order of the number of long vertical lines. In typical corridor scenarios, there are a few dozen such long near-vertical lines and the validation process usually requires less than 10 ms. The VP detection and door detection together can run at about 20 fps on a standard quad-core laptop (Sect. 4), which is fast enough for a normal speed MAV to navigate through the corridor.

3 Control Subsystem

Our control strategy combines a visual sensor for primary navigation with side-mounted sonar sensors to avoid obstacles that fall outside of the field of view. Unlike other approaches for MAV control in the prior art that incorporate optic flow to dictate MAV direction [7, 10], our approach combines feature matching from the front-facing camera with additional override inputs from the lateral sonar sensors. While computationally inexpensive, optical flow approaches to control often exhibit drift in both position and scale. This can present problems in the indoor environment, where even small errors can lead to collision and loss of control. Instead, we employ a feature-based approach wherein we receive a target and track its features across frames. The system (MAV) centers the target in the field of view and then proceeds in its direction, turning as necessary to maintain the correct heading. This straightforward approach minimizes the overall calculation required to visually navigate the MAV toward the target and avoids the need for complicated learning strategies [8] or additional external sensors [9].

A detailed block diagram of our control methodology is illustrated in Fig. 4. The system can also receive input from the front-facing camera and user input via the graphical user interface by clicking the mouse on a location in the field of view. Additionally, a sonar system is used to allow the MAV to determine its distance from obstacles to its right and left, which are out of the view of the camera. Based on the location of the target in camera coordinates relative to the center of the frame, the system will send a small control command to the MAV. The system immediately captures another frame of video and issues subsequent commands in an iterative, closed-loop manner. This continues until the MAV has arrived at the target or the system has lost the target. The sonar sensors run in a parallel control loop that has priority over the main control loop and can interrupt commands from the visual input and issue their own in order to avoid collision.

The standard operation of the system is to detect a vanishing point in a corridor and move the MAV towards it. The vanishing point is computed from the video frames of the front-facing camera using the algorithm discussed in Sect. 3. Alternatively, if the MAV is not in a corridor, the user may use the graphical user interface to choose a specific location in the scene or the location can be computed automatically based on

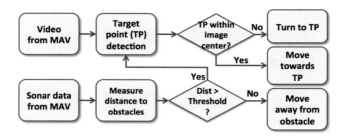

Fig. 4. Block diagram of our control strategy of indoor MAV flight.

heuristics (e.g., a landmark or a corner table) to which the MAV should fly. In either case, the input to the system is a location in camera-coordinates for a specific frame. Once a set of target coordinates has been selected, image information is extracted around its location for a given radius. Feature keypoints are extracted using a local feature algorithm (such as SIFT [14] or SURF [15]) for this image region. When a significant number of keypoints from the target region correspond with keypoints from the next frame, the target location can be imprinted on the next frame and further calculations can be made.

For each iteration of the control algorithm, the system computes the difference between the camera coordinates of the target with the center of the frame; the resulting vector indicates the direction that the MAV must move in order to acquire its target. Steering commands are issued if the horizontal difference between the target and the center of the image exceeds a given threshold. When the target is centered, the MAV will proceed forward until the originally slight discrepancy between the target and center of the image frame becomes large and another steering correction is necessary.

The control system contains sonar sensors to avoid collisions with obstacles to the sides of the MAV. Prior to each iteration of the control algorithm, the system checks the status of the sonar sensors. If the sonar detects that the MAV is too close to an obstacle, the system automatically ignores inputs from the vision system and takes immediate action to avoid the obstacle; this is a simple command to move the MAV in the direction opposite of the obstacle.

4   Experiments

The detection performance of our perception subsystem was evaluated using a number of corridor images under different lighting and viewing conditions. The perception and control subsystems described above were implemented on an offboard laptop computer and tested with a low-cost MAV quadrotor (AR.Drone2.0). We used the AR.Drone's existing front camera as our visual sensor and then added three sonar sensors for obstacle avoidance. The perception software for vanishing point and door detection was integrated with other capabilities that we have additionally developed for sensor data capturing and transmission, basic flight commands (move forward/backward, left/right, turn left/right, and others) and perception visualization, along with our control algorithm to perform autonomous corridor flying tests (see Fig. 1).

A. Evaluation of Vanishing Point and Door Detection

We tested our VP detection on 2470 test images of 720 × 480 resolution captured in different corridors using the AR.Drone camera (e.g., Figs. 5 and 6). A different set of 120 images was manually annotated with the VP to first train our MoG model. Our VP detection algorithm achieved 97.1 % true positive detection rate (recall) with 3.0 % false positive rate (alarms). In addition, we also implemented a baseline approach based on a Hidden Markov Model (HMM) by following [12] to generate the probability of VP detection through the standard Viterbi inference. This approach performed poorly compared to our approach with 88.4 % true positive detection rate and 5.1 % false positive rate. As described previously in Sect. 2, our computational complexity is lower than this baseline approach (Table 1).

Fig. 5. (Left) An example of vanishing point and door detection during a MAV flight test in a corridor. The cyan lines are the boundaries of one detected door. (Right) Another example of vanishing point and door detection. The cyan (left) and purple (right) lines are the boundaries of the detected doors. The sonar feedbacks (front, left, right) are visualized in the blue plots (Color figure online).

Table 1. Performance evaluation of vanishing point detection

	Training images	Testing images	True positive rate (recall) %	False positive rate (%)
Use MoG	120	2470	97.1	3.0
Use HMM	120	2470	88.4	5.1

We additionally evaluated our door detection algorithm on multiple MAV flights in a building corridor with multiple doors. The corridors have a total of 23 doors, about evenly distributed on the left and right sides. The average flying speed of our MAV was about 2 m/s. Table 2 summarizes the door detection performance of our algorithm with 10 runs of MAV flight in each direction for a total of 20 runs. We achieved 91.3 % true positive detection rate, with false positive and false negative rates of 4.3 % and 4.4 %, respectively. The main reasons for false positives were large windows, which look similar to doors, and strong vertical shadows on the wall.

Table 2. Performance evaluation of door detection algorithm in a corridor with 23 doors for 20 full runs of MAV flight.

Total # of doors (23 doors over 20 flights)	True positive rate (%)	False positive rate (%)	False negative rate (missed detections) (%)
460	91.3	4.3	4.4

It should also be mentioned that doors were usually hard to detect during fast flight, such as instances where the speed exceeded 5 m/s. The motion blur in such cases significantly reduced the contrast of door lines and made them harder to detect. In addition, the rolling motion of the MAV invalidated the assumption that the door lines were near-vertically oriented, causing increased missed detections and false alarms.

B. Indoor Flight Navigation

In our flight testing, the offboard laptop receives and processes video and sonar sensor data from the MAV to determine whether it should shift/turn or fly forward to the target/destination point (i.e., vanishing point, door location, or user-clicked location). If the destination point is not in the central region of the image, the MAV must to turn/shift to the left or right so that it is centered. If the destination point is not detected, the hovering command is sent to the MAV. These perception and control processes run on a laptop (with Intel® Core™ i7-2860QM CPU at 2.5 GHz running the Ubuntu Linux system). Three USB-powered sonar sensors (LV-MaxSonar-EZ1) were installed on the drone hull in the front and two sides as the proximity sensors. The transmission of sonar readings from the drone to the laptop was done by a Zigbee (XBee DigiMesh 2.4) connection. We developed custom software to interface to the AR.Drone SDK for control and camera data capture. With all sub-systems integrated and running together, we could achieve about 15–20 frames per second of visual processing on the laptop without specific optimization in a single-threaded implementation. The sonar-based update rate was even higher (e.g., >30 fps). Each time the perception module processes the sensor data, the system sends the best control command to the MAV (hover, turn left/right, move left/right). We were able to conduct multiple test flights in corridors at speeds of approximately 2 m/s speed without collision, while detecting the vanishing point and doors reliably.

Figure 5 shows two examples of our vanishing point and door detection results under different lighting conditions during MAV flight. The detected door lines are highlighted with colored lines. The probability in the upper left corner is the computed confidence of the detected vanishing point. We observed that the doors close to the camera were detected reliably. As can be seen in these examples, the

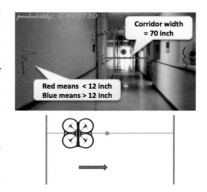

Fig. 6. Sonar-based collision avoidance. When the MAV is close to the wall, the control strategy is to move it away from the wall towards the corridor center.

illumination conditions, background, object scales and viewpoints are quite different, and yet our perception algorithms perform accurately. This is due to the fact our algorithms do not rely on any special assumptions about lightning conditions and viewpoints.

The corridors in our test environment were typically 70 inches wide, while the MAV hull is approximately 20 inches wide. This provides a 25-inch margin from each sidewall when the drone is flying in the center of the corridor. As seen in Fig. 6, if any of the side sonar distance measurements falls below a given threshold (e.g., 12 inches), then our control algorithm commands the drone to move away from the wall. This control process for moving away from the wall has a higher priority than moving forward to the destination point.

5   Conclusion and Discussion

This work describes efficient visual perception algorithms to accurately detect vanishing point and doors in indoor environments. It also describes a control and planning strategy for collision-free navigation. The key advantages of our approach are: (1) fast and efficient perception algorithms that can run onboard in power-constrained applications and platforms, e.g. MAVs; (2) robust perception algorithms that perform accurately in spite of challenging environmental conditions (e.g., changing illumination, viewpoints, appearance) due to the use of invariant geometric features; and (3) a priority-based planner and control strategy that combines outputs from the multimodal perception subsystem to navigate and avoid obstacles.

A next step for us is to port the perception-control coupled system to an onboard processing kit and evaluate its onboard performance for MAV flight. The door detection capability will also be used in a global exploration strategy to move through open doors and progressively explore the environment. We will also investigate the use of additional modalities such as 3D range sensor to improve door detection and detect moving obstacles for reactive control.

Acknowledgement. This material is based upon work supported by Defense Advanced Research Projects Agency under contract numbers W31P4Q-08-C-0264 and HR0011-09-C-0001. The views, opinions, and/or findings contained in this material are those of the authors and should not be interpreted as representing the official views or policies of the Department of Defense or the U.S. Government. Approved for Public Release, Distribution Unlimited.

References

1. Forney, G.D.: The Viterbi algorithm. Proc. IEEE **61**(3), 268–278 (1973)
2. Yang, X., Tian, Y.: Robust door detection in unfamiliar environments by combining edge and corner features. In: Proceedings of the 3rd Workshop on Computer Vision Applications for the Visually Impaired (CVAVI) (2010)

3. Hensler, J., Blaich, M., Bittel, O.: Real-time door detection based on AdaBoost learning algorithm. In: Gottscheber, A., Obdržálek, D., Schmidt, C. (eds.) EUROBOT 2009. CCIS, vol. 82, pp. 61–73. Springer, Heidelberg (2010)
4. Lee, J.-S., Doh, N.L. Chung, W.K., You, B.-J., Youm, Y.: Door detection algorithm of mobile robot in hallway using PC-camera. In: International Symposium on Automation and Robotics in Construction (2004)
5. Murillo, A.C., Košecká, J., Guerrero, J.J., Sagüés, C.: Visual door detection integrating appearance and shape cues. Robot. Auton. Syst. **56**(6), 512–521 (2008)
6. Parrot AR.Drone 2.0. http://ardrone2.parrot.com/
7. Ranft, B., Dugelay, J.-L., Apvrille, L.: 3D perception for autonomous navigation of a low-cost MAV using minimal landmarks. In: Proceedings of IMAV (2013)
8. Soundararaj, S.P., Sujeeth, A.K., Saxena, A.: Autonomous indoor helicopter flight using a single onboard camera. In: Proceedings of IROS (2009)
9. Eckert, J., German, R., Dressler, F.: On autonomous indoor flights: high-quality real-time localization using low-cost sensors. In: Proceedings of IEEE International Conference on Communications (ICC) (2012)
10. Briod, A., Zufferey, J.-C., Floreano, D.: Optic-flow based control of a 46 g quadrotor. In: Proceedings of IROS (2013)
11. Canny, J.: A computational approach to edge detection. IEEE Trans. Pattern Anal. Mach. Intell. **8**(6), 679–698 (1986)
12. Bills, C., Chen, J., Saxena, A.: Autonomous MAV flight in indoor environments using single image perspective cues. In: Proceedings of ICRA (2011)
13. Brunelli, R.: Template Matching Techniques in Computer Vision: Theory and Practice. Wiley, Hoboken (2009). ISBN 978-0-470-51706-2
14. Lowe, D.: Object recognition from local scale-invariant features. International Conference on Computer Vision, Corfu, Greece, pp. 1150–1157 (1999)
15. Bay, H., Tuytelaars, T., Van Gool, L.: SURF: speeded up robust features. In: Leonardis, A., Bischof, H., Pinz, A. (eds.) ECCV 2006, Part I. LNCS, vol. 3951, pp. 404–417. Springer, Heidelberg (2006)

Victim Detection from a Fixed-Wing UAV: Experimental Results

Anurag Sai Vempati[(✉)], Gabriel Agamennoni,
Thomas Stastny, and Roland Siegwart

Autonomous Systems Lab, ETH Zurich, Zürich, Switzerland
avempati@ethz.ch
http://www.asl.ethz.ch/

Abstract. This paper outlines a method to identify humans from a low-altitude fixed-wing UAV relying on various visual and inertial sensors including an infrared camera. The work draws inspiration from the need to detect victims in disaster scenarios in real-time, providing needed aid to rescue efforts. Such work can also be easily employed for surveillance related applications. We start by pointing out various challenges from camera imperfections, viewpoint, altitude, and synchronization. We provide a pipeline to efficiently fuse thermal and visual aerial imagery for robust real-time detections. Confident detections are tracked across various frames and the real-time GPS locations of the victims are conveyed. Performance of our detection algorithm is evaluated in a real-world victim detection scenario from an autonomous fixed-wing aircaft.

1 Introduction

Search and Rescue is a widely researched field owing to it's numerous applications in disaster scenarios. With increasing autonomy of Unmanned Aerial Vehicles (UAVs) and camera imaging technologies, it is now possible to scan large areas in a very short time and perform perception algorithms on-board at high rates with close to zero human intervention. Such technology also enables surveying in-accessible regions and hostile terrains, contributing significantly to the task of dispatching rescue efforts.

Visual spectrum cameras have been extensively used on UAVs, however, analyzing these images at high rates requires very robust algorithms to deal with various difficulties posed due to the size of objects of interest, motion blur, and viewpoint, to name a few. Detecting humans from a UAV cruising at an altitude of 50–100 m requires very high resolution cameras and the ability to quickly detect objects occupying few tens of pixels in area. On the other hand thermal cameras offer an advantage in such cases which makes it easier to narrow down the search space to hotter objects. But thermal cameras have their own limitations like low Signal-to-Noise Ratio (SNR), white-black/hot-cold polarity changes, and halos that appear around very hot or cold objects [1]. We propose a technique that best utilises the pros of either cameras using a sensor fusion technique.

© Springer International Publishing Switzerland 2015
G. Bebis et al. (Eds.): ISVC 2015, Part I, LNCS 9474, pp. 432–443, 2015.
DOI: 10.1007/978-3-319-27857-5_39

Various works leverage on hotspot techniques to quickly narrow down the potential areas to further process but they usually perform poorly in case of fast moving cameras like on UAVs. Many techniques rely on some kind of classifier to detect presence of a human in each of these potential areas. [1–4] use some kind of variant of cascade of boosted classifiers introduced by Viola and Jones [5] - which basically involves a series of weak classifiers each better than the previous one. [6] shows performance of various feature based classifiers trained on thermal data collected across wide variation in temperature, altitude and camera movement. According to their work, Histogram of Gradients (HOG) feature based classifier in conjunction with a particle filter tracker was found to perform the best.

Sensor fusion and multi-modal image registration is a well explored field. But most of the works involving registration of infrared and visual camera images like [7–10] involves image processing techniques that rely on feature extraction, edge detection, segmentation etc. Though very robust, such techniques can be quite time-consuming for real-time applications. On the other hand, [4] uses camera intrinsics and ground planarity assumption to estimate relevant part of visual camera image for stationary victims. This method has it's own limitations due to additional criteria enforced on targets' positions and the environment being scanned. In our work we make use of camera extrinsics and use techniques from multi-view geometry to get one-to-one correspondence between infrared and visual images.

In this paper we describe an algorithm that can efficiently detect stationary victims while autonomously scanning large areas using a fixed-wing UAV equipped with various sensors including visual and thermal cameras and an on-board computer to perform real-time computations. We will briefly outline individual components involved and provide results on a field experimental test.

2   Victim Detection Pipeline

Detecting humans from an altitude of 50–100 m with a camera of limited resolution poses a very challenging problem. Basic blocks of our pipeline are mentioned in the following sub-sections and in the next section we illustrate a real-case scenario.

2.1   Background Subtractor

Figure 1 shows a human as seen in false-color rendering of the thermal camera image at an altitude of about 70 m. At this scale, the humans occupy less than 50 pixels ($<0.02\,\%$) in an image of 640×512 resolution. An exhaustive search for such a tiny object of interest is very time consuming. We propose a Background Subtractor that returns regions of interest (ROI) and narrows down the search space considerably, thus enabling real-time detection. The foreground here is defined as a part of the image whose temperature differs quite significantly from it's surroundings. Since, humans are usually hotter (in winter) or colder (in summer) than the surroundings, we propose a technique that adaptively adjusts

2 threshold values (t_{low}, t_{high}) based on the surrounding pixel intensities. All the pixels with intensities less than t_{low} or greater than t_{high} are considered as foreground pixels.

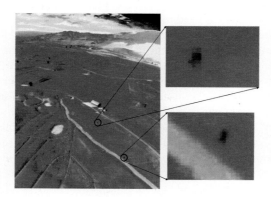

Fig. 1. A sample image recorded from the FLIR thermal camera at an approximate altitude of 70 m. Two humans are pointed out.

We employ a very simple sliding window based approach to estimate adaptive thresholds. The image is divided into various overlapping blocks and the adaptive thresholds t_{low}^b, t_{high}^b for each block b is evaluated as the higher and lower quantiles of the Gaussian models fitted to the pixel intensities of the block. Now, the threshold values at each pixel location (i, j) are chosen as weighted average of all the block thresholds t_{low}^b, t_{high}^b if the pixel belongs to block b. The weights are chosen inversely proportional to the pixel's distance from the block's center. A segmentation map (segmap) is then generated by thresholding the infrared image.

Figure 2 shows two sample images collected at different altitudes, temperature and times of the day and the corresponding segmaps. A blob detection algorithm [11] is used to find blobs of desired size, thus providing ROIs to search for humans. For the UAV scenario, we further narrow down the search space by considering only the blobs whose area lies in certain range that is calculated at every time-step based on the camera specifications, mounting, UAV's IMU pose and GPS position. This approach is computationally much cheaper and is not affected by the fast moving camera. It provides much faster and better results at real-time than many conventional approaches.

2.2   Human Classifiers

The ROIs obtained from Background Subtraction are exhaustively searched for presence of a human. For this, we use a HOG feature based learning classifier (which was found to be optimal [6]) to classify the extracted patch into human

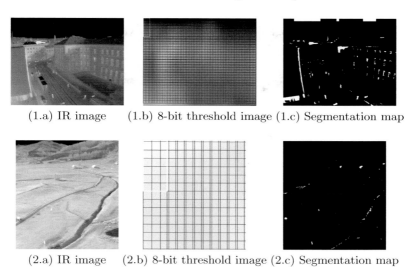

(1.a) IR image   (1.b) 8-bit threshold image (1.c) Segmentation map

(2.a) IR image   (2.b) 8-bit threshold image (2.c) Segmentation map

Fig. 2. Sequence 1 is captured in the night time at an altitude of 20 m above the ground with some crowd. Sequence 2 was captured from a UAV on a cold winter morning at an altitude of 60 m above the ground. (a) infrared image, (b) 8-bit image of pixel-wise threshold $t_{high}(i, j)$ (grid of blocks is overlaid and a sample block is marked in white on the top-left) and (c) segmentation map depicting foreground pixels.

or non-human category. The training data is obtained by manually annotating sequences from different datasets using vBBToolbox [12]. HOG features are extracted on the image patches at multiple scales to generate a descriptor. A Support Vector Machine (SVM) classifier is trained using these image patch descriptors. Type of SVM optimization problem, kernel type and other parameters are optimized using K-fold cross validation. Figure 3 shows performance of few chosen configurations.

Figure 3 shows Precision-Recall (PR) curves for the performance of the classifier on a validation dataset. The above method is then repeated to train a classifier for grayscale image patches. A C_SVC [13] optimization problem with a Linear kernel type and Radial Basis Function (RBF) kernel type are found to be two top performing configurations for the infrared classifier. The C_SVC optimization problem with a Linear kernel is found to be the optimal configuration for the grayscale classifier.

2.3   Infrared-Grayscale Image Fusion

Objects in infrared images rarely have discerning features since the temperature tends to be uniform across the objects. Therefore, using a classifier on infrared patches would be unreliable, since it only has the shape information of the object in a very small patch. Since in all our flights using the UAV we collected both infrared and grayscale images, a fusion technique is proposed to get a one to one

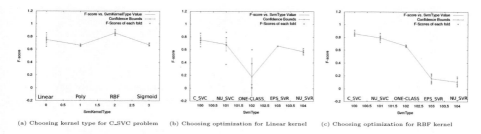

(a) Choosing kernel type for C_SVC problem   (b) Choosing optimization for Linear kernel   (c) Choosing optimization for RBF kernel

Fig. 3. Performance of infrared SVM classifier with different configurations of optimization problem and kernel type. It's found to be optimal to choose (a) RBF kernel type for C_SVC problem, (b) C_SVC problem for Linear kernel type and (c) C_SVC problem for RBF kernel type.

relationship between pixel locations of infrared and grayscale images. This will help us in extracting both infrared and grayscale image patches corresponding to the ROIs obtained from Background Subtraction which can help the classifier in producing more reliable results.

We use Kalibr [14] to calibrate the infrared and grayscale camera intrinsics and extrinsics for the sensorpod mounting. We use a radial-tangential model to estimate camera distortion parameters. An april tag pattern is used to estimate grayscale camera intrinsics. A thick matte paper with checkerboard pattern under illumination of a bright light is used to calibrate infrared camera intrinsics. Now the infrared and grayscale camera setup can be treated as a Stereo pair and standard image rectification techniques [15] can be used to evaluate the necessary projection matrices. At this point, all the epipolar lines are parallel to image edge (horizontal if horizontal rectification was employed, vertical otherwise). For our particular setup, vertical stereo was chosen since the camera centers were aligned vertically on the UAV. Now that the disparities of the image are in one axis, it's fast and easy to find pixel correspondences using robust feature matching between the images. Pseudo-code of the image fusion algorithm is shown in Algorithm 1.

Once a one-to-one correspondence is obtained between grayscale and infrared images, the patches corresponding to ROIs are passed on to individual human classifiers and pairs of patches resulting in a high cumulative classifier score are retained for probable presence of a human.

Figure 4 shows results from the fusion technique. In Sequence-1 infrared and grayscale cameras were both facing in the direction of flight and are in 25° nadir configuration with a translation of 3.5 cm between the camera centers. In Sequence-2, the infrared camera was in 25° from nadir configuration while grayscale camera was in 50° from nadir configuration with a translation of 3.5 cm between the camera centers. The results show reasonable overlap accuracy.

Figure 5 shows improvement to the PR curves after image Fusion on a dataset collected from a UAV cruising at 60 m above the ground. Close to 140 % improvement (10 % vs. 24 %) in the area under curve can be seen. Since, we have various

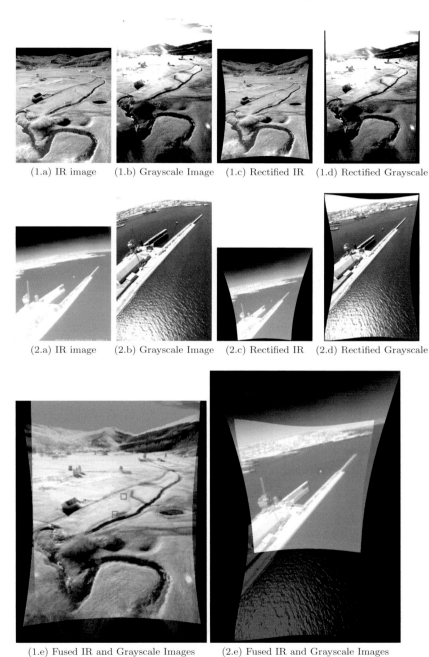

(1.a) IR image (1.b) Grayscale Image (1.c) Rectified IR (1.d) Rectified Grayscale

(2.a) IR image (2.b) Grayscale Image (2.c) Rectified IR (2.d) Rectified Grayscale

(1.e) Fused IR and Grayscale Images (2.e) Fused IR and Grayscale Images

Fig. 4. The top 2 rows show 2 sequences of original images obtained from (a) infrared, (b) grayscale cameras and (c, d) the resulting images post rectification. The last row shows fusion of both images (e) to visualize the quality of overlap. Red boxes highlight 2 humans in Sequence-1 and a boat in Sequence-2. As can be seen, the correspondence between infrared and grayscale images is reasonably good.

methods to eliminate false positives, as mentioned in next section, we operate our classifier on the higher recall point. Also the fact that very few of the false positives are of a same object recurring across several frames plays a vital role in suppressing them.

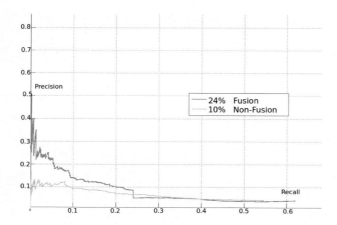

Fig. 5. Precision-Recall curves of the pipeline on a high-altitude (60 m) data collected from a UAV, before and after image fusion.

3   Experimental Results

Inspired by real-world search and rescue initiatives, we designed a representative field trial utilising our full aerial deployment and victim detection pipeline. The experiment involves the placement of multiple "victims" throughout a field over which our rapidly deployable, hand launched, fixed-wing test platform, equipped with multi-camera carrying sensorpod, must survey in a timely manner and record GPS coordinates of victims seen during the flight. In the following sections, we present the setup, enabling tools, and results.

3.1   Platform Description

All experiments are performed on a light-weight, low-altitude, and hand-launchable fixed-wing UAV, Techpod, with an integrated forward-facing sensor pod, see Fig. 6. Techpod weighs 2.65 kg, has a wingspan of 2.6 m, and flies at a nominal airspeed of 14 ms^{-1}. The UAV also consists of various telemetry, inertial modules and employs an open-source Pixhawk PX4 Autopilot [16] system. The sensorpod is equipped with rigidly connected infrared (FLIR Tau 2) and grayscale (Aptina MT9V034) cameras with oblique field-of-view, IMU (Analaog Devices ADIS16448), Skybotix VISensor [17] and a computer with an Intel Atom processor.

Algorithm 1. Image Fusion Algorithm

Require: Given a stream of ROS images I_{IR}, I_G obtained from calibrated Infrared and Grayscale camera sources with intrinsics K_{IR}, K_G and radial-tangential distortion models D_{IR}, D_G, the following algorithm specifies steps performed to un-distort images and register pixels of one image with the other. The algorithm makes use of camera rotation and translation extrinsics $[R_G^{IR} \mid t_G^{IR}]$ between the cameras to evaluate rectification transforms T_{IR} and T_G. Robust fast feature (F_{IR}, F_G) matching is further employed to evaluate post-rectification offsets O and obtain one-to-one mappings M_{IR} and M_G between rectified images I_{IR}^{new}, I_G^{new} that satisfy $I_{IR}^{new}(x,y) = I_{IR}(M_{IR}(x,y))$ and $I_G^{new}(x,y) = I_G(M_G(x,y))$

1: **procedure** IMAGEFUSION
2:     # Load pre-evaluated camera intrinsics and extrinsics.
3:     $\{K_{IR}, D_{IR}, K_G, D_G, [R_G^{IR} \mid t_G^{IR}]\} \leftarrow$ loadCalibration(**calibration.yml**)
4:     # Generate new optimal camera intrinsics that will ensure maximum possible number of pixels of both images are retained after un-distortion, in destination image size same as the original image.
5:     $K_{IR}^{new} \leftarrow$ getOptimalNewCameraMatrix(K_{IR}, D_{IR}, sizeof(I_{IR}))
6:     $K_G^{new} \leftarrow$ getOptimalNewCameraMatrix(K_G, D_G, sizeof(I_G))
7:     # Obtain single optimal camera matrix K^{new} that ensures all the pixels from both the images are retained post un-distortion.
8:     $K^{new} \leftarrow$ getOptimalNewCameraMatrix(K_{IR}^{new}, K_G^{new})
9:     # Compute rectification transforms for each camera in the calibrated stereo pair using OpenCV Stereo Rectification.
10:    $\{T_{IR}, T_G\} \leftarrow$ stereoRectify($K_{IR}, D_{IR}, K_G, D_G, [R_G^{IR} \mid t_G^{IR}]$)
11:    **while** incoming ROS messages **do**
12:        # Read ROS infrared image message.
13:        **if** incoming message topic == infrared image topic **then**
14:            $I_{IR} \leftarrow$ getImage($message$)
15:        **end if**
16:        # Read ROS grayscale image message.
17:        **if** incoming message topic == grayscale image topic **then**
18:            $I_G \leftarrow$ getImage($message$)
19:            # Evaluate transformation maps using computed destination camera matrix and the stereo rectification transforms.
20:            **if** mappings not evaluated **then**
21:                $M_{IR} \leftarrow$ initUndistortRectifyMap($K_{IR}, D_{IR}, T_{IR}, K^{new}$)
22:                $M_G \leftarrow$ initUndistortRectifyMap(K_G, D_G, T_G, K^{new})
23:            **end if**
24:            # Apply offsets to transformation maps using robust feature matching.
25:            $F_{IR} \leftarrow$ goodFeaturesToTrack(I_{IR})
26:            $F_G \leftarrow$ goodFeaturesToTrack(I_G)
27:            $O \leftarrow$ findOffsets(F_{IR}, F_G)
28:            $\{M_{IR}, M_G\} \leftarrow$ applyOffsets(O, M_{IR}, M_G)
29:            # Apply rectification transform on the images.
30:            $I_{IR}^{new} \leftarrow$ remap(I_{IR}, M_{IR})
31:            $I_G^{new} \leftarrow$ remap(I_G, M_G)
32:        **end if**
33:        # Remap the ROIs into rectified image and extract overlapping patches $(I_{IR}^{patch}, I_G^{patch})$ in both the images corresponding to the new ROIs.
34:        $ROI^{new} \leftarrow$ remapRect(ROI, M_{IR})
35:        $I_{IR}^{patch} \leftarrow I_{IR}^{new}(ROI^{new})$
36:        $I_G^{patch} \leftarrow I_G^{new}(ROI^{new})$
37:    **end while**
38: **end procedure**

Fig. 6. Techpod UAV with integrated sensing pod.

3.2 Full Coverage Aerial Inspection

To ensure victims are seen by the cameras' fields of view at some point during the flight, we employ a fast inspection planning algorithm which guarantees full coverage of a mesh-model world (obtained from GIS data) to ensure all ground is scanned for victims. The algorithm, details in [18], employs an iterative viewpoint re-sampling technique, using randomized sampling, to provide viewpoint configurations that maintain full coverage while simultaneously searching for the best route that visits them all, using the Lin-Kernighan heuristic [19]. Nonholonomic vehicle constraints and obstacle avoidance are further considered in the path search using a boundary value solver in combination with the Rapidly-Exploring Random Tree (RRT*) motion planner [20]. The path generated (see Fig. 7(a)) for this experiment ensured coverage of $52,800\,\mathrm{m}^2$ with a flight time of just under two minutes.

3.3 Estimating Victim GPS

For estimating the GPS positions of detections, we assume the ground as a plane. Using camera intrinsics and detection's pixel location, we find the directional vector in which the object might be lying. The point where this vector hits the ground plane can be found using GPS, IMU and camera-IMU extrinsics data which finally gives victim's GPS location.

Figure 7(b) shows the detected victim GPS positions, actual groundtruth victim location and the path followed by the UAV. The rest of the detections include unregistered humans in the explored area and some false positives. By enforcing a criterion - similar to [4] - that requires a particular GPS position to be detected consistently within the duration of the time that particular spot is observed, most of the false positives are eliminated and the set of true detections are grouped together and registered as victim locations. False positives are further eliminated by boosting the confidence of detections that are closer to the GPS location of the point where principal axis of camera hits the ground plane. The registered victims are marked in circles. As can be seen in the figure, we have all three successful detections and another detection whose status is not known. Figure 7(c) is a plot of GPS position estimate errors for all the three registered victims. The GPS estimates are converted to Universal Transverse Mercator (UTM) coordinates using World Geodetic System (WGS84) standard [21] to represent the error in meters. The errors obtained are a cumulative of errors in groundtruth GPS measuring device, calibration imperfections, GPS to/from UTM conversions, images' resolution, planarity assumptions, and sensor imperfections (GPS, IMU), making it quite uncertain how much of the error can be attributed to the imperfections in our algorithm. Nevertheless, the errors we obtained more than suffice to serve our goal.

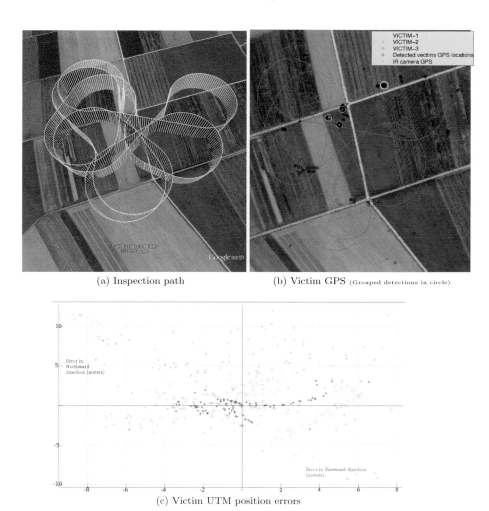

(a) Inspection path   (b) Victim GPS (Grouped detections in circle)

(c) Victim UTM position errors

Fig. 7. Full coverage field test detection results.

4   Conclusions and Future Work

In the future, we plan to have adaptive estimation of various parameters that we currently set manually. Computing HOG descriptors for each ROI is quite time consuming as well. Instead, using a weak classifier with high recall could reduce computation times, as our pipeline has the ability to deal with classifier false positives. We also plan to track the detections in UTM coordinates to have consistent detections of moving targets. Use of geographic elevation data would also be a way to remove the planar ground assumption currently implemented and improve detection location estimations. In all, we have shown, as a proof of concept, that human victim detection from upwards of 50–60 m via a fast moving fixed-wing aircraft is possible. Further, we have demonstrated the algorithm's real-time capacity for victim detection and localisation in a real-world search scenario.

Acknowledgements. This work was supported by the European Commission projects ICARUS (#285417) and SHERPA (#600958) under the 7th Framework Programme.

References

1. Wang, W., Zhang, J., Shen, C.: Improved human detection and classification in thermal images. In: 17th IEEE International Conference on Image Processing (ICIP 2010), pp. 2313–2316. IEEE (2010)
2. Davis, J.W., Keck, M.A.: A two-stage template approach to person detection in thermal imagery. In: null, pp. 364–369. IEEE (2005)
3. Treptow, A., Cielniak, G., Duckett, T.: Active people recognition using thermal and grey images on a mobile security robot. In: IEEE/RSJ International Conference on Intelligent Robots and Systems (IROS 2005), pp. 2103–2108 (2005)
4. Rudol, P., Doherty, P.: Human body detection and geolocalization for UAV search and rescue missions using color and thermal imagery. In: 2008 IEEE Aerospace Conference, pp. 1–8. IEEE (2008)
5. Viola, P., Jones, M.: Rapid object detection using a boosted cascade of simple features. In: Proceedings of the 2001 IEEE Computer Society Conference on Computer Vision and Pattern Recognition, CVPR 2001, vol. 1, pp. I-511–I-518. IEEE (2001)
6. Portmann, J., Lynen, S., Chli, M., Siegwart, R.: People detection and tracking from aerial thermal views. In: IEEE International Conference on Robotics and Automation (ICRA 2014), pp. 1794–1800. IEEE (2014)
7. Toet, A., van Ruyven, L.J., Valeton, J.M.: Merging thermal and visual images by a contrast pyramid. Opt. Eng. **28**, 287789 (1989)
8. Heo, J., Kong, S., Abidi, B., Abidi, M.: Fusion of visual and thermal signatures with eyeglass removal for robust face recognition. In: Conference on Computer Vision and Pattern Recognition Workshop, CVPRW 2004, p. 122 (2004)
9. Istenic, R., Heric, D., Ribaric, S., Zazula, D.: Thermal and visual image registration in hough parameter space. In: 14th International Workshop on Systems, Signals and Image Processing, 2007 and 6th EURASIP Conference focused on Speech and Image Processing, Multimedia Communications and Services, pp. 106–109 (2007)

10. Nandhakumar, N., Aggarwal, J.: Integrated analysis of thermal and visual images for scene interpretation. IEEE Trans. Pattern Anal. Mach. Intell. **10**, 469–481 (1988)
11. nán, C.C.L.: cvBlob. http://cvblob.googlecode.com
12. Dollár, P.: Piotr's Computer Vision Matlab Toolbox (PMT). http://vision.ucsd.edu/~pdollar/toolbox/doc/index.html
13. Cortes, C., Vapnik, V.: Support-vector networks. Mach. Learn. **20**, 273–297 (1995)
14. Furgale, P., Rehder, J., Siegwart, R.: Unified temporal and spatial calibration for multi-sensor systems. In: IEEE/RSJ International Conference on Intelligent Robots and Systems (IROS 2013), pp. 1280–1286. IEEE (2013)
15. Fusiello, A., Trucco, E., Verri, A.: A compact algorithm for rectification of stereo pairs. Mach. Vis. Appl. **12**, 16–22 (2000)
16. Meier, L., Honegger, D., Pollefeys, M.: PX4: a node-based multithreaded open source robotics framework for deeply embedded platforms. In: IEEE International Conference on Robotics and Automation (ICRA) (2015)
17. Skybotix AG (2015). http://www.skybotix.com/
18. Bircher, A., Alexis, K., Burri, M., Oettershagen, P., Omari, S., Mantel, T., Siegwart, R.: Structural inspection path planning via iterative viewpoint resampling with application to aerial robotics. In: IEEE International Conference on Robotics and Automation (ICRA 2015), pp. 6423–6430 (2015)
19. Lin, S., Kernighan, B.W.: An effective heuristic algorithm for the traveling-salesman problem. Oper. Res. **21**, 498–516 (1973)
20. Karaman, S., Frazzoli, E.: Incremental sampling-based algorithms for optimal motion planning. CoRR abs/1005.0416 (2010)
21. NGI Agency: World geodetic system (1984). http://web.archive.org/web/20120401083859/earth-info.nga.mil/GandG/wgs84/index.html

Autonomous Robotic Aerial Tracking, Avoidance, and Seeking of a Mobile Human Subject

Christos Papachristos[1]([✉]), Dimos Tzoumanikas[2],
Kostas Alexis[3], and Anthony Tzes[1]

[1] University of Patras, Patras, Greece
{papachric,tzes}@ece.upatras.gr
[2] Imperial College London, London, UK
dimosthenis.tzoumanikas14@imperial.ac.uk
[3] University of Nevada, Reno, Reno, USA
kalexis@unr.edu

Abstract. This paper presents a methodology to achieve Robotic Aerial Tracking of a mobile – human – subject within a previously-unmapped environment, potentially cluttered with unknown structures. The proposed system initially employs a high-end Unmanned Aerial Vehicle, capable of fully-autonomous estimation and flight control. This platform also carries a high-level Perception and Navigation Unit, which performs the tasks of 3D-visual perception, subject detection, segmentation, and tracking, which allows the aerial system to follow the human subject as they perform free unscripted motion, in the perceptual – and equally importantly – in the mobile sense. To this purpose, a navigation synthesis which relies on an attractive/repulsive forces-based approach and collision-free path planning algorithms is integrated into the scheme. Employing an incrementally-built map model which accounts for the ground subject's and the aerial vehicle's motion constraints, the Robotic Aerial Tracker system is capable of achieving continuous tracking and reacquisition of the mobile target.

1 Introduction

In this era of autonomous robots, Unmanned Aerial Systems (UASs) have a crucial role to play out: having fully autonomous robots capable of augmenting/surpassing the limitations of ground-based mobile or stationary systems [1,2], and additionally endowed with ample physical capabilities [3–5], perception abilities [6], and intelligence [7], holds great potential for involvement with civilian applications [8]. For the majority of case-studies UASs are evaluated in performing navigation within static maps [9]; within dynamic environments however [10] – such as economically active and/or populated areas, or in the context of dynamic missions [11] – such as search & rescue missions, and with a possibility for lack of prior knowledge regarding their structure, the

© Springer International Publishing Switzerland 2015
G. Bebis et al. (Eds.): ISVC 2015, Part I, LNCS 9474, pp. 444–454, 2015.
DOI: 10.1007/978-3-319-27857-5_40

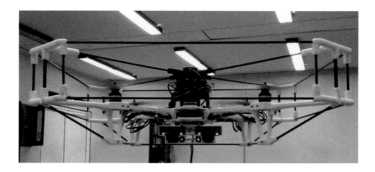

Fig. 1. The autonomous Robotic Aerial Tracker system.

autonomous robot must be capable of perceiving & monitoring dynamic entities, in order to at least avoid them, or – more interestingly – track them.

This work presents the methodology for the implementation of an Robotic Aerial Tracker system (Fig. 1), relying on the integration of several achievements in the field of small-scale robotic design, flight control, autonomous state estimation [12], and environment perception & incremental map-building and collision-free navigation. The developed system is capable of identifying human subjects, and of performing visual spatio-temporal tracking onboard the aerial robot in real-time, while having the additional capacity as a mobile agent to track the subject as it freely roams. Moreover, the robotic agent builds a representation of the map wherein the scenario takes place, enabling it to conduct collision free-navigation as it executes the tracking mission. Even more, the applied approach allows tracking of the subject even after losing visible contact – e.g. when it moves behind occlusions, as the trajectory generation is exploited in order to re-acquire the subject, accounting for the motion constraints introduced by the incrementally built environment model.

The article is structured as follows: In Sect. 2 the system framework is analyzed. In Sect. 3 the handling of the subject-tracking scenario is presented. Finally, Sect. 4 gives indicative experimental results, and Sect. 5 concludes the article.

2   Robotic Aerial Tracker – System Framework

The complete Robotic Aerial Tracker system is a high-end UAS, distinguishable into two main subsystems integrated into a single unit as illustrated in Fig. 2, namely:

2.1   Autonomous Unmanned Aerial Vehicle

The UAV subsystem is a custom-developed quadrotor aircraft, designed to accommodate all necessary sensorial and computational equipment onboard,

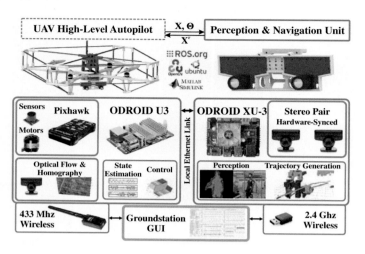

Fig. 2. Robotic Aerial Tracker UAS framework

along with a protective outer frame to prevent catastrophic contact with the environment – structures or people since it is intended to operate in a cluttered environment with dynamic human presence. It is an autonomous aerial robot, carrying a high-end suite that consists of:

(*i*) A Pixhawk ARM M4-based autopilot with Inertial Measurement Unit and Attitude Control functionality. It initially acts as a low-level bridge, interfacing with various low-level external sensors (e.g. altitude sonar), gathering internal sensor measurements (e.g. barometer, accelerometers, gyroscopes, magnetometer), interfacing with the motor Electronic Speed Controllers, and performing attitude and heading state estimation based on Extended Kalman Filtering. Secondarily, it implements a Gain-Scheduled Attitude Control scheme at a $\simeq$400 Hz rate; it hence also acts as a regulation (or remote-piloting) control fallback solution in case the top-level navigation controller hangs up.

(*ii*) An ODROID-U3 Quad-Core ARM A9-based miniature-sized computer running on Linux, which acts as a high-level autopilot. It performs 6-Degrees of Freedom (DoF) pose estimation onboard, achieved via data fusion between the measurements & state estimations collected by the low-level Pixhawk unit over a USB interface – at $\simeq$400 Hz, and a set of custom-developed Homography & Optical Flow calculation algorithms running at >100 Hz, which are conducted based on the video feed from a ground-pointed PS3 Eye camera which operates at 125 fps. Moreover, this autopilot level implements a translational Explicit Model Predictive Control structure, operating at 100 Hz as well, and drives the Pixhawk unit with the generated attitude & orientation reference commands. This framework offers agile maneuvering, control and state constraint guarantees, and an overall high-quality trajectory control performance, while it is noteworthy that the multi-core architecture provides seamless, almost real-time – multi-threaded – performance for all the necessary processes.

(iii) A 433 Mhz command and telemetry module, in order to maintain an isolated wireless link to the UAV (interfaced with the high-level autopilot) and avoid interferences with the 2.4 Ghz spectrum, and a remote direct-piloting receiver module (interfaced with the low-level autopilot with control override priority) for safety purposes.

2.2 Perception and Navigation Unit

The PNU is a high-level autonomy module, consisting of a $3D$ perception sensor and a computational module, in a compact form-factor in order to be accommodated onboard a small-scale UAS. More specifically, it is designed to be platform-agnostic, communicating over a local Ethernet connection with the UAV subsystem (the high-level autopilot unit), relying on the well-established – in the robotics community – Robot Operating System (ROS) framework. Its components are:

Fig. 3. The Hardware-Synchronized Stereo Pair. Despite the fan blade's motion – leading to blured captures – the paired images are synchronized as the CMOS chips' triggering is almost concurrent.

(i) A Stereo vision pair. It is a custom-built stereoscopic camera of a size and weight, based on two PS3 Eye cameras, which as mentioned are capable of providing video at 125 fps. The most important feature to note, is that the cameras' CMOS chips have their V^{sync}, F^{sin} pins exposed through board vias, thus allowing for hardware-synchronization of the pair, as per Fig. 3. Without this feature (which achieves latencies of the nsec-scale), and even at the highest possible capture rate, the produced results while the camera base is in motion – mounted on the UAV – are very inaccurate.

(ii) An ODROID-XU Octa-Core A15-based single board computer running a common Linux OS distribution. This module performs all the high-level computational tasks required for the Robotic Aerial Tracker's intended operation, namely stereo image pair rectification & Sum of Absolute Differences Block Matching (SAD-BM)-based disparity estimation for depth perception, implementation of the subject detection and visual tracking algorithms, incremental environment map building, and collision-free navigation & trajectory generation. As noted, the PNU communicates – onboard – bi-directionally with the UAV's high-level autopilot, allowing real-time access to the aerial vehicle's 6-DoF pose and various command parameters, and the forwarding of trajectory-reference commands.

(*iii*) A 2.4 Ghz WiFi module, which is used to forward data per request (video stream, the detected subject's visual contour, operational state of tracking scenario, generated trajectories) to a Qt-developed groundstation Graphic User Interface (GUI).

3   Perception and Navigation for Robotic Aerial Tracking

The algorithmic implementation of the perception scheme implemented for Robotic Aerial Tracking consists of two main schemes which are executed concurrently, namely a *Singular Hull Segmentation* process and a *Viewable Aspect Features* detection one. An elaboration of these can be researched at [13]; an overview of their respective functionalities is given as follows:

3.1   Subject Perception

The *Singular Hull Segmentation* (SHS) algorithm employs the $3D$-vision data as derived from the Stereo sensor, and performs spatial segmentation of the tracked subject from the rest of the environment. Its implementation as a parallelized process gives it a very fast rate of execution onboard the PNU as it essentially performs graph-connections of the surrounding – and expanding – $3D$-points based on an adaptive distance threshold, eventually finding the singular hull which determines the subject as a spatially-distinct entity within its view-field.

The *Viewable Aspect Features* (VAF) algorithm employs feature detection in order to determine if the – same – subject is being reliably tracked. It maintains a list of the key-aspects (considered in a similar sense as key-frames) of the subject that have been captured as it – or the UAS – freely moves. To this purpose, it operates on the subset of the image (contour-mask) corresponding to the tracked subject; however, if for a number of iterations a reliable percentage threshold of features belonging to one of the already captured key-aspects is not found, the tracking scheme considers that the subject is lost, and the VAF process begins searching in the entire scene for visual re-localization.

These processes operate in closed-loop; with the SHS algorithm being the fast-loop process – requiring a min/max execution time $\in [5, 15]$ ms approximately, depending on the subject's occupancy of the image. It is also noted that the SAD-BM process requires an additional 25 ms. Compared to the VAF process – which even when executed only on the subset of the image corresponding to the subject contour-mask, requires a minimum of 60 ms mainly due to the use of SURF feature detection – the SHS's execution efficiency, along with its capacity to perform $3D$-tracking on its own if the subject moves within a relatively flat environment, are noteworthy.

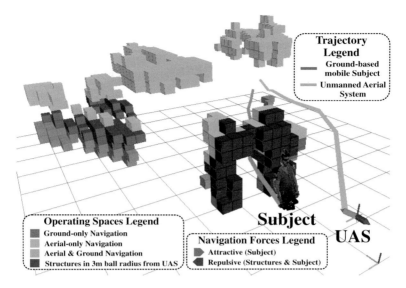

Fig. 4. The Robotic Aerial Tracking scenario Subject/UAS navigation spaces distinction, Environment/Subject Forces, and Collision-free Navigation primary concepts (Color figure online).

3.2   Navigation for Robotic Aerial Tracking

A challenging scenario for the experimental evaluation of the framework must consist of: (a) an non-flat cluttered environment populated with various static structures, and (b) a human subject, which will perform free unscripted motion at the ground level.

The PCL^{env} segmented environment-pointcloud is used to drive an Octree [14]-based incremental representation of the environment, representing the mapped occupied and free space as voxels of edge-size $d_{subj} = 0.25$ m. This sizing is proportionate to a human subject's motion abilities, i.e. a human body can fit through a hole (in the worst-case) sized as such a voxel. The occupied voxels are distinguished according to their height level, again respectively to the subject's motion abilities. For a human subject, it is considered improbable that they may navigate (jump) over obstacles higher than 1 m. Therefore, an additional encoding is employed for the environment: (a) $n^O_{subj} \in \mathbf{N}^O_{subj}$ and $n^F_{subj} \in \mathbf{N}^F_{subj}$ are mapped occupied and free voxels within the subject's available 3D navigation space $\mathbf{N}_{subj} = \mathbf{N}^O_{subj} \bigcup \mathbf{N}^F_{subj}$ ($\mathbf{N}_{subj}[z] \in [0,1]$ m), and (b) $n^O_{uas} \in \mathbf{N}^O_{uas}$ and $n^F_{uas} \in \mathbf{N}^F_{uas}$ are mapped occupied and free voxels within the Robotic Aerial Tracker's available 3D navigation space $\mathbf{N}_{uas} = \mathbf{N}^O_{uas} \bigcup \mathbf{N}^F_{uas}$ ($\mathbf{N}_{uas} \in [0.5,3]$ m). Figure 4 indicates this $2 \times 3D$-spaces distinction, with $\mathbf{N}^O_{subj}$ marked by red/orange voxels, and $\mathbf{N}^O_{uas}$ marked with green/orange voxels (orange ones are the $\mathbf{N}^O_{subj} \bigcap \mathbf{N}^O_{uas}$ overlapping region).

3.3   Navigation While Subject is Tracked

While the subject is successfully being tracked by the algorithm, the PCL^{subj} segmented subject-pointcloud's median-point $\mathbf{X}^{subj} = \{x^{med}, \ y^{med}, \ z^{med}\}$ is used to represent its $3D$-position. Additionally, the total occupied environment voxels $\mathbf{N}^O$ are iterated, and given the UAS's $3D$-position $\mathbf{X}^{uas}$ and after employing ray-casting, the first ones crossed (the outer-border of the structures in the environment) within a $3D$-ball of radius $r_{ball} = 3$ m are determined. These are denoted as $\mathbf{N}^O_{ball}$ (illustrated in Fig. 4 with purple), and the $3D$-position of each i-th voxel's centroid is denoted as $\mathbf{X}^{O,i}_{ball}$.

In order to achieve mobile tracking of the subject along with obstacle avoidance, the concept of attractive/repulsive forces [15] is employed: $\mathbf{R}^{subj}_{attr}$ and $\mathbf{R}^{subj}_{rep}$ mark the subject's attractive (towards it for following) and repulsive (away from it to avoid crashing) forces respectively, and $\mathbf{R}^{env}_{rep}$ marks the $\mathbf{N}^O_{ball}$ sum repulsive force:

$$\mathbf{R}^{subj} = \mathbf{R}^{subj}_{attr} + \mathbf{R}^{subj}_{rep} = a^{subj}\left(\frac{|\mathbf{X}^{subj} \rightarrow \mathbf{X}^{uas}|^2}{r^2_{safe}} - \frac{r^2_{safe}}{|\mathbf{X}^{subj} \rightarrow \mathbf{X}^{uas}|^2}\right) \quad (1)$$

$$\mathbf{R}^{env}_{rep} = a^{env}\frac{1}{M}\sum_{i\in\mathbf{N}^O_{ball}} -\frac{r^2_{safe}}{|\mathbf{X}^{O,i}_{ball} \rightarrow \mathbf{X}^{uas}|^2}, \quad (2)$$

with (a) $\mathbf{R}^{subj}$ in (1) behaving nonlinearly, becoming zero at a "safe" distance from the subject $r_{safe}(=1.25\,\text{m})$, and negatively increasing much faster when the UAS approaches the subject, while (b) $\mathbf{R}^{env}$ is derived from the averaged repulsion forces of each occupied voxel within the $\mathbf{N}^O_{ball}$ ball-set (M marks the total respective voxels). The parameters a^{subj} and a^{env} are used to scale the magnitude of the subject's and any obstacles' presence, and are selected as $a^{subj} > a^{env}$ to reflect the significance of the subject tracking objective.

3.4   Navigation When Subject is Lost

When the subject is lost by the tracking algorithm, an assumption is made: It is possibly attempting to evade the UAS, and has initially moved to an occluded area of the map. Furthermore, the subject may attempt to move towards an open area away from the UAS to completely escape. Consequently, the known map's furthest open point (the center of a voxel within the tracked subject's available free motion space $\mathbf{N}^F_{subj}$), for which it is quicker for the subject to reach than it is for the Robotic Aerial Tracker, is considered as the subject's target waypoint $n^{F,target}_{subj}$. This is encoded as:

$$\mathbf{n}^{F,target}_{subj} \in \mathbf{N}^F_{subj} : \max(\mathbf{X}^{uas} \rightarrow \mathbf{X}^{F,target}_{subj}) \quad (3)$$
$$\mathbf{X}^{subj} \rightarrow \mathbf{X}^{F,target}_{subj} < \mathbf{X}^{uas} \rightarrow \mathbf{X}^{F,target}_{subj},$$

where $\mathbf{X}^{F,target}_{subj}$ marks the $\mathbf{n}^{F,target}_{subj}$ voxel centroid coordinates.

For trajectory generation – as followingly elaborated – a $3D$ Rapidly-exploring Random Tree-star (RRT^*) [16] algorithm is employed to derive an obstacle collision-free path, while asymptotically converging to the path that ensures minimum cost (in this context, the cumulative Euclidian distance).

Initially, a possible trajectory for the subject is computed. The occupied voxels $\mathbf{N}^O_{subj}$ are used to construct a set of manifolds wherein the RRT^* is not allowed to navigate. If such a path is found it is considered as the subject's minimum-distance escape trajectory (otherwise another possible escape point $\mathbf{n}^{F,target}_{subj}$ is searched).

Consequently, a trajectory for the UAS is computed. The occupied voxels $\mathbf{N}^O_{uas}$ are used to construct a set of manifolds wherein the RRT^* is not allowed to navigate. This time however, the edge-size is increased to $d_{uas} = 4 * d_{subj}$, as the UAV's frame diagonal is 0.9 m. This trajectory is forwarded to the UAS, while its orientation reference slides along – scans – the subject's possible trajectory. Throughout this searching-navigation phase, the scheme of Sect. 3 is constantly attempting to visually re-locate the subject, while the $3D$-environment perception algorithm provides map updates, and the environment repulsive force of (2) is used to avoid collision with environment structures which were not present (had not been discovered) at the time that the RRT^* trajectory was generated.

The two trajectories (red for the subject, green for the UAS) are intuitively illustrated in Fig. 4. It is noted that in this illustration, the two minimum paths are very different, not only because of the different initial configurations, but also due to the differentiated motion constraints (low-elevation navigation for the subject, larger obstacles but aerial navigation capacity for the UAS).

4   Experimental Studies

The presented experimental studies were used to evaluate the level of real-time functionalities achieved with respect to the objective of Robotic Aerial Tracking of human subject. Indicative video-sequences can be found at: https://youtu.be/fwEetEWwQtw, and at: https://youtu.be/W-gX29EDRpA, while the following illustrations – visualized in Rviz-ROS based on data recorded onboard the PNU during the mission execution – are used to point out the primary operational concepts:

Initially, the attractive/repulsive forces approach in Fig. 5 is examined, with regard to its utility in achieving continuous mobile Aerial Robotic Tracking at a safe distance from the human subject. As intuitively illustrated, the segmented subject pointcloud is separately visualized from the mapped environment voxelized representation – color-coded as given in Sect. 3.2. In a) a generalized pose indicating the repulsive (purple, generated by the occupied environment voxels which form the $\mathbf{N}^O_{ball}$ set and the subject's $\mathbf{X}^{subj}$ median-point), the attractive (red, produced only by the subject's $\mathbf{X}^{subj}$), and additionally their sum force vector $\mathbf{R}^{subj} + \mathbf{R}^{env}_{rep}$ (green) are illustrated. In b) a point where those forces are in near-equilibrium – and thus the UAS remains in stationary surveillance – is shown. Illustrations c) and d) indicate points where the subject moves away-from and towards the UAS, thus generating respectively a sum attractive and repulsive force, and producing a following - or - evasive action.

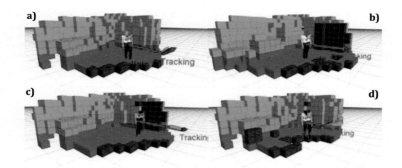

Fig. 5. Experimental results – Continuous mobile Robotic Aerial Tracking (Color figure online).

Figure 6 illustrates a sequence where the Robotic Aerial Tracker autonomously handles subject evasion – and its moving behind structural occlusions – and performs collision-free navigation until it relocates the human subject, relying on the previously elaborated framework.

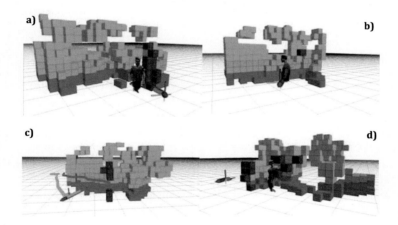

Fig. 6. Experimental results – Human subject evasion & re-localization. Experimental results – Human subject evasion & re-localization. Experimental results – Human subject evasion & re-localization (Color figure online).

In $a)$, it can be noticed that as the subject initially moves towards an environment structure, the UAS is attracted towards it. Phase $b)$ indicates the subject as it faces away from the UAS in order to distance itself and evade. At this point instant, the tracking scheme continues to achieve subject tracking, relying on the SHS and key-aspect VAF detection processes which operate in closed-loop, despite the fact that the viewed subject aspect is significantly different from the front-facing one in a. In c) and the tracking scheme has detected subject-loss –the subject has moved behind a structure within the environment – and RRT^*-based trajectory generation has been performed. It is noted that the respective trajectories are: (i) red for the subject's estimated target escape path, and (ii) green for the Aerial Robotic Tracker's collision-free $3D$-navigation. Finally in d), after following part of the reference trajectory and scanning along the target estimated path, the UAS eventually re-acquires the subject which has remained within an area previously occluded by part of the environment structure.

5 Conclusions

A methodology for autonomous Robotic Aerial Tracking of a human subject was developed and evaluated in a set of experimental case-studies. The complete framework relies on the proper integration of several achievements in the fields of robotic design and implementation, as well as the areas of autonomous perception and navigation, and is indicative of the high-level missions – with potentially dynamically evolving objectives – that can be carried out by the new generation of Unmanned Aerial Systems.

References

1. Grocholsky, B., Keller, J., Kumar, V., Pappas, G.: Cooperative air and ground surveillance. IEEE Robot. Autom. Mag. **13**, 16–25 (2006)
2. Zikou, L., Papachristos, C., Tzes, A.: The power-over-tether system for powering small UAVs: tethering-line tension control synthesis. In: Mediterranean Conference on Control Automation (MED), pp. 715–721 (2015)
3. Papachristos, C., Alexis, K., Tzes, A.: Technical activities execution with a tiltrotor uas employing explicit model predictive control. In: World Congress of the International Federation of Automatic Control (IFAC), Cape Town, South Africa, pp. 11036–11042 (2014)
4. Pounds, P.E.I., Bersak, D., Dollar, A.: Grasping from the air: hovering capture and load stability. In: IEEE International Conference on Robotics and Automation (ICRA 2011), pp. 2491–2498 (2011)
5. Alexis, K., Huerzeler, C., Siegwart, R.: Hybrid modeling and control of a coaxial unmanned rotorcraft interacting with its environment through contact. In: 2013 International Conference on Robotics and Automation (ICRA), Karlsruhe, Germany, pp. 5397–5404 (2013)
6. Weiss, S., Scaramuzza, D., Siegwart, R.: Monocular-slam based navigation for autonomous micro helicopters in GPS-denied environments. J. Field Robot. **28**, 854–874 (2011)
7. Lin, L., Goodrich, M.: UAV intelligent path planning for wilderness search and rescue. In: IEEE/RSJ International Conference on Intelligent Robots and Systems, IROS 2009, pp. 709–714 (2009)
8. Campoy, P., Correa, J., Mondragn, I., Martnez, C., Olivares, M., Mejas, L., Artieda, J.: Computer vision onboard UAVs for civilian tasks. In: Valavanis, K., Oh, P., Piegl, L. (eds.) Unmanned Aircraft Systems, pp. 105–135. Springer, Netherlands (2009)

9. Merz, T., Kendoul, F.: Dependable low-altitude obstacle avoidance for robotic helicopters operating in rural areas. J. Field Robot. **30**, 439–471 (2013)
10. Geyer, C., Singh, S., Chamberlain, L.J.: Avoiding collisions between aircraft: State of the art and requirements for UAVs operating in civilian airspace (2008)
11. Rudol, P., Doherty, P.: Human body detection and geolocalization for UAV search and rescue missions using color and thermal imagery. In: 2008 IEEE Aerospace Conference, pp. 1–8. IEEE (2008)
12. Papachristos, C., Alexis, K., Tzes, A.: Dual-authority thrust-vectoring of a tri-tiltrotor employing model predictive control. J. Intell. Robot. Syst. 1–34 (2015). http://dx.doi.org/10.1007/s10846-015-0231-1
13. Papachristos, C., Tzoumanikas, D., Tzes, A.: Aerial robotic tracking of a generalized mobile target employing visual and spatio-temporal dynamic subject perception. In: IEEE-RSJ International Conference on Intelligent Robots and Systems (IROS 2011), pp. 4319–4324 (2015)
14. Hornung, A., Wurm, K., Bennewitz, M., Stachniss, C., Burgard, W.: OctoMap: an efficient probabilistic 3D mapping framework based on octrees. Auton. Robots **34**, 189–206 (2013)
15. Nieuwenhuisen, M., Droeschel, D., Schneider, J., Holz, D., Labe, T., Behnke, S.: Multimodal obstacle detection and collision avoidance for micro aerial vehicles. In: European Conference on Mobile Robots (ECMR 2013), pp. 7–12 (2013)
16. Karaman, S., Frazzoli, E.: Incremental sampling-based algorithms for optimal motion planning. CoRR abs/1005.0416 (2010)

Inspection Operations Using an Aerial Robot Powered-over-Tether by a Ground Vehicle

Lida Zikou[1], Christos Papachristos[1], Kostas Alexis[2(✉)], and Anthony Tzes[1]

[1] University of Patras, Patras, Greece
`ece7668@upnet.gr`, {`papachric,tzes`}`@ece.upatras.gr`
[2] University of Nevada, Reno, Reno, USA
`kalexis@unr.edu`

Abstract. Within this paper, the utilization of an aerial robot that is powered-over-tether by a ground robot in order to execute long-endurance infrastructure inspection operations is described. Details on the key-concept of the Power-over-Tether system with the power bank of the aerial robot now being located on the ground vehicle are provided. To allow smooth flight of the aerial vehicle, a long power cable is wound onto a custom-developed base which is capable of releasing and retracting in a controlled manner. This design is exploited via a special controller that handles the coordinated motion of the two robots. Finally, a structural inspection planner is integrated into the system and benefitting by the on-board perception and mapping capabilities of the aerial robot autonomous 3D reconstruction of the desired objects is achieved. The overall framework is evaluated using experimental studies relevant to infrastructure inspection operations.

1 Introduction

Aerial robotics have gained great attention from the research and industrial community, with a significant number of technological and scientific breakthroughs having been achieved in order to tackle the challenges associated with their airborne nature. More specifically, the field of small-scale MAV design [1–3], the field of high-precision robust model-based control [4–6] and autonomous navigation/path-planning [7], the field of robotic manipulation [1,8–10], and the field of robotic - visual or otherwise - perception [11], accompanied by the development of small form-factor, high computational power embedded systems, have all seen major breakthroughs. This momentum has sparked a keen interest on employing small scale aerial robots in a multitude of civilian applications of a wide range, such as - and without being limited to - environment mapping [12] and precision agriculture [13], industrial infrastructure inspection [14,15], and disaster response missions [16].

Despite this progress, specific limitations remain. In the particularly widespread vehicle class of multirotor Micro Aerial Vehicles (MAVs), the challenge of achieving satisfactory levels of endurance is still not addressed. As a large

G. Bebis et al. (Eds.): ISVC 2015, Part I, LNCS 9474, pp. 455–465, 2015.
DOI: 10.1007/978-3-319-27857-5_41

Fig. 1. The combined aerial-ground robotic system that employs the power-over-tether mechanism during an infrastructure inspection test-case operation.

subset of the most critical real-life applications, such as infrastructure inspection or other camera-in-the-skye operations, require an endurance that exceeds the typically offered 15 min limitation, alternative approaches have to be proposed. This work overviews a system designed to overcome this challenge and provide a versatile framework for real-life infrastructure inspection operations. It implements and employs a method for remote powering of small-scale hovering MAVs inspired by recent advances in the field of tethered flight [17], via a ground-to-air Power-over-Tether (PoT) link which transfers power to the aerial vehicle and is carried on-board a ground robot, as illustrated in Fig. 1. A crucial requirement for such a system is to maintain the cable at non-slacking tension (to prohibit it from falling on the ground), while at the same time allowing the MAV to navigate with relative ease, and to do so in a locally-controlled way to retain operational safety in possible air-to-ground communication hang-ups or relative position estimation errors. With such a set-up at hand, a specialized structural inspection path-planning algorithm for MAVs is integrated and in combination with the perception capabilities of the aerial robot, it sets the basis for autonomous infrastructure operations. To evaluate the overall design, a relevant experimental study is conducted.

This paper is structured as follows: In Sect. 2, the PoT system is elaborated as per its design and principles of operation. In Sect. 3, the implemented control synthesis for its autonomous operation is presented. Section 4 overviews the employed inspection path planning strategy, while Sect. 5 presents results of the experimental studies. Finally, Sect. 6 concludes the article.

2   The Power-over-Tether Aerial-Ground Robotic System

The Power-over-Tether (PoT) mechanism that is carried by the ground robot and powers the aerial vehicle is presented in this Section.

2.1   Power-over-Tether System On-board the Ground Robot

As shown in Fig. 2, the main component of the PoT is the winder drum a), upon the central perimeter of which the power cable is wound. The winder drum is attached on one side on a ballbearing joint due to which, the PoT base rotates around the indicated b) Degree-of-Freedom (DoF)-1. Measurement of the winder drum's rotations is accomplished with the help of two parallel placed latching-action Hall-effect Sensors (encoder 1 in c)) which sense the presence of small neodymium magnets placed onto the perimeter of the winder-to-ballbearing support shaft. Also, it is noted that the cable is inserted into the winder drum at a hole, and exits from the support shaft at location d). As can be observed, cable exit d) supports the cable's two phases in parallel placement, hence any twisting occurs from d) and up to the external power source.

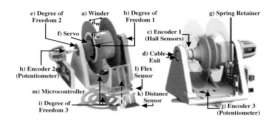

Fig. 2. The autonomous Power-over-Tether System base

The association between the free cable length L and the winder drum turns Ω can be estimated given the dimensioning of the respective elements. However, for a total cable length of 4 m, it can be approximated via a 2-nd order polynomial fitting based on measurements. This yields to the formula:

$$L \simeq L^{pol} \;=\; a_L \Omega^2 \;+\; b_L \Omega \;+\; c_L \tag{1}$$

In the proposed implementation, there is an additional DoF (marked as DoF-2) in e) , which exists due to the fact that the f) servo's shaft can rotate w.r.t. its body, which is attached onto a ballbearing joint in e). For a locked servo shaft, DoF-1 and 2 rotate coincidentally, otherwise relative rotation is possible. The servo is also attached onto the rigid PoT base body via a retainer-spring g). This allows DoF-2 to rotate, but more importantly it returns the servo and DoF-2 back to an original rotation, when the spring contracts. The rotary encoder 2 in h), implemented with a potentiometer measures the absolute deviation of DoF-2 from its original rotation (at zero spring elongation), and is used to estimate the spring's elongation. Additionally to the retainer-spring g), the servo rotation is limited up to an approximate maximum of 60° via a non-elastic retainer. This prohibits any plastic elongation of the Spring and allows "locking" of DoF-2 as well as passing the full torque of the servo onto the winder DoF-1.

The aforementioned implementation aims to achieve local "sensing" of the aerial robot's tendency to distance itself from the PoT base, or backtrack towards it: when the aerial vehicle moves away it pulls against the cable, and the mechanical tension force is transferred onto DoF-2, effectively elongating the retainer-spring. When the opposite maneuver happens, mechanical tension is decreased and the spring is compressed accordingly. This principle is used to achieve force-feedback control the PoT system via the actuation principle granted by the servo. Finally, the PoT base can also rotate w.r.t. to the world-frame **X**, as marked by DoF-3 in i) with a ballbearing joint and a fixed-base shaft guaranteering minimal friction, while the absolute Rotary Encoder 3 j) - also a potentiometer - yields the respective measurements.

A short-range infrared distance sensor marked as k) is used to measure the height of the cable at the exit point of the PoT base. It is placed at a low point to allow certain vertical distance from the cable exit, as its measurements are unreliable below a threshold of $\simeq 0.040$ m. An additional component is the Flex sensor marked as l), properly placed at the PoT base ground-to-air cable exit at a height which is marginally higher than the reliability threshold of the distance sensor k). This serves to detect cable contact which occurs before the too-short range limit of the distance sensor. Finally, an AVR-based microcontroller m) is attached onto the PoT base. This handles the acquisition of measurements from the onboard sensors, the estimation of all necessary values, and the computation of the necessary control actions. It is also responsible for the generation of the low-level control signals which are required to drive the servo actuator.

This overall PoT set-up is carried on-board the ground robot which also carries a large battery bank. Due to the high payload capacity of the ground robot, a set up 10 3-Cell LiPo batteries of 5000 mAh each is carried for the power needs of the aerial robot, therefore providing an overall endurance of approximately 2 h continuous flight.

2.2   Autonomous Quadrotor MAV

The autonomous quadrotor MAV dipicted in Fig. 3 is employed in this work. This aerial robot caries a complete sensor suite consisting of proprioceptive (Attitude and Heading Reference System, altitude-measuring sonar and LiDAR units) and exteroceptive (ground-pointed high capture rate Camera for visual odometry) sensors, and has high onboard computational power (a baseline ARM-M4 autopilot-Pixhawk, and a quad-core ARM-A9 high-level control unit-Odroid U3) allowing all state estimation processes to be performed onboard in real-time, based on a software implementation which achieves fusion of computer vision and inertial data [11,18,19]. Additionally, it incorporates a high-level perception module (the octa-core ARM-A15 Odroid XU3) which handles all non real-time critical computational tasks required for the desired operation, i.e. local communication with the Odroid U3 autopilot for access to the MAV's estimated 6-DoF pose, image acquisition from the onboard $RGB - D$ sensor (a MS Kinect), image rectification, RGB–Depth registration, 3D Point Cloud extraction, and

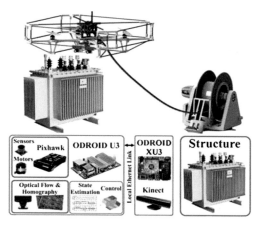

Fig. 3. The Powered-over-Tether quadrotor MAV employed for the inspection operation.

data recording. Finally, wireless communication with a groundstation for data accessing during the autonomous operation is achieved over a WiFi link.

Overall, the platform is capable of following a desired trajectory (given in the World-frame $\mathbf{X} = \{x,\ y,\ z\}$) as shown Fig. 3 without the need for remote control, as 6-DoF ego-motion estimation is achieved onboard in real-time, while concurrently capturing, processing and pose-annotating color and depth images of its environment via the use of an onboard $RGB - D$ sensor and its visual–inertial fusion pipeline), enabling the extraction of $3D$-PointCloud representations of the environment.

The aerial robot operates according to the well-established cascaded attitude and translation control methodology [20], and within the present work no knowledge of the fact that it is tethered to the ground is *intentionally* assumed in order to additionally demonstrate the plug-and-play functionality of the PoT system, i.e. a free-flight controller can be simply employed with any special adjustment.

3   Aerial-Ground Robot Motion Coordination

The power-tethered collaborative navigation methodology for the proposed autonomous inspection system is tackled by employing a leader-follower approach. Its specific implementation is mainly determined by the central objective of the Unmanned Robotic Team, namely to conduct aerial-based inspection operations. Hence, the aerial robot is considered to be the leader-robot, while the ground robot is the follower.

It should be noted, that the ground robot is a commercially-available unit, powerful enough to carry heavy payloads – and at a size that can easily accommodate large-sized ones, it is hence regarded as a realistic solution to the problem of adding mobility to the ground-located power supply for the system. As far as its motion abilities are concerned, it is considered along the lines of bi-wheeled ground robots [21]. It is capable of following nonholonomic trajectory commands [22] with precision, while its kinematics basis regarding its non-holonomic motion constraints is known and such that ensures path-following accuracy:

$$\mathbf{U}_G = \begin{bmatrix} u_G \, v_G \, r_G \end{bmatrix}^T \tag{2}$$

$$|\mathbf{U}_G| \leq \begin{bmatrix} u_G^{max} \, 0 \, r_G^{max} \end{bmatrix}^T, \tag{3}$$

where $\{u_G, \, v_G\}$ the longitudinal and lateral speed, and r_G its heading angle rate.

The aerial vehicle is considered in an isolated approach. Its trajectory-following flight control scheme does not incorporate any information about the fact that it operates in mechanical tethering to a ground mobile base; instead while conducting inspection tasks, it operates as it would in independent free-flight. It is hence important to ensure that at the ground level, the ground robot follows the aerial leader with consistency.

The Power-over-Tether system has a crucial role to play in this process. Not only does it provide prolonged endurance to the MAV, but it can also return a number of operational data as they are acquired in real-time during the autonomous mission execution. More specifically, the free cable length L, and the ψ^{PoT} PoT orientation (rotation angle) relative to the ground robot Body-Fixed Frame (onto which the PoT base is rigidly attached) are employed:

(i) The L cable length is used as a state that indicates a rough estimate of the current distance of the two vehicles, since the PoT system effectively achieves its retraction up to a point that does not impede the MAV's motion, but also ensures that cable slacking is minimal. By employing a reference value L^r, and driving the ground robot's translational velocity based on a Switched constrained Proportional Control scheme, mobile ground following of the aerial vehicle's tendency to distance itself – as it navigates towards its desired waypoints – is achieved:

$$u_G^r = K^L \, e_L, \quad e_L = \begin{cases} L^r - L, & \text{IF } |L^r - L| >= e_L^{min} \text{ AND } \overline{\mathscr{S}} \\ 0, & \text{Otherwise} \end{cases}, \tag{4}$$

where e_L^{min} a minimum error constraint that ensures that no back-and-forth ground vehicle translational motion is produced, and $\mathscr{S}$(top) a logical signal used to indicate that the ground vehicle should stop moving (potentially having reached a ground motion constraint or a minimum distance-to-be-maintained from the inspected structure). It is highlighted that even in this case, the PoT system continues to allow tethered navigation of the MAV up to its maximum cable length.

(ii) The ψ^{PoT} relative rotation-angle is used as a state that indicates a rough estimate of the current position vector between the two vehicles. In order to ensure that the ground robot follows the aerial leader in a straight line, the objective is regulation around a zero-value. A constrained Proportional Scheme is used here as well in order to drive the ground vehicle's rotation speed:

$$r_G^r = K^\psi\, e_\psi, \quad e_\psi = \begin{cases} -\psi^{PoT}, & \text{IF } |\psi^{PoT}| \geq e_\psi^{min} \\ 0, & \text{Otherwise} \end{cases}, \quad (5)$$

where e_ψ^{min} a minimum error constraint that ensures that no left-and-right ground vehicle translational motion is produced.

4   Overview of the Inspection Planner

The developed robotic system utilizes and interfaces a Structural Inspection Path-planner (SIP) that has to be capable of computing a complete coverage path for a given structure while accounting for any sensor and vehicle kinematic constraints as well as workspace constraints due to the tether limitations. Given a provided travel speed of the aerial robot, a feasible workspace, the yaw rate limitations of the robot and the Field-of-View (FoV) of the camera, the full coverage path is computed and timed at each step. Different SIP algorithms can be used, while within the framework of this work, the algorithm in [7] is employed. The algorithm considers a triangular mesh representation (open or close) of the structure and computes the path via an optimization method that alternates between two steps, namely a step that computes new viewpoints such as to reduce the cost-to-travel between itself and its neighbours and a step computes the optimal connecting tour. The specific planner, supports both rotorcraft as well as fixed-wing aerial vehicles and accounts for their motion model. Furthermore, its computational cost is particularly small due to the convexification of the viewpoint computation and the utilization of the Lin–Kernighan–Helsgaun (LKH) TSP solver for the tour computation step. Detailed formulation and computational analysis is included in [7], while demo scenarios and experimental datasets are available online.

5   Experimental Studies

The developed system was evaluated experimentally for the case of a 3-phase power-transformer structure based on a realistically-sized mockup model of an actual power transformer. The mesh model of this structure is illustrated in Fig. 4.

This mesh model was utilized by the employed SIP algorithm considering the following parameters and constraints: (a) a travel speed $V_T = 0.5\,\text{m/s}$, (b) yaw-rate constraint $\dot\psi^{max} = 0.25\,\text{rad/s}$, (c) FoV equal to $[60, 60]°$, (d) overall tether length of $4\,\text{m}$. Given these limitations, the derived inspection path is capable of covering the three main side faces as well as the top of the power transformer as shown in Fig. 5. Once this path is computed, the system commands the ground robot to keep a stationary position close to the infrastructure, while the aerial vehicle starts following the inspection trajectory. Using the pose-annotated

Fig. 4. The CAD model of the power transformer mockup used for the experimental evaluation of the coverage properties conducted by Power-over-Tether robotic system.

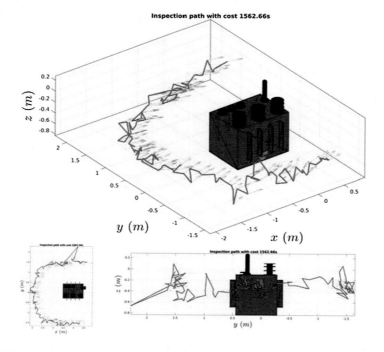

Fig. 5. The derived inspection path for the aerial robot in order to inspect the three main sides and top face of the power transformer mock-up.

images from the camera of the RGB–D sensor, the 3D reconstruction result shown in Fig. 6 is derived with the help of relevant software [23].

As shown the developed system was capable of performing high-quality inspection results due to the integration of the dedicated inspection planner and the sensing system on-board the MAV but also due to the intelligent design of the PoT mechanism which allows the vehicle to fly with limited disturbance. Furthermore, it is pointed out that although in most cases of free-flying aerial

Fig. 6. Reconstruction of the power transformer mock-up using the pose-annotated RGB images recorded with a period $T_r = 0.25\,\mathrm{s}$. This reconstruction was done using the Pix4D software.

robots of that scale, such an inspection mission is close to the battery limitations of the vehicle, this is indeed not the case for this robotic system. On the contrary, this set-up is ideal for inspecting multiple structures in an area as the ground robot can move and station in convenient multiple points, while the aerial robot can perform the corresponding subset of the inspection path while the overall endurance of the system is orders of magnitude larger compared to autonomously powered multirotor MAVs.

6   Conclusions

The development and utilization of an aerial robot that is powered-over-tether from a ground robot towards infrastructure inspection operations is overviewed and evaluated in this paper. The system is capable of long-term autonomous operation due to the prolonged endurance offered by the battery power bank carried by the ground vehicle and the advanced perception and navigation capabilities of the aerial robot. Building on top of these capabilities for autonomous navigation and self-localization, the combined robotic system was evaluated regarding its capacity to handle infrastructure inspection operations. To assist this need, a specialized structural inspection planner was integrated and the performance of the system was evaluated using experimental studies. The results indicate the capacity and potential of the overall set-up.

References

1. Alexis, K., Huerzeler, C., Siegwart, R.: Hybrid predictive control of a coaxial aerial robot for physical interaction through contact. Control Eng. Pract. **32**, 96–112 (2014)

2. Papachristos, C., Alexis, K., Tzes, A.: Design and experimental attitude control of an unmanned tilt-rotor aerial vehicle. In: 15th International Conference on Advanced Robotics, Tallin, Estonia, pp. 465–470 (2011)
3. Achtelik, M.W., Lynen, S., Chli, M., Siegwart, R.: Inversion based direct position control and trajectory following for micro aerial vehicles. In: IEEE/RSJ Conference on Intelligent Robots and Systems (IROS) (2013)
4. Alexis, K., Nikolakopoulos, G., Tzes, A.: On trajectory tracking model predictive control of an unmanned quadrotor helicopter subject to aerodynamic disturbances. Asian J. Control **16**, 209–224 (2014)
5. Alexis, K., Papachristos, C., Siegwart, R., Tzes, A.: Robust model predictive flight control of unmanned rotorcrafts. J. Intell. Robot. Syst., 1–27 (2015). http://dx.doi.org/10.1007/s10846-015-0238-7
6. Naldi, R., Gentili, L., Marconi, L., Sala, A.: Design and experimental validation of a nonlinear control law for a ducted-fan miniature aerial vehicle. Control Eng. Pract. **18**, 747–760 (2010). (Special Issue on Aerial Robotics)
7. Bircher, A., Alexis, K., Burri, M., Oettershagen, P., Omari, S., Mantel, T., Siegwart, R.: Structural inspection path planning via iterative viewpoint resampling with application to aerial robotics. In: IEEE International Conference on Robotics and Automation (ICRA), pp. 6423–6430 (2015)
8. Marconi, L., Naldi, R., Gentili, L.: Modelling and control of a flying robot interacting with the environment. Automatica **47**, 2571–2583 (2011)
9. Darivianakis, G., Alexis, K., Burri, M., Siegwart, R.: Hybrid predictive control for aerial robotic physical interaction towards inspection operations. In: IEEE International Conference on Robotics and Automation (ICRA 2014), pp. 53–58 (2014)
10. Alexis, K., Darivianakis, G., Burri, M., Siegwart, R.: Aerial robotic contact-based inspection: planning and control. Auton. Robots 1–25 (2015)
11. Papachristos, C., Tzoumanikas, D., Tzes, A.: Aerial robotic tracking of a generalized mobile target employing visual and spatio-temporal dynamic subject perception. In: IEEE/RSJ International Conference on Intelligent Robots and Systems, 2015 (IROS) (2015)
12. Oettershagen, P., Stastny, T., Mantel, T., Melzer, A., Rudin, K., Agamennoni, G., Alexis, K., Siegwart, R.: Long-endurance sensing and mapping using a hand-launchable solar-powered UAV. In: Field and Service Robotics, Toronto, Canada. Springer (2015, to appear)
13. Weiss, U., Biber, P.: Plant detection and mapping for agricultural robots using a 3d LIDAR sensor. Robot. Autonom. Syst. **59**, 265–273 (2011). Special Issue ECMR 2009
14. Burri, M., Nikolic, J., Hurzeler, C., Caprari, G., Siegwart, R.: Aerial service robots for visual inspection of thermal power plant boiler systems. In: 2012 2nd International Conference on Applied Robotics for the Power Industry (2012)
15. Metni, N., Hamel, T.: A UAV for bridge inspection: visual servoing control law with orientation limits. Autom. Constr. **17**, 3–10 (2007)
16. Rudol, P., Doherty, P.: Human body detection and localization for uav search and rescue missions using color and thermal imagery. In: 2008 IEEE Aerospace Conference, pp. 1–8. IEEE (2008)
17. Lupashin, S., D'Andrea, R.: Stabilization of a flying vehicle on a taut tether using inertial sensing. In: IEEE/RSJ International Conference on Intelligent Robots and Systems (IROS 2013), pp. 2432–2438 (2013)
18. Papachristos, C., Alexis, K., Tzes, A.: Model predictive hovering-translation control of an unmanned tri-tiltrotor. In: 2013 International Conference on Robotics and Automation, Karlsruhe, Germany, pp. 5405–5412 (2013)

19. Nikolic, J., Rehder, J., Burri, M., Gohl, P., Leutenegger, S., Furgale, P.T., Siegwart, R.: A synchronized visual-inertial sensor system with FPGA pre-processing for accurate real-time slam. In: IEEE International Conference on Robotics and Automation (ICRA). IEEE (2014)
20. Bouabdallah, S., Siegwart, R.: Full control of a quadrotor. In: 2007 IEEE/RSJ International Conference on Intelligent Robots and Systems, pp. 153–158 (2007)
21. Siegwart, R., Nourbakhsh, I.R., Scaramuzza, D.: Introduction to Autonomous Mobile Robots, 2nd edn. The MIT Press, Massachusetts Institute of Technology, Cambridge (2011)
22. Arvanitakis, I., Tzes, A.: Trajectory optimization satisfying the robot's kinodynamic constraints for obstacle avoidance. In: 20th Mediterranean Conference on Control Automation (MED 2012), pp. 128–133 (2012)
23. Pix4D: http://pix4d.com/

Autonomous Guidance for a UAS Along a Staircase

Olivier De Meyst, Thijs Goethals, Haris Balta, Geert De Cubber$^{(\boxtimes)}$, and Rob Haelterman

Royal Military Academy, Brussels, Belgium
geert.de.cubber@rma.ac.be

Abstract. In the quest for fully autonomous unmanned aerial systems (UAS), multiple challenges are faced. For enabling autonomous UAS navigation in indoor environments, one of the major bottlenecks is the capability to autonomously traverse narrow 3D - passages, like staircases. This paper presents a novel integrated system that implements a semi-autonomous navigation system for a quadcopter. The navigation system permits the UAS to detect a staircase using only the images provided by an on-board monocular camera. A 3D model of this staircase is then automatically reconstructed and this model is used to guide the UAS to the top of the detected staircase. For validating the methodology, a proof of concept is created, based on the Parrot AR.Drone 2.0 which is a cheap commercial off-the-shelf quadcopter.

1 Introduction

1.1 Problem Statement

Unmanned Aerial Systems (UAS) operating Beyond Visual Line Of Sight (BLOS) require autonomous navigation capabilities for finding their way and coping with obstacles. This is particularly true in indoor environments where the BLOS operating conditions naturally apply and where radio communication loss can easily occur. Notwithstanding these problems, there is a vast array of possible applications for UAS with autonomous indoor flight capabilities, like search and rescue [1] or indoor inspection [2]. One of the main problems these autonomous indoor UAS are confronted with are narrow passages with a highly 3-dimensional nature, like staircases. To tackle this problem, we propose in this paper a semi-autonomous navigation system which permits an UAS to detect, reconstruct and fly over a staircase. The final goal of this project is to incorporate this staircase-mounting capability into the navigation architecture of a hybrid unmanned ground + aerial system [3], enabling the vehicle to drive to a staircase and then fly over it.

1.2 Main Contributions and Relation to Previous Work

The work presented in this paper is based on the assembly of multiple state of the art algorithms in a novel system architecture geared towards an innovative

© Springer International Publishing Switzerland 2015
G. Bebis et al. (Eds.): ISVC 2015, Part I, LNCS 9474, pp. 466–475, 2015.
DOI: 10.1007/978-3-319-27857-5_42

application: autonomous stair-climbing with an unmanned aerial system (UAS). The main focus of this paper is therefore *not* on the (many) state of the art core technologies which form the basis of the navigation algorithm, but rather on the *novel* contributions in terms of system architecture (Sect. 2), staircase detection (Sect. 3) and integration of the staircase 3D model into the navigation approach (Sect. 4). All the different system modules were carefully evaluated and fine-tuned and the combination of the methods was then tested on different staircases, as described in Sect. 5.

In the literature, a lot can be found about control systems for UAS. The most popular are based on the Extended Kalman Filter (EKF). This is mainly because most robotics models are non-linear and the EKF happens to be the de facto standard in the theory of non-linear state estimation.

The authors chose to use two different Simultaneously Locating And Mapping (SLAM) algorithms in this paper. One algorithm uses visual features to help the state estimation with the localization of the quadcopter and the other algorithm generates a 3D point-cloud of the environment in near real-time. The former is based on Parallel Tracking and Mapping (PTAM) [4] while the latter is based on Large-Scale Direct Monocular SLAM (LSD-SLAM) [5] and is extensively modified to be compatible with the setup of this work.

This paper proposes a unique combination of detection algorithms in order to successfully detect a staircase. Detection of objects in general and staircases have already been subjects of many papers like [6–10]. Most of them are using depth data to detect or improve detection while visually exploiting the parallel nature of the stairs [10] or using object detection techniques [7].

This paper is based on the implementation of the algorithm in [6] where 2D edge detection is combined with 3D curvature analysis. Depth images are not obtained directly but aggregated via *LSD-SLAM* [5]. Next to 3D analysis and edge detection, visual object detection is applied on the images as described partly in [7]. Because the relatively long process times both processes are parallelized and synchronised.

In terms of application, the paper [11] is quite close to what is presented here, as it also proposes a methodology for staircase navigation. However, the methodology in [11] is based on 2D image processing. While requiring more intensive processing, the advantage of the methodology proposed in this paper (based on 3D reconstruction) allows for the potential use of the extracted 3D structural data for other state estimation, navigational or GUI processes.

An additional condition is the fact that the Parrot AR.Drone 2.0 was to be used for the development. The mandatory usage of such a small consumer drone results on the one hand in a very low cost but this type of quadcopter has on the other hand low quality sensors such as the camera which affect the performance of the algorithms and, consequently, the outcome of the staircase guidance system that is developed in this paper.

2   System Architecture

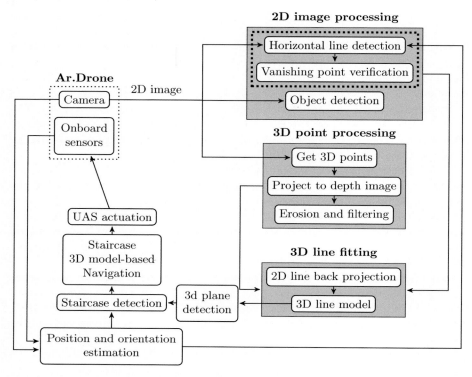

The system architecture is summarised in the diagram that is included above. The detection part is built from the ground up and can be subdivided in three parts: 2D detection, 3D detection and the 2D/3D data fusion strategy.

To detect the staircase, the authors chose to incorporate different approaches to obtain a maximal rejection of false positives (see Sect. 3). This, in the hypothesis that the different detection algorithms have very few or almost no false negatives. For the 2D detection, a Line Segment Detector (LSD) is implemented together with a self-implemented rejection criterion that is based on the vanishing points of the detected lines and the outcome of an object detector that is based on a cascade classifier, as explained in Sect. 3.1. In 3D, the generated point-cloud is projected to the drone's point of view into a depth image (as explained in Sect. 3.2) and subsequently a 3D line fitting is performed to search the 2D lines that were previously detected. The detected 3D lines, are then fitted to a 3D plane model. The size and angle of the plane with regard to the horizontal plane are calculated and should be typical for a staircase, as explained in Sects. 3.3 and 3.4. The 3D staircase model is subsequently used inside the navigation framework (which is based on the *tum_ardrone ros* navigation module) to guide the UAS towards the top of the staircase, as explained in Sect. 4. The actuation of the UAS by the navigation controller will cause new measurements of the on-board sensors, which will re-initiate the loop.

3   Staircase 3D Modeling

This paper combines the approach in [7] using common object detection techniques and the one in [10] where depth data is used to detect the stairs and improve the detection while visually exploiting the parallel nature of the stairs.

Firstly, a combination of 2D and 3D information is needed to successfully detect curbs or stairs. Secondly the basic idea behind the algorithm is based on the cascade classifier where each stage has a decent amount of rejection of false positives and and almost no false negatives.

The implementation of the 2D and 3D detection is not solely serial, as 2D detection and 3D detection can happen in parallel. The serial aspect is only in the data fusion stage and decision stage. In the following subsections, we describe the different steps of the detection algorithm.

3.1   2D Image Processing

Horizontal line detection. The only thing that is clearly visible and easily detectable is the edge of a single stairstep in an image. To detect this, the OpenCV implementation of the line segment detector developed in [12] is used on the complete image. A staircase is horizontal with respect to the ground. Using the information from the EKF, the horizon in the image can be determined. Only the lines that are parallel within tolerance to the estimated horizon are kept as shown in Fig. 1.

Object Detection. A general approach and a computational efficient technique to learn and afterwards detect distinct visual features is the cascade classifier using HAAR-like features [13]. This technique uses multiple simple classifiers to obtain a good performing classifier. The gain is found in the fact that a lot of candidates are already rejected in the early stages and do not need processing in the later more computational intensive stages. Since the first usage of cascaded classifier by [13] a lot of improvements have been proposed. A non-exhaustive overview and a new proposition can be found in [14]. In our implementation the HAAR-like features are used and the trainer is based on the ADABOOST algorithm that is already integrated in OpenCV.

Line Grouping. Only the line segments that intersect or the ones that are within the regions of interest are assumed to be stairs. This is done using an own adapted implementation of the Cohen Sutherland algorithm [15]. All the horizontal line segments within a region of interest are only meaningful if they are related to each-other. This is done by grouping the lines based on their slope or their vanishing point using an own implementation of the RANSAC algorithm.

3.2   3D Point-Cloud Processing

Collect and Projecting 3D Data. Analysis is done on a depth-image. Prior to obtaining this image, the 3D data must be gathered. This is done by projecting the data from the keyframes to the analysed camera frame.

Fig. 1. Line segment detector applied to a camera image (left) and the OpenCV object detection (right). The *blue line* is the estimated horizon, the *green lines* are within tolerance (8 deg) and the *red lines* were also detected by the line segment detector but are not within tolerance (Color figure online).

$$\mathbf{e} : \mathbb{R}^{w \times h} \to \mathbb{R}^{w \times h} : I_{i,j} \mapsto \mathbf{e}(I_{i,j}) = \min_{m=i-\mathbf{w}/2+0.5}^{i+\mathbf{w}/2+0.5} \left(\min_{n=j-\mathbf{h}/2+0.5}^{j+\mathbf{h}/2+0.5} (I_{m,n}) \right)$$

with

- $I_{m,n} = \begin{cases} I_{i,j} & \text{if } m \in [0,w] \text{ and } n \in [0,h] \\ 0 & \text{otherwise} \end{cases}$
- $w \in \mathbb{N}_0$ the width of the image and $h \in \mathbb{N}_0$ the height of the image
- $\mathbf{w}$ and $\mathbf{h}$ the width and height of the erode kernel

3.3   Combining 2D and 3D Detection

3D Line Fitting. The detected stairs have only a meaning if they can be located in 3D. Therefore, the line-segments are pointwise backprojected on the depth image. On these 3D points a robust fitting is done using built in functions from PCL. The default *SACMODEL_LINE* is used in combination with the built-in RANSAC. The inliers are kept for further processing.

3D Plane Fitting and Plane Angle. The built-in *SACMODEL_PLANE* model from PCL is used to define the detected plane. The four coefficients of the plane are its Hessian Normal form: $[n_x, n_y, n_z, d]$. A standard plane equation can be written as $ax + by + cz + d = 0$. The normal vector is defined as follows:

$$\hat{n} = (n_x, n_y, n_z) \Rightarrow \begin{cases} n_x = \frac{a}{\sqrt{a^2+b^2+c^2}} \\ n_y = \frac{b}{\sqrt{a^2+b^2+c^2}} \\ n_z = \frac{c}{\sqrt{a^2+b^2+c^2}} \end{cases} \tag{1}$$

To determine the dihedral angle between two intersecting planes, the dot product is calculated between the normalized vectors of the two planes:

$$\begin{aligned} a_1 x + b_1 y + c_1 z + d_1 &= 0 \\ a_2 x + b_2 y + c_2 z + d_2 &= 0 \end{aligned} \Rightarrow \cos\theta = \hat{n}_1 \cdot \hat{n}_2 = \frac{a_1 a_2 + b_1 b_2 + c_1 c_2}{\sqrt{a_1^2 + b_1^2 + c_1^2}\sqrt{a_2^2 + b_2^2 + c_2^2}} \tag{2}$$

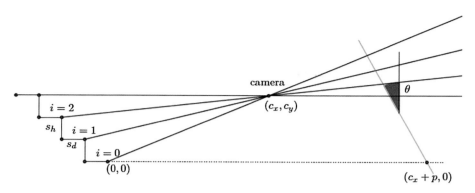

Fig. 2. Simplified staircase projection model.

3.4   Temporal Consistency by Agglomerative Clustering

To take a decision about where to find a staircase, clustering is used. Clustering is an unsupervised machine learning technique that groups relevant entities together. This can be based on any metric or derived value.

When using a clustering mechanism, the number of clusters that are needed can be set beforehand (K-mean clustering) or the clusters can be determined via a type of linkage (hierarchical clustering). Because the amount of clusters is a priori unknown, the latter was implemented.

The different detections are clustered together based on the euclidean distance in space as metric. To separate two sets from each other it must be described how they are linked. Different criteria exist but in this work the centroid linkage clustering is used $\|c_s - c_t\|$, with c_s and c_t the centers of cluster s and t.

Hierarchical clustering tries to establish a hierarchy between the different sets based on their metric and linkage. To create such an hierarchy two approaches exist: top-down and bottom-up. The top-down approach tries to iteratively divide the complete data set until every point has its proper set. The bottom-up approach iteratively groups different entities until all points are in one set. Here, the bottom-up approach is used, this is also called agglomerative clustering.

The clustering mechanism that is implemented here takes two arguments: type of cluster distance and the minimum distance between two clusters. These parameters are respectively set to the centroid cluster distance with a minimum of one meter in between two separate clusters.

3.5   Staircase Projection and Modeling

A 2D projection model of a staircase was considered here, as depicted in Fig. 2. In this model, a frontal view is assumed together with a point projection and yaw and roll are ignored.

The beginning of the staircase is situated at the origin $(0,0)$. The camera at coordinates $(c_x, c_y)[m]$ and the projection screen has a pitch angle θ. The dimensions of the staircase are characterized by the couple $(s_d, s_h)[m]$. The staircase can then be characterized by expressing the ratio between the observed distance d between the projected lines P_i of the staircase in the 2D image:

$$\frac{d_{P_i,P_{i+1}}}{d_{P_{i+1},P_{i+2}}} = \frac{c_x \cos\theta + 2s_d \cos\theta + c_y \sin\theta - 2s_h \sin\theta + s_d i \cos\theta - s_h i \sin\theta}{c_x \cos\theta + c_y \sin\theta + s_d i \cos\theta - s_h i \sin\theta} \quad (3)$$

Equation 3 is thus a function $f(c_x, c_y, s_d, s_h, i, \theta)$ that depends on six variables. This model is exploited using template matching. Template matching slides candidate samples over the image and indicates a match if the similarity is over a certain threshold. There are multiple degrees of freedom; however, c_y and θ can be measured and s_d and s_h do not vary a lot, leaving mostly c_x and i that need robust estimation by creating samples for each value.

4   Model-Based Autonomous Navigation

The state and position estimation and autonomous navigation framework used for this application is completely based on the standard *ROS tum_ardrone* package. Integration of the staircase 3D detection and 3D modeling into this navigation architecture is performed as follows:

1. When the developed staircase algorithm has enough detected points, it calculates the central point using an agglomerative clustering algorithm.
2. The top step of the staircase is identified.
3. A point 1.5 meters above the top step is calculated and chosen as a goal position.
4. A *goto x y z w* command is issued to *tum_ardrone* (w, the yaw angle, in degrees and the coordinates (x, y, z) are in meters).
5. The *tum_ardrone* navigation controller ensures the autonomous navigation of the UAS from the start to the goal position by using four built-in PID controllers.

5   Results and Validation

The goal of this paper was to implement an autonomous navigation system specifically for the Parrot AR.Drone, using the Robot Operating System (*ROS*). As the UAS features no on-board processing capability, off-board processing was used. The software in this paper was tested on normal personal computers. However, a powerful processor (e.g. intel i7 family) is recommended. During tests, an average of 5 to 10 frames could be analysed per second.

A graphical user interface was created specifically for this work to control the quadcopter and visualize what is going on. The software allows the user to manually control the drone but also to initialize the autonomous mode which is

Fig. 3. Performance monitoring. From Left: the *content of a csv-file* with KPI and the *output of the object detection*; Matlab staircase detection by calculating the angle in different parts of the depth image (Color figure online).

Fig. 4. Indoor and outdoor testing of the algorithm

supervised, meaning that the drone indicates what it wants to do and asks the user if it can continue. The guidance software listens to the output of the detection node and makes use of the agglomerative clustering technique to implement temporal consistency and filter out different detections.

To verify the results, a performance monitor was made, providing an overview of the metrics, such as the amount of horizontal lines, size and skewness of the detected plane, the processed image, etc. Possible staircase candidates are projected onto the image with red dots. Every candidate has also a unique ID during a test-run. The result is shown in Fig. 3. To evaluate the performance of the software, every processed image is manually investigated and the amount of false negatives or false positives are counted. Matlab was also used to test key properties like the angle of the detected plane. The script looks at the depth image in two different zones. The angle with respect to the ground plane is then calculated assuming the camera was perpendicular with respect to the ground.

For testing, a data recording was used of a single staircase. Afterwards, fine-tuning was done on other staircases. To facilitate testing in the field, a GUI was made to visualize and control all the required packages and events (Fig. 4). To provide a good mix of operating conditions for validation, both indoor and outdoor staircases were tested. Using the proposed autonomous navigation methodology, the following findings were obtained:

- The initialization of the *tum_ardrone* package has proven to be the most difficult part of this methodology. Planar scenes with a lot of different texture worked fine with a success rate of over 80 %. A scene with a lot of depth or similar texture (e.g. concrete wall) did not work at all.
- Outdoor staircases had a higher success rate. Mainly because of better light conditions which result in a superior performance of *lsd slam*.
- *tum_ardrone* is very sensitive to yaw movements. Sudden movements can result in losing the tracking and this can only be solved by a re-initialization.
- Once *tum_ardrone* and *lsd slam* were successful initialized, a 100 % success rate of detection was obtained and 75 % of the staircases were successfully climbed.

6   Conclusion

Multiple existing algorithms have been used in this paper in order to detect staircases as good as possible and to use this data as an input to autonomous navigation. Because a lot of the algorithms already exist, the difficulty lies not so much in the development of new methods but rather in the successful fusion of existing ones. This is what this work tried to establish.

The authors believe that a unique combination of multiple algorithms is the key to success. The methodology that is developed in this work shows great potential but requires more testing in different scenarios and especially on different platforms. The Parrot AR.Drone is a good platform for development but is not well suited for vision processing. Low frame rate and rolling shutter causes the SLAM algorithms to perform badly and this affects the 3D data quality.

This paper is a stepping stone towards a working real-time detection of staircases. However, a more general approach is recommended to be able to deal with a multitude of obstacles. The same framework could than also be used to do positive identification of objects nearby. This paper is elaborated in more in detail in a thesis that is made available online[1]. The code can be found on github[2] and visual results are posted on a YouTube channel[3].

Acknowledgment. The research leading to these results has received funding from the European Union Seventh Framework Programme (FP7/2007-2013) under grant agreement number 285417 (ICARUS).

References

1. De Cubber, G., Doroftei, D., Serrano, D., Chintamani, K., Sabino, R., Ourevitch, S.: The eu-icarus project: developing assistive robotic tools for search and rescue operations. In: IEEE International Symposium on Safety, Security, and Rescue Robotics, Sweden. IEEE, RAS (2013)

---

[1] https://github.com/olivierdm/thesis2015/raw/master/
THESIS-DE_MEYST-GOETHALS-2015.pdf.
[2] https://github.com/olivierdm/thesis2015.
[3] https://www.youtube.com/channel/UCPZkTI9SgHYo5-yVoVRwC-g.

2. Nikolic, J., Burri, M., Rehder, J., Leutenegger, S., Huerzeler, C., Siegwart, R.: A UAV system for inspection of industrial facilities. In: 2013 IEEE Aerospace Conference, pp. 1–8 (2013)
3. Yamauchi, B., Rudakevych, P.: Griffon: a man portable hybrid UGV/UAV. Ind. Robot Int. J. **31**, 443–450 (2004)
4. Klein, G., Murray, D.: Parallel tracking and mapping for small AR workspaces. In: Proceedings on Sixth IEEE and ACM International Symposium on Mixed and Augmented Reality (ISMAR 2007), Nara, Japan (2007)
5. Engel, J., Schöps, T., Cremers, D.: LSD-SLAM: Large-Scale Direct Monocular SLAM. In: Fleet, D., Pajdla, T., Schiele, B., Tuytelaars, T. (eds.) ECCV 2014, Part II. LNCS, vol. 8690, pp. 834–849. Springer, Heidelberg (2014)
6. Lu, X., Manduchi, R.: Detection and localization of curbs and stairways using stereo vision. Int. Conf. Robot. Autom. (ICRA) **4**, 4648 (2005)
7. Lee, Y.H., Leung, T.S., Medioni, G.: Real-time staircase detection from a wearable stereo system. In: International Conference on Pattern Recognition, pp. 3770–3773 (2012)
8. Delmerico, J.A., Baran, D., David, P., Ryde, J., Corso, J.J.: Ascending stairway modeling from dense depth imagery for traversability analysis, pp. 2283–2290 (2013)
9. Pérez-Yus, A., López-Nicolás, G., Guerrero, J.J.: Detection and modelling of staircases using a wearable depth sensor. In: Agapito, L., Bronstein, M.M., Rother, C. (eds.) ECCV 2014 Workshops. LNCS, vol. 8927, pp. 449–463. Springer, Heidelberg (2015)
10. Se, S., Brady, M.: Vision-based detection of staircases, vol. 1, pp. 535–540 (2000)
11. Bills, C., Chen, J., Saxena, A.: Autonomous mav flight in indoor environments using single image perspective cues. In: International Conference on Robotics and Automation (ICRA) (2011)
12. Von Gioi, R.G., Jakubowicz, J., Morel, J.M., Randall, G.: Lsd: a line segment detector. Image Process. Line **2**, 5 (2012)
13. Viola, P., Jones, M.: Robust real-time object detection. Int. J. Comput. Vis. **4**, 34–47 (2001)
14. Saberian, M., Vasconcelos, N.: Boosting algorithms for detector cascade learning. J. Mach. Learn. Res. **15**, 2569–2605 (2014)
15. Foley, J.D., Van Dam, A., Feiner, S.K., Hughes, J.F., Phillips, R.L.: Introduction to Computer Graphics, vol. 55. Addison-Wesley, Reading (1994)

Nonlinear Controller of Quadcopters for Agricultural Monitoring

Víctor H. Andaluz[1,2](✉), Edison López[2], David Manobanda[2],
Franklin Guamushig[2], Fernando Chicaiza[1], Jorge S. Sánchez[1],
David Rivas[1], Fabricio Pérez[1],
Carlos Sánchez[2], and Vicente Morales[2]

[1] Universidad de las Fuerzas Armadas ESPE, Sangolquí, Ecuador
{vhandaluz1,fachicaiza,jssanchez,
drrivas,mfperez}@espe.edu.ec
[2] Universidad Técnica de Ambato, Ambato, Ecuador
{elopez8117,amanobanda3933,fguamushig473,
carloshsanchez,jvmorales99}@uta.edu.ec

Abstract. This work presents the construction and control of an autonomous quadcopter for applications of agriculture monitoring. Mainly the navigation system is based on a Field Programmable Gate Array, FPGA, Stabilization Target and a Global Positioning System, GPS, to solve different motion problems of a mobile robot. Also, this paper shows the development of a Human Machine Interface, HMI, between the ground station and the quadcopter. The HMI allows interaction both for simulation and/or experimentally of agricultural monitoring applications through control strategies tele-operated and autonomous flight. Finally, the performance of the proposed controller is shown through real experiments.

Keywords: Unified motion control · Non-linear systems · Quadcopters · Agriculture monitoring

1 Introduction

Currently, robotics research challenge is to move from control robots designed for industrial environments to control service robots in unstructured environments [1–3]. A lot of research is performed on autonomous control of Unmanned Aerial Vehicles – UAVs-. UAVs come in all sizes, fixed wing or rotary, and are used to perform various civilian and military applications, *e.g.,* surveillance, natural risk management, infrastructure maintenance, precision agriculture and tactical missions in battlefield environments [2–5]. Precision agriculture is one of the specific applications that are resolved with this type of aircraft; e.g., photographs in oil palm plantations, fumigation, pest control, irrigation status monitoring, control soil moisture and temperature, among others, are developed with techniques of tele-operated and autonomous navigation. In this kind of application, the information about the UAV position is usually delivered by a low rate sensor, e.g. a Global Positioning System (GPS), so that a trajectory control strategy can be an extremely useful alternative, for not demanding temporal parameterizations.

© Springer International Publishing Switzerland 2015
G. Bebis et al. (Eds.): ISVC 2015, Part I, LNCS 9474, pp. 476–487, 2015.
DOI: 10.1007/978-3-319-27857-5_43

In this paper a robotic quadcopter is considered to solve the problems of trajectory tracking for the UAV within a unified structure. The controller design is based on a kinematic model of the UAV which accepts velocity inputs, as it is usual in commercial robots. This controller is designed in two parts, each one being a controller itself. The first one is a kinematic controller with saturation of velocity commands, which is based on the UAV's kinematics; and the second one is a dynamic compensation controller that is considered through of a quadcopter-inner-loop system to independently track four velocity commands: forward, lateral, up/downward, and heading rate. Furthermore, this paper shows the development of a Human Machine Interface, HMI, on MatLab between the ground station and the quadcopter. The HMI receives the information sent by the UAV, generating records of variables and analysis of the behavior of each of the monitored agricultural areas. The HMI allows interaction both for simulation and/or experimentally of agricultural monitoring applications through control strategies tele-operated and autonomous flight. To validate quadcopter construction and the proposed model, experimental tests are performed the same show good performance of it.

The paper is organized as follows: in Sect. 2 the problem formulations for agricultural monitoring through of UAVs and the kinematic modeling of the quadcopter is presented. The design of the controller, including the stability analyses is presents in Sect. 3. The Sect. 4 presents the design of software for implementing the several control strategies for agricultural monitoring applications; while that the experimental results are presented and discussed in Sect. 5. Finally, conclusions are given in Sect. 6.

2 Formulation Problem

Currently, there is a growing interest in using rotary UAVs for applications in confined spaces, more specifically in corridors, both indoor and outdoor. Potential applications include searching collapsed buildings [5], searching criminals in houses by the Police [6] and inspecting fruit orchards and vineyards from in between the tree rows [2]. The boom of the application of UAVs in precision agriculture is due to its non-invasive access in the studied area, being able to determine the state of water irrigation, temperature, humidity, the soil alkalinity and other factors that can improve crop [7, 8] without damaging to the plants. Currently, the actual climate state of the area intended for agricultural uses wireless communication nodes between sensors which detect parameters of the plant and the environment. These sensors are connected to a logical unit that processes this information to assist farmers in making decisions. In large agricultural areas, communication between the nodes and the central logic unit is based on network topologies for routing information from the sensors to the information processor [9, 10], taking in many cases, the limitation given the communication range of the devices. To cover this limiting factor of data transmission, the use of UAVs for the collection of information is added. Figure 1 shows the scene of a UAV used to acquire information from the distant agricultural areas, where data collection using wireless communication is based on the 802.15.4 protocol [11] and has modes of storage or retransmission of information.

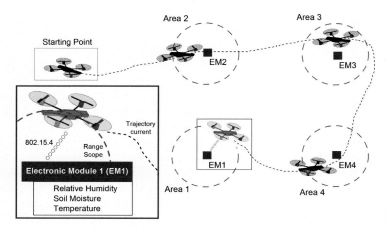

Fig. 1. Data collection from agricultural areas

Hence the quadcopter presents in this work is composed by a set of four velocities represented at the spatial frame $<Q>$. The displacement of the quadcopter is guided by the three linear velocities u_{ql}, u_{qm} and u_{qn} defined in a rotating right-handed spatial frame $<Q>$ [12–14], and the angular velocity ω_q, as shown in Fig. 2.

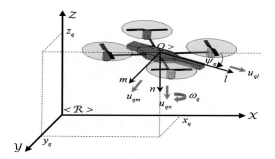

Fig. 2. Schematic of the autonomous quadcopter

Each linear velocity is directed as one of the axes of the frame $<Q>$ attached to the center of gravity of the helicopter: u_{ql} points to the frontal direction; u_{qm} points to the left-lateral direction, and u_{qn} points up [16]. The angular velocity ω_q rotates the referential system $<Q>$ counterclockwise, around the axis Z_q (considering the top view). In other words, the Cartesian motion of the quadcopter at the inertial frame <R> is defined as,

$$
\begin{bmatrix} \dot{x}_q \\ \dot{y}_q \\ \dot{z}_q \\ \dot{\psi}_q \end{bmatrix} = \begin{bmatrix} \cos\psi_q & -\sin\psi_q & 0 & 0 \\ \sin\psi_q & \cos\psi_q & 0 & 0 \\ 0 & 0 & 1 & 0 \\ 0 & 0 & 0 & 1 \end{bmatrix} \begin{bmatrix} u_{ql} \\ u_{qm} \\ u_{qn} \\ \omega_q \end{bmatrix}
$$

$$
\dot{\mathbf{h}}_q = \mathbf{J}\left(\psi_q\right)\mathbf{u}_q \tag{1}
$$

where $\dot{\mathbf{h}}_q \in \Re^n$ with $n = 4$ represents the vector of axis velocity of the <R> system; $\mathbf{J}\left(\psi_q\right) \in \Re^{n \times n}$ is a singular matrix; and the control of maneuverability of the quadcopter is defined $\mathbf{u}_q \in \Re^n$.

3   Nonlinear Controller Design

The tracking control problem of an autonomous robot consists of following a given time-varying trajectory $\mathbf{y}_{d(t)}$ and its successive derivatives $\dot{\mathbf{y}}_d(t)$ and $\ddot{\mathbf{y}}_{d(t)}$ which respectively describe the desired velocity and acceleration. That is, the desired trajectory for the UAV is defined by a vector $\mathbf{h}_{qd}(t,s) = \begin{bmatrix} \mathbf{h}_{qdpos}^T & \mathbf{h}_{qdor}^T \end{bmatrix}$ in R. The desired trajectory does not depend on the instantaneous position of the robot, but it is defined only by the time varying trajectory profile alone. The desired velocity $\mathbf{v}_{qd}$ is defined for this control problem as the time derivative of $\mathbf{h}_{qd}(t,s)$, that is $\mathbf{v}_{qd} = \dot{\mathbf{h}}_{qd}(t,s)$.

Hence, the trajectory control problem is to find the control actions for the UAV as a function of the control errors and the desired velocities of the UAV $\mathbf{u}_{ref}(t) = f\left(\tilde{\mathbf{h}}_q(t,s), \dot{\mathbf{h}}_d(t,s)\right)$ such that $\lim_{t\to\infty} \tilde{\mathbf{h}}_q(t,s) = 0$. Therefore, the desired location and velocity of the end-effector (1) for trajectory tracking can be defined as $\mathbf{h}_{qd}(t,s,h) = \mathbf{h}_{qd}(t,s)$ and $\mathbf{v}_{qd}(t,s,h) = \dot{\mathbf{h}}_{qd}(t,s)$ motion control objective.

4   Monitoring Stations and Control Software

The development of a Human Machine Interface, HMI, between the ground stations and the quadcopter is presented in this Section, in addition to the construction of the electronic stations. The HMI proposed here has been implemented in MatLab, in order to facilitate and streamline simulations experimental tests of different proposed control algorithms.

4.1   Electronic Stations

The earth station transmitter (Fig. 3) acquires sensor information relative humidity, soil moisture and air temperature, which are in the radial area of the crop. Additionally, the ground station processed this information and transmits it by wireless communication (802.15.4) to the receiving station (Fig. 3) on board the UAV. The station aboard the

a) Ground Electronic station   b) On-board Electronic station

Fig. 3. Measurement electronic stations

UAV stores or retransmits the data received by the ZigBee, where the storage of agricultural parameters uses a volatile memory programmed in the control unit for later download, while the retransmission involves a ZigBee of best features than the rest of the ground stations for reaching the main ground station. In the main ground station, the HMI programmed in MatLab receives the information stored or retransmitted by the station aboard the UAV and once processed the agricultural parameters, it shows comparative curves and tables, presented in Sect. 4.2.

4.2   Agricultural Monitoring

The HMI allows interaction both for simulation and/or experimentally of agricultural monitoring applications through control strategies tele-operated and autonomous flight, also includes buttons for storage of agricultural variables in real time. The following paragraphs describe the programming of each of the buttons on the proposed HMI.

Data Transmission Mode. The button *Retransmission* is selected to visualize agricultural variables in real time. This indicates that station is transmitting and the relationship between the number of samples and magnitude of soil moisture, temperature and relative humidity by a line graph, as shown in Fig. 4.

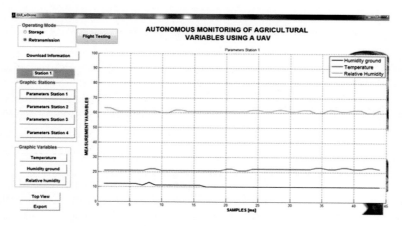

Fig. 4. HMI: data transmission mode

Download Agricultural Variables. Once the UAV performed overflight and collected data from the ground stations (sensors), this mode allowing to download the information from the sensors to visualize the variation of the parameters measured in a linear plot, as shown in Fig. 5.

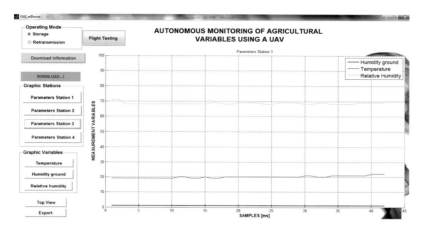

Fig. 5. HMI: download agricultural variables

Agricultural Variables Stations. This section allows graphing the record of variables stations in the two modes of operation described previously as shown in Fig. 6.

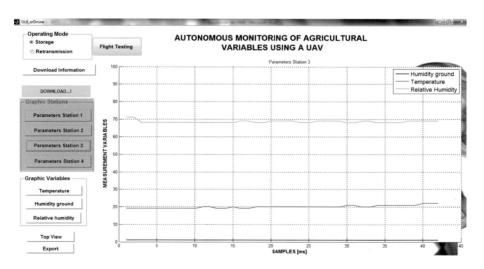

Fig. 6. HMI: agricultural variables stations

Comparing Agricultural Variables. The interface have three buttons that allow to relate a variable agricultural of each station, *i.e.*, temperature, soil moisture and relative humidity. This allow to have a preview of agricultural variables to determine the state of pre-crop land. Figure 7 shows the information for analyzing the behavior of each variable.

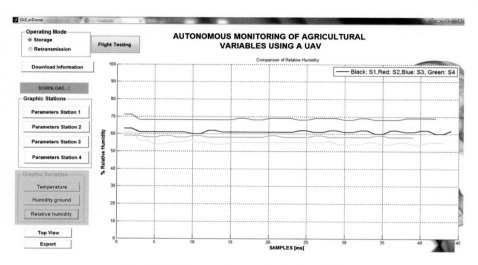

Fig. 7. HMI: comparing agricultural variables

Variables Stations. Figure 8 shows the 2D mapping of the flight by the UAV, this graphic illustrates the location of each ground station (sensor) and the values of each variable agricultural per station. Agricultural variables are represented by a range of

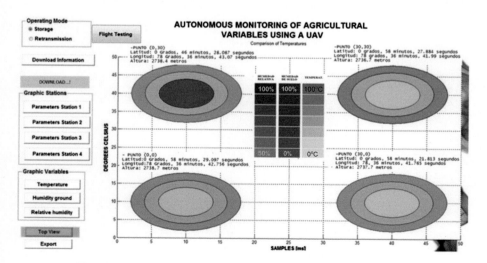

Fig. 8. HMI: agricultural variables for stations (Color figure online)

colors, *i.e.*, soil moisture, temperature and relative humidity have a shade of blue, red and green, respectively.

4.3   Tele-Operated and Autonomous Flight Control

The HMI includes a button for "Flight testing" (Fig. 8), which is implemented in order to test the construction the quadcopter and three flight strategies: point stabilization, trajectory tracking and path following, as shown in Fig. 9. While in the lower right is described by mathematical equations position or desired path or trajectory to be followed by the quadcopter independently in both axes of the X, Y, and Z of the reference system < R >, described in Sect. 2.

Simulation Mode. This button once the flight profile parameters compared to the reference system loaded in a 3D environment simulating one quadcopter whose control inputs, u_{ql}, u_{qm}, u_{qn} and ω_q respect to the frame $<Q>$ attached to the center of gravity of the helicopter, described in Sect. 2. The three-dimensional simulation is online and uses *CameraTool* to provide different views of the environment. During the simulated flight trajectory and the wisher profile quadcopter be followed is described as it is shown in Fig. 9.

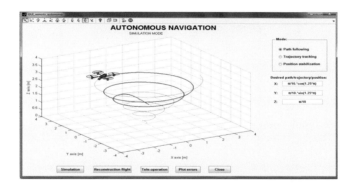

Fig. 9. HMI for autonomous and tele-operated control of a quadcopter

Flight Implementation Mode. This button allows communication from the ground station and the quadcopter through 802.15.4 protocol. The bilateral communication is in real time; the ground control station transmitted the control velocities u_{ql}, u_{qm}, u_{qn} and ω_q of the quadcopter and the quadcopter positions given by the GPS is received. It should be noted that the reconstruction of the flight is based on information provided by GPS in real time, allowing verify quadcopter construction, and performance of the implemented control algorithm. Figure 10 illustrates the three-dimensional graphical representation of quadcopter used for the reconstruction of the flight; the stroboscopic movement is based on real data quadcopter position and orientation of the axes of X, Y, and Z of the < R > system.

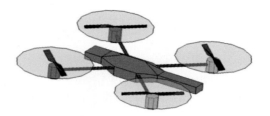

Fig. 10. 3D graphical representation of a quadcopter

Remark 2: The autonomous control algorithms are implemented in this mode.

Tele-Operation Mode. Tele-operation button toggles an autonomous navigation control to a tele-operated control. Figure 11 shows an interface where the control velocities issued by the human operator on quadcopter are determined.

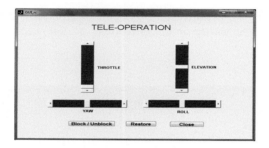

Fig. 11. Command to control tele-operated

The control signals are emitted by a remote control with ZigBee communication, Fig. 12.

Fig. 12. Remote control communication by ZigBee

Graphics Mode. *Plot errors* buttons allows determine the performance of the control algorithms implemented both in simulation and experimentally through graphs showing the evolution of control errors, $\tilde{x}(t) = x_d(t) - x_q(t)$, $\tilde{y}(t) = y_d(t) - y_q(t)$ and $\tilde{z}(t) = z_d(t) - z_q(t)$ with respect to $< \mathsf{R} >$. The subscript d determines the desired each axis position, and the subscript q represents the desired position of the quadcopter (Fig. 13). Also shows the distance between the quadcopter position $\mathbf{h}_q\left(x_q, y_q, z_q\right)$ and the desired point P_D on the trajectory $P(s)$.

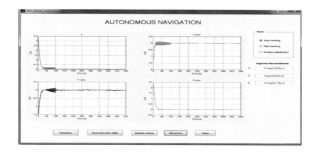

Fig. 13. Evolution of errors of the quadcopter

5   Results and Discussions

In order to illustrate the performance of the proposed controller, several experiments were carried out for trajectory tracking control of a quadcopter; the most representative simulation and results. The experiments were carried out on a quadcopter which was developed at the University of the Armed Forces ESPE [17], see Fig. 14.

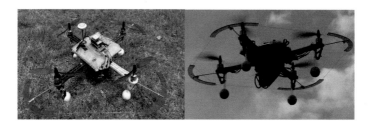

Fig. 14. Connections diagram

The hardware architecture of the quadcopter consists of a Stabilizer Flight, a Field Programmable Gate Array (FPGA), an Electronic Speed Controller (ESC), a Global Positioning System (GPS) and a ZigBee. The high-level controller is implemented in the ground station under the Windows operating system using Microsoft Visual C++ and MatLab. Several experiments are presented in order to evaluate the performance of the proposed scheme and to show the applicability of the unified algorithm to different motion control problems. The performance of the control structure for trajectory tracking is tested in the second experiment. It is important to mention that for trajectory tracking is chosen. Figures 15 and 16, represent the experimental results. Figure 15 shows the desired trajectory and the current trajectory of the UAV. It can be seen that

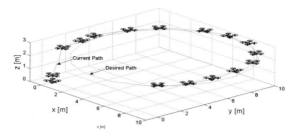

Fig. 15. Stroboscopic movement of the UAV in the trajectory following experiment.

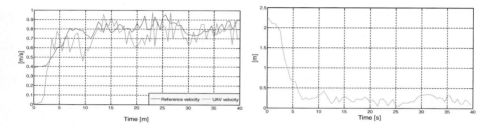

Fig. 16. End-effector velocity

Fig. 17. Distance between the UAV position and the closest point on the path

the proposed controller presents a good performance, and Fig. 17 shows the evolution of the tracking errors, which remain close to zero.

6   Conclusion

In this work a unified design of motion controllers for point stabilization, trajectory tracking and path following of a UAV has been presented. The design of the whole controller is based on two cascaded subsystems: a kinematic controller which complies with the motion control objective, and a quadcopter-inner-loop system to independently track four velocity commands: forward, lateral, up/downward, and heading rate; the kinematic controller have been designed to prevent from command saturation. Robot commands were defined in terms of reference velocities to the UAV. The performance of the proposed unified controller is shown through real experiments for the different motion control objectives. The experiments confirm the capability of the unified controller to solve different motion problems by an adequate selection of the control references.

Acknowledgment. The authors thank to the Universidad de las Fuerzas Armadas ESPE–L and to the Universidad Técnica de Ambato, for the support to develop this work.

References

1. Kelly, R., Santibañez, V.: Control de movimiento de Robots Manipuladores, p. 344. Prentice Hall, Madrid (2003)
2. Brandao, A.S., Sarcinelli-Filho, M., Carelli, R.: An analytical approach to avoid obstacles in mobile robot navigation. Int. J. Adv. Rob. Syst. (2013)
3. Ollero, A., Maza, I. (eds.): Multiple Heterogeneous Unmanned Aerial Vehicles, Springer Tracts in Advanced Robotics, vol. 37. Springer, Berlin (2007)
4. Duan, H., Liu, S.: Unmanned air/ground vehicles heterogeneous cooperative techniques: current status and prospects. Sci. China, Technol. Sci. 53(5), 1349–1355 (2010)
5. Conte, G., Duranti, S., Merz, T.: Dynamic 3d path following for an autonomous helicopter. In: IFAC Symposium on Intelligent Autonomous Vehicles, Lisbon, Portugal (2004)
6. Wang, B., Dong, X., Chen, B.M.: Cascaded control of 3d path following for an unmanned helicopter. In: IEEE International Conference on Cybernetics and Intelligent Systems, pp. 70–75. Singapore (2010)
7. Enderle, B.: Commercial Applications of UAV's in Japanese Agriculture. In: AIAA, USA (2012)
8. Saari, H., Pellikka, I., et al.: Unmanned aerial vehicle (UAV) operated spectral camera system for forest and agriculture applications. In: de Remote Sensing for Agriculture, Ecosystems, and Hydrology XIII, Czech Republic (2011)
9. Ruiz-Garciaa, L., Barreiroa, P., Roblab, J.I.: Performance of ZigBee-based wireless sensor nodes for real-time monitoring of fruit logistics. J. Food Eng. 87(3), 405–415 (2008)
10. Bender, P.: Wireless sensor/actuator network for model railroad control. In: Proceedings of the 12th ACM Conference on Embedded Network Sensor Systems, pp. 332–333 (2014)
11. Ling, T.Y., Wong, L.J., et al.: XBee wireless blood pressure monitoring system with Microsoft Visual Studio. In: 6th International Conference on Intelligent Systems, Modelling and Simulation, pp. 5–9 (2015)
12. Brandao, A.S., Sarcinelli-Filho, M., Carelli, R.: High-level under-actuated nonlinear control for rotorcraft machines. In: Proceedings of the IEEE International Conference on Mechatronics, IEEE, Vicenza, Italy, February 27–March 1, 2013
13. Raffo, G.V., Ortega, M.G., Rubio, F.R.: An integral predictive/nonlinear H control structure for a quadrotor helicopter. Automatica 46, 29–39 (2010)
14. Palunko, I., Bogdan, S.: Small helicopter control design based on model reduction and decoupling. J. Intell. Rob. Syst. 54, 201–228 (2009)
15. Brandao, A.S., Andaluz, V.H., Sarcinelli-Filho, M., Carelli, R.: 3-D path-following with a miniature helicopter using a high-level nonlinear underactuated controller. In: IEEE International Conference on Control and Automation, pp. 434–439 (2011)
16. Cai, G., Chen, B.M., Dong, X., Lee, T.H.: Design and implementation of a robust and nonlinear flight control system for an unmanned helicopter. Mechatronics 21(5), 803–820 (2011)
17. Universidad de las Fuerzas Armadas ESPE. http://www.espe.edu.ec

Medical Imaging

Groupwise Shape Correspondences on 3D Brain Structures Using Probabilistic Latent Variable Models

Hernán F. García$^{(\boxtimes)}$, Mauricio A. Álvarez, and Álvaro Orozco

Grupo de Investigación en Automática, Universidad Tecnológica de Pereira,
La Julita, Pereira, Colombia
{hernan.garcia,malvarez,aaog}@utp.edu.co

Abstract. Most of the tasks derived from shape analysis rely on the problem of finding meaningful correspondences between two or more shapes. In medical imaging analysis, this problem is a challenging topic due to the need to establish matching features in a given registration process. Besides, a similarity measure between shapes must be computed in order to obtain these correspondences. In this paper, we propose a method for 3D shape correspondences based on groupwise analysis using probabilistic latent variable models. The proposed method finds groupwise correspondences, and can handle multiple shapes with different number of objects (vertex or descriptors for every shape). By assigning a latent vector for each shape descriptor, we can cluster objects in different shapes, and find correspondences between clusters. We use a Dirichlet process prior in order to infer the number of clusters and find groupwise correspondences in an unsupervised manner. The results show that the proposed method can efficiently establish meaningful correspondences without using similarity measures between shapes.

1 Introduction

Finding correspondences between shapes, is a well known problem in medical imaging analysis. Since, processes such as registration, segmentation, parameterization, and volumetric analysis of shapes (2D or 3D), need a correct representation of the analyzed objects [1]. Generally, computing correspondences is a difficult task due to the fact that we need to understand the shape structures both at global and at local levels [2]. In order to establish a meaningful correspondence, a great variety of methods compute different similarity metrics based on texture descriptors such as bag of words features [3], geodesic contours [4], and Largest Common Pointsets [5]. However, most of these approaches only handles objects of the same size, which gives a full relation of the matched features in all of their points and builds a point-to-point correspondence problem [6].

Analyzing brain structure properties, such as shape volumetry or cortical thickness, is an important task in Parkinson's disease, due to the importance to monitoring these structures (i.e. basal ganglia area), analyze anatomic connectivity, and finding cortical disease patterns [7]. One important factor to such

© Springer International Publishing Switzerland 2015
G. Bebis et al. (Eds.): ISVC 2015, Part I, LNCS 9474, pp. 491–500, 2015.
DOI: 10.1007/978-3-319-27857-5_44

tasks is to establish meaningful cortical correspondences across a set of subject cortices (population of subjects or evolution of a given brain over time) [8]. However the high variability of these brain patterns involves a needed to compute groups of correspondences over the entire brain structures, that is why a methodology to find groupwise correspondences must be fulfilled [9].

Methods for unsupervised object matching have been proposed in the last years to modeling shape structures where there is no full correspondences of the objects to analyze or it is not possible to estimate some similarity measure between them [10]. Some works such as least squares object matching [11], variational Bayesian matching [12] and kernelized sorting [13], use probabilistic methods to modeling features between objects in order to establish matching processes.

Despite that these methods finds point-to-point correspondences, the idea to model the same information (i.e. texture descriptors in MRI data) in different shapes without using any similarity measure, is of high impact in computer vision problems [14]. However, in medical image annotation, a given label could be related to multiple images that exhibit the same information (i.e. labels maps in brain atlases), and an image that shows multiple objects could have multiple annotations such as brain structures on magnetic resonance images (MRI) [8].

In this paper, we present a method for groupwise shape correspondences based on unsupervised clustering of 3D shape descriptors over a set of brain structures using probabilistic latent variable models. Our method is based on the work proposed by Iwata *et al.* [15], in which a task for many-to-many objects matching is performed. The motivation of this method is based on the ability of modeling objects in multiple domains with different sizes (brain structures with different number of descriptors). Our contribution relies on using scale-invariant heat kernel signatures in order to extract 3D shape descriptors over a set of brain structures (i.e. thalamus, ventricles and putamen), and then use an infinite Gaussian mixture model (iGMM) to assigning a latent vector for each descriptor with the aim to cluster objects (set of shape descriptors) in different brain structures, and to finding groupwise correspondences between clusters without using any similarity measure. The paper proceeds as follows. Section 2 presents the scheme for establishing the groupwise correspondences using probabilistic latent variable models. Section 3 shows the experimental results, in which we report several experiments related to the probabilistic analysis of brain structures. The paper concludes in Sect. 4 with some conclusions of the proposed method and future works.

2   Materials and Methods

2.1   Database

In this work two MRI databases were used. First, a MRI database from Universidad Tecnológica de Pereira (DB-UTP). This database contains recordings of MRI from four patients with Parkinson's disease, and it was labeled by neurosurgegy specialists from the Institute of Parkinson and Epilepsy of the Eje Cafetero, located in the city of Pereira, Colombia. The database contains $T1$, $T2$ and CT sequences with $1\,mm \times 1\,mm \times 1\,mm$ voxel size and slices of 512×512 pixels. Second, we use the SPL-PNL[1] Brain Atlas developed by the Surgical Planning Laboratory of Hardvard Medical School, in collaboration with the Harvard Neuroscience Laboratory at Brockton VA Medical Center. The atlas was derived from a volumetric T1-weighted MR-scan of a healthy volunteer, using semi-automated image segmentation, and three-dimensional reconstruction techniques. The current version consists of: (1) the original volumetric whole brain MRI of a healthy volunteer; (2) a set of detailed label maps and (3) more than 160 three-dimensional models of the labeled anatomical structures

2.2   Scale-Invariant Heat Kernel Signature (SI-HKS)

Using a shape descriptor that maintains invariance under a wide class of transformations is convenient when it comes to shape analysis. The SI-HKS is a scale-invariant version of the heat kernel descriptor where an intrinsic local shape descriptor based on diffusion scale-space analysis is performed in order to describe a given shape [3]. Here, brain structures are modeled as Riemannian manifolds in order to compute the shape descriptors from the heat conduction properties [16]. Heat propagation on non-Euclidean domains is governed by the heat diffusion equation,

$$\left(\Delta_X + \frac{\partial}{\partial t} \right) u\left(x, t \right) = 0, \tag{1}$$

where Δ_X is the Laplace Beltrami operator and $u\left(x, t \right)$ is the heat distribution at a point x at time t. The solution $u\left(x, t \right)$ of the Eq. (1) describes the amount of heat on a given surface at point x in time t. The heat kernel signature can be described as

$$h\left(x, t \right) = K_{x,x}\left(x, z \right) = \sum_{i=0}^{\infty} e^{-\lambda_i t} \phi_i^2, \tag{2}$$

where $\lambda_0, \lambda_1, \cdots \geq 0$ are eigenvalues, and $\phi_0, \phi_1, \ldots$ are the corresponding eigenfunctions of the Laplace-Beltrami operator, satisfying $\Delta_X \phi_i = \lambda_i \phi_i$. One disadvantage of the heat kernel signatures is their sensitivity to scale. Given a shape X and its scaled version $X' = \beta X$, the new eigenvalues and eigenfunctions will satisfy $\lambda' = \beta^2 \lambda$ and $\phi' = \beta \phi$.

In order to achieve scale invariance, Broinstein proposes to remove the dependence of h from the scale factor β [16]. First, the HKS was sampled logarithmically in time ($t = \alpha^\tau$) at each shape point x to obtain a discrete function of h as h_τ. Finally, they use the discrete-time Fourier transform of h_τ to shift in time into a complex phase $H'\left(w \right) = H\left(w \right) e^{2\pi w s}$.

---

[1] This database is available on http://www.spl.harvard.edu/publications/item/view/1265.

2.3   Probabilistic Latent Variable Model for Groupwise Correspondence

In order to establish meaningful correspondences over a set of 3D shapes, the model assumes that we are given shape descriptors for D domains (set of brain structures) $\{\mathbf{X}_d\}_{d=1}^{D}$ where $\mathbf{X}_d = \{\mathbf{x}_{dn}\}_{n=1}^{N_d}$ is a set of shape descriptors in the dth domain, and $\mathbf{x}_{dn} \in \mathbb{R}_d^M$ is the feature vector of the nth shape descriptor in the dth domain. Here, we are unaware of any correspondences between shape descriptors in different domains. The number of descriptors N_d and the dimensionality M_d for each brain structure can be different from those of other structures. Therefore, our task is to match clusters of descriptors (groupwise correspondences) across multiple brain structures in an unsupervised manner [15].

As in infinite Gaussian mixture models, our approach assumes that there are a infinite number of clusters related to each correspondence, and each cluster j has a latent vector $\mathbf{z}_j \in \mathbb{R}^K$ in a latent space of dimension K. Descriptors that have the same cluster assignments s_{dn} are related by the same latent vector and considered to match (establish a groupwise correspondence).

The proposed model is based on an infinite mixture model, where the probability of descriptor $\mathbf{x}_{dn}$ is given by

$$p\left(\mathbf{x}_{dn} | \mathbf{Z}, \mathbf{W}, \boldsymbol{\theta}\right) = \sum_{j=1}^{\infty} \theta_j \mathcal{N}\left(\mathbf{x}_{dn} | \mathbf{W}_d \mathbf{z}_j, \alpha^{-1}\mathbf{I}\right), \tag{3}$$

where $\mathbf{W} = \mathbf{W}_{d=1}^{D}$ is a set of projections matrices, $\boldsymbol{\theta} = (\theta_j)_{j=1}^{\infty}$ are the mixture weights, θ_j represents the probability that the jth cluster is chosen, and $\mathcal{N}(\boldsymbol{\mu}, \boldsymbol{\Sigma})$ denotes a normal distribution with mean $\boldsymbol{\mu}$ and covariance matrix $\boldsymbol{\Sigma}$. One important contribution derived in [15], is that we can analyze multiples structures with different properties and dimensionalities, by employing projection matrices for each brain structure (domain-specific). Figure 1 shows the scheme of the proposed model, in which the relationship between shape descriptors (SI-HKS features), and latent vectors is described.

In order to draw the cluster proportions, we use a stick-breaking process to generate mixture weights for a Dirichlet process with concentration parameter γ [15] (a, b and r are the hyperparameters). The joint probability of the data $\mathbf{X}$, and the cluster assignments $\mathbf{S} = \left\{\{s_{dn}\}_{n=1}^{N_d}\right\}_{d=1}^{D}$ are given by

$$p\left(\mathbf{X}, \mathbf{S} | \mathbf{W}, a, b, r, \gamma\right) = p\left(\mathbf{S}|\gamma\right) p\left(\mathbf{X}|\mathbf{S}, \mathbf{W}, a, b, r\right). \tag{4}$$

By marginalizing out the mixture weights $\boldsymbol{\theta}$, $p\left(\mathbf{S}|\gamma\right)$ becomes in

$$p\left(\mathbf{S}|\gamma\right) = \frac{\gamma^J \prod\limits_{j=1}^{J} (N_{.j} - 1)!}{\gamma\left(\gamma + 1\right)\ldots\left(\gamma + N - 1\right)}, \tag{5}$$

where $N = \sum\limits_{d=1}^{D} N_d$ is the total number of shape descriptors, $N_{.j}$ represents the number of descriptors assigned to the cluster j, and J is the number of clusters

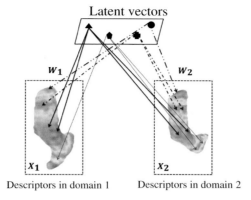

Fig. 1. Scheme for the groupwise correspondence method. The figure shows an example of establishing clusters of correspondences in two domains (left ventrals).

that satisfies $N_{\cdot j} > 0$. By marginalizing out latent vectors $\mathbf{Z}$ and the precision parameter α, the second factor of (4) is computed by

$$p\left(\mathbf{X}|\mathbf{S}, \mathbf{W}, a, b, r\right) = (2\pi)^{\frac{\Sigma_d M_d N_d}{2}} \, r^{\frac{KJ}{2}} \, \frac{b^a}{b'^{a'}} \, \frac{\Gamma\left(a'\right)}{\Gamma\left(a\right)} \prod_{j=1}^{J} |\mathbf{C}_j|^{1/2}. \tag{6}$$

Here,

$$a' = a + \frac{\Sigma_d M_d N_d}{2}, \qquad b' = b + \frac{1}{2}\sum_{d=1}^{D}\sum_{n=1}^{N_d} \mathbf{x}_{dn}^{\top}\mathbf{x}_{dn} - \frac{1}{2}\sum_{j=1}^{J} \boldsymbol{\mu}_j^{\top}\mathbf{C}_j^{-1}\boldsymbol{\mu}_j, \tag{7}$$

$$\boldsymbol{\mu}_j = \mathbf{C}_j \sum_{d=1}^{D} \mathbf{W}_d^{\top} \sum_{n:s_{dn}=j} \mathbf{x}_{dn}, \qquad \mathbf{C}_j^{-1} = \sum_{d=1}^{D} N_{dj}\mathbf{W}_d^{\top}\mathbf{W}_d + r\mathbf{I}, \tag{8}$$

where N_{dj} is the number of descriptors assigned to cluster j in the shape d (domain).

The posterior for α is $p\left(\alpha|\mathbf{X}, \mathbf{S}, \mathbf{W}, a, b\right) = \mathrm{Gamma}\left(a', b'\right)$, and for the latent vector $\mathbf{z}_j$ is given by $p\left(\mathbf{z}_j|\mathbf{X}, \mathbf{S}, \mathbf{W}, r\right) = \mathcal{N}\left(\boldsymbol{\mu}_j, \alpha^{-1}\mathbf{C}_j\right)$ [15].

2.4   Inference

We use the proposed method in [15] based on stochastic EM algorithm. Here, collapsed Gibbs sampling of cluster assignments $\mathbf{S}$, and maximum joint likelihood estimation of projection matrices $\mathbf{W}$ are alternately iterated. The main idea is to marginalize out the latent vectors $\mathbf{Z}$, and the precision parameter α.

In the E-step, a new value for s_{dn} is sampled from

$$
p\left(s_{dn} = j | \mathbf{X}, \mathbf{S}_{\backslash dn}, \mathbf{W}, a, b, r, \gamma\right) \propto \frac{p\left(s_{dn} = j, \mathbf{S}_{\backslash dn} | \gamma\right)}{\left(\mathbf{S}_{\backslash dn} | \gamma\right)} \tag{9}
$$
$$
\times \frac{p\left(\mathbf{X} | s_{dn} = j, \mathbf{S}_{\backslash dn}, \mathbf{W}, a, b, r\right)}{p\left(\mathbf{X}_{\backslash dn} | \mathbf{S}_{\backslash dn}, \mathbf{W}, a, b, r\right)},
$$

where $\backslash dn$ represents a value excluding the nth descriptor in the dth shape (domain). The first factor in the expression above is given by

$$
\frac{p\left(s_{dn} = j, \mathbf{S}_{\backslash dn} | \gamma\right)}{p\left(\mathbf{S}_{\backslash dn} | \gamma\right)} = \begin{cases} \frac{N_{.j \backslash dn}}{N - 1 + \gamma} & \text{for an existing cluster} \\ \frac{\gamma}{N - 1 + \gamma} & \text{for a new cluster.} \end{cases} \tag{10}
$$

In the M-step, we estimate $\mathbf{W}$ by maximizing the logarithm of the join likelihood (4). The gradient of the log-likelihood with respect to $\mathbf{W}_d$ is computed as

$$
\frac{\partial \log p\left(\mathbf{X}, \mathbf{S} | \mathbf{W}, a, b, r, \gamma\right)}{\partial \mathbf{W}_d} = -\mathbf{W}_d \sum_{j=1}^{J} N_{dj} \mathbf{C}_j \tag{11}
$$
$$
- \frac{a'}{b'} \sum_{j=1}^{J} \left(N_{dj} \mathbf{W}_d \boldsymbol{\mu}_j \boldsymbol{\mu}_j^\top - \sum_{n : s_{dn} = j} \mathbf{x}_{dn} \boldsymbol{\mu}_j^\top \right).
$$

3   Results

3.1   Scale-Invariant Heat Kernel Signatures

In order to construct the SI-HKS, we used a logarithmic scale-space with base $\alpha = 2$, and τ ranging between $[1 - 25]$ with increments of $1/8$. After applying the process depicted in Sect. 2.2, the first four discrete lowest frequencies were used as the local descriptor [16]. Figure 2 shows an example of computing SI-HKS descriptors over three kind of brain structures (ventricle, putamen and thalamus). The results show that spatial regions related to the same descriptor, share common information, which can be exploited to model the correspondence between brain structures.

Moreover, Fig. 3 shows the frequencies w used as scale-invariant descriptors. Here, three points were randomly chosen, in order to analyze the descriptors of each point between shapes. Since the brain structures have different scales, the results prove that the SI-HKS can efficiently compute 3D shape descriptors in a scale-invariant manner

3.2   Groupwise Correspondences

In order to compute the groupwise correspondences, we use the SI-HKS descriptors (see Sect. 2.2) as objects in the probabilistic latent variable model. Each

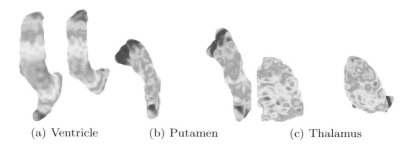

(a) Ventricle   (b) Putamen   (c) Thalamus

Fig. 2. Samples of SI-HKS descriptors for three brain structures.

brain structure of a given patient is set as a specific domain (see Sect. 2.3) in the model. Thus, we build the dataset with four domains, related to the brain structures in every subject of the database (thalamus, putamen and ventricle).

Figure 4 shows the groupwise correspondences computed from the proposed model. The results show that when more objects (number of shape descriptors) are used to build the model, the clusters related to the correspondences become more consistent. However, the number of iterations is an important factor, because the amount of clusters that are assigned, vary depending on how the model fits the data (set of brain structures).

Also, Fig. 5 shows representative results on establishing groupwise correspondences using brain structures such as thalamus and putamen. The results also show that even when the brain structures are not well defined (i.e. basal ganglia area) the model can find meaningful correspondences that cluster similar information between 3D shapes.

In order to evaluate quantitatively the efficiently of the proposed method, we used the adjusted Rand index, with the aim to quantify the similarity between the groupwise correspondences (inferred clusters and true clusters) [15]. Here, we

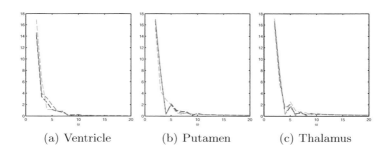

(a) Ventricle   (b) Putamen   (c) Thalamus

Fig. 3. First ten frequencies of $H(w)$ and $H(w')$ used as scale-invariant HKS. The two descriptors computed at the two different scales for the three brain structures are aligned.

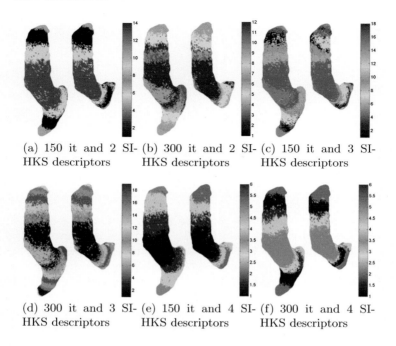

(a) 150 it and 2 SI- (b) 300 it and 2 SI- (c) 150 it and 3 SI-
HKS descriptors   HKS descriptors   HKS descriptors

(d) 300 it and 3 SI- (e) 150 it and 4 SI- (f) 300 it and 4 SI-
HKS descriptors   HKS descriptors   HKS descriptors

Fig. 4. Results of groupwise shape correspondences for ventral brain structures.

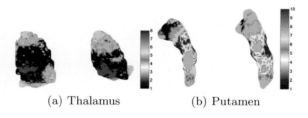

(a) Thalamus   (b) Putamen

Fig. 5. Results of groupwise shape correspondences for thalamus and putamen brain structures.

use Voronoi tessellation[2] in order to create a set of true clusters (Voronoi regions) for every brain structure. Finally, Fig. 6 show the adjusted Rand index in order to evaluate which estimated clusters (groupwise correspondence) lie on the label clusters derived by the Voronoi tessellation. The results show efficient results when two brain structures are analyzed (0.92 for ventricle, 0.86 for thalamus and 0.79 for putamen respectively). Moreover, we can notice that the index decreases as the number of brain structures (domains) to model increases, since the amount of possible groups of cluster matching increases.

---

[2] We use the fast-marching toolbox developed by Gabriel Peyre and available on https://github.com/gpeyre/matlab-toolboxes/tree/master/toolbox_fast_marching.

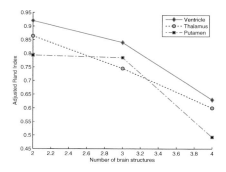

Fig. 6. Accuracy in the clusters assignment using different numbers of domains D (brain structures). Voronoi regions over 3D shapes, are used as ground-truth.

4   Conclusions

In this paper, we introduced a method for computing groupwise correspondences on 3D brain structures based on probabilistic latent variable models. The proposed approach can efficiently establish meaningful correspondences without using any similarity measure. Here, a scale-invariant heat kernel signature was used to compute 3D shape descriptors in order to extract scale-invariant features, due to the shape variability over the brain structures. Finally, we use a Bayesian nonparametric model in order to assign latent vectors for each shape descriptor. From these latent vectors we can cluster objects in different shapes, and find correspondences between clusters. By using a Dirichlet process prior, we can infer the number of clusters and find groupwise correspondences in an unsupervised manner. The results show that the proposed model can establish groupwise correspondences on multiple brain structures, and fits accurately to the Voronoi regions used as ground-truth

As future work, we propose to analyze this method on multi-atlas alignment, with the aim to capture multiple correspondences over a set of brain atlases

Acknowledgments. This research is developed under the project: *Estimación de los parámetros de neuromodulación con terapia de estimulación cerebral profunda, en pacientes con enfermedad de Parkinson a partir del volumen de tejido activo planeado*, financed by Colciencias with code $1110 - 657 - 40687$. H.F. García is funded by Colciencias under the program: *formación de alto nivel para la ciencia, la tecnología y la innovación - Convocatoria 617 de 2013*.

References

1. Lin, D., Calhoun, V.D., Wang, Y.: Correspondence between fmri and SNP data by group sparse canonical correlation analysis. Med. Image Anal. **18**, 891–902 (2014)
2. Lipman, Y., Funkhouser, T.: Mobius voting for surface correspondence. In: ACM SIGGRAPH 2009 Papers, pp. 72:1–72:12. ACM, New York(2009)

3. Bronstein, A.M., Bronstein, M.M., Guibas, L.J., Ovsjanikov, M.: Shape google: geometric words and expressions for invariant shape retrieval. ACM Trans. Graph. **30**, 1:1–1:20 (2011)

4. Liang, L., Szymczak, A., Wei, M.: Geodesic spin contour for partial near-isometric matching. Comput. Graph. **46**, 156–171 (2015)

5. Aiger, D., Mitra, N.J., Cohen-Or, D.: 4-points congruent sets for robust surface registration. ACM Trans. Graph. **27**(85), 1–10 (2008)

6. Brunton, A., Salazar, A., Bolkart, T., Wuhrer, S.: Review of statistical shape spaces for 3d data with comparative analysis for human faces. Comput. Vis. Image Underst. **128**, 1–17 (2014)

7. Hill, D.: Neuroimaging to assess safety and efficacy of ad therapies. Expert Opin. Investig. Drugs **19**, 23–26 (2010)

8. Cabezas, M., Oliver, A., Lladó, X., Freixenet, J., Cuadra, M.B.: A review of atlas-based segmentation for magnetic resonance brain images. Comput. Methods Programs Biomed. **104**, e158–e177 (2011)

9. Sidorov, K.A., Richmond, S., Marshall, D.: Efficient groupwise non-rigid registration of textured surfaces. In: Proceedings of the 2011 IEEE Conference on Computer Vision and Pattern Recognition, CVPR 2011, pp. 2401–2408. IEEE Computer Society, Washington, DC (2011)

10. Yang, X., Qiao, H., Liu, Z.Y.: Partial correspondence based on subgraph matching. Neurocomputing **122**, 193–197 (2013). Advances in cognitive and ubiquitous computingSelected papers from the Sixth International Conference on Innovative Mobile and Internet Services in Ubiquitous Computing (IMIS-2012)

11. Yamada, M., Sugiyama, M.: Cross-domain object matching with model selection. In: Gordon, G.J., Dunson, D.B. (eds.) Proceedings of the Fourteenth International Conference on Artificial Intelligence and Statistics (AISTATS 2011), Journal of Machine Learning Research - Workshop and Conference Proceedings, vol. 15, pp. 807–815 (2011)

12. Klami, A.: Variational bayesian matching. In: Proceedings of the 4th Asian Conference on Machine Learning, ACML 2012, Singapore, pp. 205–220, 4–6 November 2012

13. Quadrianto, N., Song, L., Smola, A.J.: Kernelized sorting. In: Koller, D., Schuurmans, D., Bengio, Y., Bottou, L., eds.: Advances in Neural Information Processing Systems 21, pp. 1289–1296. Curran Associates, Inc. (2009)

14. van Kaick, O., Tagliasacchi, A., Sidi, O., Zhang, H., Cohen-Or, D., Wolf, L., Hamarneh, G.: Prior knowledge for part correspondence. Comput. Graph. Forum (Proc. Eurographics) **30**, 553–562 (2011)

15. Iwata, T., Hirao, T., Ueda, N.: Unsupervised cluster matching via probabilistic latent variable models. In: desJardins, M., Littman, M.L. (eds.) AAAI. AAAI Press (2013)

16. Bronstein, M.M., Kokkinos, I.: Scale-invariant kernel signatures for non-rigid shape recognition. In: Proceedings of CVPR (2010)

Automatic Segmentation of Extraocular Muscles Using Superpixel and Normalized Cuts

Qi Xing[1], Yifan Li[2], Brendan Wiggins[3], Joseph L. Demer[4], and Qi Wei[3(✉)]

[1] Department of Computer Science, George Mason University, Fairfax, VA, USA
[2] Lake Braddock Secondary School, Burke, VA, USA
[3] Department of Bioengineering, George Mason University, Fairfax, VA, USA
[4] Department of Neurology, Jules Stein Eye Institute,
David Geffen Medical School at University of California, Los Angeles, USA
qwei2@gmu.edu

Abstract. This paper proposes a novel automatic method to segment extraocular muscles and orbital structures. Instead of conventional segmentation at the pixel level, superpixels at the structure level were used as the basic image processing unit. A region adjacency graph was built based on the neighborhood relationship among superpixels. Using Normalized Cuts on the region adjacency graph, we refined the segmentation by using a variety of features derived from the classical shape cues, including contours and continuity. To demonstrate the efficiency of the method, segmentation of Magnetic Resonance images of five healthy subjects was performed and analyzed. Three region-based image segmentation evaluation metrics were applied to quantify the automatic segmentation accuracy against manual segmentation. Our novel method could produce accurate and reproducible eye muscle segmentation.

Keywords: Automatic image segmentation · Extraocular muscle · Superpixel · Region adjacency graph · Normalized Cuts

1 Introduction

The extraocular muscles (EOMs) implement eye movements. Through Magnetic Resonance Imaging (MRI), it has been found that many forms of binocular alignment (strabismus) are associated with anatomical abnormalities of EOMs [1]. In clinical practice, EOM enlargement is a key quantity to examine in diagnosing several complex strabismus [2] including thyroid eye disease [3,4]. Therefore, how to reliably and efficiently outline the EOM boundaries from clinical MRI becomes an important practical and research question. In all published studies to date, investigators segment EOM boundaries manually [1,5,6], which is labor expensive and may introduce user dependent artifacts.

Several computer-aided semi-automatic [7–9] and automatic segmentation [10,11] methods have been developed. However, all of these methods used image pixels as the underlying representation primitive. It is known that pixels are not the most natural representation of visual scenes, since they do not take

© Springer International Publishing Switzerland 2015
G. Bebis et al. (Eds.): ISVC 2015, Part I, LNCS 9474, pp. 501–510, 2015.
DOI: 10.1007/978-3-319-27857-5_45

into account the local patterns among neighboring pixels and are subject to noise. It would be more natural and efficient to process the image with perceptually meaningful patches containing many pixels that share similar features.

We propose a fully automatic EOM segmentation method based on super-pixel, region adjacency graphing and Normalized Cuts, while integrating the prior shape information. Rather than processing images on the pixel level, the approach builds upon local feature of the EOMs. We consider small image patches obtained from superpixel over-segmentation [12–15] as the basic unit of any further image processing procedures, such as filtering, detection and segmentation. We show that by building a region adjacency graph of the superpixels, we can develop a robust method to outline the eye socket boundary and the EOMs within. The performance of our automatic segmentation method was evaluated by comparing our results to manual segmentation which showed high accuracy.

2   Related Work

Firbank et al. [8] showed the feasibility to segment EOMs using the active contours. However, this approach is sensitive to the boundary initialization, since it can be easily trapped in local minima [16]. In addition, the accuracy is influenced by the convergence criteria — higher accuracy requires tighter convergence criteria and longer computation time [17]. Souza et al. proposed a mathematical morphology method to semi-automatically segment EOMs [9, 18]. They performed an iterative grayscale closing operations to segment the orbital wall which was then used as the region of interest. The EOMs were outlined in the region of interest through Laplacian or Gaussian detector and opening operations. However, the size of flat disk used for the morphology operation was fixed and only worked on the pixel level. The number of iterations had to be carefully supervised. A more recent semi-automatic approach deformed 3D geometric template models of the EOMs to the MRI images of individual patients [19]. Image features of the EOMs were detected and filtered to guide fitting of the generic anatomical template. However, the template model has to be built by considering anatomical characteristics of the EOMs. In addition, a global registration between the image sequence and the template model had to be performed at the beginning.

3   Methodology

3.1   Superpixel Over-Segmentation

The image segmentation algorithm Superpixel [13] groups pixels with coherent intensities and spatial locations into patches of pixels. These superpixels provide a high level representation of the original image which can be used for further processing. The geometric shapes of superpixels are not restricted to rectangular. Such flexibility enables representation of features more naturally by maintaining the boundaries of the objects in the image. Accurate segmentation can then be performed by merging the local superpixels which have similar features.

We applied the k-means algorithm to group nearby pixels into superpixels in uniform sizes [12]. Unlike other superpixellization methods [20, 21], the k-means method produces a more regularized grid of superpixels, which is important for building the region adjacency graph. Figure 1(a) shows a T-1 weighted quasi-coronal MRI image perpendicular to the long axis of the orbit with 312 micron pixels and 2 mm plane thickness. Figure 1(b) illustrates the result of the superpixel over-segmentation. The boundaries of the superpixels preserve the true structure boundaries. More importantly, the shape and area characteristics of the EOMs and the eye socket are relatively consistent [9]. As the algorithm restricts, the number of pixels in each superpixel is nearly constant across the image.

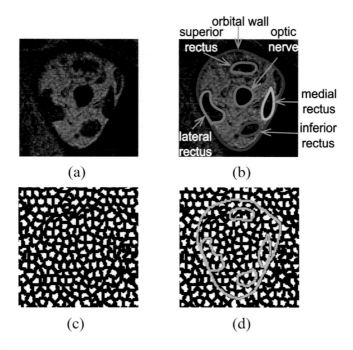

(a)   (b)

(c)   (d)

Fig. 1. (a) A representative coronal MRI image of the eye; (b) labeled ocular structures to segment; (c) superpixel over-segmentation; (d) manually segmented structured overlaid on the superpixels.

3.2   Region Adjacency Graph Construction

Human visual perception is good at recognizing individual objects even with varying intensities or textures. As an algorithm with no priors, superpixel segmentation algorithms [12, 13, 22, 23] have the tendency to over-segment the image (Fig. 1(c)(d)). Region adjacency graph (RAG) is a data structure common for many segmentation

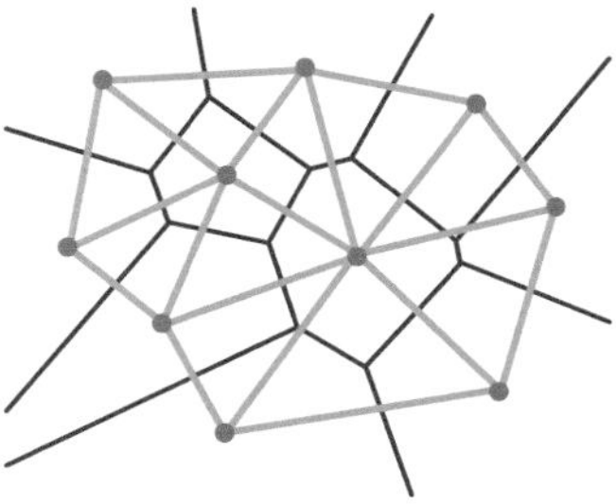

Fig. 2.   Region adjacency graph build from superpixels.

algorithms [24]. We used RAG to specify spatial connection of neighboring superpixels. Each superpixel was defined as a node in a graph (Fig. 2). Each superpixel was connected through the edges to all its neighbors. The RAG was used
to merge adjacent regions provided that these regions have similar intensity
distributions. Denote n_i to be a node in RAG with mean intensity I_i and n_j
to be one of its neighbors. The edge weight between n_i and n_j is defined as
$w_{ij} = \exp\left(\frac{-\|I_i - I_j\|^2}{\sigma^2}\right)$, where σ^2 is overall image variance. w_{ij} measures the
intensity similarity between n_i and n_j.

3.3 Normalized Cuts Segmentation

Superpixel segmentation is a bottom-up
approach as it merges individual pixels
together. Once we have the superpixels
represented by the RAG, we applied the
Normalized Cuts algorithm (Ncut) [25],
a top-down approach to partition the
graph recursively until finding the ocular structures. Applying the Normalized
Cuts other on superpixels other than pixels is more robust to image noise. In addition, it has an obvious advantage of being
computationally efficient, as the size of
the affinity matrix and the complexity of
the RAG employed for image representation are significantly reduced.

Fig. 3. Normalized Cuts segmentation
produced regions labeled in different
gray scales.

In each division, the Ncut algorithm
optimally divides one region into two subregions N_1 and N_2 by removing edges
connecting them in RAG:

$$Ncut(N_1, N_2) = \frac{cut(N_1, N_2)}{assoc(N_1, N)} + \frac{cut(N_1, N_2)}{assoc(N_2, N)}. \tag{1}$$

$cut(N_1, N_2) = \sum_{n_i \in N_1 \& n_j \in N_2} w_{ij}$ computes the degree of dissimilarity
between N_1 and N_2 as summed weight of all the removed edges. $assoc(N_1, N)$
defines the total edge weight from nodes in N_1 to all nodes in the current region.

Figure 3 shows the segmentation results after applying Normalized Cuts. The
final nodes after Normalized Cuts contain many superpixels and are colored in
grey scale. The Normalized Cuts segmentation can successfully highlight some
of the ocular structures such as the superior rectus muscle, the oblique rectus
muscle and the optic nerve. However, the lateral and medial rectus muscles had
incomplete boundaries and their boundaries are connected with the orbital wall,
making them difficult to segment. Further automatic operations are needed to
solve discontinuity issues and label each structure.

3.4   Orbital Wall and Extraocular Muscle Segmentation

Segmenting the orbital wall was studied previously. Souza et al. [9] applied an iterative grayscale mathematical morphology operation to segment the orbits. But their method required the user to specify a flat disk template and number of recursive iterations of erosions. Firbank et al. [8] manually outlined the boundary around the eye socket. We proposed an automatic method to extract the orbital wall with shape prior knowledge. The Laplacian of the Gaussian [26] and connected components labeling methods are applied to detect the connected boundaries of the orbital wall, rectus muscles and optic nerve from the segmented Normalized Cuts image shown in Fig. 4(a).

To extract the orbital wall, we considered the prior knowledge that the eye socket was always located near the image center. The center of each region produced by Normalized Cuts was calculated and shown in Fig. 4(b). The centers of the optic nerve and the orbital wall were the two closest centers near the image center. Using the k-nearest algorithm, the regions of the optic nerve and the orbital wall can be identified from the boundary map. Any region outside the orbital wall were removed (see (Fig. 4(c)). Two of the closed boundaries inside the orbital wall were identified as the superior and inferior rectus muscles (Fig. 4(c)). To segment the lateral rectus muscle and medial rectus muscle with incomplete boundaries, the convex hull around the orbital wall was calculated and shown as the red closed curve in Fig. 4(d). The generated closed orbital wall

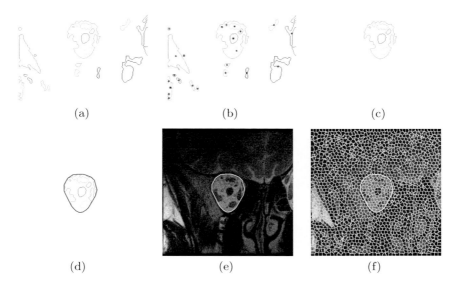

(a)   (b)   (c)

(d)   (e)   (f)

Fig. 4. Region of interest extraction. (a) Initial boundaries after Ncuts based on Fig. 3; (b) Center(*) of each boundary (c) Finding optical never and orbit regions using k-mean cluster; (d) Orbit and EOMs boundaries identified using convex hull; (e) Region of interest in the original image; (f) Region of interest in the superpixel image (Color figure online).

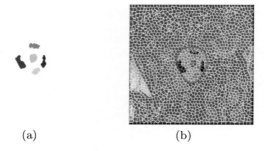

(a)   (b)

Fig. 5. (a) Segmented extraocular muscles and optic nerve. (b) Superpixels mapped with the orbital wall and the segmented ocular structures.

completed the initial discontinuous boundaries of the lateral and medial rectus muscles which are in contact with the orbital wall. The convex hull served as the region of interests and reserves the natural boundary of the eye socket in the original image (Fig. 4(e)) and the superpxiel image (Fig. 4(f)). Finally, image region and hole filling algorithms were used to complete the EOM segmentation (Fig. 5(a)). Figure 5(b) shows the segmented boundaries overlaid with the MRI and superpixel images.

4   Experiments and Results

4.1   Materials

The T1-weighted MRI images of both eyes were acquired from 5 health subjects and provided by Dr. Joseph Demer at UCLA. Eight coronal images at the slice thickness of 2 mm for each eye were segmented. All images were digitized with 256×256 pixels and 16 bits gray-level of resolution at voxel size of $0.3 \, \text{mm} \times 0.3 \, \text{mm} \times 2.0 \, \text{mm}$. We asked two operators to independently and manually trace the ocular structure boundaries, which were used as ground truth for accuracy assessment.

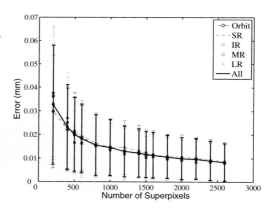

Fig. 6. The shape error between the manually segmented boundaries and the superpixel over-segmented boundaries decreases as the number of superpixels n increases for orbit, superior rectus (SR), inferior rectus (IR), medial rectus (MR), lateral rectus (LR) and all four rectus (All) muscles

4.2   Shape Error Analysis

One critical issue is to determine the number of superpixels n when applying the k-means algorithm. If there are too few superpixels, we may miss out important structures. On the other hand, if there are too many, we lose the superpixel's advantage of being representative, robust and efficient. In order to determine an appropriate n, parameter learning was performed by analyzing the influence of n on the segmentation accuracy. We first manually segmented one set of MRI images by tracing the boundaries of the ocular structures. We applied superpixel over-segmentation on the same set of images while varying $200 < n < 2600$. For each n, we overlapped the manually traced boundaries to the superpixel image (Fig. 1(c)(d)). The shape error between the manual and the superpixel segmentations was computed using the boundary-based measurement [13] to quantify how close the superpixel boundaries are to the manual segmentation (approximating ground truth). The shape error was calculated as the mean absolute distance between the superpixel boundaries and the ground truth boundaries. Figure 6 plots the shape error as a function of n for different ocular structures. Unsurprisingly, as the number of superpixels increases, the error drops monotonically to zero when each pixel becomes one superpixel. According to "elbow selection criterion" [27,28], $n = 1800$ was chosen as the number of superpixels for our MRI dataset in subsequent operations.

4.3   Performance Evaluation

Area-based metric [9] and volume-based metric [7,8] are commonly used to evaluate the accuracy of image segmentation. One drawback is that these two metrics do not consider the overlap between ground truth and computer generated segmentation results. We decided to use the region-based metrics to assess our proposed approach: Variation of Information (VI) [29], Probabilistic Rand Index (RI) [30] and Segmentation Covering Criteria (Covering) [31].

Variation of Information. The Variation of Information metric was introduced for the purpose of clustering comparison [29]. It measures the distance between two segmentations in terms of their average conditional entropy defined as

$$VI(S, S') = H(S) + H(S) - 2I(S, S'), \tag{2}$$

where H and I represent the entropies and mutual information between two clusters of data S and S'.

Rand Index. Rand Index [30] was first proposed for general clustering evaluation. It operates by comparing the compatibility of assignments between pairs of elements in the clusters. The Rand Index between the automatical segmentation and the manual segmentation X and G is given by summing the number of pairs of pixels that have the same label in $X \bigcup G$ and those with different labels in both segmentations, divided by the total number of pairs of pixels.

Table 1. Computational time (in seconds) of applying three collision detection methods.

		VI		RI		Covering	
		ODS	OIS	ODS	OIS	ODS	OIS
Left	IR	1.25	1.13	0.78	0.81	0.75	0.78
	LR	1.47	1.25	0.81	0.84	0.78	0.82
	MR	1.45	1.35	0.75	0.78	0.72	0.79
	SR	1.86	1.65	0.82	0.83	0.81	0.85
Right	IR	1.75	1.67	0.81	0.84	0.71	0.81
	LR	1.65	1.47	0.85	0.86	0.78	0.83
	MR	1.87	1.58	0.84	0.87	0.73	0.78
	SR	1.47	1.25	0.79	0.82	0.76	0.84

Segmentation Covering. We define the covering of segmentation S by segmentation S' as

$$C(S' \rightarrow S) = 1/N * \sum_{R \in S} |R| * max_{R' \in S'} O(R, R'), \qquad (3)$$

where N is the total number of pixels in the image and $O(R, R') = \frac{|R \cap R'|}{|R \cup R'|}$ is the overlap between two regions R and R' [31].

Table 1 summarized evaluation results using the region-based metrics. The IR, LR, MR and SR muscles of both left and right eyes were analyzed. We computed two scores for the Segmentation Covering for EOMs, which were segmentations at a optimal data set scale (ODS) and a optimal image scale (OIS). We also examined the Rand Index and Variation of Information quantities compared to the manual segmentation. Lower value of Variation of Information metric indicates greater similarity. Our average Variation of Information value is 1.51 which outperforms other image segmentation outcomes [32]. Rand Index metric is in the range [0 1]. Higher value indicates greater similarity between two segmentations. For different EOMs, The average value for Rand Index metric is greater than 0.82 which shows the excellent performance of our algorithm. The standard deviation of Rand Index is about 0.03 demonstrating that our method can generate consistent results for four different EOMs. With respect to Segmentation Covering, we compute the normalized overlap score in range [0 1]. The larger the value, the more accurate the algorithm. The average value for different EOMs is 0.78, which is consistent to Rand Index result. In summary, as Table 1 shows, our automatic segmentation algorithm was able to segment boundaries fairly close to the manual segmented boundaries which illustrates the accuracy and effectiveness of our approach.

5   Conclusions

Extraocular muscles segmentation from MRI is an important and challenging task for clinical diagnosis. This study has demonstrated an automatic method using Superpixel, Region Adjacency Graph and Normalized Cuts. The results were compared to manual segmentations. Region-based segmentation evaluation metrics showed that our method was able to segment boundaries fairly accurately. In the future, we plan to improve the efficiency of segmentation on superpixels using Normalized Cuts by using GPU to accelerate the Normalized Cuts. Alternatively, we will employ other more efficient graph cut algorithms to segmentation the images. To improve the reliability for the optic nerve identification, we will consider the shape prior of the optic nerve and locate it by using circle fitting method on the segmented shapes. The method will be applied to automatically reconstruct 3D patient-specific EOM models which will be used in clinical diagnosis and surgical planning.

Acknowledgments. Supported by the Jeffress Trust Awards, NIH grant EY08313 and an unrestricted grant from Research to Prevent Blindness.

References

1. Bijlsma, W.R., Mourits, M.P.: Radiologic measurement of extraocular muscle volumes in patients with Graves' orbitopathy: a review and guideline. Orbit **25**, 83–91 (2006)
2. Ben Simon, G.J., Syed, H.M., Douglas, R., McCann, J.D., Goldberg, R.A.: Extraocular muscle enlargement with tendon involvement in thyroid-associated orbitopathy. Am. J. Ophthalmol. **137**, 1145–1147 (2004)
3. Dal Canto, A.J., Crowe, S., Perry, J.D., Traboulsi, E.I.: Intraoperative relaxed muscle positioning technique for strabismus repair in thyroid eye disease. Ophthalmology **113**, 2324–2330 (2006)
4. Gupta, A., Sadeghi, P.B., Akpek, E.K.: Occult thyroid eye disease in patients presenting with dry eye symptoms. Am. J. Ophthalmol. **147**, 919–923 (2009)
5. Kono, R., Poukens, V., Demer, J.L.: Quantitative analysis of the structure of the human extraocular muscle pulley system. Invest. Ophth. Vis. Sci. **43**, 2923–2932 (2002)
6. Chaudhuri, Z., Demer, J.L.: Sagging eye syndrome: connective tissue involution as a cause of horizontal and vertical strabismus in older patients. JAMA Ophthalmol. **131**, 619–625 (2013)
7. Firbank, M.J., Coulthard, A.: Evaluation of a technique for estimation of extraocular muscle volume using 2D MRI. Brit. J. Radiol. **73**, 1282–1289 (2000)
8. Firbank, M.J., Harrison, R.M., Williams, E.D., Coulthard, A.: Measuring extraocular muscle volume using dynamic contours. Magn. Reson. Imaging **19**, 257–265 (2001)
9. Souza, A.D.A., Ruiz, E.E.S., Cruz, A.A.V.: Extraocular muscle quantification using mathematical morphology: a semi-automatic method for analyzing muscle enlargement in orbital diseases. Comput. Med. Imag. Grap. **31**, 39–45 (2007)
10. Szucs-Farkas, Z., Toth, J., Balazs, E., Galuska, L., Burman, K.D., Karanyi, Z., Leovey, A., Nagy, E.V.: Using morphologic parameters of extraocular muscles for diagnosis and follow-up of Graves' ophthalmopathy: diameters, areas, or volumes? Am. J. Roentgenol. **179**, 1005–1010 (2002)

11. Lv, B., Wu, T.N., Lu, K., Xie, Y.: Automatic segmentation of extraocular muscle using level sets methods with shape prior. In: Long, M. (ed.) IFMBE Proceedings. IFMBE, vol. 39, pp. 904–907. Springer, Heidelberg (2012)

12. Achanta, R., Shaji, A., Smith, K., Lucchi, A., Fua, P., Susstrunk, S.: SLIC superpixels compared to state-of-the-art superpixel methods. IEEE T. Pattern Anal. **34**, 2274–2282 (2012)

13. Ren, X., Malik, J.: Learning a classification model for segmentation. In: ICCV, pp. 10–17 (2003)

14. Fulkerson, B., Vedaldi, A., Soatto, S.: Class segmentation and object localization with superpixel neighborhoods. In: ICCV, pp. 670–677 (2009)

15. Levinshtein, A., Sminchisescu, C., Dickinson, S.: Optimal contour closure by superpixel grouping. In: Daniilidis, K., Maragos, P., Paragios, N. (eds.) ECCV 2010, Part II. LNCS, vol. 6312, pp. 480–493. Springer, Heidelberg (2010)

16. Chan, T., Vese, L.: Active contours without edges. IEEE T. Image Process. **10**, 266–277 (2001)

17. Kass, M., Witkin, A., Terzopoulos, D.: Snakes: active contour models. Int. J. Comput. Vision **1**, 321–331 (1988)

18. Souza, A., Ruiz, E.: Fast and accurate detection of extraocular muscle borders using mathematical morphology. In: IEMBS, pp. 1779–1782 (2000)

19. Wei, Q., Sueda, S., Miller, J., Demer, J., Pai, D.: Template-based reconstruction of human extraocular muscles from magnetic resonance images. In: ISBI, pp. 105–108 (2009)

20. Felzenszwalb, P.F., Huttenlocher, D.P.: Efficient graph-based image segmentation. Int. J. Comput. Vision **59**, 167–181 (2004)

21. Vedaldi, A., Soatto, S.: Quick shift and kernel methods for mode seeking. In: Forsyth, D., Torr, P., Zisserman, A. (eds.) ECCV 2008, Part IV. LNCS, vol. 5305, pp. 705–718. Springer, Heidelberg (2008)

22. Conrad, C., Mertz, M., Mester, R.: Contour-relaxed superpixels. In: Heyden, A., Kahl, F., Olsson, C., Oskarsson, M., Tai, X.-C. (eds.) EMMCVPR 2013. LNCS, vol. 8081, pp. 280–293. Springer, Heidelberg (2013)

23. Gould, S., Rodgers, J., Cohen, D., Elidan, G., Koller, D.: Multi-class segmentation with relative location prior. Int. J. Comput. Vision **80**, 300–316 (2008)

24. Tremeau, A., Colantoni, P.: Regions adjacency graph applied to color image segmentation. IEEE T. Image Process. **9**, 735–744 (2000)

25. Shi, J., Malik, J.: Normalized cuts and image segmentation. IEEE T. Pattern Anal. **22**, 888–905 (2000)

26. Sharifi, M., Fathy, M., Tayefeh Mahmoudi, M.: A classified and comparative study of edge detection algorithms. In: ITCC, pp. 117–120 (2002)

27. Thorndike, R.L.: Who belongs in the family? Psychometrika **18**, 267–276 (1953)

28. Tibshirani, R., Walther, G., Hastie, T.: Estimating the number of clusters in a data set via the gap statistic. J. Roy. Stat. Soc. B Met. **63**, 411–423 (2001)

29. Meila, M.: Comparing clusterings: an axiomatic view. In: ICML, pp. 577–584, New York (2005)

30. Rand, W.M.: Objective criteria for the evaluation of clustering methods. J. Am. Stat. Assoc. **66**, 846–850 (1971)

31. Malisiewicz, T., Efros, A.A.: Improving spatial support for objects via multiple segmentations. In: BMVC, pp. 1–10 (2007)

32. Arbelaez, P., Maire, M., Fowlkes, C., Malik, J.: Contour detection and hierarchical image segmentation. IEEE T. Pattern Anal. **33**, 898–916 (2011)

More Usable V-EGI for Volumetric Dataset Registration

Chun Dong[(⊠)] and Timothy S. Newman

University of Alabama in Huntsville, Huntsville, AL 35899, USA
{dongc,tnewman}@cs.uah.edu

Abstract. Enhancements to the Volume-based Extended Gaussian Image (V-EGI) registration are described. The first enhancement is the capability to recover positional difference (in additional to rotational difference) between volumetric datasets. The second, and most important, enhancement uses a multi-stage coarse-to-fine processing strategy to improve computational speed. That enhancement also incorporates an optimization scheme to enable the strategy to maintain accuracy. The third enhancement is a methodology that achieves a moderate degree of parallelism on current-generation multi-core CPUs. Results of application of these methodologies to multiple datassets are also presented.

Keywords: Image registration · Volumetric data processing · Extended Gaussian Image · Orientation histogram

1 Introduction

Many applications, such as remote sensing, medical imaging, image mosaicing, etc. [1], can require aligning image data. The process of transforming images of the same subject into a common coordinate system is called image registration. Registration can benefit a variety of domains that involve considering multiple images together. For example, medical images collected over time from one patient may need registration to aid in determining change over time. Some applications also utilize multiple modalities of images collected at a particular time. Image registration is a difficult task, in particular when images are collected by differing sensors or when images are from different viewpoints and there is noise in images. But even when these factors are not present, image registration can often be complicated by other subjects being present in some images.

One way range (or scattered point) data of objects (i.e., technically of object surfaces determined from such data) has been registered is with the Extended Gaussian Image (EGI) technique of Horn [2]. The EGI technique forms and uses a surface-based orientation histogram (that is called the EGI) that describes the distribution of the normal vectors for the object surfaces. Section 3.1 describes briefly the classical EGI.

A spin-off of the EGI, called V-EGI [3], allows for rotational registration of volume datasets. V-EGI is slow and does not recover translations, though. Here,

G. Bebis et al. (Eds.): ISVC 2015, Part I, LNCS 9474, pp. 511–520, 2015.
DOI: 10.1007/978-3-319-27857-5_46

enhancement of the V-EGI to allow for both translation and rotation recovery and multi-pronged means to improve its speed are described.

The paper is organized as follows. Section 2 describes others work on image and volume registration. The main concepts of EGI and V-EGI are described in Sect. 3. The enhancements (to recover positional difference between volumetric datasets and using multi-stage processing to determine the rotational difference between volumetric datasets) are discussed in detail in Sect. 4. Computational enhancement by multi-threading is explored in Sect. 5. Experimental results are presented in Sect. 6. Finally, the conclusion is in Sect. 7.

2   Related Work

Closely related work in registration is discussed next. The registration literature is large, although some surveys [1,4–6] provide a good overview of the field.

In image registration, the key is to align images of a common object. The Iterative Closest Points (ICP) algorithm is a widely used method to find a transformation that aligns images. The algorithm starts with two points and an estimate of the aligning rigid-body transform. Next, it refines the transform in an iterative way that considers corresponding points between the objects.

EGI-based strategies have been applied in many registrations of range images. For example, Pan et al. [7] found a plane of facial symmetry in range images using EGI.

Ikeuchi [8] used EGI as a $2\frac{1}{2}$D representation for recognizing CAD-modeled objects, such as ellipsoids and cylinders. Chibunichev and Velizhev [9] described a method that uses EGI to match terrestrial laser scans of buildings, monuments, and industrial complexes. It used an angular orientation difference, computed by searching the set of possible rotations to find the best correlation between orientation histograms.

Dold [10] used a series of EGIs to register terrestrial scan data (surface points). The method first finds an estimate of the aligning transformation using a coarse level tessellation for EGI. It then proceeds to use progressively finer tessellations. Each new, finer-level EGI uses the angular difference value of the alignment identified in the prior tessellation for its initial value. After four refinements, ICP was applied to find a final alignment.

Maes et al. [11] used an information theory measure, mutual information (MI), in a fully automatic, data independent approach for registering medical imaging datasets. It uses MI of image intensity values of matching pixel pairs as a measurement of information redundancy. It treats this MI as a similarity measure. The measure is maximal when images are geometrically aligned. It does not require any prior segmentation or preprocessing steps. They reported using the approach to align differing modalities of volume data.

Guillaume et al. [12] have described a method using surfaces of segmented kidneys to register 3D kidney CT datasets. It repeatedly traces rays from inside kidney regions towards the kidney surface to find a center of gravity. It also constructs spherical harmonics functions that are used in aligning principal axes and centers of gravity.

Ni et al. [13] have proposed a 3D SIFT-based method for generating a panorama from multiple ultrasound datasets. It combines a 3D SIFT detector with the Rohr3D detector to detect features. Then diffusion distance, which provides robustness, is used for feature matching. The method takes about one minute to register two small volumes using a 3 GHz Intel CPU.

Papazov and Burschka [14] have described a stochastic approach using global optimization to rigidly register point sets with unknown point correspondences. Their approach decomposes the rotation search space into disjoint equal-sized parts called spherical boxes to achieve fast rejection of candidates.

In addition, some registration methods have been developed for parallel computing environments [15]. Representative efforts include a MI-based 3D data registration method [16] that uses CUDA on a GPU and a client-server blackboard-based medical image registration method [17].

3 EGI and V-EGI

In this section, volumetric registration using V-EGI is described in detail. The technique is for rigid registration; it assumes there is negligible deformation.

3.1 Classic EGI Construction

First, we describe classic EGI. It is for surface-based data, such as range images. It uses a sampling of normal vectors across the surface of the object of interest, mapping each to a location (on a tesselated Gaussian sphere) that corresponds to that vector's orientation. EGI counts the normal vectors mapped to each tessel (face). The set of counts on the Gaussian sphere is a surface orientation histogram. A given rotation of the object corresponds to the same rotation of its EGI. Object translation does not affect the EGI. In EGI, the object EGI is repeatedly compared to rotations of the model's EGI in different orientations. The similarity is found for each rotation, and the one giving rise to the orientation with the highest similarity is taken as the aligning rotation.

3.2 Volumetric EGI Construction

Next, the volumetric EGI (V-EGI) technique [3] is described. V-EGI builds a volume gradient orientation histogram (henceforth, a gradient histogram), rather than a surface orientation histogram, from a target volumetric dataset to align it to a model dataset. Like in EGI, this histogram is built on a polyhedral approximation to a sphere (e.g., a 2048-tessel polyhedron [3]). The gradient histogram uses only the gradients from *significant voxels* of a dataset. A significant voxel is one with: (1) a data value, D, satisfying $M_1 \leq D \leq M_2$ for some thresholds M_1 and M_2 (M_1 is a minimal data value threshold and M_2 is a maximal one), and (2) a gradient magnitude, G, that has a value $G \geq \epsilon$, where ϵ is a threshold (set to 15 for many datasets). When V-EGI recovers orientation, only the significant voxels directly contribute to the histogram.

In uses of V-EGI here, the aligning rotation between a volumetric instance (the target) of the subject (with the model reference) uses the count of the total number of *similar vectors* between the V-EGIs of the model and the target. Similar vectors are those vectors in coincident cells of gradient histograms H_m (the model dataset's one), and H_t (the target dataset's one), that have a difference in magnitude less than a threshold λ (e.g., 1.0). First, we compute a matching criterion for each possible alignment between H_m and H_t. Then the one with the maximum criterion value is taken as the best aligning transformation. Formally, for each possible rotation of the H_m gradient histogram, we take the total number of similar vectors in the two V-EGIs as the matching criterion. This is processed as: counting the similar vectors in each cell between the two gradient histograms, and then summing these counts. Whichever rotational case has the largest number of similar vectors defines the best aligning rotation $\widetilde{R}$.

4   Pose Translation and Faster Orientation Recovery Enhancements

Next, two of the key aspects of the Enhanced V-EGI are described. One of these is making V-EGI able to recover the full rigid body transformation of a volume. Another is its major enhancement: using multi-stage processing to improve speed.

4.1   Translational Recovery

Our translation recovery uses the rotation transformation result, $\widetilde{R}$, found by the V-EGI. The basic idea of the translation recovery is to find centroids from the significant voxels used in the orientation recovery. Specifically, such centroids for the model and target datasets are found. The distance between these centroids is taken as the translation between the datasets.

Translation recovery's steps are shown in Listing 1.

Listing 1. V-EGI Translation Transformation Detection algorithm

1. Get the rotation result, $\widetilde{R}$, from the original V-EGI.
2. Apply transformation $\widetilde{R}$ on every point in point set P_m from the model dataset.
3. Compute centroid C_m for model dataset's point set P_m.
4. Compute centroid C_t for target dataset's point set P_t.
5. Compute aligning translation T_t as $T_t = C_m - C_t$.

4.2   Multi-stage Pose Orientation Recovery

Next, our multi-stage enhancement strategy is presented. It uses coarse-to-fine processing with Powell optimization. The original V-EGI only used a one stage brute force matching method without Powell optimization. The orientation recovery through our enhanced method only takes minutes compared to the hours of the brute-force matching in the original V-EGI approach.

Next, we explain our approach on gradient histogram matching, assuming a set P_m of n significant voxels $\{p_{m_1}, p_{m_2}, \cdots, p_{m_n}\}$ of the reference dataset and a set P_t of k significant voxels $\{p_{t_1}, p_{t_2}, \cdots, p_{t_k}\}$ of the target one. The approach here has three stages, each of which refines the prior one. In each stage, a rotation set R_s, which predefines all possible aligning orientations (i.e., triples of rotation angles, each triple a rotation α around the axis x, denoted as r_α , β around the axis y, denoted as r_β, and γ around the axis z, denoted as r_γ), is built. (Each triple rotation defines a transformation R that aligns the model point set and target.) The building of set R_s is based on the starting rotation angles $(r_\alpha, r_\beta, r_\gamma)$, angular range (denoted as S_r) and angular step (denoted as S_s). This building process iteratively computes each possible rotation triple by stepping through the full range of angles. Each such triple is added to R_s. In each successive stage, the orientation detection result $(r_\alpha, r_\beta, r_\gamma)$ from the last stage of V-EGI is the starting rotation angle set for a new R_s. The angular range and the angular step decrease in each stage. (Based on our experiments, we used S_r of $180\,°$ and S_s of $10\,°$ in the first stage. In the second stage, we used S_r of $10\,°$ and S_s of $5\,°$. In the last stage, we used S_r of $8\,°$ and S_s of $1\,°$).

We note if there are q rotation test cases in set R_s, the time complexity for our orientation recovery algorithm is $O(q * n * k)$.

We incorporate Powell optimization [18], which finds a local maximum of a function that counts the number of similar vectors in the multi-stage orientation detection. (Although, we have also tested the multi-stage approach without Powell Optimization.) Specifically, we utilize the Powell optimization at the end of each orientation detection stage. Utilizing Powell in this way allows optimizing the result produced by the refinement process. We perform Powell optimization using an optimization function $F(r_\alpha, r_\beta, r_\gamma)$ to compute the maximal number of similar vectors in H_t and H_m starting at an input rotation. Here angles $r_\alpha, r_\beta, r_\gamma$ represent rotation angles about the x, y and z axes, respectively.

The multi-stage strategy's steps are shown in Listing 2.

Listing 2. Multi-stage Orientation Detection Strategy

1. Based on magnitude and density thresholds, extract significant point sets P_t (from the target dataset) and P_m (from the model dataset).
2. Compute normal vectors N_t from point set P_t and N_m from point set P_m.
3. Build orientation histogram H_m from N_m.
4. Initialize *Counter*, rotation angles $r_\alpha, r_\beta, r_\gamma$, search range, S_r and step S_s.
5. Repeat until *Counter* equals 3:
   (a) Generate R_s: all rotations within S_r of $(r_\alpha, r_\beta, r_\gamma)$ when stepping by S_s.
   (b) For each rotation triple in the rotation set R_s:
      i. Retrieve a rotation triple $r_\alpha, r_\beta, r_\gamma$ from rotation set R_s and create a transformation R based on this rotation triple.
      ii. Apply this transform R on each vector in N_t and recompute the new orientation histogram H_t based on the transformed vectors.

iii. In each pair of corresponding bins (i.e., ones with the same bin index) in H_t and H_m, count the number of similar vectors, sum the number of similar vectors in all bins, and store this number and triple $r_\alpha, r_\beta, r_\gamma$ in Result List.

(c) On the Result List, find the entry with the maximal number of similar vectors, then retrieve the associated rotation triple $r_\alpha, r_\beta, r_\gamma$.

(d) Use $F()$ starting at $r_\alpha, r_\beta, r_\gamma$ to compute the local optimized (maximal) number of similar vectors (as defined in Sect. 3.2) and optimized rotation angles $\widetilde{r}_\alpha, \widetilde{r}_\beta, \widetilde{r}_\gamma$.

(e) Increment $Counter$, modify S_r and S_s, and set $r_\alpha = \widetilde{r}_\alpha$, $r_\beta = \widetilde{r}_\beta$, $r_\gamma = \widetilde{r}_\gamma$

6. Rotation $\widetilde{R}$ is the orienting transformation based on $\widetilde{r}_\alpha, \widetilde{r}_\beta, \widetilde{r}_\gamma$ between the model dataset and the target dataset.

In application, we couple this faster orientation detection with the translation recovery described in Sect. 4.1.

5   Parallelization

We next describe use of multi-threading to speedup Enhanced V-EGI. Our approach involves multi-threading of each iteration of Step 5(b) (in Listing 2) presented in Sect. 4. Specifically, we treat the iterations of Step 5(b) as independent efforts, and task one thread per effort (i.e., iteration). We create a fixed number of threads and divide the iterations of Step 5(b) among them. When any thread completes, its result, the number of the similar vectors between the reference and the target and the rotation information, is saved. Once all threads have finished, the rotation triple in the entry of Result List having the largest number of similar vectors is taken as our rotation transformation result. Figure 1 (a) illustrates the execution profile for our multi-threading registration approach.

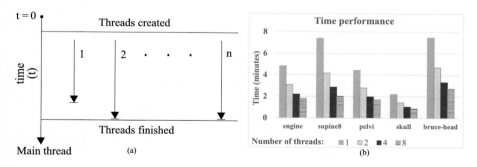

Fig. 1. Multi-threaded enhanced V-EGI registration performance. (a) Execution profile. (b) Run times.

The amenability of Enhanced V-EGI to parallelization makes it especially well-suited to current generation multi-core CPUs.

6 Experimental Results

Next, we report experiments on the computational performance and accuracy of Enhanced V-EGI. Testing was performed on five real datasets: the Engine (CT, $181 \times 181 \times 181$), Skull (CT, $181 \times 181 \times 181$), Supine8 (CT, $181 \times 181 \times 181$), Pelvi (CT, $181 \times 181 \times 181$) and BruceHead (MRI, $181 \times 181 \times 181$) datasets available from the U.S. National Library of Medicine [19] and Roettgers Volume Library [20]. For each dataset, 7 different rotated instances and four different translated instances were generated to test Enhanced V-EGI's accuracy. In addition, versions of these datasets with additive noise were generated (three levels of Gaussian noise were considered for these cases, with $\mu = 0$ and $\sigma^2 \in \{1, 4, 9\}$). Experiments of the performance were done on a PC with an Intel 3.4 GHz Dual Core i7-2600 processor with 4 GB memory on a Windows 7 system. The algorithms were coded in C++ and compiled using Microsoft Visual C/C++ 2010. Threads were implemented using the Windows thread library.

Figure 1 (b) summarizes the execution times for multi-threaded, multi-stage coarse-to-fine Enhanced V-EGI registration (i.e., with the Powell optimization) for the real datasets without any noise added. Summaries for use of 1, 2, 4 and 8 threads are shown. The speed up from multi-threading with 8 threads ranged from 2.6 to 3.6 times, with an average of 2.8 times. While the new Enhanced V-EGI only takes about 1 to 2 min to complete with 8 threads, the standard V-EGI takes about 2 h for registration with 8 threads.

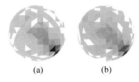

(a)   (b)

Fig. 2. Intensity-coded renderings of enhanced V-EGI gradient histograms

Figure 2 shows volumetric EGIs of the CT dataset supine8 in two poses. Cell counts are rendered using a grayscale coding, with degree of gray/black indicating counts (white indicates lower/zero counts).

Figure 3 (a)–(c) show isosurface renderings of the Engine dataset with isosurface value of 70. The reference (model) is rendered in part (a). A target dataset, which is a version of the Engine dataset rotated by $r_\alpha = 40°, r_\beta = 40°, r_\gamma = 40°$ is shown in part (b). In part (c), a rendering of the result from our Enhanced V-EGI technique (with recovered rotation of $r_\alpha = 41°, r_\beta = 41°, r_\gamma = 40°$ for the model to the target) applied to the model is shown.

Next, tests of accuracy for Enhanced V-EGI with Powell optimization are reported. These tests considered both noisy and noise-free versions of the five datasets. Summaries of results are shown in Fig. 4. Figure 4 (a) shows the average estimated orientation angle errors on the datasets. Figure 4 (b) shows the standard deviation of the error in angles. Figure 4 (c) shows the maximum errors.

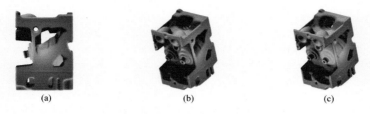

(a)   (b)   (c)

Fig. 3. Sample Renderings

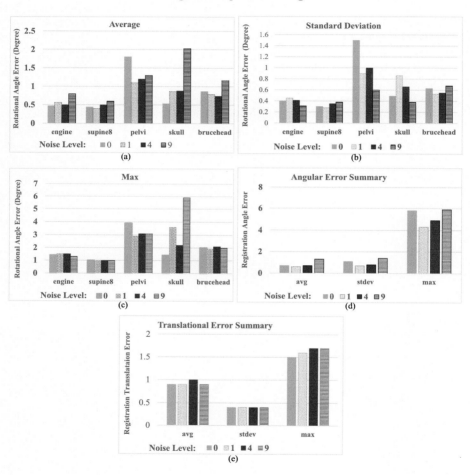

Fig. 4. Statistical summaries of accuracy

Figure 4 (d) summarizes the overall angular average error, standard deviation, and maximum error. Figure 4 (e) shows the overall average error in translation as well as its standard deviation and maximum error. Even with various noise levels (with $\mu = 0$ and $\sigma^2 \in \{1, 4, 9\}$), the average angular error is around $0.8\,°$ and standard deviation is about $0.7\,°$. The average translation error is less than 1 voxel unit and standard deviation is less than 0.5 voxel units. Best angular results were on Engine and Supine8, with average error of $0.5\,°$, standard deviation of $0.4\,°$ and max error around $1\,°$ under all noise levels. Our method achieves nearly comparable results on the Skull and Brucehead datasets under low noise ($\sigma^2 \in \{1, 4\}$), with average error of $0.8\,°$, standard deviation of $0.6\,°$ and max error of $2\,°$. The results on Pelvi were less outstanding. This could be due to more symmetry existing in this dataset that could have affected our histogram orientation matching process. The overall angular detection results for Enhanced V-EGI are more accurate and reliable than the results from angular detection method of V-EGI. With various noise levels (with $\mu = 0$ and $\sigma^2 \in \{1, 4, 9\}$), Enhanced V-EGI achieves the average angular error around $0.8\,°$, standard deviation bout $0.7\,°$ and maximal angular error about $6\,°$, while V-EGI method had the average angular error around $0.4\,°$, standard deviation of $4.4\,°$ and maximal angular error about $154.8\,°$.

7 Conclusion

In this paper, enhancements to the Volume-based Extended Gaussian Image registration have been described. These enhancements enable recovery of positional difference between volumetric datasets and speed improvement via a multi-stage coarse-to-fine process that greatly improves the time for matching. Also, we described a Powell optimization based approach that can be used in the end of each stage to improve accuracy. In addition, we have described a parallelization of the multi-stage process using multi-threading. That process enables efficient recovery of both the rotational and translational components of aligning operations. Outcomes for five datasets were also shown.

In comparison with the standard V-EGI approach, our Enhanced V-EGI method takes minutes to do the orientation detection while the standard V-EGI method takes hours and achieves similar accuracy. The time advantage of the Enhanced V-EGI method is from its multi-stage and Powell optimization strategies. They also allow high orientation and translation accuracy. In addition, the method can also be easily parallelized with multi-threading. The enhanced V-EGI proceeds in a fully automated mode and requires no prior high level features or markers to have been already found. Its computation cost is still somewhat high, although many volumetric registrations are time consuming. In future research, we hope to investigate further computational speed-ups for Enhanced V-EGI.

References

1. Zitova, B., Flussser, J.: Image registration methods: a survey. Image Vis. Comput. **21**, 977–1000 (2003)
2. Horn, B.: Extended gaussian images. Proc. IEEE **72**, 1671–1686 (1984)

3. Dong, C., Newman, T.S.: A volumetric spin-off EGI for registration of volume datasets. In: First Asian Conference on Pattern Recognition (ACPR), Beijing, pp. 470–475 (2012)
4. Oliveira, F., Tavares, J.: Medical image registration: a review. Comput. Methods Biomech. Biomed. Eng. 17, 73–93 (2014)
5. Alves, R.S., Tavares, J.M.R.S.: Computer image registration techniques applied to nuclear medicine images. In: Tavares, J.M.R.S., Jorge, R.M.N. (eds.) Computational and Experimental Biomedical Science: Methods and Applications; LNCVB. LNCVB, vol. 21, pp. 173–191. Springer, Heidelberg (2015)
6. Tavares, J.M.R.S.: Analysis of biomedical images based on automated methods of image registration. In: Bebis, G., et al. (eds.) ISVC 2014, Part I. LNCS, vol. 8887, pp. 21–30. Springer, Heidelberg (2014)
7. Pan, G., Wang, Y., Qi, Y., Wu, Z.H.: Finding symmetry plane of 3D face shape. In: Proceedings of the 18th International Conference on Pattern Recognition (ICPR06), pp. 1143–1146 (2006)
8. Ikeuchi, K.: Recognition of 3-D objects using the extended gaussian image. In: Proceedings of the 7th International Joint Conference on Artificial Intelligence, Vancouver, pp. 595–600 (1981)
9. Chibunichev, A.G., Velizhev, A.B.: Automatic matching of terrestrial scan data using orientation histogram. In: International Archives of the Photogrammetry, Remote Sensing and Spatial Information Sciences, vol. XXXVII, Beijing, pp. 601–604 (2008)
10. Dold, C.: Extented gaussian images for the registration of terrestrial scan data. In: Proceedings of ISPRS WG III/3, III/4, V/3 Workshop on Laser Scanning, Enschede, Netherlands, pp. 180–185 (2005)
11. Maes, F., Collignon, A., Vandermeulen, D., Marchal, G., Suetens, P.: Multimodality image registration by maximization of mutual information. IEEE Trans. Med. Imaging 16, 187–198 (1997)
12. Guillaume, H., Dillenseger, J.L., Patard, J.J.: Intra subject 3D/3D kidney registration/modeling using spherical harmonics applied on partial information. IEEE Trans. Pattern Anal. Mach. Intell. 14, 239–256 (1992)
13. Ni, D., Chui, Y.P., Qu, Y., Yang, X., Qin, J., Wong, T.T., Ho, S.S.H., Heng, P.A.: Reconstruction of volumetric ultrasound panorama based on improved 3D sift. Comput. Med. Imaging Graph. 33, 559–566 (2009)
14. Papazov, C., Burschka, D.: Stochastic global optimization for robust point set registration. Comp. Vis. Image Underst. 115, 1598–1609 (2011)
15. Shams, R., Sadeghi, P., Kennedy, R.A., Hartley, R.I.: A survey of medical image registration on multicore and the GPU. IEEE Signal Proc. Mag. 27, 50–60 (2010)
16. Shams, R., Sadeghi, P., Kennedy, R., Hartley, R.: Parallel computation of mutual information on the GPU with application to real-time registation of 3D medical images. Comput, Methods Progr. Biomed. 99, 133–146 (2010)
17. Tait, R.J., Schaefer, G., Hopgood, A.A., Nakeshima, T.: High performance medical image registration using a distributed blackboard architecture. In: Proceedings of 2007 IEEE Symposium on Computational Intelligence in Image and Signal Processing, Hawaii, pp. 252–257 (2007)
18. Press, W.H., Flannery, B., Teukolsky, S., Vetterling, W.T.: Numerical Recipes in C: The Art of Scientific Computing. Cambridge University Press, New York (1988)
19. U.S. National Library of Medicines. https://www.nlm.nih.gov/research/visible/animations.html. Accessed August 2015
20. Roettger, S.: The Volume Library. http://www9.informatik.uni-erlangen.de/External/vollib/. Accessed August 2015

A Robust Energy Minimization Algorithm for MS-Lesion Segmentation

Zhaoxuan Gong[1], Dazhe Zhao[1(✉)], Chunming Li[2], Wenjun Tan[1],
and Christos Davatzikos[2]

[1] Key Laboratory of Medical Image Computing of Ministry of Education,
Northeastern University, Shenyang, Liaoning 110819, China
zhaodz@neusoft.com
[2] Center of Biomedical Image Computing and Analytics,
University of PA, Philadelphia 19104, USA

Abstract. The detection of multiple sclerosis lesion is important for many neuroimaging studies. In this paper, a new automatic robust algorithm for lesion segmentation based on MR images is proposed. This method takes full advantage of the decomposition of MR images into the true image that characterizes a physical property of the tissues and the bias field that accounts for the intensity inhomogeneity. An energy function is defined in term of the property of true image and bias field. The energy minimization is proposed for seeking the optimal segmentation result of lesions and white matter. Then postprocessing operations is used to select the most plausible lesions in the obtained hyperintense signals. The experimental results show that our approach is effective and robust for the lesion segmentation.

Keywords: Multiple sclerosis lesion · MR · Bias field · True image · Energy minimization

1 Introduction

Multiple Sclerosis (MS) is a chronic and inflammatory disease, which causes morphological and structural changes to the brain. MS could cause various central nervous system dysfunctions such as numbness or weakness of a limb, in coordination, vertigo or visual dysfunction. Therefore, it is very important for radiologists to accurately detect MS lesions and follow-up the numbers, locations, and areas of MS lesions for diagnosis of each patient [1].

Automatization of MS lesion segmentation is highly desirable with regard to time and complexity and visually vague edges of anatomical borders. Their shapes are deformable, their location and area across patients may differ significantly [2–5]. There are many automatic and semiautomatic approaches for brain segmentation [6, 7], Similar to those methods, the MS lesions segmentation approaches include a variety of methods such as region partitioning, markov random field model, adaptive outlier detection and feature extraction [8–11]. Wu et al. [12] proposed an intensity-based statistical k-nearest neighbor (k-NN) classification which combined with template-driven segmentation and partial volume artifact correction (TDS+) for segmentation of

© Springer International Publishing Switzerland 2015
G. Bebis et al. (Eds.): ISVC 2015, Part I, LNCS 9474, pp. 521–530, 2015.
DOI: 10.1007/978-3-319-27857-5_47

MS lesions subtypes and brain tissue compartments. Geremia et al. [13] proposed a multi-channel MR intensities (T1, T2, FLAIR), knowledge on tissue classes and long-range spatial context to discriminate lesions from background. Abdullah [14] proposed a trained support vector machine (SVM) to discriminate between the blocks in regions of MS lesions and the blocks in non-MS lesion regions mainly based on the textural features with aid of the other features.

In this paper, we proposed a novel algorithm for automatic lesion segmentation from MR images in an energy minimization and intensity information framework. This method is an extension of Li et al.'s algorithm in [15], which only segments the normal tissues from T1 W images. There are three main steps in our proposed method shown in Fig. 1. Skull stripping methods remove non-brain voxels from the image to simplify the following lesion segmentation. The BET method [16] which employed from the FSL library is used for skull stripping first, and then an energy minimization approach is proposed to segment Gray matter(GM) class and estimate the bias field from original images, finally lesion segmentation can be compute automatically from the intensity information of the GM class.

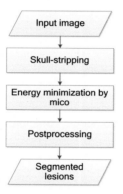

Fig. 1. The proposed approach for fully automatic segmentation of MS lesions.

2   Method

2.1   Image Model

Base on a generally accepted MR image model [17], the intensity inhomogeneity in an MR image can be modeled as a multiplicative component of an observed image described by

$$I(x) = b(x)J(x) + n(x) \tag{1}$$

where $I(x)$ is the intensity of the observed image at voxel x, $J(x)$ is the true image, $b(x)$ is the bias field that accounts for the intensity inhomogeneity in the observed image, and $n(x)$ is additive noise with zero-mean.

In this model, the true image J(x) is approximately a constant c_i for x in the i-th tissue. We denote by Ω_i the region where the i-th tissue is located. Each region (tissue) Ω_i can be represented by its membership function u_i, which satisfy $\sum_{i=1}^{N} u_i(x) = 1$. Given the membership functions u_i and constants c_i, the true image J can be approximated by

$$J(x) = \sum_{i=1}^{N} c_i u_i(x) \tag{2}$$

More generally, MS lesions are considered as the fourth type of tissue, in addition to GM, WM, and CSF. Therefore, the N is set to 4 here. Due to the partial volume effect, the fuzzy membership functions with values between 0 and 1 represent a soft segmentation result. It should satisfy

$$\begin{cases} 0 \leq u_i(x) \leq 1 & x \in \Omega_i \\ \sum_{i=1}^{N} u_i(x) = 1 \end{cases} \tag{3}$$

We represent the coefficients $w_1, \ldots, w_M$, by a column vector $w = (w_1, \ldots, w_M)^T$, where $(\cdot)^T$ is the transpose operator. The basis functions $g_1(x), \ldots g_M(x)$ are represented by a column vector valued function $G(x) = (g_1(x), \ldots g_M(x))^T$. Thus, the bias field b(x) can be expressed in the following vector form

$$b(x) = w^T G(x) \tag{4}$$

The above vector representation will be used in our proposed energy minimization method for bias field estimation, which allows us to use efficient vector and matrix computations to compute the optimal bias field derived from the energy minimization problem [15].

2.2   Energy Formulation

According to [15], the energy minimization formulation can be defined as follows:

$$F(b, J) = \int_{\Omega} |I(x) - b(x)J(x)|^2 dx \tag{5}$$

From Eqs. (2) and (4), the energy function F(u, c, w) is rewritten as:

$$F(u, c, w) = \int_{\Omega} \sum_{i=1}^{N} |I(x) - w^T G(x) c_i|^2 u_i(x) dx \tag{6}$$

In many applications, it is preferable to have fuzzy (or soft) segmentation results. To achieve fuzzy segmentation, we modify the energy function F in (6) by introducing a fuzzifier $q \geq 1$ to define the following energy:

$$F(\mathbf{u}, \mathbf{c}, \mathbf{w}) = \int_\Omega \sum_{i=1}^N |I(\mathbf{x}) - \mathbf{w}^T G(\mathbf{x}) \mathbf{c}_i|^2 u_i^q(\mathbf{x}) d\mathbf{x} \qquad (7)$$

2.3   Energy Minimization

The energy minimization can be achieved by alternately minimizing F(u, c, w, q) with respect to each of its variables given the other two fixed. The minimizer of F(u, c, w, q) in each variable is given below.

Optimization of c

For fixed w and $u = (\mathbf{u}_1, \ldots, \mathbf{u}_N)^T$, the energy $F(\mathbf{u}, \mathbf{c}, \mathbf{w})$ is minimized with respect to the variable c. Therefore, the estimated intensity mean of j-th tissue can be expressed as

$$\hat{c}_i = \frac{\int_\Omega I(\mathbf{x}) b(\mathbf{x}) u_i^q(\mathbf{x}) d\mathbf{x}}{\int_\Omega b^2(\mathbf{x}) u_i^q(\mathbf{x}) d\mathbf{x}}, \quad i = 1, \ldots, N \qquad (8)$$

Optimization of w

For fixed c and u, we minimize the energy F(u, c, w) with respect to the variable w. This can be achieved by solving the equation $\frac{\partial F}{\partial w} = 0$. It is easy to show that

$$\frac{\partial F}{\partial w} = -2r + 2Sw \qquad (9)$$

where r can be given by

$$r = \int_\Omega G(\mathbf{x}) I(\mathbf{x}) (\sum_{i=1}^N c_i u_i^q(\mathbf{x})) d\mathbf{x} \qquad (10)$$

where S is an $M \times M$ matrix,

$$S = \int_\Omega G(\mathbf{x}) G^T(\mathbf{x}) (\sum_{i=1}^N c_i^2 u_i^q(\mathbf{x})) d\mathbf{x} \qquad (11)$$

Given the solution of this equation $\hat{w} = A^{-1}v$, the vector $\hat{w}$ can be explicitly expressed as

$$\hat{w} = (\int_\Omega G(\mathbf{x}) G^T(\mathbf{x}) (\sum_{i=1}^N c_i^2 u_i^q(\mathbf{x})) d\mathbf{x})^{-1} \int_\Omega G(\mathbf{x}) I(\mathbf{x}) (\sum_{i=1}^N c_i u_i^q(\mathbf{x})) d\mathbf{x} \qquad (12)$$

With the optimal vector $\hat{w}$ given by (11), the estimated bias field is computed by

$$\hat{b}(\mathbf{x}) = \hat{w}^T G(\mathbf{x}) \qquad (13)$$

Optimization of u

By fixing c and w, the energy $F(u, c, w)$ is minimized with respect to the variable u. It can be shown that

$$\hat{u}_i(x) = \frac{(|I(x) - c_i b_i(x)|^2)^{\frac{1}{1-q}}}{(\sum\limits_{i=1}^{N} |I(x) - c_i b_i(x)|^2)^{\frac{1}{1-q}}}, \quad i = 1, \ldots, N \tag{14}$$

Each variable is updated with the other two updated in the previous iteration. The optimizations of c, w and u are performed in an iterative process for minimizing the energy F(u, c, w, q).

2.4 Lesion Extraction

Lesions are hyperintense signals on the FLAIR sequence, therefore the intensity information is used to give a preliminary segmentation of the lesions which can be compute automatically from the properties of the GM class. Let $\Omega_1 \subset \Re^2$ be the GM domain. $I : \Omega_1 \subset \Re$ be a given gray level image. Lesions can be obtained as follows:

$$\mu_{lesion} = \frac{\int_{\Omega_1} I(x) \cdot u_{lesion}(x)}{\int_{\Omega_1} u_{lesion}(x)} \tag{15}$$

$$\sigma_{lesion} = \sqrt{\frac{\int_{\Omega_1} (I(x) - \mu_{lesion}(x))^2}{\int_{\Omega_1} \mu_{lesion}(x)}} \tag{16}$$

$$S_{lesion} = \begin{cases} 1, & \mu_{lesion} + \beta \cdot \sigma_{lesion} \\ 0, & others \end{cases} \tag{17}$$

where μ_{lesion} is mean of the GM domain, and σ_{lesion} is standard deviation of the GM domain, S_{lesion} is the lesion segmentation results, β is a constant coefficient.

3   Result and Discussions

The proposed method has been tested on the images from University of Pennsylvania, section of biomedical image analysis (SBICA). There are 33 training cases which were provided with manual segmentation from different experts. The manual segmentations can be viewed as ground truth (GT) to evaluate the automatic lesion segmentations method.

3.1   Experiment Results

Our method has been validated with FLAIR image. We have tested our method on 33 real data sets. All of test data used in our method are the typical MR image with high level of intensity inhomogeneity and strong noise.

Figure 2 shows the process of lesion segmentation, lesions and white matter can be obtained by energy minimization, then lesions can be extracted from the intensity information of them. The normal tissue segmentation obtained by our method is not sufficiently accurate, as the contrast between white matter and gray matter in FLAIR images is very low. However, the errors in the normal tissue segmentation do not affect the satisfactory segmentation for MS lesions.

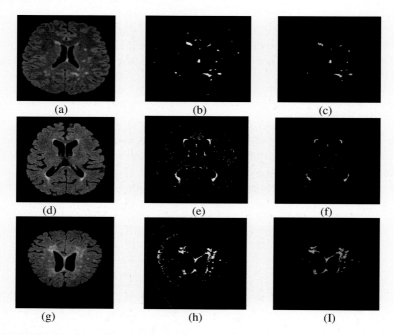

(a)   (b)   (c)

(d)   (e)   (f)

(g)   (h)   (I)

Fig. 2. Result of applying the proposed algorithm to the image of a patient with moderate lesion load:(a), (d), (g) input image, (b), (e), (h) result of energy minimization, (c), (f), (I) result of lesion extraction

As it is seen from Fig. 3, there is a good correlation between the input image and the resulted image, although the input images have the obvious intensity inhomogeneity and noise, our method carries out the desirable lesion segmentation result. Moreover, the results of our method does not rely on the type and volume of MS lesion.

3.2   Quantitative Evaluation

The specificity [18] indicates the ability of the segmentation to identify negative results. The fewer the number of false positives (FP), the greater value the specificity of the segmentation. It is defined as

$$Specificity = \frac{TN}{TN + FP} \qquad (18)$$

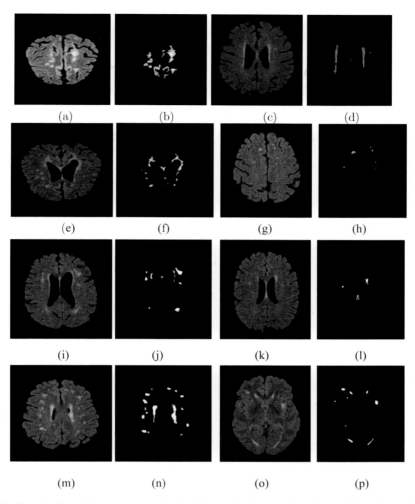

Fig. 3. Result of applying the proposed algorithm to the image of a patient with moderate lesion load: (a), (c), (e), (g), (i), (k), (m), (o) input image, (b), (d), (f), (h), (j), (l), (n), (p) result of our method.

$$TN = \left|\overline{MS} \cap \overline{Seg}\right| \qquad (19)$$

$$FP = |Seg - MS| \qquad (20)$$

where TN is true negative, which is the number of voxels marked as non-MS in both sets, and FP is the false positive, which is the number of voxels only appeared in automatic segmentation. MS is the manual segmentations, and Seg is the segmentation result by our method.

We tested specificity, false negative rate for the 33 sets of flair MR images. The results of these indexes are listed in Table 1. It can be seen that the specificity value of our method to the 33 cases approaches 1 for the manual segmentation. Therefore, the lesion segmentation results from our method are similar to the manual segmentations for most of cases.

Table 1. The detail index of our method with the manual segmentation

Data	Specificity	Data	Specificity	Data	Specificity
Case 01	0.9632	Case 12	0.9231	Case 23	0.9349
Case 02	0.9561	Case 13	0.9364	Case 24	0.9773
Case 03	0.9374	Case 14	0.9644	Case 25	0.9441
Case 04	0.9585	Case 15	0.9531	Case 26	0.9642
Case 05	0.9433	Case 16	0.9534	Case 27	0.9586
Case 06	0.9571	Case 17	0.9413	Case 28	0.9533
Case 07	0.9587	Case 18	0.9448	Case 29	0.9444
Case 08	0.9531	Case 19	0.9537	Case 30	0.9322
Case 09	0.9634	Case 20	0.9461	Case 31	0.9541
Case 10	0.9161	Case 21	0.9277	Case 32	0.9201
Case 11	0.9264	Case 22	0.9338	Case 33	0.9101

Dice's coefficient (DC) is used to measure the similarity of automatic segmentation results and ground truth. For two regions S1 and S2, Dice's coefficient is defined as twice the shared information (intersection) over the sum of cardinalities, and calculated based on the similarity of lesion regions [19]. Even though the manual segmentations cannot be considered to be the same as ground truth, they are still a good way of comparing the automatic lesion segmentation. A larger DC value suggests a better automatic segmentation result.

$$DC = \frac{2|S_1 \cap S_2|}{|S_1| + |S_2|} \tag{21}$$

From Fig. 4 we can see that the DC values between different parameter beta. It can be seen from this figure that the DC value of our method is higher with beta = 1.5 than those of other values.

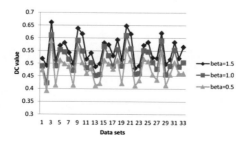

Fig. 4. The comparison of DC value based on the manual segmentation from different parameter values.

4   Conclusion

In this paper, we have presented a robust energy minimization algorithm for MS-lesion segmentation. Except the skull stripping, our method does not require other pre-processing steps for the input images. Experimental results on the real MR images have demonstrated the superior performance of our method in terms of segmentation accuracy and efficiency. Our method does not depend on the type and volume of lesions. Finally, we wish to extend our approach and apply it to other tasks and modalities in medical imaging.

Acknowledgment. This research was partly supported by National Natural Science Foundation of China(NSFC) under Grant No. 61302012 and No. 61172002, the Fundamental Research Funds for the Central Universities under Grant N130418002 and N120518001, and Liaoning Natural Science Foundation under Grant No. 2013020021.

References

1. Sivagowri, S., Jobin Christ, M.C.: Automatic lesion segmentation of multiple sclerosis in MRI images using supervised classifier. Int. J. Adv. Res. Electr. Electron. Instrum. Eng. **2**(12), 6081–6089 (2013)
2. Wallace, C.J., Seland, T.P., Fong, T.C.: Multiple sclerosis: the impact of MR imaging. Am. J. Roentgenol. **158**(1), 849–857 (1992)
3. Truyen, L.: Magnetic resonance imaging in multiple sclerosis: a review. Acta Neurol. Belg. **94**(1), 98–102 (1994)
4. Guizard, N., Coupe, P.: Rotation-invariant multi-contrast non-local means for MS lesion segmentation. NeuroImage Clinical **8**, 376–389 (2015)
5. Nyquist, P.A., Yanek, L.R.: Effect of white matter lesions on manual dexterity in healthy middle-aged persons. Neurology **84**(19), 1920–1926 (2015)
6. Yang, J., Tan, W.: Automatic MRI brain tissue extraction algorithm based on three-dimensional gray-scale transformation model. J. Med. Imaging Health Inf. **4**(6), 907–911 (2014)
7. Zhaoxuan, G., Wenjun, T.: An automatic partitioning method of CTA head-neck image. In: Proceeding of the 11th World Congress on Intelligent Control and Automation, pp. 3283–3285 (2014)
8. Akselrod-Ballin, A.: Automatic segmentation and classification of multiple sclerosis in multichannel MRI. IEEE Trans. Biomed. Eng. **56**(10), 2461–2469 (2009)
9. Khayati, R., Vafadust, M., Towhidkhah, F.: Fully automatic segmentation of multiple sclerosis lesions in brain MR FLAIR images using adaptive mixtures method and markov random field model. Comput. Biol. Med. **38**(3), 379–390 (2008)
10. Derraz, F., Pinti, A.: Multiple Sclerosis lesion segmentation using Active Contours model and adaptive outlier detection method. In: International Work-Conference on Bioinformatics and Biomedical Engineering, pp. 878–889 (2014)
11. Prastawa, M., Gerig, G.: Automatic MS lesion segmentation by outlier detection and information theoretic region partitioning. In: MICCAI 2008 Workshop (2008)
12. Wu, Y., Warfield, Simon K.: Automated segmentation of multiple sclerosis lesion subtypes with multichannel MRI. NeuroImage **32**, 1205–1215 (2006)

13. Geremia, E., Clatz, O.: Spatial decision forests for MS lesion segmentation in multi-channel magnetic resonance images. NeuroImage **57**(2), 378–390 (2011)
14. Abdullah, B.A.: Segmentation of Multiple Sclerosis Lesions in Brain MRI, University of Miami (2012)
15. Li, C., Gore, J.C., Davatzikos, C.: Multiplicative intrinsic component optimization for MRI bias field estimation and tissue segmentation. Magn. Reson. Imaging **32**, 913–923 (2014)
16. Smith, S.M.: Fast robust automated brain extraction. Hum. Brain Mapp. **17**(3), 143–155 (2002)
17. Pham, D.L., Prince, J.L.: Adaptive fuzzy segmentation of magnetic resonance images. IEEE Trans. Med. Imaging **18**(9), 737–752 (1999)
18. Richard Sims, V.G., Isambert, A.: A pre-clinical assessment of an atlas- based automatic segmentation tool for the head and neck. Radiother. Oncol. **93**, 474–478 (2009)
19. Gao, J., Li, C.: Non-locally regularized segmentation of multiple sclerosis lesion from multi-channel MRI data. Magn. Reson. Imaging **32**, 1058–1066 (2014)

Impact of the Number of Atlases in a Level Set Formulation of Multi-atlas Segmentation

Yihua Song[1,3], Zhaoxuan Gong[1], Dazhe Zhao[1], Chaolu Feng[1], and Chunming Li[2,3(✉)]

[1] No Institute Key Laboratory of Medical Image Computing Ministry of Education, Northeastern University, Shenyang, China
[2] University of Electronic Science and Technology of China, Chengdu, China
li_chunming@hotmail.com
[3] Department of Radiology, University of Pennsylvania, Philadelphia, USA

Abstract. In this paper, we present a multi-atlas segmentation method based on the level set formulation for performing label fusion that takes into account the image information and regularity of the region of interest (ROI). In the presented method, multiple atlases are first registered to a target image by deformable registration via attribute matching and mutual saliency weighting (DRAMMS) and advanced neuroimaging tools (ANTs) to get the warped labels. Then, an optimal labeling is sought by label fusion for segmentation of target image. Label fusion is achieved by seeking an optimal level set function which minimizes an energy functional in regards to three terms: label fusion term, image based term, and regularization term. In this work, we discussed the impact of subset on the accuracy of segment results. Results show that segmentation results will be much more accurate if an appropriate subset of atlases are selected for each target image than those given by non-selective combination of random atlas subsets.

1 Introduction

Due to the wide applicability and wide availability of registration tools, multi-atlas-based segmentation methods have been researched a lot in recent years in order to improve the accuracy and robustness of single atlas labeling methods [1–5]. Multi-atlas segmentation can overcome inaccurate segmentation produced by registration errors or individual propagated atlases [6]. Theoretically, deformation existing in registering the atlas images to the target image can be used to propagate segmentation of the atlas images.

Multi-atlas segmentation methods consist of two specific steps: registration and label fusion. One approach to further improve the quality of multi-atlas-based segmentation is developing more accurate and robust label fusion methods. The most common label fusion approaches include majority voting, weighted voting, and STAPLE [7–9] methods. However, this leads to a major limitation: regular shape information will be lost during the label fusion.

In this paper, the label fusion method based level set formulation for multi-atlas segmentation and atlas selection are described. This is followed by descriptions of the experiments to assess effectiveness of atlas selection on segmentation accuracy. The

© Springer International Publishing Switzerland 2015
G. Bebis et al. (Eds.): ISVC 2015, Part I, LNCS 9474, pp. 531–537, 2015.
DOI: 10.1007/978-3-319-27857-5_48

objective of this paper is searching an effective approach to select an appropriate subset of atlases for each target image provides more accurate segmentations.

2   Method

Atlas-based methods are commonly used to segment images in medical image processing. Based on this technique, templates of atlases are first registered non-rigidly to target images. The deformable labels are then considered as segmentation estimates for the target images. In this paper, DRAMMS [10] and ANTs [11] registration methods are used to compute the deformation field which are further used to generate warped labels. The top-ranked atlases are selected by comparing image similarity of the target image and each atlas [6] where correlation coefficient (CC) is used as the similarity measure. To make fusion of the transferred labels into a single label, a level set framework is adopted, which consists of label fusion term, image based term, and regularization term, as shown in Fig. 1.

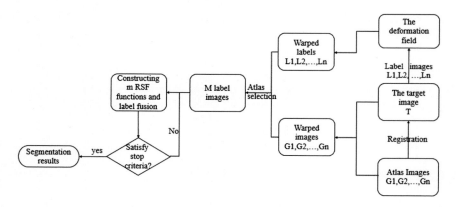

Fig. 1.   The framework of the presented segmentation method. Atlas images are registered to the target image. In order to determine the label of a voxel, level set formulation is applied to the label fusion step. RSF: Region Scalable Fitting Model [12].

2.1   Method Overview

Let image I be a function: $\Omega \rightarrow \mathfrak{R}$ defined on a continuous image domain Ω and ϕ is the level set function. The label image L corresponds with $L(x) = 1$ for x inside the label region and $L(x) = 0$ for x outside the region. Label images L can be defined as

$$\begin{aligned} \phi(x) < 0 \quad & x \in \{x{:}L(x) = 1\} \\ \phi(x) > 0 \quad & x \in \{x{:}L(x) = 0\} \end{aligned} \tag{1}$$

Therefore, the point is seeking the zero level set contour which represents the boundary of the label by minimizing an energy functional with the purpose of achieving label fusion. The following is a general variational level set formulation: formulation for label fusion, in which the energy E of the level set function ϕ is defined by:

$$E(\phi) = \alpha F(\phi; \phi_1, \dots, \phi_N) + \beta D(\phi; I) + \gamma R(\phi) \tag{2}$$

where $F(\phi; \phi_1, \dots, \phi_N)$ represents the label fusion term, $\phi_1, \dots, \phi_N$ represents each level set function with different labels, $D(\phi; I)$ represents the image based term, and $R(\phi)$ represents the regularization term with the corresponding coefficients α, β and γ.

2.2 Label Fusion and Regularization

We use weighted label fusion in which the contribution of each atlas is locally weighted by its similarity to the target image by selecting a subset of atlases. According to the definition of the label fusion and regularization above, in order to achieve optimal label fusion, these items are used in a special case of the general framework. The label fusion term $F(\phi; \phi_1, \dots, \phi_N)$ is defined as follows:

$$F(\phi; \phi_1, \dots, \phi_N) = \int_\Omega \sum_{i=1}^N \omega_i(x) \left| \phi(x) - \phi_i(x) \right|^2 dx \tag{3}$$

where $\omega_i(x)$ is a spatially varying weighting function assigned to each point x for the i-th label, which is represented by the level set function ϕ_i.

In the method, the goal is to find out the zero level contours, which represent the boundaries of the target images given by the labels. By minimizing the energy F, the zero level contour of the level set function ϕ is forced to be close to the zero level contours of the level set functions $\phi_1, \dots, \phi_N$.

In practice, a regularization mechanism is added to maintain the regularity of the contour to avoid irregular shape of the obtained contour. As in most of the active contour models, the 0-level contour is smoothed by using the regularization term

$$R(\phi) = \int |\nabla H(\phi(x))| \, dx \tag{4}$$

where H stands for the Heaviside function which is approximated by a smooth function H_ε defined by

$$H_\varepsilon(x) = \frac{1}{2}[1 + \frac{2}{\pi} \arctan(\frac{x}{\varepsilon})] \tag{5}$$

The derivative of H_ε is defined by

$$\delta(x) = H_\varepsilon'(x) = \frac{1}{\pi} \frac{\varepsilon}{\varepsilon^2 + x^2} \tag{6}$$

Traditional label fusion methods ignore the fact that different atlases may produce similar label errors, and also have a disadvantage in maintaining a regular shape of the target images. To avoid the limitation, this method proposed to take into account the image information and the regularity of the region of interest (ROI).

2.3   Level Set Formulation

A region-scalable fitting (RSF) [12,13] energy functional is defined in terms of the 0-level contour and two fitting functions that locally approximate image intensities on the both sides of the contour. The region-scalability of the RSF energy functional contains a scale parameter of kernel function which allows the use of intensity information in regions at a controllable scale, from small neighborhoods to the entire domain.

Consider a given vector valued image $I:\Omega \to \Re$, where $\Omega \to \Re$ is the image domain. Let C be a closed contour in the image domain Ω, which divides Ω into two regions: $\Omega_1 = outside(C)$ and $\Omega_2 = inside(C)$. For a given image I, the following energy is defined to seek an optimal contour and fitting functions f_1 and f_2

$$E_x(C, f_1(x), f_2(x);I) = \lambda_1 \int_{\Omega_1} |I(y) - f_1(x)|^2 \, dy$$
$$+ \lambda_2 \int_{\Omega_2} |I(y) - f_2(x)|^2 \, dy \tag{7}$$

where λ_1 and λ_2 are the weighting coefficients, and $f_1(x)$ and $f_2(x)$ are two values that approximate image intensities in Ω_1 and Ω_2. The energy is minimized for all $x \subset \Omega$ for the image I. Minimization of $E_x(C,f_1(x),f_2(x);I)$ for all x can be achieved by minimizing the integration of E_x and the energy D, defined by

$$D(C, f_1, f_2;I) = \int_{\Omega} E_x(C, f_1(x), f_2(x);I)dx \tag{8}$$

A contour $C \subset \Omega$ is represented by the zero level set of a Lipschitz function $\phi:\Omega \to \Re$, which is called a level set function. In this paper, the level set function ϕ takes positive and negative values outside and inside the contour C, respectively.

The local intensity fitting energy can be considered as a region-scalable fitting (RSF) energy

$$D(\phi, f_1, f_2;I) = \sum_{i=1}^{2} \int (\int K_\rho(x - y) |I(y) - f_i(x)|^2 M_i(\phi(y))dy)dx \tag{9}$$

where kernel function $K_\rho:\Re^n \to [0, +\infty)$ is defined by $K_\rho(u) = a$ if $|u| \le \rho$ an $K_\rho(u) = 0$, and a is a normalization factor such that $\int_{|u|\le\rho} K_\rho(u) = 1$, and also, $M_1(\phi(y)) = 1 - H(\phi(y))$, $M_2(\phi(y)) = 1 - H(\phi(y))$.

Although only intensity in the definition is used in the paper, previous research [1, 2, 6] shows various image based information, such as edges and texture, can be used in the image based term in the Eq. (9). With respect to the functions f_1 and f_2, the function $D(\phi,f_1,f_2;I)$ is minimized. By calculus of variations, it can be shown that the functions f_1 and f_2 which minimize $D(\phi,f_1,f_2;I)$ with ϕ fixed also satisfy the following Euler-Lagrange equations:

$$\int K_\rho(x - y)M_i^\epsilon(\phi(y))(I(y) - f_i(x))dy = 0, \quad i = 1, 2. \tag{10}$$

From Eq. (10), the function f is written as:

$$f_i(x) = \frac{K_\rho(x) * [M_i^\epsilon(\phi(x))I(x)]}{K_\rho(x) * M_i^\epsilon(\phi(x))}, i = 1, 2 \tag{11}$$

3   Experimental Results

The performance of the presented method was evaluated on data set from MICCAI 2012 Multi-Atlas Labeling Challenge, which is publicly available from the following website: https://masi.vuse.vanderbilt.edu/workshop2012/. The data set has 15 testing T1 MR images and 35 training T1 MR images with manually segmented ROI labels being provided as ground-truth.

Taking the manual segmentation as the ground truth, we evaluated the agreement between segmentation results of the presented method B and ground truth using Dice coefficient, given by

$$DC = \frac{2|A \cap B|}{|A| + |B|} \tag{12}$$

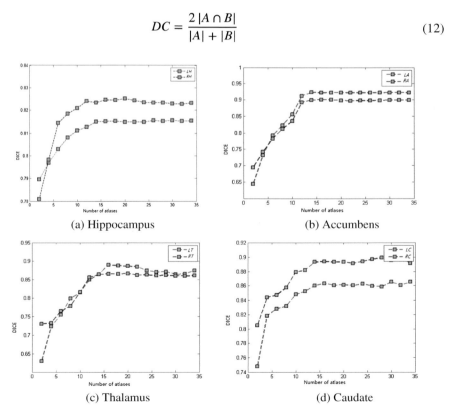

(a) Hippocampus

(b) Accumbens

(c) Thalamus

(d) Caudate

Fig. 2. Segmentation accuracy for various structures after fusing increasing numbers of ranked atlases. (Red: left ROI; blue: right ROI) (Color figure online)

DC demonstrates similarity measure between the presented segmentation method and the ground truth. The higher the *DC* value is, the more similar between segmentation results of the presented method and the ground truth.

In this work, automated segmentation is carried out based on ranking and selecting atlases from the database of atlases. The accuracy of each segmentation estimate was assessed using its Dice overlap with manual segmentation of the target image. The segmentation results of all the left and right ROIs selecting k atlases are the same after $k \geq 15$. The preliminary results indicate that after the k reaching a value, the Dice score tend to be stable. In this experiments, 15 selected atlases are reasonable. Data for 4 ROIs (Hippocampus, Accumbens, Thalamus, and Caudate) of the results are shown. The horizontal axis shows the number of the selected fusing atlases and the vertical axis shows the Dice score. Each point plotted in Fig. 2 shows the average Dice overlap for all atlases at a given rank in segmenting the target images for which the rank is calculated. Some of the graphs are randomly superimposed. In all cases, this effect of varying the number of atlases selected upon the final segmentation Dice score is exceeded.

4   Conclusion

We have presented an investigation of strategies for atlas selection prior for multi-atlas segmentation based on the level set formulation. This method has an intrinsic mechanism of maintaining shape regularity while exploiting image information within the label fusion framework. Warped labels from different registration approaches with complementary advantages were combined in the label fusion step. The suitability of images similarity is considered as the atlas selection criterion. Experimental results demonstrate that accuracy varied with the changing of the number of atlases but the fusion of approximately 15–20 atlases generates the maximal accuracy. Results suggest that the choice of 15 atlases after selection is reasonable for the efficiency and accuracy.

Acknowledgment. This research was partly supported by National Natural Science Foundation of China (NSFC) under Grant No. 61302012 and No. 61172002, the Fundamental Research Funds for the Central Universities under Grant N130418002, N120518001, N140403006 and N140402003. The National Key Technology support Program 2014BAI17B02, and Liaoning Natural Science Foundation under Grant No. 2013020021.

References

1. Isgum, I., Staring, M., Rutten, A., Prokop, M., Viergever, M.A., van Ginneken, B.: Multi-atlas-based segmentation with local decision fusion-application to cardiac and aortic segmentation in CT scans. IEEE Trans. Imag. Proc. **28**(7), 1000–1010 (2009)
2. Lötjönen, J.M.P., Wolz, R., Koikkalainen, J.R., et al.: Fast and robust multi-atlas segmentation of brain magnetic resonance images. Neuroimage **49**(3), 2352–2365 (2010)
3. Van Rikxoort, E.M., et al.: Adaptive local multi-atlas segmentation: application to the heart and the caudate nucleus. Med. Image Anal. **14**(1), 39–49 (2010)
4. Wang, H., et al.: Regression-based label fusion for multi-atlas segmentation. In: 2011 IEEE Conference on Computer Vision and Pattern Recognition (CVPR). IEEE (2011)

5. Gholipour, A., Akhondi-Asl, A., et al.: Multi-atlas multi-shape segmentation of fetal brain MRI for volumetric and morphometric analysis of ventriculomegaly. Neuroimage **60**(3), 1819–1831 (2012)
6. Aljabar, P., Heckemann, R.A., Hammers, A., Hajnal, J.V., Rueckert, D.: Multi-atlas based segmentation of brain images: atlas selection and its effect on accuracy. Neuroimage **46**(3), 726–738 (2009)
7. Asman, A.J., Landman, B.A.: Non-local statistical label fusion for multi-atlas segmentation. Med. Image Anal. **17**(2), 194–208 (2013)
8. Artaechevarria, X., Arrate, M.B., Ortiz-de-Solórzano, C.: Combination strategies in multi-atlas image segmentation: application to brain MR data. IEEE Trans. Med. Imaging **28**(8), 1266–1277 (2009)
9. Warfield, S.K., Zou, K.H., Wells, W.M.: Simultaneous truth and performance level estimation (STAPLE): an algorithm for the validation of image segmentation. IEEE Trans. Med. Imaging **23**(7), 903–921 (2004)
10. Ou, Y., et al.: DRAMMS: deformable registration via attribute matching and mutual-saliency weighting. Med. Image Anal. **15**(4), 622–639 (2011)
11. Avants, B.B., Tustison, N.J., Song, G., Cook, P.A., Klein, A., Gee, J.C.: A reproducible evaluation of ants similarity metric performance in brain image registration. Neuroimage **54**(3), 2033–2044 (2011)
12. Li, C., Kao, C., Gore, J.C., et al.: Minimization of region-scalable fitting energy for image segmentation. IEEE Trans. Image Process. **17**(10), 1940–1949 (2008)
13. Feng, C., Li, C., Zhao, D., Davatzikos, C., Litt, H.: Segmentation of the left ventricle using distance regularized two-layer level set approach. In: Mori, K., Sakuma, I., Sato, Y., Barillot, C., Navab, N. (eds.) MICCAI 2013, Part I. LNCS, vol. 8149, pp. 477–484. Springer, Heidelberg (2013)

Probabilistic Labeling of Cerebral Vasculature on MR Angiography

Benjamin Quachtran, Sunil Sheth, Jeffrey L. Saver,
David S. Liebeskind, and Fabien Scalzo[⊠]

Stroke Program, Department of Neurology,
University of California, Los Angeles (UCLA), Los Angeles, USA
fscalzo@mednet.ucla.edu

Abstract. Automatic labeling of cerebrovascular territories would greatly advance our ability to systematically study large datasets and also provide rapid decision support during the assessment of stroke patients, for example. Previous attempts have been challenged by the wide inter-subject variation in vascular topography. We investigate the use of a probabilistic model that learns the configurational characteristics of vascular territories to better annotate the cerebrovasculature. In the George Mason Brain Vasculature database, we identified patients with MRA reconstructions segmented into seven major regions (left and right MCA, PCA, and ACA and Circle of Willis). We then augmented these labels by manually segmenting the MCA territory into an additional eight regions. Among 54 patients that met the inclusion criteria, 39 reconstructions were used as training input to the MCA, ACA, and PCA model among the 61 digital reconstructions of human brain arterial structures available. The model was then validated on an independent cohort of 15 patients. The MCA segmentation algorithm was trained and tested using leave-one-out crossvalidation. The algorithm was found to be $94 \pm 5.2\,\%$ accurate in annotating the seven major regions and $88 \pm 9.3\,\%$ accurate in annotating the MCA subterritories.

1 Introduction

Advancements in medical imaging and computational methods have led to a new wave of algorithmic-based techniques to analyze patient data. In the past, cerebral scans have been studied in a more case-by-case basis, leaving the interpretations subject to debate and reducing analysis efficiency. The inherent lack of quantification limits the recognition of correlations and patterns, obscuring meaningful clinical findings within the noise. Digitizing data and automating interpretation paves the way for deeper and more objective analysis, and in the case of cerebral angiography, moves towards more unbiased assessments that can directly improve clinical decision-making.

Previous attempts at automated 3D vessel segmentation [14] and labeling [1,3,15] have been challenged by the difficulties related to the quality of the images and the wide inter-subject variability that normally occurs. Image processing

© Springer International Publishing Switzerland 2015
G. Bebis et al. (Eds.): ISVC 2015, Part I, LNCS 9474, pp. 538–548, 2015.
DOI: 10.1007/978-3-319-27857-5_49

and computer vision methods can play a major role in solving these issues if the algorithms are well trained to handle highly variable data. A study [2] conducted at the University of Toronto has investigated the use of Bayesian-based inference to label cerebral vasculature of mice. An arterial set is initialized with random labels and then a relaxation algorithm iteratively re-labels segments in manner that minimizes the data's energy function, a negated summation of each label's posterior probability. Despite being $\geq 75\%$ accurate in recognizing each of the sample's 54 territories, the method took an average of $100 \pm 18\,h$ to segment each image. Recently, promising advances [13] have also been achieved for the automatic segmentation and labeling of the arterial tree from whole-body MRA data. Labeling of the vascular tree was performed by using a combination of graph-based and atlas-based approaches.

The characteristics chosen as inputs in the training data have proven to be a critical element in determining the robustness of an algorithm. Shape recognition, through use of principal component analysis, for example, provides a reasonably effective and computationally efficient way to quantify cerebral territories in segmenting the majority of vasculature sets [5]. The dependence on consistent arterial structure made by this technique is not resilient to the wide variation in morphological structure present in cerebral anatomy. Instead, selecting spatial features as inputs seem to lead to better results when performing vascular segmentation because of the visual nature of territorial labeling [2]. Labeling studies based on other modalities such as CT [6] have successfully used various features such as diameter, curvature, direction, and running vectors of a branch to infer the labels on abdominal arteries.

In this paper, we propose and test a method to automatically label the major vascular territories of cerebral reconstructions through use of kernel density estimation (KDE) and Bayesian inference. Through this form of probabilistic estimation, we are able to create accurate approximations of multi-dimensional density functions describing the cartesian coordinates of the right and left MCA, PCA, and ACA regions. From these probabilistic distributions, each vascular segment is evaluated as a function of the likelihoods of each sub-segment. A maximum a posteriori algorithm decides the appropriate label based on the optimal characterization of the evaluation.

2 Methods

2.1 Dataset Acquisition and Properties

The 3D arterial models were collected from the George Mason Brain Vasculature database [17] which is freely available at http://cng.gmu.edu/brava/ and contains 61 digital reconstructions created from magnetic resonance angiography (MRA) scans of healthy adult subjects (mean age, 31 years; age range, 19 to 64; 36 women). The dataset was created using Neuron_Morpho, a plug-in of ImageJ [5]. Through Neuron_Morpho, users are able to trace neurons contained in image stacks and export them in swc format, representing the cerebral vasculature as a series of interconnected cylindrical segments (segments have a specified x, y, and z coordinate, type, radius, and parent segment). Following the use of this software, the reconstructions were additionally verified for accuracy through juxtaposition with 3D renderings of the MRA scans. In the Brain Vasculature database, each reconstruction is available both unlabeled and labeled, with the labeled models identifying seven major regions (Fig. 1): Circle of Willis and the left and right middle cerebral artery (MCA), posterior cerebral artery (PCA), and anterior cerebral artery (ACA). Vaa3D [9], an open source tool for 3D bio-image visualization and editing, was used to render and view these models.

While the arterial data from BraVa is pre-segmented into the seven aforementioned regions, we are interested in further annotating the MCA territory, a site of interest in stroke patients. To create the necessary training data, we used Vaa3D's built-in neuron utilities to expand the MCA territory into eight additional regions [7] (Fig. 2): (1) Posterior Temporal, (2) Temporo-Occipital, (3) Angular, (4) Posterior Parietal, (5) Rolandic, (6) Precental, (7) Prefrontal, and (8) Orbitofrontal. A neurologist from UCLA manually labeled the MCA regions of training images. Vasculatures in the database were co-registered using landmark points placed manually so that the relative location of the vessel segments could be used. Data registration is necessary to properly characterize the vasculatures anatomical locations with the respective territories.

2.2   Bayesian Labeling Framework

We assume that the labeling framework is presented with a set of unlabeled vessel segments $S_{i=1...n}$ that can be represented as a tree-structured graph. Each segment S_i is characterized by a state x_i in the model that represents the label probabilities of vascular territories. Because of the natural variations occurring in the vasculature across subjects, the number of states N for a specific model varies $[x_1, \ldots, x_N]$. A state x_i is a M-dimensional discrete vector that associates a probability to each of the M possible labels of the segment.

To each state x_i is associated an observation y_i, directly obtained from the normalized location of the segment and its radius. It differs from the state x_i in the sense that it comes vessel detectors that can be affected by noise and transient artifacts present in the image, whereas the value x_i is obtained through inference, thus believed to be more robust.

The labeling model is assumed to follow the general properties of a tree-structured Markov model where each state x_i is connected to at most one parent state x_i^p and can have several children states $x_{i,1...N_c}^c$ where $N_c \geq 0$. This means that the probability of a segment x_i given all the states available $x_{1...N}$ depends only its parent and children states,

$$p(x_i | x_{\{1...N\}}) = p(x_i) \, p(x_i | x_i^p) \, p(x_i | x_{i,1...N_c}^c) \tag{1}$$

where $p(x_i)$ is the prior distribution and $p(x_i | x_j)$ represents the conditional dependency between two connected vessel segments. By introducing observations

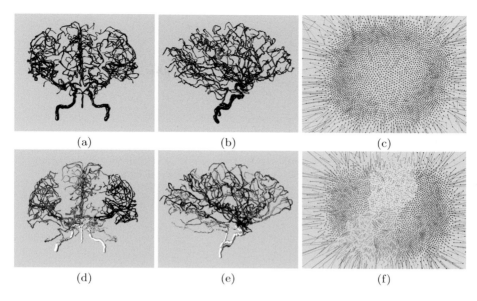

Fig. 1. Figures (a) and (b) show the unlabeled reconstructions of a cerebral vasculature. Figures (d) and (e) show their respective labels using color mapping. The annotations were chosen as Pink = LMCA, Blue = RMCA, Cyan = LACA, Red = RACA, Green = LPCA, Yellow = RPCA, White = Circle of Willis and ICA. The corresponding graphical representations of the unlabeled and labeled vessel segments are shown in (c) and (f). The vascular model shown in this figure was composed of 3514 vessel segments (Color figure online).

y_i in the model, the posterior marginal of the state is defined as follows,

$$p(x_i|y_{\{1...N\}}) \propto p(y_i|x_i)\, p(x_i) \int p(x_i|x_i^p)\, p(x_i^p|y_{\{1...N\},1}) dx_i^p \,\cdots$$
$$\prod_j^{N_c} \int p(x_i|x_{i,j}^c)\, p(x_{i,j}^c|y_{\{1...N\}})\, dx_{i,j}^c \qquad (2)$$

where $p(y_i|x_i)$ is the likelihood. We propose to use a graphical model to represent this recursive problem, and Belief Propagation to perform the labeling process.

Graphical Model. The graphical model used in our labeling framework defines relations between pairs of nodes only. It is usually referred to as Pairwise Markov Random Field (PMRF) in the literature. As illustrated in Fig. 3, states $x_i \in x$ and observations $y_i \in y$ are represented in the graphical model by white, and shaded nodes, respectively. Edges represent dependencies between states by two types of functions: observation potentials $\phi(x_i, y_i)$ that are the equivalent of the likelihood part $p(y_i|x_i)$, and compatibility potentials $\psi_{ij}(x_i, x_j)$ that embed the conditional parts $p(x_i|x_j)$, $p(x_j|x_i)$ of the Bayesian formulation and can be used

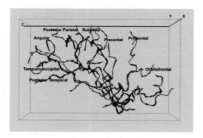

Fig. 2. Illustration of the additional annotations made by a neurologist on the middle cerebral artery (MCA) territory indicating the eight sub-territories of interest.

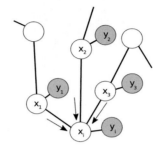

Fig. 3. In this graphical representation, y_1, y_2, y_3 indicate each node's observational features and x_1, x_2, x_3 represent their respective states. Messages received by state x_i are denoted by directed arrows.

by conditioning them in either directions during inference. In addition, the prior distribution over the labels is denoted $\psi_i(x_i)$.

Observation Model. An observation y_i represents the label information about the ith vessel segment that is directly extracted from the image. An observation $y_i \in R^M$ assigns a likelihood to each possible label with respect to the position of the segment a, b, c and its radius r using a previously learned kernel density estimation (KDE) $\hat{f}(S_i; \Theta_l)$ (Eq. 3) that is constructed by collecting a total of N labeled vessel segments $\{a_k, b_k, c_k, r_k\}_{1...N}$ across the training set,

$$\hat{f}(S_i; \Theta_l) = \frac{1}{N} \sum_{k=1}^{N} \mathcal{G}(S_i; \{a_k, b_k, c_k, r_k\}, \Sigma_i) \tag{3}$$

where $\mathcal{G}(S_i; \{a_k, b_k, c_k, r_k\}, \Sigma_i)$ is a Gaussian kernel centered at $\{a_k, b_k, c_k, r_k\}$ with standard deviation $\Sigma_{i,l}$ which is common to all the components k of a given label l. $\Theta_l = \{\{a_i^k, b_j^k, c_j^k, r_j^k\}_{1...N_c}, \Sigma_i\}$ denotes the parameter set of the KDE associated with a given label l.

Observations y_i are linked probabilistically to their state x_i through a Gaussian observation potential $\phi(x_i, y_i)$,

$$\phi(x_i, y_i) = \exp(-|y_i - x_i|^2/\sigma_o^2) \tag{4}$$

where σ_o is a smoothing parameter.

Compatibility Potentials. Compatibility potentials $\psi(x_i, x_j)$ define the relationship between two connected states. They are defined as a Gaussian difference between their arguments,

$$\psi(x_i, x_j) = \exp(-|x_i - x_j|^2/\sigma_t^2) \tag{5}$$

where the standard deviation σ_t of the model can be estimated using maximum likelihood (ML) on training data.

2.3   Labeling Using Belief Propagation

Labeling vessel segments in the 3D vasculature amounts to estimating $p(x_i|y_{\{1...N\}})$, the posterior belief associated with the state x_i given all observations $y_{\{1...N\}}$ accumulated. Thus, labeling is achieved through inference in our graphical model. One way to do this efficiently is to use Belief Propagation (BP) [8], a method implemented successfully in numerous computer vision applications such as for image segmentation [16] and object recognition [10–12].

It is a message passing algorithm for graphical models where messages are repeatedly exchanged between nodes to perform inference. Following the notation of BP, a message m_{ij} sent from node i to j is written,

$$m_{i,j}(x_j) \leftarrow \int \psi_{i,j}(x_i, x_j)\, \psi_i(x_i)\, \phi_i(x_i, y_i) \prod_{k \in \mathcal{N}_i \backslash j} m_{k,i}(x_i)\, \mathrm{d}x_i \tag{6}$$

where $\mathcal{N}_{i \backslash j}$ is the set of neighbors of state i where j is excluded, $\psi_{i,j}(x_i, x_j)$ is the pairwise potential between nodes i, j, and $\phi_i(x_i, y_i)$ is the observation potential.

After any iteration of message exchanges, each state can compute an approximation $\hat{p}(x_i|y_{\{1...N\}})$, called belief, to the marginal distribution $p(x_i|y_{\{1...N\}})$ by combining the incoming messages with the local observation:

$$\hat{p}(x_i|y_{\{1...N\}}) \leftarrow \psi_i(x_i)\, \phi_i(x_i, y_i) \prod_{k \in \mathcal{N}_i} m_{k,i}(x_i) \tag{7}$$

For tree-structured graphs like ours, the beliefs (Eq. 7) will converge to the exact solution $p(x_i|y_{\{1...N\}})$ [4].

3   Experiments

The goal of the experiments presented in this section is to demonstrate the effectiveness of the developed probabilistic labeling algorithm. To evaluate the model,

Table 1. Results of the probabilistic labeling model on 15 test subjects.

Cerebral territory	Accuracy
LACA	$88.9 \pm 14.3\%$
RACA	$93.8 \pm 10.3\%$
LPCA	$93.6 \pm 10.3\%$
RPCA	$97.3 \pm 3.2\%$
LMCA	$98.7 \pm 1.7\%$
RMCA	$98.7 \pm 1.4\%$
Circle of Willis	$84.1 \pm 14.4\%$
Average	**$94 \pm 5.2\%$**

two sets of experiments were ran; the first one using the 7 major vascular territories provided as part of BraVa, and the second one based on the 8 additional MCA sub-territories established in this study (as described in Sect. 2.1). In both experiments, we evaluate the accuracy by finding the percentage of segment labels in the testing data consistent with the manual annotation.

To quantify the performance of the ACA, MCA, and PCA labeling, we used 39 annotated reconstructions from BraVa as training to build and optimize the graphical model, utilizing the cartesian data and radii as inputs. The algorithm was then tested on the data of 15 remaining subjects. The split between the training and testing set was done randomly. For the MCA subdivision, we tested the algorithm using leave-one-out crossvalidation. In this method, we train the model using all subjects except one, and test on the remaining subject, repeating the process for each set.

Table 1 summarizes the accuracy of the ACA, MCA, and PCA segmentation over the 15 tested data sets was $94 \pm 5.2\%$ (as illustrated in Fig. 5). For the

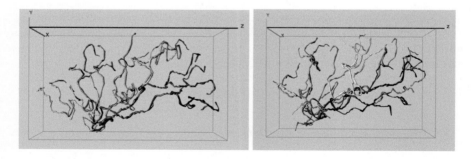

Fig. 4. Results of the MCA segmentation algorithm on two vascular models. Black = Posterior Temporal, Red = Temporo-Occipital, Blue = Angular, Cyan = Posterior Parietal, Yellow = Rolandic, Green = Precental, Brown = Prefrontal, Beige = Orbifrontal (Color figure online).

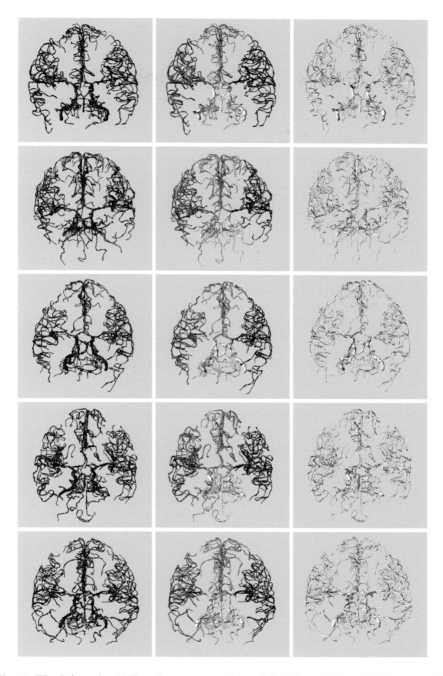

Fig. 5. The left and middle columns show the unlabeled models and their respective probabilistic segmentation results for five subjects obtained from BraVa [17]. The right column indicates the automated labeling errors in red (Color figure online).

MCA subdivision, we observed an accuracy of $88 \pm 9.3\%$ for the 7 tested sets (Fig. 4). These results demonstrate that our model is effective in predicting vascular regions in arterial reconstructions. The larger MCA subdivision deviation is a result of one experiment producing an accuracy of 67%, indicating that our model is not totally impervious to the wide variation in vascular topography. The major artery segmentation algorithm proved to be more consistent in its results; the accuracy for any test never fell below 79%.

4   Discussion

The results obtained during our experiments have demonstrated that the use of a graphical model that combines kernel density estimation (KDE) in conjunction with belief propagation inference is capable to learn and label vascular territories automatically with high accuracy. We believe this performance can be attributed to both the robustness of nonparametric estimation of the likelihood model and the optimal sharing of information by the message passing algorithm. Training and labeling times allow for processing of a large cohort of subject data; reconstructions can be labeled in a matter of seconds, and complete training and optimization in a few minutes.

In order to fully evaluate how well the model performs in real clinical conditions, a larger, more diverse pool of subjects needs to be tested. Because all training and test sets were retrieved from a single database, we are exposed to sampling bias, perhaps overly estimating the accuracy of the results of the framework on a new dataset where the input features would be extracted with a vessel segmentation technique different from Neuron_Morpho.

More importantly, the ability of our framework to study brain vasculatures automatically is hindered by pre-processing tasks that require manual input, preventing large data sets from being analyzed. Being able to automatically label vascular territories solves a piece of that puzzle, however, more efforts should be invested in the automatic vessel segmentation process.

In terms of clinical relevance of the developed tool for stroke patients, the automatic labeling allows us to quantify arterial regions missing due to occlusion and assess brain health. The MCA territory is a common site of occlusion in ischemic stroke and, through automated labeling, we can identify areas of the MCA that are frequently targeted and recognize features symptomatic of stroke. The proposed algorithm provides a framework for the learning and labeling of other vascular territories as well; whole-body and abdominal vessel segmentation are also very relevant clinical applications that could benefit from the probabilistic framework presented in this paper.

5   Conclusion

The results indicate that the developed framework, which is based on Bayesian formulation of the labeling problem, is effective in labeling vascular features in 3D reconstructions using belief propagation algorithm. We anticipate that this approach will prove useful for learning arterial characteristics in any setting, particularly neurovascular, and circumventing the need for manual annotation. In order to fully automate the labeling process, an effective registration method should be considered. This requirement will be addressed in future work.

References

1. Dufour, A., Ronse, C., Baruthio, J., Tankyevych, O., Talbot, H., Passat, N.: Morphology-based cerebrovascular atlas. In: ISBI, pp. 1210–1214 (2013)
2. Ghanavati, S., Lerch, J.P., Sled, J.G.: Automatic anatomical labeling of the complete cerebral vasculature in mouse models. Neuroimage **95**, 117–128 (2014)
3. Ghanavati, S., Lerch, J.P., Sled, J.G.: Improved method for automatic cerebrovascular labelling using stochastic tunnelling. In: International Workshop on Pattern Recognition in Neuroimaging, pp. 1–4 (2014)
4. Ihler, A.T., Fisher, J.W., Willsky, A.S.: Message errors in belief propagation. In NIPS, no. 17, pp. 609–616. MIT Press (2005)
5. Mut, F., Wright, S., Ascoli, G.A., Cebral, J.R.: Morphometric, geographic, and territorial characterization of brain arterial trees. Int. J. Numer. Method Biomed. Eng. **30**(7), 755–766 (2014)
6. Oda, M., Hoang, B.H., Kitasaka, T., Misawa, K., Fujiwara, M., Mori, K.: Automated anatomical labeling method for abdominal arteries extracted from 3D abdominal CT images (2012)
7. Osborn, A.: Osborn's Brain: Imaging, Pathology, and Anatomy. Amirsys Pub., Salt Lake City (2013)
8. Pearl, J.: Probabilistic Reasoning in Intelligent Systems: Networks of Plausible Inference. Morgan Kaufmann Publishers Inc., San Francisco (1988)
9. Peng, H., Ruan, Z., Long, F., Simpson, J.H., Myers, E.W.: V3D enables real-time 3D visualization and quantitative analysis of large-scale biological image data sets. Nat. Biotechnol. **28**(4), 348–353 (2010)
10. Scalzo, F., Piater, J.H.: Statistical learning of visual feature hierarchies. In: CVPR, p. 44 (2005)
11. Scalzo, F., Piater, J.H.: Unsupervised learning of visual feature hierarchies. In: Perner, P., Imiya, A. (eds.) MLDM 2005. LNCS (LNAI), vol. 3587, pp. 243–252. Springer, Heidelberg (2005)
12. Scalzo, F., Piater, J.H.: Adaptive patch features for object class recognition with learned hierarchical models. In: CVPR, pp. 1–8 (2007)
13. Shahzad, R., Dzyubachyk, O., Staring, M., Kullberg, J., Johansson, L., Ahlstrom, H., Lelieveldt, B.P., van der Geest, R.J.: Automated extraction and labelling of the arterial tree from whole-body MRA data. Med. Image Anal. **24**(1), 28–40 (2015)
14. Stefancik, R., Sonka, M.: Highly automated segmentation of arterial and venous trees from three-dimensional magnetic resonance angiography (MRA). Int. J. Cardiovasc. Imaging **17**(1), 37–47 (2001)
15. Uchiyama, Y., Yamauchi, M., Ando, H., Yokoyama, R., Hara, T., Fujita, H., Iwama, T., Hoshi, H.: Automated classification of cerebral arteries in MRA images and its application to maximum intensity projection. IEEE Eng. Med. Biol. Soc. **1**, 4865–4868 (2006)
16. Wang, J., Cohen, M.F.: An iterative optimization approach for unified image segmentation and matting. In: ICCV, vol. 2, pp. 936–943 (2005)

17. Wright, S.N., Kochunov, P., Mut, F., Bergamino, M., Brown, K.M., Mazziotta, J.C., Toga, A.W., Cebral, J.R., Ascoli, G.A.: Digital reconstruction and morphometric analysis of human brain arterial vasculature from magnetic resonance angiography. Neuroimage **82**, 170–181 (2013)

Virtual Reality

Lateral Touch Detection and Localization for Interactive, Augmented Planar Surfaces

A. Ntelidakis[(✉)], X. Zabulis, D. Grammenos, and P. Koutlemanis

Institute for Computer Science,
Foundation for Research and Technology - Hellas (FORTH),
Heraklion, Crete, Greece
{ntelidak,zabulis,gramenos,koutle}@ics.forth.gr

Abstract. This work regards fingertip contact detection and localization upon planar surfaces to provide interactivity in augmented displays implemented upon these surfaces, by projector-camera systems. In contrast to the widely employed approach where user hands are observed from above, lateral camera placement avails increased sensitivity to touch detection. An algorithmic approach for the treatment of the laterally acquired visual input is proposed and is comparatively evaluated against the conventional.

1 Introduction

Interactive surfaces are key elements in Augmented, Mixed Reality and Ambient Intelligence environments, which endorse the notion of enhancing physical surfaces with interactivity. A widely used approach employs a projector that creates a "display" upon the, typically planar, surface. At the same time, it utilizes sensing to detect and localize the contact of fingertips upon the surface and generate touch events. Motivation stems from natural interaction and avoidance of instrumentation of the surface.

This work focuses on the detection and localization of touch events upon planar surfaces, using a depth camera. As reviewed in Sect. 2, the norm in recent approaches to this problem is the placement the sensor above the interactive surface, so that contact is detected from depth differences between fingers and the surface. In Sect. 3, the lateral placement of the sensor is proposed, along with a method that detects touch based on the location of fingers in the depth image. As shown in the same section, the proposed approach is scalable to multiple sensor for creating larger interactive surfaces.

When viewing fully laterally, the workspace is imaged as a line, or a "horizon". Fingers in contact with the surface are detected within a thin region of interest, adjacent to this line. In Fig. 1 *left*, two approaches are illustrated. In Sect. 4, the approaches are compared showing that the proposed approach exhibits increased sensitivity to touch detection and, furthermore, covers more area per sensor with the same or better accuracy. A summary is provided in Sect. 5.

© Springer International Publishing Switzerland 2015
G. Bebis et al. (Eds.): ISVC 2015, Part I, LNCS 9474, pp. 551–560, 2015.
DOI: 10.1007/978-3-319-27857-5_50

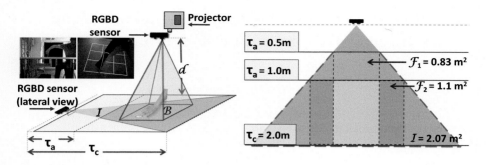

Fig. 1. *Left:* Top and lateral sensor placement and acquired images (although not employed by the method, their RGB component is shown in the thumbnails, for illustration purposes). For lateral placement, $[\tau_a, \tau_c]$ is the utilized depth range and $\mathcal{I}$ the area that fingertips can be detected. For top placement, d is the distance of the sensor to the surface and $\mathcal{B}$ the area that fingertips can be detected. *Right:* Geometry of lateral sensor placement. Sensor FOV is shown in light blue and area $\mathcal{I}$ with dashed orange lines. Rectangular subregions in $\mathcal{I}$ (i.e. $\mathcal{F}_1$, $\mathcal{F}_2$) can be defined to avail "portrait" or "landscape" interaction areas (Color figure online).

2   Related Work

Spatial Augmented Reality (SAR) [1] is related to smart environments, as it envisages the utilization of any physical surface as an interactive display. A prominent category of surfaces are planar ones, due to ease of projection, contact detection, and intuitiveness of use. Distortion-free projection upon planar surfaces can be achieved despite a relative slant between the projector and surface [2,3]. A special case of planar interactive surfaces, utilizes semi-transparent backprojected surfaces, where IR illumination is cast to provide contact detection and localization [4–7]. Front projection systems are more appealing, as they can be applied to virtually any surface without the need of instrumentation.

In [8], a seminal method for touch detection using a depth camera is proposed, where the camera is above a planar interaction surface. In setup, depth data are used to approximate the planar surface. In runtime, only 3D points close to this plane are considered. Touch events are implicitly detected, as only the top face of the finger is imaged, rather than the bottom which comes in contact with the surface. Detection utilizes an upper threshold to isolate candidate pixels close to the surface. A lower one, τ_X (≈ 1 cm), is used to select pixels imaging fingertips, from pixels imaging the interaction surface. Consequently, fingertips closer than τ_X to the surface, but not in contact with it will, still trigger a touch event, thus reducing interaction intuitiveness. The proposed work, increases sensitivity to fingertip contact, by reducing this distance where touch is spuriously triggered.

In [9,10], the same principle to [8] is adopted, but the shape of imaged fingers is analyzed to increase localization accuracy. In [11], the work in [8] is extended for multiple surfaces. In [2,3] the approach in [8] is extended for a steerable planar surface: instead of a priori modeling the planar surface, it continuously estimates its orientation at run-time, excluding user hands with a robust plane fitting method. In [12] the principle in [8] is adopted, but palms instead of fingers are detected. Spatial consistency is better exploited this way, as only large blobs trigger touch events. Thus, τ_X is lower ($\approx$3 mm), but at the cost of spatial granularity (palms instead of fingertips).

The requirement for planarity of the interaction surface was relaxed in [13,14], which enable touch detection on arbitrary surfaces. The depth camera is employed to model the interaction surface as background. In [13] the same principle as in [8] is adopted to detect touch, while [14] employs a stylus. The methods in [15–17], employ top camera placement and extend interaction affinity using collision detection to detect touch with the interactive surface. As above, a 3D representation of the surface is captured off-line.

3   The Proposed Approach

In the proposed approach, the camera is static and observing the scene laterally. Ideally, the camera is placed so that the interaction plane is perpendicular to the image plane and appears as a 2D line $\mathcal{L}$, or a "horizon", in the middle row of the acquired image. As this is practically difficult to achieve and the sensor may be placed slightly higher (see Fig. 2). Thresholding depth values, constrains the search for fingertips within the $[\tau_a, \tau_c]$ range of depth, where the sensor is reliable. To collect pixel support, a rectangular region of interest $\mathcal{Z}$, oriented parallel to the horizon and whose height is τ_h pixels, is considered in the image. The proposed approach is scalable to the case of combining multiple sensors to increase the interaction area. Further technical details of the techniques employed for calibration are presented in [18], along with analogous information for setup of the experiments presented in the next section.

3.1   Projector-Camera System Calibration

A difficulty in the calibration of the projector-camera system is that the camera does not image the projection area, due to its high slant. Typically null values are returned from such surfaces in depth cameras (i.e. see Fig. 2, bottom-right). Thus to associate coordinates on the interaction area with projector coordinates, the proposed touch detection method in Sect. 3.2 is utilized. The following, in this subsection, are quantities and geometrical entities determined during the calibration.

Region of interest $\mathcal{Z}$ in the depth image. $\mathcal{L}$ is determined by selection of two points upon the horizon of the interaction plane by the operator and of the computer mouse. $\mathcal{Z}$ is a 2D rectangle oriented as $\mathcal{L}$ and with a small height τ_h (i.e. 5 pixels). In the case of "approximate" sensor placement, τ_h is larger to support touch detection at close ranges (see Fig. 2).

Interaction plane $\mathcal{P}$ is estimated using 3D touch points, of the operator purposefully tracing the surface. $\mathcal{P}$ is computed by robustly fitting a plane to these points. The transformation $\{R, \mathbf{t}\}$ that maps $\mathcal{P}$ to plane $z = 0$ is computed from $\mathcal{P}$ through SVD decomposition.

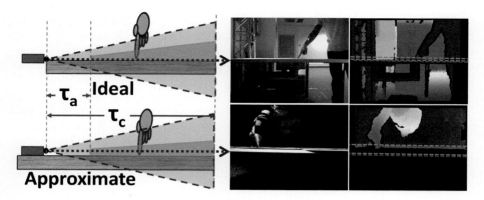

Fig. 2. *Left:* Side views of ideal (top) and approximate (bottom) lateral sensor placement. Dashed blue arrows plot sensor principal axes. The green area represents the volume within which 3D data is collected. Brown rectangles represent the interaction plane. *Right:* Images acquired for the ideal (top) and approximate (bottom) sensor placement. Superimposed, dashed green rectangles show $\mathcal{Z}$, whose lower edge occurs upon the horizon $\mathcal{L}$ (*red dashed line*) (Color figure online).

Homography H, a 3×3 matrix, maps coordinates from $\mathcal{P}$ to $\mathcal{F}$, the image frame of the projector. To estimate H, small dots are projected on the surface and the operator touches them. The 3D contact points, establish correspondences between the projector and the touch detection, 2D, reference frames. Points $\mathbf{c}_j$, are converted to 2D coordinates first by bringing them in $\mathcal{P}$'s reference frame, as $[x_j \, y_j \, z_j]^T = R \, \mathbf{c}_j + \mathbf{t}$. Truncation of the z dimension, projects these points on $z = 0$, converting them to 2D points $[x_j \, y_j \, 1]^T$. H is estimated from the correspondences between $\mathcal{P}$ and $\mathcal{F}$, via least squares optimization. The 2D polygon $\mathcal{Q}$, in $\mathcal{P}$, that outlines the interactive surface is computed from points $\mathbf{c}_j$.

Multiple sensors are enumerated by k and calibration is independently performed, for each. Corresponding results are denoted as $\mathcal{Z}_k$, P_k, H_k, and $\mathcal{Q}_k$. Let $\mathcal{O}_n$ be the n intersections of polygons $\mathcal{Q}_k$. These intersections are utilized in Sect. 3.3, where touch events occurring in $\mathcal{O}_n$ are treated specially.

In all experiments, depth sensors with FOV ($58°, 45°$) were employed and $\tau_a = 0.5$ m, $\tau_c = 2.0$ m. In an ideal sensor placement, this range forms an isosceles trapezoid $\mathcal{I}$ of 2.07 m^2 area (see Fig. 1 *right*). To create rectangular displays within $\mathcal{I}$, different regions can be designated. When comparing with the top view approach [8], in Sect. 4, the sensor is placed at $d = 1.3$ m where depth measurement is still reliable for finger touch detection [19]. For $d = 1.3$ m, $\mathcal{B} = 1.55$ m^2 compared to the 2.07 m of the proposed approach.

3.2   Contact Detection and Localization

Depth values in $\mathcal{Z}$ are interpreted in 3D. In particular, only points in $\mathcal{Q}$ are considered, ensuring that touch events are triggered only by events in the defined workspace. The result is a set of 3D points, $\mathbf{p}_i$. Then a simple 3D clustering technique, based on Connected Component Labeling, is employed upon $\mathbf{p}_i$ to find fingertips in contact with the interaction surface. The benefit over conventional, 2D blob detection on "foreground" pixels is that it resolves cases where two fingers at different depths are imaged as joined. The resultant clusters are fingertip detection candidates.

These candidates are filtered based on their spatial extent in the direction of $\mathcal{L}$. Clusters that are larger than the typical spatial extent of a fingertip (i.e. $1-3\,\mathrm{cm}$) are rejected. The result is a set of clusters, with centroids $\mathbf{c}_j$. Finally, points $\mathbf{c}_j$ are projected on $z = 0$ and mapped to display coordinates $\mathbf{q}_j$, as in Sect. 3.1: $\mathbf{q}_j=(u_j, v_j)$, as $[u_j\, v_j\, 1]^T = H\,[x_j\, y_j\, 1]^T$.

3.3 Multiple Sensors

Multiple sensors are combined in a horizontal placement, resulting in a display with "landscape" aspect ratio (see. Fig. 3). This placement maximizes interaction area, while retaining minimum depth interference between the sensors. The opposite case of stacking multiple sensors in a "portrait" configuration (one on top of the other) is not practical, as the users cannot reach all areas of the workspace. In addition, self-occlusion effects are more frequent when imaging hand from the side (see Sect. 4.3).

$\mathcal{O}_n$ are the n intersections of polygons $\mathcal{Q}_k$. Touch estimates in $\mathcal{O}_n$ are imaged by more than one sensor and treated specially, to avoid multiple touch detections for the same fingertip. A point in polygon test upon $\mathbf{q}_j$ groups cluster centers in and nearby $\mathcal{O}_n$, which are closer than distance τ_o.

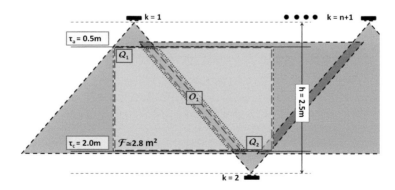

Fig. 3. Illustration of horizontal placement of k sensors. Sensors are placed equidistantly and against each other with a small overlap, minimizing interference between them. Sensor distance is $h = 2.5\,\mathrm{m}$ and $[\tau_a, \tau_c] = [0.5\,\mathrm{m}, 2.0\,\mathrm{m}]$. This allows for finger touch detection, in a rectangular $(2.8\,\mathrm{m}^2$, for $k = 2)$ interaction area on the surface (marked in yellow). Black dots indicate possible geometry with more sensors (Color figure online).

3.4   Fingertip Tracking and Touch Event Creation

A 2D tracker assigns unique ids to $\mathbf{q}_j$, tracks them in consecutive frames and generates touch events to the operating system for multiple finger interaction. The tracker also compensates for transient detection failures and spurious detections.

Touch points $\mathbf{q}_j$ are assigned unique ids the first time detected. Temporal correspondences are established with the closest touch locus at the previous frame, if its distance is below threshold τ_t; otherwise a new id is generated.

A 2D Kalman filter [20] is employed to better estimate touch location and compensate for transient failures in estimates. The latter is achieved by inputting to the tracker, upon disappearance of a 2D touch point, the predicted state for a small sequence of frames (i.e. 5). Contact points are also tracked for a few frames (i.e. 4) before deeming them valid, to mitigate spurious detections.

Tracking of touch point ids is required, to avoid confusion in time-lasting events. The implemented touch events are *touchstart*, *touchmove* and *touchend*, of the *Windows 8 Touch Injection API*, which transparently provides the emulated events to applications run by the operating system.

4   Experiments

Experiments were performed using the *Asus Xtion Pro* RGB-D sensors (480×640, 30 fps), on a conventional PC (*Intel i7* CPU, 2.93 GHz, 8 GB RAM). The proposed method is computationally lightweight, as only subset $\mathcal{Z}$ of the depth image is considered. The implementation runs at 4.6 msec for a frame. The top view method runs at 7.3 msec for frames of the same resolution.

4.1   Contact Localization Accuracy and Limitations

To assess the accuracy of the proposed method to the top-view approach two experiments were conducted.

Experiment 1. The proposed method was installed in a wall configuration. The top view sensor placement is impractical for this configuration, as it interferes with the user's position. Range $[\tau_a, \tau_c]$ was $[0.6\,\mathrm{m}, 1.9\,\mathrm{m}]$ and $\mathcal{I}$ was $\approx 1.2\,\mathrm{m}^2$. The projector displayed a grid of 160 dots, radius 1 cm, and 3 users touched their centers. Error was the distance of the estimated touch location to the center of the dot, in $\mathcal{F}$. The mean localization error was 3.02 (std: 2.19) pixels. The reported errors are the effective errors as, besides localization error, they also contain the calibration error due to the estimation of H (Sect. 3.1). Error increases with distance, indicating localization accuracy < 2 pixels in short range and no more than 6 pixels at $\approx 2\,\mathrm{m}$ distance (see Fig. 4).

Experiment 2. Two configurations, a lateral and top-view, were implemented to image the same planar surface simultaneously, as in Fig. 1. The common workspace covered by both sensors was $\approx 0.7\,\mathrm{m}^2$ ($0.9 \times 0.8\,\mathrm{m}^2$). The task and error measurement were the same as in the first experiment. The projector displayed

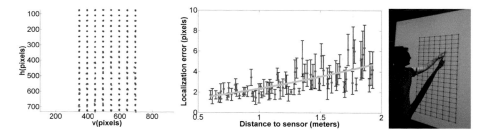

Fig. 4. Accuracy of touch detection. *Left:* Plot illustrates the projected 160 target dots (blue) and the estimated contact locations (red) for 3 users. *Middle:* Mean localization error (blue) with standard deviation (gray) as a function of distance. A second degree polyonym (green) is fit. *Right:* A user touching a dot in the target grid (Color figure online).

a grid of 16 dots and a user touched the dots. At the same time, activation of the two sensors alternated, so the active illumination systems of the two sensors would not interfere. Thus, the two systems were compared on the same physical events, in images acquired (approximately) simultaneously. The mean fingertip localization error (and standard deviation) were 4.60 (2.80) for the lateral and 4.59 (3.14) for the top view in pixels, indicating that the proposed approach is at least equivalent, in terms of accuracy, with the conventional one.

Discussion. The results of these experiments indicate a similar contact localization accuracy for both methods. However, the lateral placement approach is computationally lightweight and exhibits an advantage in terms of area covered by a sensor (see Sect. 3.1), as well as, suitability for wall configurations. The higher accuracy for the proposed approach in the first experiment is attributed to the closer distance ($\tau_a = 0.6\,\text{m}$) of fingers compared to the second experiment ($\tau_a = 1.0\,\text{m}$). However, for $\tau_b = 2.0\,\text{m}$ the mean error is ≈ 4 pixels, comparable to mean accuracy of top view approach in the second experiment.

4.2   Sensitivity to Height Threshold

Sensitivity s of contact detection is defined as the maximum distance from the surface that contact is detected. Intuitively, sensitivity is the minimum distance at which the system reliably discriminates when the fingertip is in contact to the surface, or not. The smaller s is, the more sensitive contact detection becomes.

The setup was as in Sect. 4.1, Exp. 2. A grid of 16 dots, covering [780 × 780 mm^2] was displayed. On each dot, 4 *Lego* blocks were stacked comprising 4 different height configurations (3.16, 6.33, 9.5 and 19 mm). For each configuration, depth data of a user placing its finger at the top of the stack were acquired from both sensors. Acquired data were post-processed to remove depth pixels imaging the Lego blocks. For each method and at each location, sensitivity s was measured as the maximum height that a touch detection occurred out of the 4

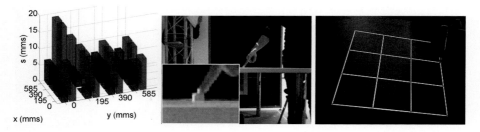

Fig. 5. Touch sensitivity. *Left:* s values for the lateral (blue) and top-view approach (red) on each of the 16 positions of the $[780 \times 780\,\text{mm}^2]$ grid. *Middle-Right:* Two simultaneous input images from the experiment, one from each sensor (Color figure online).

configurations (see Fig. 5). The mean of s was 4.55 (std: 1.99) for the lateral and 11.08 (std: 4.01) for the top view, in mm.

As the mean values of s for the lateral approach are smaller than for the top-view, it is concluded that the former yields greater sensitivity to touch.

4.3   Multiple Finger Interaction

Multiple finger interaction is subject to distance limitations and self-occlusions. In the experiments, interaction area was $1.08\,\text{m}^2$ $(1.2 \times 0.9\,\text{m}^2)$ and range $[\tau_a, \tau_c]$ was $[0.7\,\text{m}, 1.9\,\text{m}]$. The projector displayed 3 parallel lines with in-between distance 5 cm sequentially, in 3 regions (total 9 lines). The lines had either parallel or perpendicular orientation to the principal axis of the sensor. They approximately covered the interaction area, as illustrated in the plots of Fig. 6.

The user traced each set of lines with three fingers. As distance from the sensor increases, fingers may appear merged due to the reduced resolution; loss of reliability occurs at distances of $2\approx$ m where 3D clustering cannot discriminate between fingers (see Fig. 6).

Self-occlusions (i.e. a finger occluding another finger) may occur even at close range. Ergonomy indicates that self-occlusions are limited by construction when the sensor is placed "opposite" to the user rather than the side. In Fig. 6, it is shown that when the sensor is placed on the side of the user, self-occlusions are more likely to occur than when placed opposite to the user.

4.4   Use Cases

The proposed method was employed in two interactive applications, which were presented in a public installation during the TEDx Heraklion 2015 event. The same configuration for both, employed two Xtion sensors mounted on a wall that covered interaction area $\mathcal{F} = 2.8\,\text{m}^2$. In the first application, users interact conventionally with multimedia shown by large displays via multi-touch events. The application presents collections of multimedia information, where users interact

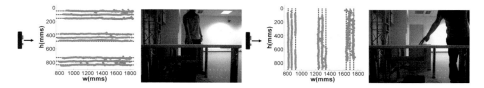

Fig. 6. Multitouch detection results, with the sensor placed opposite (left) and on the side of the user (right). RGB images are shown for illustration only. Red dots plot the estimated coordinates in $\mathcal{F}$, while blue dashed dots the projected pivot line pattern. Approximate sensor placement is shown on the left of plots (Color figure online).

Fig. 7. Pilot applications. *Left:* Multimedia application. *Right:* Shooting game.

via multi-touch input to open/close windows, drag images etc. Generation of native multi-touch events also allows utilization of multi-touch gesture detection provided by the *Windows* 8 operating system (zoom in/out and scroll). The second application is a game where users shoot toy balls at digital targets on the wall. It demonstrates system robustness for cases of brief contact. Targets are "hit" upon contact detection of the ball to the wall at the target's location. The applications received positive feedback, as interaction was intuitive and robust, indicating the suitability of the method for augmented, interactive displays. Figure 7 illustrates users interacting with the applications.

5 Summary

An approach is presented for touch detection and localization upon planar surfaces using a depth camera, placed laterally. Correspondingly, an algorithmic approach for the treatment of this input is proposed. Experimental evaluation shows that it provides greater sensitivity to touch than the conventional top view approach, while being computationally lightweight. Increased sensitivity is due to the detection of contact based on xy pixel coordinates, rather than depth values which are less reliable. Furthermore, the proposed approach exhibits greater accuracy of touch localization, allows for greater interactive areas per sensor, and supports multiple sensors to create large interactive displays.

Acknowledgments. This work has been supported by the FORTH-ICS internal RTD Programme "Ambient Intelligence and Smart Environments".

References

1. Bimber, O., Raskar, R.: Spatial Augmented Reality: Merging Real and Virtual Worlds. A.K. Peters Ltd., Natick (2005)
2. Zabulis, X., Koutlemanis, P., Grammenos, D.: Augmented multitouch interaction upon a 2-DOF rotating disk. In: Bebis, G., et al. (eds.) ISVC 2012, Part I. LNCS, vol. 7431, pp. 642–653. Springer, Heidelberg (2012)
3. Koutlemanis, P., et al.: A steerable multitouch display for surface computing and its evaluation. IJAIT **22**, 1–29 (2013). doi:10.1142/S0218213013600166
4. Matsushita, N., Rekimoto, J.: Holowall: designing a finger, hand, body, and object sensitive wall. In: ACM UIST, pp. 209–210 (1997)
5. Han, J.: Low-cost multi-touch sensing through frustrated total internal reflection. In: ACM UIST, pp. 115–118 (2005)
6. Michel, D., et al.: Building a multi-touch display based on computer vision techniques. In: MVA, pp. 74–77 (2009)
7. Leibe, B., et al.: Toward spontaneous interaction with the perceptive workbench. IEEE CG&A **20**, 54–65 (2000)
8. Wilson, A.: Using a depth camera as a touch sensor. In: ACM ITS, New York, NY, USA, pp. 69–72 (2010)
9. Klompmaker, F., et al.: Authenticated tangible interaction using RFID and depth-sensing cameras. In: ACHI, pp. 141–144 (2012)
10. Klompmaker, F., et al.: dSensingNI: a framework for advanced tangible interaction using a depth camera. In: TEI, pp. 217–224 (2012)
11. Wilson, A., Benko, H.: Combining multiple depth cameras and projectors for interactions on, above and between surfaces. In: ACM UIST, pp. 273–282 (2010)
12. Xiao, R., et al.: WorldKit: rapid and easy creation of ad-hoc interactive applications on everyday surfaces. In: CHI, pp. 879–888 (2013)
13. Harrison, C., et al.: OmniTouch: wearable multitouch interaction everywhere. In: UIST, pp. 441–450 (2011)
14. Jones, B., et al.: Build your world and play in it: Interacting with surface particles on complex objects. In: ISMAR, pp. 165–174 (2010)
15. Hilliges, O., et al.: HoloDesk: direct 3D interactions with a situated see-through display. In: ACM CHI, pp. 2421–2430 (2012)
16. Benko, H., Jota, R., Wilson, A.: MirageTable: freehand interaction on a projected augmented reality tabletop. In: ACM SIGCHI, pp. 199–208 (2012)
17. Jones, B., et al.: Roomalive: Magical experiences enabled by scalable, adaptive projector-camera units. In: ACM UIST, pp. 637–644 (2014)
18. Ntelidakis, A., Zabulis, X., Grammenos, D., Koutlemanis, P.: Lateral touch detection based on depth cameras. Technical report FORTH-ICS-459, Foundation for Research and Technology - Hellas, Institute for Computer Science (2015)
19. Smisek, J., et al.: 3D with Kinect. In: ICCV Workshops, pp. 1154–1160. IEEE (2011)
20. Bishop, C.M.: Pattern Recognition and Machine Learning (Information Science and Statistics). Springer, New York (2006)

A Hybrid Real-Time Visual Tracking Using Compressive RGB-D Features

Mengyuan Zhao[1], Heng Luo[2], Ahmad P. Tafti[3], Yuanchang Lin[4], and Guotian He[4(✉)]

[1] College of Computer Science and Technology, Chongqing University of Posts
and Telecommunications, Chongqing, China
[2] College of Communication and Information Engineering, Nanjing University of Posts
and Telecommunications, Nanjing, China
[3] Department of Computer Science, University of Wisconsin-Milwaukee, Milwaukee,
WI 53211, USA
[4] Chongqing Institute of Green and Intelligent Technology, Chinese Academy of Sciences,
Chongqing, China
heguotian@cigit.ac.cn

Abstract. The online multi-instance learning tracking (MIL) algorithm is known for its ability of alleviating tracking drift by training classifiers with positive and negative bag. However, the increased computational complexity results in time consuming due to the lack of consideration of sampling importance when collecting training samples. Additionally, the MIL method, as a 2D feature-based tracking algorithm, performs unsteadily when the object changes poses or rotates seriously. In this paper, a histogram-based feature similarity measurement is employed as a weighting strategy to select positive samples. Benefited from profitable depth information, the tracking algorithm we proposed achieves higher tracking performance. For computational efficiency, a compressive sensing method is adopted to extract features and reduce dimensionality. Experimental results demonstrate that our algorithm is better in robustness, accuracy, efficiency than three state-of-the-art methods on challenging video sequences.

Keywords: Visual tracking · MIL · Histogram feature similarity · Depth feature · Compressive sensing

1 Introduction

Target tracking plays an important role in compute vision, with a wide range of applications in intelligent monitoring, intelligent transportation, human interaction, machine vision and many other fields [1]. However, due to issues such as shape change, brightness variation and occlusion, it is still a great challenge to design robust object tracking algorithms.

In recent years, object tracking algorithm based on online learning and training classifier [2–10] has drawn wide attention. The key lies in separating foreground objects from the background in a sequence of images. Tracking is viewed as a special binary classification problem in this method. For example, the MIL algorithm [8] proposed a

© Springer International Publishing Switzerland 2015
G. Bebis et al. (Eds.): ISVC 2015, Part I, LNCS 9474, pp. 561–573, 2015.
DOI: 10.1007/978-3-319-27857-5_51

multi-instance learning method to alleviate the problem of target-drifting in the online training classifiers from a single positive sample and negative sample. On account of paying no attention to sampling importance, the MIL algorithm is, to some extent, time-consuming. Zhang et al. presented a novel online weighted MIL (WMIL) tracker [9] to improve the timeliness, which integrates the sampling importance into an efficient online learning procedure by assuming some most important samples when training the classifier. The WMIL algorithm proposed the strategy that gives larger weights to the samples near the target and smaller weights to the ones far away to speed up the samples selection thereby improving the time performance of algorithm. However, these positive samples stained by sudden occlusion contributes little to the following procedure in that they are selected according to the Euclidean distance between samples and destinations. In this paper, we exploit a weighting principle of histogram feature similarity strategy [11, 12] to select positive samples. Samples similar to the feature of object is picked to form positive sample bags so as to handle the occlusion.

The MIL algorithm is based on 2D features with single feature species, thus the tracking often fails when the target changes poses or rotates. The proposed tracking algorithm utilizes features in the depth map and achieves higher tracking accuracy by combining the color, texture feature and depth information.

2   The Proposed Tracking Method

Motivated by the framework of MIL algorithm, we proposed an improved tracking algorithm which is showed in Fig. 1. For higher efficiency and precision, the depth information is used to find the primary location of the object. In order to reduce computational cost in the case of few information loss, the compressive sensing theory [13, 14] is adopted to extract Haar-like features in R, G and B channel. A weak classifier is constructed with compressive Haar features. Then the boosting framework is employed to generate a strong classifier. In the step of training classifier, the training samples are selected by a weighting principle of histogram-based feature similarity. Meanwhile, undesirable samples are rejected to alleviate error accumulation. The parameters of classifier are updated using online multiple instance learning to ensure perfect tracking performance.

2.1   Image Representation

Compressive Haar-C Feature. In order to describe the visual characteristics, salient features of the object need to be extracted and effectively represented. To balance the performance and efficiency, the Haar-like feature is also adopted to build the object's surface model [15]. Firstly, the target is located by cropping several patches, i.e., the candidate samples in the i-th frame I_i. For each patch p_i, the Harr-like features is obtained by the little boxes whose size and position are determined randomly from the three channels, i.e., R, G, B. Then, a strong feature C_i, namely, the Haar-C feature is acquired by the combination of feature vector $\{R_i, G_i, B_i\}$. Considering that there exists more than millions of boxes, a random measurement matrix [16–23] $R \in R^{n \times m}$ is introduced to

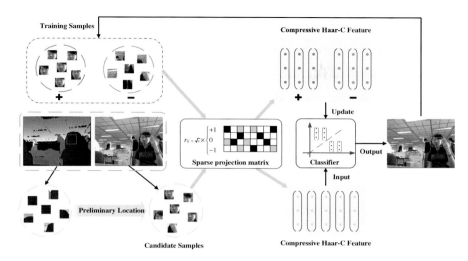

Fig. 1. The outline of proposed tracking algorithm

projecting data from the high-dimensional image space $v \in R^m$ to a lower-dimensional space $x \in R^n$, formulated as below

$$x = Rv, \tag{1}$$

where the vector v represents weak feature vectors in R, G, B channel respectively. Thus, we obtain a strong feature vector, which is called compressive Haar-C feature $C_i * (x)$, by combining the low dimensional vector v in three channels. This procedure of compressing Haar-C feature from single channel is vividly showed in Fig. 1. Baraniuk [24] proved that the random matrix satisfy the Johnson-Lindenstrauss lemma [19], which ensure that x preserves almost all the information in v. Note that the matrix R is generated randomly in the initialization of the tracker and remain unchanged during subsequent tracking process.

Depth Information. Several algorithms [2–4] which extract Haar features based on gray pictures are witnessed not good enough during the tracking due to insufficiency discriminative features provided by the intension of pixels. Therefore, by utilizing the depth information of the picture scene that can be real-timely captured by a Kinect device, more abundant description about the object and the background can be obtained. This can benefit the identification of the disturbance extremely similar to the truth target from the background, which is a huge challenge for the common algorithms. As the gray value indicates the distance between the target and camera in the depth map, all the pixels representing the same object should have the same or similar gray value. Accordingly, if the gray value of each pixel is regarded as a random variable x (ranging from 0 to 255), some statistic numerical properties can be utilized to formulate useful information contained in the depth picture. The mean and variance of all pixels in the depth map's corresponding region are calculated as

$$\mu_i = E\{X_K\} = \frac{1}{N} \sum_{K=1}^{N} X_K, \tag{2}$$

$$\sigma_i = E\left\{[X_K - \mu_K]^2\right\} = \frac{1}{N} \sum_{K=1}^{N} (X_K - \mu_i)^2, \tag{3}$$

where N is the number of pixels in p_i and $D_i = (\mu_i, \sigma_i)$ is the depth feature of p_i. Ideally, if patch p_i exactly represents an object, then μ_i indicates the depth of object in current scene and σ_i approximates to 0. Generally, distinct boundary exists between target and background in depth map as their gray level are extremely dissimilar. The variance is used for evaluating how many pixels from background are mixed in p_i and the mean recognizes the potential disruptor with disparate depth level to the real object. Based on the analysis above, the depth information can be applied to identify the approximate target location in the new frame, which improves the accuracy of the algorithm and reduces the calculation amount of the follow-up procedure.

2.2 Selection of Training Samples

The MIL algorithm trains the classifier online using the positive and negative samples, which solves the problems of tracking failure or model drifting due to the positive samples caused by noise when target detection in the current frame is not accurate. As this approach ignores the importance of selecting samples, the real-time performance is not good. The WMIL algorithm, however, considers the location relationship between the sample and the target, giving larger weight coefficient to samples near the target and lower weight to samples far from the target, so as to speed up the selection of the samples in real-time.

The WMIL tracker defines two bag probabilities as follows

$$p(y = 1|X^+) = \sum_{j=0}^{N-1} w_{j0} p(y = 1|x_{1j}), \tag{4}$$

$$p(y = 0|X^-) = \sum_{j=N}^{N+L-1} w p(y_0 = 0|x_{0j}) = w \sum_{j=N}^{N+L-1} (1 - p(y_0 = 1|x_{0j})), \tag{5}$$

where $p(y = 1|x_{1j})$ is the prior probability of the samples x_{1j}. The weight w_{j0} is a monotone decreasing function. In WMIL algorithm, a higher value is assigned to the samples near the target, assuming that all negative instances contribute equally to the negative bag.

The WMIL algorithm works well in samples selection according to Euclidean distance between sample and destination. However, the samples near the tracking location may not always contribute much to the subsequent tracking process. For example, when sudden occlusion occurs, the error is potential to be accumulated if samples near occlusion region are all added to positive bag. An advisable strategy is adopted to eliminate these error-prone samples and also reject them in negative bag. In the purpose of noise restraint, an isolation strip between the positive and negative sample selection area is built to avoid degrading the contrast of these two classes. In order to enhance the resistance to sudden occlusion, samples are filtrated based on the weight which is determined by the similarity of feature histogram between target and samples. The weight is calculated by Bhattacharyya coefficient

$$w_{j0} = \sum_{u=1}^{m} \sqrt{p_u(y)q_u},$$
(6)

where $p_u(y)$ is the feature histogram of the target, and q_u is the feature histogram of the sample. For simplicity, undesired samples is discarded by evaluating Bhattacharyya coefficient. The selection strategy is described in Algorithm 1. In general, the more similarity between the sample and the target, the greater contribution to the subsequent tracking process. We use this strategy to choose samples based on the similarity of histogram of features to solve the sudden occlusion problem.

Algorithm 1. Selection of Training Samples

Input: the object location $l_t(x^*)$, the feature vector of object $C_t^*(x)$;

1. Sample a set of patches $D^\alpha = \{x \mid\mid\mid l_t(x) - l_t(x^*) \mid\mid < \alpha\}$ to form positive bag;
2. Sample a set of patches $D^{\beta,\gamma} = \{x \mid \beta \mid\mid l_t(x) - l_t(x^*) \mid\mid < \gamma\}$ to form negative bag;
3. Compute each w_{j0} according to Eq.13;
4. Find the maximum and the minimum among w_{j0}, $j = 0,1,\ldots,$N-1;
5. Reject samples with weight ranking the last 10%.

Output: positive bag, weight of each positive sample, negative bag.

2.3 Classifier Construction and Selection

In the MIL algorithm, the posterior probability of labeling sample x, which is positive, is computed with the Bayesian theorem.

$$p(y = 1|x) = \frac{p(x|y = 1)p(y = 1)}{\sum_{y \in \{0,1\}} p(x|y)p(y)} = \sigma \left(\ln \left(\frac{p(x|y = 1)p(y = 1)}{p(x|y = 0)p(y = 0)} \right) \right),$$
(7)

where $\sigma(z) = 1/(1 + e^{-z})$ is a sigmoid function, x is the labeled sample, which is represented by a feature vector function $f(x) = (f1(x), \ldots, fk(x))^T$. We assume that the features in $f(x)$ are independently distributed and $p(y = 0) = p(y = 1)$. The discriminative appearance model is a classifier $H_K(x)$ defined as

$$H_K(x) = \ln \left(\frac{p(x|y = 1)p(y = 1)}{p(x|y = 0)p(y = 0)} \right) = \sum_{k=1}^{K} h_k(x).$$
(8)

Then Eq. (7) can be rewritten as

$$p(y = 1|x) = \sigma(H_k(x)),$$
(9)

where $H_k(.)$ is a strong classifier, which is built with k weak classifiers $h_k(.)$, related to each Haar-C feature $f_k(.)$. The conditional distributions are modeled as Gaussian function

$$p(f_k(x_{ij})|y_i = 1) \sim N(\mu_1, \delta_1), \tag{10}$$

$$p(f_k(x_{ij})|y_i = 0) \sim N(\mu_0, \delta_0), \tag{11}$$

where μ_1 and δ_1 are the mean and standard deviation of the positive class. μ_0 and δ_0 are the mean and standard deviation of the negative class. The update schemes for the parameters and are computed as follows

$$\mu_1 \leftarrow \eta\mu_1 + (1 - \eta)\frac{1}{N}\sum_{j|y_i=1} f_k(x_{ij}), \tag{12}$$

$$\delta_1 \leftarrow \eta\delta_1 + (1 - \eta)\sqrt{\frac{1}{N}\sum_{j|y_i=1} \left(f_k(x_{ij}) - \mu_1\right)^2}, \tag{13}$$

where N is the number of positive samples and η is the learning rate. The μ_0, δ_0 are updated with similar rules.

2.4   Tracking Process

To initialize our tracker, we set an initial bounding box through artificial annotation. The depth feature D_0 is calculated by Eqs. 15 and 16, the training samples is collected in line with Algorithm 1 and the compressive Haar-C feature Ct*(x) is extracted complying with Fig. 2. Then K weak classifiers are selected by solving the following equation

$$h_K = \arg\max_{h \in \Phi} <h, \nabla l(H)> |_{H=H_{k-1}}, \tag{14}$$

where $\nabla l(.)$ is the gradient of possibility function described and computed in [9]. When a new frame arrives, candidate samples are collected in depth map around the object location in previous frame for determining primary search as below

$$D^\zeta = \left\{x| \left\|l_t(x) - l_{t-1}(x^*)\right\| < \zeta\right\}, \tag{15}$$

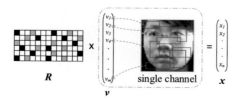

Fig. 2. Illustration of compressing Haar-C feature

Depth information is firstly used to detect the approximate location of the target. Since the object depth is likely to change between two successive frames (i.e., the distance between object and camera may get farther or closer) and the variance of depth is expected to be as low as possible, the approximate location $l_t(x')$ can be obtained by optimizing the following formula

$$D'_t = \arg \min_{i \in S} f(D_i), \tag{16}$$

where D'_t is the depth feature corresponding to $l_t(x')$, S is the amount of collected samples, and function $f(.)$ is defined as

$$f(D_i) = (\sigma_i + 1) e^{|\mu_i - \mu_t^*|}. \tag{17}$$

Considering that μ_i has more powerful influence upon the output of $f(.)$ than σ_i, the purpose of solving Eq. 16 is to find the one with minimum variance in sample set which have similar depth value to object. In color map, candidate samples is cropped around $l_t(x')$ with similar operating manner as Algorithm 1, thus a new samples set is formed as

$$D^\xi = \left\{ x | \left\| l_t(x) - l_t(x') \right\| < \xi \right\}. \tag{18}$$

Then, the strong classifier trained in previous frame is used to find the final position of object, i.e., the maximum confidence position of object which is achieved by solving

$$l_t(x^*) = \arg \max_{x \in D^\xi} P(y|x). \tag{19}$$

Algorithm 2. Tracking Using Compressive RGB-D Hybrid Features

---

Input: the object location $l_{t-1}(x^*)$, the depth feature of object D_{t-1}

1. Crop candidate samples D^ζ according to Eq.15;
2. Compute each D_i according to Eq.2-3;
3. Obtain $l_t(x')$ according to Eq.16-17;
4. Crop candidate samples D^ξ according to Eq.18;
5. Extract and compress Haar-C feature C_i^* according to Eq.1;
6. Use strong classifier H_k to estimate $P(y=1|x)$ for $x \in D^\eta$;
7. Locate object $l_t(x^*)$ via Eq.19;
8. Update the classifier parameters according to Eqs. 12-13

Output: $l_t(x^*)$, D_t

The summary of proposed tracking process is presented in Algorithm 2.

3   Experiments

3.1   Experimental Data and Settings

In the following experiments, we test our algorithms on a PC with Intel Core CPU (2.81 GHz) and 4G memory. The dataset used in experiments is mainly collected from the following publicly available website: http://vision.cs.princeton.edu/projects/2013/tracking/dataset.html/.

We have compared ours algorithm with three state-of-the-art trackers on three video sequences, which contain challenging factors, pose and scale variations and heavy occlusions. The existing excellent algorithms we compared our algorithms with include: Visual tracking with online multiple instance learning (MIL) [8], real-time compressive tracker (CT) [7], real-time multi-scale tracking based on compressive sensing (MSCT) [25].

Given a target location at the current frame, the primary search radius ζ and the secondary search radius ξ is set as 5 pixels and 10 pixels respectively. In Algorithm 1, the inner and outer radius is set as $\alpha = 5$ and $\beta = 16$ respectively. By adjusting appropriate search step-length, 30 positive samples and 40 negative samples are randomly selected from the corresponding region respectively. Our tracker sets $K = 8 \sim 10$ features to design the weak classifier, and the learning parameter is set as $\eta = 0.95 \sim 0.96$. A smaller learning rate can make the tracker quickly adapted to the fast appearance and larger rate can reduce the likelihood that the tracker drifts off the target. The best parameter we have tested is 0.96.

3.2   Experimental Results

In this paper, we use two evaluation criteria: center location error and tracking success rates [26], which are both computed against manually labeled ground truth, to measure the four target tracking algorithms. The center location error is defined as the Euclidean distance between the central locations of the tracked objects and the manually labeled ground truth. It is calculated as follows

$$error(i) = \sqrt{\left(x_i - x_{gi}\right)^2 + \left(y_i - y_{gi}\right)^2}, \tag{20}$$

where (x_i, y_i) is the tracking result of the i-th frame and (x_{gi}, y_{gi}) is the position of ground truth. The metric success rate is defined as

$$score = \frac{area(B_T \cap B_G)}{area(B_T \cup B_G)}, \tag{21}$$

Where B_T is the tracking bounding box and B_G is the ground truth bounding box. If the score is larger than 0.5 in one frame, the tracking results are considered as a success.

Table 1 shows the experimental results in terms of average center location error. Table 2 presents the tracking results in terms of success rate. Table 3 shows the frame rates of all methods. Tables 1 and 2 tell that our tracking algorithm achieves the best or second best performance in most sequences based on both average center location error and success rate. Furthermore, the proposed tracker is obviously much better than the original MIL algorithm in real-time, performing well in terms of speed among all the evaluated algorithms on the same machine.

Table 1. The average tracking center positioning error (pixels)

Sequence	Ours	MIL	CT	MSCT
two_people_1.1	20.0	158.4	84.1	60.6
basketballnew	16.1	80.5	71.0	74.4
face_occ2	28.9	107.1	117.7	59.8
basketball1	9.19	70.0	68.7	77.9

Table 2. The success rate of the algorithm (SR %)

Sequence	Ours	MIL	CT	MSCT
two_people_1.1	64	11	52	25
basketballnew	67	25	27	30
face_occ2	53	8	11	28
basketball1	78	19	16	15

Table 3. Frame rate comparison (frames/second)

Sequence	Ours	MIL	CT	MSCT
two_people_1.1	7.1	2.6	9.5	7.0
basketballnew	8.7	2.7	9.6	7.1
face_occ2	8.6	2.6	10.1	7.6
basketball1	8.3	2.5	9.1	6.9

Sudden occlusion: The target object in the two_people_1.1 and face_occ2 sequences in Fig. 3 undergoes sudden occlusion (#31, #212 in Fig. 3(a) and #17, #207 in Fig. 3(b)). From Fig. 4, Tables 1 and 2, we can see that our algorithm outperforms the other methods on those sequences in terms of success rate and center location error. The main reason is that our algorithm has a good mechanism, choosing samples based on the similarity of histogram of features, to solve the sudden occlusion problem. As shown in #264 in the Fig. 3(a) and #330 in the Fig. 3(b), all trackers except ours fail to work properly with occlusion. Furthermore, the proposed algorithm outperforms most of the other methods in the speed as shown in Table 3, because our algorithm uses space compressive sensing matrix to reduce the time complexity.

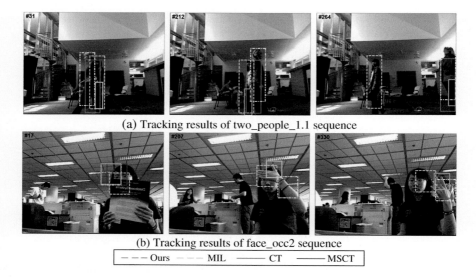

(a) Tracking results of two_people_1.1 sequence

(b) Tracking results of face_occ2 sequence

— — — Ours   — — — MIL   ——— CT   ——— MSCT

Fig. 3. Tracking results of four trackers on two_people_1.1 and face_occ2 sequences

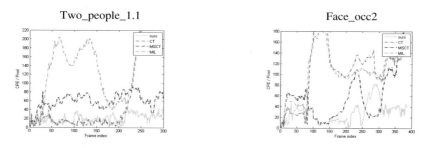

Fig. 4. Quantitative comparison of the four trackers with the center location error on the two_people_1.1 and the face_occ2 video sequences

Abrupt motion and pose variation: For the basketballnew sequence shown in Fig. 5(a) and the basketball1 sequence in Fig. 5(b), the object undergoes abrupt motion and pose variation. We can see that only our algorithm performs well and is able to track the object reliably in that our algorithm utilizes more robust 3D information. By using complementary features and feature similarities as the weights to update online samples, which can easily distinguish between the background and foreground compared with other algorithms. Therefore, our algorithm can adapt to the changes of target appearance model in the procedure of the abrupt motion and pose variation of target. Furthermore, from Table 3, we can see that space compressive sensing matrix is used to reduce the time complexity to perform well when the object undergoes abrupt motion and pose variation (Fig. 6).

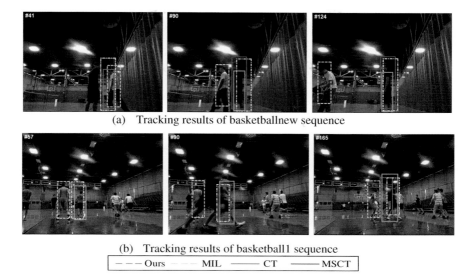

(a)   Tracking results of basketballnew sequence

(b)   Tracking results of basketball1 sequence

| − − − Ours | − − − MIL | ——— CT | ——— MSCT |

Fig. 5. Tracking results of the four trackers on basketballnew and basketball1 sequence

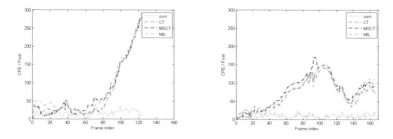

Fig. 6. Quantitative comparison of the four trackers with the center location error on the basketballnew and basketball1 video sequence

4   Conclusions

The present work is based on the MIL algorithm, with additional depth information and the strategy of choosing samples based on the similarity of histogram of features to solve the sudden occlusion problem. The sparse compressive sensing method is used to compress features, improving the robustness of the algorithm and reducing the time complexity of the algorithm at the same time. As can be seen from the experimental results and analysis, the proposed algorithm can solve the tracking failure owing to sudden occlusion, abrupt motion and pose variation. Furthermore, from Table 3, the time-consuming of our algorithm is also low. Overall, our tracking algorithm can achieve a unity of good robustness and real-time performance.

References

1. Yilmaz, A., Javed, O., Shah, M.: Object tracking: a survey. ACM Comput. Surv. **38**, 13 (2006)
2. Collins, R.T., Liu, Y., Leordeanu, M.: Online selection of discriminative tracking features. IEEE Trans. Pattern Anal. Mach. Intell. **27**, 1631–1643 (2005)
3. Grabner, H., Bischof, H.: On-line boosting and vision. In: IEEE Computer Society Conference on Computer Vision and Pattern Recognition, vol. 1, pp. 260–267 (2006)
4. Grabner, H., Grabner, M., Bischof, H.: Real-time tracking via on-line boosting. BMVC **1**, 6 (2006)
5. Avidan, S.: Ensemble tracking. IEEE Trans. Pattern Anal. Mach. Intell.**29**, 261–271 (2007)
6. Babenko, B., Yang, M.H., Belongie, S.: Robust object tracking with online multiple instance learning. IEEE Trans. Pattern Anal. Mach. Intell.**38**, 1619–1632 (2011)
7. Zhang, K., Zhang, L., Yang, M.-H.: Real-time compressive tracking. In: Fitzgibbon, A., Lazebnik, S., Perona, P., Sato, Y., Schmid, C. (eds.) ECCV 2012, Part III. LNCS, vol. 7574, pp. 864–877. Springer, Heidelberg (2012)
8. Babenko, B., Yang, M.H., Belongie, S.: Visual tracking with online multiple instances learning. In: IEEE Conference on Computer Vision and Pattern Recognition 2009, CVPR 2009, pp. 983–990. IEEE (2009)
9. Zhang, K., Song, H.: Real-time visual tracking via online weighted multiple instance learning. Pattern Recogn. **46**(January), 397–411 (2013)
10. Zhang, K., Zhang, L., Yang, M.-H.: Fast compressive tracking. IEEE Trans. Pattern Anal. Mach. Intell. **36**(10), 2002–2015 (2014)
11. Guo, C.: Research on Online Video Object Tracking Algorithm in Presence of Its Zoom and Occlusions. Chongqing University, Chongqing (2014)
12. Smeaton, A.F., O'Connor, N.E.: An improved spatiogram similarity measure for robust object localization. In: Proceedings of ICASSP, pp. 1067–1072. IEEE, Honolulu (2007)
13. Candes, E., Tao, T.: Decoding by linear programming. IEEE Trans. Inf. Theor. **51**(12), 4203–4215 (2005)
14. Candes, E., Tao, T.: Near-optimal signal recovery from random projections: universal encoding strategies. IEEE Trans. Inf. Theor. **52**(12), 5406–5425 (2006)
15. Viola, P., Jones, M.: Rapid object detection using a boosted cascade of simple features. In: Proceedings of IEEE Conference on Computer Vision and Pattern Recognition, vol. 1, pp. 511–518 (2001)
16. Li, H., Shen, C., Shi, Q.: Real-time visual tracking using compressive sensing. In: CVPR, pp. 1305–1312 (2011)
17. Wright, J., Yang, A., Ganesh, A., Sastry, S., Ma, Y.: Robust face recognition viasparse representation. PAMI **31**, 210–227 (2009)
18. Liu, L., Fieguth, P.: Texture classic cation from random features. PAMI **34**, 574–586 (2012)
19. Achlioptas, D.: Database-friendly random projections: Johnson-Lindenstrauss with binary coins. J. Comput. Syst. Sci. **66**(4), 671–687 (2003)
20. Bingham, E., Mannila, H.: Random projection in dimensionality reduction: applications to image and text data. In: International Conference on Knowledge Discovery and Data Mining, pp. 245–250 (2001)
21. Baraniuk, R., Davenport, M., DeVore, R., Wakin, M.: A simple proof of the restricted isometry property for random matrices. Constr. Approx. **28**, 253–263 (2008)
22. Achlioptas, D.: Database-friendly random projections: Johnson-Lindenstrauss with binary coins. J. Comput. Syst. Sci. **66**, 671–687 (2003)
23. Li, P., Hastie, T., Church, K.: Very sparse random projections. In: KDD, pp. 287–296 (2006)

24. Baraniuk, R.: Compressive sensing. IEEE Sig. Process. Mag. **24**(4), 118–121 (2007)
25. Wu, Y., Jia, N., Sun, J.: Real-time multi-scale tracking based on compressive sensing. Original Article **30**(4) (2014)
26. Wu, Y., Lim, J., Yang, M.-H.: Online object tracking: a benchmark. In: Proceedings of IEEE Conference on Computer Vision and Pattern Recognition, pp. 2411–2418 (2013)

High-Quality Consistent Illumination in Mobile Augmented Reality by Radiance Convolution on the GPU

Peter Kán[✉], Johannes Unterguggenberger, and Hannes Kaufmann

Institute of Software Technology and Interactive Systems,
Vienna University of Technology, Vienna, Austria
peterkan@peterkan.com

Abstract. Consistent illumination of virtual and real objects in augmented reality (AR) is essential to achieve visual coherence. This paper presents a practical method for rendering with consistent illumination in AR in two steps. In the first step, a user scans the surrounding environment by rotational motion of the mobile device and the real illumination is captured. We capture the real light in high dynamic range (HDR) to preserve its high contrast. In the second step, the captured environment map is used to precalculate a set of reflection maps on the mobile GPU which are then used for real-time rendering with consistent illumination. Our method achieves high quality of the reflection maps because the convolution of the environment map by the BRDF is calculated accurately per each pixel of the output map. Moreover, we utilize multiple render targets to calculate reflection maps for multiple materials simultaneously. The presented method for consistent illumination in AR is beneficial for increasing visual coherence between virtual and real objects. Additionally, it is highly practical for mobile AR as it uses only a commodity mobile device.

1 Introduction

Augmented reality is the technology which superimposes virtual objects into the real world while accurate spatial and visual registration has to be achieved [1]. In order to achieve accurate visual registration and visual coherence, a consistent illumination of real and virtual objects is essential. Therefore, illumination from the real world has to be captured and used to illuminate the virtual objects. However, previous methods for consistent illumination were either limited to light with low angular frequency [2,3] or required a complex hardware setup [4–6] which is not suitable for mobile AR.

In this paper, we propose a method for rendering with consistent illumination in mobile AR which does not require special hardware and utilizes both low-frequency and high-frequency illumination in the angular domain. Our method overcomes the limitation of low-frequency lighting by reconstructing environment map, representing real illumination, from images captured by mobile device. The

© Springer International Publishing Switzerland 2015
G. Bebis et al. (Eds.): ISVC 2015, Part I, LNCS 9474, pp. 574–585, 2015.
DOI: 10.1007/978-3-319-27857-5_52

overlapping image data are accumulated and a high dynamic range (HDR) environment map is reconstructed interactively. This map is then processed on the mobile GPU and used for rendering of virtual objects. In order to calculate high-quality light reflections on the virtual surfaces, our method generates reflection maps by convolving the input map by the bidirectional reflectance distribution functions (BRDFs) of materials in the scene. Previous methods approximated this convolution either by MIP-mapping or by the Gaussian blur which leads to inaccurate light reflections. We overcome this limitation by calculating high-quality reflection maps on the mobile GPU which operates on output pixels in parallel. The reflection maps are calculated at the beginning of the rendering step and then they are used to render high-quality reflections in real-time. All calculations are done in HDR. Our method utilizes multiple render targets to simultaneously generate reflection maps for multiple shininess factors of Phong BRDF. Additionally, we use lightmaps to precalculate ambient occlusion and simulate a local visibility occlusion in high quality. The process of HDR environment map creation, reflection maps calculation, and rendering with consistent illumination is depicted in Fig. 1. This figure shows the tonemapped environment maps by exposure and gamma correction.

The results of this paper demonstrate the quality improvement of the rendering with our method in comparison to the common MIP-mapping approach. Moreover, we compare our results to the reference solution, rendered by offline path tracing. Our method produces results which are close to the reference solution. Both, light capturing and rendering are done on site and require only a commodity mobile device. Therefore, our method is highly usable to render high-quality consistent illumination in mobile AR.

The main contributions of this paper are:

- A two step method for rendering with consistent illumination in AR which uses only a mobile device.
- A method for efficient calculation of high-quality irradiance environment maps on a mobile GPU.
- Parallel generation of reflection maps for multiple materials.

2   Related Work

Rendering of virtual objects with consistent illumination has been a challenging problem in AR. In order to solve this problem, the illumination of the real world has to be captured or estimated and then used for the rendering of virtual objects. Previous methods which addressed this problem can be divided into two categories based on how the real light is acquired: Methods using a light probe (either passive reflective sphere or active camera with fish eye lens) and methods estimating the light from a main camera image. Some of these methods estimate the light in real-time and the others need captured illumination as an input.

Light Probe Methods. Methods based on light probes use a special hardware (usually a camera with a fish eye lens) to capture the illumination in high quality.

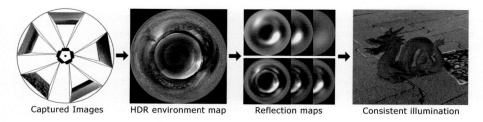

Captured Images   HDR environment map   Reflection maps   Consistent illumination

Fig. 1. The process of HDR environment map capturing and rendering with consistent illumination in AR. From left to right: Multiple input images with different exposures and different device orientations are combined into one HDR spherical map. This environment map is used to calculate reflection maps on mobile GPU. Finally, the reflection maps are used for real-time rendering with consistent illumination in AR.

Approaches which capture the illumination in real-time and use it for rendering in AR were presented in [5–10]. These approaches also calculate global illumination in an AR scene which introduces the light reflections between virtual and real objects. The advantage of these methods is a high visual fidelity of the rendered result. However, they need a high computational power and therefore are not suitable for mobile devices. A method for consistent near-field illumination which runs on a mobile device was presented by Rohmer et al. [11]. This method renders virtual objects lit consistently with the real world and it calculates global illumination. Its limitation is the requirement of multiple external cameras connected to a computer which sends the data to the mobile device via Wi-Fi. In contrast to that our method requires only a mobile device and does not need any additional hardware.

Other methods use the captured illumination as an input for the rendering and then render the virtual objects with natural light. Pessoa et al. [12] used the environment map reconstructed from multiple exposure images of a reflective sphere. The authors also generate the first glossy, second glossy, and diffuse map. Then these are interpolated to approximate a specific material. Our method calculates a glossy reflection map specifically for each material. Therefore, a high accuracy of the rendering is achieved. Meilland et al. [4] proposed a method for dense reconstruction of HDR light field in a real scene by registering multiple low dynamic range images with different exposures, captured by the RGB-D camera. The reconstructed light field is then used to generate multiple light probes which illuminate virtual objects. The similarity with our method is that we also firstly capture an HDR environment map and then use it for rendering with consistent illumination. An approach for fast light factorization for AR rendering based on spherical harmonics was presented in [13].

Light Estimation Methods. Several methods for light source estimation from a main camera image were proposed. The advantage of these methods is that they do not require additional cameras or light probes. The estimation of real illumination from shadows was proposed by Sato et al. [14]. The authors estimate the distribution of illumination by analyzing the relationships between the image brightness and the occlusions of incoming light. Gruber et al. [3] presented a method for real-time estimation of diffuse illumination from arbitrary geometry, captured by an RGB-D camera. This method reconstructs the real geometry and surrounding illumination which is used for rendering of the virtual content in AR with consistent illumination. Recently, a method for illumination estimation from a user's face, based on trained set of radiance transfer functions, was presented in [2]. The limitation of the approaches for light estimation from the main camera image is that they do not reconstruct high-frequency illumination.

Rendering with Natural Illumination. Once the real light is reconstructed, rendering with natural illumination plays an important role for achieving a consistent appearance of virtual and real objects. Precalculation of reflection maps for the rendering of diffuse and specular reflection was discussed in [15, 16]. Ramamoorthi and Hanrahan [17] proposed an efficient method for irradiance environment maps calculation by utilizing spherical harmonics. This method is well suitable to calculate the diffuse illumination. A fast approximation of environment map convolution can be achieved by MIP-mapping [18–20]. Our method calculates accurate convolution of illumination by the BRDF per each pixel. Therefore, it can be used for diffuse as well as glossy, and specular materials. The methods which used real-time rendering with natural illumination in AR by reflection maps were presented in [12, 21–23]. Recently, Mehta et al. [24] presented a method for physically-based rendering in AR based on GPU path tracing and real-time filtering. Their method illuminate the scene by a captured light probe. While producing high-quality rendering results, their method is not suitable for mobile AR due to the requirement of high computational power which can be provided only by a high-performance desktop GPU.

3   Consistent Illumination in AR

The proposed method for rendering with consistent illumination in AR works in two main steps. In the first step, the user scans the environment by a mobile device. A rotational motion of the mobile device is assumed while the images of the camera are accumulated to the spherical environment map. HDR reconstruction is performed to capture the wide range of light intensities from the real world. In the second step, this captured map is convolved by the BRDF of each material in the virtual scene. The convolved reflection maps are then used for real-time rendering. We achieve consistent illumination of virtual and real objects by using the real world's light to illuminate the virtual scene. The captured environment map contains both high and low frequencies and its convolution is calculated per each pixel on the GPU. Therefore, our method achieves high quality of the final rendering. The details of each step are revealed in the next sections.

3.1   Illumination Capturing

In order to render the virtual objects lit consistently with the real world, real illumination needs to be captured. For this purpose, the user scans the surrounding environment by a mobile device performing 360° rotational motion in yaw and pitch angles to cover the full sphere of directions. An HDR environment map is reconstructed interactively during scanning. The environment light scanning utilizes an approach presented in [25] to capture HDR environment map.

Each camera image is projected to the spherical environment map according to the orientation of the device, estimated by the inertial measurement unit (IMU) of the mobile device which includes gyroscope, accelerometer and magnetometer. We use the mapping to the sphere similar to Debevec [26]. Distinctly, our map is oriented to have positive z axis (up vector) projected to the image center because we prefer to have a consistent image deformation along the horizontal direction. Geometric and radiometric calibration is performed to accurately project the captured data.

High dynamic range of the reconstructed radiance data is essential to accurately represent the high contrast of real light. Therefore, low dynamic range data, captured by the image sensor, have to be converted to the HDR radiance. Our approach for HDR reconstruction is based on the work of Robertson et al. [27]. The inverse camera response function is reconstructed for each color channel in the calibration process. Then, we use these functions for the HDR reconstruction from multiple overlapping images with known exposure times [25,27]. A mobile camera is set to the auto-exposure mode to correctly adapt exposure time to the local portion of the scene where the camera is pointing. If two images overlap, they usually have slightly different exposure time due to the adapting auto-exposure. Therefore, we can merge the overlapping data and calculate incoming radiance in HDR. The HDR reconstruction runs on the mobile GPU and the environment map is accumulated to the floating point framebuffer to achieve interactive speed. Additionally, we employ an alignment correction based on feature matching to compensate the drift of IMU.

3.2   Environment Map Convolution

The captured HDR environment map is used to render virtual objects lit consistently with the real world lighting. At the beginning of the rendering phase, the environment map is convolved by the BRDF of each material to create a reflection map which will allow fast high quality lighting. We developed a new method for the convolution of the environment map on the GPU which is capable of convolving the map by multiple BRDFs simultaneously. Multiple render targets are utilized for this purpose. High quality of the resulting reflection map is achieved by taking each pixel of the input map into the calculation.

In order to enable the precalculation of reflected light we need to reduce the number of parameters in the reflectance calculation. We assume a distant light, represented by the captured environment map. Moreover, we calculate only single light reflection on the surface and scene materials are assumed to be non-emissive. Based on these assumptions, the rendering equation [28] can be simplified to:

$$L_o \approx \frac{2\pi}{N} \sum_{i=1}^{N} f_r(x, \boldsymbol{\omega}_i, \boldsymbol{\omega}_o) L_i(x, \boldsymbol{\omega}_i) \mid \boldsymbol{n} \cdot \boldsymbol{\omega}_i \mid \tag{1}$$

which calculates the reflected radiance L_o by taking into account each incoming radiance L_i from the environment map and the BRDF function denoted by f_r. The BRDF function depends on the direction of incoming light $\boldsymbol{\omega}_i$, direction of reflected right $\boldsymbol{\omega}_o$ and the surface position x. N stands for the number of pixels in the environment map and $\boldsymbol{n}$ is the local surface normal. In the following we loosely write $=$ instead of $\approx$ also when referring to the approximation of the incoming light by the environment map. Equation 1 represents the convolution of the environment map by the BRDF function f_r. However, the BRDF function still varies with too many parameters. Therefore, we split Eq. 1 into diffuse and specular convolutions based on the assumption that the BRDF can be separated to the diffuse and specular components.

Our rendering utilizes the Phong BRDF model [29]. In our approach, the specular component of the Phong BRDF is divided by $\mid \boldsymbol{n}{\cdot}\boldsymbol{\omega}_i \mid$. This modification makes the specular reflection dependent only on the reflection of view direction (perfect specular reflection direction) and the shininess n. Then, the term $(n+1)$ has to be used in the normalization factor to obey the energy conservation for the Phong BRDF. The used BRDF model is described by the following equation:

$$f_r = k_d \frac{1}{\pi} + k_s \frac{(n+1)\,\cos^n{(\alpha)}}{2\pi\,\mid \boldsymbol{n} \cdot \boldsymbol{\omega}_i \mid} \tag{2}$$

α is the angle between the incoming light direction and the reflected view direction. k_d and k_s are diffuse and specular coefficients of the material. By adding the Phong BRDF, we can split the Eq. 1 to diffuse and specular components and it can be rewritten to:

$$L_o = \frac{2\pi}{N} \sum_{i=1}^{N} \frac{k_d}{\pi} L_i(x, \boldsymbol{\omega}_i) \mid \boldsymbol{n} \cdot \boldsymbol{\omega}_i \mid$$

$$+ \frac{2\pi}{N} \sum_{i=1}^{N} \frac{k_s(n+1)\,\cos^n{(\alpha)}}{2\pi\,\mid \boldsymbol{n} \cdot \boldsymbol{\omega}_i \mid}\, L_i(x, \boldsymbol{\omega}_i) \mid \boldsymbol{n} \cdot \boldsymbol{\omega}_i \mid \tag{3}$$

The diffuse (k_d) and specular (k_s) coefficients can be taken out of the sums. Then, these sums can be precalculated to the diffuse and specular reflection maps. Due to the modified Phong BRDF, the radiance reflected specularly varies only with the mirrored view direction and shininess n. Thus, the convolution of the specular BRDF component and the input environment map can be precalculated to the reflection map for a specific shininess n. In order to calculate the reflected radiance in rendering, the specular reflection map is addressed by the reflected ray direction. Additionally, we precalculate the irradiance environment map (diffuse reflection map) by convolving the input environment map by the diffuse component of Eq. 3. This map is the same for each material and it is accessed by the surface normal because it only depends on the orientation of the surface. The calculation of reflection maps by convolving the input map by a variable BRDF kernel is depicted in Fig. 2. Finally, the reflected radiance is calculated by summing the diffuse and specular reflected radiances (obtained

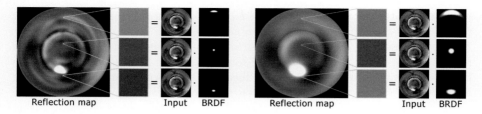

Fig. 2. Calculation of reflection maps by convolution of the input environment map by varying BRDF kernel. The materials with shininess 160 (left) and 20 (right) are shown.

from the precalculated reflection maps), multiplied by their corresponding BRDF coefficients k_d, k_s:

$$L_o = k_d E_d(\boldsymbol{n}) + k_s E_s^n(\boldsymbol{\omega}_r) \qquad (4)$$

The terms E_d and E_s^n represent the calculated diffuse and specular reflection maps. n is the shininess of the material and $\boldsymbol{\omega}_r$ stands for the mirror reflection of the view direction.

All reflection maps are calculated on the mobile GPU to increase the efficiency of the calculation. We use multiple render targets to calculate the maps for multiple materials simultaneously. This efficiency enhancement is of high benefit because most of the shader code is the same for all shininess factors except the power of $\cos^n(\alpha)$. The algorithm for reflection maps calculation goes through all pixels of the input map to calculate the result for a specific pixel of the output map. The pixels of the output map are calculated in parallel by the OpenGL rasterization pipeline. If a high output resolution is requested, our method subdivides the 2D output space into tiles and processes them sequentially. By this procedure the latency caused by the reflection maps calculation is reduced and they can be recalculated even during rendering if needed. The reflection maps, calculated by our method are shown in Fig. 5.

3.3   Rendering with Natural Light

When the reflection maps are calculated, we use them to consistently illuminate the virtual objects in real-time. This step is very efficient because only two texture lookups and color multiplications are required. The resulting radiance, reflected towards the camera, is calculated by Eq. 4. The preconvolved diffuse reflected radiance is obtained from the diffuse irradiance environment map $E_d(\boldsymbol{n})$ which is accessed by the surface normal. Radiance, reflected specularly is obtained from the specular reflection map E_s^n by using the reflected view direction $\boldsymbol{\omega}_r$. The rendering of the virtual dragon with calculated reflection maps is shown in Fig. 1.

3.4 Lightmapping

If the virtual object is simply superimposed onto the real image, the light interaction between the virtual and the real scene is missing. In order to address this problem, we precalculate the visibility by an offline high-quality ambient occlusion calculation and store it to the lightmaps. Then the diffuse reflected radiance is multiplied by the ambient occlusion factor for both real and virtual objects. This enables the self-shadowing of the virtual objects as well as the local proximity shadows on the real scene cast by the virtual objects. In order to render the virtual shadows on real objects we need to know the geometry of the real scene. Currently, we use a proxy geometry which models the real world. The advantage of multiplication by the shadowing factor is that we can avoid compositing by differential rendering [26] which requires two rendering solutions. A comparison of rendering without and with ambient occlusion can be seen in Fig. 3. The image without ambient occlusion does not correctly indicate the spatial relationships of virtual and real objects because of missing contact shadows. In contrast to that, the image with ambient occlusion contains contact shadows which indicate the proximity of the dragon to the real table.

3.5 Implementation

The mobile implementation of our method for consistent illumination in AR has the advantage of high usability. We implemented the shaders for camera image projection, HDR reconstruction, reflection maps calculation, and rendering with natural illumination in OpenGL Shading Language and we run them in OpenGL ES 3.0. Vuforia SDK was used in our implementation to track the real camera and to achieve a correct spatial registration of virtual objects in the real scene. Our experiments, described in Sect. 4, were performed on an NVIDIA Shield tablet. Rendering of virtual objects is done in HDR which ensures proper reflections of light on the surfaces. Finally, a tonemapping is applied to show the AR image on the mobile display. We use the photographic tone reproduction method, presented in [30], to convert computed HDR radiances to low dynamic range intensities.

Fig. 3. Rendering in AR without (left) and with (right) ambient occlusion. The image shows a virtual dragon, a real cup, and a real cube.

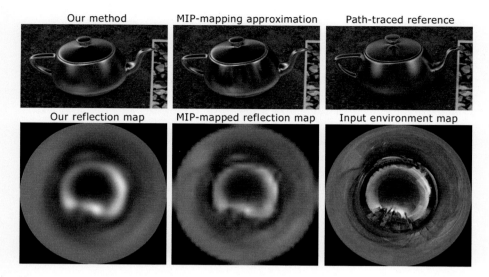

Fig. 4. Comparison of rending with our method, rendering with MIP-mapped reflection maps, and path-traced reference. Bottom row shows the reflection map generated by our method and one generated by MIP-mapping. The material shininess is 60.

4   Results

The results of rendering with consistent illumination by the presented method are shown in Figs. 1, 3, and 4. These figures show that the high dynamic range of the captured environment map ensures proper shiny light reflections on the virtual objects. Moreover, precalculated lightmaps create consistent local proximity shadows which increase visual coherence.

In order to evaluate the quality of the presented method we compared it to common calculation of reflection maps by the MIP-mapping approach and to the reference solution, calculated by offline path tracing. The results of this comparison can be seen in Fig. 4. Glossy material with shininess 60 was used to render a teapot lit by natural illumination. As we can see, our method produces higher quality than the MIP-mapping based calculation of reflection maps and we achieve results similar to the reference. The reflections in the MIP-mapped result are sharper than the reference. Our method correctly calculates the reflections of the real world on the virtual content. The reference solution also features the self-reflections which are not simulated by our method.

We analyzed the performance of our method by measuring the computational time of each particular step. First, we measured the computational time of reflection maps calculation. The time measurements for different sizes of the reflection maps can be seen in Table 1. 8 reflection maps were calculated in each measurement in parallel. The computational time of reflection maps for lower resolutions ($\leq 512 \times 512$) is acceptable for AR applications which can capture and use the real illumination on site. Additionally, the frame rate of rendering different virtual objects in AR with our method was measured. The results

Table 1. Computational time of reflection maps calculation for different output resolutions. Eight reflection maps were calculated simultaneously for each evaluated resolution. The computational time is stated in format minutes:seconds:milliseconds.

Resolution of reflection maps	Reflection maps calculation time
32×32	00:00:040
64×64	00:01:450
128×128	00:06:360
256×256	00:28:090
512×512	02:38:370
1024×1024	23:01:650

of this measurement are shown in Table 2. Our rendering achieves high-quality results with interactive performance which is essential for AR.

In our evaluation, we also measured the average time in which a non-skilled user was able to capture the environment map by the presented method. 10 users (5 men and 5 women) participated in a short experiment in which their performance of environment map capturing was measured. None of the participants used our capturing method before. The average capturing time was 52 seconds in this experiment. This result demonstrates usability of our method and its applicability to on-site AR scenarios.

5   Limitations and Future Work

Currently, our implementation uses Phong BRDF to represent the materials. In future, it can be extended to more complex BRDFs. If additional parameter is required in BRDF convolution (e.g. in case of BRDFs with Fresnel term), 3D textures can be used to store the convolved reflection map.

Ambient occlusion, baked into the lightmaps, limits the presented method to static geometry. This limitation can be solved by using real-time screen-space ambient occlusion calculation instead of pre-calculated one.

Table 2. Frame rates of rendering with consistent illumination in AR by our method. The rendering was done in Full HD resolution.

3D model	Number of triangles	Frame rate
Reflective cup	25 600	29 fps
Teapot	15 704	30 fps
Dragon	229 236	13 fps

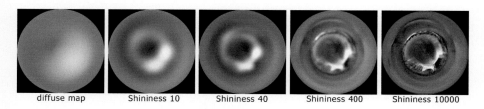

| diffuse map | Shininess 10 | Shininess 40 | Shininess 400 | Shininess 10000 |

Fig. 5. HDR reflection maps calculated by our method.

6   Conclusion

This paper presents a two-step method for rendering with consistent illumination in AR. In the first step, the surrounding illumination is captured into an HDR environment map. In the second step a set of reflection maps is precalculated and used for real-time rendering with natural light. Our method uses only a commodity mobile device and does not require any special hardware. Therefore, it is highly usable for mobile AR applications. We utilize lightmaps to simulate the local occlusion of real world objects by virtual geometry. The results show that the presented method increases the quality of lighting in AR in comparison to MIP-mapping reflection maps calculation. Finally, multiple presented AR images demonstrate the quality of lighting by the presented method.

Acknowledgements. The dragon model is the courtesy of Stanford Computer Graphics Laboratory. The teapot model is the courtesy of Martin Newell. This research was funded by Austrian project FFG-BRIDGE 843484.

References

1. Azuma, R.T.: A survey of augmented reality. Presence Teleoperators Virtual Environ. **6**, 355–385 (1997)
2. Knorr, S., Kurz, D.: Real-time illumination estimation from faces for coherent rendering. In: IEEE ISMAR 2014, pp. 113–122 (2014)
3. Gruber, L., Richter-Trummer, T., Schmalstieg, D.: Real-time photometric registration from arbitrary geometry. In: IEEE ISMAR, pp. 119–128 (2012)
4. Meilland, M., Barat, C., Comport, A.: 3D high dynamic range dense visual slam and its application to real-time object re-lighting. In: IEEE ISMAR (2013)
5. Kán, P., Kaufmann, H.: High-quality reflections, refractions, and caustics in augmented reality and their contribution to visual coherence. In: IEEE ISMAR, pp. 99–108. IEEE Computer Society (2012)
6. Knecht, M., Traxler, C., Mattausch, O., Wimmer, M.: Reciprocal shading for mixed reality. Comput. Graph. **36**, 846–856 (2012)
7. Grosch, T., Eble, T., Mueller, S.: Consistent interactive augmentation of live camera images with correct near-field illumination. In: ACM Symposium on Virtual Reality Software and Technology, pp. 125–132. ACM, New York (2007)
8. Kán, P., Kaufmann, H.: Differential irradiance caching for fast high-quality light transport between virtual and real worlds. In: IEEE ISMAR, pp. 133–141 (2013)

9. Franke, T.: Delta voxel cone tracing. In: IEEE ISMAR, pp. 39–44 (2014)
10. Franke, T.: Delta light propagation volumes for mixed reality. In: IEEE ISMAR, pp. 125–132 (2013)
11. Rohmer, K., Buschel, W., Dachselt, R., Grosch, T.: Interactive near-field illumination for photorealistic augmented reality on mobile devices. In: IEEE ISMAR, pp. 29–38 (2014)
12. Pessoa, S., Moura, G., Lima, J., Teichrieb, V., Kelner, J.: Photorealistic rendering for augmented reality: a global illumination and brdf solution. In: 2010 IEEE Virtual Reality Conference (VR), pp. 3–10 (2010)
13. Nowrouzezahrai, D., Geiger, S., Mitchell, K., Sumner, R., Jarosz, W., Gross, M.: Light factorization for mixed-frequency shadows in augmented reality. In: IEEE ISMAR, pp. 173–179 (2011)
14. Sato, I., Sato, Y., Ikeuchi, K.: Illumination from shadows. IEEE Trans. Pattern Anal. Mach. Intell. **25**, 290–300 (2003)
15. Miller, G.S., Hoffman, C.R.: Illumination and reflection maps: simulated objects in simulated and real environments. In: SIGGRAPH 1984 (1984)
16. Kautz, J., Vzquez, P.P., Heidrich, W., Seidel, H.P.: A unified approach to prefiltered environment maps. In: Péroche, B., Rushmeier, H. (eds.) Rendering Techniques 2000, pp. 185–196. Springer, Vienna (2000)
17. Ramamoorthi, R., Hanrahan, P.: An efficient representation for irradiance environment maps. In: SIGGRAPH, pp. 497–500. ACM, New York (2001)
18. McGuire, M., Evangelakos, D., Wilcox, J., Donow, S., Mara, M.: Plausible Blinn-Phong reflection of standard cube MIP-maps. Technical report CSTR201301, Department of Computer Science, Williams College, USA (2013)
19. Scherzer, D., Nguyen, C.H., Ritschel, T., Seidel, H.P.: Pre-convolved Radiance Caching. Comput. Graph. Forum **4**, 1391–1397 (2012)
20. Kautz, J., Daubert, K., Seidel, H.P.: Advanced environment mapping in VR applications. Comput. Graph. **28**, 99–104 (2004)
21. Agusanto, K., Li, L., Chuangui, Z., Sing, N.W.: Photorealistic rendering for augmented reality using environment illumination. In: IEEE ISMAR, pp. 208–218 (2003)
22. Supan, P., Stuppacher, I., Haller, M.: Image based shadowing in real-time augmented reality. IJVR **5**, 1–7 (2006)
23. Franke, T., Jung, Y.: Real-time mixed reality with GPU techniques. In: GRAPP, pp. 249–252 (2008)
24. Mehta, S.U., Kim, K., Pajak, D., Pulli, K., Kautz, J., Ramamoorthi, R.: Filtering environment illumination for interactive physically-based rendering in mixed reality. In: Eurographics Symposium on Rendering (2015)
25. Kán, P.: Interactive HDR environment map capturing on mobile devices. In: Eurographics 2015 - Short Papers, pp. 29–32. The Eurographics Association (2015)
26. Debevec, P.: Rendering synthetic objects into real scenes: Bridging traditional and image-based graphics with global illumination and high dynamic range photography. In: SIGGRAPH 1998, pp. 189–198. ACM, New York (1998)
27. Robertson, M., Borman, S., Stevenson, R.: Dynamic range improvement through multiple exposures. ICIP **3**, 159–163 (1999)
28. Kajiya, J.T.: The rendering equation. In: SIGGRAPH, pp. 143–150 (1986)
29. Lafortune, E.P., Willems, Y.D.: Using the modified phong reflectance model for physically based rendering. Technical report, K.U. Leuven (1994)
30. Reinhard, E., Stark, M., Shirley, P., Ferwerda, J.: Photographic tone reproduction for digital images. ACM Trans. Graph. **21**, 267–276 (2002)

Efficient Hand Articulations Tracking
Using Adaptive Hand Model and Depth Map

Byeongkeun Kang[✉], Yeejin Lee, and Truong Q. Nguyen

Department of Electrical and Computer Engineering,
UC San Diego, La Jolla, CA, USA
bkkang@ucsd.edu

Abstract. Real-time hand articulations tracking is important for many applications such as interacting with virtual/augmented reality devices. However, most of existing algorithms highly rely on expensive and high power-consuming GPUs to achieve real-time processing. Consequently, these systems are inappropriate for mobile and wearable devices. In this paper, we propose an efficient hand tracking system which does not require high performance GPUs.

In our system, we track hand articulations by minimizing discrepancy between depth map from sensor and computer-generated hand model. We also re-initialize hand pose at each frame using finger detection and classification. Our contributions are: (a) propose adaptive hand model to consider different hand shapes of users without generating personalized hand model; (b) improve the highly efficient re-initialization for robust tracking and automatic initialization; (c) propose hierarchical random sampling of pixels from each depth map to improve tracking accuracy while limiting required computations. To the best of our knowledge, it is the first system that achieves both automatic hand model adjustment and real-time tracking without using GPUs.

1 Introduction

Hands are used in daily lives to handle objects and to better communicate with others. Especially, hands are almost the only way to control electronic devices except limited usage of speech. It is limited since speech is hard to protect privacy, and understanding of speech is difficult in noisy environment. Recent advancements in mobile devices and wearable devices demand better communication methods rather than touch screens which limit physical space. Due to the demand of more natural and convenient interacting methods, interaction using hand gestures has received lots of attention for human-computer interactions, virtual/augmented reality, and robot controls.

1.1 Related Work

Previously, hand pose estimation methods are classified into single frame-based methods and model-based tracking methods [1]. Single frame-based methods

© Springer International Publishing Switzerland 2015
G. Bebis et al. (Eds.): ISVC 2015, Part I, LNCS 9474, pp. 586–598, 2015.
DOI: 10.1007/978-3-319-27857-5_53

estimate hand pose by searching huge databases or by recovering hand pose from hand joint classification. Athitsos *et al.* and Wang *et al.* used a color image to retrieve hand pose from large databases [2,3]. The method in [3] used a color glove for better searching from a database. However, since the database has limited number of hand pose images, it can only estimate the poses in the database. Recently, Tang *et al.* and Tompson *et al.* estimated hand pose by applying hand joint classification using random forest and convolutional neural networks (CNNs) respectively [4,5]. Tang *et al.* proposed the semi-supervised transductive regression (STR) forest method with joint refinement procedure [4]. Tompson *et al.* employed CNNs and pose recovery to achieve continuous pose estimation [5]. These methods require high performance GPUs to achieve real-time processing and also require large real and synthetic database for training.

Model-based tracking methods estimate hand pose by finding optimal parameters of computer-generated hand model using both current input image and previous results [6]. Rehg *et al.* and Oikonomidis *et al.* used multiple RGB cameras to reduce occlusions and to increase visual features [7,8]. The method in [9] tracks a full DOF hand motion by minimizing discrepancy between input RGB-D image and computer-generated model using particle swarm optimization (PSO). Generating many possible hand pose images for each frame using computer graphics is computationally expensive and requires high performance GPUs to achieve about $15 \sim 20$ frames per second (fps) performance. Also, this method requires a user to place a hand on pre-determined position and pose to initialize tracking.

Recently, Sridhar *et al.* proposed the combined method of single frame-based method and model-based tracking method to blend advantages of each method. They used multiple color cameras for model-based tracking and a depth sensor to search their database, then a voting algorithm is applied to combine the results. Although multiple camera system helps to achieve better accuracy, it requires setup and calibration processes. Moreover, it requires GPUs to process multiple inputs at each frame. Qian *et al.* proposed another combined method using a depth sensor [10]. They combined an efficient initialization method and a tracking method using PSO and iterative closest point (ICP). Their hand model is designed using only spheres to simplify objective function. Sharp *et al.* also proposed the combined method of two-layer re-initialization using random forest and model fitting using PSO and genetic algorithm [11].

Most of existing hand tracking systems are rely on expensive, high power-consuming, and high performance GPUs since hand tracking is challenging because of complex articulations, self-occlusions, deformation, and rapid motions. Consequently, these systems are inappropriate for portable and wearable devices. Hand tracking systems for those devices do not need to consider huge viewpoint changes since in general, a user's hand is relatively close to camera. In this paper, we focus on efficient hand articulations tracking system that does not require GPUs and considers mainly the situation when a user's hand is close to camera. Even though we mainly consider small viewpoint changes, it is very challenging because of very limited computational power. Although

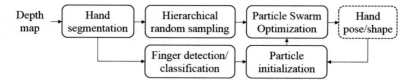

Fig. 1. Block diagram of the proposed method.

we design and test our system without using GPUs, it can be implemented with inexpensive and low power-consuming GPUs for better accuracy while still being able to be used for mobile devices. To the best of our knowledge, the proposed system is the first one that can automatically adjust hand shapes which we call adaptive hand model while other methods manually and experimentally decided hand model size for each user. Moreover, we focus on real-time system without using GPUs for mobile and wearable devices. We also propose hierarchical random sampling of pixels on each depth map to achieve better performance with limited computations. Lastly, we improve an efficient re-initialization method at each frame using finger detection and classification.

Figure 1 shows the entire process of proposed method. We first process hand segmentation from depth map and choose partial pixels from hand region using hierarchical random sampling. We also extract fingertip positions and finger joint rotations using an efficient finger detection and classification algorithm for re-initialization at each frame. Then hand pose and shape are estimated by minimizing discrepancy between the selected partial pixels and computer-generated hand model using PSO. Particles are initialized by previous frame's result and finger detection/classification result.

2   Method

2.1   Hand Segmentation

We process a very simple and effective segmentation by using a black wrist band and by assuming that a user's hand is the closest object from camera. This assumption is valid in general environments where one interacts with mobile and wearable devices. The black wrist band is to get depth voids around the wrist since depth sensor cannot capture depth from black object well. The segmentation is processed by finding the connected components from the closest point. For details, we refer the reader to [12].

2.2   Hierarchical Random Sampling

We propose hierarchical random sampling of pixels on each depth map for efficient comparison between depth map from sensor and computer-generated hand model. It is computationally expensive to draw computer-generated hand model on image plane and compare entire image to input depth map. To reduce required computations, this process is replaced by comparing subset of pixels on input image to computer-generated hand model with only spheres. Thus we do not need to draw computer-generated hand model on image plane since the difference can be computed without drawing the model. Although tracking accuracy is improved with more subset points, required computation for comparison is also increased. Therefore, we focus on improving the selection of pixels to process from each depth map by applying hierarchical random sampling. The sampling aims to include more pixels on the region which has large depth variations since the region can be interpreted as more informative on depth map.

First, initial samples S_1 are randomly sampled. Then, hierarchical sampling S_2 is employed to include more points on large depth variation regions. To find large depth variation regions, the gradient matrix G is computed as the sum of absolute x- and y- directional gradient of depth map D. The gradients of x- and y- direction are computed by 3×3 Sobel operators (O_x and O_y) in hand segment region:

$$G = |O_x * D| + |O_y * D|, \tag{1}$$

where $|\cdot|$ indicates component-wise absolute value and $*$ represents convolution.

For initial samples with large gradient, random samples S_R are selected around initial samples by adding random values u_1, u_2 from discrete uniform distribution to x- and y- coordinates respectively:

$$S_R = \{S_1 + (u_1, u_2) \mid G(S_1) > t_1\}. \tag{2}$$

The random samples in S_R are included in hierarchical sample set S_2 if the depth difference between initial sample and random sample is greater than a threshold t_2:

$$S_2 = \{S_R \mid |D(S_R) - D(S_1)| > t_2\}. \tag{3}$$

The final sample set S includes both initial samples S_1 and hierarchical samples S_2. Sampled points $S \in \mathbb{Z}^{2 \times N_s}$ are converted to $X \in \mathbb{R}^{3 \times N_s}$ with (x, y, z) in millimeter, where N_s is the total number of samples (Fig. 2).

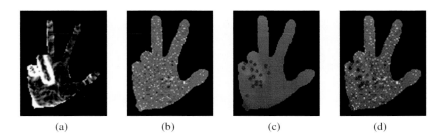

(a)	(b)	(c)	(d)

Fig. 2. Hierarchical random sampling of pixels from each depth map. This sampling aims to include more pixels on the region which has large depth variations. (a) gradient matrix G of depth map; (b) random sampling S_1; (c) hierarchical sampling S_2; (d) hierarchical random sampling S.

2.3   Adaptive Hand Model

Tracking accuracy increases as the hand model becomes more similar to each user's hand. Although a personalized model can be generated by scanning the user's hand, it requires a pre-processing step for each user. Therefore, we propose the adaptive hand model to consider different hand size and shape while avoiding to scan each user's hand.

Our adaptive hand model consists of a hand size parameter vector $\mathbf{l} \in \mathbb{R}^6$ for palm size and finger lengths, and a hand pose parameter vector $\mathbf{p} \in \mathbb{R}^{26}$ for hand position and joint rotations. For hand size parameters, one parameter l_0 is for palm size and five parameters $\{l_1, ..., l_5\}$ are for finger lengths from thumb finger to pinky finger. Finger width parameter is not considered since finger width is relatively less important and also computation power is limited. Hand pose parameter vector is defined to estimate translations and rotations of hand joints as in Fig. 3(a).

Since hand shape at each frame is dependent on both size and pose parameters, two parameter vectors should be optimized simultaneously. However, the combination of two parameter vectors is 32 parameters, which is really complex to optimize even without considering correlation between parameters. Therefore, size parameters are considered only when five fingers are detected and classified by the re-initialization method in Sect. 2.5 since in that case, the optimization of pose parameters is relatively accurate and robust.

Hand model is designed using only spheres to reduce the computational complexity of objective function [10]. The main reason is that in comparison between input depth map and the hand model, the length from a point to the surface of a sphere is simply the length from the point to the center of sphere subtracted by the radius.

2.4   Optimization

Objective Function. Our objective function is designed mainly to minimize Euclidean distance between sampled pixels from input depth map and computer-

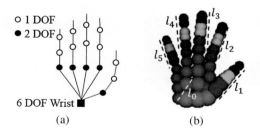

(a)   (b)

Fig. 3. Hand model. (a) skeletal structure of hand model with 26 DOFs. Each DOF represents a parameter in hand pose parameter vector; (b) an example of hand model with 48 spheres. Two spheres are hidden at connection between palm and thumb. Hand size parameters are visualized.

generated hand model. The function consists of two discrepancy terms between hand model and depth map and one validity term. Two discrepancy terms are the Euclidean distance from depth map to hand model and from the model to depth map. The validity term is to check invalid overlapping between parts of hand model. The total cost is the weighted summation of these three terms. For details of general objective function, we refer the reader to the paper [10]. For our system, since hand shape at each frame is determined by both hand pose and hand size parameters, cost is also determined by both parameters. We conducted experiment with another objective function which incorporates additional cost term to minimize temporal hand size parameter changes. However, tracking accuracy is not improved since the variation of hand size parameters are already regularized by the normal distribution.

Particle Swarm Optimization. Modified Particle Swarm Optimization (PSO) is used to find the best hand pose parameters and hand size parameters by minimizing an objective function. Each particle represents one state of hand pose and hand size parameters in our algorithm. The optimization method first initializes particles over possible solution range, and then finds the best solution among particles using objective function. Then particles are moved from current state to the direction of the best solution. The algorithm iterates finding the best solution and moving particles to the direction of best solution until it reaches maximum generation or termination condition.

Entire particles are initialized as the sum of optimized parameters $(\mathbf{p}_o^{(t-1)}, \mathbf{l}_o^{(t-1)})$ at previous frame and random values $(\mathbf{r}_p, \mathbf{r}_l)$ from normal distribution if corresponding finger is not detected and classified by the method in Sect. 2.5. Otherwise, 75 % of the corresponding parameters are initialized with the same method, and 25 % of them are initialized with the sum of measured parameters from Sect. 2.5 and random values. The distribution of random values are $\mathbf{r}_p \sim \mathcal{N}(0, \sigma_1^2)$ and $\mathbf{r}_l \sim \mathcal{N}(0, \sigma_2^2)$. However, at the first generation, hand size parameters are not considered to focus on the optimization of pose parameters since inaccurate pose parameters lead to wrong size parameters and the size parameters from previous frame are relatively reliable.

At each generation, for pose parameters, particles are updated using global best particle $\mathbf{g}$ and personal best particle $\mathbf{b}$. Personal best particle is the state when the particle has the lowest cost until current generation. Global best particle is the lowest cost state among all particles and all generations until current generation. Particles are updated to the direction of personal best particle and global best particle using the following rules:

$$
\begin{aligned}
\mathbf{b}_{i,j} &= \{\mathbf{p}_{i,\tilde{k}} | \tilde{k} = \arg \min_k C(X^{(t)}, \mathbf{l}_{i,1}^{(t)}, \mathbf{p}_{i,k}^{(t)})\}, \\
\mathbf{g}_j &= \{\mathbf{p}_{\tilde{i},\tilde{k}} | (\tilde{i}, \tilde{k}) = \arg \min_{i,k} C(X^{(t)}, \mathbf{l}_{i,1}^{(t)}, \mathbf{p}_{i,k}^{(t)})\}, \\
\mathbf{p}_{i,j} &= \mathbf{p}_{i,j-1} + \alpha_1(\mathbf{b}_{i,j-1} - \mathbf{p}_{i,j-1}) + \alpha_2(\mathbf{g}_{j-1} - \mathbf{p}_{i,j-1}),
\end{aligned}
\tag{4}
$$

where $C(\cdot)$ is the objective function, weight $\alpha_1 \sim U[0.5, 1.5]$, weight $\alpha_2 = 2-\alpha_1$, i is the index of each particle, j and k denote the generation index, and t is current frame index.

For size parameters, particles are not updated to avoid misleading caused by the dependency between size parameters and pose parameters. Although size parameters are not updated at each generation, the particles with better size parameters are likely to have lower cost after many generations since pose parameters will become similar.

After reaching maximum generation, both size parameters and pose parameters are updated with the global best particle.

$$(\mathbf{p}_o^{(t)}, \mathbf{l}_o^{(t)}) = \{(\mathbf{p}_{i,k}^{(t)}, \mathbf{l}_{i,1}^{(t)})|(\tilde{i}, \tilde{k}) = \arg \min_{i,k} C(X^{(t)}, \mathbf{l}_{i,1}^{(t)}, \mathbf{p}_{i,k}^{(t)})\}. \tag{5}$$

2.5   Re-initialization at Each Frame

Re-initialization at each frame is important to avoid manual initialization and error accumulation. However, general re-initialization methods using random forest or CNNs require large computation load which needs high performance GPUs to achieve real-time. Therefore, we improved the efficient finger detection and classification method proposed by [10]. Although this re-initialization method works in limited cases, it initializes hand pose automatically, improves tracking accuracy, and is incorporated in real-time tracking without GPUs. We improve the re-initialization by estimating palm orientation using both current measurement and prior knowledge from previous frames, which is inspired by Kalman filtering [13].

Finger Detection. A simple finger detection algorithm is employed to detect planar fingers and orthogonal fingers. We define planar fingers as the fingers which are parallel to image plane and orthogonal fingers which are orthogonal to image plane. First, palm center is measured as the maximum of distance transform of hand segment. Then a planar finger candidate is detected by finding connected component from extreme distance point from palm center until the component reaches finger length. The detected finger candidate is classified into either a finger or a non-finger based on the component size. This process is iterated until it detects five fingers or the segment does not have any extreme distance point. After detecting planar fingers, an orthogonal finger candidate is detected by finding connected component from the closest point from a camera on both depth map and hand segment within a small window. It is also classified to either a finger or a non-finger based on the size of region. This process is also iterated until it reaches same condition as planar finger case. For planar fingers, principal component analysis (PCA) is applied to each detected region in order to calculate the orientation of each detected finger. For orthogonal fingers, the orientation is assumed that it is orthogonal to image plane.

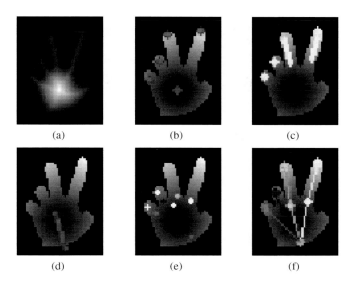

Fig. 4. Finger detection and classification. (a) distance transformation; (b) detection of fingertips (red) and palm center (blue); (c) computed finger orientation using PCA; (d) estimated palm orientation; (e) joint junctions (white: detected, colored: predicted using hand model and palm orientation); (f) classification result (Color figure online).

Finger Classification. Finger classification is to use the result from finger detection for particle initialization process in Sect. 2.4. The algorithm classifies each detected finger to one of five finger classes. First, palm orientation is measured by applying PCA to palm segment. However, the measured orientation θ_m is not robust and accurate enough to use directly for classification. Therefore, palm orientation $\theta_{p,t}$ is predicted by the weighted summation of previously estimated palm orientation $\theta_{o,t-1}$ from PSO and currently measured palm orientation $\theta_{m,t}$. The weights in summation are decided by the measurement error e_m and priori estimation error e_o. This is inspired by Kalman filtering [13].

$$\theta_{p,t} = \frac{e_{m,t}}{e_{m,t} + e_{o,t}}(\theta_{o,t-1} + c_{t-1}) + \frac{e_{o,t}}{e_{m,t} + e_{o,t}}\theta_{m,t},$$
$$e_{m,t} = |\theta_{o,t-1} - \theta_{m,t-1}| + a_1 e_{m,t-1},$$
$$e_{o,t} = |\theta_{o,t-1} - (\theta_{o,t-2} + c_{t-2})| + a_2 e_{o,t-1},$$

(6)

where t is current frame index and a_i is chosen as a constant for simplicity. In our experiments, both a_1 and a_2 are set to 0.5. After each frame, priori estimation constant c is updated as $c_t = \theta_{o,t} - \theta_{o,t-1}$.

A set F of junction points of fingers and palm is computed using detected fingertips, finger orientations, and finger length. Another set Q of junction points is calculated using hand model and predicted palm orientation. Each junction point $f_i \in F$ from an input image is matched to the closest junction point $q_j \in Q$ from hand model.

$$J(f_i) = \arg\min_j ||f_i - q_j||_2. \tag{7}$$

If more than two of detected junctions are classified into the same class, optimization is employed to find the result with minimum cost. The algorithm first finds possible combinations B that have minimum changes in initial classification.

$$B_k = \{(J(f_1) + v_1, ..., J(f_n) + v_n) \mid \min \sum_{i=1}^{n} |v_i|^2, \forall i \neq j, J(f_i) + v_i \neq J(f_j) + v_j\}. \tag{8}$$

The final class L is classified as follows:

$$L = \{B_{\tilde{k}} \mid \tilde{k} = \arg\min_k \sum_{i=1}^{n} (||f_i - q_{B_k(i)}||_2)\}, \tag{9}$$

where i is the index of detected finger and n is the number of detected fingers. Figure 4(d), (e), and (f) illustrate the procedure of finger classification. In Fig. 4(e), white circles represent detected finger junctions and colored circles from red to purple indicate hand model junctions from thumb finger to pinky finger. A yellow circle which corresponds to predicted joint junction of index finger is overlapped with a white circle. Although this re-initialization method cannot detect all the fingers at every frame, it improves tracking accuracy. It can also initialize at each frame including the very first frame, and takes only a few milliseconds using only CPU.

3   Experimental Results

The algorithm is tested using a Creative Senz3D camera and a computer with Intel Core i7-3770 3.4 GHz CPU, 16 GB RAM, and without GPUs. Although the machine has 16 GB RAM, this algorithm only uses about 60MB memory. Although a 3.4 GHz CPU is used, we believe similar computation power can be obtained using the combination of mobile CPU and mobile GPUs.

We captured 500 frames for each subject and labeled wrist and five fingertips of last 400 frames. The dataset is available on our repository[1]. Initial 100 frames contain open-hand pose that camera can capture at least some fingers using the algorithm in Sect. 2.5 for automatic initialization. The initial frames are not used to compute accuracy. Error in accuracy is computed using 3D Euclidean distance in millimeter.

Unless specifically mentioned, we sample 256 points from each depth map and optimize with 256 particles and 6 generations. Table 1 shows the processing time at this setting. Even though the algorithm is not fully optimized, it achieves about 16 FPS using eight threads on CPU.

---

[1] https://github.com/byeongkeun-kang/HandTracking.

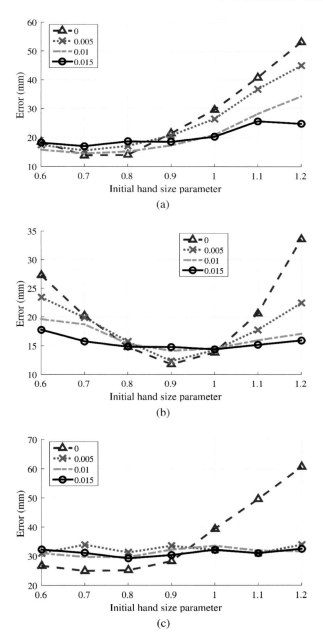

Fig. 5. Comparison of adaptive hand model and fixed scale hand model for three subjects (top: subject 1, middle: subject 2, bottom: subject 3). The legend represents standard deviations σ_2 of hand size parameter randomness. Standard deviation of 0 means that the scaling factor of hand model is always the same with initial scaling factor. Larger standard deviation means more possibility of large update of scaling factor at each frame. The result shows that the performance of adaptive hand model is better than the performance of fixed scaled hand model in general.

Table 1. Processing time.

Process	Time (in ms)
Finger detection/classification	2.5
Optimization	53.5
Others	6.5
Total	62.5

3.1 Adaptive Hand Model

The performance comparison of hand model with adaptive scaling and fixed scaling is demonstrated in Fig. 5 for different subjects, initial scaling factors, and hand size parameter randomness. Scaling factors are chosen from 0.6 to 1.2 with 0.1 step size and standard deviation of randomness σ_2 is chosen from 0 to 0.015 with 0.005 step size. Also, Table 2 clearly shows that adaptive hand model reduces error about 5 mm. The overall results show that adaptive hand model adjusts hand scaling factor automatically to minimize discrepancy between pre-defined hand model and user's hand, and users do not need to manually and experimentally select scaling factor of hand model. Moreover, to consider the case that the user is changed after starting tracking, we keep update hand scaling factor.

Table 2. Average error for different standard deviations σ_2 of hand size parameter randomness. To compute average error, we consider three subjects and seven initial scaling factors chosen from 0.6 to 1.2. The detailed explanation of standard deviation is on Fig. 5.

σ_2 in Sect. 2.5	0	0.005	0.01	0.015
Error (in mm)	28.00	25.31	22.90	22.38

3.2 Finger Classification

Correct classification rate (CCR) is calculated for each finger with and without the proposed palm orientation prediction in Table 3. The result shows that by using the proposed prediction, the average CCR is improved from 70.6 % to 84.8 %.

Table 3. Correct classification rate (in %) of finger classification.

CCR	Thumb	Index	Middle	Ring	Pinky	Average
Without prediction	87.1	65.5	69.4	59.0	71.8	70.6
With prediction	97.2	86.3	78.1	70.0	92.1	84.8

Table 4. Performance comparison of random sampling [10] and the proposed hierarchical random sampling.

Sampling	Random [10]	Hierarchical random
Error (mm)	17.01	16.11

3.3   Hierarchical Random Sampling

The performance of random sampling and hierarchical random sampling is compared in Table 4. The average error is computed using the hand model scaled by the best performance fixed scaling factor, 0.7, 0.9, and 0.7 for subject 1, 2, and 3 respectively. The average accuracy is improved from 17.01 mm to 16.11 mm. The computational cost of calculating gradient is much smaller than increasing the number of generations, particles, or samples to achieve the same improvement. However, if the number of sampling pixels is too small, random sampling might be better since hierarchical random sampling prevents that sampled pixels are more globally distributed.

4   Conclusion

We present an efficient hand articulations tracking system for mobile and wearable devices which do not have high performance GPUs. We show that the proposed system achieves both automatic hand model adjustment using adaptive hand model and real-time tracking without using GPUs. We also achieve improved accuracy using hierarchical random sampling and improved efficient re-initialization at each frame while limiting required computations.

References

1. Erol, A., Bebis, G., Nicolescu, M., Boyle, R.D., Twombly, X.: Vision-based hand pose estimation: a review. Comput. Vis. Image Underst. **108**, 52–73 (2007)
2. Athitsos, V., Sclaroff, S.: Estimating 3D hand pose from a cluttered image. In: 2014 IEEE Conference on Computer Vision and Pattern Recognition (CVPR) (2003)
3. Wang, R.Y., Popović, J.: Real-time hand-tracking with a color glove. ACM Trans. Graph. **28**, 63 (2009)
4. Tang, D., Yu, T.H., Kim, T.K.: Real-time articulated hand pose estimation using semi-supervised transductive regression forests. In: 2013 IEEE International Conference on Computer Vision (ICCV) (2013)
5. Tompson, J., Stein, M., Lecun, Y., Perlin, K.: Real-time continuous pose recovery of human hands using convolutional networks. ACM Trans. Graph. **33**, 63 (2014)
6. Stenger, B., Thayananthan, A., Torr, P., Cipolla, R.: Model-based hand tracking using a hierarchical Bayesian filter. IEEE Trans. Pattern Anal. Mach. Intell. **28**, 1372–1384 (2006)
7. Rehg, J.M., Kanade, T.: Visual tracking of high dof articulated structures: an application to human hand tracking. In: Computer Vision ECCV 1994 (1994)

8. Oikonomidis, I., Kyriazis, N., Argyros, A.A.: Markerless and efficient 26-dof hand pose recovery. In: Proceedings of the 10th Asian Conference on Computer Vision (2011)

9. Oikonomidis, I., Kyriazis, N., Argyros, A.: Efficient model-based 3D tracking of hand articulations using kinect. In: Proceedings of the British Machine Vision Conference (2011)

10. Qian, C., Sun, X., Wei, Y., Tang, X., Sun, J.: Realtime and robust hand tracking from depth. In: 2014 IEEE Conference on Computer Vision and Pattern Recognition (CVPR) (2014)

11. Sharp, T., Keskin, C., Robertson, D., Taylor, J., Shotton, J., Kim, D., Rhemann, C., Leichter, I., Vinnikov, A., Wei, Y., Freedman, D., Kohli, P., Krupka, E., Fitzgibbon, A., Izadi, S.: Accurate, robust, and flexible real-time hand tracking. In: Proceedings of the 33rd Annual ACM Conference on Human Factors in Computing Systems, CHI 2015. ACM (2015)

12. Kang, B., Tripathi, S., Nguyen, T.: Real-time sign language fingerspelling recognition using convolutional neural networks from depth map. In: 2015 3rd IAPR Asian Conference on Pattern Recognition (ACPR) (2015)

13. Kalman, R.E.: A new approach to linear filtering and prediction problems. J. Fluids Eng. **82**, 35–45 (1960)

Eye Gaze Correction with a Single Webcam Based on Eye-Replacement

Yalun Qin$^{(\boxtimes)}$, Kuo-Chin Lien, Matthew Turk, and Tobias Höllerer

University of California, Santa Barbara, USA
allenchin1990@gmail.com

Abstract. In traditional video conferencing systems, it is impossible for users to have eye contact when looking at the conversation partner's face displayed on the screen, due to the disparity between the locations of the camera and the screen. In this work, we implemented a gaze correction system that can automatically maintain eye contact by replacing the eyes of the user with the direct looking eyes (looking directly into the camera) captured in the initialization stage. Our real-time system has good robustness against different lighting conditions and head poses, and it provides visually convincing and natural results while relying only on a single webcam that can be positioned almost anywhere around the screen.

1 Introduction

With the introduction of both consumer and enterprise video communication technologies, videoconferencing is becoming more and more widely used, allowing people in different locations to meet face-to-face with each other without having to travel to a common location. However, most current videoconferencing systems are not able to maintain eye contact between the conversation participants due to the disparity between the locations of the videoconferencing window and the camera. This problem has been addressed in some high-end level videoconferencing systems by using special hardware, such as semitransparent mirrors/screens [1,2], cameras with depth sensors [3,4], or stereo cameras [5,6]. However due to the extra hardware dependence, these solutions cannot be applied to consumer level videoconferencing systems, which typically only use webcams.

In this paper, we present a gaze correction system that relies only on a single webcam, and thus can be integrated in current videoconferencing systems running on various consumer level devices, ranging from desktop computers to smartphones. Our system corrects the gaze by replacing the eyes of a person in the video with images of eyes looking directly at the camera; these are captured during an initialization session. By further applying several computer vision and computer graphics techniques we make the result look very natural under a range of lighting conditions and head poses. The system supports cameras placed above, below, or to either side of the display screen. The speed of our

© Springer International Publishing Switzerland 2015
G. Bebis et al. (Eds.): ISVC 2015, Part I, LNCS 9474, pp. 599–609, 2015.
DOI: 10.1007/978-3-319-27857-5_54

system is about 30 fps on 640 × 480 streams from a laptop camera. With minor changes this approach can be easily extended to correct the gaze of multiple users simultaneously.

2   Related Work

Previous research using a single webcam for gaze correction has taken two main approaches. The first category of approaches aims to perform the gaze correction by synthesizing a new view of the user's face from a virtual point behind the screen in the center, which is equivalent to performing a rotation of head/face or the entire image by some degrees [7–11]. However these approaches exhibit distortions on the contour of the head, occluded areas and glasses frames. In addition, most of these methods do not perform any explicit changes in the shapes of the eyes, which we believe is very critical to gaze correction and might not be modified sufficiently by merely rotating the head model.

The second category of approaches performs gaze correction by only processing the eyes [12–16]. Most of these methods aim to reposition the iris using computer vision techniques like segmenting and warping [12–14], which all suffer some robustness issues under various lighting conditions and do not have very satisfactory real-time performance. The eye-replacement method proposed by Wolf et al. [15] can establish very natural looking eye contact and is quite robust under various lighting conditions; however due to the quality of the tracker, their system will fail to work when there are large head motions. Their method is also unable to detect whether the user is looking sideways or not and simply replaces the eyes regardless of the actual gaze direction. A recent approach proposed by Kononenko et al. [16] aims to synthesize eyes looking up just by modifying the original image, which turns out to have a good real-time performance, and the result looks natural even when the user is looking sideways. However, due to the limitation of training data, the system only supports vertical gaze redirection by 10 to 15 degrees.

3   Eye-Gaze Correction Method

3.1   System Overview

Our work is based on the eye-replacement method proposed by Wolf et al. [15]. We improved it to make it more robust to changes in lighting conditions and head poses. In contrast to Wolf's approach, ours does not require training and the quick initialization could potentially be done automatically. To ensure the results look good, we also defined the *zone of correction*, which is the range of head poses and eye openness when the gaze correction should be applied. We made the results look natural not only inside the zone but also in the transition areas between inside the zone of correction and outside. Ideally the actual gaze direction should also be taken into account, so that gaze correction is performed

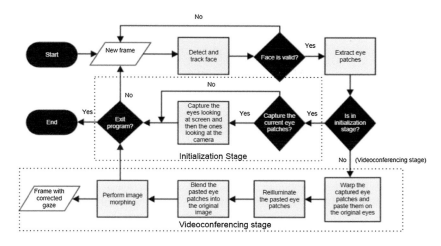

Fig. 1. Work flow of our system

only when the user is looking at the video participant. We further extended our system to support the case when the camera is positioned lateral to the screen.

Our system consists of two stages (Fig. 1):

1. Initialization stage, in which the user is asked to look at the screen first(this step is used for setting the ratio of vertical eye enlargement, which is detailed in Sect. 3.3) and then look straight at the camera. The system captures a pair of eyes that are looking at the camera, used for replacement. Eye contact detection method like [17] can be applied in this stage to make the capture of direct looking eyes more automatic.
2. Videoconferencing stage, in which the user's eyes are tracked and the previous captured eyes are warped and pasted on the original eyes. Reillumination, blending and morphing techniques are then applied to improve the gaze correction results, which are displayed in real time.

Our system is implemented in C++ using OpenCV and OpenGL libraries.

3.2 Face Tracking

The face tracker used in our work is the IntraFace tracker [18]. As is shown in Fig. 2, the IntraFace tracker outputs 49 facial feature points in total, of which the 12 eye points are used by our system. We compute the bounding rotated box for each eye, and enlarge the width of it with a scale factor 3.5 and the height of it with a scale factor 2 to create the bounding rotated box of the eye patch. The areas within the outer rotated rectangles are what we refer as "eye patches". The purpose of this enlargement step is to make sure the pasted eyes completely cover the original eyes. The two rectangles also define the area iterated in the reillumination step, which will be explained in Sect. 3.3. The two scale factors are selected through experimentation to achieve the best visual result.

Fig. 2. Face and eye tracking result. The dots represent the feature points returned by the IntraFace tracker. Inner rectangles around the eyes are the rotated rectangles of the minimum area enclosing the eye points. Outer rectangles are the enlarged rectangles. The coordinate system at the top left represents the head pose direction.

| (a) | (b) | (c) | (d) |

Fig. 3. Comparison of intermediate results: (a) original uncorrected eyes, (b) pasted eyes without any adjustment, (c) reilluminated eyes, (d) final result after Laplacian blending

3.3   Eye Replacement

Warping and Pasting of Eyes. In order to replace the eyes with the direct looking ones captured in the initialization stage, we need to first warp the captured eyes so that they have similar shapes to the original eyes, which we approximate by an affine transformation. To achieve this we use four pairs of points (four corners of the enlarged bounding rotated box) in the source image and destination image to establish the mapping. Note that in the destination image we enlarge the eyes by some ratio in the vertical direction if the gaze direction needs to be corrected vertically, which is equal to the ratio of width of the direct looking eyes to the width of the eyes looking at the screen. The enlargement step is important to make sure the direct looking eyes look real after warping.

Reillumination. The pasted eyes look very unnatural due to the fixed illumination of the pasted eye images. Ideally the illumination of the pasted eye patches should be very similar to the original eye patches, changing with the actual lighting condition. We use Wolf's [15] reillumination method due to its low computational complexity and good results. The idea is that assuming the user's face is Lambertian, then the intensity of a pixel in each of the RGB channels can be approximated using a third order polynomial [19,20]. Unlike Wolf's

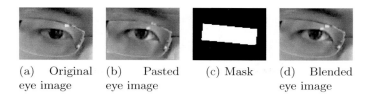

(a) Original   (b)   Pasted   (c) Mask   (d)   Blended
eye image   eye image   eye image

Fig. 4. Input and output of Laplacian pyramid blending

method, we do not iterate through every pixel in the eye patch. Instead, we only iterate through the pixels that are in between the original rotated bounding box and the enlarged one (the area between the two rectangles around each eye in Fig. 2). This is because we found that the pixel intensities in the eye area introduce some unwanted noise to the solution, which as a result makes the reilluminated eye image noisy, having some unwanted slim shadows in the eyes. The reilluminated result is shown in Fig. 3(c).

Laplacian Pyramid Blending. After reillumination, the contrast between the pasted eye patch and the surrounding area is greatly reduced. However the border of the eye patch is still noticeable (Fig. 4(b)). To smooth these artifacts out, we apply Laplacian pyramid blending on the eye images, which is a simple and efficient algorithm to blend one image seamlessly into another [21].

3.4   Smoothing of Transition

Due to the limitation of the face tracker, face tracking fails for some head poses. And since our eye-replacement method doesn't work well when the rotation angle of the user's head is too large, it is necessary to define a zone of correction so that gaze correction is performed only when both processes work. Specifically, the zone of correction is defined as follows:

1. The pitch and yaw angles of head rotation are within some ranges when both face tracking and eye replacement processes work well (Fig. 5). According to our tests, they are $\pm 10°$ for the pitch angle and $\pm 19°$ for the yaw angle.
2. Correction should be applied only when the eyes are open. The degree of openness is measured by the ratio of the height of eye box to the width of eye box. If it is larger than a threshold, a blink is detected and the correction will not be applied.

In order to provide a good user experience, we also define a transition zone (Fig. 5) in which alpha blending (cross-dissolving) is performed to make the transition smooth. To achieve this, the alpha should be changed gradually from 1.0 inside to 0.0 outside the transition zone. The algorithm that determines the value of alpha works as follows:

1. Check if the head pose direction is within the transition range of pitch angles. If it is, then alphaPitch = (maxOuterPitch - pitchAngle)/ (maxOuterPitch - maxInnerPitch) (or (pitchAngle - minOuterPitch)/ (minInnerPitch - minOuterPitch) depending on the side). If it is outside both the correction and transition zones, then alphaPitch = 0; Otherwise(inside the correction zone), alphaPitch = 1.
2. Compute alphaYaw for the yaw angles in a similar manner.
3. The initial value of alpha is the minimum of alphaPitch and alphaYaw.
4. The transition zone of eye openness is determined by minEllipseRatio and maxEllipseRatio, which are minimum and maximum ratios of the height of eye ellipse to the width of eye ellipse. Compute ratioEye as (maxEllipseRatio - curEyeRatio)/(maxEllipseRatio - minEllipseRatio) if the eye openness is inside the transition zone, 1 if it is inside the correction zone and 0 if otherwise.
5. The final alpha value is equal to the initial alpha value times ratioEye.

After we determine the value of alpha, the final image is computed as:

$$finalImage(x, y) = correctedImage(x, y) \cdot \alpha + originalImage(x, y)(1 - \alpha) \quad (1)$$

According to our experiments, when the transition happens during a blink, it does not make much difference whether we apply this blending or not since the corrected eye image is very similar to the original one in appearance when the eye is nearly closed (first row in Fig. 6). However when the transition occurs during a change of head pose, we can see some small artifacts around the eye contours (last three rows in Fig. 6). This is mainly because of the small error of eye point locations returned by the face tracker; given the 22+ fps of the video sequence, these small artifacts are not easily noticed.

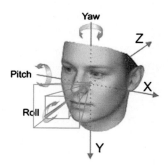

Fig. 5. Correction zone (indicated by the inner square pyramid) and transition zone (indicated by the space between the inner and outer square pyramids) of head pose. Each square pyramid, which is normal to the face plane, denotes the range of possible head pose vectors. The boundary of the outer square pyramid is determined by four angles: maxOuterPitch, minOuterPitch, maxOuterYaw, minOuterYaw. Similarly, the angles maxInnerPitch, minInnerPitch, maxInnerYaw, minInnerYaw determine the boundary of the inner square pyramid. These values are set as constants.

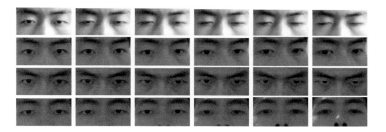

Fig. 6. Transition from open eyes to closed eyes (first row from left to right). Transition when the user's head is turning to one side(second row from left to right). Transition when the user is lowering down his head(third row from left to right). Transition when the user is raising up his head(fourth row from left to right)

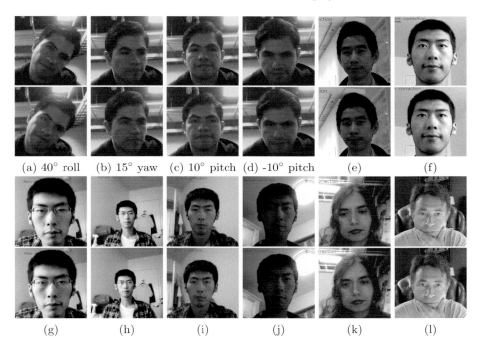

  (a) 40° roll   (b) 15° yaw (c) 10° pitch (d) -10° pitch   (e)   (f)

    (g)   (h)   (i)   (j)   (k)   (l)

Fig. 7. Good results. Top rows are original images and bottom rows are corrected ones. Please refer to our supplementary video for more results.

4　Results

We tested our system on a laptop equipped with an Intel(R) Core(TM) i7-4710HQ CPU, 16 GB RAM, and a NVIDIA GeForce GTX 860M (GPU acceleration not used). The frame rate of the built-in webcam is 22 fps for 800×600 input videos and 30 fps for 640×480 input videos. In typical usage, the face height in the camera image is half the height of the image. About one third of

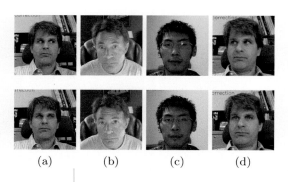

(a)   (b)   (c)   (d)

Fig. 8. Unsatisfactory results. Top rows are original images and bottom rows are corrected ones. Please refer to our supplementary video for more results.

the time is spent on reading the video sequence from the webcam, which might vary if the video resolution changes. Face tracking takes up a very small portion of the processing time, about 5 to 10 ms, which is not very dependent on the video resolution. The remaining time is spent on the eye-replacement process. Since the processing time of warping and pasting of eyes and the reillumination algorithm is dependent on the resolution of eyes and video frame, the entire processing of each frame takes longer when the face is closer or when the image resolution is higher, but for the common use settings like above our system does support real-time interaction.

We conducted several experiments to test the robustness and temporal consistency of our system. The hardware settings are similar to the ones listed above but the resolution might vary a bit since we also used an external webcam with a different resolution setting, which does not affect the gaze correction result.

Figure 7 shows screenshots from testing the robustness against head pose changes. The system gives good correction results as long as the head pose is within the zone of correction, which is set to be $\pm 10°$ of vertical rotation and $\pm 19°$ of horizontal rotation. If the user's head pose is in the transition zone, morphing will be applied and the results are similar to those in Fig. 6, which still look natural. If the head pose is outside the transition zone, then no correction is applied. The test results shown in Fig. 7 and the supplementary video also demonstrate the robustness of our system against facial expressions. Figure 7(g) and (h) illustrate the results when the user is close to and far from the webcam. As long as the user's face is within the working range of the face tracker, the gaze correction process will also work. Figure 7(i) and (j) show the results when the lighting and background change. They belong to the same video sequence, in which the user is moving from one place to another while holding the laptop (along with the built-in webcam). Good results are achieved when the facial features are visible and the image quality doesn't change much during the movement. Figure 7(l) and (k) show the case when the face is partially occluded by hand or hair. As long as the facial features like eyes and mouth are visible, the tracker will work and so will our gaze correction system. Our system also supports the cases when the camera is positioned to the side of the screen (Fig. 7(e)) or below the screen (Fig. 7(f)). According to our test results, our system works for almost all the camera positions as long as the head pose in the image is within the correction zone, which means that the disparity angle between the locations of the camera and the videoconferencing window should not be larger than 19 degrees horizontally and 10 degrees vertically.

The supplementary video shows how our system performs in real time. Overall the temporal consistency is good, i.e., the corrected eyes do not flicker much from frame to frame. It still flickers a little bit from time to time due to the noise of the image. We find that if a noise reduction algorithm is applied to each frame beforehand the results become much more stable.

The results shown in Fig. 8 demonstrate some of the major limitations of our system. When the user is looking extremely sideways (Fig. 8(a)), the large area of eye white may whiten the reilluminated eye image due to the limitation of our reillumination method. When the user's eyes are widely opened (Fig. 8(b)), the enlarged replacement eyes may differ from the actual eyes quite a lot, because of the difference between the captured direct looking eyes and the actual eyes. Moreover, if there are strong reflection of light on the glasses, the tracked eye points may be very inaccurate and unstable, and therefore the final correction result will also look very bad (Fig. 8(c)).

One drawback of our gaze correction approach is that by simply replacing the original eyes with the direct looking eyes, it ignores the actual gaze direction, i.e., the gaze direction in the corrected image does not change with the actual gaze direction (Fig. 8(d)). It would be preferable to perform gaze correction only when the user is looking directly at the screen, or design a gaze correction approach that takes into account the actual gaze correction.

5   Conclusion

In this paper we presented our gaze correction system based on eye-replacement, similar to Wolf et al. [15]. Compared to their approach, ours does not require training, and we explicitly define the correction zone and handle the gaze in the transition zone. Our system also supports more camera positions around the display. According to our test results, the system has good robustness against different common lighting conditions and head poses. Our system is also shown to have a good run-time speed and provide visually convincing and natural results.

Acknowledgements. This work was supported in part by Citrix Online and by NSF Grants IIS-1423676 and IIS-1219261.

References

1. Jones, A., Lang, M., Fyffe, G., Yu, X., Busch, J., McDowall, I., Bolas, M., Debevec, P.: Achieving eye contact in a one-to-many 3d video teleconferencing system. ACM Trans. Graph. (TOG) **28**, 64:1–64:8 (2009)

2. Okada, K.I., Maeda, F., Ichikawaa, Y., Matsushita, Y.: Multiparty videoconferencing at virtual social distance: Majic design. In: Proceedings of the 1994 ACM Conference on Computer Supported Cooperative Work, pp. 385–393. ACM (1994)
3. Kuster, C., Popa, T., Bazin, J.C., Gotsman, C., Gross, M.: Gaze correction for home video conferencing. ACM Trans. Graph. (TOG) **31**, 174 (2012)
4. Zhu, J., Yang, R., Xiang, X.: Eye contact in video conference via fusion of time-of-flight depth sensor and stereo. 3D Res. **2**, 1–10 (2011)
5. Criminisi, A., Shotton, J., Blake, A., Torr, P.H.: Gaze manipulation for one-to-one teleconferencing. In: Ninth IEEE International Conference on Computer Vision, 2003, Proceedings, pp. 191–198. IEEE (2003)
6. Yang, R., Zhang, Z.: Eye gaze correction with stereovision for video-teleconferencing. In: Heyden, A., Sparr, G., Nielsen, M., Johansen, P. (eds.) ECCV 2002, Part II. LNCS, vol. 2351, pp. 479–494. Springer, Heidelberg (2002)
7. Cham, T.J., Krishnamoorthy, S., Jones, M.: Analogous view transfer for gaze correction in video sequences. In: 7th International Conference on Control, Automation, Robotics and Vision, 2002. ICARCV 2002, vol. 3, pp. 1415–1420. IEEE (2002)
8. Yip, B., Jin, J.S.: An effective eye gaze correction operation for video conference using antirotation formulas. In: Proceedings of the 2003 Joint Conference of the Fourth International Conference on Information, Communications and Signal Processing and Fourth Pacific Rim Conference on Multimedia, vol. 2, pp. 699–703. IEEE (2003)
9. Gemmell, J., Toyama, K., Zitnick, C.L., Kang, T., Seitz, S.: Gaze awareness for video-conferencing: a software approach. IEEE Multimedia **7**, 26–35 (2000)
10. Gemmell, J., Zhu, D.: Implementing gaze-corrected videoconferencing. In: Communications, Internet, and Information Technology, pp. 382–387 (2002)
11. Giger, D., Bazin, J.C., Kuster, C., Popa, T., Gross, M.: Gaze correction with a single webcam. In: 2014 IEEE International Conference on Multimedia and Expo (ICME), pp. 1–6. IEEE (2014)
12. Hundt, E., Riegel, T., Schwartzel, H., Ziegler, M.: Correction of the gaze direction for a videophone, US Patent 5,499,303 (1996)
13. Jerald, J., Daily, M.: Eye gaze correction for videoconferencing. In: Proceedings of the 2002 Symposium on Eye Tracking Research & Applications, pp. 77–81. ACM (2002)
14. Weiner, D., Kiryati, N.: Virtual gaze redirection in face images. In: 12th International Conference on Image Analysis and Processing, 2003, Proceedings, pp. 76–81. IEEE (2003)
15. Wolf, L., Freund, Z., Avidan, S.: An eye for an eye: a single camera gaze-replacement method. In: 2010 IEEE Conference on Computer Vision and Pattern Recognition (CVPR), pp. 817–824. IEEE (2010)
16. Kononenko, D., Lempitsky, V.: Learning to look up: realtime monocular gaze correction using machine learning. In: Computer Vision and Pattern Recognition (CVPR). IEEE (2015)
17. Smith, B.A., Yin, Q., Feiner, S.K., Nayar, S.K.: Gaze locking: passive eye contact detection for human-object interaction. In: Proceedings of the 26th Annual ACM Symposium on User Interface Software and Technology, pp. 271–280. ACM (2013)
18. Xiong, X., De la Torre, F.: Supervised descent method and its applications to face alignment. In: 2013 IEEE Conference on Computer Vision and Pattern Recognition (CVPR), pp. 532–539. IEEE (2013)
19. Basri, R., Jacobs, D.W.: Lambertian reflectance and linear subspaces. IEEE Trans. Pattern Anal. Mach. Intell. **25**, 218–233 (2003)

20. Bitouk, D., Kumar, N., Dhillon, S., Belhumeur, P., Nayar, S.K.: Face swapping: automatically replacing faces in photographs. ACM Trans. Graph. (TOG) **27**, 39 (2008)
21. Burt, P.J., Adelson, E.H.: A multiresolution spline with application to image mosaics. ACM Trans. Graph. (TOG) **2**, 217–236 (1983)

ST: Observing Humans

Gradient Local Auto-Correlations and Extreme Learning Machine for Depth-Based Activity Recognition

Chen Chen[1], Zhenjie Hou[2(✉)], Baochang Zhang[3], Junjun Jiang[4], and Yun Yang[3]

[1] Department of Electrical Engineering,
University of Texas at Dallas, Richardson, TX, USA
chenchen870713@gmail.com
[2] School of Information Science and Engineering,
Changzhou University, Changzhou, China
houzj@cczu.edu.cn
[3] School of Automation Science and Electrical Engineering,
Beihang University, Beijing, China
bczhang@buaa.edu.cn, bhu_yunyang@163.com
[4] School of Computer Science, China University of Geosciences, Wuhan, China
junjun0595@163.com

Abstract. This paper presents a new method for human activity recognition using depth sequences. Each depth sequence is represented by three depth motion maps (DMMs) from three projection views (front, side and top) to capture motion cues. A feature extraction method utilizing spatial and orientational auto-correlations of image local gradients is introduced to extract features from DMMs. The gradient local auto-correlations (GLAC) method employs second order statistics (i.e., auto-correlations) to capture richer information from images than the histogram-based methods (e.g., histogram of oriented gradients) which use first order statistics (i.e., histograms). Based on the extreme learning machine, a fusion framework that incorporates feature-level fusion into decision-level fusion is proposed to effectively combine the GLAC features from DMMs. Experiments on the MSRAction3D and MSRGesture3D datasets demonstrate the effectiveness of the proposed activity recognition algorithm.

Keywords: Gradient local auto-correlations · Extreme learning machine · Activity recognition · Depth images · Depth motion map

1 Introduction

Human activity recognition is one of the important areas of computer vision research today. It has a wide range of applications including intelligent video surveillance, video analysis, assistive living, robotics, telemedicine, and human computer interaction (e.g., [1–4]). Research on human activity recognition has initially focused on learning and recognizing activities from video sequences captured by conventional RGB cameras.

Since the recent release of cost-effective 3D depth cameras using structured light or time-of-flight sensors, there has been great interest in solving the problem of human

© Springer International Publishing Switzerland 2015
G. Bebis et al. (Eds.): ISVC 2015, Part I, LNCS 9474, pp. 613–623, 2015.
DOI: 10.1007/978-3-319-27857-5_55

activity recognition by using 3D data. Compared with traditional color images, depth images are insensitive to changes in lighting conditions and provide body shape and structure information for activity recognition. Color and texture are precluded in the depth images, which makes the tasks of human detection and segmentation easier [5]. Moreover, human skeleton information can be estimated from depth images providing additional information for activity recognition [6].

Research on activity recognition has explored various representations (e.g., 3D point cloud [7], projection depth maps [8], spatio-temporal interest points [9], and skeleton joints [10]) of depth sequences. In [7], a bag of 3D points was sampled from depth images to characterize the 3D shapes of salient postures and Gaussian mixture model (GMM) was used to robustly capture the statistical distribution of the points. A filtering method to extract spatio-temporal interest points (STIPs) from depth videos (called DSTIP) was introduced in [9] to localize activity related interest points by effectively suppressing the noise in the depth videos. Depth cuboid similarity feature (DCSF) built around the DSTIPs was proposed to describe the local 3D depth cuboid. Inspired by motion energy images (MEI) of motion history images (MHI) [11], depth images in a depth video sequence were projected onto three orthogonal planes and differences between projected depth maps were stacked to form depth motion maps (DMMs) [8]. Histogram of oriented gradients (HOG) [12] features were then extracted from DMMs as global representations of a depth video. DMMs effectively transform the problem in 3D to 2D. In [13], the procedure of generating DMMs was modified to reduce the computational complexity in order to achieve real-time action recognition. Later in [14], local binary pattern [15] operator was applied to the overlapped blocks in DMMs to enhance the discriminative power for action recognition. Skeleton information has also been explored for activity recognition, for example [10]. Reviews of skeleton based activity recognition methods are referred to [10, 16].

Motivated by the success of DMMs in depth-based activity recognition, our method proceeds along with this direction. Specifically, we introduce the gradient local auto-correlations (GLAC) [17] descriptor and present a new feature extraction method using GLAC and DMMs. A fusion framework based on extreme learning machine (ELM) [18] is proposed to effectively combine the GLAC features from DMMs for activity recognition. The main contributions of this paper are summarized as follows:

1. We introduce a new feature descriptor, GLAC, to extract features from DMMs of depth sequences. The GLAC descriptor, which is based on the second order of statistics of gradients (spatial and orientational auto-correlations of local image gradients), can effectively capture rich information from images.
2. We present a unified fusion framework which incorporates feature-level fusion into decision-level fusion for activity recognition.
3. We demonstrate that ELM has better performance than support vector machine (SVM) in our proposed method for depth-based activity recognition.

The rest of this paper is organized as follows. Section 2 describes the proposed feature extraction method. Section 3 overviews ELM and provides details of the unified fusion framework for activity recognition. Experimental results and discussions are presented in Sect. 4. Finally, Sect. 5 concludes the paper.

2 Feature Extraction from Depth Sequences

2.1 Depth Motion Map

To extract features from depth images, depth motion maps (DMMs) discussed in [13] are used due to their computational efficiency. More specifically, each 3D depth image in a depth video sequence is first projected onto three orthogonal Cartesian planes to generate three 2D projected maps corresponding to front, side, and top views, denoted by map_f, map_s, and map_t, respectively. For a depth video sequence with N frames, the DMMs are obtained as follows:

$$DMM_{\{f,s,t\}} = \sum_{i=1}^{N-1} \left| map_{\{f,s,t\}}^{i+1} - map_{\{f,s,t\}}^{i} \right|, \tag{1}$$

where i represents frame index. A bounding box is considered to extract the foreground in each DMM. Since foreground DMMs of different video sequences may have different sizes, bicubic interpolation is applied to resize all such DMMs to a fixed size and thus to reduce the intra-class variability. Figure 1 shows two example sets of DMMs.

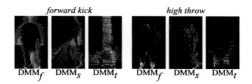

Fig. 1. DMMs for the *forward kick* and *high throw* depth action video sequences.

2.2 Gradient Local Auto-Correlations

GLAC [17] descriptor is an effective tool for extracting shift-invariant image features. Let I be an image region and $\mathbf{r} = (x, y)^t$ be a position vector in I. The magnitude and the orientation angle of the image gradient at each pixel can be represented by $n = \sqrt{\frac{\partial I^2}{\partial x} + \frac{\partial I^2}{\partial y}}$ and $\theta = \arctan\left(\frac{\partial I}{\partial x}, \frac{\partial I}{\partial y}\right)$, respectively. The orientation θ is then coded into D orientation bins by voting weights to the nearest bins to form a gradient orientation vector $\mathbf{f} \in \mathbb{R}^D$. With the gradient orientation vector $\mathbf{f}$ and the gradient magnitude n, the Nth order auto-correlation function of local gradients can be expressed as follows:

$$R(d_0, \ldots, d_N, \mathbf{a}_1, \ldots, \mathbf{a}_N) = \int_I \omega[n(\mathbf{r}), n(\mathbf{r}+\mathbf{a}_1), \ldots, n(\mathbf{r}+\mathbf{a}_N)] f_{d_0}(\mathbf{r}) f_{d_1}(\mathbf{r}+\mathbf{a}_1) \cdots f_{d_N}(\mathbf{r}+\mathbf{a}_N) d\mathbf{r}, \tag{2}$$

where $\mathbf{a}_i$ are displacement vectors from the reference point $\mathbf{r}$, f_d is the dth element of $\mathbf{f}$, and $\omega(\cdot)$ indicates a weighting function. In the experiments reported later, $N \in \{0, 1\}$,

$a_{1x,y} \in \{\pm \Delta r, 0\}$, and $\omega(\cdot) \equiv \min(\cdot)$ were considered as suggested in [17], where Δr represents the displacement interval in both horizontal and vertical directions. For $N \in \{0, 1\}$, the formulation of GLAC is given by

$$
\begin{aligned}
\mathbf{F}_0 &: R_{N=0}(d_0) = \sum_{\mathbf{r} \in I} n(\mathbf{r}) f_{d_0}(\mathbf{r}) \\
\mathbf{F}_1 &: R_{N=1}(d_0, d_1, \mathbf{a}_1) = \sum_{\mathbf{r} \in I} \min[n(\mathbf{r}), n(\mathbf{r} + \mathbf{a}_1)] f_{d_0}(\mathbf{r}) f_{d_1}(\mathbf{r} + \mathbf{a}_1).
\end{aligned}
\tag{3}
$$

The spatial auto-correlation patterns of $(\mathbf{r}, \mathbf{r} + \mathbf{a}_1)$ are shown in Fig. 2.

Fig. 2. Configuration patterns of $(\mathbf{r}, \mathbf{r} + \mathbf{a}_1)$.

The dimensionality of the above GLAC features ($\mathbf{F}_0$ and $\mathbf{F}_1$) is $D + 4D^2$. Although the dimensionality of the GLAC features is high, the computational cost is low due to the sparseness of $\mathbf{f}$. It is worth noting that the computational cost is invariant to the number of bins, D, since the sparseness of $\mathbf{f}$ doesn't depend on D.

2.3   DMMs-Based GLAC Features

DMMs generated from a depth sequence are pixel-level features. To enhance the discriminative power and gain a compact representation, we adopt the method in [14] to extract GLAC features from DMMs. Specifically, DMMs are divided into several overlapped blocks and the GLAC descriptor is applied to each block to compute GLAC features (i.e., $\mathbf{F}_0$ and $\mathbf{F}_1$). For each DMM, GLAC features from all the blocks are concatenated to form a single composite feature vector. Therefore, three feature vectors $\mathbf{g}_1$, $\mathbf{g}_2$ and $\mathbf{g}_3$ corresponding to three DMMs are obtained for a depth sequence.

3   Classification Fusion Based on ELM

3.1   Elm

ELM [18] is an efficient learning algorithm for single hidden layer feed-forward neural networks (SLFNs) and has been applied in various applications (e.g., [19, 20]).

Let $\mathbf{y} = [y_1, \ldots, y_k, \ldots, y_C]^T \in \mathbb{R}^C$ be the class to which a sample belongs, where $y_k \in \{1, -1\}$ ($1 \leq k \leq C$) and C is the number of classes. Given n training samples $\{\mathbf{x}_i, \mathbf{y}_i\}_{i=1}^n$, where $\mathbf{x}_i \in \mathbb{R}^M$ and $\mathbf{y}_i \in \mathbb{R}^C$, a single hidden layer neural network having L hidden nodes can be expressed as

$$\sum_{j=1}^{L} \boldsymbol{\beta}_j h\big(\mathbf{w}_j \cdot \mathbf{x}_i + e_j\big) = \mathbf{y}_i, \quad i = 1, \ldots, n, \tag{4}$$

where $h(\cdot)$ is a nonlinear activation function, $\boldsymbol{\beta}_j \in \mathbb{R}^C$ denotes the weight vector connecting the jth hidden node to the output nodes, $\mathbf{w}_j \in \mathbb{R}^M$ denotes the weight vector connecting the jth hidden node to the input nodes, and e_j is the bias of the jth hidden node. (4) can be written compactly as:

$$\mathbf{H}\boldsymbol{\beta} = \mathbf{Y}, \tag{5}$$

where $\boldsymbol{\beta} = [\boldsymbol{\beta}_1^T; \ldots; \boldsymbol{\beta}_L^T] \in \mathbb{R}^{L \times C}$, $\mathbf{Y} = [\mathbf{y}_1^T; \ldots; \mathbf{y}_n^T] \in \mathbb{R}^{n \times C}$, and $\mathbf{H}$ is the hidden layer output matrix. A least-squares solution $\hat{\beta}$ to (5) is

$$\hat{\boldsymbol{\beta}} = \mathbf{H}^\dagger \mathbf{Y}, \tag{6}$$

where $\mathbf{H}^\dagger$ is the *Moore-Penrose inverse* of $\mathbf{H}$. The output function of the ELM classifier is

$$\mathbf{f}_L(\mathbf{x}_i) = \mathbf{h}(\mathbf{x}_i)\boldsymbol{\beta} = \mathbf{h}(\mathbf{x}_i)\mathbf{H}^T \left(\frac{\mathbf{I}}{\rho} + \mathbf{H}\mathbf{H}^T \right)^{-1} \mathbf{Y}, \tag{7}$$

where $1/\rho$ is a regularization term. The label of a test sample is assigned to the index of the output nodes with the largest value. In our experiments, we use a kernel-based ELM (KELM) with a radial basis function (RBF) kernel.

3.2   Proposed Fusion Framework

In [14], both feature-level fusion and decision-level fusion were examined for action recognition. It was demonstrated that decision-level fusion had better performance than feature-level fusion. To further improve the performance of decision-level fusion, we propose a unified fusion framework that incorporates feature-level fusion into decision-level fusion.

In decision-level fusion, each feature (e.g., $\mathbf{g}_1$, $\mathbf{g}_2$, and $\mathbf{g}_3$) is used individually as input to an ELM classifier. The probability outputs of each individual classifier are merged to generate the final outcome. The posterior probabilities are estimated using the decision function of ELM (i.e., $\mathbf{f}_L$ in (7)) since it estimates the accuracy of the output label. $\mathbf{f}_L$ is normalized to $[0, 1]$ and Platt's empirical analysis [21] using a Sigmoid function is utilized to approximate the posterior probabilities,

$$p(y_k|\mathbf{x}) = \frac{1}{1 + \exp\big(Af_L(\mathbf{x})_k + B\big)}, \tag{8}$$

where $f_L(\mathbf{x})_k$ is the kth output of the decision function $\mathbf{f}_L(\mathbf{x})$. In our experiments, $A = -1$ and $B = 0$. Logarithmic opinion pool (LOGP) [20] is used to estimate a global membership function:

$$\log P(y_k|\mathbf{x}) = \sum_{q=1}^{Q} \alpha_q p_q(y_k|\mathbf{x}), \tag{9}$$

where Q is the number of classifiers and $\{\alpha_q\}_{q=1}^{Q}$ are uniformly distributed classifier weights. The final class label y^* is determined according to

$$y^* = \arg\max_{k=1,\ldots,C} P(y_k|\mathbf{x}). \tag{10}$$

To incorporate feature-level fusion into decision-level fusion, we stack $\mathbf{g}_1$, $\mathbf{g}_2$, and $\mathbf{g}_3$ (feature-level fusion) as the fourth feature $\mathbf{g}_4 = [\mathbf{g}_1, \mathbf{g}_2, \mathbf{g}_3]$. Then these four feature vectors are used individually as inputs to four ELM classifiers. Note that principal component analysis (PCA) is employed for dimensionality reduction of the feature vectors. For a testing sample $\mathbf{x}$, five sets of probability outputs including four sets of probability outputs $\{p_q(y_k|\mathbf{x})\}_{q=1}^{4}$ corresponding to the four classifiers and a set of fusion probability outputs $P(y_k|\mathbf{x})$ by using (9) can be obtained. The class label of the testing sample $\mathbf{x}$ is assigned based on $P(y_k|\mathbf{x})$.

Since each set of probability outputs is able to make a decision on the class label of the testing sample $\mathbf{x}$, we could use the label information (five class labels) from the five sets of probability outputs to reach a more robust classification decision. Here, we employ the majority voting strategy on the five labels. If at least three class labels are the same, we consider there is a major certainty among the five sets of probability outputs and use the majority voted label as the final class label; otherwise, we use the label given by $P(y_k|\mathbf{x})$. By introducing a majority voting step in the decision-level fusion, we not only consider the classification probability but also the classification certainty.

4   Experiments

In this section, we evaluate our proposed activity recognition method on the public MSRAction3D [7] and MSRGesture3D [22] datasets which consist of depth sequences captured by RGBD cameras. Some example depth images from these two datasets are presented in Fig. 3. Our method is then compared with the existing methods. The source code of our method will be available on our website.

4.1   MSRAction3D Dataset

The MSRAction3D dataset [7] includes 20 actions performed by 10 subjects. Each subject performs each action 2 or 3 times. To facilitate a fair comparison, the same

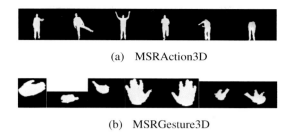

(a)   MSRAction3D

(b)   MSRGesture3D

Fig. 3. Sample depth images of different actions/gestures.

experimental setup in [23] is used. A total of 20 actions are employed and one half of the subjects (1, 3, 5, 7, 9) are used for training and the rest subjects are used for testing.

First of all, we estimate the optimal parameter set $(D, \Delta r)$ for the GLAC descriptor using the training data via five-fold cross validation. The recognition results with various parameter sets for the MSRAction3D dataset are shown in Table 1. Therefore, $D = 10$ and $\Delta r = 8$ are chosen in terms of the activity recognition accuracy.

Table 1. Recognition accuracy (%) of GLAC with different parameter sets $(D, \Delta r)$ for the MSRAcion3D dataset using training data

D / Δr	1	2	3	4	5	6	7	8	9	10	11	12
1	79.0	84.5	87.8	90.7	90.4	90.0	90.4	91.5	92.2	92.2	92.6	91.8
2	78.3	84.2	87.1	89.6	91.1	90.7	91.5	91.1	91.1	90.7	91.8	91.5
3	79.4	82.7	86.7	89.6	90.7	90.4	91.8	91.5	91.5	91.8	91.1	91.8
4	79.8	82.0	84.9	88.9	88.9	88.5	90.7	91.1	91.1	90.0	91.5	91.1
5	81.2	84.5	85.6	90.4	90.4	91.8	92.2	92.2	92.9	92.9	92.6	92.6
6	83.1	86.0	86.3	90.7	91.5	91.1	92.6	91.8	92.9	92.9	92.9	92.9
7	83.8	86.7	86.3	90.4	90.4	92.6	91.8	92.9	92.9	93.7	92.9	92.6
8	84.2	86.0	86.3	90.4	91.5	92.6	92.9	93.3	93.7	93.7	93.7	92.9
9	84.2	83.8	85.6	89.6	90.0	90.7	91.8	91.8	92.6	92.9	92.9	92.9
10	82.3	85.3	85.6	89.3	90.0	90.7	91.5	91.8	92.6	92.6	92.6	92.2

A comparison of our method with the existing methods is carried out. The outcome of the comparison is listed in Table 2. We also report the results of feature-level fusion and decision-level fusion using only $\mathbf{g}_1$, $\mathbf{g}_2$, and $\mathbf{g}_3$. These two methods are denoted by DMM-GLAC-DF and DMM-GLAC-FF. As we can see that our method achieves an accuracy of 92.31 %, only 0.78 % inferior to the state-of-the-art accuracy (93.09 %) of SNV [5]. Moreover, the proposed unified fusion method outperforms DMM-GLAC-FF by 1.83 %, which demonstrates the benefit of incorporating the concatenated feature $\mathbf{g}_4$ into decision-level fusion. The confusion matrix of our method for the MSRAction3D dataset is shown in Fig. 4(a). The recognition errors concentrate on similar actions, e.g., *draw circle* and *draw tick*. This is mainly because the DMMs of these actions are similar.

Table 2. Comparison of recognition accuracy on the MSRAction3D dataset

Method	Accuracy
Bag of 3D points [7]	74.70 %
EigenJoints [10]	82.30 %
STOP [24]	84.80 %
Random Occupancy Pattern [23]	86.50 %
Actionlet Ensemble [25]	88.20 %
DMM-HOG [8]	88.73 %
Histograms of Depth Gradients [27]	88.80 %
HON4D [26]	88.89 %
DSTIP [9]	89.30 %
DMM-LBP-FF [14]	91.90 %
DMM-LBP-DF [14]	93.00 %
SNV [5]	**93.09 %**
DMM-GLAC-FF	89.38 %
Proposed	90.48 %
DMM-GLAC-DF	92.31 %

4.2   MSRGesture3D Dataset

The MSRGesture3D dataset [22] consists of 12 gestures defined by American Sign Language (ASL). It contains 333 depth sequences. The leave one subject out cross-validation test [23] is utilized in our evaluation. $D = 12$ and $\Delta r = 1$ are selected for GLAC based on the parameter tuning experiment for this dataset. Our method obtains the state-of-the-art accuracy of 95.5 % which outperforms all previous methods as shown in Table 3. The confusion matrix of our method for the MSRGesture3D dataset is demonstrated in Fig. 4(b).

Table 3. Comparison of recognition accuracy on the MSRGesture3D dataset

Method	Accuracy
Random Occupancy Pattern [23]	88.50 %
DMM-HOG [8]	89.20 %
Histograms of Depth Gradients [27]	93.60 %
HON4D [26]	92.45 %
Action Graph on Silhouett [22]	87.70 %
DMM-LBP-FF [14]	93.40 %
DMM-LBP-DF [14]	94.60 %
SNV [5]	94.74 %
Proposed	**95.50 %**

4.3   ELM vs. SVM

We also conduct comparison between ELM and SVM for activity recognition. For our method using SVM as the classifier, LIBSVM [28] toolbox is utilized to provide

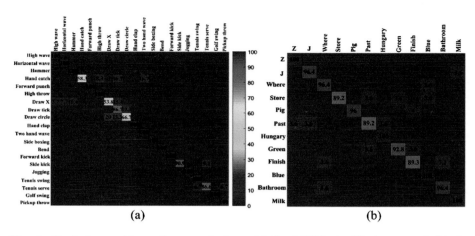

Fig. 4. Confusion matrices of our method on (a) the MSRAction3D dataset and (b) the MSRGesture3D dataset. This figure is best seen on screen.

probability estimates for multi-class classification. The comparison results in terms of recognition accuracy are presented in Table 4. It is easy to see that ELM has superior performance over SVM for both datasets. The standardized McNemar's test [20] is employed to verify the statistical significance in accuracy improvement of our ELM-based method. A value of $|Z| > 1.96$ in the McNemar's test indicates there is a significant difference in accuracy between two classification methods. The sign of Z indicates whether classifier 1 statistically outperforms classifier 2 ($Z > 0$) or vice versa. As we can see that the ELM-based method statistically outperforms SVM-based method.

Table 4. Performance comparison between ELM and SVM in our method

MSRAction3D		MSRGesture3D	
Accuracy		Accuracy	
ELM	92.31 %	ELM	95.50 %
SVM	86.80 %	SVM	92.80 %
Z/significant?		Z/significant?	
ELM (classifier 1) *vs.* SVM (classifier 2)		ELM (classifier 1) *vs.* SVM (classifier 2)	
3.13/yes		2.04/yes	

5   Conclusions

We have presented a novel framework for activity recognition from depth sequences. The gradient local auto-correlations (GLAC) features utilize spatial and orientational auto-collections of local gradients to describe the rich texture information of the depth motion maps generated from a depth sequence. A unified fusion scheme that combines feature-level fusion and decision-level fusion is proposed based on extreme learning machine for activity recognition. Our method is evaluated on two public benchmark datasets and the experimental results demonstrate that the proposed method can achieve competitive or better performance compared to a number of state-of-the-art methods.

Acknowledgement. We acknowledge the support of the Industry, Teaching and Research Prospective Project of Jiangsu Province (grant No. BY2015027-12), the Natural Science Foundation of China, under contracts 61063021, 61272052 and 61473086, and the Program for New Century Excellent Talents of the University of Ministry of Education of China.

References

1. Chen, C., Jafari, R., Kehtarnavaz, N.: Improving human action recognition using fusion of depth camera and inertial sensors. IEEE Trans. Hum.-Mach. Syst. **45**(1), 51–61 (2015)
2. Chen, C., Liu, K., Jafari, R., Kehtarnavaz, N.: Home-based senior fitness test measurement system using collaborative inertial and depth sensors. In: EMBC, pp. 4135–4138 (2014)
3. Theodoridis, T., Agapitos, A., Hu, H., Lucas, S.M.: Ubiquitous robotics in physical human action recognition: a comparison between dynamic ANNs and GP. In: ICRA, pp. 3064–3069 (2008)
4. Chen, C., Kehtarnavaz, N., Jafari, R.: A medication adherence monitoring system for pill bottles based on a wearable inertial sensor. In: EMBC, pp. 4983–4986 (2014)
5. Yang, X., Tian, Y.: Super normal vector for activity recognition using depth sequences. In: CVPR, pp. 804–811 (2014)
6. Shotton, J., Fitzgibbon, A., Cook, M., Sharp, T., Finocchio, M., Moore, R., Blake, A.: Real-time human pose recognition in parts from single depth images. In: CVPR, pp. 1297–1304 (2011)
7. Li, W., Zhang, Z., Liu, Z.: Action recognition based on a bag of 3D points. In: CVPRW, pp. 9–14 (2010)
8. Yang, X., Zhang, C., Tian, Y.: Recognizing actions using depth motion maps-based histograms of oriented gradients. In: ACM Multimedia, pp. 1057–1060 (2012)
9. Xia, L., Aggarwal, J.K.: Spatio-temporal depth cuboid similarity feature for activity recognition using depth camera. In: CVPR, pp. 2834–2841 (2013)
10. Yang, X., Tian, Y.: Effective 3d action recognition using eigenjoints. J. Vis. Commun. Image Represent. **25**(1), 2–11 (2014)
11. Bobick, A.F., Davis, J.W.: The recognition of human movement using temporal templates. IEEE Trans. Pattern Anal. Mach. Intell. **23**(3), 257–267 (2001)
12. Dalal, N., Triggs, B.: Histograms of oriented gradients for human detection. In: CVPR, pp. 886–893 (2005)
13. Chen, C., Liu, K., Kehtarnavaz, N.: Real-time human action recognition based on depth motion maps. J. Real-Time Image Process., 1–9 (2013). doi:10.1007/s11554-013-0370-1
14. Chen, C., Jafari, R., Kehtarnavaz, N.: Action recognition from depth sequences using depth motion maps-based local binary patterns. In: WACV, pp. 1092–1099 (2015)
15. Ojala, T., Pietikäinen, M., Mäenpää, T.: Multiresolution gray-scale and rotation invariant texture classification with local binary patterns. IEEE Trans. Pattern Anal. Mach. Intell. **24** (7), 971–987 (2002)
16. Vemulapalli, R., Arrate, F., Chellappa, R.: Human action recognition by representing 3D skeletons as points in a lie group. In: CVPR, pp. 588–595 (2014)

17. Kobayashi, T., Otsu, N.: Image feature extraction using gradient local auto-correlations. In: Forsyth, D., Torr, P., Zisserman, A. (eds.) ECCV 2008, Part I. LNCS, vol. 5302, pp. 346–358. Springer, Heidelberg (2008)
18. Huang, G.B., Zhu, Q.Y., Siew, C.K.: Extreme learning machine: theory and applications. Neurocomputing **70**(1), 489–501 (2006)
19. Chen, C., Zhou, L., Guo, J., Li, W., Su, H., Guo, F.: Gabor-filtering-based completed local binary patterns for land-use scene classification. In: 2015 IEEE International Conference on Multimedia Big Data, pp. 324–329 (2015)
20. Li, W., Chen, C., Su, H., Du, Q.: Local binary patterns and extreme learning machine for hyperspectral imagery classification. IEEE Trans. Geosci. Remote Sens. **53**(7), 3681–3693 (2015)
21. Platt, J.: Probabilistic outputs for support vector machines and comparisons to regularized likelihood methods. Adv. Large Margin Classifiers **10**(3), 61–74 (1999)
22. Kurakin, A., Zhang, Z., Liu, Z.: A real time system for dynamic hand gesture recognition with a depth sensor. In: EUSIPCO, pp. 1975–1979 (2012)
23. Wang, J., Liu, Z., Chorowski, J., Chen, Z., Wu, Y.: Robust 3D action recognition with random occupancy patterns. In: Fitzgibbon, A., Lazebnik, S., Perona, P., Sato, Y., Schmid, C. (eds.) ECCV 2012, Part II. LNCS, vol. 7573, pp. 872–885. Springer, Heidelberg (2012)
24. Vieira, A.W., Nascimento, E.R., Oliveira, G.L., Liu, Z., Campos, M.F.: STOP: space-time occupancy patterns for 3D action recognition from depth map sequences. In: Alvarez, L., Mejail, M., Gomez, L., Jacobo, J. (eds.) CIARP 2012. LNCS, vol. 7441, pp. 252–259. Springer, Heidelberg (2012)
25. Wang, J., Liu, Z., Wu, Y., Yuan, J.: Mining actionlet ensemble for action recognition with depth cameras. In: CVPR, pp. 1290–1297 (2012)
26. Oreifej, O., Liu, Z.: HON4D: Histogram of oriented 4D normals for activity recognition from depth sequences. In: CVPR, pp. 716–723 (2013)
27. Rahmani, H., Mahmood, A., Huynh, D.Q., Mian, A.: Real time action recognition using histograms of depth gradients and random decision forests. In: WACV, pp. 626–633 (2014)
28. Chang, C.C., Lin, C.J.: LIBSVM: a library for support vector machines. ACM Trans. Intell. Syst. Technol. **2**(3), 1–27 (2011)

An RGB-D Camera Based Walking Pattern Detection Method for Smart Rollators

He Zhang and Cang Ye[✉]

Department of Systems Engineering, University of Arkansas at Little Rock,
2801 S. University Ave, Little Rock, AR 72204, USA
{hxzhang1,cxye}@ualr.edu

Abstract. This paper presents a walking pattern detection method for a smart rollator. The method detects the rollator user's lower extremities from the depth data of an RGB-D camera. It then segments the 3D point data of the lower extremities into the leg and foot data points, from which a skeletal system with 6 skeletal points and 4 rods is extracted and used to represent a walking gait. A gait feature, comprising the parameters of the gait shape and gait motion, is then constructed to describe a walking state. K-means clustering is employed to cluster all gait features obtained from a number of walking videos into 6 key gait features. Using these key gait features, a walking video sequence is modeled as a Markov chain. The stationary distribution of the Markov chain represents the walking pattern. Five Support Vector Machines (SVMs) are trained for walking pattern detection. Each SVM detects one of the five walking patterns. Experimental results demonstrate that the proposed method has a better performance in detecting walking patterns than three existing methods.

1 Introduction

Walking therapy is a particular physical therapy (or physiotherapy) that assists a motor-impaired patient to recover his/her walking ability. This treatment requires interaction and cooperation between a therapist and a patient. The patient is offered instructions to perform the physiotherapy exercises in a monitored manner that provides feedbacks to the therapists for evaluating the effectiveness of the exercises and adjusting the therapy parameters. However, due to the lengthy recovery process and the need of travel, one-to-one in-clinic treatment is prohibitively expensive. As a result, the patient is taught in clinic about the therapy exercises and performs the exercises at home. While it is cost-effective and save the patient time in travel, at-home physiotherapy does not provide the therapist with feedback in a timely fashion for evaluation and adjustment of the exercises. Often, a patient uses a rolling walker (aka rollator) [2, 6–8] as a walking aid and to support the therapy exercises during the recovery process. Our work is therefore to develop a computer vision method for automatic detection of walking patterns and devise a smart rollator system that is able to provide persistent monitor on the user's walking patterns for at-home walking therapy. The system can be used to score a physiotherapy exercise by monitoring the change in the user's walking patterns during the course of recovery.

© Springer International Publishing Switzerland 2015
G. Bebis et al. (Eds.): ISVC 2015, Part I, LNCS 9474, pp. 624–633, 2015.
DOI: 10.1007/978-3-319-27857-5_56

We define a walking pattern as a sequence of walking postures and speeds. In the course of a walking therapy, a patient undergoes changes in both walking postures and speeds. If the prescribed exercises are effective, his/her walking gait will change from abnormal to normal and the speed from slow to normal. Otherwise, there will be no noticeable change of walking pattern. In other words, detection of walking pattern change is critical for the therapist to judge the effectiveness of the at-home walking therapy. Walking pattern recognition by computer vision involves lower limb detection, gait feature (gait shape and gait motion parameters) extraction, walking pattern representation (as a sequence of gait features), machine learning for pattern detection.

In the literature, force and moment sensors [6] have been used on a smart rollator to estimate the step count, pace and stride time of the rollator user. However, these sensors cannot measure the walking posture. In [7], a video camera is mounted on the front bar of a rollator to monitor the lower limb behavior of the user for balance control. The system measures the displacements and velocities of the feet and it requires the user to wear markers on the shoes for foot detection. Recently, RGB-D camera has been employed to measure a person's walking postures and speeds [1, 2]. An RGB-D camera provides reliable depth data for lower limb detection. Gritti, *et al.* [1] propose a histogram based lower limb detection algorithm that extracts a person's feet and legs from an RGB-D camera's depth data and tracks them over time. However, it does not measure the walking postures. Joly, *et al.* [2] propose a model-fitting method to detect bare legs and bare feet from the depth data of a Kinect sensor. The method models a bare leg as a cylinder and a bare foot as a plane and fit the parametric models to the Kinect data to detect the leg and foot. Although a skeleton representation of the lower limb can be created from the detected leg and foot, the work in [2] mainly focuses on determining foot orientation and ankle angle. However, the cylinder model fitting approach cannot be used in our case where the human legs are covered by the deformable pant. It is not possible to use any parametric model to describe the motion-induced deformation of the pant. In this paper, we propose a new method to determine the leg skeleton by least square plane fitting. Based on the skeletons of the lower limbs, we introduce a new gait feature to describe the gait shape and motion for walking pattern recognition. The gait feature resembles the action feature, consisting of shape and motion parameters of a full skeletal system of human body, that has been used in [4] for human action recognition.

Existing methods [4, 5] for human action recognition may be applied to walking pattern recognition. In [4], an image-to-image difference of the action features between a test video and a class—a video representing a particular class of action—is computed and the sum of the differences for all image frames is used for action detection. The sum does not take into account the transitions between action states. The method in [5] allows comparison of one image frame against multiple image frames. However, transitions between action states are not considered. In this paper, we propose a Markov chain model to capture the characteristics of a gait feature sequence. A gait feature consists of the shape and motion parameters of the walking gait. The transition matrix of the Markov chain model records both the state and state transition information of a walking sequence. If a walking sequence has a fixed pattern, the transition matrix should converge and thus the stationary distribution represents the walking pattern. To automatically identify the walking pattern, we used Support Vector Machine (SVM) [3] because it has been proved efficient in recognizing human actions with discriminative feature descriptors.

This paper is organized as follows. Section 2 briefly describes the smart rollator system. Section 3 introduces the data processing pipeline of the proposed method. Section 4 presents leg and foot extraction and gait feature construction. Sections 5 and 6 details the Markov chain modelling of a gait feature sequence and the SVMs for walking pattern recognition, respectively. Section 7 presents the experimental results. The paper is concluded in Sect. 8.

2   Smart Rollator Setup

As depicted in Fig. 1(a), an RGB-D camera (ASUS Xtion PRO LIVE) is installed on a rollator, facing towards the lower-extremity of the user with a 20° tilt-down angle. This view angle ensures that the feet and the lower parts of the legs are inside the camera's field of view when the user is walking. The RGB-D camera provides a color video and a depth video with 640×480 pixels at 30 fps. Given a depth image, the 3D point cloud of the user' lower body can be obtained in real time. Figure 1(b) shows the point cloud of the user' legs and feet from a depth image frame. To analyze the lower extremity motion, an absolute coordinate system $X_f Y_f Z_f$ is fixed on the floor plane. Point cloud data captured from the camera are then transformed from the camera coordinate system $X_c Y_c Z_c$ into. $X_f Y_f Z_f$ is initialized by using the first frame of point cloud data. First, the floor plane is extracted from the point cloud using a RANSAC plane segmentation method [10]. Second, randomly select a point from the extracted floor plane and fix the origin O_f on this point. Third, align Y_f with the normal of the plane and X_f with X_c and Z_f is determined to form a right-handed coordinate system. The transformation matrix T that transforms a point from $X_c Y_c Z_c$ into $X_f Y_f Z_f$ is computed by letting $Y_f = (0,1,0)$.

(a) Rollator with Xtion camera   (b) Floor plane (purple) and point cloud of lower-extremity

Fig. 1. Smart rollator and its coordinate systems

In this paper, a simulated walking therapy case is used for the development and validation of the proposed method. The rollator user imitates the walking patterns of a patient with knee injury during the course of recovery. Both normal walk and abnormal walk will be performed and video data (stream of RGB and depth data) will be captured by the RGB-D camera for training and testing the walking pattern recognition method. A Normal Walk (NW) is one with a sequence of normal walking gait at a regular speed.

An abnormal walk is one with a sequence of normal/abnormal walking gaits at a slow speed or zero speed (i.e., halt). The abnormal gait is a lame walking gait. Abnormal walks include Slow Walk (SW), Slow Walk with Halt (SWH), Slow Lame Walk (SLW), and Slow Lame Walk with Halt (SLWH) in this work.

3   Data Processing Pipeline of the Smart Rollator System

As depicted in Fig. 2, the data processing pipeline of the smart rollator system consists of three main modules: Gait Feature Extraction (GFE), Markov Chain Modeling (MCM), and SVM-based Walking Pattern Recognition (SWPR). The GFE extracts the point cloud of the user's lower-extremity from a depth data frame. After locating the foot and leg, the GFE models the lower-extremity as a skeletal system consisting of skeletons and skeletal points. It then computes the position and motion parameters of the left and right skeletal systems' skeletons and skeletal points. Using these parameters, the GFE constructs a gait feature for the frame. The MCM first clusters the gait features extracted from a number of walking videos into six classes, each of which is a key gait feature. It then describes each frame by one of the six key gait features. This turns the walking video into a Markov chain whose stationary distribution represents a certain walking pattern. The SWPR maps the stationary distribution to a walking pattern by a trained SVM. The technical details of the three modules will be given in the following sections.

Fig. 2. Data processing pipeline

4   Gait Feature Extraction

In this paper, a gait feature contains parameters describing gait shape and gait motion. The gait shape parameters encode the current information of the user's lower extremity posture while the gait motion parameters describe how one gait shape evolves into another. The gait shape parameters are the positions of skeletal points of each lower extremity and the gait motion parameters are the velocities of these skeletal points. Collectively, these parameters describe a walking state of the rollator user. The process of gait feature extraction is divided into four steps: lower limb detection, leg and foot segmentation, leg and foot skeletons extraction, and gait feature construction.

4.1   Lower Limb Detection

Considering the case of walking on a flat ground with the rollator, we use the RANSAC plane segmentation method [10] to extract the floor plane from the first frame of the camera's point cloud data. We then initialize the rollator's coordinate systems as mentioned in Sect. 2. Data points (above the floor) within the view volume clipped by the rollator chassis and the camera's depth limit (0.8 m $\sim$ 3.5 m) are identified, out of which we select the clusters within the first 70 cm above the floor plane as lower limb cluster P containing feet and legs.

4.2   Leg and Foot Segmentation

A 2-stage processing is employed to find the foot and leg segments from each lower limb cluster P. In the first stage, the minimum y coordinate y_{min} of the lower limb cluster's data points is obtained and the foot and leg segments P'_f and P'_l, are roughly located based on the data points' y-coordinates. Assuming the y-span between the toe and ankle of a human's feet is smaller than 0.2 meter, we locate points within [y_{min}, y_{min} + 0.2 m] as the foot segment P'_f and the rest the leg segment P'_l. In the second stage, the normal vector of each point in P'_f is first computed. Then, taking P'_f as input, the normal vector based region growing segmentation algorithm [11] is implemented to extract the foot segment P_f. The leg segment is determined by $P_l = (P - P_f) \cap P'_l$.

Figure 3(b) depicts the segmentation result of a depth data frame. The points of the leg segment are shown in green while the points of the foot segment are shown in red.

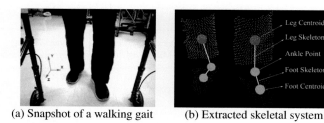

Leg Centroid
Leg Skeleton
Ankle Point
Foot Skeleton
Foot Centroid

(a) Snapshot of a walking gait   (b) Extracted skeletal system

Fig. 3.   Skeletal systems extracted from the point cloud data of a normal walking gait.

4.3   Leg and Foot Skeleton Extraction

Three skeletal points are computed and used to form the leg and foot skeletons. The first two skeletal points are the centroids of the leg and foot segments. And the third point—the ankle point—is determined as the point where the leg intersects the foot-plane. Similar to [2], a Least-Square Plane (LSP) to the data points of the foot segment P_f is first computed and its normal vector $\bar{n}$ is used to describe the foot orientation. The LSP is called a foot-plane.

The skeleton of the leg can be extracted from the data points p_j (for $j = 1, \ldots, N$) of the leg segment P_l. In an ideal case where the pant leg is pleat-free and the data points are noise-free, the orientation of the leg skeleton, denoted μ, is orthogonal with the surface normal of p_j, denoted $\mathbf{n_j} = (n_{jx}, n_{jx}, n_{jx})$. If we treat $\mathbf{n_j}$ as a data point, μ is the normal of the LSP to point set $\mathbf{n_j}$ for $j = 1, \ldots, N$. The least-square problem is equivalent to find the normal of the LSP to a point set $q_j = \left(k_j n_{jx}, k_j n_{jy}, k_j n_{jz}\right)$ for $j = 1, \ldots, N$, where k_j is a randomly generated non-zero value for $\mathbf{n_j}$. By applying k_j, we spread the data points in a larger area without changing each point's vector direction. This treatment avoids the case where all data points locate in a narrow area, making the LSP sensitive to noise. Using all data points to compute μ minimizes the effects of noise and pant-pleats. The LSP problem is solved by the SVD method. The centroid of P_l and μ are then used to describe the leg skeleton.

The ankle point is determined as the intersection of the leg skeleton and the foot-plane. A lower limb skeletal system, consisting of 2 skeletons and 3 skeletal points, is then formed as shown in Fig. 3(b).

4.4 Gait Feature Representation

The extraction of the two skeletal systems results in 6 skeletal points $sp_i (i = 1, \ldots 6)$. The skeletal points' positions determine the gait shape. Using the 6 skeletal points' centroids $sp_c = \{x_c, y_c, z_c\}$ as the reference point, the skeletal points' coordinates are re-computed by $sp_i' = sp_i - sp_c$, from which a bounding box $[x_{min}, x_{max}, y_{min}, y_{max}, z_{min}, z_{max}]$ is created. Finally, a 18-dimensional vector f^s representing the gait shape is computed from $sp_i' = \{x_i', y_i', z_i'\}$ by:

$$f_j^s = \begin{cases} \left(x_i' - x_{min}\right)/\left(x_{max} - x_{min}\right) & for\ j = 1 \ldots 6, i = 1 \ldots 6 \\ \left(y_i' - y_{min}\right)/\left(y_{max} - y_{min}\right) & for\ j = 7 \ldots 12, i = 1 \ldots 6 \\ \left(z_i' - z_{min}\right)/\left(z_{max} - z_{min}\right) & for\ j = 13 \ldots 18, i = 1 \ldots 6 \end{cases} \qquad (1)$$

The velocity of each skeletal point is computed from its positional change between two consecutive frames. Since the ankle point is indirectly computed from the point cloud (as the intersection between the leg skeleton and foot-plane), its position may incur a larger error than the other two skeletal points. This means that its velocity computed from two consecutive frames is not reliable. However, we observe that the angle between the foot and leg skeletons is relatively more accurate, indicating the angular velocity ω_p may be used as a reliable motion parameter. Therefore, we use the velocities of the foot and the leg centroids, denoted by $v_f = \left(v_x, v_y, v_z\right)$ and $v_l = \left(v_x, v_y, v_z\right)$, and the angular velocity ω_p to form a 14-dimensional vector f^m:

$$f^m = \left[v_f^1, v_l^1, \omega_p^1, v_f^2, v_l^2, \omega_p^2\right] \qquad (2)$$

The superscripts 1 and 2 denote the left and the right legs, respectively. By concatenating (1) and (2), a 32-dimensional feature is constructed as $[f^s, f^m]$. In order to rule out correlation between the elements of the feature vector, the principal component analysis is employed to reduce the feature dimension from 32 to 18. We observe that the first 18 eigenvalues weight over 95 %.

5 Markov Chain Modeling

The gait feature extraction process can produce a large number of raw gait features from a walking video. They may be classified into to a few representative gait features, called Key Gait Features (KGFs), to simplify the representation. After the KGFs are determined, each gait feature of a walking video is represented by one of the KGFs and the sequence of KGFs forms a Markov chain whose stationary distribution is used to detect the rollator user's walking pattern.

In this paper, we use K-means classification algorithm to partition the raw gait features into k clusters and the centroid of each cluster is computed as a KGF [3, 5]. The value of k is determined by the Bayesian Information Criterion (BIC) [9]. The BIC is a criterion for selecting a model out of a finite set of models. In our case, we choose k with the lowest BIC. In addition, if the number of gait features belonging to a cluster is smaller than a threshold $\tau = 50$, this cluster is treated as an outlier and thus deleted. Using this scheme, we extract 6 KGFs from a number of walking videos.

A gait feature extracted from a walking video is then represented by a specific KGF if the norm of the difference between the gait feature and the KGF is below a threshold. By treating a KGF as a state, we denote the gait sequence of a walking video by a Markov chain S. The transition matrix P of the Markov chain is of 6×6 dimensions. Each entry of the matrix p_{ij} represents the probability, with which state i evolves into state j. p_{ij} can be obtained from the state sequence. Therefore, P can be computed from S. The following is a Markov chain sample obtained from a portion of a walking video:

$$S : 111323233\underline{34}444446565232\underline{33}44446656232\underline{33}\underline{344}$$

S describes how long a gait is held and what gait it transforms into. For this sequence, p_{34} can be computed by $p_{34} = n_{34}/n_3 = 0.27$, where $n_{34} = 3$ is the number of times state 3 transforms into state 4, and $n_3 = 11$ is the number of times state 3 appears in the sequence. The other entries can be computed in a similar way to obtain P.

If a walking video has a fixed pattern and a sufficiently large number of frames, the transition matrix P of the Markov chain should converge. In this case, the stationary distribution π of the Markov chain holds the inherent property of the walking pattern. π, a row vector whose entries are non-negative and sum to 1, is defined by:

$$\pi P = \pi \tag{3}$$

It can be seen that π is a left eigenvector of P with an eigenvalue of 1. Therefore, it can be computed from P. In this work, it represents the pattern of a walking video.

6 SVM for Walking Pattern Recognition

As indicated earlier, there are five types of walking patterns to be detected. Therefore, a multi-class classifier is required for pattern recognition. In this paper, we use the one-vs-all strategy to train five SVMs to detect the walking patterns. One SVM will be trained to recognize a particular type of walking patterns by using the relevant training data (π_i, y_i); $i = 1, \ldots, N$, where N is the number of walking videos used for training the SVM while y_i is the SVM output for π_i. y_i is manually labeled. Taking the training of the 4th SVM (for SLW detection) as an example, feature vector π_i is computed for the ith video. If the walking pattern of this video is SLW, $y_i = +1$; Otherwise $y_i = -1$. The kernel function of the SVM is Gaussian kernel ($\sigma = 0.03$) and the regularization parameter is 0.01.

7   Walking Pattern Detection

7.1   Data Collection

Nine human subjects participated in data collection. They were instructed to perform the five types of walks. For each walk, the image and depth data streams were recorded from the Xtion. The video for each walk is 12–17 s long, containing 360–500 data frames. 7–9 video clips were recorded for the experiments performed by each human subject, resulting in 72 video clips. These video clips form a dataset with 9 NWs, 9 SWs, 9 SWHs, 27 SLWs and 18 SLWHs.

7.2   Experimental Results

We classify the walking pattern detection results into three categories: correct, acceptable and incorrect. The rationale for the category of acceptable detection is that the Markov chain model captures the probabilistic property of the gait states and state transitions for a walking video that may mix two or more walking patterns. For instance, it may have both SW and SWH or both SW and NW. Given that human subjects may interpret a slow walk/normal walk differently and measurement noise can make it difficult to tell the difference between a slow walk and a halt, there is no clear cut between two close walking patterns. Therefore, a "misdetection" of a walking pattern as its closest pattern is regarded as acceptable. When we tabulate our experimental results, we use a green, blue and red cells to indicate a correct, acceptable and incorrect detection results, respectively. We compute the rate of incorrect detection by $e = n_r/N$ and the rate of correct detection by $c = n_{bg}/N$. Here, n_r is the sum of the numbers in the red cells; n_{bg} is the sum of the numbers in the green and blue cells; and $N = 72$ is the number of the test videos.

In our experiments, leave-one-out cross validation is adopted to compare the performance of the following four methods: the Bag-of-Words (BoW) based one-vs-all SVM [3], Naive-Bayes-Nearest-Neighbor (NBNN) classifier [4], key pose based Dynamic Time Warping (DTW) [5], and Markov chain based one-vs-all SVM (the proposed) methods. The results are tabulated in Tables 1, 2, 3, and 4. In each table, the 1st row shows the five types of walking videos with the number of video clips for each type in the parenthesis. The values in each column are the number of video clips detected as NW, SW, SWH, SWL, and SWLH, respectively. Taking the numbers in the first column of Table 1 as an example, there are 9 NW video clips, out of which the method detects 2 as NW, 5 as SW and 2 as SWH. For each table, the incorrect and correct detection rates, e and c, are also computed.

Table 1. Detection result of the BoW based one-vs-all SVM: $e = 0.125,\ c = 0.875$

True \ Result	NW (9)	SW (9)	SWH (9)	SWL (27)	SWLH (18)
NW	2	2	1	2	1
SW	5	4	5	3	0
SWH	2	3	3	3	0
SWL	0	0	0	13	5
SWLH	0	0	0	6	12

Table 2. Detection result of the NBNN classifier: $e = 0.222,\ c = 0.778$

True \ Result	NW (9)	SW (9)	SWH (9)	SWL (27)	SWLH (18)
NW	2	0	1	1	0
SW	1	0	1	0	0
SWH	0	1	0	0	0
SWL	4	5	7	18	12
SWLH	2	3	0	8	6

Table 3. Detection result of key pose based DTW classifier: $e = 0.250,\ c = 0.750$

True \ Result	NW (9)	SW (9)	SWH (9)	SWL (27)	SWLH (18)
NW	2	1	0	2	3
SW	1	2	3	2	0
SWH	2	3	3	2	1
SWL	3	2	2	14	9
SWLH	1	1	1	7	5

Table 4. Detection result of the Markov chain based one-vs-all SVM: $e = 0.097,\ c = 0.903$

True \ Result	NW (9)	SW (9)	SWH (9)	SWL (27)	SWLH (18)
NW	3	3	0	1	0
SW	3	4	3	2	0
SWH	2	3	6	1	1
SWL	1	0	0	14	6
SWLH	0	0	0	9	11

From the results in these tables, it is apparent that the proposed method has the best overall performance ($e = 0.097,\ c = 0.903$), followed by the BoW based one-vs-all SVM, the NBNN classifier and the DTW classifier. The NBNN and DTW classifiers have a similar overall performance.

8   Conclusion

This paper presents an RGB-D camera based walking pattern detection method for a smart rollator system. The method extracts the user's lower limbs from the camera's depth data to obtain the gait information represented by a skeletal system with 6 skeletal points and 4 skeletons. By combining the parameters of the gait shape and gait motion, a gait feature is constructed to describe a walking state. K-means is employed to cluster all gait features extracted from a number of walking videos into six key gait features. Using the key gait features, a walking video is modeled as a Markov chain, of which the stationary distribution represents the walking pattern. Five SVMs are trained and used to detect five walking patterns. Experimental results validate that the proposed method outperforms three existing methods in detecting walking patterns.

In term of future research, we will use video data collected from real patients' to test the method and compare its performance with that of the other methods.

Acknowledgment. This work was supported in part by the NICHD, NINR and NIBIB of the National Institutes of Health under Award R01NR016151, and in part by NASA under Award NNX13AD32A. The content is solely the responsibility of the authors and does not necessarily represent the official views of the funding agencies.

References

1. Gritti, A., Tarabini, O., Guzzi, J.: Kinect-based people detection and tracking from small-footprint ground robots. In: IEEE/RSJ International Conference on Intelligent Robots and Systems, Chicago, IL (2014)
2. Joly, C., Dune, C.: Feet and legs tracking using a smart rollator equipped with a kinect. In: IEEE/RSJ International Conference on Intelligent Robots and Systems. Tokyo, Japan. IEEE (2013)
3. Laptev, I., Caputo, B., Schüldt, C., Lindeberg, T.: Local velocity-adapted motion events for spatio-temporal recognition. Comput. Vis. Image Underst. **108**, 207–229 (2007)
4. Yang, X., Tian, Y.: Eigenjoints-based action recognition using naive-bayes-nearest-neighbor. In: IEEE Computer Vision and Pattern Recognition Workshops, Providence, RI (2012)
5. Chaaraoui, A.A., Padilla-López, J.R., Climent-Pére, P., Flórez-Revuelta, F.: Evolutionary joint selection to improve human action recognition with RGB-D devices. Expert Syst. Appl. **41**, 786–794 (2014)
6. Alwan, M., Ledoux, A., Wasson, G., Sheth, P., Huang, C.: Basic walker-assisted gait characteristics derived from forces and moments exerted on the walker's handles: results on normal subjects. Med. Eng. Phys. **29**, 380–389 (2007)
7. Tung, J.: Development and evaluation of the iWalker: an instrumented rolling walker to assess balance and mobility in everyday activities. Ph.D. dissertation, University of Toronto (2010)
8. Dune, C., Gorce, P., Merlet, J.P.: Can smart rollators be used for gait monitoring and fall prevention? In: IEEE/RSJ International Conference on Intelligent Robots and Systems (2012)
9. Pelleg, D., Moore, A.W.: X-means: extending k-means with efficient estimation of the number of clusters. In: International Conference on Machine Learning (ICML). IEEE (2000)
10. Qian, X., Ye, C.: NCC-RANSAC: a fast plane extraction method for 3D range data segmentation. IEEE Trans. Cybern. **44**, 2771–2783 (2014)
11. http://pointclouds.org/documentation/tutorials/region_growing_segmentation.php

Evaluation of Vision-Based Human Activity Recognition in Dense Trajectory Framework

Hirokatsu Kataoka[1]([✉]), Yoshimitsu Aoki[2], Kenji Iwata[1], and Yutaka Satoh[1]

[1] National Institute of Advanced Industrial Science and Technology (AIST),
Tskuba, Japan
[2] Keio University, Tokyo, Japan
hirokatsu.kataoka@aist.go.jp

Abstract. Activity recognition has been an active research topic in computer vision. Recently, the most successful approaches use dense trajectories that extract a large number of trajectories and features on the trajectories into a codeword. In this paper, we evaluate various features in the framework of dense trajectories on several types of datasets. We implement 13 features in total by including five different types of descriptor, namely motion-, shape-, texture- trajectory- and co-occurrence-based feature descriptors. The experimental results show a relationship between feature descriptors and performance rate at each dataset. Different scenes of traffic, surgery, daily living and sports are used to analyze the feature characteristics. Moreover, we test how much the performance rate of concatenated vectors depends on the type, top-ranked in experiment and all 13 feature descriptors on fine-grained datasets. Feature evaluation is beneficial not only in the activity recognition problem, but also in other domains in spatio-temporal recognition.

1 Introduction

Recently, activity recognition has become one of the most active topics in the field of computer vision. Since space-time interest points (STIP) [1] were proposed, many researchers have studied activity recognition. Several survey papers have been published in activity recognition such as Moeslund et al. [2] and Aggarwal et al. [3]. Moeslund et al. [2] introduced a large number of approaches, not only in activity recognition, but also in human detection and tracking in their paper, and Aggarwal et al. [3] listed several recognition styles such as single person's activity and interaction recognition.

In their study of activity representation, Wang et al. [4] evaluated several space-time features for activity recognition, e.g., STIP [5], cuboid [6], Hessian [7] and dense [4] features with more detailed experimental settings. This evaluation has led to the idea of dense trajectories (DT) [8], which outperform other space-time features. In follow-up work with improved dense trajectories (iDT) [9], they improved their idea by implementing estimating camera motion with speeded-up robust features (SURF) [10] and a homography matrix, human rectangles and Fisher vector [11]. The improvements induced outstanding performance rates

© Springer International Publishing Switzerland 2015
G. Bebis et al. (Eds.): ISVC 2015, Part I, LNCS 9474, pp. 634–646, 2015.
DOI: 10.1007/978-3-319-27857-5_57

such as UCF50 [12] (91.2 %), and Hollywood2 [13] (64.3 %). The current state-of-the-art approach on the side of accuracy is the combination of iDT and per frame deep net features (6,7,8-layers) [14]. According to the THUMOS challenge, which consists of activity classification in a large-scale database [14], the iDT should be used to more completely understand all human activity, and not only deep net features.

Benenson *et al.* [15] cited and implemented over 40 approaches including various features and classifiers so as to detect a pedestrian in traffic scenes. The results of three familiar frameworks (random forests [16], deformable part model (DPM) [17] and deep learning [18]) are close if there are enough fine-tuned parameters. Thus, the comparison of various approaches will be a significant test to determine how much to change and how to apply the feature descriptors. In activity recognition, feature evaluation is important to gain knowledge of a more practical use of a space-time feature descriptor for activity recognition.

In this paper, we execute efficient evaluations with various dense trajectory-based feature descriptors on multiple types of datasets including traffic (NTSEL–self-collected), surgery (INRIA surgery [19]), daily living (MSR daily activity 3D [20]) and sports (UCF50 [12]) scenes. Moreover, the 13 features are assigned and divided into five feature properties: (i) trajectory (ii) shape (iii) motion (iv) texture and (v) co-occurrence. The performance rate of activity classification depends on the computational environment, i.e., activity codewords, trajectory patterns, classifier settings and cross-validation task. We furthermore evaluate various features in a fair experimental setting.

The rest of the paper is organized as follows. In the next section we describe the dense feature and 13 feature descriptors used in this paper. In Sect. 3, we show the effectiveness of the 13 feature descriptors and their concatenated vectors in our experimental results by means of four datasets. Finally, in the last section we conclude the paper.

2   Feature Evaluation Strategy

Fig. 1 shows the framework of the 13 feature descriptors in the dense trajectories framework. We applied 13 features – trajectory feature (traj.) [8], histograms of oriented gradients (HOG) [21], scale invariant feature transform (SIFT) [22], histograms of optical flow (HOF) [23], motion boundary histogram (MBHx &MBHy) [24], motion interchange patterns (MIP) [25], higher-order local auto correlation (HLAC) [26], local binary patterns (LBP) [27], improved LBP (iLBP) [28], local trinary patterns (LTP) [29], Co-occurrence HOG (CoHOG) [30], and Extended CoHOG (ECoHOG) [31] in this evaluation. We categorized the 13 features into five topics, namely: (i) trajectory – traj. (ii) shape – HOG and SIFT (iii) motion – HOF, MBH, and MIP (iv) texture – HLAC, LBP, iLBP and LTP, and (v) co-occurrence – CoHOG and ECoHOG. Moreover, the dense trajectories (DT) [8] + bag-of-words (BoW) model (not improved DT + Fisher vector [9]) was used in this evaluation because we evaluated the performance ability of the feature descriptors themselves. We used a support vector machine (SVM) as a multi-class classifier following [8].

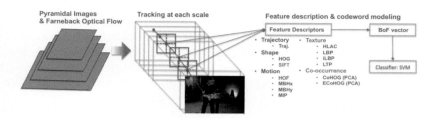

Fig. 1. The framework of the 13 feature descriptors in the dense feature framework.

2.1 Dense Trajectories

We used Wang's dense trajectories [8] (DT) to create bag-of-words (BoW) vectors [32] for activity recognition. The idea of DT includes dense sampling and space-time feature extraction at the sampled points. Feature points at each grid cell are computed with Farneback optical flow. To take care of scale changes, the DT extracts dense flows in multiple image scales, where the image size increases by a scale factor $1/\sqrt{2}$. A large number of DT flows among the multiple scales are integrated into a feature vector based on BoW. This setup allows us to obtain a detailed motion at the specified patch. The length of the trajectory was set as 15 frames. Therefore, we recorded 0.5 s activities from a 30 fps video. Moreover, we set all BoW vectors to 4000 dimensions at each feature descriptor following [8].

2.2 Thirteen Feature Descriptors

Trajectory Feature [8]. In activity analysis, the trajectory feature (traj.) [8] was extracted at each image patch. The size of the patch was 32×32 pixels, which is divided into 2×2 blocks. Here, the trajectory feature (T) is calculated as below:

$$T = \frac{(\delta P_t, ..., \delta P_{t+L-1})}{\Sigma_{j=t}^{t+L-1} ||P_j||} \tag{1}$$

$$\delta P_t = (\delta P_{t+1} - \delta P_t) = (x_{t+1} - x_t, y_{t+1} - y_t) \tag{2}$$

where L is the trajectory length. The feature represents a shape of connected optical flow.

Histograms of Oriented Gradients (HOG) [21]. HOG describes a feature vector with accumulating edge-magnitude into a quantized edge direction histogram. The process of feature extraction consists of edge calculation and normalization. Edge magnitude is accumulated into the quantized histogram by edge direction with $m(u,v) = \sqrt{f_u^2 + f_v^2}$ and $\theta(u,v) = arctan(f_v/f_u)$, where the magnitude and direction are $m(u,v)$ and $\theta(u,v)$, $f(x,y)$ is the differences between two pixels in the x and y directions. Feature extraction is executed on overlapping blocks, and the feature vector is normalized every block with a norm.

Scale Invariant Feature Transform (SIFT) [22]. This approach has characteristics of scale- and rotation-invariant features. SIFT contains keypoint detection and feature description; however, we mainly apply feature description to evaluate as a descriptor. To describe a feature vector, SIFT takes care of the image rotation by deciding a maximum direction. SIFT extracts 8 orientations divided into 4×4 blocks, giving 128 dimensions from an image patch.

Histograms of Optical Flow (HOF) [23]. The captured optical flows are quantized into nine directions. Wang *et al.* implemented HOF with a 4-divided image patch in his paper [8], therefore, a 36-dimension feature is extracted in an image patch. The feature represents normalized optical flow on a human motion area.

Motion Boundary Histograms (MBH) [24]. The motion boundary calculates the difference between two temporal frames. Therefore, it is less susceptible to capturing background noise when the camera motion is stable. Usually MBH features include the x- and y-directions together. However, we separate MBH into each direction MBHx and MBHy to analyze the properties of the feature descriptor at each scene.

Motion Interchange Patterns (MIP) [25]. This feature basically extracts a feature vector with trinary encoded pattern changes from a noticed area. Three temporal frames are applied to construct a motion interchange pattern.

Higher-Order Local Auto Correlation (HLAC) [26]. HLAC describes a feature by counting 25 significant mask patterns. The patterns indicate the displacement of a human in an image patch. The 25 pattern count allows us to capture a high-level movement, and we capture the patterns from the edge and the binarized image.

Local Binary Patterns (LBP) [27]. The process of LBP is constructed using a binarization step and an encoding step. In the binarization step, we process each 3×3 pixel patch to compare two pixels at the center of patch. The values are binarized with magnitude correlation in the patch. We capture eight binarized values, then the values are translated into 0–255 as a feature (this is an encoding step).

Improved LBP (iLBP) [28]. The basic idea of binarization is close to the normal LBP. The iLBP compares the eight nearest pixels with the averaged value of the nine pixels in a 3×3 patch. The feature emphasizes an edge element compared with LBP.

Local Trinary Patterns (LTP) [29]. The improved feature descriptor with trinary patterns has the same description as LBP. The feature instinctively captures by preparing an additional neutral class from two binarized classes. Because of the third class, it has more powerful representation as a texture-based descriptor.

CoHOG [30]. Co-occurrence Histogram of Oriented Gradient (CoHOG) is able to describe more complex shapes by pairing the brightness gradient directions in HOG. The brightness gradient direction of the pair is calculated using the co-occurrence matrix. The co-occurrence matrix is calculated by counting the number of brightness gradient directions of the pair that are a target pixel, and the specific positional relationship from the target pixel in the block.

$$C_{x,y}(i,j) = \sum_{p=1}^{n} \sum_{q=1}^{m} \begin{cases} 1 \\ (if\ d(p,q) = i\ and \\ \quad d(p+x, q+y) = j) \\ \\ 0 \\ (otherwise) \end{cases} \tag{3}$$

where $C(i,j)$ is the co-occurrence histogram that accumulates pairs of the pixel of interest and an objective pixel. Coordinates (p,q) indicate the pixel of interest (center of window) and coordinates $(p+x, p+y)$ indicate the objective pixel. m and n are the width and height of the feature extraction window. $d(p,q)$ is a function that quantizes the edge direction as an integer from 0 to 7 at pixel (p,q).

ECoHOG] [31]. Extended Co-occurrence Histogram of Oriented Gradient (ECoHOG) enables a more efficient feature description by deleting the feature dimensions in CoHOG and extracting only the valid features. ECoHOG makes improvements over CoHOG via the accumulation of edge strength, the step acquisition of edge pairs and time series feature representation. We describe each of these processes below. CoHOG generates a histogram by counting the number of pairs of brightness gradients. However, ECoHOG describes not only the shape but also the intensity of light and shade and the condition of the change by accumulating edge intensity.

$$m_1(x_1, y_1) = \sqrt{f_{x1}(x_1, y_1)^2 + f_{y1}(x_1, y_1)^2} \tag{4}$$

$$m_2(x_2, y_2) = \sqrt{f_{x2}(x_2, y_2)^2 + f_{y2}(x_2, y_2)^2} \tag{5}$$

$$C_{x,y}(i,j) = \sum_{p=1}^{n} \sum_{q=1}^{m} \begin{cases} m_1(x_1, y_1) + m_2(x_2, y_2) \\ (if\ d(p,q) = i\ and \\ \quad d(p+x, q+y) = j) \\ \\ 0 \\ (otherwise) \end{cases} \tag{6}$$

$m_1(x,y)$ and $m_2(x,y)$ are the magnitudes of the pixel of interest (at the center of the window) and the magnitudes in the objective window, respectively.

3   Experiments

We carried out evaluations of feature descriptors in the framework of dense trajectories. Figure 2 shows the performance rates and Table 1 shows the top three features with dense trajectories on the four different types of datasets, namely in traffic (NTSEL traffic), surgery (INRIA surgery), daily living (MSR daily activity 3D) and sports (UCF50) scenes. Moreover, the classification with concatenated vectors is shown in Table 2. Here, Fig. 3 shows examples of the datasets used in the experiments.

3.1   Results

NTSEL Traffic Dataset. We collected 100 videos with four pedestrian activities in traffic scenes (see Fig. 3 top left). The activities include *walking, crossing, turning*, and *riding a bicycle*, where all of the activities indicate fine-grained pedestrian motion with three people. The dataset contains a cluttered background in small areas, making it difficult to capture optical flows. Presented activities are also fine grained as there are only small variations between *walking, crossing* and *turning* that is, they have a very few appearance and motion differences. We evaluated this dataset using 5-fold cross validation.

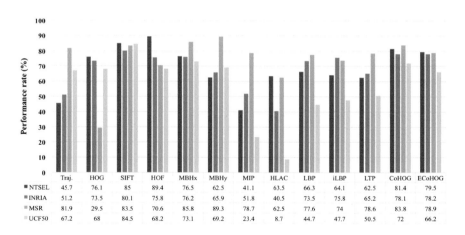

	Traj.	HOG	SIFT	HOF	MBHx	MBHy	MIP	HLAC	LBP	iLBP	LTP	CoHOG	ECoHOG
■ NTSEL	45.7	76.1	85	89.4	76.5	62.5	41.1	63.5	66.3	64.1	62.5	81.4	79.5
■ INRIA	51.2	73.5	80.1	75.8	76.2	65.9	51.8	40.5	73.5	75.8	65.2	78.1	78.2
■ MSR	81.9	29.5	83.5	70.6	85.8	89.3	78.7	62.5	77.6	74	78.6	83.8	78.9
▫ UCF50	67.2	68	84.5	68.2	73.1	69.2	23.4	8.7	44.7	47.7	50.5	72	66.2

Fig. 2. Overall rate of 13 features in the framework of dense trajectories [8] on the four datasets (NTSEL, INRIA surgery, MSR, and UCF50) (Color figure online).

According to Table 1, HOF (89.4 %), SIFT (85.0 %) and CoHOG (81.4 %) are the top three features in the NTSEL traffic dataset. The activities included in the NTSEL dataset are fine-grained walking activities. From these results, the optical flow vectorization (HOF) and detailed shape descriptors (SIFT, CoHOG) are effective for the classification of walking activities. HOF calculates 9 (dim.) × 4

Fig. 3. Four different datasets: NTSEL traffic dataset (top left), INRIA surgery dataset (top right), MSR daily 3D activity dataset (middle), and a part of the UCF50 dataset (bottom)

(blocks) × 3 (frames) optical flow-based features, therefore it can significantly represent activities, e.g., *walking* and *crossing*, that show subtle differences when walking at a vertical or horizontal angle to a camera. The quantized optical flow vectors are able to classify fine-grained walking activities on the dataset. SIFT (4 × 4 block division) and CoHOG (co-occurrence feature with extraction window) are the detailed shape descriptors and extract temporal differences on a trajectory.

INRIA Surgery Dataset [19]. This dataset includes four activities performed by 10 different people with occlusions; e.g., people are occluded by a table or a chair (see Fig. 3 top right). The activities include *cutting, hammering, repositioning,* and *sitting.* Each person performs the same activity twice, one for training and another for testing in this experiment.

The top three features are SIFT (80.1 %), ECoHOG (78.2 %) and CoHOG (78.2 %) from Table 1. Near to the NTSEL traffic dataset, the INRIA surgery dataset contains fine-grained activities in experimental surgery scenes. The camera angle is fixed and the arm swing activities are confusing, so the detailed shape features such as SIFT, ECoHOG and CoHOG are important for surgery activity classification. The difference between ECoHOG and CoHOG is edge-magnitude extraction and edge-pair counting. In this situation, ECoHOG fetches co-occurrence features with magnitude accumulation, which is slightly better than CoHOG.

MSR Daily Activity 3D Dataset [20]. The dataset is basically used as a depth-based activity recognition with a Kinect sensor. Depth and 3D-posture are given in this dataset, and at the same time we can access RGB videos (see Fig. 3, middle). In this experiment, *only* RGB information is assigned as input data to calculate feature vectors. There are 16 activities in the dataset: *drink, eat, read book, call cellphone, write on a paper, use laptop, use vacuum cleaner, cheer up, sit still, toss paper, play game, lie down on sofa, walk, play guitar, stand up, sit down.* The experiment is executed with leave-one-person-out cross-validation. Moreover, the trajectory length (L) is 50 because the normal setting of dense trajectories (15 frame accumulation) is extremely short so as to capture feature elements from an activity in a daily living activity.

MBHx (89.3 %), MBHy (85.8 %) and CoHOG (83.8 %) are listed as the top three feature descriptors. The original paper on dense trajectories [8] reported that the MBH effectively extracts motions for human activity. The motion boundary is a well-classified feature that is not dependent on horizontal and vertical direction. Large variations exist in understanding activities of daily living. For example, *sit still &lie down on sofa* are stable activities, and *cheer up &walk* are whole body motions. Although the dense trajectories accumulate over 50 frames to calculate a feature vector on the dataset, both MBH-based features perform better than the other features. The motion boundary feature enables subtle motions such as *sit still* and *lie down on sofa* to be distinguished, not only whole body activities. We believe that CoHOG contains a similar process to MBH in terms of feature description. CoHOG also extracts a detailed feature from a human area; however, CoHOG captures a co-occurrence shape with edge-pair counting.

UCF50 Dataset [12]. The UCF50 dataset comprises 1168 videos in 50 categories collected from YouTube (see Fig. 3 bottom for examples of the dataset). There are many categories in this dataset, for example, *Baseball Pitch*, *Breaststroke*, *Playing Guitar*, *Jumping Jack*, *Punch*, *Tennis Swing* and *Walking with a dog*. The dataset also includes several elements that computer vision has difficulty with, such as camera motion, complicated backgrounds, occlusions and personal variations. Performance rate is calculated with leave-one-group-out cross-validation in 25 groups for each activity.

Table 1 shows that SIFT (84.5 %), MBHx (73.1 %) and CoHOG (72.0 %) are better descriptors than the other 10 features on the UCF50 dataset. SIFT is an advanced feature descriptor itself in recognition tasks. Here, UCF50 is made large-scale enough by including an element of object recognition, and therefore SIFT outperforms other features on the dataset. The MBHx achieved the second highest score because this feature is likely to be robust to background noise. Only motion boundary is recorded as a feature. The co-occurrence feature stably accomplishes a good result on activity recognition by combining dense feature representations. The relationship between CoHOG and ECoHOG is competitive; however, CoHOG generally outputs a better score.

3.2 Concatenated Features for Activity Classification

The several-feature concatenation generally performs with a higher percentage on a vision-based classification. Here, we carried out the evaluations as to how to combine the 13 features based on the experiments in Sect. 3.1. Table 2 describes how much the score changes with feature combinations. We prepared a variety of feature combination approaches: (i) original dense trajectories [8] (as baseline), (ii) five feature categories, in belief trajectory, shape, motion, texture and co-occurrence, and (iii) top-ranked shown in Sect. 3.1 and all 13 feature combinations. We experimentally used two fine-grained datasets (NTSEL traffic and INRIA surgery dataset) to evaluate feature combination approaches.

Table 1. The top three performance rates with dense trajectories on the four datasets.

Dataset	Outline	Top three features
NTSEL	Fine-grained pedestrian activities.	HOF (89.4 %)
		SIFT (85.0 %)
		CoHOG (81.4 %)
INRIA	Fine-grained surgery activities.	SIFT (80.1 %)
		ECoHOG (78.2 %)
		CoHOG (78.1 %)
MSR	Daily living activities.	MBHy (89.3 %)
		MBHx (85.8 %)
		CoHOG (83.8 %)
UCF50	Large-scale sports activities.	SIFT (84.5 %)
		MBHx (73.1 %)
		CoHOG (72.0 %)

Baseline. The original dense trajectories achieved an outstanding score as they are well-organized structures in activity classification. The four features (traj., HOG, HOF, MBH) are combined with kernel SVM for fine-grained recognition. The performance rates were 89.6 % and 87.5 % on the NTSEL traffic and INRIA surgery datasets, respectively.

Several Types of Feature. We categorized 13 features into five topics based on feature characteristics. The statements of several features are itemized here: trajectory – traj. feature, shape – HOG, SIFT, motion – HOF, MBHx, MBHy and MIP, texture – HLAC, LBP, iLBP and LTP, co-occurrence [31] – CoHOG, ECoHOG. The recognition scores were 45.7 % (traj.), 85.2 % (shape), 85.9 % (motion), 60.7 % (texture), 85.0 % (co-occurrence) on the NTSEL, and 51.2 % (traj.), 85.7 % (shape), 82.5 % (motion), 76.4 % (texture) and 89.6 % (co-occurrence) on the INRIA surgery dataset. The motion feature gave the best rate (85.9 %), and the rates of the shape (85.2 %) and co-occurrence (85.0 %) features came next on the NTSEL traffic dataset. The motion feature included HOF and MBH, which are highly-accurate descriptors in the dataset. In the INRIA surgery dataset, the co-occurrence feature indicated a significant rate for classifying fine-grained activities with a precise description. The co-occurrence feature that contained CoHOG and ECoHOG showed stable vectorization in both datasets. The combination (co-occurrence feature &dense trajectories) of detailed description and dense representation is sophisticated in terms of fine-grained recognition.

Top-Ranked and All 13 Feature Combinations. The top 3, 5, 7 and all 13 features should be combined to measure the ability of concatenated vectors. The scores of the concatenated vectors are in a narrow margin; however, the top seven (90.9 %–HOF, SIFT, CoHOG, ECoHOG, MBHx, HOG and LBP) on

Table 2. The performance rate with concatenated features on the NTSEL traffic and INRIA surgery dataset.

Baseline	(%) on NTSEL	(%) on INRIA
Wang *et al.* [8]	89.6 %	87.5 %
Feature Type	(%) on NTSEL	(%) on INRIA
Trajectory	45.7 %	51.2 %
Shape	85.2 %	85.7 %
Motion	**85.9 %**	82.5 %
Texture	60.7 %	76.4 %
Co-occurrence [31]	85.0 %	**89.6 %**
Concatenated Vector	(%) on NTSEL	(%) on INRIA
Top 3 feature concatenation	89.6 %	90.4 %
Top 5 feature concatenation	90.2 %	**91.5 %**
Top 7 feature concatenation	**90.9 %**	90.8 %
All 13 features	90.7 %	90.7 %

the NTSEL traffic dataset and the top five (91.5 % – SIFT, ECoHOG, CoHOG, MBHx and HOF) on the INRIA surgery dataset showed the best rates. The results demonstrate that all-feature concatenation does not always give the best accuracy on an activity recognition dataset. At this point, the 13-feature concatenation achieved 90.7 % on both datasets. We believe that both datasets contain fine-grained activities, therefore the top-ranked features should be combined for high-accuracy fine-grained activity classification. Although the number of top-ranked features is an ad-hoc problem, the accuracy is easy to investigate in a classification experiment. In every case, we obtained a lower processing speed with a selected concatenated vector using dense trajectory-based feature extraction and activity classification. We therefore improve both performance rate and processing speed by using a concatenated vector for fine-grained recognition.

4   Conclusion

In this paper, we evaluated 13 features in the framework of dense trajectories for more effective activity recognition. We carried out all experiments at fair settings in terms of activity codewords, captured trajectories, classifier settings and cross-validation. The four scenes are included in the experiments using traffic, surgery, daily living and sports datasets. The results describe the best performance rate at each dataset – HOF (89.4 %) on the NTSEL traffic dataset, SIFT (80.1 %) on the INRIA surgery dataset, MBHy (89.3 %) on the MSR daily activity 3D dataset and SIFT (84.5 %) on the UCF50 dataset. Detailed analysis indicated that the co-occurrence feature containing CoHOG and ECoHOG would be a stable descriptor in activity recognition. The co-occurrence feature becomes significant representation by using dense trajectories. In feature concatenation, we found that the combination of highly-selected features tends to give a better rate than the integration of all 13 features. Therefore, we experimentally chose sophisticated features with top-ranked listed features. Particularly in a fine-grained activity dataset, an extra feature descriptor should NOT be included for a fine-grained classification by means of only effective features. Moreover, selected feature concatenation allows us to improve both processing speed and accuracy.

For more detailed analysis, we visualize how to execute classification with various features. The subspace representation is one of the most important tasks in pattern recognition. Moreover, we try to correct for various durations of human activity.

Acknowledgements. This work was partially supported by JSPS KAKENHI Grant Number 24300078.

References

1. Viola, P., Jones, M.: Rapid object detection using a boosted cascaded of simple features. In: IEEE Conference on Computer Vision and Pattern Recognition (CVPR) (2001)
2. Moeslund, T.B., Hilton, A., Kruger, V.: A survey of advances in vision-based human motion capture and analysis. Comput. Vis. Image Underst. (CVIU) **104**, 90–126 (2006)
3. Aggarwal, J.K., Ryoo, M.S.: Human activity analysis: a review. ACM Comput. Surv. **43**, 16 (2011)
4. Wang, H., Ullah, M.M., Klaser, A., Laptev, I., Schmid, C.: Evaluation of local spatio-temporal features for action recognition to cite this version. In: British Machine Vision Conference (BMVC) (2009)
5. Laptev, I.: On space-time interest points. Int. J. Comput. Vis. (IJCV) **64**, 107–123 (2005)
6. Dollar, P., Rabaud, V., Cottrell, G., Belongie, S.: Behavior recognition via sparse spatio-temporal features. In: Visual Surveillance and Performance Evaluation of Tracking and Surveillance (PETS), pp. 65–72 (2005)
7. Willems, G., Tuytelaars, T., Van Gool, L.: An efficient dense and scale-invariant spatio-temporal interest point detector. In: Forsyth, D., Torr, P., Zisserman, A. (eds.) ECCV 2008, Part II. LNCS, vol. 5303, pp. 650–663. Springer, Heidelberg (2008)
8. Wang, H., Klaser, A., Schmid, C.: Dense trajectories and motion boundary descriptors for action recognition. Int. J. Comput. Vis. (IJCV) **103**, 60–79 (2013)
9. Wang, H., Schmid, C.: Action recognition with improved trajectories. In: IEEE International Conference on Computer Vision (ICCV) (2013)
10. Bay, H., Tuytelaars, T., Van Gool, L.: SURF: speeded up robust features. In: Leonardis, A., Bischof, H., Pinz, A. (eds.) ECCV 2006, Part I. LNCS, vol. 3951, pp. 404–417. Springer, Heidelberg (2006)
11. Perronnin, F., Sánchez, J., Mensink, T.: Improving the Fisher kernel for large-scale image classification. In: Daniilidis, K., Maragos, P., Paragios, N. (eds.) ECCV 2010, Part IV. LNCS, vol. 6314, pp. 143–156. Springer, Heidelberg (2010)

12. Reddy, K.K., Shah, M.: Recognizing 50 human action categories of web videos. Mach. Vis. Appl. (MVA) **24**, 971–981 (2012)
13. Marszalek, M., Laptev, I., Schmid, C.: Actions in context. In: IEEE Conference on Computer Vision and Pattern Recognition (CVPR) (2009)
14. Jiang, Y.G., Liu, J., Roshan Zamir, A., Toderici, G., Laptev, I., Shah, M., Sukthankar, R.: THUMOS challenge: action recognition with a large number of classes (2014). http://crcv.ucf.edu/THUMOS14/
15. Benenson, R., Omran, M., Hosang, J., Schiele, B.: Ten years of pedestrian detection, what have we learned? In: Agapito, L., Bronstein, M.M., Rother, C. (eds.) ECCV 2014 Workshops. LNCS, vol. 8926, pp. 613–627. Springer, Heidelberg (2015)
16. Breiman, L.: Random forests. Mach. Learn. **45**, 5–32 (2001)
17. Felzenszwalb, P., Girshick, R., McAllester, D., Ramanan, D.: Object detection with discriminatively trained part based models. IEEE Trans. Pattern Anal. Mach. Intell. **32**, 1627–1645 (2010)
18. Sermanet, P., Kavukcuoglu, K., Chintala, S., LeCun, Y.: Pedestrian detection with unsupervised multi-stage feature learning. In: IEEE Conference on Computer Vision and Pattern Recognition (CVPR) (2013)
19. Huang, C.H., Boyer, E., Navab, N., Ilic, S.: Human shape and pose tracking using keyframes. In: IEEE Conference on Computer Vision and Pattern Recognition (CVPR) (2014)
20. Wang, J., Liu, Z., Wu, Y., Yuan, J.: Mining actionlet ensemble for action recognition with depth cameras. In: IEEE Conference on Computer Vision and Pattern Recognition (CVPR) (2012)
21. Dalal, N., Triggs, B.: Histograms of oriented gradients for human detection. In: IEEE Conference on Computer Vision and Pattern Recognition (CVPR) (2005)
22. Lowe, D.G.: Distinctive image features from scale-invariant keypoints. Int. J. Comput. Vis. **60**, 91–110 (2004)
23. Laptev, I., Marszalek, M., Schmid, C., Rozenfeld, B.: Learning realistic human actions from movies. In: IEEE Conference on Computer Vision and Pattern Recognition (CVPR) (2008)
24. Dalal, N., Triggs, B., Schmid, C.: Human detection using oriented histograms of flow and appearance. In: Leonardis, A., Bischof, H., Pinz, A. (eds.) ECCV 2006. LNCS, vol. 3952, pp. 428–441. Springer, Heidelberg (2006)
25. Kliper-Gross, O., Gurovich, Y., Hassner, T., Wolf, L.: Motion interchange patterns for action recognition in unconstrained videos. In: Fitzgibbon, A., Lazebnik, S., Perona, P., Sato, Y., Schmid, C. (eds.) ECCV 2012, Part VI. LNCS, vol. 7577, pp. 256–269. Springer, Heidelberg (2012)
26. Kobayashi, T., Otsu, N.: Image feature extraction using gradient local autocorrelations. In: Forsyth, D., Torr, P., Zisserman, A. (eds.) ECCV 2008, Part I. LNCS, vol. 5302, pp. 346–358. Springer, Heidelberg (2008)
27. Ojala, T., Pietikainen, M., Maenpaa, T.: Multiresolution grayscale and rotation invariant texture classification with local binary patterns. IEEE Trans. Pattern Anal. Mach. Intell. **24**, 971–987 (2002)
28. Mohamed, A.A., Yampolskily, R.V.: An improved lbp algorithm for avatar face recognition. In: International Symposium on Information, Communication and Automation Technologies (ICAT) (2011)
29. Yeffet, L., Wolf, L.: Local trinary patterns for human action recognition. In: International Conference on Computer Vision (ICCV) (2009)
30. Watanabe, T., Ito, S., Yokoi, K.: Co-occurrence histograms of oriented gradients for pedestrian detection. In: Wada, T., Huang, F., Lin, S. (eds.) PSIVT 2009. LNCS, vol. 5414, pp. 37–47. Springer, Heidelberg (2009)

31. Kataoka, H., Hashimoto, K., Iwata, K., Satoh, Y., Navab, N., Ilic, S., Aoki, Y.: Extended co-occurrence HOG with dense trajectories for fine-grained activity recognition. In: Cremers, D., Reid, I., Saito, H., Yang, M.-H. (eds.) ACCV 2014. LNCS, vol. 9007, pp. 336–349. Springer, Heidelberg (2015)
32. Csurka, G., Dance, C.R., Fan, L., Willamowski, J., Bray, C.: Visual categorization with bags of keypoints. In: European Conference on Computer Vision Workshop (ECCVW) (2004)

Analyzing Activities in Videos Using Latent Dirichlet Allocation and Granger Causality

Dalwinder Kular[✉] and Eraldo Ribeiro

Computer Vision Laboratory, Department of Computer Sciences,
Florida Institute of Technology, Melbourne, FL, USA
dkular2009@my.fit.edu

Abstract. We propose an unsupervised method for analyzing motion activities from videos. Our method combines Latent Dirichlet Allocation with Granger Causality to discover the main motions composing the activity as well as to detect how these motions relate to one another in time and space. We tested our method on synthetic and real-world datasets. Our method compares favorably with state-of-the-art methods.

1 Introduction

In this paper, we address the problem of automatically analyzing motion activities in videos, using as input just local motion measurements such as optical flow. The activities that we target in this paper are those consisting of sequences of basic motions or co-occurring basic motions. Examples of such activities include sports players interactions, car driving through traffic intersections, and guests walking in a hotel lobby (Fig. 1). Analyzing these activities is a simple task for our natural vision system. Indeed, as we watch an activity for some time, we begin to notice the emergence of basic patterns of motions, which have various degrees of spatial and temporal coherence. The development of algorithms that can analyze these types of activities is a main goal of computer vision.

Some recent methods for analyzing activities use statistical approaches such as probabilistic topic modeling. Li et al. [1] analyzed activities by using a two-layer probabilistic semantic analysis. The first layer discovers basic actions, and the second discovers activities. Wang et al. [2] grouped motion events into activities using hierarchical Bayesian models. These methods do not model temporal relationships among motions. Hospedales et al. [12] proposed a Markov Clustering Topic Model based on the Latent Dirichlet Allocation [3] for clustering spatial visual events into activities and identify the temporal dynamics of the visual events. Here, the use of a single Markov chain for learning occurrence cycles may result in incomplete activities.

In this paper, we propose an unsupervised method for analyzing motion activities. Our method combines Latent Dirichlet Allocation [3] with partial-Granger Causality [4] to discover the main motions composing the activity as well as detect how these motions relate to one another in time and space. For example, given a video of a traffic-intersection scene (Fig. 1), we want to determine the

© Springer International Publishing Switzerland 2015
G. Bebis et al. (Eds.): ISVC 2015, Part I, LNCS 9474, pp. 647–656, 2015.
DOI: 10.1007/978-3-319-27857-5_58

Fig. 1. Motion activities. Superimposed curves show examples of motion patterns.

following: the main traffic lanes through which vehicles drive, the sequences of movement that take place in these lanes as vehicles go from one place to another (e.g., moving forward, moving forward then turning right, turning about). In addition to finding lanes and major flow directions, we also want to learn how these movements are temporally synchronized (i.e., motions that flow sequentially and in co-occurrence). Finally, our method takes as input just local velocity measurements (i.e., optical flow) of the moving objects.

Our method is based on the concept of motion patterns [5], which are spatially and temporally coherent clusters of moving pixels. These clusters form a spatio-temporal region describing the motion of objects from one place to another in the scene (e.g., curves shown in Fig. 1). In our method, we analyze an activity using three main steps. First, we detect the actions that make up the activity's motion patterns. Here, actions are the hidden topics of a Latent Dirichlet Allocation (LDA) [3,12]. We fit a LDA model to the motion data from multiple optical-flow vector fields, and solve it for the actions. Secondly, we use partial-Granger causality [4] to find temporal connections between spatially coherent actions. This step creates the motion patterns. The method's final step uses Granger causality again to learn the dynamic relationships among the motion patterns. The activity encompasses the motion patterns themselves and the motion patterns' dynamic interplay. In our activity model, this dynamic interplay is represented by the motion patterns' sequences and co-occurrences. Figure 2 illustrates the feature-extraction process in our method. Figure 3 shows an overview of the activity analysis method.

We tested our method on the same datasets as in [6,12] (Sect. 3). Our method compares favorably with the methods in [6,12].

2   Method

2.1   Activities, Motion Patterns, and Actions

We assume that an activity is composed by sequences of motion patterns as well as sets of co-occurring motion patterns. *Motion patterns* are spatio-temporal clusters of moving pixels describing the flow of objects along a spatial path [5].

The top-left side of the diagram in Fig. 2 illustrates an idealized activity composed of two motion patterns. The pink curve is a vertical forward motion pattern and the orange curves is a vertical right turn, respectively.

To form motion patterns, our method clusters optical-flow data, without information about individual objects. As a result, different motion patterns that share a spatial path in the scene may be split into parts. These parts, when temporally connected together along a spatial path, form a motion pattern. The motion patterns in Fig. 2 are composed by three simpler motion parts, which are shown as arrows in red, green, and blue colors (also labeled by 1, 2, and 3). The motion pattern of the longer vertical-forward motion (i.e., pink curve) is formed by two shorter vertical forward motions shown by arrows 1 (red) and 3 (blue). The motion pattern of to the vertical right turn is formed by a short vertical-forward motion (i.e., action 1 in red) followed by a horizontal forward motion (i.e., action 2 in green). We call these parts *actions*.

2.2 Basic Actions as Topics of a Latent Dirichlet Allocation

We approach the detection of actions as a probabilistic inference. We begin by assuming that motion events (i.e., optical flow) happening at a certain time interval are generated by a Latent Dirichlet Allocation (LDA). LDA is a generative probabilistic model originally designed for finding topics in large collection of documents using as input only the words in the documents [3]. LDA sees documents as a mixture of topics, where each topic is a probability distribution over words, and the words are elements of a fixed-size vocabulary. In LDA, each document has its own unique mixture of topics. But, the topics (i.e., distribution over the same words) are shared among documents.

When using LDA for analyzing motion activities in videos, documents and topics have a different interpretation than that used for document analysis. The documents are short fixed-duration video clips into which the main video is divided. The topics are the actions that compose the motion patterns. Finally, the words in each document are the basic motion events that occur during the clip. These events occur with certain frequency that depends on the topics. They are the optical-flow data. Figure 3 summarizes the main steps of our method.

We call a *motion event* as any sufficiently fast motion happening in the scene at a given time instant. These motions are the instantaneous velocities of objects moving in the scene. The detection of motion events is done as follows. First, we divide the video into a set of non-overlapping clips, $\mathcal{C} = \{c_1, \ldots, c_M\}$, where each clip has a fixed number of image frames, i.e., $c_i = \{I_1, \ldots, I_N\}$. Then, we compute the optical flow between pairs of all consecutive frames within the clip. This calculation produces a set of 2-D vector fields of motion events $\mathbf{v}(\mathbf{x}) = (u, v)^\mathsf{T}$ for each clip, where u and v are the components of the velocity vector at pixel $\mathbf{x}$. To create a suitable statistical analogy from a video clip to a document in LDA, we summarize the clip's motion information as a two-dimensional matrix E whose elements contain the frequency of all sufficiently fast motion events occurring at each pixel location during the clip. For a clip c of N image frames, we have:

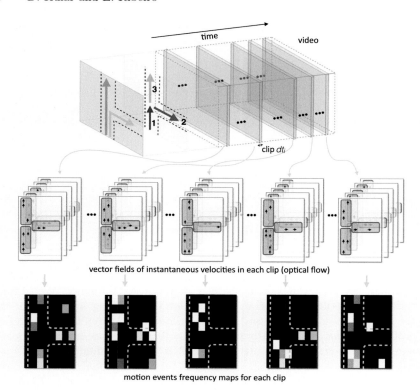

Fig. 2. Feature extraction. An activity formed by two motion patterns: a forward motion (pink) and a right-turn motion (orange). Motion patterns are connected short-duration motion parts (red, green, and blue arrows). Video is divided into consecutive clips. For each clip, we calculate the optical flow for pairs of consecutive frames. Pixel locations with high-magnitude motion are marked in each vector field. The frequency of these pixels is recorded into a motion-event map created for each clip. Both the velocity fields and the motion-event maps are the input to our method (Color figure online).

$$E_c\left(\mathbf{x}\right) = \frac{1}{N} \sum_{n=1}^{N} \mathbb{1}\left(\|\mathbf{v}\left(\mathbf{x}\right)\| > \tau\right), \tag{1}$$

where the indicator function $\mathbb{1}(\cdot)$ returns 1 when its argument is true and 0 otherwise, and τ is a pre-defined threshold value. By considering only motions above certain magnitude, we filter out noise and less-relevant motions. Figure 2 (bottom row) illustrates the motion-frequency maps summarizing the motion in each video clip. These maps are the input to our LDA model for discovering actions (i.e., motion-pattern parts).

LDA is a generative probabilistic model, which can be thought as a sequence of simple statistical steps that can be used for generating the data. The joint distribution reflects the multiple dependencies of the model's random variables.

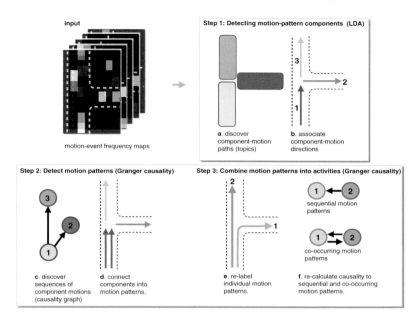

Fig. 3. Discovering activities. Given a set of maps of motion-event frequency and corresponding velocity fields as input, our method first uses LDA to discover the three motion-pattern components, which are shown in red, green, and blue (Row 1: a, b). These motion components are then connected into motion patterns using Granger causality (Row 2: c, d). Finally, we re-calculate the causality maps to recover both sequential and co-occurring motion patterns, which forms the activity (Row 2: e, f) (Color figure online).

In our LDA model, the action label, $a_{n,c}$, associated with motion event $e_{n,c}$ in clip c_i depends on the action proportion θ_c. The motion event $e_{n,c}$ depends on both the action label and the categorical distribution β_k over motion events. The motion events are observed variables and the actions are hidden variables. The complete LDA process for generating the motion events in a clip is as follows:

1. For each clip c,
   (a) Sample a per-clip action proportion, $(\theta_c|\alpha) \sim Dirichlet(\alpha)$. The action proportion is described by the Dirichlet distribution with parameter α,
   (b) For each visual event in the clip,
      i. Sample the action assignment $a_{n,c}$ from the per-clip action proportion from (a), $p(a_{n,c}|\theta_c)$,
      ii. Sample a visual event $e_{n,c}$ for the clip c from action assignment (i), $(\beta_k, a_{n,c}, e_{n,c}) \sim p(e_{n,c}|a_{n,c}, \beta_{1:K})$.

The joint distribution of variables $\{e_{n,c}, a_{n,c}\}_1^M$ and parameters $\{\beta, \theta\}$ given the hyper-parameters $\{\phi, \alpha\}$ for LDA is given by:

$$p(\{e_{n,c}, a_{n,c}\}_1^M, \beta, \theta, |\phi, \alpha) = \prod_k \underbrace{p(\beta|\phi)}_{\text{actions}} \prod_c \underbrace{p(\theta|\alpha)}_{\text{action proportion}}$$

$$\times \prod_n \underbrace{p(a_{n,c}|\theta_c)}_{\text{action}} \underbrace{p(e_{n,c}|a_{n,c}, \beta_k)}_{\text{visual event}}, \qquad (2)$$

where:

$$p(\beta_k|\phi) \sim Dirichlet(\phi), \qquad (3)$$
$$p(\theta_c|\alpha) \sim Dirichlet(\alpha), \qquad (4)$$
$$p(a_{n,c}|\theta_c) \sim Multinomial(\theta_c), \qquad (5)$$
$$p(e_{n,c}|a_{n,c}, \beta_{1:K}) \sim Multinomial(\beta_{a_{n,c}}). \qquad (6)$$

We need to estimate the action assignments, the per-clip action proportions, and the actions based on the observed variables. To obtain the actions in the clips, we evaluate the posterior distribution over the action labels. Let $\mathbf{a} = \{a_{c,n}\}_{c=1,n=1}^{M,N}$ be unknown action labels, and $\mathbf{e} = \{e_{c,n}\}_{c=1,n=1}^{M,N}$ be the observed motion events, the posterior distribution over the action labels is given by:

$$p(\mathbf{a}|\mathbf{e}) = \frac{p(\mathbf{e}|\mathbf{a})p(\mathbf{a})}{\sum_{\mathbf{e}} p(\mathbf{e}|\mathbf{a})p(\mathbf{a})}. \qquad (7)$$

Calculating Eq. 7 is impractical because its denominator sums over all possible assignments of motion events $\mathbf{e}$. Thus, we use Gibbs sampling [7] to obtain an approximate solution. However, the actions discovered by the LDA model are only parts of motion patterns. To find the motion patterns, we connect the actions together using temporal connections obtained by calculating Granger causality. This procedure is described next.

2.3   Hierarchical-Partial Granger Causality

This step of our method connects the actions produced by LDA into motion patterns. The motion patterns are formed by connecting nearby actions that are consecutive. Here, we use partial-Granger causality (pGC) to identify a causal influence occurring between actions. According to Granger causality [8,9], an observed time series t_x Granger-causes another series t_y, if addition of past data from t_x to past data of t_y decreases the prediction error of t_y. In computer vision, pairwise Granger causality has been applied to analyzing activities such as hand shaking, patty-cake game [10], and also in the analysis of dynamic scenes [11].

However, standard pairwise Granger causality is applicable only to bivariate time series. A multi-series extension can be done by means of partial-Granger causality [4]. Our method uses a hierarchical partial-Granger Causality model (H-pGC). In the first layer, we use Granger causality for discovering which

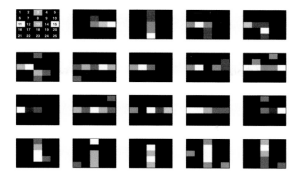

Fig. 4. Twenty event-frequency maps for a synthetic 4-way intersection. Maps are 5×5 matrices, whose cells store the frequency of motion events occurring at that pixel. Brighter pixels indicate higher frequency. Matrix elements are indexed from 1 to 25.

actions can be connected into motion patterns. Once the motion patterns are connected, we repeat the causality analysis to discover temporal relationships among motion patterns. Our H-pGC discovers temporally co-occurring motion patterns as well as motion patterns that follow each other in time. The first layer requires the action and clip co-occurrence matrix as input. This matrix is the action-proportion θ for every clip learned by LDA. The action-proportion matrix is a k-dimensional multivariate action-proportion series with M clips. In the second layer of our hierarchical-pGC, we create the motion pattern and clip co-occurrence matrix. This matrix has the motion-pattern proportion for each clip discovered during the first layer of the hierarchical-pGC. The motion-proportion matrix is a l-dimensional multivariate motion-proportion series with M clips. The influences observed by the hierarchical-pGC identify the activities.

3   Experiments

3.1   Synthetic Data

We began by testing our method on a synthetic dataset [12]. The dataset represents a scene observed at a 4-way traffic light intersection consisting of four kinds of traffic flows that are horizontal crossing, vertical crossing, vertical turn and horizontal turn (Fig. 4). In Fig. 4, the 5×5 matrices representing video clips c_m. Each cell contains the frequency of traffic that was observed at that pixel location, i.e., the motion events ($e_n \in \{e_1, e_2, \ldots, e_{25}\}$). Clips are placed according to their occurrence in the scene (i.e., row-wise).

Our model explained the dataset shown in Fig. 4 in terms of actions, motion patterns, and activities. We generated $1,000$ synthetic clips. We provided the occurrence of visual events in every clip as input (Fig. 4). In the first stage of our method, LDA returns spatially co-occurring visual events (i.e., actions). The learned actions correctly represent the four lanes, i.e., vertical top, vertical bottom, horizontal right, and horizontal left, formed on the roads (as shown in

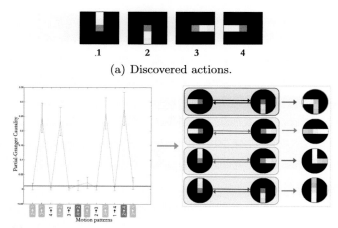

(a) Discovered actions.

(b) Discovered motion patterns. Left: Granger causality graph. Right: interpretation of the graph. For example, actions 2 and 4 causing each other. This relationship generated vertical turn motion pattern.

Fig. 5. Actions and motion patterns discovered by our approach. Discovered motion patterns are vertical turn, horizontal crossing, horizontal turn, and vertical crossing.

Fig. 5.a). In the second stage, the two-layer hierarchical-pGC discovers motion patterns and activities. In the first layer of the two-layer hierarchical-pGC, the action distribution per clip returned by LDA is used as input for pGC to produce motion patterns, i.e., vertical turn, horizontal crossing, horizontal turn, and vertical crossing (as shown in Fig. 5.b). The output of first layer becomes input for the second layer and the second layer of pGC generates the causal influence shared among motion patterns. The causal information represents the order in which motion patterns occurred (Fig. 6).

3.2   Real-World Data

Datasets and Setting. QMUL Street Intersection Dataset: Traffic at a four-way traffic intersection. **Pedestrian-Crossing Dataset:** Traffic of vehicles moving in vertical direction and pedestrians walking in horizontal direction. **Roundabout Dataset:** Traffic regulated by traffic lights observed at roundabout. **MIT Traffic Dataset:** A street view with less traffic flow as compared to the QMUL Street intersection. **Roundabout-India Traffic Dataset:** This video contains a roundabout in India.

We used 5-minute videos from each dataset. Each video was divided into 10-second video clips as done in [6, 12]. Then, optical flow was computed to generate visual events. To reduce the computation time, video frames were downsized to 72×90 pixels. Next, the bag of visual events (i.e., thresholded optical flow) was input to LDA to generate actions. We ran LDA for 500 iterations. In each case,

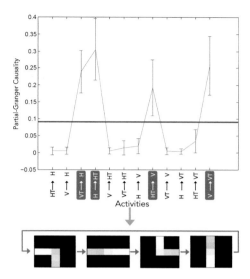

Fig. 6. Activities discovered by our approach. Partial Granger causality graph for activities (top) and graphs interpretation (bottom). Activities: vertical turn, horizontal crossing, horizontal turn, vertical crossing, and back to vertical turn.

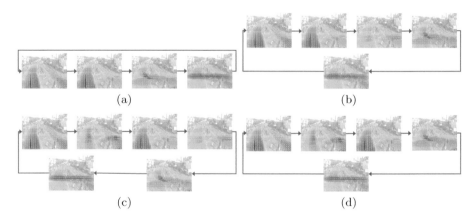

Fig. 7. Activities for the QMUL street-intersection dataset. (a) Vertical traffic from opposite direction with interleaving traffic followed by horizontal right and left traffic. (b) Vertical traffic with interleaving traffic from one side followed by right turns and horizontal right and left turn. (c) Vertical traffic with interleaving traffic from both directions followed by horizontal traffic. (d) Vertical traffic followed by horizontal traffic.

we selected the number of actions $K = 8$ as in [12]. Dirichlet hyper-parameters were fixed at $\alpha = .25, \beta = 0.05$ for all experiments to encourage composition of specific visual events into general actions.

For all datasets, our method discovered the same activities as in [6,12], but also discovered additional activities in the QMUL street-intersection dataset (Fig. 7), and in the pedestrian-crossing dataset.

4   Conclusion

To analyze videos, we combined Latent Dirichlet Allocation with hierarchical-partial Granger Causality to automatically learn co-occurring visual events and their occurrence sequence. Our model discovers spatially co-occurring visual events using LDA and learn temporally co-occurring visual events using two layer hierarchical-pGC. We tested our method on five datasets. Our method successfully extracted actions, motion patterns, and activities. In future work, we plan to work with tracklets (i.e., trajectory fragments).

References

1. Li, J., Gong, S., Xiang, T.: Global behaviour inference using probabilistic latent semantic analysis. In: BMVC, vol. 3231, p. 3232 (2008)
2. Wang, X., Ma, X., Grimson, W.E.L.: Unsupervised activity perception in crowded and complicated scenes using hierarchical Bayesian models. IEEE Trans. Pattern Anal. Mach. Intell. **31**(2009), 539–555 (2009)
3. Blei, D.M., Ng, A.Y., Jordan, M.I.: Latent Dirichlet allocation. J. Mach. Learn. Res. **3**, 993–1022 (2003)
4. Guo, S., Seth, A.K., Kendrick, K.M., Zhou, C., Feng, J.: Partial Granger causality - eliminating exogenous inputs and latent variables. J. Neurosci. Meth. **172**, 79–93 (2008)
5. Saleemi, I., Shafique, K., Shah, M.: Probabilistic modeling of scene dynamics for applications in visual surveillance. IEEE Trans. Pattern Anal. Mach. Intell. **31**, 1472–1485 (2009)
6. Kuettel, D., Breitenstein, M.D., Van Gool, L., Ferrari, V.: What's going on? discovering spatio-temporal dependencies in dynamic scenes. In: IEEE CVPR, pp. 1951–1958 (2010)
7. Griffiths, T.: Gibbs sampling in the generative model of latent Dirichlet allocation. **518**(11), 1–3 (2002). Standford University
8. Wiener, N.: The theory of prediction. In: Modern Mathematics for the Engineer, pp. 165–190. McGraw-Hill, New York (1956)
9. Granger, C.W.: Investigating causal relations by econometric models and cross-spectral methods. Econom. J. Econom. Soci. **33**, 424–438 (1969)
10. Prabhakar, K., Oh, S., Wang, P., Abowd, G.D., Rehg, J.M.: Temporal causality for the analysis of visual events. In: IEEE CVPR, 1967–1974 (2010)
11. Fan, Y., Yang, H., Zheng, S., Su, H., Wu, S.: Video sensor-based complex scene analysis with Granger causality. Sensors **13**, 13685–13707 (2013)
12. Hospedales, T., Gong, S., Xiang, T.: Video behaviour mining using a dynamic topic model. Int. J. Comput. Vis. **98**, 303–323 (2012)

Statistical Adaptive Metric Learning for Action Feature Set Recognition in the Wild

Shuanglu Dai[✉] and Hong Man

Department of Electrical and Computer Engineering,
Stevens Institute of Technology,
1 Castle Point on Hudson, Hoboken, NJ 07030, USA
{sdai1,hong.man}@stevens.edu

Abstract. This paper proposes a statistical adaptive metric learning method by exploring various selections and combinations of multiple statistics in a unified metric learning framework. Most statistics have certain advantages in specific controlled environments, and systematic selections and combinations can adapt them to more realistic "in the wild" scenarios. In the proposed method, multiple statistics, include means, covariance matrices and Gaussian distributions, are explicitly mapped or generated in the Riemannian manifolds. Subsequently, by embedding the heterogeneous manifolds in their tangent Hilbert space, the deviation of principle elements is analyzed. Hilbert subspaces with minimal principle elements deviation are then selected from multiple statistical manifolds. After that, Mahalanobis metrics are introduced to map the selected subspaces back into the Euclidean space. A uniformed optimization framework is finally performed based on the Euclidean distances. Such a framework enables us to explore different metric combinations. Therefore our final learning becomes more representative and effective than exhaustively learning from all the hybrid metrics. Experiments in both static and dynamic scenarios show that the proposed method performs effectively in the wild scenarios.

1 Introduction

Set classification has been studied many years with application to video sequences and image sets recognition. The task of feature set classification is to classify an input feature set to one of the sets in the training gallery [1]. Compared to image sets, feature sets are more diverse. They can not to be assumed to follow certain distribution or lie in some scale and affine invariant linear subspace. One of the effective techniques handling such problem is by using statistical representations for the original feature samples. For action recognition in the wild scenarios, combinations of statistics from lower-order to higher-order have shown promising representation capabilities, while how to combine these multiple statistics in a near optimal way remains a technical challenge [1].

Generally, three types of statistics have been applied on set modeling: sample-based statistics (SAS) [2–4], subspace-based statistics (SUS) [5–7] and

© Springer International Publishing Switzerland 2015
G. Bebis et al. (Eds.): ISVC 2015, Part I, LNCS 9474, pp. 657–667, 2015.
DOI: 10.1007/978-3-319-27857-5_59

distribution-based statistics (DIS) [8,9]. Utilizing affine transformation and centroid of samples, sample-based statistics represent d-dimension feature sets with first-order statistics in the R^d space. Well-known sample-based methods include Minimum Mean Discrepancy (MMD) [10], Affine (Convex) Hull based Image Set Distance (AHISD, CHISD) [11] and Information Theoretic Metric Learning (ITML) [12]. Differently from sample-based statistics, subspace-based statistics analyze sets lying on a specific Riemannian manifold. By embedding manifolds into their Hilbert subspaces, Distance discriminant functions can then performed with unified measurements.Second-order statistics based methods have better representation of the data, but it is hard to design a discriminant function directly with a distance measure for the manifolds. Typical subspace-based methods include Manifold Discriminant Analysis (MDA) [6], Grassmann Discriminant analysis (GDA) [5], Covariance Discriminative Learning (CDL) [7] and etc. Distribution based statistics model each sample in the feature set with a distribution, which can be expressed as an expansion of the Riemannian manifold from the 2nd-order statistic space Sym_d^+ to Sym_{d+1}^+. Such methods are often with 3rd-order statistics and may lead to complex parametric distribution comparison. Typical examples include Single Gaussian Models (SGM) [8], Gaussian Mixture Models (GMM) [9] and kernel version of ITML with DIS-based set model (DIS-ITML). Although 3rd-order statistics model sets with more consolidated representations, the hypothesis tests often require significant amount of computations in distribution comparisons. More adaptive forms of set modeling methods have been proposed by combining multiple statistical metrics in certain heuristic ways. Some of the recent hybrid statistical models include Projection Metric Learning (PML) on Grassmann manifold and hybrid Euclidean-and-Riemannian Metric Learning (HERML) [13]. The main idea of multiple statistics combination is to project measurements from multiple heterogeneous spaces into high-dimensional Hilbert spaces. The key issue then becomes the learning of unified discriminant functions from the training sets.

Focusing on complex feature set classification, this paper proposes an adaptive selection method for hybrid statistical metrics in Hilbert space embeddings. The selected single or multiple statistics can be suitable for a wide range of feature sets. Inspired by the discriminant function design in the second-order based methods, LogDet divergence is introduced as a unified discriminant function to unify different statistics into a common measurement. The unifying enables nearest neighbor (NN) method easily to be performed for classification. The whole process of modeling and learning consists of several steps. Firstly, heterogeneous statistics including mean, covariance matrix and Gaussian distribution are introduced to project data into high-dimensional Hilbert spaces. Typically, d-dimensional mean vectors represent samples from R^d to Sym_d^+ expanded by a Point-to-Set projection; covariance matrices lie in Riemannian manifold Sym_d^+ and multivariate Gaussian distributions expand the second order statistics into Riemannian manifold Sym_{d+1}^+ expressed by relative entropy. Secondly, by embedding the heterogeneous spaces into high-dimensional Hilbert spaces, the Mahalanobis distance is introduced as our discriminant metric. Then, the Hilbert spaces selection is conducted based on the minimum Hilbert subspaces. The hybrid statistics are then reduced to single or multiple statistics combination. Finally, with LogDet divergence mapping all the Hilbert space points into R^d, a constrained metric learning is performed. Recognitions are mainly conducted on video sequences in both static and dynamic scenarios using spatial image features such as edges, SIFT, HOG, and texture features.

2 Statistical Feature Set Modeling

2.1 Set Description and Discriminant Function

Let $X = [X_1, ..., X_N]$ denotes the training set formed by N feature sets, where $X_i = [x_1, x_2, ..., x_M] \in R^{n_i \times d}$ indicates the i-th feature set, $1 \leq i \leq N$, and n_i is the number of samples in this set. Considering $\phi : R^d \rightarrow F$ or $Sym_d^+ \rightarrow F$ as an explicit mapping with the statistical kernels, Φ_i^r denotes the high dimensional feature of r-th statistics extracted from the feature set X_i, $1 < r < R$. R is the number of statistics being used.

A typical metric, the Log-Euclidean Distance (LED) is used as a measure of the true geodesic distance on Sym^+. As a geodesic distance on the Riemannian manifold, logarithm directions are computed as tangent vectors. In our case, the geodesic distance between two kernelized points X and Y is computed as the inner product: $\kappa_{XY}(X, Y) = tr(log(X) \cdot log(Y))$.

Since our set model is a kernelized space lying on Sym^+ Riemannian manifold, a true geodesic distance between the kernelized i-th and j-th points is computed by

$$d_{A_r}(\Phi_i^r, \Phi_j^r) = tr(A_r(\Phi_i^r - \Phi_j^r)(\Phi_i^r - \Phi_j^r)^T) \tag{1}$$

As a distance metric and discriminant function between Φ_i and Φ_j, $d_{A_r}(\Phi_i^r, \Phi_j^r)$ shows outstanding shift-rotation invariant properties. A_r is the learned Mahalanobis distance representing the r-th statistics in the Hilbert space. Once the Mahalanobis distances have been learned, the classification can be simply completed by nearest neighbor (NN) with $d_{A_r}(\Phi_i^r, \Phi_j^r)$.

2.2 Statistical Modeling

This section discusses multiple statistics as Φ. Before adaptive selection, we uniformly map feature set $X_i, 1 \leq i \leq N$ with following three statistics: sample-based, subspace-based and distribution-based statistics.

Sample-Based Statistics (SAS). Given sample $x_k \in X_i$, $1 \leq k \leq M$, the mean vector μ_i of X_i is computed as: $u_i = \frac{1}{M} \sum_{k=1}^{M} x_k$.

Subspace-Based Statistics (SUS). Given sample $x_k \in X_i$, $1 \leq k \leq M$, the covariant matrix C_i of X_i is computed as: $C_i = \frac{1}{M-1} \sum_{k=1}^{M} (x_k - \mu_i)(x_k - \mu_i)^T$.

Distribution-Based Statistics (DIS). The d-dimensional distribution of set X_i is modeled as a Single Gaussian Model (SGM) with an estimated d-dimensional mean vector $\widehat{m}_{k,i}, 1 \leq k \leq d$ and a covariance matrix $\widehat{C}_i$: $x \sim N(\widehat{m}_i, \widehat{C}_i)$.

2.3   Minimum Hilbert Subspace Selection

For a point μ in the Euclidean space R^d, map it into a symmetrical positive definite matrix $Sym_d^+ = \{\mu\mu^T | R^{d \times d}, |\mu\mu^T| > 0\}$. For DIS, the space of Gaussian distribution is able to be embedded into a Riemannian manifold Sym_{d+1}^+. According to the information geometry [14], If random vector x normally distributed, then its affine transformation $Qx + \widehat{m}$ follows $N(\widehat{m}, Q)$, where Q is a semi-positive definite square matrix of covariant matrix $\widehat{C} = QQ^T$. Define the affine group $Aff_d^+ = \{(\widehat{m}, Q) | \widehat{m} \in R^d, Q \in R^{d \times d}, |Q| > 0\}$ to the simple Lie group $Sl_{d+1} = \{V | V \in R^{(d+1) \times (d+1)}, |V| > 0\}$:

$$Aff^+ \to Sl_{d+1}, (m, \widehat{Q}) \to V, (m, \widehat{Q}) \to |Q|^{-\frac{1}{d+1}} \begin{pmatrix} Q & \widehat{m} \\ \widehat{m}^T & 1 \end{pmatrix} \tag{2}$$

Then we map the simple Lie group back to symmetric positive defined (SPD) matrices $Sym_{d+1}^+ = \{VV^T | VV^T \in R^{(d+1) \times (d+1)}, |VV^T| > 0\}$ of $\widehat{C}$ by

$$Sl_{d+1} \to Sym_{d+1}^+, V \to VV^T, (m, \widehat{Q}) \to |Q|^{-\frac{2}{d+1}} \begin{pmatrix} QQ^T + \widehat{m}\widehat{m}^T & \widehat{m} \\ \widehat{m}^T & 1 \end{pmatrix} \tag{3}$$

where $QQ^T = \widehat{C}$. Therefore, after embedding, we have 3 statistical representations:

$$\Phi_{SAS} = \mu\mu^T \in Sym_d^+ \quad \Phi_{SUS} = C \in Sym_d^+ \quad \Phi_{DIS} = \begin{pmatrix} \widehat{C} + \widehat{m}\widehat{m}^T & \widehat{m} \\ \widehat{m}^T & 1 \end{pmatrix} \in Sym_{d+1}^+ \tag{4}$$

The SPD matrices lie in a specific Riemannian manifold, where tangent vector of Sym^+ is often utilized as a projection to R^d. For the three different embedded statistics and all their possible combinations, we examine their logarithm Hilbert subspace through an eigen-decomposition: $log(\boldsymbol{\Phi}) = U(log(\boldsymbol{\Lambda}))U^T$ where $\boldsymbol{\Phi} = \{\Phi_{SAS}, \Phi_{SUS}, \Phi_{DIS}, \Phi_{SAS}\Phi_{SUS}, \Phi_{SAS}\Phi_{DIS}, \Phi_{SUS}\Phi_{DIS}, \Phi_{SAS}\Phi_{SUS}\Phi_{DIS}\}$ indicates the set of all possible statistic combinations; $\boldsymbol{\Lambda} = \{\Lambda_{SAS}, \Lambda_{SUS}, \Lambda_{DIS}, \Lambda_{SAS}\Lambda_{SUS}, \Lambda_{SAS}\Lambda_{DIS}, \Lambda_{SUS}\Lambda_{DIS}, \Lambda_{SAS}\Lambda_{SUS}\Lambda_{DIS}\}$ indicates the set of all the possible eigen-value matrices of $\boldsymbol{\Phi}$, respectively; U is the decomposed projection matrices. Set $\boldsymbol{\Phi}$ describes eigen-spaces $\boldsymbol{\Lambda}$ of all the possible statistics combinations. The last element utilizes all the statistics, while the first three have only single statistic.

As a kernel representation of the feature sets, a subspace with smaller inner deviation on its principal elements generally gives a better representation. Thus, kernels are selected by $min\{var(diag(log\boldsymbol{\Lambda}))\}$ where $var(\cdot)$ calculates the correlation and $diag(\cdot)$ denotes the diagonal matrix.

2.4   Learning Problem and Its Optimization

As defined in Eq. (1), our learning problem is to learn the Mahalanobis distances A for all the selected statistics. The learned distance should have maximized intra-class and minimized inter-class. While in the Sym^+ Hilbert spaces, one way to optimize A_r is to approximate it to an initial Mahalanobis matrix.

$$\min_{A_1 \leq 0,...,A_r \leq 0,\epsilon} \frac{1}{R}\sum_{r=1}^{R} D_{ld}(A_r, A_0) + \gamma D_{ld}(diag(\epsilon), diag(\epsilon_0))$$

$$s.t. \quad \frac{\delta_{ij}}{R}\sum_{r=1}^{R} d_{A_r}(\Phi_i^r, \Phi_j^r) \leq \epsilon_{ij} \forall (i,j)$$

$$(5)$$

δ_{ij} denotes the similarity between i-th and j-th kernel. If they are similar, $\delta_{ij} = 1$, otherwise $\delta_{ij} = -1$. $D_{ld}(A_r, A_0)$ is the LogDet divergence measuring the correlation between matrix A_r and A_0 where $D_{ld}(A_r, A_0) = tr(A_r, A_0^{-1}) - logdet(A_r A_0^{-1}) - d$. d is the dimension of A. ϵ_{ij} denotes the boundaries of similarity and dissimilarity, which is introduced as an upper/lower bound for the i-th and j-th sample covariance [15]. γ is a parameter arbitrarily set before the computation. In Eq. (5), the minimization leads to maximumal inter-class and the constrain leads to minimal intra-class. The form of Eq. (5) can be considered as Bregman optimizations for R selected statistics [16]. With Bregman projection, Eq. (5) is iteratively updated by

$$\begin{cases} A_r^{t+1} = A_r^t + \frac{\delta_{ij}\alpha}{1-\delta_{ij}\alpha d_{A_r^t}(\Phi_i^r, \Phi_j^r)} A_r^t(\Phi_i, \Phi_j)(\Phi_i, \Phi_j)^T A_r^t \\ \epsilon_{ij}^{t+1} = \gamma\epsilon_{ij}^t / (\gamma + \alpha\epsilon_{ij}^t) \end{cases}$$

$$(6)$$

The Lagrange multiplier α for R statistics can be solved by

$$d_{A_r^t}(\Phi_i^r, \Phi_j^r) + \frac{\delta_{ij}}{R}\sum_{r=1}^{R} \frac{\alpha d_{A_r^t}(\Phi_i^r, \Phi_j^r)}{1 - \alpha d_{A_r^t}(\Phi_i^r, \Phi_j^r)} = \frac{\gamma\epsilon_{ij}^t}{\gamma - \alpha\delta_{ij}\epsilon_{ij}^t} r = 1,...,R \quad (7)$$

According to (6) and (7), statistical kernels $A_r (1 \leq r \leq R)$ are learned by Algorithm 1.

3   Relations with HERML and ITML

It is essential to point out that our method, Minimum Hilbert adaptive metric learning (MHAML) is an adaptive version of hybrid statistical metric learning, e.g. Hybrid Euclidean-and-Riemannian Metric Learning (HERML). HERML individually fixes the number of learning Mahalanobis Matrices as 3 for all kinds of feature sets. In HERML, $min\{var(diag(log\Lambda))\} = log\Phi_{SAS}\Phi_{SUS}\Phi_{DIS} = U(log\Lambda_{SAS}\Lambda_{SUS}\Lambda_{DIS})U^T$ is arbitrarily set as an implicit condition. Different from HERML, MHAML selects all the statistic combinations of Eq. (4). handling feature sets. Minor variations will usually not lead to large fluctuations in

---

Algorithm 1. Minimum Hilbert Adaptive Metric Learning

Input: Training pairs $\{(\Phi_i^r, \Phi_j^r), \delta_{ij}\}$, input Initial Mahalabonis matrix A_0 , self-covariance co-efficient γ, distance thresholds ρ , margin parameter τ and tuning scale ς

Output:
1: $t \leftarrow 1$, $A_r^1 \leftarrow A_0$, for r=1,,R, $\lambda_{ij} \leftarrow 0$, $\epsilon_{ij} \leftarrow \delta_{ij}\rho - \varsigma\tau, \forall (i,j)$
2: *Repeat*
3: Pick a constraint (i,j), and compute the distances $d_{A_r}(\Phi_i^r, \Phi_j^r)$
4: Solve α with Eq. (16)
5: $\epsilon_{ij} \leftarrow \gamma\epsilon_{ij}/(\gamma + \alpha\epsilon_{ij})$
6: $\lambda \leftarrow \lambda_{ij} - \alpha$
7: $A_r \leftarrow A_r + \frac{\lambda_{ij}\alpha}{1-\lambda_{ij}\alpha d_{A_r}(\Phi_i^r, \Phi_j^r)} A_r(\Phi_i^r - \Phi_j^r)(\Phi_i^r - \Phi_j^r)^T A_r$, for r=1,,R
8: Until convergence
9: **return** Mahalanobis Matrices $A_1, ..., A_R$

---

images, but they may likely lead to large fluctuations in features, e.g. HOG and SIFT. HERML individually combines all three metrics, while MHAML explores the correlation among these metrics. Thus effective combination of the metrics can be achieved on various feature sets.

As a low-rank kernel learning, this paper utilizes methods of ITML to optimize the selected kernel combinations. We can observe if the selected combination includes third-order statistic, the solution to our problem becomes a KL-divergence minimization derived a metric learning; otherwise, it is a Mahalabonis metric learning.

4   Experiments

4.1   Methods in the Comparative Study and Experimental Settings

In the experiments, the proposed Minimum Hilbert Adaptive Metric Learning (MHAML) method is mainly compared with statistical metric learning algorithms in the following four categories, respectively, includes two state-of-art methods. These methods are:

1. Sample-based methods: Information theoretic metric learning (ITML) [12], Maximum Mean Discrepancy (MMD) [10]
2. Subspace-based methods: Manifold Discriminant Analysis (MDA) [6], Covariance Discriminant Learning (CDL) [7]
3. Distribution-based methods: Gaussian Mixture Model (GMM) [9], Single Gaussian Model (SGM) [8]
4. Hybrid statistic methods: Hybrid Euclidean-and-Riemannian Metric Learning (HERML) [1] and Distribution-based ITML (DIS-ITML)

In addition, In order to obtain a comprehensive view, we also compare our method with some methods using spatio-temporal feature analysis. Static computer vision features are extracted frame by frame. In case of set modeling, common scale invariant image features are employed, including Histogram of Oriented Gradients (HOG), Shift-invariant Feature Transformation (SIFT), Harris corners and Textures. All the features are clustered into 128 dimensions, while SIFT features are clustered into 128 points for all the frames in use. Thus each frame generates a $128 \times k$ feature set. For videos with different resolutions, we uniformly resize every frame into 320×240 at QVGA resolution. When compared with other methods, feature combination of SIFT+HOG+Corners are uniformly used. Average recognition rate is used to measure the accuracy.

The task of our action classification is to classify input feature sets of one testing video clip to one of action class sets in the training gallery. Recognitions are performed repeatedly over 10 trials for each database. In each trial, 10 video clips are randomly picked out for every action video class in the dataset and feature sets are extracted from the video clips uniformly. Feature sets of the i-th action are all labelled as "i". Average recognition rate is calculated from the 10 trials.

All the source codes of the compared methods are obtained from their original authors, except for SGM and GMM. We implemented SGM and GMM meticulously. For methods using Bregman optimization like ours, distance thresholds ρ is set as the mean distances, margin parameter τ is set as the standard deviations and tuning scale $\varsigma \in [0.01 \; 10]$. Furthermore, since our experiments are tested on feature sets, we carefully tuned each models of all the methods to achieve the best possible recognition performance. For MMD, the number of iteration is set to 10. For ITML, we set the parameter of Bregman optimization as stated above. The parameter settings of MDA is according to [6]. For CDL we adopt the KLDA for the discriminant analysis and k-NN for classification.

Five famous human action data sets are used in our experiments: Weizmann human action, UCF sports, KTH human action, YouTube and HMDB51. Among those five video data sets, Weizmann and KTH have very static backgrounds with different human actions, while YouTube and HMDB51 both have complex and varying backgrounds and dynamic scenarios with different humans. Experiments are conducted both on human verification and action recognition.

4.2 Results and Analysis

Performance Evaluation and Methods Comparisons. First of all, Table 1 shows the performances of different feature set classification methods in terms of average recognition rate. Three benchmark video analysis datasets are in use: UCF101, YouTube and HMDB51. Note that the videos sequences in those datasets are not just with human action scenes, but also with more common and complicated individual shapes, streets and dynamic scenarios.

From Table 1, we observe that our MHAML has better recognition performance than that of either single or hybrid statistic methods on UCF and HMDB51. On YouTube data, MHAML performs better than other methods, except HERML. All the results of Table 1 comprehensively show that our method is able to adaptively take advantage of sample-based, subspace-based, distribution-based and hybrid statistic, and consequently boost their performance to some extent.

Table 1. Average recognition rate (%) of different image set classification methods on UCF101, YouTube and HMDB51 datasets.

Methods		UCF	YouTube	HMDB51
Sample-based	MMD [10]	73.5	51.2	32.4
	ITML [12]	76.2	67.6	51.2
Subspace-based	MDA [6]	88.9	65.3	63.3
	CDL [15]	93.5	67.7	70.3
Distribution-based	GMM [9]	87.5	61.2	32.7
	SGM [8]	82.3	50.3	26.4
Hybrid statistics	HERML [1]	92.5	74.6	68.1
	DIS-ITML [12]	86.1	70.3	47.8
Proposed	MHAML	95.5	70.83	71.3

Table 2. Average recognition rate (%) compared with well-known human action classification methods on KTH, Weizmann and UCF human action benchmark datasets.

Methods	KTH	Weizmann	UCF
Campos et al. [17]	91.5	96.7	80
Fathi et al. [18]	90.5	100	-
Jhuang et al. [19]	91.7	98.8	-
Wu et al. [20]	94.5	-	91.3
MHAML	100	100	95.5

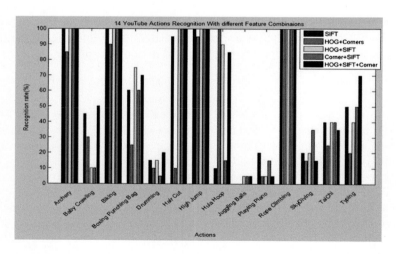

Fig. 1. Robustness test of MHAML with different feature combinations on YouTube database in terms of recognition rate(%). Feature combinations: SIFT, HOG+Corners, HOG+SIFT, Corner+SIFT, HOG+SIFT+Corner

Table 3. Training and Testing time (in seconds) of HERML and MHAML in the experiments on five video databases: UCF, KTH, Weizmann, YouTube and HMDB51.

Training time	UCF101	KTH	Weizmann	YouTube	HMDB51
HERML	227.2	765.737	280.57	366.99	567.36
MHAML	77.43	123.676	71.35	100.77	83.49
Testing time	UCF101	KTH	Weizmann	YouTube	HMDB51
HERML	0.03	0.037	0.0351	0.037	0.0342
MHAML	0.0277	0.031	0.029	0.032	0.0335

Secondly, our methods are further compared with a few well-known action recognition methods on three benchmark human action video datasets: UCF101, KTH and Weizmann. From Table 2, we observe that the recognition performance of MHAML is better than other well-known methods.

Performance Analysis. Furthermore, our method is tested on different feature combinations to evaluate the feature robustness. Figure 1 shows an example of recognition rate with 14 randomly picked actions from Youtube database. With 5 available static features, the figure shows that the recognition rates of MHAML are nearly the same among different combinations for each of the actions.

Computational Costs. Last but not least, computational costs of our method on different dataset are also evaluated in Table 3. In Table 3, both the training time and testing time of MHAML are faster than those of HERML. Such acceleration is mainly contributed by the statistic kernel reduction introduced by our selection. Classifications with only necessary statistics combination efficiently enhance the computation and recognition.In summary, our method performs well in human action recognition. It generally outperforms other hybrid statistic metric learning methods in terms of accuracy and computational efficiency.

5 Conclusions

In this paper, we proposed a statistical adaptive metric learning method for feature set modeling and classification. The extensive experiments have shown that our proposed method outperforms several state-of-the-art methods in human action recognition over a variety of public datasets. To our best knowledge, the optimal selection of metric learning combination in the Hilbert spaces has not been investigated before this work. In the future, it would be interesting to expand our method for other possible hybrid metric selections with more robust statistics modeling feature sets in different structures and real-world classifications.

Acknowledgments. This work was supported in part by the National Science Foundation under Award Number IIS-1350763.

References

1. Huang, Z., Wang, R., Shan, S., Chen, X.: Hybrid euclidean-and-riemannian metric learning for image set classification. In: Cremers, D., Reid, I., Saito, H., Yang, M.-H. (eds.) ACCV 2014. LNCS, vol. 9005, pp. 562–577. Springer, Heidelberg (2015)
2. Huang, Z., Zhao, X., Shan, S., Wang, R., Chen, X.: Coupling alignments with recognition for still-to-video face recognition. In: 2013 IEEE International Conference on Computer Vision (ICCV), pp. 3296–3303. IEEE (2013)
3. Yang, M., Zhu, P., Van Gool, L., Zhang, L.: Face recognition based on regularized nearest points between image sets. In: 2013 10th IEEE International Conference and Workshops on Automatic Face and Gesture Recognition (FG), pp. 1–7. IEEE (2013)
4. Zhu, P., Zhang, L., Zuo, W., Zhang, D.: From point to set: extend the learning of distance metrics. In: 2013 IEEE International Conference on Computer Vision (ICCV), pp. 2664–2671. IEEE (2013)
5. Hamm, J., Lee, D.D.: Grassmann discriminant analysis: a unifying view on subspace-based learning. In: Proceedings of the 25th International Conference on Machine Learning, pp. 376–383. ACM (2008)
6. Wang, R., Chen, X.: Manifold discriminant analysis. In: IEEE Conference on Computer Vision and Pattern Recognition, CVPR 2009, pp. 429–436. IEEE (2009)
7. Wang, R., Guo, H., Davis, L.S., Dai, Q.: Covariance discriminative learning: a natural and efficient approach to image set classification. In: 2012 IEEE Conference on Computer Vision and Pattern Recognition (CVPR), pp. 2496–2503. IEEE (2012)
8. Tistarelli, M., Grosso, E.: Identity management in face recognition systems. In: Schouten, B., Juul, N.C., Drygajlo, A., Tistarelli, M. (eds.) BIOID 2008. LNCS, vol. 5372, pp. 67–81. Springer, Heidelberg (2008)
9. Arandjelović, O., Shakhnarovich, G., Fisher, J., Cipolla, R., Darrell, T.: Face recognition with image sets using manifold density divergence. In: IEEE Computer Society Conference on Computer Vision and Pattern Recognition, CVPR 2005, vol. 1, pp. 581–588. IEEE (2005)
10. Gretton, A., Borgwardt, K.M., Rasch, M.J., Schölkopf, B., Smola, A.: A kernel two-sample test. J. Mach. Learn. Res. **13**, 723–773 (2012)
11. Cevikalp, H., Triggs, B.: Face recognition based on image sets. In: 2010 IEEE Conference on Computer Vision and Pattern Recognition (CVPR), pp. 2567–2573. IEEE (2010)
12. Davis, J.V., Kulis, B., Jain, P., Sra, S., Dhillon, I.S.: Information-theoretic metric learning. In: Proceedings of the 24th International Conference on Machine Learning, pp. 209–216. ACM (2007)
13. Huang, Z., Wang, R., Shan, S., Chen, X.: Face recognition on large-scale video in the wild with hybrid euclidean-and-riemannian metric learning. Pattern Recognit. **48**, 3113–3124 (2015)
14. Amari, S.I., Nagaoka, H.: Methods of Information Geometry, vol. 191. American Mathematical Soc, New York (2007)
15. Bregman, L.M.: The relaxation method of finding the common point of convex sets and its application to the solution of problems in convex programming. USSR Comput. Math. Math. Phys. **7**, 200–217 (1967)

16. Kulis, B., Sustik, M.A., Dhillon, I.S.: Low-rank kernel learning with bregman matrix divergences. J. Mach. Learn. Res. **10**, 341–376 (2009)
17. De Campos, T., Barnard, M., Mikolajczyk, K., Kittler, J., Yan, F., Christmas, W., Windridge, D.: An evaluation of bags-of-words and spatio-temporal shapes for action recognition. In: 2011 IEEE Workshop on Applications of Computer Vision (WACV), pp. 344–351. IEEE (2011)
18. Fathi, A., Mori, G.: Action recognition by learning mid-level motion features. In: IEEE Conference on Computer Vision and Pattern Recognition, CVPR 2008, pp. 1–8. IEEE (2008)
19. Jhuang, H., Serre, T., Wolf, L., Poggio, T.: A biologically inspired system for action recognition. In: IEEE 11th International Conference on Computer Vision, ICCV 2007, pp. 1–8. IEEE (2007)
20. Wu, X., Xu, D., Duan, L., Luo, J.: Action recognition using context and appearance distribution features. In: 2011 IEEE Conference on Computer Vision and Pattern Recognition (CVPR), pp. 489–496. IEEE (2011)

ST: Spectral Imaging Processing

Learning Discriminative Spectral Bands
for Material Classification

Chao Liu, Sandra Skaff$^{(\boxtimes)}$, and Manuel Martinello

Innovation Center, Canon USA, San Jose, CA, USA
sskaff@cusa.canon.com

Abstract. This paper describes a novel setup to capture images of the spectral response of different materials to improve their classification. The proposed system involves a Liquid Crystal Tunable Filter (LCTF) that, placed in front of the camera, allows the capture of narrow spectral band images for each material from different illumination directions. We analyze the captured spectral images and propose a learning based method to select a subset of bands (or filters), the corresponding images of which can be used without compromising on material classification performance. Results on both binary and multi-class classification tasks are reported in the experimental section.

1 Introduction

The classification of raw materials, especially metal, is a long-standing and challenging task in some industrial applications, such as recycling and quality inspection. Recently, Jehle *et al.* [1] and Gu *et al.* [2] have proposed two very interesting systems for raw material classification based on the physical properties of a sample related to its appearance. These capturing systems achieve good results but are quite expensive and complicated to build and calibrate. Being able to have simpler and inexpensive hardware would allow to reduce the cost and the time spent in the capturing process.

We propose a setup which captures images over a large number of spectral bands and a few variations in the illumination (Fig. 1). With capturing the spectral components of different materials, we explore which component carries the largest amount of information when discriminating materials. The main advantage of the proposed system is that, once the learning process is completed, the hardware can be very simple and inexpensive: the few spectral bands that will be selected for a specific material classification task can be reproduced by a small set of filters or by designing an *ad hoc* color filter array for material sensing. Such a setup would provide portability for robotics or mobile imaging systems.

The main contributions of this paper can be summarized as follows. First, we propose a novel image capture setup to build a spectral image database of materials corresponding to multiple illumination directions. Next, we describe a learning based method to optimally select a subset of spectral bands which

---

C. Liu—Currently Ph.D. candidate at Carnegie Mellon University, PA, USA.

© Springer International Publishing Switzerland 2015
G. Bebis et al. (Eds.): ISVC 2015, Part I, LNCS 9474, pp. 671–681, 2015.
DOI: 10.1007/978-3-319-27857-5_60

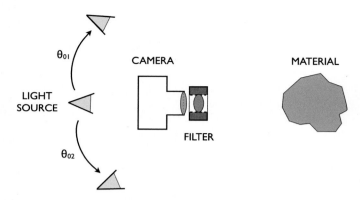

Fig. 1. Image capture setup. A liquid crystal tunable filter is placed in front of the camera lens in order to capture different spectral responses of materials under different illumination orientations.

provide the most discriminability among materials. Finally, we show that material classification can be improved when using images corresponding to only the selected spectral bands, as compared with all bands.

2   Prior Work

An increasing amount of work in material classification is using spectral information, since the spectral characteristics of imaged materials are better modulated in the hyperspectral domain than with the standard trichromatic red, blue, and green data [3]. The main issue with this type of data is the large amount of information needed for processing. Therefore a key challenge is to optimally reduce the number of features or the spectral bands.

Towards this goal, several approaches have been proposed: Bajcsy and Groves [4] rank spectral bands using different unsupervised and supervised methods, but results strongly depend on suitable parameter selection; Chang and Wang [5] add a linear constraint to each band of signal of interest, while minimizing the band correlation or band dependence on other bands. In [6,7] higher classification accuracies are achieved by clustering spectral bands: Martinez-Uso *et al.* [6] cluster the bands by similarity measures (such as the Kullback-Leibler divergence), thus reducing the computational complexity, while Qian *et al.* [7] utilize affinity propagation and also use wavelet shrinkage to remove noisy bands. Similarly Kumar *et al.* [8] propose a clustering based band selection algorithm to group similar bands of the hyperspectral images into the same cluster, using Principal Component Analysis (PCA). Finally, Fu and Robles-Kelly [9] use geometric properties of spectra to model absorption in addition to statistical techniques for band selection.

The above approaches focus on remote sensing images, which are captured at very high altitudes. In terms of material classification in indoor environments targeted mainly for industrial use, the following approaches have been proposed. Picon *et al.* [3] combine spectral and spatial information to increase efficiency and accuracy of the sorting of nonferrous material. Other researchers use polarization [10–12], or near infrared reflectance [13] for material classification, while Ibrahim *et al.* [14] use spectral reflectance for printed circuit board inspection.

Recently novel imaging systems have been presented in order to use illumination from different angles [1], or spectral BRDFs (Bidirectional Reflectance Distribution Functions) captured using illumination from different angles and spectral bands [2]. These capture systems are complex: Jehle *et al.* [1] consider illumination directions produced by illuminating a parabolic mirror using parallel light and therefore require a very accurate illumination process. Gu and Liu *et al.* [2] build a dome of 25 clusters corresponding to 25 illumination directions, with each cluster comprising 6 LEDs of different spectral bands. Ali *et al.* [15] measure reflectance of isotropic materials by carefully choosing a set of sampling directions such that the corresponding BRDF provides similar representation as the exhaustively captured set of BRDFs. Inspired by this line of work, we propose a simpler imaging system where the light directions are reduced to 3 and the tunable filter in front of the camera allows the capture of the effects of 32 narrow spectral bands. Imaging materials with narrow bands would provide images which are more discriminative in terms of subtle changes in the spectral reflectance of different materials.

3 Image Capturing System

The goal of our image capturing system is to be able to capture characteristics in the spectral reflectance, such that we can learn the most discriminative spectral bands for different materials.

3.1 System Components

To be able to record as much spectral information as possible, we place a Liquid Crystal Tunable Filter (LCTF) in front of the camera lens, as illustrated in Fig. 1. In this configuration, the filter selects a material response in a specific wavelength range, which will be recorded as an image in the camera sensor.

The LCTF that we use in this imaging system works in the range of 400–720 nm with 10 nm intervals between adjacent peaks, and therefore provides the capability of imaging using 32 narrow spectral bands. The plot of these bands are represented by the differently colored curves in the plot provided by the manufacturer (Fig. 2). To make sure that all the spectral components being measured by the LCTF are available, we illuminate the material using the white light of a conventional projector. As illustrated in Fig. 3, by projecting white light onto the scene we use the broadest available band range and therefore maximize the range of wavelengths which can be selected by the filter. In this work we consider 16 categories of flat material samples, which are metals.

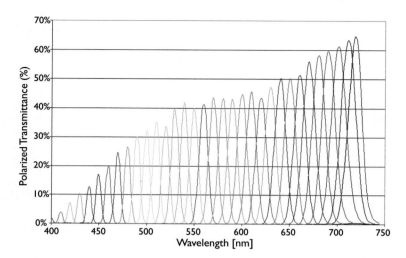

Fig. 2. Bandpass transmittance spectra as a function of wavelength for our LCTF. Plot provided by the manufacturer.

3.2   Illumination Directions

The appearance of some materials, especially metals, changes with the light direction. To include this variation in our images, we move the light source to 3 different positions during capture. As illustrated in Fig. 1, the reference position is aligned with the camera, while the other two are obtained by a rotation of θ_{01} and θ_{02}, with $\theta_{01} \neq \theta_{02}$, from the reference position. Experimentally we found that by setting $\theta_{01} = 45°$ and $\theta_{02} = 20°$ we can capture the most discriminative images of materials.

Therefore, for every material we consider, 3 sets of 32 images are captured: every set corresponds to a different position of the light source, and the images forming each set correspond to the spectral bands of the LCTF. These 3 sets of images are then merged into a single set of 32 images, by weighing the contributions given by the different illuminations in order to maximize material discriminability (Sect. 4.1). Consequently, we use the combined set to learn the optimal subset of spectral bands for binary and multi-class classification tasks (Sect. 4.2).

4   Learning-Based Band Selection

In this section we describe an approach to learn the weights of spectral bands for a given material classification task. The weights of the bands can be used to select the most discriminative bands.

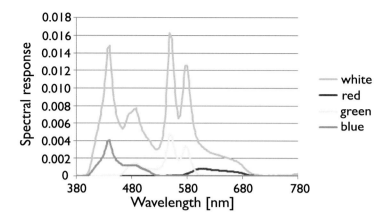

Fig. 3. The spectra for the four channels of the projector used as a light source: spectral response measured from the PR-650 spectrometer at each wavelength [nm] (Color figure online).

4.1   Illumination Weights

Before learning the spectral bands, we merge the contributions of the 3 different illuminations to the image corresponding to each band in order to increase the Signal-to-Noise-Ratio (SNR).

Let $\boldsymbol{I}_{\lambda L}$ be an image captured using the filter corresponding to the spectral band λ and under a light source placed at location L. For a given material, at each band λ, with $\lambda \in \{1, \ldots, 32\}$, the resulting image is a linear combination of the images captured at 3 different illuminations:

$$\boldsymbol{I}_\lambda = w_{\lambda 1}\boldsymbol{I}_{\lambda 1} + w_{\lambda 2}\boldsymbol{I}_{\lambda 2} + w_{\lambda 3}\boldsymbol{I}_{\lambda 3}. \tag{1}$$

The weights $w_{\lambda 1}$, $w_{\lambda 2}$, and $w_{\lambda 3}$ are estimated using a Support Vector Machine (SVM): the optimal weights should generate the most discriminative image $\boldsymbol{I}_\lambda$ compared to the images corresponding to the other bands.

4.2   Learning Spectral Bands

We use the Adaboost algorithm [16], where each spectral band is considered as a weak classifier $\boldsymbol{h}_\lambda(\mathbf{x})$:

$$H(x) = \sum_{\lambda=1}^{32} \alpha_\lambda \boldsymbol{h}_\lambda(\mathbf{x}). \tag{2}$$

The weights α_λ indicate the discriminative ability of each spectral band and are estimated for two-class material classification tasks. The feature $\mathbf{x}$ corresponds to a pixel in the image $\boldsymbol{I}_\lambda$, obtained from Eq. 1.

For a two-class classification task, the most discriminative spectral bands would be the first N bands with the largest weights, where N can be set by the user. For a multi-class task, instead, the classification algorithm is typically broken down into a set of binary classification tasks. The ensemble of the sets will eventually constitute more than N spectral bands if sets of selected bands are not exactly the same. In order to combine the N selected filters from the individual binary classification tasks, the weights α_λ calculated by the Adaboost algorithm are combined to obtain a new set of weights $W = \{W_1, \ldots, W_{32}\}$. W_λ is the combined weight for the spectral band λ and is estimated as

$$W_\lambda = \frac{1}{Z} \sum_{t \in T} e^{-\left(\frac{1 - \tilde{\alpha}_{\lambda,t}}{\tilde{\alpha}_{\lambda,t}}\right)^p}, \tag{3}$$

where $\tilde{\alpha}_{\lambda,t}$ is the normalized weight of spectral band λ corresponding to the binary classification task t, and Z is a normalization term. Finally the selected spectral bands are taken to be the first N bands with the largest W_λ. The shape of the mapping function in Eq. 3 is determined by the parameter p. The mapping function maps the weights of spectral bands as estimated in the two-class classification tasks to the set of weights W for the multi-class classification tasks. In our experiments we set $p = 0.5$.

5   Experiments

To evaluate the proposed approach we built the capturing system described in Fig. 1, and tested material classification performance on both metals and non-metals.

5.1   System Setup

In our capturing system, the camera is a Grasshopper GRASS-50S5M-C, which is a monochromatic camera with maximum resolution at 2448×2048 pixels and 16-bit depth of data. The lens used by the camera is a Canon 50 mm EF lens with an $f/1.8$ aperture size. The LCTF is a VariSpec filter manufactured by Cambridge Research &Instrumentation, Inc. (CRi), and now part of PerkinElmer; the working aperture size of the filter is 20 mm. The LCTF used in our setup has filters in the wavelength range of 400–720 nm with 10 nm intervals between adjacent peaks, as displayed in Fig. 2. A picture of the capturing imaging system is shown in Fig. 4.

The Grasshopper camera provides pixel binning where blocks of pixels, such as 2×2 or 3×3 adjacent pixels are combined into a single pixel to achieve an increased sensitivity and reduced SNR.

The light source is generated by a conventional InFocus LP-600 projector, whose spectrum has been measured using a PR-650 spectrometer from Photo Research (Fig. 3).

The material samples used in this work can be divided into 16 types of metals: Zinc (2 types), Mild Steel, Nickel, Lead, Stainless Steel, Copper, Brass, Steel, Titanium, Tin, Aluminum (5 types). In Fig. 5 we show the two groups of metals as used in our experiments (Sect. 5.2).

Fig. 4. Camera setup. The liquid filter is placed in front of the camera lens.

(a)   (b)

Fig. 5. Material samples used in experiment. There are two groups of material categories, which from left to right are: (a) Zinc, Steel, Lead, Aluminum, Brass and (b) Nickel, Titanium, Tin, Aluminum, Copper. This figure is better viewed on a monitor than in print.

5.2   Material Classification

We tested our algorithm on several material classification tasks. We report how the material classification performance changes when using only images corresponding to the most discriminative spectral bands given as outputted by our method.

Given a five-class classification task (Nickel, Titanium, Tin, Aluminum, and Copper), the algorithm described in Sect. 4 ranks the spectral bands for each binary task. Table 1 shows the wavelengths of the 5 bands with the largest weights, in descending order, for each pair of materials. Using Eq. 3 we can then compute the weights and obtain the wavelengths of the spectral bands selected for this five-class classification task: 550, 580, 440, 690, and 420 nm. The images corresponding to the N optimal bands selected for each binary task are then used to train and test linear SVM classifiers, with 50 % of the pixels used for training and 50 % used for testing. In total, four images per spectral band were captured for each material, each of which has an approximate size of 1000×800 pixels. The classification accuracies in % are reported in Table 1. Also shown in the table are accuracies computed using the images from all bands, when no learning is performed, for comparison.

Results show that only the images corresponding to the selected subset of spectral bands are sufficient to provide competitive performance. Also note that for a few binary classification tasks there is an improvement when using only the

Table 1. Binary classification tasks within a multi-class material classification framework: Nickel, Titanium, Tin, Aluminum, and Copper. The first column shows the materials used in each binary classification task. The second column shows the selected bands for the task arranged in descending order of their weights. The third column shows the accuracy given the N selected bands. The last column shows the accuracy given all bands (32 in this case). The selected 5 bands for the framework are: 550, 580, 440, 690, and 420 nm.

	Selected bands [nm] ($N = 5$)	% Accuracy (N bands)	% Accuracy (all bands)
Nickel-Titanium	550, 580, 560, 670, 680	99.9	99.9
Nickel-Tin	580, 550, 570, 710, 600	99.9	100.0
Nickel-Aluminum	440, 690, 490, 430, 700	70.6	77.4
Nickel-Copper	420, 450, 440, 710, 430	98.8	99.6
Titanium-Tin	690, 700, 530, 430, 570	97.1	94.8
Titanium-Aluminum	550, 580, 500, 480, 570	99.9	100.0
Titanium-Copper	550, 560, 570, 710, 600	99.9	100.0
Tin-Aluminum	550, 580, 480, 440, 420	99.9	100.0
Tin-Copper	550, 580, 530, 430, 570	99.9	100.0
Aluminum-Copper	440, 420, 430, 690, 700	99.8	99.8

N images corresponding to the most discriminative bands. This is due to the fact that the most discriminative features, corresponding to the selected bands, are used for classification.

We also plot the spectral bands selected for each binary classification task as listed in Table 1. Since the spectral bands are too many to show in one figure, we group them into three as shown in Fig. 6. The grouping is performed by taking all the binary tasks including (a) Nickel, (b) Titanium, and (c) Copper. In each plot, the spectral band curves for one binary classification task are shown in one color. For example, the 5 spectral bands selected for discriminating Nickel and Titanium are shown in green in Fig. 6(a). The peaks of the spectra are proportional to the weights of the bands (Eq. 3), meaning the larger the weight, the higher the peak. By examining the plots we can observe that the spectral bands selected for each binary classification task generally span two regions of the wavelength range, except for when Copper is one of the materials. In the latter case, the most selected bands for one binary task lie generally in one region of the wavelength range. Such a result can be explained by the fact that Copper is bronze in color and therefore can be more easily discriminated from the remainder of materials which are gray. The classification accuracies depicted for the binary tasks including Copper are in line with this explanation as shown in Table 1 (rows 4, 7, 9, and 10).

In order to more closely examine the impact of using different numbers of filters on the multi-class classification algorithm, we consider different numbers

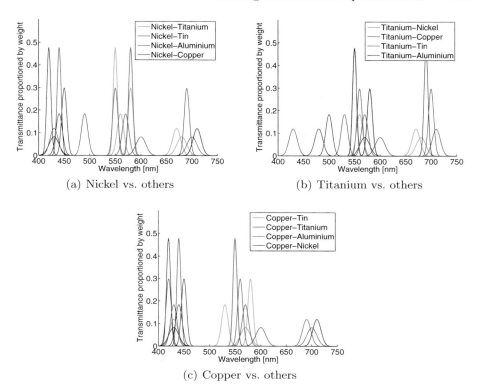

(a) Nickel vs. others

(b) Titanium vs. others

(c) Copper vs. others

Fig. 6. Visualization of the selected spectral bands for the binary material classification tasks corresponding to Table 1 when (a) Nickel, (b) Titanium, and (c) Copper is one of the materials. The spectra corresponding to the selected bands for one binary classification task are represented in one color. The peaks of the spectral band curves are proportional to their weights as computed by our algorithm (Color figure online).

of N selected filters: 3, 5, 7, 11, and 32. Table 2 lists the classification accuracies for the multi-class classification framework when images corresponding to each number of filters are used. The material categories used are Nickel, Titanium, Tin, Aluminum, and Copper (shown in the top row), and Zinc, Steel, Lead, Aluminum and Brass (shown in the bottom row). Once again, results show that increasing N does not necessarily lead to an increase in classification accuracy.

Our results show that we can reduce the number of spectral bands needed for capture and consequently the dimensionality of the feature vectors needed to represent the materials in the classification algorithm. For a binary material classification case, the capture time for one image given one illumination direction would be 32 bands × 5 s = 160 s. Using only 5 bands, the capture time would be 5 bands × 5 s = 25 s. Additionally, the sizes of the feature vectors for the 32 band case and the 5 band case are about 100 Mb and 16 Mb respectively. This implies that the sizes of the feature vectors are reduced by a factor of 6 upon

Table 2. Overall classification accuracies using different numbers of spectral bands for two multi-class classification scenarios. Top row (NTTAC): Nickel, Titanium, Tin, Aluminum, and Copper; Bottom row (ZSLAB): Zinc, Steel, Lead, Aluminum, and Brass.

	Spectral bands used (N)				
	3	5	7	11	ALL
% Acc. (NTTAC)	90.3	96.6	92.8	94.2	97.2
% Acc. (ZSLAB)	85.5	86.9	82.4	82.7	82.8

using a subset of filters. Note that while in this example case very few images are used, the number of images can scale considerably in a factory or industrial application setting.

Given that our method demonstrates good performance in the visible range, it would be interesting to extend it to beyond the visible, especially given the increase in popularity of RGB-IR sensors in mobile devices.

6   Conclusion

In this paper we described a novel setup to capture the spectral characteristics of different materials while considering multiple directions of illumination. We then proposed a learning algorithm to select the most discriminative spectral bands and showed that a small set of bands can be used without compromising on material classification performance. Selecting spectral bands implies reducing the sizes of the feature vectors, consequently reducing training and test computation time and storage. Decreasing computation time and storage allows for better portability in real-time and mobile devices.

References

1. Jehle, M., Sommer, C., Jähne, B.: Learning of optimal illumination for material classification. In: Goesele, M., Roth, S., Kuijper, A., Schiele, B., Schindler, K. (eds.) Pattern Recognition. LNCS, vol. 6376, pp. 563–572. Springer, Heidelberg (2010)
2. Gu, J., Liu, C.: Discriminative illumination: per-pixel classification of raw materials based on optimal projections of spectral BRDF. In: IEEE Conference on Computer Vision and Pattern Recognition (2012)
3. Picón, A., Ghita, O., Bereciartua, A., Echazarra, J., Whelan, P., Iriondo, P.: Real-time hyperspectral processing for automatic nonferrous material sorting. J. Electron. Imaging **21**(1), 1–8 (2012)
4. Bajcsy, P., Groves, P.: Methodology for hyperspectral band selection. Photogram. Eng. Remote Sens. J. **70**, 793–802 (2004)
5. Chang, C.I., Wang, S.: Constrained band selection for hyperspectral imagery. IEEE Trans. Geosci. Remote Sens. **44**, 1575–1585 (2006)
6. Martínez-Usó, A., Pla, F., Sotoca, J., Garcia-Sevilla, P.: Clustering-based hyperspectral band selection using information measures. IEEE Trans. Geosci. Remote Sens. **45**, 4158–4171 (2007)

7. Qian, Y., Yao, F., Jia, S.: Band selection for hyperspectral imagery using affinity propagation. IET Comput. Vision **3**, 213–222 (2009)
8. Kumar, V., Hahn, J., Zoubir, A.: Band selection for hyperspectral images based on self-tuning spectral clustering. In: European Signal Processing Conference (2013)
9. Fu, Z., Robles-Kelly, A.: Discriminant absorption feature learning for material classification. IEEE Trans. Geosci. Remote Sens. **49**, 1536–1556 (2010)
10. Tominaga, S., Kimachi, A.: Polarization imaging for material classification. Opt. Eng. **47** (2008)
11. Chen, H., Wolff, L.: Polarization phase-based method for material classification in computer vision. Int. J. Comput. Vision **28**, 73–83 (1998)
12. Wolff, L.: Polarization-based material classification from specular reflection. IEEE Trans. Pattern Anal. Mach. Intell. **12**, 1059–1071 (1990)
13. Salamati, N., Fredembach, C., Süsstrunk, S.: Material classification using color and NIR images. In: IS&T Color Imaging Conference (2009)
14. Ibrahim, A., Tominaga, S., Horiuchi, T.: Spectral imaging method for material classification and inspection of printed circuit boards. Opt. Eng. **49**, 057201–057210 (2010)
15. Ali, M., Sato, I., Okabe, T., Sato, Y.: Toward efficient acquisition of BRDFs with fewer samples. In: IEEE Asian Conference on Computer Vision (2012)
16. Freund, Y., Schapire, R.: Experiments with a new boosting algorithm. In: International Conference on Machine Learning (1996)

A Deep Belief Network for Classifying Remotely-Sensed Hyperspectral Data

Justin H. Le, Ali Pour Yazdanpanah[(✉)], Emma E. Regentova,
and Venkatesan Muthukumar

Department of Electrical and Computer Engineering,
University of Nevada, Las Vegas, Las Vegas, USA
`pouryazd@unlv.nevada.edu`

Abstract. Improving the classification accuracy of remotely sensed data is of paramount interest for science and defense applications. In this paper, we investigate deep learning architectures (DLAs), whose popularity has grown recently due to the discovery of efficient algorithms to train them, one of which, unsupervised pre-training, seeks to initialize the learned model in a way that greatly encourages efficient supervised learning. We propose a structure for a DLA, the deep belief network (DBN), suitable for the classification of remotely-sensed hyperspectral data. To arrive at this structure, we first study the role of the DBN's width and the duration of pre-training in the learning of features used for the multiclass discrimination of spectral data. We then study the effect of exploiting joint spectral-spatial information. The support vector machine (SVM) is used as a baseline to determine that the proposed method is feasible, offering consistently high classification accuracies in comparison.

1 Introduction

Research in hyperspectral data classification has focused primarily on either feature selection (FS) or feature extraction (FE) as a way to prime the data for use by a classifier [1–3, 22, 23]. FS involved the limiting of data to an appropriate subset of spectral bands, and FE involved the transformation of the data into a space of lesser dimensionality [2]. Being especially well-suited for the classification of high-dimensional data with few training samples, support vector machines (SVMs) were a prime candidate for the classification of hyperspectral data and could have potentially removed the need for further research in dimensionality reduction [3–5]. Interest in feature extraction remains strong, however, as methods that exploit joint spectral-spatial information have been shown to outperform SVMs that use untransformed data [3, 6–9].

We are primarily interested in deep learning architectures (DLA) that perform FE effectively for hyperspectral data. DLAs are a recent development in neural networks that have achieved historically high classification accuracies and, in many cases, have yet to be outperformed by other methods [10–12, 24]. It has been suggested that their ability to generalize comes closer to hard artificial

© Springer International Publishing Switzerland 2015
G. Bebis et al. (Eds.): ISVC 2015, Part I, LNCS 9474, pp. 682–692, 2015.
DOI: 10.1007/978-3-319-27857-5_61

intelligence than do other state-of-the-art learning machines [13–15]. Two similar types of DLA, the deep belief network (DBN) and the stacked autoencoder, have been shown to outperform SVMs in the classification of hyperspectral data [16–18]. These experiments sought to determine how the depth of the networks and the number of principal components of the data would affect classification accuracy and running time. In this paper, we seek to further study how the structure of a DBN affects its performance by observing how the accuracy is affected by the width of the network and the duration of training. Based on the results of our study, we propose a structure for a DBN that allows it to achieve high classification accuracies on remotely-sensed hyperspectral data, using a support vector machine (SVM) as a reference for comparison.

1.1 Why Deep Architectures?

We define a local estimator to be a learning machine that classifies an input x under the assumption that the target function $f(x)$ is smooth. That is, after learning from a data point (x_i, y_i), the estimator outputs a y similar to y_i if x is similar to x_i. Examples of local estimators are kernel machines such as the support vector machine and nonparametric methods such as the k-nearest neighbor algorithm. Local estimators have two major disadvantages [13–15]. Firstly, under the smoothness assumption, each new input x can only be predicted if the estimator has previously seen training data x_i in the neighborhood of x, and the estimator would thus require many training examples in order to generalize well. For target functions that have many complex variations, the required amount of training data becomes unmanageably large. Secondly, good local estimators employ either well-designed features or an appropriate kernel for a particular task, which requires prior knowledge of the task that is often unavailable.

These issues can be addressed by learning representations. Particularly, the learned representations must be compact and distributed [13–15,19]. Here, "distributed" refers to the effect of representing an input x by features that are not mutually exclusive, i.e., that encode information about x through their collective pattern of activation rather than through the local, independent activation of each feature. Distributed representations can be compact in the sense that they may require exponentially fewer values to encode information about x than would be required by one-hot representations such as those learned by local estimators. For example, the local (one-hot) representation of $x = 7$ in binary digits is 1000000, whereas its distributed (base-2) representation is 111. This shows that a distributed representation can encode information in fewer digits, or features, than a local one.

Furthermore, the expressive power of distributed representations allows them to be reused in forming more abstract representations [13–15,19,20]. It is expected that, at higher levels of abstraction, the learned representation can capture any essential factors that explain the data and shed any inessential details that would otherwise complicate it. A machine that learns such high-level representations can generalize well, even when the target function varies greatly in the input space, as the prediction does not rely solely on the target function's smoothness but also on the assumption that descriptive patterns in the data can be discovered and used in explaining unseen data. The reuse of learned representations can be achieved by learning a representation at each layer of a deep neural network, passing the representation learned at each layer as the input to the next layer. With the advent of efficient training algorithms and highly-parallel computing hardware, this deep approach to learning has become feasible [10, 20].

2   Method

To explain the structure and theory of the DBN, we first describe its main component, the Restricted Boltzmann Machine (RBM) and the algorithm used to train it. We then describe how training is performed for the DBN as a whole and the parameters involved in the construction and training of the DBN. We refer to these parameters henceforth as hyper-parameters to distinguish them from the parameters of the learned model (the weights and biases that connect the layers of the DBN).

2.1   Restricted Boltzmann Machines

A distributed representation can be formed by a restricted Boltzmann machine (RBM), an undirected graphical model composed of two layers of units, one visible $\mathbf{v}$ and one hidden $\mathbf{h}$, with full connections between the two layers. The RBM is an energy-based model; the joint distribution of the layers is expressed as the Boltzmann distribution, a function of the energy of the joint configuration of units:

$$P(\mathbf{h}|\mathbf{v}) = \frac{\exp(-\text{Energy}(\mathbf{v}, \mathbf{h}; \theta))}{\sum_{\mathbf{h}} \exp(-\text{Energy}(\mathbf{v}, \mathbf{h}; \theta))}. \tag{1}$$

θ is the model to be updated during learning and includes weights $\mathbf{w}$ between pairs of connected units, visible biases $\mathbf{b}$, and hidden biases $\mathbf{c}$. As in the definition borrowed from statistical physics, states of high energy have a low probability of occurrence. Units within the same layer are not connected (hence the name "Restricted"), which allows the energy function to be linear in both $\mathbf{v}$ and $\mathbf{h}$ and as a result, to factorize well, leading to an expression for the conditional probability that can be efficiently computed [14]. For the case where $v_i = \{0, 1\}^I$ and $h_j = \{0, 1\}^J$, where I and J are the number of visible and hidden units, respectively:

$$P(h_j|\mathbf{v}) = \text{sigmoid}(c_j + \sum_{i \in I} W_{ij} v_i), \tag{2}$$

where W_{ij} is the matrix of weights between visible and hidden units. Symmetry across the two layers allows the derivation of Eq. 2 to be repeated for the reverse conditional case [14]:

$$P(v_i|\mathbf{h}) = \text{sigmoid}(c_i + \sum_{j \in J} W_{ij} h_j), \tag{3}$$

which enables the RBM to reconstruct its visible units. Reconstruction is an efficient way to estimate the log-probability gradient of the visible units with respect to the model and thus leads to learning rules that can be efficiently computed [20]:

$$\Delta w = \epsilon(\mathbf{h_0}\mathbf{v_0'} - P(\mathbf{h_1.} = \mathbf{1}|\mathbf{v_1})\mathbf{v_1'}) \tag{4}$$

$$\Delta b = \epsilon(\mathbf{h_0} - P(\mathbf{h_1.} = \mathbf{1}|\mathbf{v_1})) \tag{5}$$

$$\Delta c = \epsilon(\mathbf{v_0} - \mathbf{v_1}), \tag{6}$$

where Δ is the change in a parameter per update, ϵ is the learning rate (the scaling factor of the change per update), and $\mathbf{v_k}$ and $\mathbf{h_k}$ are the k-th sample of the visible and hidden vectors' probability distributions as computed by Eqs. 3 and 2, respectively. The symbol $'$ denotes the transpose of a vector. This method of approximation is known as contrastive divergence [10,19].

2.2   Deep Belief Networks

RBMs can be stacked by feeding the hidden vector of one RBM as the input (visible vector) of another RBM. We use the term "layer" to refer to a layer of model parameters between two vectors of visible or hidden units. In the layers beyond the first one, the conditional probabilities given by Eqs. 2 and 3 can be rewritten in the general form

$$P(h_j^{(\ell+1)}|\mathbf{h}^{(\ell)}) = \text{sigmoid}(\mathbf{c_i}^{(\ell+1)} + \sum_{j \in \mathbf{J}} \mathbf{W_{ij}}^{(\ell+1)} \mathbf{h}_j^{(\ell)}) \tag{7}$$

where $\mathbf{h}^{(\ell)}$ denotes the input vector of the RBM at the ℓ-th layer of the network, and $\mathbf{h^{(0)}}$ represents the input vector of the DBN [10,20]. When the network is constructed in this way and trained layerwise in an unsupervised fashion using Eq. 7, it is known as a Deep Belief Network (DBN). Although a DBN can be used as a generative model due to the RBM's ability to reconstruct data, we are only interested here in the discriminative case, as our intention is to classify data rather than to reconstruct it. To use a DBN discriminatively, the output of the top-layer RBM is fed as input to a classifier such as a logistic regression, at which point, the model is then trained in a supervised fashion [10,20], as shown in Fig. 1.

The unsupervised training stage is referred to as pre-training. Its purpose is to initialize the model in such a way as to improve the efficiency of supervised training. The supervised training stage is referred to as fine-tuning, as it adjusts the classifier's prediction to match the ground truth of the data. Each iteration of pre-training or fine-tuning is referred to as an epoch [10,21].

In order for accurate predictions to be made, the DBN must learn to produce "good" representations. A good representation is one that captures only the most essential underlying patterns of the data and discards the rest. The best representation is one which not only captures these patterns, but disentangles them from each other, just as the human mind learns to recognize an object

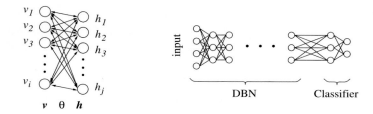

Fig. 1. Left: An RBM with i visible units, j hidden units, and model parameters θ. Right: A DBN with arbitrary width and a classifier, e.g., softmax, at the output

by disentangling it from the shadows cast upon it [13]. These definitions give rise to the term "representation learning". The term "deep learning" refers to the implementation of this theory through a learning machine that consist of many layers, such as a DBN. In this paper, the DBN is chosen over other deep architectures as we are interested in studying one particular technique used by the DBN, pre-training, whose nature has yet to be fully understood [13].

3   Experiments

To study the performance of the DBN across multiple settings and ensure that it yields consistently high accuracies, we conduct experiments in classification with spectral signatures, in which we vary the hyper-parameters whose effect on performance would be unpredictable without empirical guidance. We then repeat the experiments using joint spectral-spatial information in an attempt to support or dispute the evidence of any trends observed in spectral classification. In this section, we describe the dataset, our choice of hyper-parameters and their justification, and the process of learning using spectral and spectral-spatial data.

3.1   Data

Pavia Centre is a hyperspectral dataset with 102 bands and 1096-by-715 pixels, depicting a scene from the town of Pavia in Italy. Figure 4 depicts a false-color map of the data. The data was collected by the ROSIS sensor at a geometric resolution of 1.3 m. The sensor has a bandwidth of 4 nm and a wavelength range of 400 to 900 nm.

The nine materials present in Pavia Centre are labelled according to the color scheme shown in Fig. 4. Pixels that belong to the zero class are those that do not belong to any of these nine classes. Zero-class pixels have their own spectral signatures, but they must be excluded from the training process, as they are so numerous (relative to other classes) that they would skew the learning process in a way that causes the classifier to predict only zero-class pixels. Training was thus performed using only the pixels that belong to one of the nine labelled classes. Particularly, training was performed using a randomly selected half of

Table 1. The labels, names, and number of training and validation samples of each land cover class in the spectral and spectral-spatial datasets

Label	Class	Number of samples					
		Spectral data			Spectral-spatial data		
		Training	Validation	Total	Training	Validation	Total
1	Water	32986	32985	65971	32662	32661	65323
2	Trees	3799	3799	7598	3767	3767	7534
3	Asphalt	1545	1545	3090	1470	1470	2940
4	Bricks	1343	1342	2685	1343	1342	2685
5	Bitumen	3292	3292	6584	3285	3285	6570
6	Tiles	4624	4624	9248	4590	4589	9179
7	Shadows	3644	3643	7287	3644	3643	7287
8	Meadows	21413	21413	42826	21190	21189	42379
9	Bare soil	1432	1431	2863	1429	1428	2857

this set, with the other half of the pixels reserved for validation. Table 1 shows how many pixels from each class are present in the dataset.

Each pixel of the Pavia Centre dataset is treated as a single data point. The values at the input (visible) layer are normalized with respect to the maximum value across all pixels and all bands (a value of 8000).

3.2   Variation of Hyper-Parameters

We define width to be the number of units in a hidden vector $\mathbf{h}^{(\ell)}$ for $\ell > 0$, as described by Eq. 7, and depth to be the number of layers of the model. It is expected that accuracy will either improve or remain roughly the same as the number of fine-tuning epochs increases, and so we fix the fine-tuning duration at 20,000 epochs. In this paper, we are only interested in studying the width of the DBN and the pre-training duration, and so, we wish to fix the depth. We choose a depth of 4 for spectral classification and 3 for spectral-spatial, as suggested by the experimental results in [18].

Given that the fine-tuning duration and depth are fixed, the hyperparameters that chiefly interest our study are the width of the layers and the pre-training duration. We perform a grid-search of several configurations of these two hyper-parameters as given in Tables 2 and 3. The widths vary in increments of 5 from 20 to 35. These values are chosen because they correspond to a "valley" in the classification error; we found that widths beyond this range tend toward higher errors regardless of other hyper-parameter settings. For the same reason, we vary the pre-training duration between 1000 and 3000 epochs.

For all experiments, we use binary RBM units and apply the equations given in Sects. 2.1 and 2.2.

3.3 Learning from Spectral Features

The input vector is defined as $\mathbf{h}^{(\ell)}$ in Eq. 7 for $\ell = 0$ and represents the visible vector of the first-layer RBM. To learn from spectral signatures, we construct the input vector as a pixel of the hyperspectral image, as shown in Fig. 2, and its length is thus equal to the number of spectral bands (102). The first RBM undergoes pre-training by first updating the binary values and probabilities of activation of its hidden units using Eq. 2, then reconstructing the visible units (input vector) using Eq. 3. To obtain the conditional probability used in Eqs. 4 and 5 for updating the weights and visible biases, respectively, the probabilities of the hidden units are then computed again using the reconstructed visible vector as the prior. This process is repeated for each data point (pixel). An epoch is defined as the time taken for this process to take place over all data points in a layer. Once the specified number of epochs have taken place in the first layer, the process is repeated for all subsequent layers using the hidden vector from the previous layer as the visible (input) vector of the current layer, as described by Eq. 7.

Table 2 compares the DBN's performance to that of a radial basis function kernel support vector machine (RBF-SVM). "Best Epoch" indicates the epoch that produced the highest overall accuracy (OA) for a certain configuration. OA is measured as the total number of correctly classified samples to the total number of samples, whereas average accuracy (AA) is measured as the average across all classes of the number of correctly classified samples in a class to the number of samples in that class. The Kappa statistic (κ) measures the agreement between these two accuracies. The DBN outperforms the SVM when constructed with a width of 25 and pre-trained for 1000 epochs, yielding an OA of 97.928 %, an improvement of 2.528 %. Although the SVM yielded a higher AA than the DBN, the DBN yields a higher κ in the case when it achieves the highest OA.

3.4 Learning from Spectral-Spatial Features

Spatial information conveys the probability that the given pixel belongs to a particular, e.g., a pixel is likely to depict soil if it is surrounded by other soil pixels, as soil often appears in patches that are spanned by multiple pixels. To extract spatial information, a neighborhood of 7-by-7 pixels is chosen as a

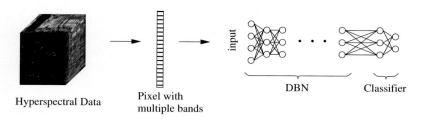

Fig. 2. Training the DBN pixelwise with spectral signatures

data point. The center pixel of this neighborhood represents this data point's coordinate. Hence, the label of the center pixel is used as the label for the data vector extracted in this manner. The values are extracted in row major fashion. Due to the immensity of these data points, the number of bands for each pixel in a neighborhood must be reduced to 5 using principal component analysis, as suggested by [18]. Each reduced data point obtained this way is then stacked onto the corresponding data point used in the spectral classification method to form a combined spectral-spatial feature vector, which is then fed as input to the DBN, as shown in Fig. 3.

Unlike the previous experiment, a width of 35 yields the lowest accuracy when the duration of pre-training is fixed, and the best case is achieved with more pre-training epochs rather than less. The other three cases show that fewer epochs yield higher accuracy with fixed width.

Table 2. Overall accuracy, average accuracy, Kappa statistic, and the epoch at which the highest accuracy was reached for spectral classification

Pre-training epochs	Width	OA (%)	AA (%)	κ	Best epoch
1000	20	92.2828	89.6493	0.2544	19888
	25	**97.9285**	89.7680	**0.7975**	19737
	30	94.3058	89.7029	0.4470	4492
	35	97.9224	89.7644	0.7970	17954
3000	20	92.5076	89.6689	0.2748	15736
	25	93.3237	89.7156	0.3508	18099
	30	95.2784	89.7092	0.5412	19881
	35	94.0290	89.7113	0.4197	7776
	SVM	95.4054	**89.8087**	0.5492	

The results of the two experiments show that a greater number of pre-training epochs does not necessarily yield higher accuracy, although in some cases it does. It is possible that further experiments would show no correlation between pre-training duration and accuracy. In contrast, the results suggest that a width of 25 is ideal for learning the features of the Pavia Centre dataset, as both spectral and spectral-spatial classification reach a peak in accuracy when using this width.

3.5   Logistic Regression

The final hidden vector of the DBN is fed as input to a logistic regression stage in order to learn a model for predicting the class to which a data point belongs, as depicted by the Classifier in Fig. 1. Learning in this layer is referred to as fine-tuning and is performed by stochastic gradient descent. As in pre-training, an epoch is defined as the time taken for fine-tuning to take place over all data points.

Table 3. Overall accuracy, average accuracy, Kappa statistic, and the epoch at which the highest accuracy was reached for spectral-spatial classification

Pre-training epochs	Width	OA (%)	AA (%)	κ	Best epoch
1000	20	97.3268	83.5989	0.8370	18897
	25	97.3431	83.8029	0.8360	18247
	30	97.8474	83.7758	0.8673	18543
	35	96.3612	83.1496	0.7841	19832
3000	20	95.0835	83.6071	0.7001	12906
	25	**98.1233**	83.7873	**0.8842**	19921
	30	95.6395	83.6134	0.7339	17591
	35	93.7642	83.3949	0.6245	6415
	SVM	89.3068	**83.8555**	0.3377	

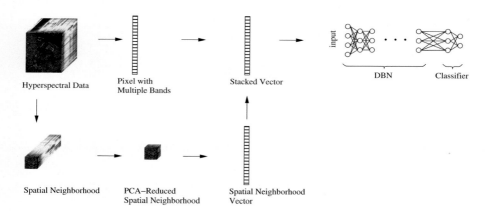

Fig. 3. Stacking spectral data onto spatial data that is constructed by flattening PCA-reduced neighborhoods

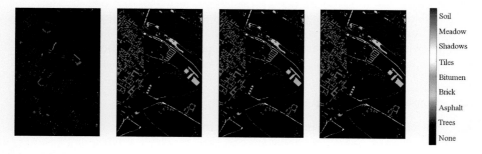

Fig. 4. From left to right: the Pavia Centre data in grayscale, the color maps for the ground truth and labels predicted by spectral and spectral-spatial classification, and the color scale (Color figure online)

In the experiments that produced the highest accuracies, the best epoch occurred near the final epoch of fine-tuning, which shows that in these cases, the accuracy improved even toward the end of training, suggesting that it would continue to improve had training been extended beyond the maximum number of epochs used.

4   Conclusion

We have presented a new DBN structure for the classification of remotely-sensed hyperspectral images. To achieve the best classification performance, we investigated the effect of the DBN's width and pre-training time on multi-class discrimination by spectral signatures. Then, we incorporated spatial information to form a spectral-spatial input, which improved the accuracy for five out of eight cases and produced the best accuracy. The performance of the achieved architecture was compared with that of the radial basis function kernel support vector machine.

Acknowledgements. This research was supported by NASA EPSCoR under cooperative agreement No. NNX10AR89A.

References

1. Bruce, L., Koger, C., Li, J.: Dimensionality reduction of hyperspectral data using discrete wavelet transform feature extraction. IEEE Trans. Geosci. Remote Sens. **40**, 2331–2338 (2002)
2. Harsanyi, J., Chang, C.I.: Hyperspectral image classification and dimensionality reduction: an orthogonal subspace projection approach. IEEE Trans. Geosci. Remote Sens. **32**, 779–785 (1994)
3. Kang, X., Li, S., Benediktsson, J.: Spatial hyperspectral image classification with edge-preserving filtering. IEEE Trans. Geosci. Remote Sens. **52**, 2666–2677 (2014)
4. Melgani, F., Bruzzone, L.: Classification of hyperspectral remote sensing images with support vector machines. IEEE Trans. Geosci. Remote Sens. **42**, 1778–1790 (2004)
5. Pal, M., Foody, G.: Feature selection for classification of hyperspectral data by SVM. IEEE Trans. Geosci. Remote Sens. **48**, 2297–2307 (2010)
6. Li, J., Bioucas-Dias, J., Plaza, A.: Hyperspectral image segmentation using a new bayesian approach with active learning. IEEE Trans. Geosci. Remote Sens. **49**, 3947–3960 (2011)
7. Li, J., Bioucas-Dias, J.M., Plaza, A.: Semisupervised hyperspectral image classification using soft sparse multinomial logistic regression. IEEE Geosci. Remote Sens. Lett. **10**, 318–322 (2013)
8. Li, J., Bioucas-Dias, J., Plaza, A.: Spatial classification of hyperspectral data using loopy belief propagation and active learning. IEEE Trans. Geosci. Remote Sens. **51**, 844–856 (2013)
9. Bernard, K., Tarabalka, Y., Angulo, J., Chanussot, J., Benediktsson, J.: Spatial classification of hyperspectral data based on a stochastic minimum spanning forest approach. IEEE Trans. Image Process. **21**, 2008–2021 (2012)

10. Hinton, G.E., Osindero, S., Teh, Y.W.: A fast learning algorithm for deep belief nets. Neural Comput. **18**, 1527–1554 (2006)
11. Larochelle, H., Erhan, D., Courville, A., Bergstra, J., Bengio, Y.: An empirical evaluation of deep architectures on problems with many factors of variation. In: Proceedings of the 24th International Conference on Machine Learning, ICML 2007, pp. 473–480. ACM, New York (2007)
12. Hinton, G.E., Salakhutdinov, R.R.: Reducing the dimensionality of data with neural networks. Science **313**, 504–507 (2006)
13. Bengio, Y., Courville, A.C., Vincent, P.: Unsupervised feature learning and deep learning: a review and new perspectives. CoRR abs/1206.5538 (2012)
14. Bengio, Y.: Learning deep architectures for AI. Found. Trends Mach. Learn. **2**, 1–127 (2009)
15. Bengio, Y., Lecun, Y., Operationnelle, D.D.E.R., Montreal, U.D.: Scaling learning algorithms towards AI. In: Bottou, L., Chapelle, O., DeCoste, D., Weston, J. (eds.) Large-Scale Kernel Machines. MIT Press, Cambridge (2007)
16. Lin, Z., Chen, Y., Zhao, X., Wang, G.: Spectral-spatial classification of hyperspectral image using autoencoders. In: 2013 9th International Conference on Information, Communications and Signal Processing (ICICS), pp. 1–5 (2013)
17. Chen, Y., Lin, Z., Zhao, X., Wang, G., Gu, Y.: Deep learning-based classification of hyperspectral data. IEEE J. Sel. Top. Appl. Earth Obs. Remote Sens. **7**, 2094–2107 (2014)
18. Chen, Y., Zhao, X., Jia, X.: Spectral-spatial classification of hyperspectral data based on deep belief network. IEEE J. Sel. Top. Appl. Earth Obs. Remote Sens. **8**, 2381–2392 (2015)
19. Le Roux, N., Bengio, Y.: Representational power of restricted boltzmann machines and deep belief networks. Neural Comput. **20**, 1631–1649 (2008)
20. Bengio, Y., Lamblin, P., Popovici, D., Larochelle, H., Montral, U.D., Qubec, M.: Greedy layer-wise training of deep networks. In: Schölkopf, B., Platt, J., Hoffman, T. (eds.) NIPS. MIT Press, Cambridge (2007)
21. Bengio, Y.: Practical recommendations for gradient-based training of deep architectures. CoRR abs/1206.5533 (2012)
22. Serpico, S., Bruzzone, L.: A new search algorithm for feature selection in hyperspectral remote sensing images. IEEE Trans. Geosci. Remote Sens. **39**, 1360–1367 (2001)
23. Chang, C.I., Du, Q., Sun, T.L., Althouse, M.: A joint band prioritization and band-decorrelation approach to band selection for hyperspectral image classification. IEEE Trans. Geosci. Remote Sens. **37**, 2631–2641 (1999)
24. Makantasis, K., Karantzalos, K., Doulamis, A., Doulamis, N.: Deep supervised learning for hyperspectral data classification through convolutional neural networks. In: IGARSS, pp. 1771–1800 (2015)

Variational Inference for Background Subtraction in Infrared Imagery

Konstantinos Makantasis[1][✉], Anastasios Doulamis[2],
and Konstantinos Loupos[3]

[1] Technical University of Crete, University Campus,
Kounoupidiana, 73100 Chania, Greece
kmakantasis@isc.tuc.gr
[2] National Technical University of Athens,
Zografou Campus, 15780 Athens, Greece
adoulam@cs.ntua.gr
[3] Institute of Communication and Computer Systems, Athens, Greece
kloupos@iccs.gr

Abstract. We propose a Gaussian mixture model with fixed but unknown number of components for background subtraction in infrared imagery. Following a Bayesian approach, our method automatically estimates the number of components as well as their parameters, while simultaneously it avoids over/under fitting. The equations for estimating model parameters are analytically derived and thus our method does not require any sampling algorithm that is computationally and memory inefficient. The pixel density estimate is followed by an efficient and highly accurate updating mechanism, which permits our system to be automatically adapted to dynamically changing visual conditions. Experimental results and comparisons with other methods indicate the high potential of the proposed method while keeping computational cost suitable for real-time applications.

1 Introduction

Pixels values of infrared frames correspond to the relative differences in the amount of thermal energy emitted or reflected from objects in the scene. Due to this fact, infrared cameras are equally applicable for both day and night scenarios, while at the same time, compared to visual-optical cameras, are less affected by changing illumination or background texture. Furthermore, infrared imagery eliminates any privacy issues as people being depicted in the scene can not be identified. These features make infrared cameras prime candidate for persistent video surveillance systems.

Although, infrared imagery can alleviate several problems associated with visual-optical videos, it has its own unique challenges such as (a) low signal-to-noise ratio (noisy data) and (b) almost continuous pixel values that model objects' temperature. Both issues complicate pixel responses modeling and most of conventional computer vision techniques, that successfully used for visual-optical data, can not be applied straightforward on thermal imagery.

© Springer International Publishing Switzerland 2015
G. Bebis et al. (Eds.): ISVC 2015, Part I, LNCS 9474, pp. 693–705, 2015.
DOI: 10.1007/978-3-319-27857-5_62

For many high-level vision based applications, either they use visual-optical videos [1,2] or infrared data [3–5], the task of background subtraction constitutes a key component, as this is one of the most common methods for locating moving objects, facilitating search space reduction and visual attention modeling [6–8]. Background subtraction methods applied on visual-optical videos model the color properties of depicted objects [9,10] and can be classified into three main categories [11]: basic modeling [12], statistical modeling [13,14] and background estimation [15,16].

The most used methods are the statistical ones due to their robustness to critical situations. In order to statistically represent the background, a probability distribution is used to model pixel intensities over time. Towards this direction, the work of Stauffer and Grimson [17], is one of the best known approaches. It uses a Gaussian mixture model, with fixed number of components, for a per-pixel density estimate. Similar to this approach, Makantasis et al. in [18] propose a Student-t mixture model, taking advantage of Student-t distribution compactness and robustness to noise and outliers. The works of [19,20] extend the method of [17] by introducing a rule based on a user defined threshold to estimate the number of components. However, this rule is application dependent and not directly derived from the data. All these techniques present the drawback that objects' color properties are highly affected by scene illumination, making the same object to look completely different under different lighting or weather conditions.

Although, thermal imagery can provide a challenging alternative for addressing the aforementioned difficulty, there exist few works for thermal data. The authors of [21,22] exploit contour saliency to extract foreground objects. Initially, they utilize a unimodal background modeling technique to detect regions of interest and then exploit the halo effect of thermal data for extracting foreground objects. However, unimodal background modeling is not usually capable of capturing background dynamics. Baf et al. in [11] present a fuzzy statistical method for background subtraction to incorporate uncertainty into the mixture of Gaussians. This method requires a predefined number of components and thus is application dependent. Elguebaly and Bouguila in [23] propose a finite asymmetric generalized Gaussian mixture model for object detection. Again this method requires a predefined maximum number of components, presenting therefore limitations when this technique is applied on uncontrolled environments. Dai et al. in [24] propose a method for pedestrian detection and tracking using infrared imagery. This method consists of a background subtraction technique that exploits a two-layer representation (one for foreground and one for background) of infrared frame sequences. However, the assumption made is that the foreground is restricted to moving objects, a consideration which is not sufficient for dynamically changing environments.

One way to handle the aforementioned difficulties is to introduce a background model, the parameters and the structure of which are directly estimated from the data, while at the same time it takes into account the special properties of infrared imagery.

1.1   Our Contribution

This work presents background modeling able to provide a per pixel density estimate, taking into account the special characteristics of infrared imagery, such as low signal-to-noise ratio. Our method exploits a Gaussian mixture model with unknown number of components. The advantage of such a model is that its own parameters and structure can be directly estimated from data distribution, allowing dynamic model adaptation to uncontrolled and changing environments.

An important issue in the proposed Gaussian mixture modeling concerns learning the model parameters. In our method, this is addressed using a variational inference framework to associate the functional structure of the model with real data distributions obtained from the infrared images. Then, the Expectation-Maximization (EM) algorithm is adopted to fit the outcome of variational inference to real measurements. Updating procedures are incorporated to allow dynamic model adaptation to the forthcoming infrared data. Our updating method avoids of using heuristics by considering existing knowledge accumulated from previous data distributions and then it compensates this knowledge with current measurements. Our overall strategy exploits a Bayesian framework to estimate all the parameters of the mixture model and thus avoids over/under fitting issues. To compensate computational challenges arising from the non a priori known nature of the mixture model, we utilize conjugate priors and thus we derive analytical equations for model estimation. In this way, we avoid the need of any sampling method, which are computationally and memory inefficient.

2   Gaussian Mixture Modeling

In this section we formulate the Bayesian framework adopted in this paper to estimate all the parameters of the proposed mixture model. In Sect. 2.1 we briefly describe the basic theory behind Gaussian mixtures, while in Sect. 2.2 we describe the introduction of conjugate priors that assist us in yielding analytical model estimations.

2.1   Model Fundamentals

The Gaussian mixture distribution can be seen as a linear superposition of Gaussian functional components,

$$p(x|\boldsymbol{\varpi}, \boldsymbol{\mu}, \boldsymbol{\tau}) = \sum_{k=1}^{K} \varpi_k \mathcal{N}(x|\mu_k, \tau_k^{-1}) \tag{1}$$

where the parameters $\{\varpi_k\}_{k=1}^{K}$ must satisfy $0 \leq \varpi_k \leq 1$ for every k and $\sum_{k=1}^{K} \varpi_k = 1$ and K is the number of Gaussian components. In the proposed mixture modeling, variable K can take any natural value up to infinity. However, it is highly recommended to set the upper bound for K less than the cardinality of the dataset, i.e. the number of observed samples. By introducing a K-dimensional binary latent variable z, such as $\sum_{k=1}^{K} z_k = 1$ and $p(z_k = 1) = \varpi_k$, the distribution $p(x)$ can be defined in terms of a marginal distribution $p(z)$ and a conditional distribution $p(x|z)$ as follows

$$p(x|\boldsymbol{\varpi},\boldsymbol{\mu},\boldsymbol{\tau}) = \sum_z p(z|\boldsymbol{\varpi})p(x|z,\boldsymbol{\mu},\boldsymbol{\tau}) \tag{2}$$

where $p(z|\boldsymbol{\varpi})$ and $p(x|z)$ are in the form of

$$p(z|\boldsymbol{\varpi}) = \prod_{k=1}^{K} \varpi_k^{z_k} \quad \text{and} \quad p(x|z,\boldsymbol{\mu},\boldsymbol{\tau}) = \prod_{k=1}^{K} \mathcal{N}(x|\mu_k,\tau_k^{-1})^{z_k} \tag{3}$$

where $\boldsymbol{\mu} = \{\mu_k\}_{k=1}^{K}$ and $\boldsymbol{\tau} = \{\tau_k\}_{k=1}^{K}$, correspond to the mean values and precisions of Gaussian components. By introducing latent variables and transforming the Gaussian mixture distribution into the form of (2), we are able to exploit the EM algorithm for fitting our model to the observed data, as shown in Sect. 4.

If we have in our disposal a set $\boldsymbol{X} = \{x_1,...,x_N\}$ of observed data we will also have a set $\boldsymbol{Z} = \{z_1,...,z_N\}$ of latent variables. Each z_n will be a K-dimensional binary vector, such as $\sum_{k=1}^{K} z_{nk} = 1$, and, in order to take into consideration the whole dataset of N samples, the distributions of (3) will be transformed to

$$p(\boldsymbol{Z}|\boldsymbol{\varpi}) = \prod_{n=1}^{N} \prod_{k=1}^{K} \varpi_k^{z_{nk}} \tag{4}$$

$$p(\boldsymbol{X}|\boldsymbol{Z},\boldsymbol{\mu},\boldsymbol{\tau}) = \prod_{n=1}^{N} \prod_{k=1}^{K} \mathcal{N}(x_n|\mu_k,\tau_k^{-1})^{z_{nk}} \tag{5}$$

2.2   Conjugate Priors

To avoid computational problems in estimating the parameters and the structure of the proposed Gaussian mixture, we introduce conjugate priors, over the model parameters $\boldsymbol{\mu}$, $\boldsymbol{\tau}$ and $\boldsymbol{\varpi}$, that allow us to yield analytical solutions. Introduction of priors implies the use of a Bayesian framework for the analysis.

Let us denote as $\boldsymbol{Y} = \{\boldsymbol{Z},\boldsymbol{\varpi},\boldsymbol{\mu},\boldsymbol{\tau}\}$ the set which contains all model latent variables and parameters and as $q(\boldsymbol{Y})$ its distribution. Then, our goal is to estimate $q(\boldsymbol{Y})$ which maximizes model evidence $p(\boldsymbol{X})$.

$$q(\boldsymbol{Y}) : \max \ln p(\boldsymbol{X}) \tag{6}$$

where in (6) we used the logarithm of $p(\boldsymbol{X})$ for calculus purposes. For maximizing (6) we need to define the distribution over $\boldsymbol{Y}$, that is, $p(\boldsymbol{Z}|\boldsymbol{\varpi})$ from (4), $p(\boldsymbol{\varpi})$ and $p(\boldsymbol{\mu},\boldsymbol{\tau})$.

Due to the fact that $p(\boldsymbol{Z}|\boldsymbol{\varpi})$ is a Multinomial distribution, its conjugate prior is a Dirichlet distribution over the mixing coefficients $\boldsymbol{\varpi}$

$$p(\boldsymbol{\varpi}) = \frac{\Gamma(K\lambda_0)}{\Gamma(\lambda_0)^K} \prod_{k=1}^{K} \varpi_k^{\lambda_0-1} \tag{7}$$

where $\Gamma(\cdot)$ is the Gamma function. Parameter λ_0 has a physical interpretation; the smaller the value of this parameter is, the larger is the influence of the data rather than the prior on the posterior distribution $p(\boldsymbol{Z}|\boldsymbol{\varpi})$. In order to introduce uninformative priors and not prefer a specific component against the other, we choose to use a single parameter λ_0 for the Dirichlet distribution, instead of a vector with different values for each mixing coefficient.

Similarly, the conjugate prior of (3) takes the form of a Gaussian-Gamma distribution, because both $\boldsymbol{\mu}$ and $\boldsymbol{\tau}$ are unknown. Subsequently, the joint distribution of parameters $\boldsymbol{\mu}$ and $\boldsymbol{\tau}$ can be modeled as

$$p(\boldsymbol{\mu},\boldsymbol{\tau}) = p(\boldsymbol{\mu}|\boldsymbol{\tau})p(\boldsymbol{\tau}) = \prod_{k=1}^{K} \mathcal{N}(\mu_k|m_0,(\beta_0\tau_k)^{-1})Gam(\tau_k|a_0,b_0) \tag{8}$$

where $Gam(\cdot)$ denotes the Gamma distribution. In order to not express any specific preference about the form of the Gaussian components through the introduction of priors, we use uninformative priors by setting the values of hyperparameters m_0, β_0, a_0 and b_0 to appropriate values as shown in Sect. 4.

Having defined the parametric form of observed data, latent variables and parameters distributions, our goal is to approximate the posterior distribution $p(\boldsymbol{Y}|\boldsymbol{X})$ and the model evidence $p(\boldsymbol{X})$, where $\boldsymbol{Y} = \{\boldsymbol{Z},\boldsymbol{\varpi},\boldsymbol{\mu},\boldsymbol{\tau}\}$ is the set with distribution $q(\boldsymbol{Y})$, which contains all model latent variables and parameters. Based on the Bayes rule, the logarithm of distribution $p(\boldsymbol{X})$ can be expressed as

$$\ln p(\boldsymbol{X}) = \int q(\boldsymbol{Y}) \ln \frac{p(\boldsymbol{X},\boldsymbol{Y})}{q(\boldsymbol{Y})} d\boldsymbol{Y} - \int q(\boldsymbol{Y}) \ln \frac{p(\boldsymbol{Y}|\boldsymbol{X})}{q(\boldsymbol{Y})} d\boldsymbol{Y} \tag{9a}$$

$$= \mathcal{L}(q) + KL(q||p) \tag{9b}$$

where $KL(q||p)$ is the Kullback-Leibler divergence between $q(\boldsymbol{Y})$ and $p(\boldsymbol{Y}|\boldsymbol{X})$ distributions and $\mathcal{L}(q)$ is the lower bound of $\ln p(\boldsymbol{X})$. Since $KL(q||p)$ is a non negative quantity, equals to zero only if $q(\boldsymbol{Y})$ is equal to $p(\boldsymbol{Y}|\boldsymbol{X})$, maximization of $\ln p(\boldsymbol{X})$ is equivalent to minimizing of $KL(q||p)$. For minimizing $KL(q||p)$ and estimating $p(\boldsymbol{X})$ we exploit the EM algorithm, as shown in Sect. 4.

By making the assumption, based on variational inference, that the distribution $q(\boldsymbol{Y})$ can be factorized over M disjoint sets such as $q(\boldsymbol{Y}) = \prod_{i=1}^{M} q_i(\boldsymbol{Y}_i)$, as shown in [25], the optimal solution $q_j^*(Y_j)$ corresponds to the minimization of $KL(q||p)$ is

$$\ln q_j^*(\boldsymbol{Y}_j) = \mathbb{E}_{i \neq j}[\ln p(\boldsymbol{X},\boldsymbol{Y})] + \mathcal{C} \tag{10}$$

where $\mathbb{E}_{i \neq j}[\ln p(\boldsymbol{X},\boldsymbol{Y})]$ is the expectation of the joint distribution over all variables $\boldsymbol{Y}_j$ for $j \neq i$ and $\mathcal{C}$ is a constant. $P(\boldsymbol{X},\boldsymbol{Y})$ is modeled through (11). In the following, we present the analytical solution for the optimal distributions $q_j^*(Y_j)$ for model parameters and latent variables, i.e. the optimal distributions $q^*(\boldsymbol{Z}), q^*(\boldsymbol{\varpi}), q^*(\boldsymbol{\tau})$ and $q^*(\boldsymbol{\mu}|\boldsymbol{\tau})$.

3   Optimal Model Parameter Distributions

According to (4), (5), (7) and (8), the joint distribution of all random variables can be factorized as follows

$$p(\boldsymbol{X}, \boldsymbol{Z}, \boldsymbol{\varpi}, \boldsymbol{\mu}, \boldsymbol{\tau}) = p(\boldsymbol{X}|\boldsymbol{Z}, \boldsymbol{\mu}, \boldsymbol{\tau})p(\boldsymbol{Z}|\boldsymbol{\varpi})p(\boldsymbol{\varpi})p(\boldsymbol{\mu}|\boldsymbol{\tau})p(\boldsymbol{\tau}) \qquad (11)$$

where $\boldsymbol{X}$ are the observed variables.

Using (10) and the factorized form of (11) the distribution of the optimized factor $q^*(\boldsymbol{Z})$ is given by a Multinomial distribution of the form

$$q^*(\boldsymbol{Z}) = \prod_{n=1}^{N} \prod_{k=1}^{K} \left(\frac{\rho_{nk}}{\sum_{j=1}^{K} \rho_{nj}} \right)^{z_{nk}} = \prod_{n=1}^{N} \prod_{k=1}^{K} r_{nk}^{z_{nk}} \qquad (12)$$

as ρ_{nk} we have denote the quantity

$$\rho_{nk} = \exp \left(\mathbb{E}\big[\ln \varpi_k\big] + \frac{1}{2}\mathbb{E}\big[\ln \tau_k\big] - \frac{1}{2}\ln 2\pi - \frac{1}{2}\mathbb{E}_{\boldsymbol{\mu}, \boldsymbol{\tau}}\big[(x_n - \mu_k)^2 \tau_k\big] \right) \qquad (13)$$

The expected value $\mathbb{E}[z_{nk}]$ of $q^*(\boldsymbol{Z})$ is equal to r_{nk}. Using (10) and (11) the distribution of the optimized factor $q^*(\boldsymbol{\varpi})$ is given a Dirichlet distribution of the form

$$q^*(\boldsymbol{\varpi}) = \frac{\Gamma(\sum_{i=1}^{K} \lambda_i)}{\prod_{j=1}^{K} \Gamma(\lambda_j)} \prod_{k=1}^{K} \varpi_k^{\lambda_k - 1} \qquad (14)$$

λ_k is equal to $N_k + \lambda_0$, where $N_k = \sum_{n=1}^{N} r_{nk}$ represents the proportion of data that belong to the k-th component. Similarly, the distribution of the optimized factor $q^*(\mu_k, \tau_k)$ is given by a Gaussian distribution of the form

$$q^*(\mu_k|\tau_k) = \mathcal{N}(\mu_k|m_k, (\beta_k\tau)^{-1}) \qquad (15)$$

where the parameters m_k and β_k are given by

$$\beta_k = \beta_0 + N_k \qquad (16a)$$

$$m_k = \frac{1}{\beta_k}\left(\beta_0 m_0 + N_k \bar{x}_k\right) \qquad (16b)$$

where $\bar{x}_k$ is equal to $\frac{1}{N_k}\sum_{n=1}^{N} r_{nk}x_n$ represents the centroid of the data that belong to the k-th component. After the estimation of $q^*(\mu_k|\tau_k)$, distribution of the optimized factor $q^*(\tau_k)$ is given by a Gamma distribution of the following form

$$q^*(\tau_k) = Gam(\tau_k|a_k, b_k) \qquad (17)$$

while the parameters a_k and b_k are given by the following relations

$$a_k = a_0 + \frac{N_k}{2} \qquad (18a)$$

$$b_k = b_0 + \frac{1}{2}\left(N_k \sigma_k + \frac{\beta_0 N_k}{\beta_0 + N_k}\left(\bar{x}_k - m_0\right)^2\right) \qquad (18b)$$

where $\sigma_k = \frac{1}{N_k}\sum_{n=1}^{N}(x_n - \bar{x}_k)^2$.

4 Distribution Parameters Optimization

After the approximation of random variables distributions, we will use the EM algorithm in order to find optimal values for model parameters, i.e. maximize (9). In order to use the EM algorithm, we have to initialize priors hyperparameters λ_0, a_0, b_0, m_0 and β_0 and the model parameters ϖ_k, μ_k, τ_k, β_k, a_k, b_k and λ_k (see Sect. 3).

The parameter λ_0 can be interpreted as the effective prior number of observations associated with each component. We introduce an uninformative prior for ϖ by setting the parameter λ_0 equal to N/K (the same number of observations is associated to each component). Parameters a_0 and b_0 were set to the value of 10^{-3}. When the values for a_0 and b_0 are close to zero the results of updating Eqs. (18a) and (18b) are primarily affected by the data and not by the prior. The mean values of the components are described by conditional Normal distribution with means m_0 and precisions $\beta_0\tau_k$. We introduce an uninformative prior by setting the value for m_0 to the mean of the observed data and $\beta_0 = \frac{b_0}{a_0 v_0}$, where v_0 is the variance of the observed data.

The convergence of EM algorithm is facilitated by initializing the parameters ϖ_k, μ_k, τ_k and β_k using the k-means. To utilize k-means, the number of clusters, i.e. the Gaussian components, should be a priori known. Although the number of components is unknown, it should be less or equal to the number of observed data. If we have no clue about the number of classes K_{max} we can set it to equal N. If we denote as $\hat{N}_k$ the number of observation that belong to k-th cluster, then we can set the value of parameter μ_k to equal the centroid of k-th cluster, the parameter ϖ_k to equal the proportion of observations for the k-th cluster, the parameter τ_k to equal $\hat{v}_k^{-1}$, where v_k stands for the variance of the k-th cluster and the parameter β_k to equal $\hat{N}_k^{-1}$. Then, we can exploit the formula for the expected value of a Gamma distribution to initialize the parameters a_k and b_k to values τ_k and one respectively. Finally, the initialization of ϖ_k allows us to initialize the parameter λ_k, which can be interpreted as the effective number of observations associated with each Gaussian component, to the value $N\varpi_k$.

After the initialization of model parameters and priors hyperparameters, the EM algorithm can be used to minimize $KL(q\|p)$ of (9). During the E step, r_{nk} is calculated given the initial/current values of all the parameters of the model. Using (12) r_{nk} is

$$r_{nk} \propto \tilde{\varpi}_k \tilde{\tau}_k^{1/2} \exp\left(- \frac{a_k}{2b_k}\left(x_n - m_k\right)^2 - \frac{1}{2\beta_k} \right) \tag{19}$$

Due to the fact that $q^*(\varpi)$ is a Dirichlet distribution and $q^*(\tau_k)$ is a Gamma distribution, $\tilde{\varpi}_k$ and $\tilde{\tau}_k$ will be given by

$$\ln \tilde{\varpi}_k \equiv \mathbb{E}\left[\ln \varpi_k \right] = \Psi(\lambda_k) - \Psi\left(\sum_{k=1}^{K} \lambda_k \right) \tag{20a}$$

$$\ln \tilde{\tau}_k \equiv \mathbb{E}\left[\ln \tau_k \right] = \Psi(a_k) - \ln b_k \tag{20b}$$

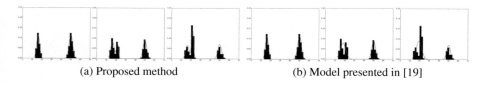

(a) Proposed method   (b) Model presented in [19]

Fig. 1. Updating to new observed data.

where $\Psi(\cdot)$ is the digamma function.

During the M step, we keep fixed the value of r_{nk} (calculated during the E step), and we re-calculate the values for model parameters using (14), (16) and (18). The steps E and M are repeated sequentially untill the values for model parameters are not changing anymore. As shown in [26] convergence of EM algorithm is guaranteed because bound is convex with respect to each of the factors $q(\mathbf{Z}), q(\boldsymbol{\varpi}), q(\boldsymbol{\mu}|\boldsymbol{\tau})$ and $q(\boldsymbol{\tau})$.

During model training the mixing coefficient for some of the components takes value very close to zero. Components with mixing coefficient less than $1/N$ are removed (we require each component to model at least one observed sample) and thus after training the model has automatically determined the right number of components.

5   Online Updating Mechanism

Having described how our model fits to N observed data, in this section we present the mechanism that permits our model to automatically adapt to new visual conditions. Contrary to [19] where heuristic rules are used during the adaptation of the model, we exploit statistics based on the observed data.

Let us denote as x_{new} a new observed sample. Then, there are two cases; either this sample is successfully modeled by our trained model, or not. To estimate if a new sample is successfully modeled, we find its closest component the Mahalanobis distance, since this is reliable distance measure between a point and a distribution.

The closest component c to the new sample is the one that presents the minimum Mahalanobis distance D_k

$$c = \arg\min_k D_k = \arg\min_k \sqrt{(x_{new} - \mu_k)^2 \tau_k} \tag{21}$$

The probability of the new sample to belong to c is

$$p(x_{new}|\mu_c, \tau_c) = \mathcal{N}(x_{new}|\mu_c, \tau_c^{-1}) \tag{22}$$

where μ_c and τ_c stand for the closest component mean value and precision respectively.

Let us denote as Ω the initially observed dataset. Then, we can assume that the probability to observe the new sample x_{new} is given by

$$p(x_{new}|e) = \frac{N_e}{N}\mathcal{U}(x_{new}|x_{new} - e, x_{new} + e) \tag{23}$$

where $N_e = \left| \{x_i \in \Omega : x_{new} - e \leq x_i \leq x_{new} + e\} \right|$ and $\mathcal{U}(x_{new}|x_{new} - e, x_{new} + e)$ is a Uniform distribution with lower and upper bounds to equal $x_{new} - e$ and $x_{new} + e$ respectively. Eq. (23) suggests that the probability to observe x_{new} is related to the proportion of data that have already been observed around x_{new}. By increasing the neighborhood around x_{new}, i.e. increasing the value of e, the quantity $\mathcal{U}(x_{new}|x_{new} - e, x_{new} + e)$ is decreasing, while the value of N_e is increasing. Upon the arrival of a new sample, we can estimate the optimal range ϵ around it that maximizes (23) as

$$\epsilon = \arg\max_e p(x_{new}|e) \tag{24}$$

Then, if $p(x_{new}|\mu_c, \tau_c) \geq p(x_{new}|\epsilon)$ the new observed sample can sufficiently represented by our model. Otherwise, a new Gaussian component must be created.

For model updating, we need to define the binary variable o, called ownership and associated with the Gaussian components, as

$$o_k = \begin{cases} 1, & \text{if } k = c \\ 0, & \text{otherwise} \end{cases} \tag{25}$$

where c represents the index of the closest component and k is the index of k-th component. When the new observed sample is successfully modeled, the parameters for the Gaussian components are updated using the *following the leader* [27] approach

$$\varpi_k \leftarrow \varpi_k + \frac{1}{N}\left(o_k - \varpi_k\right) \tag{26a}$$

$$\mu_k \leftarrow \mu_k + o_k\left(\frac{x_{new} - \mu_k}{\varpi_k N + 1}\right) \tag{26b}$$

$$\sigma_k^2 \leftarrow \sigma_k^2 + o_k\left(\frac{\varpi_k N(x_{new} - \mu_k)^2}{(\varpi_k N + 1)^2} - \frac{\sigma_k^2}{\varpi_k N + 1}\right) \tag{26c}$$

where σ_k^2 is equal to τ_k^{-1}.

When the new observed sample cannot be modeled by the existing components, a new component is created with mixing coefficient ϖ_{new}, mean value μ_{new} and standard deviation σ_{new}, the parameters of which are given as

$$\varpi_{new} = \frac{1}{N} \ , \ \mu_{new} = x_{new} \ , \ \sigma_{new}^2 = \frac{(2\epsilon)^2 - 1}{12} \tag{27}$$

From (27), we see that the mixing coefficient for the new component is equal to $1/N$ since it models only one sample (the new observed one), its mean value equals the value of the new sample and its variance the variance of the Uniform distribution, whose the lower and upper bounds are $x_{new} - \epsilon$ and $x_{new} + \epsilon$ respectively. When a new component is created the values for the parameters for all the other components remain unchanged. After each adaptation of the system to new observed samples, either they modeled or not, the mixing coefficients of the components are normalized to sum to one, while components with mixing coefficients less than $1/N$ are removed.

Figure 1 presents the adaptation of our model and the model presented in [19] to new observed data. To evaluate the quality of the adaptation of the models, we used a toy dataset with 100 observations, which were generated from two Normal distributions with mean values 16 and 50 and standard deviations 1.5 and 2.0 respectively. The initially trained models are presented in the left column. Then, we generated 25 new samples form a Normal distribution with mean value 21 and standard deviation 1.0. Our model creates a new component and successfully fits the data. On the contrary, the model of [19] is not able to capture the statistical relations of the new observations and fails to separate the data generated from distributions with mean values 16 and 21 (middle column). The quality of our adaptation mechanism becomes more clear in the right column, which presents the adaptation of both models after 50 new observations.

6   Background Subtraction

In this section we utilize our model for background subtraction. We initially capture N frames used to create a history of infrared responses for each pixel. These histories act as observed data and used to train one model for each pixel. To classify a pixel of a new captured frame as background or foreground, we compute the probability its value to be represented by the mixture model. If this value is larger than a threshold the pixel is classified as background, otherwise it is classified as foreground. The threshold is defined in relation to the parameters of the mixture, in order to be dynamically adapted.

7   Experimental Results

For evaluating our algorithm, we used the Ohio State University (OSU) thermal datasets and an dataset captured at Athens International Airport (AIA) during eVACUATE fp7 project OSU datasets contain frames that have been captured using a thermal camera and have been converted to grayscale images. In contrast, the AIA dataset contains raw thermal frames whose pixel values correspond to the real temperature of objects. OSU datasets [21, 22] are widely used for benchmarking algorithms for pedestrian detection and tracking in infrared imagery. Videos were captured under different illumination and weather conditions. AIA dataset was captured using a Flir A315 camera at different Airside Corridors and the Departure Level. Totally, 10 video sequences were captured, with frame dimensions 320×240 pixels of total duration 32051 frames, at 7.5 fps.

We compared our method with method presented by Zivkovic in [19] (MOG), which is one of the most robust and widely used background subtraction technique, and with the method for extracting the regions of interest presented in [22] (SBG). To conduct the comparison we utilized the objective metrics of *recall*, *precision* and *F1 score* on a pixel wise manner. Figure 2 visually present the performance of the three methods. Our method outperforms both MOG and SBG

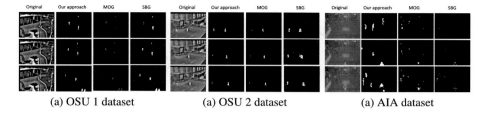

(a) OSU 1 dataset   (a) OSU 2 dataset   (a) AIA dataset

Fig. 2. Visual results for all datasets.

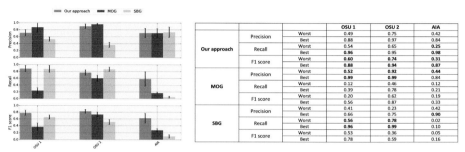

(a) Precision, recall and F1 score   (b) Best/worst case for precision, recall and F1 score

			OSU 1	OSU 2	AIA
Our approach	Precision	Worst	0.49	0.75	0.42
		Best	0.88	0.97	0.84
	Recall	Worst	0.54	0.65	0.25
		Best	0.96	0.95	0.98
	F1 score	Worst	0.60	0.74	0.31
		Best	0.88	0.94	0.87
MOG	Precision	Worst	0.52	0.92	0.44
		Best	0.99	0.99	0.84
	Recall	Worst	0.12	0.46	0.12
		Best	0.39	0.78	0.21
	F1 score	Worst	0.20	0.62	0.19
		Best	0.56	0.87	0.33
SBG	Precision	Worst	0.41	0.23	0.42
		Best	0.66	0.75	0.90
	Recall	Worst	0.56	0.78	0.02
		Best	0.96	0.99	0.10
	F1 score	Worst	0.53	0.36	0.05
		Best	0.78	0.59	0.16

Fig. 3. Algorithms performance per dataset (Color figure online).

on all datasets. While MOG and SBG perform satisfactory on grayscale frames of OSU datasets, their performance collapses when they applied on AIA dataset, which contains actual thermal responses. Regarding OSU datasets, although MOG algorithm presents high precision it yields very low recall values, i.e. the pixels that have been classified as foreground are indeed belong to the foreground class, but a lot of pixels that in fact belong to background have been misclassified. SBG algorithm seems to suffer by the opposite problem. Regarding AIA dataset, our method significantly outperforms both methods. In particular, MOG and SBG algorithms present high precision, but their recall values are under 0.2. Figure 3(a) presents average precision, recall and F1 score per dataset and per algorithm for all frames examined to give an objective evaluation. In Fig. 3(b) presents the best and worst case in terms of precision, recall and F1 score among all frames examined.

Regarding computational cost, the main load of our algorithm is in the implementation of EM optimization. In all experiments conducted, the EM optimization converges within 10 iterations. Practically, the time required to apply our method is similar to the time requirements of Zivkovic's method making it suitable for real-time applications.

8   Conclusions

This paper presents a background subtraction method applicable to thermal imagery, based on Gaussian mixtures with unknown number of components. We analytically derive the solutions that describe the parameters of the model and we use the EM optimization to estimate their values, avoiding this way sampling algorithms and high computational cost. Due to its low computational cost and the real-time operation, our method is suitable for real-world applications.

Acknowledgements. This research was funded from European Unions FP7 under grant agreement n.313161, eVACUATE Project (www.evacuate.eu).

References

1. Porikli, F.: Achieving real-time object detection and tracking under extreme conditions. J. Real-Time Image Proc. **1**, 33–40 (2006)
2. Tuzel, O., Porikli, F., Meer, P.: Pedestrian detection via classification on riemannian manifolds. IEEE Trans. Pattern Anal. Mach. Intell. **30**, 1713–1727 (2008)
3. Jungling, K., Arens, M.: Feature based person detection beyond the visible spectrum. In: IEEE Computer Vision and Pattern Recognition Workshops, CVPR 2009, pp. 30–37 (2009)
4. Latecki, L., Miezianko, R., Pokrajac, D.: Tracking motion objects in infrared videos. In: IEEE Conference on Advanced Video and Signal Based Surveillance, AVSS 2005, pp. 99–104 (2005)
5. Wang, W., Zhang, J., Shen, C.: Improved human detection and classification in thermal images. In: 2010 17th IEEE International Conference on Image Processing (ICIP), pp. 2313–2316 (2010)
6. Doulamis, N.D.: Coupled multi-object tracking and labeling for vehicle trajectory estimation and matching. Multimedia Tools Appl. **50**, 173–198 (2010)
7. Kosmopoulos, D.I., Doulamis, N.D., Voulodimos, A.S.: Bayesian filter based behavior recognition in workflows allowing for user feedback. Comput. Vis. Image Underst. **116**, 422–434 (2012)
8. Voulodimos, A.S., Doulamis, N.D., Kosmopoulos, D.I., Varvarigou, T.A.: Improving multi-camera activity recognition by employing neural network based readjustment. Appl. Artif. Intell. **26**, 97–118 (2012)
9. Brutzer, S., Hoferlin, B., Heidemann, G.: Evaluation of background subtraction techniques for video surveillance. In: 2011 IEEE Conference on Computer Vision and Pattern Recognition (CVPR), pp. 1937–1944 (2011)
10. Herrero, S., Bescós, J.: Background subtraction techniques: systematic evaluation and comparative analysis. In: Blanc-Talon, J., Philips, W., Popescu, D., Scheunders, P. (eds.) ACIVS 2009. LNCS, vol. 5807, pp. 33–42. Springer, Heidelberg (2009)
11. El Baf, F., Bouwmans, T., Vachon, B.: Fuzzy statistical modeling of dynamic backgrounds for moving object detection in infrared videos. In: IEEE Conference on Computer Vision and Pattern Recognition Workshops, CVPR Workshops 2009, pp. 60–65 (2009)
12. Zheng, J., Wang, Y., Nihan, N., Hallenbeck, M.: Extracting roadway background image: mode-based approach. Transp. Res. Rec. J. Transp. Res. Board **2006**, 82–88 (1944)

13. Elgammal, A., Harwood, D., Davis, L.: Non-parametric model for background subtraction. In: Vernon, D. (ed.) ECCV 2000. LNCS, vol. 1843, pp. 751–767. Springer, Heidelberg (2000)
14. Wren, C., Azarbayejani, A., Darrell, T., Pentland, A.: Pfinder: real-time tracking of the human body. IEEE Trans. Pattern Anal. Mach. Intell. **19**, 780–785 (1997)
15. Messelodi, S., Modena, C.M., Segata, N., Zanin, M.: A Kalman filter based background updating algorithm robust to sharp illumination changes. In: Roli, F., Vitulano, S. (eds.) ICIAP 2005. LNCS, vol. 3617, pp. 163–170. Springer, Heidelberg (2005)
16. Toyama, K., Krumm, J., Brumitt, B., Meyers, B.: Wallflower: principles and practice of background maintenance. In: The Proceedings of the Seventh IEEE International Conference on Computer Vision, vol. 1, pp. 255–261 (1999)
17. Stauffer, C., Grimson, W.: Adaptive background mixture models for real-time tracking. In: IEEE Conference on Computer Vision and Pattern Recognition, vol. 2, p. 252 (1999)
18. Makantasis, K., Doulamis, A., Matsatsinis, N.: Student-t background modeling for persons' fall detection through visual cues. In: 2012 13th International Workshop on Image Analysis for Multimedia Interactive Services (WIAMIS), pp. 1–4 (2012)
19. Zivkovic, Z.: Improved adaptive gaussian mixture model for background subtraction. In: Proceedings of the 17th International Conference on Pattern Recognition, ICPR 2004, vol. 2, pp. 28–31 (2004)
20. Zivkovic, Z., van der Heijden, F.: Efficient adaptive density estimation per image pixel for the task of background subtraction. Pattern Recogn. Lett. **27**, 773–780 (2006)
21. Davis, J., Sharma, V.: Fusion-based background-subtraction using contour saliency. In: IEEE Computer Society Conference on Computer Vision and Pattern Recognition - Workshops, CVPR Workshops 2005, pp. 11–11 (2005)
22. Davis, J.W., Sharma, V.: Background-subtraction in thermal imagery using contour saliency. Int. J. Comput. Vis. **71**, 161–181 (2007)
23. Elguebaly, T., Bouguila, N.: Finite asymmetric generalized gaussian mixture models learning for infrared object detection. Comput. Vis. Image Underst. **117**, 1659–1671 (2013)
24. Dai, C., Zheng, Y., Li, X.: Pedestrian detection and tracking in infrared imagery using shape and appearance. Comput. Vis. Image Underst. **106**, 288–299 (2007)
25. Bishop, C.: Pattern Recognition and Machine Learning. Information Science and Statistics. Springer, New York (2007)
26. Boyd, S., Vandenberghe, L.: Convex Optimization. Cambridge University Press, Cambridge (2004)
27. Dasgupta, S., Hsu, D.: On-line estimation with the multivariate gaussian distribution. In: Bshouty, N.H., Gentile, C. (eds.) COLT. LNCS (LNAI), vol. 4539, pp. 278–292. Springer, Heidelberg (2007)

Image Based Approaches for Tunnels' Defects Recognition via Robotic Inspectors

Eftychios Protopapadakis[1,3]([⊠]) and Nikolaos Doulamis[2,3]

[1] Technical University of Crete, 73100 Chania, Greece
eprotopapadakis@isc.tuc.gr
[2] National Technical University of Athens,
9 Iroon Polytechneiou, 15780 Athens, Greece
ndoulam@cs.ntua.gr
[3] Institute of Communication and Computer Systems, Athens, Greece

Abstract. In this paper we present a visual based approach, utilized for the detection of concrete defects in tunnels. The detection mechanism is a hybrid approach, based on both image processing and deep learning models. Initial detections are validated by an expert, in order to create a robust data set in short time, saving resources during annotation process. Then, a deep-learning classifier is trained and applied for the inspection. The fully automated system, performs well, in various environments, and can be, easily, implemented with most robotic systems.

1 Introduction

Civil infrastructures are progressively deteriorating (ageing, environmental factors), calling for inspection and assessment. Presently, structural tunnel inspection is predominantly performed through tunnel-wide visual observations by inspectors who identify structural defects, rate them and then, based on their severity, categorize the liner. This visual inspection (VI) process is slow, labour intensive and subjective (depending on the experience and fatigue), while working in an unpleasant environment and uncomfortable conditions [1]. Moreover, the classification of the liner condition is empirical and incomplete and lacks any engineering analysis. It is therefore, unreliable.

Approaches that utilize automated procedures for VI of concrete infrastructures aim specifically to defects detection for structure evaluation. Towards this direction, such methods exploit image processing and machine learning techniques. Initially, low-level image features are used towards the construction of complex handcrafted features, which are used to train learning models; *i.e.* the detection methods. Automated approaches have been applied in practical settings including roads, bridges, fatigues, and sewer-pipes [2–5].

Recent research in the robotics sector and the relevant sectors such as computer vision and sensors, have significantly increased the competitiveness of components needed in automated systems that can perform quick and robust inspection/assessment, in general transportation and tunnel infrastructures. However, such an integrated and automated system is not yet available.

© Springer International Publishing Switzerland 2015
G. Bebis et al. (Eds.): ISVC 2015, Part I, LNCS 9474, pp. 706–716, 2015.
DOI: 10.1007/978-3-319-27857-5_63

The work that will be presented in this paper, involves the core mechanism of such system; a computer vision scheme, easy to integrate with all the required components for the inspection and assessment of tunnels in one pass.

1.1 Related Work

Intensity features and Support Vector Machines (SVMs) for crack detections on tunnel surfaces where used in [6]. Color properties, different non-RGB color spaces and various machine learning algorithms are also investigated in [7]. Edge detection techniques are applied in [8] for detecting concrete defects. Edge detection algorithms (i.e. Sobel and Laplacian operators) and graph based search algorithms are also utilized in [1] to extract crack information.

An image mosaic technology for detecting tunnels surface defects was further extended in [9]. A pothole detection system [10], based on histogram shape-based thresholding and low level texture features, has been used in asphalt pavement images. A concrete spalling measurement system for post-earthquake safety assessments, using template matching techniques and morphological operations, has been proposed by [11].

The exploitation of more sophisticated features has also been proposed. Histograms of Oriented Gradient features and SVMs are utilized in the work of [12], to support automated detection and classification of pipe defects. Shape-based filtering is exploited in the work of [13] for crack detection and quantification. The constructed features are fed as input to ANN or SVM classifiers in order to discriminate crack from non-crack patterns.

1.2 Our Contribution

The conventional paradigm of pattern recognition consists of two steps: complex handcrafted feature construction and classifiers training. However, variety in defect types makes difficult the feature construction/selection task [12]. Deep learning models [14,15] are a class of machines that can learn a hierarchy of features by building complex, high-level features from low-level ones, automating the process of feature construction for the problem at hand.

The work of [16] exploit a Convolutional Neural Network (CNN) to hierarchically construct high-level features, describing the defects, and a Multi-Layer Perceptron (MLP) that carries out the defect detection task in tunnels. Such an approach offers an automated feature extraction, adaptability to the defect type(s), and has no need for special set-up for the image acquisition. Nevertheless, there is a major drawback regarding the applicability in real life scenarios: resources spend for data annotation. Data annotation is a time consuming job that requires a human expert; it is therefore prone to segmentation errors.

In this paper we enrich the approach of [16] by incorporating a prior, image processing, detection mechanism. Such mechanism stands as a simple detector and is only used at the beginning of the inspection. Possible defects are annotated and then validated by an expert; after validating few samples, the required training dataset for the deep learning approach has been formed. From this point onwards, a CNN is trained and, then, utilized for the rest of the inspection process.

2   Data Acquisition and Processing Challenges

Tunnel inspection is a tedious task; extreme temperatures, low luminosity, short spaces, floods, fires, are few among the dangers. Even if we employ robots for the inspection, optical information comes at a great cost; occlusions (e.g. bugs, webs, graffiti) and noise (e.g. dust, steam). Ideally, during image acquisition, no special setup should take place since we try to develop a generic optic inspection method. In other words, images can be taken from any angle and distance from the tunnel surfaces. Even if we have the ideal conditions, during acquisition, the defect types make the problem difficult enough. The term "defect" can be interpreted in many ways; deformations, cracks, surface disintegration, and other defects are widely known and commonly appear.

Discreet, parallel cracks that look like tearing of the surface are caused by shrinkage while the concrete is still fresh, called plastic shrinkage cracks. Fine random cracks or fissures that may only be seen when the concrete is drying after being moistened are called crazing cracks. Cracking that occurs in a three-point pattern is generally caused by drying shrinkage. Large pattern cracking, called map-cracking, can be caused by alkali-silica reaction within the concrete. Structural failure cracking may look like many other types of cracking; however, in slabs they are often associated with subsequent elevation changes, where one side of the crack is be lower than the other.

Disintegration of the surface is generally caused by three types of distress: (a) dusting, due to carbonation of the surface by unventilated heaters or by applying water during finishing, (b) ravelling or spalling at joints, when pieces of concrete from the joint edges are dislodged and, (c) breaking of pieces from the surface of the concrete, generally caused by delaminations and blistering. Popouts are conical fragments that come off the surface, typically leaving a broken aggregate at the bottom of the hole. Popoffs, or mortar flaking, is similar to popouts, except that the aggregate is not broken and the broken piece is generally smaller. Flaking of the concrete surface over a widespread area is called scaling.

Other defects include discolouration of the concrete; bugholes, which are small voids in the surface of vertical concrete placements, and honeycombing, which is the presence of large voids in concrete caused by inadequate consolidation. At this point, we understand the defect identification problem: it is extremely difficult to extract features suitable for the accurate description of such a large number of defect alternatives, simultaneously.

3   Image Analysis

Computer vision (CV) algorithms [17,18] are being designed to perform tunnel inspection and assessment, by detecting structural defects. Assuming that the CV system operates at a rate of about 1 m/sec and acquires 2D images of

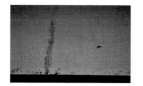

Fig. 1. Illustration of various images acquired and require further investigation

the tunnel lining at a coarse level of detail, we need fast object recognition techniques to identify areas of interest. Then, at a slower rate, concentrate the image acquisition on details of interest.

Hierarchical CV schemes are applied to make the recognition process just-in-time, and thus significantly reduce the time and effort needed for visual inspections. The system will apply recent advances in active continuous learning to tunnels' inspection mechanisms, so as to achieve on-line understanding of the cracks while surveying the tunnels.

Cracks are the most common defect type. They appear in concrete for many reasons; *usually as secondary symptoms of other defects*, such as a long rounded crack following the structural failure of a warped slab, caused by volume changes and structural failure. As such, the identification of a crack should be the first step, prior to an extensive analysis in the surrounding area. Additionally, given the magnitude of the investigated area, crack identification should be as fast as possible.

The detection of defects can be seen as an image segmentation problem, which entails to the classification of each one of the pixels in the image into one of two classes; defects class and non-defects class. Such a task requires the description of pixels by a set of highly discriminative features that fuse visual and spatial information. However, the features extraction is depended on the problem at hand. Such drawback can be eliminated following the deep learning paradigm.

At first low-level features are extracted over RGB tunnel's surfaces images. These features consist the CNN's input, which fuses them in order to hierarchically construct complex, high-level features, which in turn are fed to a MLP that conducts the classification task. Visual information about a specific pixel is related to its responses to different low-level feature extraction algorithms, while its spatial information is obtained by taking into consideration its neighbors.

After the extraction of low-level image features the raw image has been transformed to a 3D tensor of dimensions $h \times w \times c$, where h and w correspond to the height and width of the image and c to its channels (number of extracted features). In order to be aligned with the specific nature of CNNs (global features construction), we have to decompose the transformed image into *patches*, each one of which contains visual and spatial information for a specific pixel.

Concretely, in order to classify a pixel p_{xy}, located at (x, y) point on image plane, and successfully fuse visual and spatial information, we use a square patch of size $s \times s$ centered at pixel p_{xy}. If we denote as l_{xy} the class label of the pixel at location (x, y) and as b_{xy} the patch centered at pixel p_{xy}, then, we can form a dataset $D = \{(b_{xy}, l_{xy})\}$ for $x = 1, 2, \cdots, w$ and $y = 1, 2, \cdots, h$. Patch b_{xy} is also 3D tensor, which is divided into c matrices of dimensions $s \times s$ (2D inputs). These matrices are fed as input into the CNN. Then, the CNN hierarchically builds complex, high-level features that encode visual and spatial characteristics of pixel p_{xy}. The output of the CNN is sequentially connected with the MLP. Therefore, obtained features are used as input by the MLP classifier, which is responsible for detecting defects.

3.1   Crack Identification

In order to detect cracks, we employ, at first, image processing techniques. Such techniques exploit: the intensity of pixels and their spatial relations, morphological operations, filtering schemes. Additionally, we perform simple shape analysis on detection results.

Shape properties describe the ratio length to width for the detected edges. Further, we are looking for curves. Regarding intensity, pixels corresponding to cracks are expected to be darker than their neighbouring pixels. Thus, based on cracks characteristics our approach consists of the following steps: (a) Lines enhancing, (b) Noise removal, (c) Straight lines removal, (e) Shape filtering, and (f) Morphological reconstruction.

Line enhancement occurs by comparing the intensity of the specific pixel to its neighbours. Such an approach result in "salt and pepper" noise. The next step focuses on noise removal, exploiting a traditional median filter. Straight lines, is something common, and correspond to man made crafts (e.g. wiring) are located according to Hough transform by thresholding the detected outputs.

Shape filtering using appropriate moments is another crucial step. By locating minimum enclosing circles we are able to exclude symmetrical areas. Finally, we perform a classical morphological operation called "opening by reconstruction". Reconstruction starts from a set of starting pixels and then grows in flood-fill fashion to include complete connected components.

3.2   Deep Learning Defect Recognition

CNNs apply trainable filters and pooling operations on their input resulting in a hierarchy of increasingly complex features. Convolutional layers consist of a rectangular grid of neurons (filters), each of which takes inputs from rectangular sections of the previous layer. Each convolution layer is followed by a pooling layer that subsamples block-wise the output of the precedent convolutional layer and produces a scalar output for each block. Formally, if we denote the k-th output of a given convolutional layer as h^k whose filters are determined by the weights W^k and bias b^k then the h^k is obtained as:

$$h_{ij}^k = g((W^k * x)_{ij} + b^k) \tag{1}$$

where x stands for the input of the convolutional layer and indices i and j correspond to the location of the input where the filter is applied. Star symbol (*) stands for the convolution operator and $g(\cdot)$ is a non-linear function. Max pooling layers simply take some $k \times k$ region and output the maximum value in that region.

Max pooling layers introduce scale invariance to the constructed features, which is a very important property for object detection/recognition tasks, where scale variability problems may occur. However, for the problem of tunnel defects detection, we involve CNNs to construct features that encode spatio-visual information, which indicates the presence or absence of a defect to a specific pixel. Thus scale invariance, which is addressed through the use of a Gaussian pyramid, does not consist a significant property for our learning model. Due to this fact, we do not involve pooling layers into our learning architecture.

4   Core System Evaluation

The proposed system was developed on a conventional laptop with i7 CPU, 8GB RAM, using Theano library [19] in Python. The proposed CNN is compared against well known techniques in pattern recognition, which rely on the exploitation of handcrafted features. Compatibility with robotic parts has been also verified using YARP [20].

All the images originate from Metsovo motorway tunnel in Greece, which is a 3.5 km long twin tunnel. In a distance of 20 m parallel and north to this bore, runs the ventilation tunnel. The main tunnel suffered a significant deformation due to water inflow. Image data were captured at this part of the tunnel, using a hand held DSLR camera. Figure 1 illustrates various tunnel images during data acquisition process. Regions depicting defects, for each of the captured images, were manually annotated, by experts (i.e. about 100 images).

All algorithms are applied on the raw image data and their performance is evaluated in regard to the ground truth data. The unbalanced nature of classes, defects span very few areas, may deteriorate the performance of the classifiers. Due to this fact, we truncate the non-defects class to contain the same number of samples as the class that represents defects. The final dataset that is used for training and testing, is created by concatenating the elements of the two classes.

4.1   Crack Detection

Line enhancement is performed on 13×13 windows by thresholding the 0.99 % of the mean intensity value. Then, areas spanning less than 550 pixels are considered noise and, thus, excluded. Hough transform distance and angle resolution were set to 5 pixels and 0 radians respectively. Finally, areas of defects should span at least 30 % of the minimum enclosing circle. An illustration of the process outcome can be found in Fig. 2.

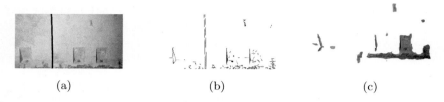

(a)   (b)   (c)

Fig. 2. Illustration of the annotation steps at the beginning of the inspection, the crack identification. (a) Depicts the original image, (b) the areas of interest after the median filtering, and (c) demonstrates the final image presented to the expert for annotation corrections

4.2 Multiple Defects Detection

The input of the CNN are patches of dimensions $s \times s \times c$ (Sect. 3). After the construction of feature maps, the parameter c corresponds to the dimension of feature vector $\boldsymbol{f}$, in our case it equals 17, while the parameter s determines the number of neighbors of each pixel that will be taken into consideration during classification task.

During experimentation process we set the parameter s to be equal to 5, in order to take into consideration the closest 24 neighbors of each pixel. By increasing the value of s, the number of neighbors that are taken into consideration is increased and thus the computational cost of classification is increased, also. However, setting the parameter s to a value larger than 5, no further improvement on classification accuracy was reported. On the contrary, increasing the value of s over 13, deteriorates classification accuracy.

Having estimate the values s and c, we can proceed with the CNN architecture design. The first layer of the proposed CNN is a convolutional layer with $C_1 = 30$ trainable filters of dimensions 3×3. This layer delivers C_1 matrices of dimensions 3×3 (during convolution we do not take into consideration the border of the patch). Due to the fact that we do not employ a max pooling layer, the output of the first convolutional layer is fed to the second convolutional layer.

Preprocessing. In this section we describe the process for encoding visual information. This process takes place exploiting low-level features. There are two main reasons we used such features:similar features were used by many researchers (e.g. [7,8,10–12]). and, such low-level features are calculated over raw-data and are computationally less-expensive than other high-level features. It has to be mentioned that these features are used by the CNN to hierarchically construct high-level features. They are not used directly for classification purposes.

Using low-level feature extraction techniques, each pixel p_{xy} is described by a feature vector $\boldsymbol{f}_{xy} = [f_{1,xy}, \cdots, f_{k,xy}]^T$, where $f_{1,xy}, \cdots, f_{k,xy}$ are scalars correspond to the presence and magnitude of the low level features detected at location (x, y). Feature vectors along with the class labels of every pixel are used to form a dataset for training, validating and testing our learning model. In the following we describe, which features are used to form feature vector $\boldsymbol{f}$.

In order to successfully exploit image **edges**, the system must be able to detect them in a very accurate way and it must preserve their magnitude. For this reason we combined the Canny and Sobel operators. **Frequency** feature is utilized to emphasize regions of high frequency in the image and at the same time suppress low frequency regions. Frequency components of an I are computed as: $\mathcal{F}_I = \nabla^2 \cdot I$. Image **entropy** quantifies the information coded in an image, i.e. the amount of information which can be coded by a compression algorithm. Images that depict large homogeneous regions, present low entropy, while highly textured images will present high entropy. **Histogram of Oriented Gradients** (HOG) is a popular dense feature descriptor used for the task of object detection. It exploits image gradients to capture contour, silhouette and texture information, producing an encoding that is sensitive to local image content while remaining resistant to small changes in pose or appearance. Finally, for **texture** identification we used Gabor filters.

Following the aforementioned procedure we construct 16 low level features for an image. By combining these features with the raw pixels intensity, feature vector $\boldsymbol{f}$ takes the form of an 1×17 vector containing visual information that characterizes each one of the image's pixels.

Comparison Against Other Techniques. We use a variety of conventional pattern recognition techniques; all the techniques utilize a variety of parameters (usually user defined). The proposed methods are: Artificial neural networks (ANNs), Support Vector Machines (SVMs), k-nearest neighbors (kNNs) and Classification trees (Ctree).

kNNs and Ctrees are rather intuitively methods. The former is a non-parametric method, which exploits a similarity metric in order to identify the k more similar instances (among the labelled ones) to a specific datum i. The latter is a tree shape structure, in which every node investigates the values of a specific feature. As we propagate through the tree brunches, by evaluating each of the datum features, we will end to a terminal node, indicating the corresponding class.

More sophisticated methods are ANNs and SVMs. ANNs are universal function aproximators, which, however, have multiple local minima, i.e. solutions, due to their structure. ANNs are composed from multiple hierarchical layers of interconnected nodes/neurons. SVMs select a subset among the labeled data in order to create appropriate separation hyperplane(s) among the classes. SVMs can efficiently perform a non-linear classification using what is called the "kernel trick", implicitly mapping their inputs into high-dimensional feature spaces, where the different classes are linear separable.

4.3   Performance Evaluation

In this paper we have two possible classes; defect or non-defect, named positive (P) and negative (N) class, respectively. Given the outputs, we form the

Table 1. Comparison of state-of-the-art techniques for various performance metrics. These metrics represent the ability to distinguish defect areas over a given image.

	Quantitative performance metrics (test set)								
	TPR	SPC	PPV	NPV	FPR	FDR	FNR	ACC	F1 score
CNN	**0.890**	**0.883**	**0.883**	**0.890**	**0.117**	**0.117**	**0.110**	**0.886**	**0.886**
Ctree	0.721	0.591	0.751	0.553	0.409	0.249	0.279	0.673	0.736
kNN	0.845	0.575	0.773	0.685	0.425	0.227	0.155	0.746	0.807
FFNN	0.854	0.554	0.766	0.689	0.446	0.234	0.146	0.743	0.808
linSVMs	0.833	0.514	0.746	0.643	0.486	0.254	0.167	0.716	0.787
polySVMs	0.877	0.036	0.609	0.146	0.964	0.391	0.123	0.567	0.719
rbfSVMs	0.864	0.470	0.736	0.669	0.530	0.264	0.136	0.719	0.795

confusion table, which is a 2×2 matrix that reports the number of false positives (FP), false negatives (FN), true positives (TP), and true negatives (TN). Given these values we are able to calculate various performance metrics regarding the defect detection performance. Calculated metrics are: sensitivity (TPR), specificity (SPC), precision (PPV), negative predicted value (NPV), false possitive rate (FPR), false discovery rate (FDR), miss rate (FNR), accuracy (ACC) and F1 score (F1).

A quantitative evaluation of our method compared to traditional pattern recognition methods is shown in Table 1 over the test set. The proposed methodology outperforms all competitive techniques. It is worth mentioning the lowest rate in false negative detection (FNR), without sacrificing the classification accuracy for the non-defect class (ACC and FDR). That suggest a well adapted classifier, able to distinguish between the two classes.

5   Conclusions

In this paper, we point the suitability for deep learning architectures for the tunnel defect inspection problem. The proposed approach surpass a variety of well known approaches without making any assumptions on the given images. In order to facilitate the system initialization, an image processing scheme has been developed. Hierarchical construction of complex high-level features in an automated way result in better classification than the conventional handcrafted features, while at the same time it minimized the feature construction effort, compared to the traditional approaches.

Acknowledgments. The research leading to these results has received funding from the EC FP7 project ROBO-SPECT (Contract N.611145). Authors wish to thank all partners within the ROBO-SPECT consortium. The work has been, also, partially supported by IKY Fellowships of excellence for postgraduate studies in Greece-Siemens program.

References

1. Yu, S.N., Jang, J.H., Han, C.S.: Auto inspection system using a mobile robot for detecting concrete cracks in a tunnel. Autom. Constr. **16**, 255–261 (2007)
2. Pynn, J., Wright, A., Lodge, R.: Automatic identification of cracks in road surfaces. In: Seventh International Conference on Image Processing and Its Applications, 1999 (Conf. Publ. No. 465), vol. 2, pp. 671–675 (1999)
3. Kim, Y.S., Haas, C.T.: A model for automation of infrastructure maintenance using representational forms. Autom. Constr. **10**, 57–68 (2000)
4. Tung, P.C., Hwang, Y.R., Wu, M.C.: The development of a mobile manipulator imaging system for bridge crack inspection. Autom. Constr. **11**, 717–729 (2002)
5. Sinha, S.K., Fieguth, P.W.: Automated detection of cracks in buried concrete pipe images. Autom. Constr. **15**, 58–72 (2006)
6. Liu, Z., Suandi, S.A., Ohashi, T., Ejima, T.: Tunnel crack detection and classification system based on image processing. Proc. SPIE Int. Soc. Opt. Eng. **4664**, 145–152 (2002)
7. Son, H., Kim, C., Kim, C.: Automated color ModelBased concrete detection in construction-site images by using machine learning algorithms. J. Comput. Civ. Eng. **26**, 421–433 (2012)
8. Abdel-Qader, I., Abudayyeh, O., Kelly, M.E.: Analysis of edge-detection techniques for crack identification in bridges. J. Comput. Civ. Eng. **17**, 255–263 (2003)
9. Mohanty, A., Wang, T.T.: Image mosaicking of a section of a tunnel lining and the detection of cracks through the frequency histogram of connected elements concept, vol. 8335 (2012) 83351P–83351P-9
10. Koch, C., Brilakis, I.: Pothole detection in asphalt pavement images. Adv. Eng. Inform. **25**, 507–515 (2011)
11. German, S., Brilakis, I., DesRoches, R.: Rapid entropy-based detection and properties measurement of concrete spalling with machine vision for post-earthquake safety assessments. Adv. Eng. Inform. **26**, 846–858 (2012)
12. Halfawy, M.R., Hengmeechai, J.: Automated defect detection in sewer closed circuit television images using histograms of oriented gradients and support vector machine. Autom. Constr. **38**, 1–13 (2014)
13. Jahanshahi, M.R., Masri, S.F., Padgett, C.W., Sukhatme, G.S.: An innovative methodology for detection and quantification of cracks through incorporation of depth perception. Mach. Vis. Appl. **24**, 227–241 (2011)
14. Hinton, G.E., Salakhutdinov, R.R.: Reducing the dimensionality of data with neural networks. Science **313**, 504–507 (2006)
15. Hinton, G.E., Osindero, S., Teh, Y.W.: A fast learning algorithm for deep belief nets. Neural Comput. **18**, 1527–1554 (2006)
16. Makantasis, K., Protopapadakis, E., Doulamis, A.D., Doulamis, N.D., Loupos, C.: Deep convolutional neural networks for efficient vision based tunnel inspection. In: 2015 IEEE International Conference on Intelligent Computer Communication and Processing (ICCP), pp. 335–342. IEEE, Cluj-Napoca, Romania (2015)
17. Doulamis, A.: Event-driven video adaptation: a powerful tool for industrial video supervision. Multimedia Tools Appl. **69**, 339–358 (2012)
18. Doulamis, A., Matsatsinis, N.: Visual understanding industrial workflows under uncertainty on distributed service oriented architectures. Future Gener. Comput. Syst. **28**, 605–617 (2012)

19. Bastien, F., Lamblin, P., Pascanu, R., Bergstra, J., Goodfellow, I., Bergeron, A., Bouchard, N., Warde-Farley, D., Bengio, Y.: Theano: new features and speed improvements. arXiv:1211.5590 [cs] (2012)
20. Fitzpatrick, P., Metta, G., Natale, L.: Towards long-lived robot genes. Robot. Auton. Syst. **56**, 29–45 (2008)

Deep Learning-Based Man-Made Object Detection from Hyperspectral Data

Konstantinos Makantasis[1]([✉]), Konstantinos Karantzalos[2],
Anastasios Doulamis[2], and Konstantinos Loupos[3]

[1] Technical University of Crete, University Campus,
73100 Kounoupidiana, Chania, Greece
`kmakantasis@isc.tuc.gr`
[2] National Technical University of Athens, Zografou Campus, 15780 Athens, Greece
`karank@central.ntua.gr, adoulam@cs.ntua.gr`
[3] Institute of Communication and Computer Systems, Athens, Greece
`kloupos@iccs.gr`

Abstract. Hyperspectral sensing, due to its intrinsic ability to capture the spectral responses of depicted materials, provides unique capabilities towards object detection and identification. In this paper, we tackle the problem of man-made object detection from hyperspectral data through a deep learning classification framework. By the effective exploitation of a Convolutional Neural Network we encode pixels' spectral and spatial information and employ a Multi-Layer Perceptron to conduct the classification task. Experimental results and the performed quantitative validation on widely used hyperspectral datasets demonstrating the great potentials of the developed approach towards accurate and automated man-made object detection.

1 Introduction

Automatic extraction of man-made objects, such as buildings, building blocks or roads, is of major importance for supporting several government activities, such as urban planning, cadastre, monitoring of protected nature areas, environmental monitoring and various GIS applications like map generation and update [1–4]. To this end, the accurate extraction and recognition of man-made objects from remote sensing data has been an important topic in remote sensing, photogrammetry and computer vision for more than two decades [5,6]. Today, man-made object extraction is, still, an active research field, with the focus shifting to object detailed representation, the use of data from multiple sensors and the design of novel, generic, spatially accurate algorithms. Recent quantitative results from the International Society for Photogrammetry and Remote Sensing (Working Group III/4) benchmark on urban object detection and 3D building reconstruction [7] indicated that, in 2D, buildings can be recognized and separated from the other terrain objects, however, there is room for improvement towards the detection of small building structures and the precise delineation of building boundaries.

© Springer International Publishing Switzerland 2015
G. Bebis et al. (Eds.): ISVC 2015, Part I, LNCS 9474, pp. 717–727, 2015.
DOI: 10.1007/978-3-319-27857-5_64

Most research and development efforts do consider a classification approach at the core of their man-made detection framework [3], while the detection is based mainly on satellite multispectral datasets [8,9]. For multi-temporal data their efficient registration is a prerequisite [10]. However, due to recent advances in photonics, optics and nanotechnology hyperspectral satellite and airborne data are becoming more (openly) available and cost-effective while new object detection algorithms are introduced and validated [11]. In particular, for remote sensing applications and the classification of hyperspectral data, deep learning algorithms have recently indicated their very promising potentials with relative low classification errors in comparison with other state-of-the-art methodologies [12,13].

In contrast to the approaches that follow the conventional paradigm of pattern recognition, which consists of the construction of complex handcrafted features from the raw data input [14,15], deep learning models [16–18] are a class of machines that can learn a hierarchy of features by building high-level features from low-level ones, thereby automating the process of feature construction for the problem at hand. Furthermore, human brains perform well in object recognition tasks because of its multiple stages of information processing from retina to cortex [19]. Similarly, machine learning architecture with multiple layers of information processing construct more abstract and invariant representations of data, and thus are believed to have the ability of yielding higher classification accuracy than swallow architectures [20]. Convolutional Neural Networks (CNN), Stacked Auto-Encoders, Deep Belief Networks, etc. consist examples of deep learning models.

We tackle the problem of man-made object detection through a deep learning classification framework using on hyperspectral data. In particular, through the exploitation of a CNN we encode pixels' spectral and spatial information and a Multi-Layer Perceptron (MLP) is employed to perform the classification task. Experimental results and quantitative validation on widely used datasets showcasing the potential of the developed approach for accurate man-made object detection.

The rest of this paper is organized as follows; Sect. 2 presents our approach overview by describing the fundamentals of the deep learning architecture and its application on hyperspectral data classification. Section 3 describes the architecture of the proposed system, experimental results and comparisons are presented in Sects. 4 and 5 concludes this work.

2   Approach Overview

We consider the exploitation of a deep learning architecture for the detection of man-made objects using hyperspectral data, *i.e* the classification of each pixel to *man-made* or *non man-made* classes based on their spectral signatures and spatial properties. Spectral signatures are associated with the reflectance properties at every pixel for every spectral band, while spatial information is derived by taking into consideration its neighbors. The exploitation of spatial information is justified by the fact that, due to the nature of the problem, neighboring pixels is very probable to belong to the same class.

Towards this direction, high-level features that encode pixels' spectral and spatial information, are hierarchically constructed in an automated way using a CNN [16]. CNNs consist a type of deep models, which apply trainable filters and pooling operations on the raw input, resulting in a hierarchy of increasingly complex features.

2.1 Convolutional Neural Networks Fundamentals

A CNN consists of a number of convolutional and sub-sampling layers. The input to a convolutional layer is a 3D tensor of dimensions $h \times w \times c$, where h, w and c correspond to input's height, width and channels respectively. The convolutional layer contains C trainable filters of dimensions $m \times m \times q$, where m is smaller than h and w and q is usually equal to c. Each filter is small spatially, but it extends through all channels of the input. By convolving each filter across the width and height of the input, 2D activation maps (feature maps) of that filter are produced.

Intuitively, the network learns filters that activate when they see some specific type of feature at some spatial position in the input. Stacking these activation maps for all filters along the depth dimension forms the full output volume, which is a 3D tensor of dimensions $h - m + 1 \times w - m + 1 \times C$ (convolution does not take into consideration the border of the input). The final output of a convolutional layer incorporates non-linearities, which are modeled through the application of non linear functions on the full output volume (*e.g.* sigmoid, tanh) and the addition of a bias term.

The output of the convolutional layers is fed to a sub-sampling layer, where each activation map is sub-sampled typically using the max polling operator over $k \times k$ contiguous regions. The sub-sampling layer incorporates scale and translation invariance to constructed activation maps. Again the output of the sub-sampling layer is a 3D tensor, whose dimensions are $\frac{h-m+1}{k} \times \frac{w-m+1}{k} \times C$.

A deep learning architecture may consist of many convolutional and sub-sampling layers. The last sub-sampling layer typically is sequentially connected with a fully-connected MLP, which is responsible for conducting the classification or regression task. The whole deep learning architecture is trained using the well known back propagation algorithm.

2.2 Convolutional Neural Networks for Hyperspectral Data

A hyperspectral image is represented as a 3D tensor of dimensions $h \times w \times c$, where h and w correspond to the height and width of the image and c to its channels (spectral bands). As mentioned before, CNNs produce global image features. However, man-made object detection can be seen a pixel-based classification problem. In order to be able to exploit CNNs for man-made object detection, we have to decompose the captured hyperspectral image into *patches*, each one of which contains spectral and spatial information for a specific pixel.

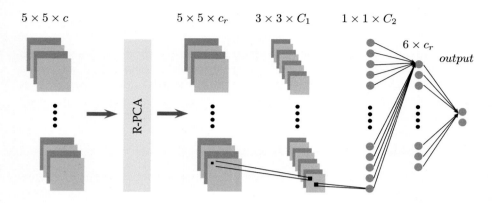

Fig. 1. Overall system architecture.

More specifically, in order to classify a pixel $p_{x,y}$ at location (x, y) on image plane and successfully fuse spectral and spatial information, we use a square patch of size $s \times s$ centered at the same location. Let us denote as $l_{x,y}$ the class label of the pixel at location (x, y) and as $t_{x,y}$ the patch centered at pixel $p_{x,y}$. Then, we can form a dataset $D = \{(t_{x,y}, l_{x,y})\}$ for $x = 1, 2, \cdots, w$ and $y = 1, 2, \cdots, h$. Patch $t_{x,y}$ is also a 3D tensor with dimension $s \times s \times c$, which contains spectral and spatial information for the pixel located at (x, y) on image plane.

Moreover, tensor $t_{x,y}$ is divided into c matrices of dimensions $s \times s$ which are fed as input into a CNN, which hierarchically builds high-level features that encode spectral and spatial characteristics of pixel $p_{x,y}$. These features are fed to a MLP, which is responsible for the classification task. At this point, it has to be mentioned that after training the deep learning architecture is capable of classifying patches and not pixels. We assume that the label of the patch centered at location (x, y) on image plane must be the same with the pixel at the same location. Although, it is a strong assumption, for this specific problem at hand, it is valid for the vast majority of the pixels.

3   System Architecture

In this section the developed system architecture is described. Firstly, the proposed approach for the dimensionality reduction of the input raw data is presented and then the structures of CNN and MLP are given.

3.1   Raw Input Data Dimensionality Reduction

CNNs are capable of hierarchically constructing high-level features in an automated way. Constructed features are the outcome of the convolution between the trainable filters and 2D network's input, which takes place during the training process. In contrast to RGB images that consist of three color channels, the *hundreds* of channels (network inputs) along the spectral dimension of a hyperspectral image may increase the computational cost of training and prediction phases to non acceptable levels.

However, the spectral signature of a material is very specific. Thus, pixels that depict the same material is expected to present very similar spectral responses [13]. Indeed, through a simple statistical analysis of pixels' spectral responses we observed two different things. Firstly, the variance of spectral responses of pixels that depict the same material is very small, and secondly, pixels that depict different materials, either they respond to different spectral bands or when they respond to the same spectral bands the values of their responses is very divergent. These two observations suggest that redundant information is present along the spectral dimension of the hyperspectral image. Therefore, a dimensionality reduction technique can be employed to reduce the dimensionality of the raw input data in order to speed up the training and prediction processes.

For dimensionality reduction, Randomized Principal Component Analysis (R-PCA) [21] is introduced along the spectral dimension to condense the whole image. Principal Component Analysis projects data to a lower dimensional space that preserves most of the variance by dropping components associated with lower eigenvalues. R-PCA limits the computation to an approximate estimate of principal components to perform data transformation. Thus, it is much more computationally efficient than PCA and suitable for large scale datasets, like hyperspectral images.

It should be noted that this step does cast away spectral information, but since R-PCA is applied along the spectral dimension, the spatial information remains intact. The number of principal components that are retained after the application of R-PCA, is appropriately set, in order to keep at least 99.9 % of initial information. This is very important, since dimensionality reduction is conducted by taking into consideration the maximum allowed information loss and not a fixed number of principal components.

During the experimentation process on widely used hyperspectral datasets, 99.9 % of initial information is preserved by using the first 10 to 20 principal components, reducing this way up to 20 times the dimensionality of the raw input.

3.2   Detection Structure

After dimensionality reduction, each patch is a 3D tensor of dimensions $s \times s \times c_r$. Parameter c_r corresponds to the number of principal components that preserve at least 99.9 % of initial information, while the parameter s determines the number of neighbors of each pixel that will be taken into consideration during classification task. The neighbors of a pixels are utilized to represent its spatial information.

During experimentation process we set the parameter s to be equal to 5, in order to take into consideration the closest 24 neighbors of each pixel. By increasing the value of s, the number of neighbors that are taken into consideration is increased and thus the computational cost of classification is increased, also.

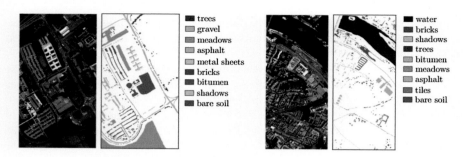

Fig. 2. Pavia University (left) and Pavia Centre (right) datasets along with their ground truths and labels of depicted materials (figures taken from [22]).

However, setting the parameter s to a value larger than 5, no further improvement on classification accuracy was reported in all experiments. On the contrary, increasing the value of s over 13, deteriorates classification accuracy. This is justified by the fact that our previous assumption, which states that the label of the patch centered at location (x, y) on image plane must be the same with the label of the pixel at the same location, is not valid for large s.

Having estimate the values of the parameters s and c_r, we can proceed with the CNN structure design. The first layer of the proposed CNN is a convolutional layer with $C_1 = 3 \times c_r$ trainable filters of dimension 3×3. This layer delivers C_1 matrices of dimensions 3×3 (during convolution we don't take into consideration the border of the patch). In contrast to conventional CNNs, we do not use a subsampling layer after the convolution layer, since we don't take into account any translation and scale invariance. For this reason the first convolutional layer is followed by a second convolutional layer with $C_2 = 3 \times C_1$ trainable filters. Again, the filters are 3×3 matrices.

The second convolutional layer delivers a vector with C_2 elements, which is fed as input to the MLP classifier. The number of MLP hidden units is smaller than the dimensionality of its input. In particular, we set the number of hidden units to equal $6 \times c_r$. For training the deep learning architecture the standard back propagation algorithm was employed, in order to learn the optimal model parameters, *i.e.* minimize the negative log-likelihood of the data sets under the model parameterized by MLP weights and filters elements. The overall system architecture is presented in Fig. 1.

4   Experimental Results and Validation

In this section we present the experimentation framework. Specifically, we present the dataset description, comparisons with other techniques and prediction capabilities of the proposed method.

4.1 Datasets for Validating the Proposed Method

In our study we experimented and validated the developed framework with widely known and publicly available ROSIS hyperspectral datasets. In particular, we utilized (i) the Pavia Centre and (ii) Pavia University datasets, whose number of spectral bands are 102 and 103 respectively. Pavia Centre is a 1096×1096 pixels image, and Pavia University is 610×610 pixels. Some of the samples in both images contain no information and discarded before the analysis. The geometric resolution is 1.3 meters. Ground truths for both images contain 9 classes. Pixels of ground truth images that are depicted in white color are not annotated. Both datasets along with their ground truths and the labels of depicted materials are presented in Fig. 2.

In this paper we focus on the detection of man-made objects. Thus, we grouped together pixels that depict man-made objects and discriminated them than the rest of the pixels. Pixels that are not annotated were not taken into consideration for classification purposes. For the Pavia University dataset, pixels that depict man-made objects are labeled as *asphalt, metal sheets, bricks* and *bitumen*, while for the Pavia Centre dataset pixels that depict man-made objects are labeled as *asphalt, tiles, bricks* and *bitumen*. Pixels that are labeled as *shadows* were not taken into consideration for classification purposes because it is doubtful whether they represent man-made objects.

Supervised training was conducted using pixels that depict man-made and not man-made objects. In particular, we split the tagged parts of the aforementioned datasets into three sets, *i.e.* training, validation and testing data. During experimental validation we tested different split ratios in order to evaluate the classification accuracy of the proposed system when different amounts of data are available. For quantifying the classification accuracy of the proposed system we used the percentage of misclassification on testing set. Finally, validation set is used to enable early stopping criteria during the training process.

4.2 Quantitative Evaluation

Our method was compared against Support Vector Machines (SVM) approaches that use Radial Basis Function (RBF) and linear kernels. In order to conduct

Table 1. Quantitative evaluation results for Pavia University dataset. Split ratio ranges from 5 % to 75 % of the size of the whole dataset and corresponds to the size of the training set.

	Pavia University - misclassification error (%)					
Split ratio	5 %	10 %	20 %	30 %	50 %	75 %
Our approach	**2.5773**	**2.1455**	**1.3203**	**0.7729**	**0.2645**	**0.1861**
RBF-SVM	3.6213	3.2620	3.0960	2.8825	2.7685	2.6976
Linear SVM	4.4065	4.1332	3.9775	3.9693	3.9339	3.9166

Table 2. Quantitative evaluation results for Pavia Centre dataset. Split ratio ranges from 5 % to 75 % of the size of the whole dataset and corresponds to the size of the training set.

	Pavia Centre - misclassification error (%)					
Split ratio	5 %	10 %	20 %	30 %	50 %	75 %
Our approach	**0.2281**	**0.1223**	**0.0759**	**0.0663**	**0.0329**	**0.0086**
RBF-SVM	0.6598	0.5553	0.4730	0.4057	0.3876	0.3578
Linear SVM	1.1418	1.0988	1.0323	1.0810	1.1004	1.0563

a fair comparison SVM should be able to exploit pixels spectral and spatial information during training. For this reason each pixel was represented by its own spectral responses as well as the responses of its $(s \times s) - 1$ closest neighbors. In other words, a pixels at location (x, y) on image plane was represented by the spectral responses of a patch centered at the same location. Although this representation is a 3D tensor, it was flattened to form a 1D vector, in order to be utilized for training the SVM.

We conducted the experiments separately for each one of the datasets. During the experiments we formed six varying size training datasets, whose size ranges from 5 % to 75 % of the size of the whole dataset. Due to the fact that training, validation and testing set were formed by *randomly* selecting samples according to a pre-specified splitting ratio, we replicated each one of the experiments 25 times. Therefore, misclassification error corresponds to average error and classification accuracy to average accuracy.

Tables 1 and 2 present the quantitative evaluation of our proposed method against SVM-based approaches. Our method outperforms SVM-based methods in both datasets and for all training set sizes. It is capable of achieving high classification accuracy scores for very small training datasets, while at the same time it avoids over-fitting when larger training datasets are used. Figure 3 visualizes the outcomes of the evaluation in terms of classification accuracy.

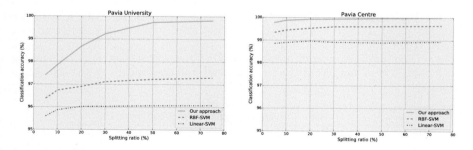

Fig. 3. Methods evaluation and comparison in terms of classification accuracy for both datasets.

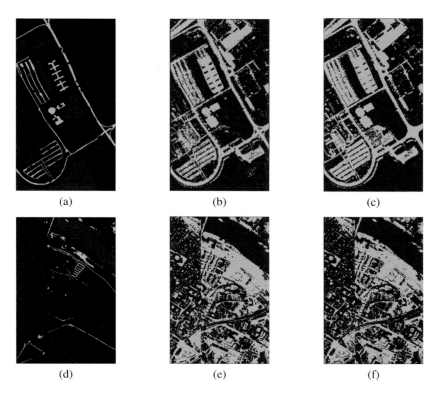

(a)   (b)   (c)

(d)   (e)   (f)

Fig. 4. Classification accuracy of the proposed method using annotated and non-annotated pixels of each hyperspectral image. (a) and (d) Ground truth images for Pavia University and Pavia Centre datasets, (b) and (e) classification results when splitting ratio equals 5 % and (c) and (f) classification results when splitting ratio equals 75 %.

4.3 Prediction Capabilities

Furthermore, we examine the classification accuracy from a visual perspective using all the pixels of each image. In other words, pixels, corresponding to annotated and not-annotated regions, for each one of the hyperspectral images where classified using our deep learning approach. The classification results were obtained by setting the splitting ration equal to 5 % and 75 %.

The classification results after the application of the developed framework are presented in Fig. 4. The resulted classification maps along with the ground truths are shown for both datasets. Green and red colored pixels correspond to man-made objects and non man-made object, respectively. Blue colored pixels in ground truth images correspond to non annotated pixels. As we can see, by fusing spectral and spatial information for each pixel, classification process results to the formation of compact areas, avoiding the presence of noisy scatter points, while at the same time it retains the shape and details of the depicted objects. Finally, the compactness of the areas is getting stronger as the size of the training set is increased.

5 Conclusions

In this paper, we propose a deep learning based approach for man-made object detection using hyperspectral data. Through the deep learning paradigm, our approach hierarchically constructs high-level features that encode pixels spectral and spatial information. Experimental validation of the proposed method and comparisons against SVM based methods showcase the high potential of the developed man-made objects detection system. Finally, among the future perspectives is the application of the developed framework for the detection of human behavior from hyperspectral video sequences.

Acknowledgements. This research was funded from European Unions FP7 under grant agreement n.313161, eVACUATE Project (www.evacuate.eu)

References

1. Vescoukis, V., Doulamis, N., Karagiorgou, S.: A service oriented architecture for decision support systems in environmental crisis management. Future Gener. Comput. Syst. **28**, 593–604 (2012)
2. Pesaresi, M., Huadong, G., Blaes, X., Ehrlich, D., Ferri, S., Gueguen, L., Halkia, M., Kauffmann, M., Kemper, T., Lu, L., Marin-Herrera, M., Ouzounis, G., Scavazzon, M., Soille, P., Syrris, V., Zanchetta, L.: A global human settlement layer from optical HR/VHR RS data: concept and first results. IEEE J. Sel. Top. Appl. Earth Obs. Remote Sens. **6**, 2102–2131 (2013)
3. Karantzalos, K.: Recent advances on 2D and 3D change detection in urban environments from remote sensing data. In: Helbich, M., Arsanjani, J.J., Leitner, M. (eds.) Computational Approaches for Urban Environments, vol. 13, pp. 237–272. Geotechnologies and the Environment, Springer, Switzerland (2015)
4. Florczyk, A., Ferri, S., Syrris, V., Kemper, T., Halkia, M., Soille, P.: A new european settlement map from optical fine scale remote sensed data. IEEE J. Sel. Top. Appl. Earth Obs. Remot. Sens. **PP**(99), 1–15 (2015)
5. Gruen, A., Kuebler, O., Agouris, P.: Automatic Extraction of Man-made Objects from Aerial and Space Images I. Birkhaeuser, Basel (1995)
6. Karantzalos, K., Argialas, D.: A region-based level set segmentation for automatic detection of man-made objects from aerial and satellite images. Photogram. Eng. Remote Sens. **75**, 667–677 (2009)
7. Rottensteiner, F., Sohn, G., Gerke, M., Wegner, J.D., Breitkopf, U., Jung, J.: Results of the ISPRS benchmark on urban object detection and 3d building reconstruction. ISPRS J. Photogram. Remote Sens. **93**, 256–271 (2014)
8. Vakalopoulou, M., Karantzalos, K., Komodakis, N., Paragios, N.: Simultaneous registration and change detection in multitemporal, very high resolution remote sensing data. In: IEEE Computer Vision and Pattern Recognition Workshops (CVPRW 2015) (2015)

9. Vakalopoulou, M., Karantzalos, K., Komodakis, N., Paragios, N.: Building detection in very high resolution multispectral data with deep learning features. In: IEEE International Geoscience and Remote Sensing Symposium (IGARSS 2015) (2015)
10. Karantzalos, K., Sotiras, A., Paragios, N.: Efficient and automated multi-modal satellite data registration through mrfs and linear programming. In: IEEE Computer Vision and Pattern Recognition Workshops, pp. 1–8 (2014)
11. Manolakis, D., Truslow, E., Pieper, M., Cooley, T., Brueggeman, M.: Detection algorithms in hyperspectral imaging systems: an overview of practical algorithms. IEEE Signal Process. Mag. 31, 24–33 (2014)
12. Chen, Y., Lin, Z., Zhao, X., Wang, G., Gu, Y.: Deep learning-based classification of hyperspectral data. IEEE J. Sel. Top. Appl. Earth Obs. Remote Sens. 7, 2094–2107 (2014)
13. Makantasis, K., Karantzalos, K., Doulamis, A., Doulamis, N.: Deep supervised learning for hyperspectral data classification through convolutional neural networks. In: IEEE International Geoscience and Remote Sensing Symposium (IGARSS 2015) (2015)
14. Camps-Valls, G., Bruzzone, L.: Kernel Methods for Remote Sensing Data Analysis. J. Wiley and Sons, NJ, USA (2009)
15. Camps-Valls, G., Tuia, D., Bruzzone, L., Benediktsson, J.A.: Advances in hyperspectral image classification: earth monitoring with statistical learning methods. IEEE Signal Process. Mag. 31, 45–54 (2014)
16. Lecun, Y., Bottou, L., Bengio, Y., Haffner, P.: Gradient-based learning applied to document recognition. Proc. IEEE 86, 2278–2324 (1998)
17. Hinton, G.E., Salakhutdinov, R.R.: Reducing the dimensionality of data with neural networks. Science 313, 504–507 (2006)
18. Bengio, Y., Lamblin, P., Popovici, D., Larochelle, H.: Greedy layer-wise training of deep networks. Adv. NIPS 19, 153–160 (2007)
19. Kruger, N., Janssen, P., Kalkan, S., Lappe, M., Leonardis, A., Piater, J., Rodriguez-Sanchez, A.J., Wiskott, L.: Deep hierarchies in the primate visual cortex: What can we learn for computer vision? IEEE Trans. PAMI 35, 1847–1871 (2013)
20. Bengio, Y., Courville, A., Vincent, P.: Representation learning: a review and new perspectives. IEEE Trans. Pattern Anal. Mach. Intell. 35, 1798–1828 (2013)
21. Halko, N., Martinsson, P., Tropp, J.: Finding structure with randomness: stochastic algorithms for constructing approximate matrix decompositions (2009). http://arxiv.org/abs/0909 (4061)
22. Mura, M.D., Villa, A., Benediktsson, J.A., Chanussot, J., Bruzzone, L.: Classification of hyperspectral images by using extended morphological attribute profiles and independent component analysis. IEEE Geosci. Remote Sens. Lett. 8, 542–546 (2011)

Hyperspectral Scene Analysis
via Structure from Motion

Corey A. Miller$^{(\boxtimes)}$ and Thomas J. Walls

Naval Research Laboratory, 4555 Overlook Avenue, Washington, DC 20375, USA
corey.miller@nrl.navy.mil

Abstract. We present an overview of a structure from motion (SFM) pipeline for processing hyperspectral imagery (HSI), and demonstrate the data exploitation advantages associated with post-processing HSI data in a 3D environment. Using only raw HSI datacubes as input, we leverage HSI anomaly detection and spectral matching to create a 3D spatial model of the scene being imaged. The resulting 3D space provides an intuitive basis for all forms of HSI analysis. We demonstrate the usefulness of the proposed HSI SFM pipeline through an experimental data set collected using an aerial hyperspectral sensor.

1 Introduction

Structure from motion (SFM) algorithms are frequently used to transform motion-based 2D imagery into 3D reconstructions [1], generating both sparse [2] and dense [3] point cloud representations of a scene using 2D imagery alone. Recent progress in hyperspectral imaging (HSI) technology, which adds a third, spectral dimension to collected imagery, has resulted in these non-traditional image sensors being integrated in a wide variety of applications including surveillance/reconnaissance [4], geological surveying [5,6], and structural inspection [7]. As these sensors continue to become smaller and lighter, their integration into motion-based platforms, both aerially and on the ground, is becoming more commonplace. In such applications, it is advantageous to process the additional spectral content in an intuitive 3D environment. Processing HSI data in a 3D environment is quickly finding its way into active areas of research. For example, Nieto et al. fuse hyperspectral classifications with laser-based range data to classify 3D geological maps [8]. Similarly, Kim et al. integrate a hyperspectral data with a 3D scanner to study the spectral reflectance of objects [9]. These techniques, however, rely on the combination of HSI data with a secondary data source in order to generate the 3D content.

Here we describe an extension of state-of-the-art SFM algorithms to build 3D reconstructions of spectral scene content using only the hyperspectral data cubes. This concept of using SFM algorithms on raw hyperspectral data was previously explored by Miller et al. [10], who characterized the advantages of using multiple spectral bands to merge several SFM reconstructions into a single high-accuracy 3D reconstruction [10]. Here we build on their work, refining

© Springer International Publishing Switzerland 2015
G. Bebis et al. (Eds.): ISVC 2015, Part I, LNCS 9474, pp. 728–738, 2015.
DOI: 10.1007/978-3-319-27857-5_65

their algorithms and adding additional steps of spectral analysis, and have developed a complete and deployable HSI SFM processing system. The rest of paper is organized as follows. Section 2 describes the hyperspectral imaging system used in our analysis. Section 3 introduces the proposed 3D pipeline algorithms. Section 4 presents experimental results using our pipeline and explores several data exploitation techniques, with conclusions and future work given in Sect. 5.

2   Hyperspectral Imaging

Hyperspectral sensors work by collecting densely sampled spectral information for each pixel in an image, producing a three-dimensional (x, y, λ) datacube as output. Compared to visible cameras which only record data at red, blue, and green wavelengths, hyperspectral sensors typically capture dozens if not hundreds of individual wavelengths in the visible and non-visible spectrum. This collection allows for the pixel-level detection of anomalous spectra, known target spectra, and common elemental spectral signatures in the scene of interest.

Hyperspectral scanners typically use either a spatial or a spectral scanning technique to generate the third dimension of the data cube. The hardware used in this discussion is of the former variety, where the two-dimensional focal-plane array is representative of a full slit spectrum (x, λ) and the third dimension (y) is manually generated through a line-scanning motion. The sensor used for this analysis contains a 190-band short-wave infrared focal plane, housed in a fully-stabilized ball gimbal designed to be used for aerially-based, cued operation at long standoff distances [11]. As the second spatial dimension is manually introduced through mechanical motion, high-fidelity stabilization is essential in producing spatially-accurate imagery.

3   Structure from Motion Pipeline

The structure from motion (SFM) processing chain aims to estimate three-dimensional structure from imagery collected along an axis of motion. Typically, the pipeline consists of four steps: feature extraction, feature matching, triangulation, and bundle adjustment/optimization [1]. SFM algorithms were originally developed under the assumption of using standard 2D panchromatic imagery, so we have introduced variations in several of the algorithms in order to incorporate and leverage the spectral content within HSI data [10].

The first step in our pipeline is to run the Reed-Xiaoli Detector (RXD) [12] algorithm on each data cube, an anomaly detection routine that measures the spectral differences between spatial regions of the image. By detecting subtle variations in the spectra across the image, we look to establish a set of baseline features to reliably track between all images. We threshold the raw output values of the RXD algorithm to identify only those points which are the most unique within each image. The full spectrum of these RXD anomaly points are extracted and used as their descriptor vectors in the latter matching step.

In order to leverage existing state-of-the-art SFM feature detector algorithms, we next flatten each HSI data cube into a false-color RGB representation. We select three individual bands to use to represent each of the three RGB channels ($\lambda = 1043\,nm$, $1230\,nm$, and $1577\,nm$), however our selection is specific to our sensor and should be adjusted appropriately for other devices and/or materials. For example, an average over a range of bands could be used to generate the RGB channels as opposed to single band selection. We take this false-color image, along with monochromatic images from each of the three bands individually, and run them through several feature detection and extraction routines. The output of this step is a set of feature points and their respective descriptor vectors for each data cube in our set.

Once we have a set of RXD and standard feature points, we match them between views to generate a basis for extracting the camera motion. Brute-force feature matcher routines common to SFM are combined with additional spectral verification in our pipeline, providing a two-factor verification that reduces mismatched features between views that can easily skew model estimation in later steps. We begin with the RXD features first, provided there are enough that have been reliably identified and are common to several views. Their feature descriptors are matched using a nearest-neighbor routine and their spectral similarity is computed. The cosine similarity between spectrum S_i^v and S_i^w of points x_i^v and x_i^w in match i and cameras v, w is defined by according to

$$\text{similarity}_i = \frac{\sum_\lambda S_i^v \times S_i^w}{\sqrt{\sum_\lambda (S_i^v)^2} \times \sqrt{\sum_\lambda (S_i^w)^2}}, \tag{1}$$

where similarity values range from -1, meaning exactly opposite, to +1, meaning exactly the same, with a value of 0 indicating independence. Once we are confident the same real-world point is being matched between views, we use these matches to build a set of feature tracks that follow a given point across several views in the data set. That is, each real-world point of interest corresponds to a single feature track that contains all of the image locations of that point as the sensor passes over it. Once all the RXD feature tracks have been established, we repeat the matching routine with the standard RGB and single band imagery feature points, extending our feature track set.

Once the set of feature tracks is built we must relate the motion of the feature tracks across views to the respective motion of the camera. In this analysis we make no assumptions about the camera's intrinsic parameters (focal length (f_x, f_y), principal point (p_x, p_y), and skew coefficient (γ), and therefore can only relate camera positions up to a projective transformation via the fundamental matrix F constraint

$$x^v F_{v,w} x^w = 0, \tag{2}$$

for pixel x and a pair of cameras $v, w \in 1, \dots, N$ out of N total cameras. The fundamental matrix is first computed between all possible pairs of camera views via a RANSAC routine using only the baseline set of RXD feature tracks. By providing a baseline estimate for the camera's intrinsic parameters, we can transform the fundamental matrix to its essential matrix equivalent, which is a metric relation between scenes. Our baseline estimate for these parameters only needs to be sufficiently accurate to extract a baseline camera pose from the feature tracks, as these values will be included in our bundle adjustment step and will therefore be optimized as the reconstruction builds itself. These parameters are represented by a camera calibration matrix (K) which relates the fundamental matrix to the calibrated essential matrix by

$$E_{v,w} = K^t F_{v,w} K, \tag{3}$$

where ^{t} indicates matrix transpose. We can then extract estimated rotation (R) and translation (T) extrinsic camera parameters from our estimated essential matrix. We take the singular value decomposition of E and extract R and T by defining matrix W such that

$$E = USV^t \tag{4}$$

$$T = VWSV^t \tag{5}$$

$$R = UW^{-1}V^t, \tag{6}$$

where the subscripts v, w are withheld for simplicity. Determining the extrinsic rotation and translation camera parameters between views allows us to triangulate the coordinates of the real-world points associated with each of the respective feature tracks. The image point x_i^j, which is the ith point when viewed from camera $j \in 1, \ldots, N$, can be related to its 3D world coordinate point X_i according to

$$x_i^j \propto K_j \left[R_j | T_j \right] X_i. \tag{7}$$

Our algorithms step through each pair of cameras and determine which has the highest percentage of accurately-projected feature points [13]. We measure accuracy here through the percent of points with positive z-axis projections relative to the camera plane (i.e. in front of the camera) combined with the average error of the 3D points as they are reprojected back onto the various camera planes. Once a baseline camera pair has been established with their corresponding 3D points, the initial bundle adjustment (BA) routine is run to optimize the triangulation and initial camera parameters [14].

The 3D reconstruction is then iteratively built by adding camera views to the existing reconstruction. For each additional camera, feature tracks that span into the established 3D space are used to compute the camera's extrinsic camera motion through an iterative 3D-2D perspective-n-point (PnP) RANSAC routine based on Levenberg-Marquardt optimization. As this system builds upon itself, the BA routine is called after each update to the 3D space. The resulting 3D point cloud space is a sparse representation of the global scene being imaged.

4 Experimental Results

We demonstrate the output of our full HSI SFM pipeline here. Our testing data set is composed of 26 individual HSI data cubes collected aerially as the

HSI system completed a full orbit of an industrial complex at a large standoff distance. Each data cube is 1280 pixels × 966 pixels × 190 bands in dimension and has had atmospheric corrections applied. It should be noted that while the discussion within this paper relates to aerially-based data, these techniques and methodologies extend to any data set that contains a changing perspective that fits the projective geometry requirements.

4.1 HSI SFM

An example portion of the industrial complex being imaged in this example can be seen in Fig. 1. The 26 HSI data cubes that make up our data set generated 86,958 feature tracks when processed in our HSI SFM pipeline, or just over 3,300 per camera view. Of these, 80,902 were determined to triangulate accurately and therefore make up the resulting 3D point cloud which can be seen in Fig. 2.

Fig. 1. A false-color RGB image of the industrial complex. Three individual wavelengths from our data cube, $\lambda = 1043\,nm$, $1230\,nm$, and $1577\,nm$, are mapped to the RGB channels shown here.

4.2 3D Exploitation

As hyperspectral imaging becomes more commonplace, analysis techniques will be developed in order to exploit the additional spectral content. The benefit of generating a 3D representation of a scene using hyperspectral data is the wealth of intuitive post-processing techniques that can be utilized. From the generation of digital elevation maps to match filtering visualization tools, performing scene analysis in a 3D environment provides the operator with an intuitive understanding of location and situational awareness that is lost in standard 2D processing. We provide several examples of this capability here.

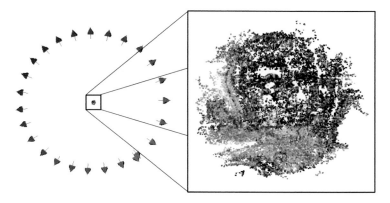

Fig. 2. Top-down view of the 3D reconstruction shows the image-extracted camera positions circling the 3D pointcloud. Close-up (on right) of the pointcloud.

False-color representation can be painted onto the sparse point cloud in any manner in which the user desires. Our 3D SFM pipeline selects three specific bands early in the processing to generate a single false-color representation of each HSI data cube, and these RGB values automatically get pushed onto the 3D space as can be seen in Fig. 3. It is trivial to paint the scene with alternate spectral content, which provides a scene-wide visualization of spectral variations across all views simultaneously. Results from the RX anomaly algorithm can also be viewed in the 3D space, giving localization awareness of spectral irregularities relative to the global view.

A main focus of hyperspectral imaging exploitation is imaging spectroscopy, which is the identification of materials based on their absorption and reflection characteristics. Numerous algorithms have been developed for the detection of targets with known spectral signatures in unknown or unstructured backgrounds [15]. The output of these algorithms is often some measure of target abundance relative to the sensor's local field of view. Detection values that meet the identification threshold can be localized in the global coordinate frame and viewed as an overlay in the 3D space relative to the entire global scene. Because the 3D space builds upon itself as additional data is collected, this 3D representation allows the operator to keep awareness of all identified targets simultaneously, regardless of their presence in the current field of view or not.

As previously discussed, the default output of the 3D SFM algorithms is a sparse point cloud of feature points matched between scenes. Dense 3D reconstructions require additional processing, and can be calculated once the user has identified specific areas of interest within the global scene. We now demonstrate this feature within our interface, where the only user interaction is to highlight a region of the 3D space that requires additional detail. Algorithms are then automatically run which utilize optical flow techniques to match the highlighted subset of the global scene on an individual pixel level. These matches are triangulated using the previously calculated camera positions, and the dense points

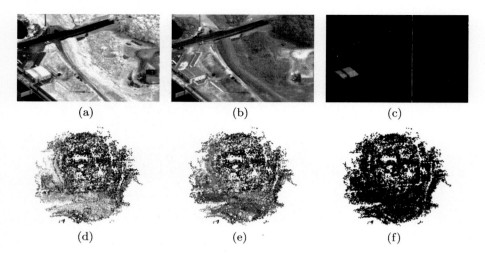

Fig. 3. Single-band imagery (top) and 3D point cloud overlaid with the single-band data (bottom) for $\lambda_1 = 1230\,nm$ [(a), (d)], $\lambda_1 = 1577\,nm$ [(b), (e)], as well as RX anomaly results where white indicates higher anomalous value [(c), (f)].

are pushed into the 3D scene. All of the previously discussed analysis capabilities are still available to the user on the dense points, and can be exploited similarly. An example of this sparse-to-dense transition can be seen in Fig. 4.

In addition to exploiting the spectral content of the scene, traditional 3D analysis can also be performed on the reconstruction. With aerial data, a common request for output is a digital elevation map (DEM) of the scene. Our hyperspectral SFM pipeline can output a DEM of the scene by flattening the z-coordinates of the global scene onto the $x - y$ axis. These values are scaled and missing pixels are filled in with interpolation, producing an HSI-derived elevation map. An example of this can be seen in Fig. 5.

4.3   Discussion on RXD

Our initial analysis has shown that the contribution of using RXD features is highly dependent on the scene content being studied. Our industrial complex data set, for example, contained enough different materials present with spectral variation that the anomaly detector was able to identify and use these points to build RXD feature tracks which established baseline camera estimates. A secondary HSI data set we have studied is of a country estate, an example of which is shown in Fig. 6, where the scene is of a large white house surrounded mainly by trees and grass. This scene lacks spectrally anomalous features, and therefore we were unable to leverage the RXD algorithm to build camera estimates from. In such a case, we rely solely on the RGB and single-band image feature points to extract camera pose information. One approach that could remove this dependence on scene content would be to place spectral targets throughout the

(a) Industrial Building - RGB false color imagery

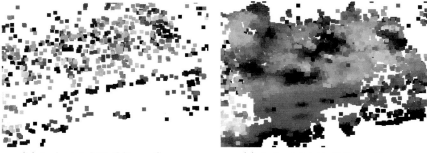

(b) Industrial Building - Sparse   (c) Industrial Building - Dense

Fig. 4. (a) Raw HSI false-color RGB image of a building in the industrial complex. (b) Sparse representation of the building generated as output from the SFM algorithms. (c) Dense reconstruction applied locally to this building reveals 3D features found in the original image.

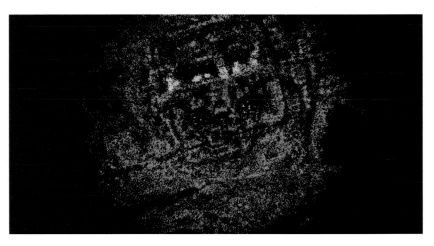

Fig. 5. Sparse HSI reconstruction with a false-color map corresponding to elevation applied. These elevation values are the source for generating digital elevation maps (DEMs).

scene being imaged. Anomaly detection or matched filter detection could easily identify and match these targets from view to view, automatically creating a set of baseline feature correspondences to build feature tracks on.

(a)   (b)

Fig. 6. (a) A false-color RGB image of the country estate. Three individual wavelengths from our data cube, $\lambda = 1043\,nm$, $1230\,nm$, and $1577\,nm$, are mapped to the RGB channels shown here. (b) RXD output on this image. The lack of high anomaly output values can be seen here, and is a result of the minimal spectral variation in the scene.

5   Conclusions and Future Work

We have presented a 3D reconstruction pipeline for processing hyperspectral imagery that can be exploited for a variety of spectral analysis techniques. We have shown that hyperspectral data can be translated into an intuitive, three-dimensional environment which can be generated with a single pass or orbit of an aerially-based HSI sensor. This domain provides new visualization techniques for a variety HSI analysis tools, including content-based surveying, digital elevation mapping, and matched-filter target detection, adding a level of intuitiveness not found in standard 2D image processing. An example reconstruction generated from 26 individual hyperspectral images as the sensor orbited an industrial complex was provided. Merging traditional SFM techniques with the spectral content of the data proved successful, as the SFM pipeline converged on accurate camera parameters for our system, with correct camera position and intrinsic values determined solely from pixel movement between views. Anomaly detection results were integrated into the pipeline, but their contribution was shown to depend heavily on the type of scene being imaged.

Future work includes further integration of additional spectral analysis techniques into the interface that has been created. We are looking to include additional adaptive matched filter detection algorithms into our processing arsenal, as well as fine-tuning the existing capabilities. As we continue to collect additional data, we are looking to distribute targets with known spectra throughout the scene to act as baseline tie-points for the reconstruction algorithm.

References

1. Hartley, R., Zisserman, A.: Multiple View Geometry in Computer Vision. Cambridge University Press, New York (2003)
2. Snavely, N., Seitz, S.M., Szeliski, R.: Photo tourism: exploring photo collections in 3d. In: ACM Transactions on Graphics (TOG), vol. 25, pp. 835–846. ACM (2006)
3. Geiger, A., Ziegler, J., Stiller, C.: Stereoscan: dense 3d reconstruction in real-time. In: Intelligent Vehicles Symposium (IV), 2011 IEEE, pp. 963–968. IEEE (2011)
4. Yuen, P.W., Richardson, M.: An introduction to hyperspectral imaging and its application for security, surveillance and target acquisition. Imaging Sci. J. $58(5)$, 241–253 (2010)
5. Van der Meer, F.D., van der Werff, H., van Ruitenbeek, F.J., Hecker, C.A., Bakker, W.H., Noomen, M.F., van der Meijde, M., Carranza, E.J.M., Smeth, J., Woldai, T.: Multi-and hyperspectral geologic remote sensing: a review. Int. J. Appl. Earth Obs. Geoinf. $14(1)$, 112–128 (2012)
6. Tochon, G., Féret, J., Valero, S., Martin, R., Knapp, D., Salembier, P., Chanussot, J., Asner, G.: On the use of binary partition trees for the tree crown segmentation of tropical rainforest hyperspectral images. Remote Sens. Environ. $\mathbf{159}$, 318–331 (2015)
7. Resende, M.R., Bernucci, L.L.B., Quintanilha, J.A.: Monitoring the condition of roads pavement surfaces: proposal of methodology using hyperspectral images. J. Transp. Lit. $8(2)$, 201–220 (2014)
8. Nieto, J.I., Monteiro, S.T., Viejo, D.: 3D geological modelling using laser and hyperspectral data. In: 2010 IEEE International Geoscience and Remote Sensing Symposium (IGARSS), pp. 4568–4571. IEEE (2010)
9. Kim, M.H., Harvey, T.A., Kittle, D.S., Rushmeier, H., Dorsey, J., Prum, R.O., Brady, D.J.: 3D imaging spectroscopy for measuring hyperspectral patterns on solid objects. ACM Trans. Graph. (TOG) $31(4)$, 38 (2012)
10. Miller, C.A., Walls, T.J.: Passive 3D scene reconstruction via hyperspectral imagery. In: Bebis, G., Boyle, R., Parvin, B., Koracin, D., McMahan, R., Jerald, J., Zhang, H., Drucker, S.M., Kambhamettu, C., Choubassi, M., Deng, Z., Carlson, M. (eds.) ISVC 2014, Part I. LNCS, vol. 8887, pp. 413–422. Springer, Heidelberg (2014)
11. Neumann, J., Allman, E.C., Downes, T., Howard, J., Kruer, M., Lee, J., Linne von Berg, D., Leathers, R., Murray-Krezan, J., Nezis, N.: Demonstration of the MX-20SW standoff SWIR hyperspectral imaging ball gimbal system. MSS, Passive Sensors (2008)
12. Reed, I.S., Yu, X.: Adaptive multiple-band cfar detection of an optical pattern with unknown spectral distribution. IEEE Trans. Acoust. Speech Signal Process. $38(10)$, 1760–1770 (1990)
13. Hartley, R.I., Sturm, P.: Triangulation. Comput. Vis. Image Underst. $68(2)$, 146–157 (1997)

14. Zach, C.: Simple Sparse Bundle Adjustment (SSBA) (2011) http://www.inf.ethz. ch/personal/chzach/opensource.html. Accessed on October 2013
15. Manolakis, D.G., Shaw, G.A., Keshava, N.: Comparative analysis of hyperspectral adaptive matched filter detectors. In: AeroSense 2000, International Society for Optics and Photonics, pp. 2–17 (2000)

ST: Intelligent Transportation Systems

Detecting Road Users at Intersections Through Changing Weather Using RGB-Thermal Video

Chris Bahnsen$^{(\boxtimes)}$ and Thomas B. Moeslund

Visual Analysis of People Laboratory, Aalborg University, Aalborg, Denmark
cb@create.aau.dk

Abstract. This paper compares the performance of a watch-dog system that detects road user actions in urban intersections to a KLT-based tracking system used in traffic surveillance. The two approaches are evaluated on 16 h of video data captured by RGB and thermal cameras under challenging light and weather conditions. On this dataset, the detection performance of right turning vehicles, left turning vehicles, and straight going cyclists are evaluated. Results from both systems show good performance when detecting turning vehicles with a precision of 0.90 and above depending on environmental conditions. The detection performance of cyclists shows that further work on both systems is needed in order to obtain acceptable recall rates.

1 Introduction

Road safety is a subject of high interest amongst governments and the research community. In the European Union, for instance, it is the goal of the European Commission to halve the number of road deaths by 2020 [1]. One of the ways to increase the passive safety of a road is to improve the layout of the road based on historical data of traffic events such as police and hospital records. However, not all accidents are reported and when conflict data is sparse, it might be difficult to assess the safety of a particular road. Video data, on the other hand, allows the generation of detailed information about the road users such as trajectories, speed profiles, and road user types.

However, when conflict data is sparse, one must look through thousands of hours of video to extract events of interest. Another approach is *surrogate safety analysis* which studies potential conflicts as a surrogate for real conflicts. The foundation of surrogate safety analysis is the existence of a continuous relationship between the severity of the interactions and the volume which forms a so-called conflict pyramid [2]. The fatal injuries reside on the top of the pyramid and resembles a low volume of the traffic data while normal traffic with no conflicts make up the majority of the traffic and resides in the lower part of the pyramid. Surrogate safety analysis builds on the claim that the analysis of near-conflicts gives a surrogate measure of the number of fatal interactions between road users [3].

The long-time goal of this work is to obtain a surrogate measure of the safety of bicyclists, pedestrians, and other vulnerable road users at urban intersections.

© Springer International Publishing Switzerland 2015
G. Bebis et al. (Eds.): ISVC 2015, Part I, LNCS 9474, pp. 741–751, 2015.
DOI: 10.1007/978-3-319-27857-5_66

In order to do this, one must reliably detect and track all road users in intersections by automatic analysis of video data. This paper presents a thorough evaluation of the block-based road user action detection technique presented in [4] on 16 h of thermal and RGB video in three urban intersections. The data set includes a variety of sub-optimal conditions for visual algorithms, including rain, hard shadows, reflections, low lighting, and occlusion. We compare the mentioned approach to a feature-based Kanade-Lucas-Tomasi (KLT) tracker for traffic analysis presented by Saunier et al. [5] which is made available through the open-source *trafficintelligence* project [6]. To the best of our knowledge, this is the first cross-evaluation of tracking algorithms for infrastructure-side monitoring on real-world, non-optimal thermal-visible video data.

The following section of this paper contains an overview of related work in infrastructure-side traffic surveillance. Section 3 outlines the main methods used for the block-based road user action detection system used in [4] and Sect. 4 explains the KLT-based feature tracker used for comparison. In Sect. 5, the thermal-visible dataset is described, including context and weather information of the data. Section 6 contains the experimental results of using the mentioned algorithms for cyclist and vehicle detection and Sect. 7 concludes the work.

2   Related Work

Traditional computer-vision based methods on traffic surveillance concerns the monitoring of motorized vehicles at highways [7]. In highways, the detection and tracking of vehicles is easier because they are usually well separated, run in separate lanes, and follow certain routes. A comprehensive survey of traffic surveillance in highway applications is found in [8].

In the past decade, researchers have explored the more complex task of monitoring vehicles at urban areas which includes monitoring of intersections [5,9] and pedestrians and two-wheelers such as mopeds and cyclists [10–12]. Monitoring urban traffic is challenging due to the density of the traffic, variable types of road users, and lower camera orientations which aggravates occlusion. Due to the vast amount of challenges, the field of traffic analysis in computer vision is very diverse and includes a broad range of approaches. In their extensive review of urban traffic analysis with computer vision, Buch et al. divides the field into two main approaches; top-down and bottom-up surveillance systems which eventually are combined with a tracking system [13].

The foundation of *top-down surveillance* is the segmentation of the foreground which is accomplished by using a variety of classic techniques, including frame differencing, background averaging, Kalman filtering, and the Gaussian mixture model (GMM). Foreground segmentation is followed by grouping and vehicle classification which includes region- and contour-based features, and advanced machine learning. Examples of top-down approaches are found in [9,14,15] where the authors use a background model to detect vehicles and [16] which is based on frame differencing.

In *bottom-up surveillance*, the foreground segmentation is replaced by patch detectors and classifiers. Examples in traffic surveillance include Hessian corners [5], SIFT [17], and boosting [18].

Tracking is used to connect observations of road users in consecutive frames into spatio-temporal trajectories. The classic Kalman filter is used in a variety of applications, including [10]. Trackers based on the Kalman filter assumes a Gaussian process and measurement noise, which is not fulfilled in the general case of tracking urban traffic. The Particle Filter removes these assumptions at the cost of computational simplicity and is used for tracking motorcycles in [19]. Saunier and Sayed use the KLT tracker to track keypoints of vehicles in intersections [5]. Tracked features are grouped over time according to the spatial distance of the tracks. The work of Saunier and Sayed has been used in [20] to predict collision amongst vehicles in intersections and extended in [12] to include the classification of road users, including pedestrians and bicycles.

3   Watch-Dog Detection of Road User Actions

Detection, tracking, and classification of urban traffic in unconstrained scenarios pose a substantial challenge to computer vision algorithms. Existing methods are typically evaluated either at short intervals or under ideal conditions. An automated system which is able to accomplish these tasks in unconstrained scenarios does currently not exist [13]. On the other hand, manual monitoring of vast amounts of video is an expensive and tedious task which indeed does not scale well to analysis of complex transport networks. In the recent work of [4], the authors take steps to close the gap between automated and manual analysis by introducing a semi-automated watch-dog system, whose aim is to reduce the amount of video data for inspection by the traffic analysts. This semi-automated system is specialized for the detection of interactions between Right Turning Vehicles (RTV), Left Turning Vehicles (LTV), and Straight Going Cyclists (SGC) at intersections. The goal of the watch-dog is not to perform perfect tracking of road users but to obtain a reasonable data reduction.

The watch-dog system contains a cascade of two fundamental detector types that registers either presence or movement in a region of interest (ROI). Each fundamental detector is laid out in a predefined ROI where the road users of interest may be observed. In the watch-dog system, a detector is either triggered or non-triggered, and it is a combination of this binary logic that lays out the RTV, LTV, and SGC detectors.

The *presence detector* uses a background subtraction technique based on reference images. The reference images are compared to the current frame by computing the Canny edges [21] in the ROI of the frame. If the difference of the two edge images is greater than a specified threshold, the detector is marked as triggered. If the difference of the edge images is below 80 % of the threshold, and the background has not been updated for τ consecutive frames, the background image is updated with the current image.

The *movement detector* estimates the movement in a certain ROI of the intersection by computing the dense optical flow [22] between two frames. Only the flow vectors within a desired direction of movement and above a certain magnitude are kept. If the number of remaining flow vectors surpass a threshold, the detector is triggered. A detailed description of the movement and presence detectors is found in [4].

The movement and presence detectors are overlaid on specific parts of the intersection and several detectors are chained to detect RTV, LTV, or SGC. For instance, if we want to detect RTV, we know that vehicles of interest must enter the intersection, perform a right turn in a designated area, and eventually exit the intersection. In the watch-dog framework, this translates to three sequential detections; detecting presence, detecting movement, and detecting presence. Prototype layouts of the RTV, LTV and SGC detectors are shown in Fig. 1. The presence detector is abbreviated E (edge), the movement detector F (flow), and a new S (stationary) detector is introduced which detects if something is present, but not moving. The stationary detector is a combination of the presence and movement detectors configured on the same ROI. Activity diagrams which describe the sequential logic of the RTV and SGC detectors, are shown in Fig. 2. The LTV and RTV detectors contain one or more modules (F2, F3, F4) which are used to prevent the detection of road users from other directions.

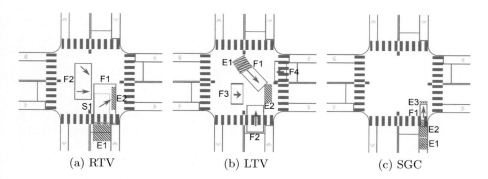

(a) RTV   (b) LTV   (c) SGC

Fig. 1. Prototype layouts for detecting road user actions. The arrows of the F detectors indicates the direction of movement for which the detector is configured

3.1   Fusing Modalities

The watch-dog operates on RGB, thermal, and combined RGBT video data. In RGBT mode, the fundamental detectors of the watch-dog run in parallel on the synchronized RGB and thermal video data. The fundamental detectors output a confidence value between 0 and 1 and a value ≥ 0.5 indicates that the detector is triggered. If the averaged confidence value of the RGB and thermal detectors is above 0.5, the multi-modal detector is triggered.

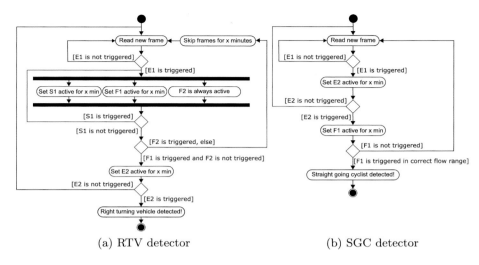

(a) RTV detector   (b) SGC detector

Fig. 2. Activity diagrams of RTV and SGC watch-dog detectors. The LTV detector is similar to the RTV detector. In the LTV detector, F3 and F4 shares the behaviour of F2 in the RTV detector

4   Feature-Based Tracking of Road Users

We use the feature-based tracker of Saunier et al. [5] to compare the results of the watch-dog and assess the method in less-than-ideal weather conditions. The feature-based tracking algorithm is an extension of the method by Beymer et al. [7] which was used to track vehicles at highways. The cornerstone of the tracking algorithm is the KLT-tracker [23]. The extracted features are tracked over time and in order to mimic the physical constraints of the movement of the road users, the features are kept only if the spatio-temporal displacement is small and the averaged motion of the features is smooth. The displacement and motion constraints also entail that only objects in motion are tracked, i.e. if road users stop at any point in the intersection, the tracking of the road user is suspended and will be initiated under a different ID once the road user starts moving.

The features from the tracking stage are grouped in a subsequent, offline step. The grouping algorithm operates in world coordinates and groups feature tracks with similar motion. A feature is grouped to nearby features if the distance is below the maximum distance threshold, $D_{\text{connection}}$. A new feature is easily grouped to several feature groups through this connection stage. However, new features are only added to existing groups if the feature and the group are similar for a minimum number of frames.

For each frame, it is checked if the available pairs of connected features are still belonging together. The distance $d_{i,j}$ between the connected features is computed and it is checked if the relative motion between the two feature tracks

Table 1. Environmental conditions of the proposed data set. The sequences are distributed over three days for each intersection

Location	Seq.	Time of day	Weather	Temp.	Lighting
A	1	07:00–08:00	Partly cloudy	12 °C	Full daylight
A	2	07:00–08:00	Light rain	17 °C	Overcast
A	3	07:00–08:00	Mostly cloudy	15 °C	Overcast
A	4	12:00–13:00	Clear	19 °C	Full daylight
A	5	15:00–16:00	Light rain	19 °C	Overcast
A	6	16:00–17:00	Mostly cloudy	19 °C	Full daylight
B	1	06:00–07:00	Rain	12 °C	Twilight
B	2	07:00–08:00	Rain	12 °C	Overcast
B	3	07:00–08:00	Shallow fog, partly cloudy	6 °C	Overcast
B	4	12:00–13:00	Mostly cloudy	13 °C	Full daylight
B	5	16:00–17:00	Partly cloudy	17 °C	Full daylight
C	1	07:00–08:00	Light rain showers	13 °C	Overcast
C	2	07:00–08:00	Mostly cloudy	13 °C	Overcast
C	3	12:00–13:00	Mostly cloudy	16 °C	Overcast
C	4	16:00–17:00	Light rain showers	14 °C	Overcast
C	5	21:00–22:00	Rain showers	12 °C	Deep twilight

is within a segmentation threshold, $D_{\text{segmentation}}$. If their relative motion is above $d_{\text{segmentation}}$, the features are disconnected.

The $D_{\text{connection}}$ and $D_{\text{segmentation}}$ thresholds are tuned to obtain a balance between overgrouping and oversegmentation. If the thresholds are set too low, road users will be oversegmented, i.e. a single road user will be represented as multiple tracks. If the thresholds are set too high, adjacent road users are detected as one. Finding the right thresholds is a challenge, however, and one has to choose the road user type for which the thresholds should be optimized. For instance, the algorithm might correctly detect car-sized objects while smaller road user types, such as cyclists, are prone to overgrouping and larger objects, such as lorries, are oversegmented.

4.1 Fusing Modalities

Although the feature-based tracker of [5] is built to operate on RGB video, it also translates well to video in the thermal domain. As described in the following section, objects in thermal video generally contain less information than their RGB representation. This is taken into account by adjusting the KLT-tracking parameters and allowing the formation of trajectory groups with fewer trajectories. Additional grouping parameters need not be changed because the grouping is performed in world coordinates in both domains. In RGBT mode, features

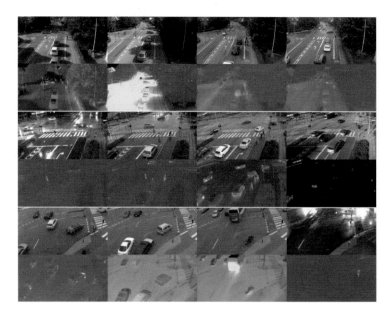

Fig. 3. Sample images of intersection A (top), B (middle), and C (bottom). The appearance of the intersections in both modalities vary greatly due to changing environmental conditions

are extracted separately in the RGB and thermal modality and mapped to a common world coordinate system. Grouping is then performed on the combined RGB and thermal trajectories to produce one single output.

In order to compare the performance of the watch-dog and feature-based tracking, we need to obtain measures of RTV, LTV, and SGC from the latter. Because the entry and exit points of these road users are well-defined in intersections, we define entry and exit masks for each road user action type. We thus define a RTV from an object trajectory if the trajectory passes through entry and exit masks defined for the intersection in question.

5   Thermal-Visible Intersection Data Set

In order to track road users in a diverse range of weather and environmental conditions, the road users themselves must be visible to the algorithms. This is difficult during the night if artificial illumination is sparse. Vehicles might be detected by their headlights - but what about pedestrians and bicyclists?

Thermal cameras are independent of the availability of visible light and only depends on the emitted radiation from objects. Thus, thermal cameras allows to see objects through the night, as long as the temperature of the objects is different from the temperature of the surroundings. Contrary to RGB cameras, thermal cameras are not susceptible to shadows. However, features are sparse in

Table 2. RTV, LTV, and SGC detection performance of the watch-dog and feature-based trackers. The number of manually annotated SGC, RTV, and LTV road user actions is marked in italics in the right side of the table

Seq.		Block-based Watch-dog Precision SGC	RTV	LTV	Recall SGC	RTV	LTV	Feature-based tracker Precision SGC	RTV	LTV	Recall SGC	RTV	LTV	GT
	RGB	0.20	0.52	0.93	**0.71**	0.50	0.84	0.48	0.72	**0.91**	0.27	**0.80**	0.62	*146*
A1	T	0.77	**0.98**	0.94	0.70	0.75	0.65	**0.96**	**0.96**	0.77	**0.32**	0.79	**0.97**	*122*
	RGBT	**0.80**	0.97	**0.96**	**0.71**	**0.80**	**0.92**	0.57	0.74	0.86	0.24	0.75	0.92	*77*
	RGB	0.29	0.98	**0.99**	**0.64**	0.78	**0.99**	0.95	0.96	0.88	0.43	0.81	0.96	*131*
A2	T	0.57	**0.99**	0.91	0.56	0.61	0.55	**1.00**	**0.99**	0.93	0.19	0.69	0.96	*130*
	RGBT	**0.76**	0.97	0.97	0.60	**0.84**	0.89	0.95	0.95	**0.94**	**0.44**	**0.82**	**0.97**	*74*
	RGB	0.68	0.98	0.91	0.65	**0.90**	**0.97**	0.93	**0.99**	**0.87**	0.47	**0.88**	0.87	*141*
A3	T	0.85	0.98	0.94	0.65	0.85	0.16	**0.96**	**0.99**	0.83	0.33	0.70	**0.95**	*120*
	RGBT	**0.90**	**1.00**	**0.97**	**0.69**	0.90	0.89	0.95	0.97	0.85	**0.52**	0.87	0.94	*93*
	RGB	0.11	**1.00**	0.91	**0.90**	0.50	0.87	0.75	0.98	**0.93**	0.10	0.88	0.68	*29*
A4	T	0.73	0.97	0.90	0.55	**0.83**	0.87	**1.00**	**1.00**	0.84	0.10	0.77	**0.99**	*101*
	RGBT	**0.77**	0.99	0.90	0.79	0.79	**0.96**	0.83	0.99	0.91	**0.17**	0.87	**0.99**	*82*
	RGB	0.54	**0.98**	0.95	**0.70**	0.78	**0.91**	0.95	**0.98**	**0.93**	**0.41**	**0.88**	0.85	*44*
A5	T	0.63	0.94	0.90	**0.70**	0.78	0.27	**1.00**	0.96	0.91	0.07	0.51	**0.90**	*156*
	RGBT	**0.66**	0.97	**0.99**	**0.70**	**0.91**	0.73	0.92	0.96	0.91	0.27	0.77	0.88	*139*
	RGB	**1.00**	0.96	**0.98**	0.08	0.80	0.88	**1.00**	0.97	**0.91**	0.18	**0.90**	0.86	*39*
A6	T	0.27	0.73	0.96	**0.90**	0.27	0.88	**1.00**	0.98	0.89	0.15	0.82	**0.99**	*137*
	RGBT	**1.00**	0.95	0.97	0.00	0.42	**0.96**	**1.00**	0.98	0.89	0.15	0.82	**0.99**	*155*
	RGB	0.19	**0.97**	**0.98**	0.82	0.55	0.48	**1.00**	0.94	0.92	0.71	**0.85**	**1.00**	*28*
B1	T	**0.78**	0.91	0.97	**0.89**	0.71	0.69	**1.00**	0.93	**0.95**	0.64	0.67	0.70	*195*
	RGBT	0.63	0.94	0.96	0.86	**0.78**	**0.82**	0.96	0.92	0.62	**0.79**	0.81	0.65	*89*
	RGB	0.34	**0.97**	**0.99**	**0.73**	0.67	0.83	**1.00**	0.97	**0.93**	**0.72**	0.88	**0.97**	*71*
B2	T	0.65	0.84	0.98	0.68	0.78	0.69	0.98	**0.97**	0.92	0.61	0.75	0.91	*353*
	RGBT	**0.67**	0.95	0.98	0.68	**0.85**	**0.95**	**1.00**	**0.97**	0.83	0.69	0.83	0.79	*210*
	RGB	0.43	**0.99**	0.92	**0.65**	0.89	0.93	**1.00**	**1.00**	0.90	0.63	0.84	**0.98**	*92*
B3	T	0.19	0.94	0.92	0.64	0.80	0.83	0.98	0.99	0.71	**0.68**	**0.87**	0.81	*377*
	RGBT	**0.49**	**0.99**	**0.93**	0.63	**0.90**	**0.99**	**1.00**	**1.00**	**0.91**	0.61	0.81	0.88	*177*
	RGB	**0.08**	0.98	0.96	**0.82**	0.67	0.89	**1.00**	0.96	0.93	**0.71**	0.81	0.99	*28*
B4	T	0.03	**1.00**	**1.00**	0.79	0.00	0.02	**1.00**	0.99	0.86	0.64	0.78	**0.99**	*205*
	RGBT	0.04	**1.00**	**1.00**	0.79	0.00	0.05	**1.00**	0.98	**0.95**	**0.71**	0.75	**0.99**	*87*
	RGB	0.09	**0.99**	0.97	**0.83**	0.88	**0.94**	0.83	**0.99**	0.95	0.60	0.83	**0.99**	*48*
B5	T	0.26	**0.99**	**1.00**	0.60	0.91	0.87	**0.95**	0.96	0.87	**0.77**	**0.89**	0.91	*347*
	RGBT	0.07	**0.99**	0.99	0.81	**0.93**	**0.94**	0.86	**0.99**	**0.98**	0.63	0.80	0.95	*102*
	RGB	0.90	**0.94**	**0.97**	0.73	**0.94**	**0.95**	**1.00**	0.96	0.96	0.51	0.94	0.69	*74*
C1	T	0.89	0.88	0.96	0.66	0.91	0.81	**1.00**	0.93	0.93	**0.54**	**0.97**	**0.96**	*116*
	RGBT	**0.96**	**0.94**	0.96	0.70	0.93	0.94	**1.00**	**0.97**	**0.97**	0.53	**0.97**	0.70	*99*
	RGB	0.80	0.97	0.94	**0.75**	0.94	0.81	**0.98**	0.97	0.98	**0.52**	0.97	0.72	*88*
C2	T	0.93	0.42	**0.97**	0.74	0.96	0.88	**0.98**	0.94	0.94	0.51	**1.00**	**0.96**	*120*
	RGBT	**0.94**	**0.98**	0.95	**0.75**	**0.97**	**0.92**	**0.98**	**0.99**	**0.99**	**0.52**	0.97	0.73	*113*
	RGB	**0.79**	**0.99**	0.94	0.92	**0.94**	**0.94**	0.91	**1.00**	0.98	**0.83**	0.97	0.82	*12*
C3	T	0.03	0.55	0.81	**1.00**	0.46	0.48	0.98	0.98	0.98	**0.83**	**1.00**	**0.91**	*128*
	RGBT	0.14	0.98	**0.94**	0.75	0.48	0.61	0.91	**1.00**	**0.99**	**0.83**	0.99	0.77	*109*
	RGB	**0.84**	0.98	**0.99**	0.79	0.35	**0.93**	**1.00**	**1.00**	**0.99**	0.56	0.96	0.82	*34*
C4	T	0.31	0.93	0.88	**0.79**	**0.74**	0.81	**1.00**	0.97	0.98	0.65	**0.98**	**0.93**	*155*
	RGBT	0.33	0.96	**0.99**	0.76	0.61	0.81	**1.00**	0.99	**0.99**	**0.62**	0.97	0.80	*103*
	RGB	0.01	0.53	**1.00**	**0.67**	0.24	0.16	**1.00**	0.97	**1.00**	**0.67**	0.94	0.84	*3*
C5	T	**0.67**	0.94	**1.00**	**0.67**	0.88	0.68	0.67	0.92	**1.00**	**0.67**	**1.00**	**0.95**	*33*
	RGBT	**0.67**	**1.00**	0.93	**0.67**	**1.00**	0.74	**1.00**	**1.00**	**1.00**	**0.67**	**1.00**	**0.95**	*19*

thermal images, and it might thus be more difficult to determine identities or distinguish between objects. An extensive review of thermal cameras in computer vision is found in [24]. When combined, RGB and thermal cameras extend the visibility of the road users and improves the robustness of traffic surveillance algorithms.

The proposed data set is an extension of the data set used in [4]. We extend the original four hours of video data to 16 h and include a broader variety of weather and lighting conditions such as rain, wind, twilight, overcast, and full daylight. Further details regarding the contextual information of the data set is found in Table 1. Samples of each of the three locations are shown in Fig. 3. For each of the three locations, RTV, LTV, and SGC have been manually annotated and assigned a time stamp whenever the desired road user type enters the area of the intersection corresponding to the E2 or E3 module of the watch-dog detectors shown in Fig. 1.

6   Experimental Results

The watch-dog and the feature-based tracker are applied on the 16 h of video described in Table 1. In order to mitigate the oversegmentation of vehicles caused by the feature-based tracker, duplicate tracks are filtered in a post-processing step. Only one object trajectory per second is allowed to pass through what corresponds to the E2 and E3 module of the watch-dog detectors and other tracks within the second are filtered out. Because oversegmentation is not a problem amongst cyclists, the 1 s filter is only applied to the detection of RTV and LTV. For detection of SGC, the threshold is relaxed to 0.3 s. A detection is marked as a true positive if is within ± 2 s of the ground truth.

The results of the test are shown in Table 2. It is seen that the watch-dog based detection of RTV and LTV generally is robust with precision and recall rates above 0.90 in several sequences. The RGBT mode of the watch-dog is shown to be stable whenever each individual modality gives acceptable results. Whenever the watch-dog fail to detect road users in a single modality, the RGBT results will suffer accordingly. Whenever results are stable, the RGBT mode performs best or trails behind the best performing modality by few percentage points. The results of the feature-based tracker shows remarkable precision with few or none false positives. Recall of RTV and LTV shows to be comparable or slightly better than the watch-dog performance. However, the detection of SGC by either the watch-dog or feature-based approaches shows considerable room for improvement. The smaller size and the irregular motion of the SGC are still challenges that need to be solved.

7   Conclusion

This work evaluates the detection performance of left and right turning vehicles and straight going cyclists at urban intersections. Two detection approaches are evaluated at 16 h of RGB and thermal video data featuring challenging weather and light levels. The first approach, the watch-dog, detects road user actions by using a chained set of basic detectors and spatial constraints of the intersection. The second approach, the feature-based detector, uses a KLT-tracker and additional grouping to track moving objects in the intersection. Both approaches show promising results when detecting vehicles while the detection of cyclists shows room for further improvement. The use of RGB and thermal modalities generally results in more stable performance for both detection approaches. However, more sophisticated weighting of modalities is needed to filter out false negatives whenever a detection algorithm breaks down in one modality.

Future work includes more persistent tracking of road users at all speeds in the intersection and further road user classification. Once full trajectories are found, trajectory classification techniques will be investigated to gather more detailed information of the road user actions [25].

Acknowledgements. The authors thank Tanja Kidmann Osmann Madsen for acquiring the data as well as assistance on the ground truth. This project has received funding from the European Unions Horizon 2020 research and innovation programme under grant agreement No. 635895. This publication reflects only the author's view. The European Commission is not responsible for any use that may be made of the information it contains.

References

1. EC European Commission et al.: WHITE PAPER roadmap to a single European transport area towards a competitive and resource efficient transport system. COM (2011) 144 (2011)
2. Hydén, C.: The development of a method for traffic safety evaluation: The swedish traffic conflicts technique. Bulletin Lund Institute of Technology, Department (1987)
3. Svensson, Å., Hydén, C.: Estimating the severity of safety related behaviour. Accid. Anal. Prev. **38**, 379–385 (2006)
4. Bahnsen, C., Moeslund, T.B.: Detecting road user actions in traffic intersections using RGB and thermal video. In: Advanced Video and Signal Based Surveillance, AVSS 2015. IEEE (2015)
5. Saunier, N., Sayed, T.: A feature-based tracking algorithm for vehicles in intersections. In: The 3rd Canadian Conference on Computer and Robot Vision, pp. 59–59. IEEE (2006)
6. Saunier, N.: Trafficintelligence (2015). https://bitbucket.org/Nicolas/trafficintelligence. Accessed 29 Jul 2015
7. Beymer, D., McLauchlan, P., Coifman, B., Malik, J.: A real-time computer vision system for measuring traffic parameters. In: Proceedings of the 1997 IEEE Computer Society Conference on Computer Vision and Pattern Recognition, pp. 495–501. IEEE (1997)
8. Kastrinaki, V., Zervakis, M., Kalaitzakis, K.: A survey of video processing techniques for traffic applications. Image Vis. Comput. **21**, 359–381 (2003)
9. Veeraraghavan, H., Masoud, O., Papanikolopoulos, N.P.: Computer vision algorithms for intersection monitoring. IEEE Trans. Intell. Transp. Syst. **4**, 78–89 (2003)

10. Maurin, B., Masoud, O., Papanikolopoulos, N.P.: Tracking all traffic: computer vision algorithms for monitoring vehicles, individuals, and crowds. IEEE Robot. Autom. Mag. **12**, 29–36 (2005)

11. Messelodi, S., Modena, C.M., Zanin, M.: A computer vision system for the detection and classification of vehicles at urban road intersections. Pattern Anal. Appl. **8**, 17–31 (2005)

12. Zangenehpour, S., Miranda-Moreno, L.F., Saunier, N.: Automated classification in traffic video at intersections with heavy pedestrian and bicycle traffic. In: Transportation Research Board 93rd Annual Meeting, No. 14–4337 (2014)

13. Buch, N., Velastin, S., Orwell, J., et al.: A review of computer vision techniques for the analysis of urban traffic. IEEE Trans. Intell. Transp. Syst. **12**, 920–939 (2011)

14. Aköz, Ö., Karsligil, M.E.: Traffic event classification at intersections based on the severity of abnormality. Machine Vis. Appl. **25**, 613–632 (2014)

15. Kamijo, S., Matsushita, Y., Ikeuchi, K., Sakauchi, M.: Traffic monitoring and accident detection at intersections. IEEE Trans. Intell. Transp. Syst. **1**, 108–118 (2000)

16. Ki, Y.K., Lee, D.Y.: A traffic accident recording and reporting model at intersections. IEEE Trans. Intell. Transp. Syst. **8**, 188–194 (2007)

17. Zhang, W., Yu, B., Zelinsky, G.J., Samaras, D.: Object class recognition using multiple layer boosting with heterogeneous features. In: IEEE Computer Society Conference on Computer Vision and Pattern Recognition, CVPR 2005, vol. 2, pp. 323–330. IEEE (2005)

18. Khammari, A., Nashashibi, F., Abramson, Y., Laurgeau, C.: Vehicle detection combining gradient analysis and adaboost classification. In: Proceedings of the 2005 IEEE Intelligent Transportation Systems, pp. 66–71. IEEE (2005)

19. Nguyen, P.-V., Le, H.-B.: A multi-modal particle filter based motorcycle tracking system. In: Ho, T.-B., Zhou, Z.-H. (eds.) PRICAI 2008. LNCS (LNAI), vol. 5351, pp. 819–828. Springer, Heidelberg (2008)

20. Saunier, N., Sayed, T., Ismail, K.: Large-scale automated analysis of vehicle interactions and collisions. Transp. Res. Rec. J. Transp. Res. Board **2147**, 42–50 (2010)

21. Canny, J.: A computational approach to edge detection. IEEE Trans. Pattern Anal. Mach. Intell. **6**, 679–698 (1986)

22. Farnebäck, G.: Two-frame motion estimation based on polynomial expansion. In: Bigun, J., Gustavsson, T. (eds.) SCIA 2003. LNCS, vol. 2749, pp. 363–370. Springer, Heidelberg (2003)

23. Baker, S., Matthews, I.: Lucas-kanade 20 years on: a unifying framework. Int. J. Comput. Vis. **56**, 221–255 (2004)

24. Gade, R., Moeslund, T.B.: Thermal cameras and applications: a survey. Mach. Vis. Appl. **25**, 245–262 (2014)

25. Morris, B.T., Trivedi, M.M.: Learning, modeling, and classification of vehicle track patterns from live video. IEEE Trans. Intell. Transp. Syst. **9**, 425–437 (2008)

Safety Quantification of Intersections Using Computer Vision Techniques

Mohammad Shokrolah Shirazi[(✉)] and Brendan Morris

University of Nevada, Las Vegas, Las Vegas, USA
shirazi@unlv.nevada.edu, brendan.morris@unlv.edu

Abstract. Vision-based safety analysis is a difficult task since traditional motion-based techniques work poorly when pedestrians and vehicles stop due to traffic signals. This work presents a tracking method in order to provide a robust tracking of pedestrians and vehicles, and quantify safety through investigating the tracks. Surrogate safety measurements are estimated including TTC and DTI values for a highly cluttered video of Las Vegas intersection and the performance of the tracking system is evaluated at detection and tracking steps separately.

1 Introduction

Intersection safety is one of the most important transportation concerns due to complex behavior of vehicles and pedestrians and their interactions which might lead to accidents. Around 2 million accidents and 6,700 fatalities in the United States occur at intersections every year which constitutes 26 % of all collisions [1, 2].

Although safety is emerging as an area of increased attention and awareness, it is difficult to assess due to the lack of good predictive models of accident potentials and lack of consensus on what constitutes a safe or unsafe facility. There are two major methods for intersection safety analysis. The basic method uses data mining techniques on real accident datasets to find the contributing reasons. Since there are availability and quality problems regarding collision data, some studies rely on traffic conflict analysis as an alternative or complementary approach to analyze traffic safety.

The correlation between accidents and conflict-based safety measurements [3] encourage researches to use surrogate safety measurements for safety quantification. Surrogate safety measurements are reliable and consistent in definition, and they are proven to be a practical metric for safety analysis [3]. The surrogate safety measurements such as time to collision (TTC), distance to intersection (DTI), and time to intersection (TTI) are usually estimated through the video frames. TTC is defined as a time for two vehicles (or a vehicle and pedestrian) to collide if they continue at their present speeds on their paths [3].

Vision-based techniques are used for automatic detection, tracking, and safety analyses of vehicles and pedestrians from the sequence of images (i.e. videos) [4]. For instance, Sayed et el. [5] assessed vehicle-vehicle conflicts by calculating TTC values. Low TTC value indicates the severity of the near-accident

© Springer International Publishing Switzerland 2015
G. Bebis et al. (Eds.): ISVC 2015, Part I, LNCS 9474, pp. 752–761, 2015.
DOI: 10.1007/978-3-319-27857-5_67

events. Further, computer vision techniques are the useful tools to analyzing vehicle-pedestrian conflicts since the collisions involving pedestrians are less frequent than other collision types. As an example, Zaki et al. [6] showed the maturity of computer vision techniques to estimate pedestrians' conflict and violation by calculating TTC values.

The performance of a vision-based safety quantification method is directly affected by the underlying detection and tracking algorithm. Vision based methods usually use motion as cue such as optical flow and background subtraction; however, they are not robust to track the temporarily stopped vehicles and pedestrians at intersections due to traffic signals [7]. As a result, DTI and TTI values can not be efficiently estimated and appearance-based methods are required to improve safety analyses results.

This work presents a robust tracking system in order to provide the reliable vehicle and pedestrian trajectories at intersections. The system benefits contextual fusion of appearance and motion at detection step followed by improved version of optical flow tracking for pedestrians and vehicles. Since stopped participants are tracked, more accurate estimation is provided in comparison with traditional methods (i.e., optical flow). The proposed system was evaluated separately at detection and tracking steps, and DTI, TTI, and TTC values for one of the Las Vegas intersections were estimated through a video. The remainder of the paper presents more details regarding tracking system and safety quantification process.

The paper is organized as follows: Sect. 2 presents the proposed system and Sect. 3 shows the scene preparation process. Section 4 shows experimental evaluations, and finally Sect. 5 concludes the paper.

2   Detection and Tracking System

Vehicle and pedestrian detection systems are performed through the two different modules since pedestrians require enhanced method to provide the robust trajectories. The enhancement is conducted through the detection and tracking steps.

Gaussian mixture model (GMM) [8] is used to create an adaptive background model for background subtraction method. Moving vehicles are detected in the motion area to initialize the tracks (see Fig. 1a). Since detection by motion is prone to occlusion or blob merging [7], the motion area is placed in a location in which vehicles do not usually stop even when the queue line is created behind the red signal. In addition, motion area should be close enough to the camera as it ensures stable moving objects (i.e. blobs) obtained by background subtraction.

Pedestrian detection is performed separately through another module since it is more challenging than vehicles, and it worsens for intersection videos. Pure motion-based method for pedestrian detection fails for different scenarios:

1. Although motion is widely used for highway scenarios, it is not consistent at intersection since traffic signals force pedestrians to stop. It also affects waiting time estimation of pedestrians.

(a)   (b)

Fig. 1. Defining areas (a) Motion area for vehicle detection (b) Mix area for pedestrian detection through contextual fusion

2. Pedestrians usually cross together in group and using motion is prone to detect a moving object. Detecting individuals is more appropriate for behavior analysis of individuals [7].

Appearance-based detection methods can significantly improve the pedestrian detection performance in video surveillance [7]. They recognize a pedestrian directly from an image by evaluating pixel values. Therefore, they do not need sequence of frames; instead, they use positive and negative samples to train the classifiers.

The local binary pattern (LBP) is used as an appearance-based detection method to improve the detection performance. LBP is the particular case of the texture spectrum and it has been found to be a powerful feature for texture classification [9]. LBP value of each pixel is computed by constructing 3×3 sub-window thresholded by the center value. Histograms of decimal numbers for each pixel are concatenated to provide a feature vector.

The key role of the proposed system is the contextual fusion which pools the best detections from several positives. The proposed contextual fusion [7] works at the decision level to combine the outputs from the GMM, and LBP detections in special mix areas where both detectors are active. In this method, appearance-based detections are limited to smaller processing regions for speed and reliability. The mix area is defined around signals and crosswalks and appearance-based detections out of this area are removed. Bounding box detections are grouped if they have more than 50 % overlapping and the bigger one is removed for each two detections. The process continues until there are no multiple detections with more than 50 % overlapping. More details are found in [7] regarding contextual fusion.

Optical flow is enhanced for vehicle and pedestrian tracking. Further, pedestrian tracking system benefits cooperation with bipartite graph to handle temporarily miss detected pedestrians. Figure 2 shows the tracking system. Final detected pedestrians (i.e., through contextual fusion in mix area) and vehicles

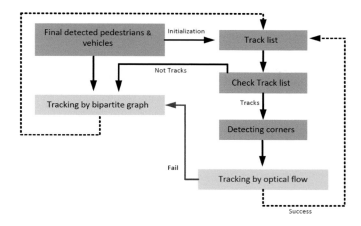

Fig. 2. Pedestrian and vehicle tracking system, red arrow only exists for pedestrian tracking (Color figure online)

(i.e. through GMM in motion area) are given to the tracking system which uses bipartite graph at first to initialize the tracks. The initialized tracks use enhanced optical flow which relies on detected corner and texture-based features inside tracks. The initialized tracks use optical flow as long as the tracking process is successful. The likelihood of successful tracking is determined based on the quality of the detected matches and the estimated bounding box around the features. If these values are less than predefined thresholds, which means the optical flow tracker failed, detected pedestrians are used by bipartite graph tracker to handle the problem. This is due to running appearance-based classifiers at each frame. If contextual fusion does not provide any detection for the bipartite tracker at the time of optical flow failure, the track of a pedestrian is finally lost.

2.1 Enhanced Optical Flow

Optical flow is a default tracking method used for initialized tracks since it is robust against partial occlusion and its effectiveness has been shown for behavior analysis of vehicles and pedestrians [5,6]. The optical flow is enhanced in this work through the three steps of enriching feature points, filtering features, and bounding box estimation.

The reduction problem of feature points is prevalent for small size objects with low quality. This worsens during the optical flow tracking for stopped or slow moving vehicles and pedestrians. The idea is to sample each vehicle and pedestrian with more feature points, called enriching feature points, to tackle this problem. Filtering features is a process of rectifying by removing falsely detected features inside a track. The filtering process is performed by determining the state of the track and filtering the opposing features. For example, when a vehicle or pedestrian state is waiting, the static features are predominant and moving features are removed. The way of determining the state of the track is based on the average of the displacement vector. Bounding box estimation is a crucial part of the tracking process since it helps to keep the track of stopped vehicles in a queue. The bounding box estimation of stopped vehicles and pedestrians is a challenging task which leads to a tracking drift. As a result, the fixed bounding box around the fixed position is leveraged when the waiting state is determined.

2.2   Bipartite Graph

Bipartite Graph uses a greedy approach to find a nearest detection for a track [10]. The nodes of the graph are tracks and frame's detections and a cost between each two nodes is the difference between appearance measurements as calculated in (1),

$$A(v_1, v_2) = A_{pos}(p_1, p_2) A_{size}(s_1, s_2) A_{appr}(h_1, h_2) \tag{1}$$

where A_{pos}, A_{size} and A_{appr} are affinities based on position, size and appearance defined as follows.

$$A_{pos}(p_1, p_2) = \frac{1}{2\pi\sigma_x\sigma_y} exp\left(-\frac{(x_1 - x_2)^2}{2\sigma_x^2}\right) exp\left(-\frac{(y_1 - y_2)^2}{2\sigma_y^2}\right) \tag{2}$$

$$A_{size}(s_1, s_2) = \frac{1}{\sqrt{2\pi}\sigma_s} exp\left(-\frac{(s_1 - s_2)^2}{2\sigma_s^2}\right) \tag{3}$$

$$A_{appr}(h_1, h_2) = \sum_{i=1}^{m} \sqrt{h_1^u h_2^u} \tag{4}$$

Difference of distance and size plugs into Gaussian kernel for A_{pos}, A_{size} and A_{appr} is Bhattacharyya distance calculated separately for two histograms of colors and edges. When a detection does not find any match any tracks in the track-list, a new track is created. If an existing track does not find a detection, it is marked for deletion.

3   Scene Configuration

Scene configuration includes the essential steps required to feed the tracking system by defining mix area, typical paths and training appearance based classifier. Mix area, where LBP appearance detections are also performed, are defined in regions where stopped pedestrians are expected such as around the signal and crosswalk. The GMM is used across the entire scene to account for any visible motion (see Fig. 1b). Typical paths are defined to recognize vehicles and pedestrians regarding different lanes and crosswalk. This can be used for lane-based

Table 1. Pedestrian detection performance during traffic phases

Type	Method	Green		Red		Total	
		TPR	Jaccard	TPR	Jaccard	TPR	Jaccard
Motion	Optical flow	0.02	0.01	0.04	0.02	0.03	0.01
	GMM	0.08	0.06	0.18	0.15	0.13	0.11
Appearance	LBP	0.64	0.18	0.61	0.20	0.64	0.19
Contextual fusion	GMM+LBP	0.52	0.41	0.58	0.41	0.55	0.41

behavior analysis of vehicles (e.g., flow, speed) and observing pedestrian crossing behavior (e.g., crossing speed, crossing count).

Datasets are required to effectively train appearance-based classifier. A large image dataset was created by collecting public available datasets listed in [7]. Finally, 60726 and 89798 positive and negative samples were collected and prepared to train LBP classifiers. The OpenCV implementation of the LBP cascaded classifier is used by Adaboost procedure to learns weak classifiers which are combined to form a strong classifier.

4   Experimental Results

Experimental results include two steps. Detection and tracking performance of the proposed system is evaluated which is followed by the safety quantification results. The vision-based tracking system was implemented by C++ using OpenCV 2.3 and it was run on Intel i7 quad core.

4.1   Detection and Tracking System Evaluation

Detection results are evaluated using true positive rate (TPR) and $Jaccard$ coefficient shown in (5). $Jaccard$ coefficient is way of accounting both false positive (FP) and false negative (FN) values with one indicator. $Jaccard$ value is always less or equal to TPR and its value is increased when the wrongly detected pedestrians (FP) and missed detected pedestrians (FN) are reduced.

$$TPR = \frac{TP}{TP + FN}, \quad Jaccard = \frac{TP}{TP + FP + FN} \tag{5}$$

Table 1 shows the performance of the detected methods for 1000 frames. The higher values are shown with blue and red colors for TPR and $Jaccard$ values respectively. Motion-based techniques indicate the lower detection rates for green and red traffic signals. Contextual fusion always have higher $Jaccard$ values since appearance-based detections are constrained to the mix area. The GMM+LBP is finally used in this work since it has the higher $Jaccard$ values for all traffic phases.

Five criteria are defined to evaluate the performance of the tracking system quantitatively [11]:

Table 2. Comparison of the optical flow with the proposed tracking methods.

Tracker	Tracking method	GT	MT	ML	FG	FT	IS	Duration
Vehicle	Optical flow [5,12]	42	18	5	12	0	6	2000 frames
	Proposed	42	31	3	4	0	1	
Pedestrian	Optical flow [6,13]	14	4	7	19	30	3	5000 frames
	Proposed	14	8	4	5	17	2	

1. Number of mostly tracked (MT) trajectories: more than 80 % of the trajectory is tracked. Value should be high.
2. Number of mostly lost (ML) trajectories: more than 80 % of the trajectory is lost. Value should be low.
3. Number of fragments (FG) of trajectories: the generated trajectory is between 80 % and 20 % of the ground truth.
4. Number of false trajectories (FT): trajectories corresponding to no real object. Value should be low.
5. Frequency of identity switches (IS): identify exchanges between a pair of result trajectories. Value should be low.

The tracking system is compared against our implementation of the pure optical flow used in [5,6,12,13]. The pure optical flow detects moving objects by clustering the features using motion magnitude and direction. Each moving object is initialized by features, and features find their match using KLT algorithm. The main difference in comparison with the proposed method is the lack of three introduced steps (i.e., enriching features, filtering features, bounding box estimation) by enhanced optical flow. Moreover, when pure optical flow fails due to dramatic reduction in the number of matched features, it loses its track since there is no assistance by another tracker through the detections by contextual fusion for pedestrians.

Table 2 shows the comparison of the pure optical flow with proposed tracking method. The proposed tracking method outperforms all criteria values for vehicles and pedestrians; it provides higher MT value and lower ML, FG, FT and IS values than optical flow.

4.2   Safety Evaluation

The intersection safety was evaluated by the proposed system for 18318 frames (i.e., 12:04 p.m–12:56 p.m). DTI and TTC values were calculated in this work for safety analysis. TTI and DTI are naturally conducted by drivers before taking turns to assess the level of threat posed by opposing traffic. These important safety measurements are used in safety systems including advanced driver assistant systems (ADAS) and decision support systems.

DTI is calculated based on the distance of the vehicle from the stop bar. Figure 3 shows the DTI plots of 4 vehicles. DTI value has a decreasing trend until a vehicle stops which causes DTI remains unchanged (i.e., vehicles 1 and 2).

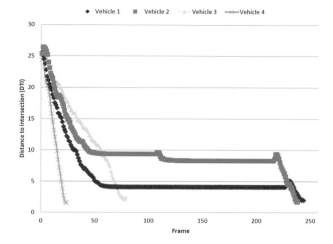

Fig. 3. DTI of four typical vehicles

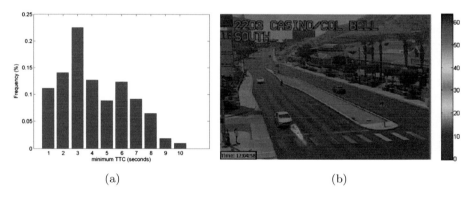

(a)   (b)

Fig. 4. TTC evaluation (a) Probability density of TTC values (b) Vehicle-pedestrian conflict heatmap frequency

DTI values of vehicles 1 and 2 indicate different position of waiting vehicles in a queue behind the red signal. Vehicles 3 and 4 never face with the red signal and they continue their moving during their track life time. Vehicle 3 takes brake at some time since its slope is reduced but vehicle 4 moves with the same speed during its travel time. DTI plots can not be easily utilized with traditional vision-based frameworks since the trajectories of stopped vehicles are lost after a while.

TTC is calculated based on the predicted arrival time and it is defined as the time for two objects to collide if they continue with their present speed on their paths. Each partially observed trajectory of vehicles is compared against typical paths to find its most probable path and its associated conflict point. The time to conflict point for a vehicle is compared to those pedestrians that are moving toward it and the minimum TTC value is counted if both timings are in a same window.

Figure 4a shows the probability density of TTC values for vehicle-pedestrian conflicts. The intersection has a peak at 3 s which classifies intersection as a medium hazardous level since TTC values of less than 2 s indicate the high severity of conflicts. Figure 4b shows the high frequency of conflict on the cross-walk between pedestrians and vehicles on the right lane. The major reason is the higher flow of vehicles on this lane and higher number of pedestrians crossing from the right to left.

5   Conclusion

This paper presents a vision-based tracking system to provide the reliable safety measurements including DTI and TTC. The proposed system benefits contextual fusion (i.e. for pedestrians) at detection level and enhanced optical flow to provide long term track of moving and stopping vehicles and pedestrians at intersections. The system was evaluated at detection and tracking levels, and it analyzed safety of a Las Vegas intersection by estimating the surrogate safety measurements. The work can be further improved by running the system for longer time periods during the peak hours and finding the contributing factors (i.e. vehicle flow, pedestrian waiting time) to vehicle-pedestrian conflicts.

Acknowledgement. The authors acknowledge the Nevada Department of Transportation for their support of this research.

References

1. Liu, Y., Ozguner, U., Ekici, E.: Performance evaluation of intersection warning system using a vehicle traffic and wireless simulator. In: Proceedings of IEEE Intelligent Vehicles Symposium, pp. 171–176 (2005)
2. Shirazi, M., Morris, B.: Observing behaviors at intersections: a review of recent studies and developments. In: 2015 IEEE Intelligent Vehicles Symposium (IV), pp. 1258–1263 (2015)
3. Chin, H.C., Quek, S.T.: Measurement of traffic conflicts. J. Comput. Sci. **26**(3), 169–185 (1997)
4. Shirazi, M.S., Morris, B.: A typical video-based framework for counting, behavior and safety analysis at intersections. In: 2015 IEEE Intelligent Vehicles Symposium (IV), pp. 1264–1269 (2015)
5. Sayed, T., Zaki, M.H., Autey, J.: A novel approach for diagnosing cycling safety issues using automated computer vision techniques. Transportation Research Board Annual Meeting Compendium of Papers (2013)
6. Zaki, M.H., Tarek, S., Tageldin, A., Hussein, M.: Application of computer vision to diagnosis of pedestrian safety issues. Transp. Res. Rec. J. Transp. Res. Board **2393**, 75–84 (2013)
7. Shirazi, M.S., Morris, B.: Contextual combination of appearance and motion for intersection videos with vehicles and pedestrians. In: Bebis, G., et al. (eds.) ISVC 2014, Part I. LNCS, vol. 8887, pp. 708–717. Springer, Heidelberg (2014)

8. Stauffer, C., Grimson, W.E.L.: Adaptive background mixture models for real-time tracking, pp. 246–252 (1999)
9. Ojala, T., Pietikainen, M., Harwood, D.: A comparative study of texture measures with classification based on feature distributions. Trans. Pattern Recogn. **29**, 51–59 (1996)
10. Shirazi, M.S., Morris, B.: Vision-based turning movement counting at intersections by cooperating zone and trajectory comparison modules. In: Proceedings of 17th International IEEE Conference on Intelligent Transportation Systems, Qingdao, China, pp. 3100–3105 (2014)
11. Wu, B., Nevatia, R.: Detection and tracking of multiple, partially occluded humans by bayesian combination of edgelet based part detectors. Intern. J. Comput. Vis. **75**, 247–266 (2007)
12. Saunier, N., Sayed, T.: A feature-based tracking algorithm for vehicles in intersections. Proceedings of 3rd Canadian Conference on Computer and Robot Vision, Quebec, Canada, p. 59 (2006)
13. Ismail, K., Sayed, T., Saunier, N.: Automated analysis of pedestrian-vehicle: conflicts context for before-and-after studies. Transp. Res. Rec. J. Trans. Res. Board **2198**, 52–64 (2010)

Vehicles Detection in Stereo Vision Based on Disparity Map Segmentation and Objects Classification

Djamila Dekkiche[1,2]([✉]), Bastien Vincke[2], and Alain Mérigot[2]

[1] IRT SystemX Palaiseau, Palaiseau, France
[2] Institut d'Electronique Fondamentale, Université Paris Sud, Orsay, France
djamila.dekkiche@irt-systemx.fr

Abstract. This paper presents a coarse to fine approach of on-road vehicles detection and distance estimation based on the disparity map segmentation supervised by stereo vision. Scene segmentation is first performed relying on the robustness of the UV-disparity maps to generate free space and obstacles space. This last is investigated for on-road vehicles detection. The detection process starts with off-road objects substraction based on the connected component labeling algorithm which is also used for on-road segments extraction instead of the traditional hough transform for more robust, precise and fast detection. Objects classification is then applied to the on-road segments by using some cues describing the geometry of vehicles like width and height. However, these latter have been measured not in meter but rather in pixels in function of the disparity. The whole approach is presented and the experimental results of evaluation are shown.

1 Introduction

Stereo vision systems have recently emerged in the domain of robotics and autonomous cars. These systems provide the 3D perception of the environment which is employed in Advanced Driver Assistance Systems (ADAS) to support a variety of functions including obstacles detection [1], lane departure warning [2] and collision warning systems [3]. While the depth measurement precision of stereo vision systems is not as high as with active sensors such as RADAR and LIDAR, the stereo camera can compete with these active technologies due to its low cost in one hand and the amount of traffic scence information it provides in the other hand.

The litterature describes several works on stereo vision based vehicles detection. The majority of the proposed approaches rely on a depth map obtained usually from a disparity map through stereo matching. The study of the state of the art shows two major axes in this field. The first one includes motion based approaches [4]. The idea is to perform features tracking in the monocular image plane of one of the stereo cameras and 3-D localization in the disparity and depth maps. The second axis deals with the transformation of the disparity

G. Bebis et al. (Eds.): ISVC 2015, Part I, LNCS 9474, pp. 762–773, 2015.
DOI: 10.1007/978-3-319-27857-5_68

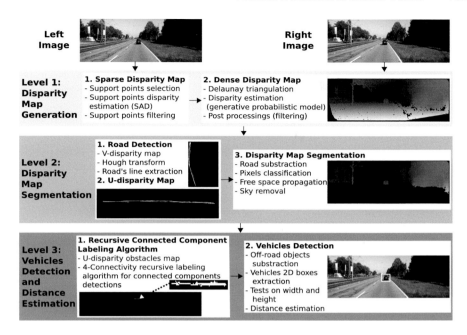

Fig. 1. Algorithm's functional diagram

map in a more represenrative and compact form including occupancy grid [5] and ground surface modeling [6]. These different transformations aim mainly to facilitate scene segmentation and reduce computation time.

The V-disparity [6] is a popular approach for ground plane estimation and road scene analysis [7]. It is a compact representation of the disparity map in a new space more robust and more representative for obstacles detection. This transformation models the road surface as a slanted line and vertical obstacles as vertical lines. Using curve fitting techniques such as Hough transform [8], lines can be detected. The same principle has been used to generate the U-disparity map which has been used for free space estimation [9] as well as for obstacles detection [7].

In this paper, we propose a coarse to fine approach for vehicles detection and distance estimation based on stereo vision. The approach can be divided into three processing levels as shown in Fig. 1(to better view the different images in this article, readers are encouraged to view them on a computer). The first level deals with the generation of a dense disparity map, the approach proposed in [10] has been used for this purpose. The second level performs scene segmentation relying on the V-disparity map and a set of post processings. Finally, vehicles detection is processed in the third level based on the U-disparity map generated from the obstacles disparity map and a set of post processings.

One of the common issues in on-road obstacles detection are the off-road information such as high walls, buildings and trees along the road which may affect the precision of the detection and increase the false alarms. The important contribution of this paper is the approach proposed to remove the off-road objects based on the connected segments labeling algorithm. While this approach has been used in some articles [11] mainly for on-road obstacles detection, we propose to use it also for off-road objects substraction. The second contribution of the paper is the proposed approach for obstacles classification. We rely on a set of tests based on the geometry of the obsatcles like the width and height which are not measured in meters but in pixels as functions of disparity. The idea is based on the fact that close objects are projected with many pixels and far objects with few pixels. This solution reduces the errors of disparity quantification and allows more accurate detection.

In the remainder of this paper, the next section describes the complete on-road vehicles detection algorithm. Section 3 presents the experimental results. Finally, the last section sums up the contributions and concludes with future works.

2 Vehicles Detection Algorithm

The proposed algorithm performs vehicles detection as well as distance estimation based on stereo vision. The disparity map is first processed for scene segmentation to get a confident map of obstacles that simplifies the process of vehicles detection. Then, on-road vehicles detection is performed starting from removing off-road objects belonging to the static obstacles space like trees and sidewalks panels. For this purpose, a recursive connected labeling algorithm is applied to the U-disparity map. Finally, a set of post processings is applied for selecting vehicles and identifying their corresponding distances.

2.1 Level 1: Disparity Map Generation

In this paper, we used the Efficient LArge scale Stereo matching (ELAS) algorithm [10]. It is a Bayesian approach for dense stereo matching (Fig. 2(b)). It starts by selecting a set of supporting points and identifying their corresponding disparity. These support points are then used to build a 2D mesh via Delaunay triangulation which is processed to find the disparity of the remaining pixels based on a generative probabilistic model.

2.2 Level 2: Scene Segmentation

The dense disparity map is segmented into two distinctive spaces; free space and obstacles space (Fig. 2). The former one includes road, sidewalks and sky, the later one covers static obstales (trees, building, panels) and dynamic ones (moving vehicles and pedestrians). The obstacles map can be viewed as a confident map since the probability of detecting vehicles is higher in this map compared to the complete dense disparity map.

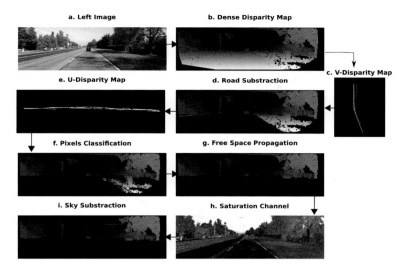

Fig. 2. Scene segmentation approach

Road Detection. The detection of the road profile is based on the V-disparity space [6] where the x axis plots the disparity d and the y axis plots the image row number. The intensity value of a pixel $p_{vdisp}(d, v)$ on this map is identified According to Eq. (1).

$$p_{vdisp}(d, v) = \sum_{u=0}^{cols} \lambda, \ \lambda = \begin{cases} 1 & \text{if } p_{disp}(u, v) = d. \\ 0 & \text{otherwise.} \end{cases} \tag{1}$$

The ground plane is projected as a slanted line under the hypothesis that the road's surface is plane and horizontal. The hough transform [8] has been used to detect this line and to identify its equation $ad + b$. Then, for each pixel $p_{disp}(u, v)$ in the disparity map, we calculate its distance $dist$ (Eq. (2)) with respect to the road's line which should be less than a threshold ϵ (few pixels) to classify this pixel as road's pixel. Through this process, road's pixels are substracted from the disparity map. Figure 2(b–d) shows results of road segmentation (d) based on the dense disparity map (b) and the V-disparity map (c).

$$p_{disp}(u, v) = \begin{cases} 0 & \text{if } dist < \epsilon \\ p_{disp}(u, v) & \text{otherwise} \end{cases}, \ dist = \frac{|au - vb|}{\sqrt{1 + a^2}}. \tag{2}$$

Pixels Classification. The classification phase aims to recover the non-road free space pixels by using the U-disparity map. The intensity of a pixel $p_{udisp}(u, d)$ in the U-disparity map is determined according to (Eq. (3)).

$$p_{udisp}(u, d) = \sum_{v=0}^{rows} \lambda, \ \lambda = \begin{cases} 1 & \text{if } p_{disp}(u, v) = d. \\ 0 & \text{otherwise.} \end{cases} \tag{3}$$

If the intensity of a pixel on the U-disparity map is higher than a certain threshold τ, this means that in a certain column u of the disparity map, there are too many pixels with the same distance to the camera and these points belong to potential obstacles. Based on this observation, pixels with high intensity are kept and the others are set to 0 (Eq. (4)). The threshold τ refers to the height of obstacles measured in pixels, in our case, it has been set to 40 pixels. Figure 2(d–f) shows results of pixels classification (f) based on the disparity map after road substraction (d) and by using the U-disparity map (e).

$$p_{disp}(u, v) = \begin{cases} p_{disp}(u, v) & \text{if } p_{udisp}(u, d) > \tau, d = p_{disp}(u, v). \\ 0 & \text{otherwise.} \end{cases} \tag{4}$$

Propagation of Free Space. Pixels classification has been performed to get two sets of pixels candidates belonging to free space and obstacles space. These classified pixels are taken as initial seed points to determine the class of the non-classified pixels. The idea is to count the contribution of classified free space and obstacles space pixels on the neighbour of each pixel. An accumulator $accum$ has been used to count this contribution. If the neighbour's pixel belongs to free space, $accum$ is decremented, otherwise it is incremented (Eq. (5)). Figure 2(g) shows the results of free space propagation.

$$d(u, v) = \begin{cases} d(u, v) & \text{(free space pixel) if } accum < 0. \\ 0 & \text{(obstacles space pixel) otherwise.} \end{cases} - \tag{5}$$

Sky Substraction. While many free space pixels have been recovered through free space propagation, wrong classification may happen concerning the sky's pixels classified as obstacles pixels (Fig. 2(g)). The reason is that, the propagation task relies on the analysis of the surrounding pixels, since the sky's pixels are on the top part of the image far from the free space and close to obstacles space, the contribution of obstacles pixels is higher. To remove sky from disparity map, the saturation (S) channel (Fig. 2(h)) of the HSL color space is used and applied as a mask on the last segmented disparity map by using Eq. (6).

$$d(u, v) = \begin{cases} 0 & \text{if } s(u, v) \text{ is (black(0)} \vee \text{white(255))} \wedge (v < b - \epsilon). \\ d(u, v) & \text{otherwise.} \end{cases} \tag{6}$$

The saturation channel has been used because the sky's pixel $s(u, v)$ on this channel is either white or black. It is white in case the sky is blue or grey on the RGB color space and it is black when the sky is white. We may then apply the S channel as a mask to each pixel $d(u, v)$. However, to avoid removing pixels belonging to potential vehicles, the mask is applied only for the part above the horizon line b identified previously (see Sect. 2.2) with a tolerance range ϵ (30 pixels). The result is shown on Fig. 2(i).

2.3   Level 3: Vehicles Detection Approach

Vehicles detection task relies on the U-disparity map which is less noisy compared to the V-disparity map. This choice is based on the fact that, in the U-disparity map, obstacles in general are represented by separate horizontal lines even side to side vehicles which is not the case on the V-disparity map where obstacles at the same distance are ovelayed and represented by the same line. However, the crucial point with the U-disparity map is how to remove the off-road features. To cope with this issue, connected segments labeling algorithm has been used. Finally, on-road vehicles are detected and recognised based on their geometry. The whole process can be divided into two phases: an off-road features substraction phase and an on-road vehicles detection phase.

Phase 1: Off-Road Features Substraction. Figure 3 shows an example of the complete off-road substraction phase. First, the U-disparity map is generated from the obstacles disparity map. Then, connected segments algorithm is applied based on the 4-connected neighborhood approach (Fig. 5) as follows:

1. Scan the U-disparity map from left to right and from top to bottom
2. The first non zero pixel is taken as the seed point
3. check for the horizontal (left and right) and the vertical (top and bottom) neighbours
4. Join each non-zero neighbour to the seed point
5. Apply recursively the 4-connected neighborhood for each new joined pixel (steps 3 and 4)

Once the whole U-disparity map is processed, a list of segments is recovered and each one is treated according to its width. To take into consideration the fact that an obstacle looks smaller when it is far and bigger when it is close, the width is measured in function of the disparity. Figure 4 illustrates the principle.

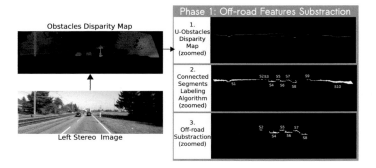

Fig. 3. Vehicles detection Algorithm: *Phase 1* off-road features substraction

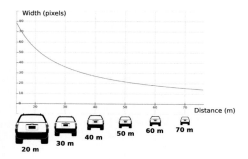

Fig. 4. Vehicle's width variation in pixels with distance for a 1.5 m vehicle's width

Let the pixel $P_i(u,v)$ be a projection on the image plane of a point $P_w(X,Y,Z)$ in the real world plane. To recover X based on the image coordinate system we rely on Eq. (7). C_u is the projection of the x coordinate of the camera's optical center in the image plane, Z is the distance and f is the focal distance.

$$X = (u - C_u) \times Z/f. \tag{7}$$

Let suppose u_{min} and u_{max} are the minimum and maximum vertical limits of a vehicle in the image plane. By using Eq. (7), we can determine the width W in meter as shown in Eq.(8). Figure 4 shows the variation of the width in pixels $(u_{max} - u_{min})$ with distance for a fixed vehicle's width of 1.5 m.

$$\begin{cases} X_{min} = (u_{min} - C_u) \times Z/f \\ X_{max} = (u_{max} - C_u) \times Z/f \end{cases} \Rightarrow W = (X_{max} - X_{min}) = (u_{max} - u_{min}) \times Z/f. \tag{8}$$

To remove off-road segments represented by long segments, we have fixed the width interval from 1.5 m to 3 m and the detection distance from 30 m to 70 m. Based on this, we have determined the variation interval of the width in pixel to take into consideration. Figure 3 shows the result of applying this to substract the off-road features.

Phase 2: On-Road Vehicles Detection

Hypothesis Generation. To detect on-road vehicles, the list of on-road segments generated is investigated. To distinguish between on-road vehicles and other on-road objects we rely on the geometry features of vehicles. From the previous phase, the vertical position of each segment on the image plane (u_{min}, u_{max}) is recovered (Fig. 5) and hence the width in meter is deduced according to Eq. (8). Also, the disparity range is determined (Fig. 5). Then, for each on-road segment, in the disparity map region limited by u_{min} and u_{max}, we recover all pixels having disparity in the disparity range $[disp_{min}, disp_{max}]$ (Fig. 6). The horizontal position is then determined (v_{min}, v_{max}) and the height is found. Also we refine the vertical position (U'_{min}, U'_{max}) as shown in Fig. 6. Figure 8 shows these different steps at the hypothesis generation level.

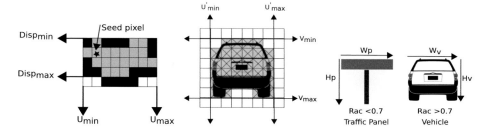

Fig. 5. 4-Connected neighborhood approach

Fig. 6. Vertical and horizontal vehicle's position identification

Fig. 7. 2D Box area/external contour ratio principle

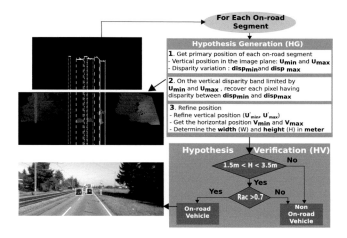

Fig. 8. Vehicles detection Algorithm: *phase 2* on-road vehicles detection

Hypothesis Verification. To select vehicles among other on-road obstacles, two tests are applied (Fig. 8). First,the height in pixels is checked (Eq. 8). Vehicles have height between 1.5 m and 3.5 m, any obstacle with height outside this range is rejected. If the test is verified, the ratio between the 2D box area and the external contour is computed. The external contour is limited by the pixels which have been recovered (Fig. 6). The reason of using this ratio as a metric is that, for vehicles, this ratio is supposed to be high which is not the case for panels as shown in Fig. 7.

3   Experimental Results

The algorithm has been evaluated on KITTI datasets. For each dataset, a list of tracklets is available as the ground truth representing the different objects available on each frame. Each tracklet is presented in velodyne coordinates as a 3D box with its corresponding width, height and length.

The algorithm is implemented on a standard PC with an Intel CPU (i5) of 1.8 GHz. The operating system is Ubuntu 14.04. The disparity map as mentionned previously is generated from LibELAS library based on the approach explained on Sect. 2.1.

3.1   Experimental Design

Before using the tracklets for evaluation, there are some points to take into consideration. The first one deals with the obstacles type; since our algorithm deals only with vehicles, we need first to filter all non-vehicles tracklets. The second point concerns the off-road objects which are not taken into consideration in our algorithm like parked vehicles. Finally, the detection distance range of our algorithm is set up from 30 m to 70 m, hence, we have to take into consideration only tracklets having distance within this range. To cope with these issues the position and motion history data of the tracklets have been used, for instance, to remove non-vehicles objects, the object's type has been employed. For evaluation, we determine the rate of correct detections or missed ones and false alarms. For better evaluation, these criteria are checked manually when the data provided by the tracklets is not sufficient. To test, we need first to perform a 2D projection of the 3D boxes representing the tracklets. Then, we determine the percentage of intersection between the 2D box of each tracklet taken as a ground truth and the 2D box generated by our algorithm. A detection is then validated once the percentage of intersection is higher than 70 %. For evaluation we have selected 3 datasets; Dataset 1: High way, Dataset 2: Urban road and Dataset 3: Rural road.

3.2   On-Road Vehicles Detection Results

Table 1 shows the evaluation results of the algorithm whitout the aid of a tracking module on the 3 different datasets. Two hundreds frames have been selected for each dataset. Dataset 1 contains 270 on-road detections, 241 of them have been well detected. In dataset 2 there exists 236 detections, 211 have been well detected, while the rest have been missed. Dataset 3 contains 128 on-road vehicles associated with tracklets, 211 have been correctely detected.

For evaluation, the three criteria previously presented have been determined for each dataset. During the evaluation, we noticed the absence of redundant detections due to the use of connected component algorithm to recover the segments on the U-disparity map. This solution also increases the accuracy of distance estimation. The results show that high successful detection rate can be achieved as shown in the top left image in Fig. 9 and the estimated distance reaches 70 m. Also the classification task works well since different types of vehicles have been detected like cars and trucks, the bottom left image in Fig. 9 illustrates this situation.

Fig. 9. Some detections results obtained from KITTI datasets

Table 1. On-road vehicles detection evaluation

Dataset	Correct detections	Missed detections	False alarms
Dataset 1	89.26 %	10.74 %	8.88 %
Dataset 2	89.40 %	10.59 %	2.96 %
Dataset 3	92.19 %	7.81 %	11.71 %

Table 2. Comparaison of our Algorithm with other on-road vehicles detection Algorithms

Article	FDD	CDR	FD	Detection type
Sun et al. [12]	32 × 32 image region	98.5 %	2 %	Rear
Alonso et al. [13]		92.63 %	3.63 %	Rear and front
Bergmiler et al. [14]		83.12 %	16.7 %	Rear
Southall et al. [15]	40 m	99 %	1.7 %	Single lane rear
Chang et Cho [16]	32 × 32 image region	99 %	12 %	Rear
Kowsari et al. [17]	120 m	98.6 %	13 %	Multi view
Our algorithm	70 m	90.28 %	7.85 %	Multi view

Although the algorithm gives sufficient results, it has still some deficiencies. The first point concerns the use of the connected component algorithm to remove off-road features, although this techniques works well it fails when we deal with on-road vehicles close to high off-road features, in this case, we will loose the vehicles, also this approach is highly sensitive to the stereo matching errors since it is related to the U-disparity map generated from the projection of the disparity map which may increase the false alarms and the rate of missed detections as shown in the top right image of Fig. 9. The false alarms are generally sidewalks panels and road markers as illustrated on the top and bottom right images in Fig. 9. They are usually generated because of the stereo matching errors that affect directely the scene segmentation task and hence the on-road vehicles detection phase. This can be solved by merging the proposed algorithm with a lane marking detection module. We have compared the performances of our algorithm with some contributions cited in Table 2 in terms of the Farthest

Detection Distance (FDD), the Correct Detections Rate (CDR) and the False Detections FD).

4   Conclusion

In this paper, a stereo vision based scene segmentation and on-road vehicles detection algorithm has been proposed. While a variety of vehicles detection algorithms exist in the litterature, the proposed algorithm provides improuvements to cope with some crucial issues. The first one concerns the off-road features substraction. It is an important point to deal with when on-road objects are targeted for detection. We proposed to use the connected segments labeling algorithm which has been also used to recover the on-road segments instead of the traditional hough transform for better and fast obstacles detection. For the second improvement, width and height of objects are not measured in meter but rather in pixels in funtion of the disparity. This solution increases the detection precision and distance estimation. Also, some cues have been presented for objects classification like the ratio between the 2D box area and the external contour of the object. The algorithm has been evaluated on the KITTI vision dataset and the experimental results show that it can detect the most on-road vehicles and determine their distance up to 70 m.

For futur works and to improve the current algorithm, a tracking component will be added to increase the rate of correct detections while decreasing the rate of false and missed detections. Also, since ADAS applications are real time, we intend to integrate this algorithm in a dedicated hardware architecture.

Acknowledgments. This research work has been carried out in the framework of the Technological Research Institute SystemX, and therefore granted with public funds within the scope of the French Program "Investissements d'Avenir".

References

1. Yu, Q., Araújo, H., Wang, H.: A stereovision method for obstacle detection and tracking in non-flat urban environments. Auton. Robots **19**, 141–157 (2005)
2. Jiang, Y., Gao, F., Xu, G.: Computer vision-based multiple-lane detection on straight road and in a curve. In: Proceedings of the International Conference on Image Analysis and Signal Processing (IASP), pp. 114–117. IEEE (2010)
3. Khan, W., Klette, R.: Stereo accuracy for collision avoidance for varying collision trajectories. In: Intelligent Vehicles Symposium (IV), pp. 1259–1264. IEEE (2013)
4. Rabe, C., Franke, U., Gehrig, S.: Fast detection of moving objects in complex scenarios. In: Intelligent Vehicles Symposium, pp. 398–403. IEEE (2007)
5. Perrollaz, M., Yoder, J.D., Negre, A., Spalanzani, A., Laugier, C.: A Visibility-based approach for occupancy grid computation in disparity space. IEEE Trans. Intell. Transp. Syst. **13**, 1383–1393 (2012)
6. Labayrade, R., Aubert, D., Tarel, J.P.: Real time obstacle detection in stereovision on non flat road geometry through "v-disparity" representation. In: Intelligent Vehicle Symposium, Vol. 2, pp. 646–651. IEEE (2002)

7. Gao, Y., Ai, X., Rarity, J., Dahnoun, N.: Obstacle detection with 3D camera using UV-Disparity. In: 7th International Workshop on Systems, Signal Processing and their Applications (WOSSPA), pp. 239–242. IEEE (2011)
8. Duda, R.O., Hart, P.E.: Use of the Hough transformation to detect lines and curves in pictures. Commun. ACM **15**, 11–15 (1972)
9. Perrollaz, M., Spalanzani, A., Aubert, D.: Probabilistic representation of the uncertainty of stereo-vision and application to obstacle detection. In: Intelligent Vehicles Symposium (IV), pp. 313–318 (2010)
10. Geiger, A., Roser, M., Urtasun, R.: Efficient large-scale stereo matching. In: Kimmel, R., Klette, R., Sugimoto, A. (eds.) ACCV 2010, Part I. LNCS, vol. 6492, pp. 25–38. Springer, Heidelberg (2011)
11. Wang, B., Rodriguez Florez, S.A., Frémont, V.: Multiple obstacle detection and tracking using stereo vision: application and analysis. In: 13th International Conference on Control, Automation, Robotics & Vision (ICARCV), Singapore, Singapore, pp. 1074–1079 (2014)
12. Sun, Z., Bebis, G., Miller, R.: Monocular precrash vehicle detection: features and classifiers. IEEE Trans. Image Process. **15**, 2019–2034 (2006)
13. Alonso, D., Salgado, L., Nieto, M.: Robust vehicle detection through multidimensional classification for on board video based systems. In: IEEE International Conference on Image Processing, vol. 4, pp. IV–321. IEEE (2007)
14. Bergmiller, P., Botsch, M., Speth, J., Hofmann, U.: Vehicle rear detection in images with generalized radial-basis-function classifiers. In: Intelligent Vehicles Symposium, pp. 226–233. IEEE (2008)
15. Southall, B., Bansal, M., Eledath, J.: Real-time vehicle detection for highway driving. In: IEEE Conference on Computer Vision and Pattern Recognition, pp. 541–548. IEEE (2009)
16. Chang, W.-C., Cho, C.-W.: Online boosting for vehicle detection. IEEE Trans. Syst. Man Cybern. B (Cybernetics) **40**(3), 892–902 (2010)
17. Kowsari, T., Beauchemin, S.S., Cho, J.: Real-time vehicle detection and tracking using stereo vision and multi-view AdaBoost. In: 14th International Conference on Intelligent Transportation Systems (ITSC), pp. 1255–1260. IEEE (2011)

Traffic Light Detection at Night: Comparison of a Learning-Based Detector and Three Model-Based Detectors

Morten B. Jensen[1,2]($\boxtimes$), Mark P. Philipsen[1,2], Chris Bahnsen[1],
Andreas Møgelmose[1,2], Thomas B. Moeslund[1], and Mohan M. Trivedi[2]

[1] Visual Analysis of People Laboratory, Aalborg University, Aalborg, Denmark
mboj@create.aau.dk
[2] Computer Vision and Robotics Research Laboratory, UC San Diego,
La Jolla, USA

Abstract. Traffic light recognition (TLR) is an integral part of any intelligent vehicle, it must function both at day and at night. However, the majority of TLR research is focused on day-time scenarios. In this paper we will focus on detection of traffic lights at night and evaluate the performance of three detectors based on heuristic models and one learning-based detector. Evaluation is done on night-time data from the public LISA Traffic Light Dataset. The learning-based detector outperforms the model-based detectors in both precision and recall. The learning-based detector achieves an average AUC of 51.4 % for the two night test sequences. The heuristic model-based detectors achieves AUCs ranging from 13.5 % to 15.0 %.

1 Introduction

Traffic lights are used to safely regulate the traffic flow in the current infrastructure, they are therefore a vital part of any intelligent vehicle, whether it is fully autonomous or employ Advanced Driver Assistance Systems (ADAS). In either application, TLR must be able to perform during both day and night. TLR for night-time scenarios is especially important as more than 40 % of accidents at intersections occur during the late-night/early-morning hours, in fact a crash is 3 times more probable during the night than during the day [1]. For more introduction to TLR in general we refer to [2] where an overview is given of the current state of TLR. In the same paper, the lack of a large public dataset is addressed with the introduction of the LISA Traffic Light Dataset, which contains challenging conditions and both day- and night-time data.

Before the state of traffic lights (TLs) can be determined they must first be detected. Traffic light detection (TLD) has proven to be very challenging under sub-optimal and changing conditions. The purpose of this paper is therefore to evaluate the night-time TLD performance of three heuristic TL detectors and compare this to a state-of-the-art learning based detector relying on Aggregated Channel Features (ACF). The same learning-based detection framework has previously been applied for day-time TLD in [3]. This makes it possible to compare

© Springer International Publishing Switzerland 2015
G. Bebis et al. (Eds.): ISVC 2015, Part I, LNCS 9474, pp. 774–783, 2015.
DOI: 10.1007/978-3-319-27857-5_69

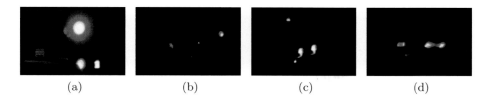

(a)   (b)   (c)   (d)

Fig. 1. Challenges samples from the LISA Traffic Light Dataset.

the detector's performance at night and day. Evaluation is done on night-time sequences from the extensive and difficult LISA Traffic Light Database. The contributions are thus threefold:

1. First successful application of a state-of-the-art learning-based detector for TLD at night.
2. Comparison of three model-based TLD approaches and a learning-based detector using ACF.
3. Clarification of the challenges for night-time TLD.

The paper is organized as follows: Challenges specific to night-time TLD are clarified in Sect. 2. Relevant research is summarized in Sect. 3. In Sect. 4 we present the detectors, followed by the evaluation in Sect. 5. Finally, Sect. 6 rounds of with our concluding remarks.

2   Traffic Lights and Their Variations

In this section we present some challenges particular to night-time TLD.

1. Lights may seem larger than the actual source [4], see Fig. 1a.
2. Colors saturate to white [4], see Fig. 1a.
3. Lack of legal standards for tail-lights in the USA, tail-lights may therefore resemble TLs [5], see Fig. 1d.
4. TL may be reflected in reflective surfaces, e.g. storefronts, see Fig. 1b.
5. Street lamps and other light sources may look similar to TLs, see Fig. 1c.

Type 1 and 2 can be reduced by increasing the shutter speed at the risk of getting underexposed frames. One solution to this problem is seen in [6], where frames are captured by alternating between slow and fast shutter speed. Generally, it is hard to cope with the remaining issues from a detection point of view. One solution to removing type 3, 4 and 5 false positives could be the introduction of prior maps with information of where TLs are located in relation to the ego-vehicles location, as e.g. seen in [5].

3   Related Work

Most research on TLD and TLR has been focused on day-time, only a handful of publications evaluate their systems on night-time data. One is [4] where a fuzzy clustering approach is used for detection. Gaussian distributions are calculated based on the red, amber, green, and black clusters in a large number of combinations of the RGB and RGB-N image channels. In [7] the work from [4] is expanded, by the introduction of an adaptive shutter and gain system, advanced tracking, distance estimation, and evaluate on a large and varied dataset with both day-time and night-time frames. Because of the differences in light conditions between night and day, they use one fuzzy clustering process for day conditions and another for night conditions. [8] finds TL candidates by applying the color transform proposed in [9]. The color transform determines the dominant color of each pixel based on the RGB values. Dominant color images are only generated for red and green, since no transform is presented for yellow. After thresholding of the dominant color images, BLOBs are filtered based on the width to height ratio and the ratio between the area of the BLOB and the area of the bounding box. The remaining TL candidates are then classified using SVM on a wide range of BLOB features.

When looking at TL detectors which have been applied to day-time data, two recent papers have employed learning-based detectors. [10] is combining occurrence priors from a probabilistic prior map and detection scores based on SVM classification of Histogram of Oriented Gradients (HoG) features to detect TLs. [3] uses the ACF framework provided by [11]. Here features are extracted as summed blocks of pixels in 10 channels created from transformations of the original RGB frames. The extracted features are classified using depth-2 learning trees. Spotlight detection using the white top hat operation on intensity images is seen in [12–15]. In [16], the V channel from the HSV color space is used with the same effect. A high proportion of publications use simple thresholding of color channels in some form. [6] is a recent example where traffic light candidates are found by setting fixed thresholds for red and green TL lamps in the HSV color space.

For a more extensive overview of the TLR domain, we refer to [2].

4   Methods

In this section we present the used methods. In the first subsection the learning-based detector is described. The second describes each of the three model-based detectors and how the confidence scores are calculated for the TL candidates found by these model-based detectors.

4.1   Learning-Based Detection

In this subsection we describe how the successful ACF detector has been applied to the night-time TL detection problem. The learning-based detector is provided

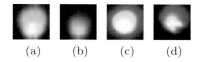

(a)   (b)   (c)   (d)

Fig. 2. Positive samples for training the learning-based detector.

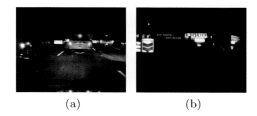

(a)   (b)

Fig. 3. Negative samples for training the learning-based detector.

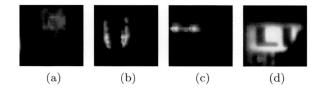

(a)   (b)   (c)   (d)

Fig. 4. Hard negative samples for training the learning-based detector.

as part of the Matlab toolbox from [11]. It is similar to the detectors seen in [17] for traffic signs and [3] for day-time TLs, except for few differences in the configuration and training which are described below.

Features. The learning-based detector is based on features from 10 channels as described in [18]. A channel is a representation of the input image, which is obtained by various transformations. The 10 different channels include 6 gradient histogram channels, 1 for unoriented gradient magnitude, and 3 for each channels in the CIE-LUV color space. In each channel, the sums of small blocks are used as features. These features are evaluated using a modified AdaBoost classifier with depth-4 decision trees as weak learners.

Training. Training is done using 7,456 positive TL samples with a resolution of 25×25 and 163,523 negative samples from 5,772 selected frames without TLs. Figure 2 shows four examples of the positive samples used for training the detector. Similarly, Fig. 3 shows two examples of frames used for negative samples. Finally, Fig. 4 shows four hard negative samples extracted using false positives from the training dataset.

AdaBoost is used to train 3 cascade stages, 1st stage consists of 10 weak learners, 2nd stages of 100, and 3rd stage is set to 4,000 but converges at 480.

In order to detect TLs at a greater interval of scales, the octave up parameter is set to 1 instead of the default 0. The octave up parameters defines the number of octaves to compute above the original scale.

Detection. A 18×18 sliding window is used across each of the 10 aggregated channels in the frames from the test sequences.

4.2   Heuristic Model-Based Detection

We want to compare the learning-based detector to more conventional detector types which are based on heuristic models. For each of the three model-based detectors, a short description is given along with output showing central parts of the detectors. The sample shown in Fig. 1a is used as input.

Detection by Thresholding. The detector which uses thresholding is mainly based on the work presented in [6]. Thresholds are found for each TL color in the HSV color space by looking at values of individual pixels from TL bulbs sampled from the training clips in the LISA Traffic Light dataset. Figure 5(a) shows the input sample and Fig. 5(b) shows output after thresholding input. Pixels that fall inside the thresholds for one of the three colors are labeled green, yellow or red in Fig. 5. For the input sample only pixels which fell within the yellow and red thresholds were present.

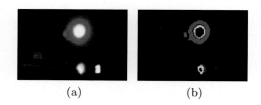

(a)   (b)

Fig. 5. Thresholded TL (Color figure online).

Detection by Back Projection. Back projection begins with the generation of color distribution histograms. The histograms are two-dimensional and are created for each of the TL colors using 20 training samples for each of the TL colors, green, yellow, and red. From the training samples the U and V channels of the LUV color space are used. The histograms are normalized and used to generate a back projection which is thresholded to remove low probability pixels from the TL candidate image. The implementation is similar to our previous work in [3]. Figure 6a shows the back projected TL candidate image. Figure 6b shows the processed back projected TL candidate image after removal of low probability pixels and some typical morphology operations.

(a)   (b)

Fig. 6. Back projected TL (Color figure online).

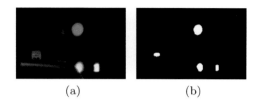

(a)   (b)

Fig. 7. Spotlights found using the white top-hat operation.

Detection by Spotlight Detection. Spotlights are found in the intensity channel L from the LUV colorspace using the white top-hat morphology operation. The implementation is similar to our previous work in [3]. This method has been used in a many recent TLR papers [12–16]. Figure 7a shows the output of the white top-hat operation. Figure 7b shows the binarized TL candidate image after thresholding and some typical morphology operations.

Confidence Scores for TL Candidates. Confidence scores are calculated for all TL candidates found by the three model-based detectors. The TL BLOB characteristics used in this work have seen use in earlier work, such as [8,9]. Scores from individual characteristics are generated ranging from [0–1], with 1 being the best. These are summed for each TL candidate, resulting in a combined score ranging from [0–5].

Bounding box ratio: The bulbs of TLs are circular, therefore under ideal conditions the bounding box will be quadratic. The bounding box ratio is calculated as the ratio between width and height of the bounding box.

Solidity ratio: Since TL bulbs are captured as circular and solid under ideal conditions, a BLOBs solidity is a characteristic feature for a TL. The solidity is calculated as the ratio between the convex area of detected BLOBs and the area of a perfect circle, with a radius approximated from the dimensions of the BLOB.

Mean BLOB intensity: Each of the three detectors produce an intensity channel which can be interpreted as a confidence map of TL pixels. The best example is from detection by back projection, where the result of the back projection

Table 1. Overview of night test sequences from the LISA Traffic Light Dataset.

Sequence name	Description	# Frames	# Annotations	# TLs	Length
Night seq. 1	night, urban	4,993	18,984	25	5 min 12 s
Night seq. 2	night, urban	6,534	23,734	62	6 min 48 s
		11,527	42,718	87	12 min

is an intensity channel with normalized probabilities of each pixel being a TL pixel. The intensity channel employed from the spotlight detector is less informative, since it describes the strength of the spotlight. From the threshold based detector, we simply use the intensity channel from the LUV color space.

Flood-filled area ratio: The bulbs of TLs are surrounded by darker regions, by applying flood filling from a seed inside the found BLOBs, it can be confirmed that this contrast exists. We use the ratio between the area of the bounding box and the area of the bounding box from the flood filled area as a measure for this.

Color confidence: Using basic heuristics based thresholding we find the most prominent color inside the TL candidates' bounding boxes. The confidence is calculated based on the number of pixels belonging to that color and the total number of pixels within the bounding box. Pixels with very low saturation are not included in the confidence calculation.

5   Evaluation

Most TL detectors have been evaluated on datasets which are unavailable to the public. This makes it difficult to determine the quality of the published results and compare competing approaches. We strongly advocate that evaluation is done on public datasets such as the LISA Traffic Light Dataset[1].

5.1   LISA Dataset

The four detectors presented in this paper are evaluated on the two night test sequences from the LISA Traffic Light Dataset. This provides a total of 11,527 frames, and a total ground truth of 42,718 annotated TL bulbs. Additional information of the video sequences can be found in Table 1. The resolution of the LISA Traffic Light Database is 1280×960, however only the upper 1280×580 pixels are used in this paper.

---

[1] Freely available at http://cvrr.ucsd.edu/LISA/datasets.html for educational, research, and non-profit purposes.

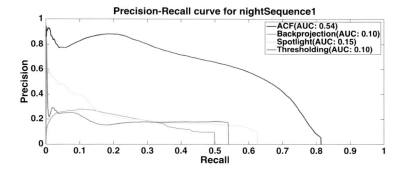

Fig. 8. Precision-Recall curve of night sequence 1 using 50 % overlap criteria.

5.2 Evaluation Criteria

In order to insure that the evaluation of TL detectors provide a comprehensive insight into the detectors performance, it is important to use descriptive and comparable evaluation criteria. The presented detectors are evaluated based upon the following four criteria:

PASCAL overlap criterion defines a true positive (TP) to be a detection with more than 50 % overlap over ground truth (GT).

Precision is defined in Eq. (1).

$$Precision = \frac{TP}{TP + FP} \tag{1}$$

Recall is defined in Eq. (2).

$$Recall = \frac{TP}{TP + FN} \tag{2}$$

Area-under-curve (AUC) for a precision-recall (PR) curve is used as a measure for the overall system performance. A high AUC indicates good performance, an AUC of 100 % indicates perfect performance for the testset.

5.3 Results

We present the final results according to the original PASCAL overlap criteria of 50 % in Figs. 8 and 9.

By examining Figs. 8 and 9, it is clear that the learning-based detector outperforms the other detectors in both precision and recall on both night sequences. The odd slopes of the PR curves for the back projection detectors are a result of problems with getting filled and representative BLOBs. The learning-based

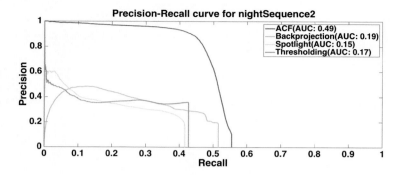

Fig. 9. Precision-Recall curve of night sequence 2 using 50 % overlap criteria.

detector is able to differentiate well between TLs and other light sources, leading to a great precision and smooth precision-recall curve. The main problems with the learning-based detector's PR curves are the false negatives caused by detections not meeting the PASCAL criteria but still reaching a very high score, and problems with detecting TLs from far away. These detections causes some instability in the precision especially around 0.05 recall in Fig. 8.

6   Concluding Remarks

We have compared three detectors based on heuristic models to a learning-based detector based on aggregated channel features. The learning-based detector reached the best AUC because of the significantly higher precision and good recall. Recall is generally seen as the most important performance metric for detectors since precision can be improved in later stages, whereas false negatives are lost for good. The learning-based detector achieves an average AUC of 51.4 % for the two night test sequences. The heuristic model-based detectors achieved average AUCs ranging from 13.5 % to 15.0 %, with detection by back projection and spotlight detection achieving the highest AUCs.

Interesting future TLD work could involve applying and comparing the performance of deep learning methods on the LISA TL Dataset with the results presented in this paper.

References

1. Federal Highway Administration: Reducing late-night/early-morning intersection crashes by providing lighting (2009)
2. Jensen, M.B., Philipsen, M.P., Trivedi, M.M., Møgelmose, A., Moeslund, T.B.: Vision for looking at traffic lights: Issues, survey, and perspectives. IEEE Trans. Intell. Transport. Syst. IEEE (2015, in submission)

3. Philipsen, M.P., Jensen, M.B., Møgelmose, A., Moeslund, T.B., Trivedi, M.M.: Traffic light detection: a learning algorithm and evaluations on challenging dataset. In: 18th International Conference on IEEE Intelligent Transportation Systems Conference, pp. 2341–2345 (2015)

4. Diaz-Cabrera, M., Cerri, P.: Traffic light recognition during the night based on fuzzy logic clustering. In: Moreno-Díaz, R., Pichler, F., Quesada-Arencibia, A. (eds.) EUROCAST. LNCS, vol. 8112, pp. 93–100. Springer, Heidelberg (2013)

5. Fairfield, N., Urmson, C.: Traffic light mapping and detection. In: Proceedings of ICRA 2011, pp. 5421–5426 (2011)

6. Jang, C., Kim, C., Kim, D., Lee, M., Sunwoo, M.: Multiple exposure images based traffic light recognition. In: IEEE Intelligent Vehicles Symposium Proceedings, pp. 1313–1318 (2014)

7. Diaz-Cabrera, M., Cerri, P., Medici, P.: Robust real-time traffic light detection and distance estimation using a single camera. Expert Syst. Appl. **42**(8), 3911–3923 (2015)

8. Kim, H.K., Shin, Y.N., Kuk, S.G., Park, J.H., Jung, H.Y.: Night-time traffic light detection based on SVM with geometric moment features. In: 76th World Academy of Science, Engineering and Technology, pp. 571–574 (2013)

9. Ruta, A., Li, Y., Liu, X.: Real-time traffic sign recognition from video by class-specific discriminative features. Pattern Recogn. **43**, 416–430 (2010)

10. Barnes, D., Maddern, W., Posner, I.: Exploiting 3D semantic scene priors for online traffic light interpretation. In: Proceedings of the IEEE Intelligent Vehicles Symposium (IV), Seoul, South Korea (2015)

11. Dollár, P.: Piotr's Computer Vision Matlab Toolbox (PMT) (2015). http://vision.ucsd.edu/~pdollar/toolbox/doc/index.html

12. Trehard, G., Pollard, E., Bradai, B., Nashashibi, F.: Tracking both pose and status of a traffic light via an interacting multiple model filter. In: 17th International Conference on Information Fusion (FUSION), pp. 1–7. IEEE (2014)

13. Charette, R., Nashashibi, F.: Traffic light recognition using image processing compared to learning processes. In: IEEE/RSJ International Conference on Intelligent Robots and Systems, pp. 333–338 (2009)

14. de Charette, R., Nashashibi, F.: Real time visual traffic lights recognition based on spot light detection and adaptive traffic lights templates, In: IEEE Intelligent Vehicles Symposium, pp. 358–363 (2009)

15. Nienhuser, D., Drescher, M., Zollner, J.: Visual state estimation of traffic lights using hidden markov models. In: 13th International IEEE Conference on Intelligent Transportation Systems, pp. 1705–1710 (2010)

16. Zhang, Y., Xue, J., Zhang, G., Zhang, Y., Zheng, N.: A multi-feature fusion based traffic light recognition algorithm for intelligent vehicles. In: 33rd Chinese Control Conference (CCC), pp. 4924–4929 (2014)

17. Mogelmose, A., Liu, D., Trivedi, M.M.: Traffic sign detection for us roads: Remaining challenges and a case for tracking. In: Intelligent Transportation Systems, pp. 1394–1399. IEEE (2014)

18. Dollár, P., Tu, Z., Perona, P., Belongie, S.: Integral channel features. In: BMVC, vol. 2, p. 5 (2009)

Modelling and Experimental Study for Automated Congestion Driving

Joseph A. Urhahne[1(✉)], Patrick Piastowski[1], and Mascha C. van der Voort[2]

[1] Ford Motor Company, Cologne, Germany
jurhahne@ford.com
[2] University of Twente, Enschede, The Netherlands

Abstract. Taking a collaborative approach in automated congestion driving with a Traffic Jam Assist system requires the driver to take over control in certain traffic situations. In order to warn the driver appropriately, warnings are issued ("pay attention" vs. "take action") due to a control transition strategy that reacts to lane change manoeuvres by surrounding traffic. This paper presents the outcome of a driving simulator study regarding the evaluation of a control transition strategy. The strategy was found to provide adequate support to drivers. However, driver acceptance can be increased. A refined model is proposed.

1 Introduction

Media frequently report on the progress made regarding automated driving, but it will expectedly take another few years before fully automated vehicles or robot taxis will appear on public roads. In the meantime car manufactures opt for collaborative driving, i.e. vehicles equipped with a combined longitudinal and lateral support system [1] still request human supervision and control take-over for driving situations. For instance while driving with TJA Traffic Jam Assist, the driver has to observe the traffic in parallel to the intelligent sensors of the TJA system, in particular during surrounding lane change manoeuvers. Automated driving with TJA requires a novel control transition strategy because it goes beyond studies [14, 15] that investigate congestion assistant systems supporting drivers by warning for upcoming traffic jams and by using pedal counterforce feedback.

Drivers are commonly alerted and engaged by means of so called 'soft' and 'hard' warnings [13]. Soft warnings are intended to ask drivers for attention – but do not require immediate intervention. They create awareness of the driver that a take-over request can happen at any time. However, if soft warnings are provided too often or are perceived inappropriate, this is likely to reduce the acceptance of a warning system. A clear transition with handover of control to the driver is demanded with 'hard' warnings: This type of transition requires a real and immediate action from the driver because the system ramps down its support within a defined time span.

The aim of this study is to create a control transition strategy that has a high level of usability. According to ISO 9241-11 usability is defined as: *The extent to which a product can be used by specified users to achieve specified goals with effectiveness, efficiency and satisfaction in a specified context of use.*

© Springer International Publishing Switzerland 2015
G. Bebis et al. (Eds.): ISVC 2015, Part I, LNCS 9474, pp. 784–794, 2015.
DOI: 10.1007/978-3-319-27857-5_70

Here high usability means that the control transition strategy of the Traffic Jam Assist (1) effectively alerts or warns the driver in case control take-over, (2) at the same time limiting the number of warnings with compromising provided support (efficiency) and (3) ensuring that circumstances under which warning are issued are in line with drivers' acceptance (satisfaction).

A previous paper [11] revealed a list of physical indicators that are recommended for modelling the control transition strategy, among them the velocity of a TJA vehicle itself and the distance to the leading vehicle (primary object). This paper consisted of a naturalistic driving study NDS with traffic jam data of approximately 30 h and 900 lane change manoeuvers. Post-processing of the collected data allowed to identify and categorise relevant lane changes in the surrounding of the TJA vehicle. Video material of traffic scenes was selected that provide insight in indicators for lane change manoeuvres that would lead to hard warnings (i.e. transition of control required), so called "close cut-ins". Lane change manoeuvres that requested soft warnings (i.e. driver's attention) were labelled as "normal cut-ins" and "cut-outs".

This paper presents the outcome of a driving simulator study regarding the evaluation of a control transition strategy that reacts to lane change manoeuvres by surrounding traffic.

2   Modelling of Control Transition Strategy

The scope of modelling of the control transition strategy is to generate warnings effectively and efficiently in order to reach a high customer acceptance and maintain the comfort and benefit of a TJA system. Any lane change manoeuver can be a potential candidate for a warning. Based on the previous study, the velocity of a TJA vehicle (vehicle_speed) and the distance to the leading vehicle (PO_range) are introduced as indicators to lane change manoeuvers that require a warning. Complementary, the indicator Q is introduced, being the quotient of speed and range signals:

$$Q = \frac{\text{vehicle\_speed}}{\text{PO\_range}}$$

The lower the quotient Q, the less worrying and safety relevant a lane change manoeuver is likely to be for a TJA driver. A minimal threshold for Q might therefore be appropriate to limit the number of warnings provided to drivers. Q presents the inverse value of the time headway THW that is used e.g. in [3] to describe traffic flow. In addition drivers are obliged by traffic regulations like [4] to keep a safety margin at all times. The applied rule of thumb for city traffic and traffic jam speed below 50 kph is a time headway of 1 s, equivalent to a quotient Q = 3,6 kph/m. On rural roads and motorways this rule changes to a German byword "halber Tacho" that represents a quotient Q = 2,0 kph/m.

In contrast to cut-ins a completed cut-out manoeuver extends the gap to the next leading vehicle. As a consequence the TJA vehicle will accelerate in order to readjust to a standard safety distance. This can directly be perceived by the TJA driver. A warning interest in this situation could be assumed because the traffic flow is dynamic. A bigger gap might inspire some other drivers from the neighboring lanes to use this space for a cutting-in lane change.

It is unknown how many and which soft warnings are requested and accepted by TJA drivers and which threshold for indicator Q is desired. Figure 1 presents exemplarily the distribution of Q for 433 normal cut-in manoeuvers identified within the NDS data of the previous study. Approximately 50 % of warnings can be reduced with a threshold quotient $Q_{cut-in} < 1,0$ kph/m. Dependent on the chosen threshold the model for the control transition strategy defines whether a lane change generates a soft warning. It is expected that the chosen threshold will directly affect the warning acceptance by individual TJA drivers. This study will therefore investigate the validity of the model.

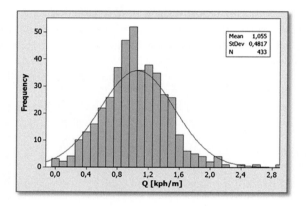

Fig. 1. Distribution of Q for cut-in manoeuvers

A draft model is represented by Fig. 2. It outlines a strategy how the quantity of warnings can be influenced by freely varying and combining thresholds for the indicators Q, Speed and Range with an emphasis on Q. In this phase of the study the number of hard warnings is considered non-reductional, as they are necessary to ensure traffic safety and their acceptance can be tested rather than their efficiency.

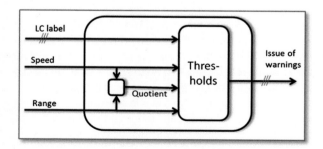

Fig. 2. Preliminary warning model

In short, the draft model for reduction of warnings is led by the hypothesis that TJA drivers will appreciate a warning system that warns for relevant lane change manoeuvers only. This hypothesis stresses that limiting the pure quantity of warnings should not be the target while designing collaborative automated driving but that issuing correct, relevant warnings is essential for the acceptance of the TJA driver. To verify this hypothesis a driving simulator experiment is conducted.

3   Simulator Experiment

Basic aim of the driving simulator study is to explore and understand the acceptance of drivers regarding the proposed control transition strategy for a Traffic Jam Assist, and learn more about drivers' preferences. Within the study, participants will be confronted with video material of traffic scenarios including lane change manoeuvers of surrounding traffic, selected from the available NDS data. As part of the presented traffic scenarios TJA warnings are issued in correspondence to the proposed control transition strategy. Participants are asked to indicate for each presented scenario whether the issued warning or non-warning is relevant, efficient and satisfactory. Furthermore drivers are asked to indicate whether the type of warning issued (i.e. soft or hard) is considered appropriate. Finally participants are asked to fill out a questionnaire that aims to detect further indicators for a transition strategy independent of the preliminary model.

3.1   Setup

A fundamental constraint of the experiment lies in using a driving simulator in lieu of a field study test in real life traffic. Due to the laboratory setting, one should be careful in generalising of the research findings. The application of driving simulators that uses a video method is however known in the field of automotive research. Exemplarily [12] describes a simulator experiment with traffic videos and labels. The reasoning to select a laboratory study as experimental research is founded in the high controllability as well as the assurance of participant safety at current stage of TJA development.

To ensure internal validity of the simulator study the experimental set-up is constructed consciously in accordance with [5, 10]: The simulator is located in a quiet environment that avoids acoustic and visual distraction of the participants. The acoustic scene contains only a constant low driving noise in order to concentrate on the chimes for soft and hard warnings. A stop button is used to interrupt the video when scene separators appear to fill in the assessment sheet. Intentionally it is not possible for all participants to replay a scene a second time. These environmental conditions are applied constantly and equally to all test persons.

3.2   Participants and Procedure

Participants were selected on a voluntary basis. All 25 participants understood the subject of automated driving well and had at least 5 years of driving experience. This quantity of participants is aligned with the recognitions from related literature [8] to confirm significance of the chosen video experiment setup. The complete conduct of the experiment is presented in Fig. 3. It was reproduced in the same way for all participants, starting with a briefing and guided examples prior to the independent video sequences. In order to exclude a fatigue effect for the participants on a particular video scene the sequence of the traffic scene videos is changed by the rules of partial counterbalancing as proposed by [7].

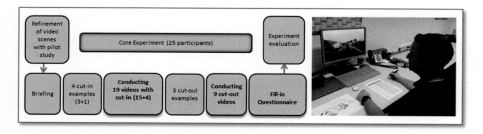

Fig. 3. Test sequence and environment

Prior to the core experiment a pilot-study was conducted [2] with six randomly selected participants. The results turned out to proof a successful setup of the experiment with a manageable structure for the data analysis. To be more representative for potential users of a TJA system, it is aimed that ca. 50 % of participants should have experience with ACC systems (Adaptive Cruise Control) in the core experiment to understand whether they distinguish from non-ACC drivers.

The procedure supports the research question: Does the preliminary model for a control transition strategy satisfy TJA drivers to bring the warnings to a high level of usability? Based on the results of the experiment the model shall be confirmed, refined or to redefined consequently.

3.3   Stimulus Material

In alignment with the proposed model a list of independent variables is provided as chosen stimuli:

(1) Type of lane change manoeuvre
(2) Speed, Range and Quotient
(3) ACC driving experience

As additional indicator the near-field distance to the cutting-in object is available for some cut-in manoeuvers. The dependent variables of the experiment are the percentile agreement to hard transitions due to close cut-ins and the percentile agreement to soft transitions due to normal cut-ins and cut-outs.

Table 1 presents the selection of scenes that are divided into three categories: normal cut-ins, close cut-ins and cut-outs. It shows the compilation of the selected lane changes with characteristic indicators. One can perceive that a wide variation of indicators was chosen. This serves to determine if the initial threshold is sufficient to satisfy the participants and if more indicator thresholds shall be used.

Table 1. Experiment stimulus material

NORMAL CUT INS			CLOSE CUT INS			CUT OUTS		
Scenes	Speed	Range	Quotient	Scenes	Speed	Range	Quotient	Scenes	Speed	Range	Quotient

Scenes	Speed	Range	Quotient	Scenes	Speed	Range	Quotient	Scenes	Speed	Range	Quotient
A1	31	23	1,35	A5	18	27	0,67	E1	12	15	0,80
A2	25	36	0,69	B5	20	12	1,67	E2	30	20	1,50
A3	37	26	1,42	C5	35	20	1,75	E3	36	13	2,77
A4	26	14	1,86	D4	12	12	1,00	F1	20	20	1,00
B1	21	18	1,17					F2	31	21	1,48
B2	41	34	1,21					F3	27	19	1,42
B3	10	14	0,71					G1	23	20	1,15
B4	31	28	1,11					G2	6	11	0,55
C1	27	21	1,29					G3	35	15	2,33
C2	31	37	0,84								
C3	30	21	1,43								
C4	38	28	1,36								
D1	24	26	0,92								
D2	40	26	1,54								
D3	24	16	1,50								

As a starting point it was selected to apply thresholds for Q that reduce the quantity of soft warnings by half of the occurrence of lane changes (e.g. $Q_{cut-in} > 1,0$ kph/m).

For those scenes that exceed the thresholds soft warnings were issued to the driver by means of an according warning chime. In Table 1 these scenes are shadowed like A1. There are exceptions to this rule in cases that show an underlined scene number, e.g. A2 or C3. For the exceptions the participants have actively to contradict to the actual warning. The participants were explicitly encouraged in the preparation phase that a dislike of a warning proposal is possible.

Four close cut-in scenarios are selected, each accompanied by a hard warning with an according warning chime. A reduction of quantity is not intended so far. However - with scene A5 the experiment also offers a manoeuver with a low quotient $Q = 0,67$ kph/m to investigate whether the issued warning is still accepted by the participants.

After the video sessions the participants are asked for further details in a questionnaire. The questions serve

– to describe the profile of chosen participants.
– to explore whether (separately from the chosen model indicators) participants apply their own indicators, e.g. the influence of weather or lighting conditions
– to evaluate the appreciation of TJA as automated driving with cooperative approach. What is the acceptance of automated driving with cooperative approach in general?

3.4   Experiment Results and Discussion

In Table 2 the percentile agreement for the proposed control transition strategy is listed. The second column presents results separately for ACC drivers and non-ACC drivers. This is reflected by showing percentages in three ways: An overall percentage in bold numbers and the distributed percentages for ACC drivers and non-ACC drivers.

Table 2. Agreement to warning model (ACClnon-ACC drivers)

	General Agreement to Model-proposed warning type	Agreement to Model-proposed warnings	Agreement to Model-proposed non-warnings	Optimization	Theoretical max due to voting
Close cut-ins	84% 80% 89%	-	-	-	+ 0%
Normal cut-ins	66% 65% 67%	70%	56%	+ 3%	+ 5%
Cut-outs	76% 72% 80%	65%	79%	+ 0%	+ 2%

Concerning the quantity of warnings it can be stated that ACC drivers are more critical toward the issue of warnings than non-ACC drivers. This specially applies in direct comparison for close cut-in and cut-out manoeuvers. The significance of the difference between ACC drivers and non-ACC drivers is investigated by a Chi-Square test. The null hypothesis H_0 states that there are no differences between the two types of participants in the core experiment. The results of Chi-Square testing [6] in all three types of lane changes cannot reject the null hypothesis H_0 ($P > 0{,}05$) and thus there are no significant differences between ACC drivers and non-ACC drivers to consider in further experiments.

For all three lane change manoeuver types a review of the proposed warning model is provided below. Furthermore an analysis was performed to identify individual traffic scenes with peculiarities. To support the discussion of the warning model a scatterplot with linear fit in Fig. 4 is used. It shows the participants' agreement as a function of Q.

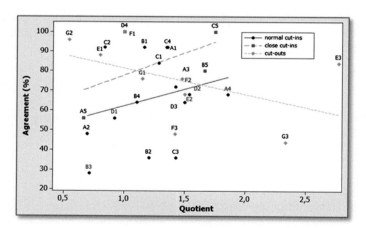

Fig. 4. Agreement vs. Quotient by lane change type

Close Cut-ins. The agreement of 84 % is on a significant high level. At first instance optimization is not possible due to the binding relation of close cut-in labels to hard warnings.

Peculiarities: Scene A5 has a remarkably low agreement factor of 55 %. This can be connected to an eye-catching low value of Q = 0,67 kph/m. It shows that indicator Q would be feasible to confirm or reject close cut-in manoeuvers and consequently hard warnings. Since the model uses Q with a minimal threshold it can be stated from Fig. 4 that the agreement rises with higher values of Q. The Pearson correlation of the graph is 0,567 (P = 0,433). Further it is recommendable to replace the subjective close cut-in labels by objective measurements. A preliminary study of this recommendation was presented by [11] and outlines a significant correlation of close cut-in labels with the indicators of near-field sensor values and values of Q.

Normal Cut-ins. For the traffic scenes including normal cut-in manoeuvers a percentile agreement of 66 % was found while using solely the threshold of Q_{cut-in}. This result is seen as a moderate success only. By combining the dominant threshold of Q with those of indicators speed (e.g. >20 kph) or range (e.g. <33 m) only offers a limited optimization potential of 3 % higher agreement.

Figure 4: The Pearson correlation of the graph is 0,253 (P = 0,364) and shows that with higher values of Q there is increasing agreement. Thus the model threshold of Q supports the results of the simulator experiment and validates parts of the model.

Peculiarities: A consensus to the proposed warning model is more difficult to achieve when the participants have to contradict actively to the model: scenes A2 and C3 need an active dissent to comply with the model. This might be the cause for the lower percentage of agreement.

Scene B2 also had a very low agreement of 36 %. A peculiarity in the known indicators could be identified that provides a plausible cause: The scene deals with a rather high range value (34 m). Scene A4 had strong demand to change to hard warning with a high value of Q = 1,86 kph/m which supports the importance of Q as an indicator for soft and hard warnings. An underlying near field sensor value confirms the tendency to convert to the desired close cut-in type manoeuver. Another aspect is that the lane changing vehicle was a police car. In the participants' comments there were hints that it has influenced the subjective ratings of video scene A4. Likewise in scene D3 several participants demanded a hard warning - with Q = 1,5 kph/m. Surprisingly for scene B3 as a non-warning scene with Q = 0,7 kph/m they also demanded a hard warning. An in-depth investigation unveiled that in both cases – D3 and B3 – the TJA prototype vehicle has failed to keep the normal distance after a lane change. This is a sacrifice for using real video scenes in lieu of composing an artificial simulator experiment. In retrospect these two scenes can be declared as invalid for the experiment. All these peculiarities have led to the unsatisfactory results in acceptance (66 %) for normal cut-in warnings. Therefore the results of the questionnaire shall specifically be considered in order to identify model improvements.

Cut-outs. A moderate model acceptance of 76 % was achieved. It is worth to mention that only two participants (8 %) reject all warnings for cut-out manoeuvers, all other accepted or demanded at least one scene for a soft warning.

Peculiarities: Scenes F3 and G3 have gained low level of agreement to the model. In scene F3 we find a special behaviour of cut-out because the vehicle that leaves the lane drives over a shaded area, i.e. the participants regard this as a driving rule violation of the leading vehicle and would have expected a warning due to this fact although the model identifies this as a non-warning scene.

An explanation for the evaluation of scene G3 could not be found as the high value for Q = 2,33 kph/m fully contradicts the model to suppress a warning.

Figure 4: The Pearson correlation of the graph is -0,485 (P = 0,185). The negative value is contra-productive for the validity of the proposed model that uses a higher Q for warnings in cut-out manoeuvers. The result can lead to reject the use of Q further-on for the warning strategy of cut-out manoeuvers.

Another attempt was made to identify thresholds and a correlation between the agreement and the model indicators Speed and Range. Initially the Speed indicator with a lower threshold was considered, e.g. proposing warnings only for Speed <25 kph. But a well agreed warning for scene E3 at higher speed discards the proposal. Despite the fact that only 8 % of participants do not agree with any warning the strategy for cut-outs could be redesigned not to consider any warnings for theses manoeuvers at all because a suitable model with indicator thresholds cannot be identified.

Questionnaire Findings. The profile of participants shows a fair gender distribution and a wide age profile (20–60 years). With 56 % the envisioned target of experienced ACC drivers was reached. The question about the subjective reasons to agree or disagree to a lane change warning is treated in Fig. 5.

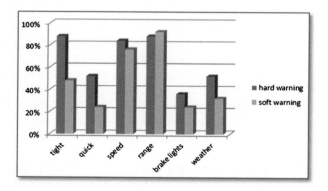

Fig. 5. Rationales to accept a warning

It is worth to discuss the levels above 50 %: Hard warnings are expected if the lane changing vehicle is in tight distance (88 %) to the TJA vehicle. This recognition can serve again as argument to use near-field sensor measurements to identify acceptance for close cut-ins and the related hard warning. Two further indicators above 50 % for close cut-ins are to be mentioned: Rapidity ("quick") and weather conditions could be other motivating factors to postulate a hard warning.

High relevancy was found for the indicators *Speed* and *Range* for both hard and soft warnings (76–92 %). The choice of indicators from the data analysis is in accordance with the subjective rating of the participants. Another relevant indicator for soft warnings is the attribute "tight" showing ca. 45 % of relevance. This attribute deals with the distance measurement to the cutting-in vehicles: Independent from the mentioned near-field measurement which was partially available in the NDS data this implies to use a new indicator that represent the distance of the TJA vehicle to the relevant secondary object for every cut-in manoeuvers.

The pie charts in Fig. 6 show the results of two questions that deal with an overall judgment and efficiency of the tested TJA system with warnings and transitions in control. It is represented on the Likert scales from x = 1 (no agreement) to 5 (full agreement). A simplified level of agreement is presented by a percentage summary factor P. Due to [9] this factor is only applicable if the 5 steps of the Likert scale are considered equidistant. The arithmetic mean values for the questions are calculated for ACC and non-ACC drivers:

$$P = \frac{1}{n} \sum_1^n 0,25(x-1)$$

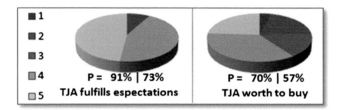

Fig. 6. TJA judgment (ACC divers ǀ non-ACC drivers)

ACC drivers turn out at a more enthusiastic level with the proposed TJA design as a collaborative system: in the answer of TJA worthiness they gave higher credits to the proposed TJA system. In several cases the participants mentioned that the actual added value would depend on the cost of ownership.

4   Conclusions and Outlook

In the framework of automated driving with cooperative approach this study presents a control transition strategy that is built on surrounding lane change manoeuvers. A model is developed with the target to provide a high level of usability for hard and soft warnings. As core indicator for warnings quotient Q was identified that takes vehicle speed and distance into account. A simulator experiment based on video scenes was conducted to assess model validity. This study revealed an overall positive attitude of drivers towards an automated driving technology with cooperative approach. Previous experience with ACC driving proved to be of no significant effect on the acceptance of the proposed warnings.

The results of the experiment showed overall a moderate and adequate perceived usability of the current model and led to the following recommendations for further development of the strategy: For hard warnings it will be beneficial to consider objective indicators like near-field distance and Q. For soft warnings a redesign and refinement is proposed. Normal cut-in manoeuvers have a significant correlation with indicator Q and can potentially be enhanced by an indicator that represents the distance to the cutting-in vehicle. Cut-out manoeuvers showed a rather uncontrollable behaviour with indicators. A revised strategy could consider neglecting any warning for cut-outs.

A follow-up driving simulator experiment will be conducted to assess driver acceptance with respect to this revised control transition strategy as well as the proposed refined warning model.

References

1. Schaller, T.: Congestion assistance – supporting the driver in lateral and longitudinal guidance in congestion situations, dissertation TU Munich, Garching (2009)
2. Henel, C.: Analyse und experimentelle Nutzung von naturalistischen Fahrdaten für automatisiertes Fahren, RWTH Aachen (2014)
3. Gilchrist, R., et al.: Three-dimensional relationships among traffic flow theory variables. Transp. Res. Rec. **1225**, 99–108 (1992)
4. BMJ: Straßenverkehrsordnung. German Road Traffic Regulation, Ministry of Justice, Berlin (2013)
5. Krauth, J.: Experimental Design. Elsevier/Saunders, München (2000)
6. Minitab Inc.: Minitab Handbook (2007)
7. Pelham, B., Blanton, H.: Conducting Research in Psychology, Measuring the Weight of Smoke, 4th edn. Wadsworth/Cengage Learning, Belmont (2013)
8. International Telecommunication Union: Methodology for the subjective assessment of the quality of television pictures, Geneva (2012)
9. Rost, J.: Lehrbuch Testtheorie, Testkonstruktion Huber, Bern (1996)
10. Sadish, W.: Experimental and Quasi-Experimental Designs for Generalized Causal Inference. Houghton Mifflin, Boston (2002)
11. Urhahne, J.: Analysis of Traffic Congestions for Automated Driving with Cooperative Approach, FISITA World Congress, Maastricht NL (2014)
12. Stutts, J.: Distractions in Everyday Driving, AAA Foundation for Traffic Safety, Washington (2003)
13. Dambӧck, D., Kienle, M., Bengler, K., Bubb, H.: The H-metaphor as an example for cooperative vehicle driving. In: Jacko, J.A. (ed.) Human-Computer Interaction, Part III, HCII 2011. LNCS, vol. 6763, pp. 376–385. Springer, Heidelberg (2011)
14. Brookhuis, K. et al.: Driving with a congestion assistant; mental workload and acceptance. Appl. Ergon. **40**, 1019–1025 (2008). (Elsevier)
15. Van Driel, C., et al.: Impacts of a congestion assistant on driving behaviour and acceptance using a driving simulator. Transp. Res. Part F **2**, 139–152 (2007). (Elsevier)

Visualization

Aperio: A System for Visualizing 3D Anatomy Data Using Virtual Mechanical Tools

T. McInerney[(⊠)] and D. Tran

Department of Computer Science, Ryerson University,
Toronto, ON M5B 2K3, Canada
tmcinern@ryerson.ca

Abstract. We present Aperio - an interactive real-time system for visualizing complex organic-shaped 3D models such as anatomy data or medical data. Aperio employs an interaction model based on a mechanical tool analogy via a small set of virtual "metal" tools, such as rods, rings, cutters, and scalpels. The familiar and well-differentiated tool shapes, combined with their initial pose and metallic appearance, suggest a tool's function to the user. Cutter tools are designed to create easily-understood cutaway views, and rings and rods provide simple oriented path constraints that support rigid transformations of models via "sliding", including interactive exploded view capabilities. GPU rendering provides realistic real-time "solid cut" previewing of surface-mesh models. We demonstrate Aperio using a human anatomy data set and present user studies to provide supporting evidence of Aperio's interaction simplicity and its effectiveness for visualizing model spatial interrelationships.

1 Introduction

Perceiving the spatial relationships between 3D geometric models and between model subsystems, and perceiving the hidden internal structure of complex geometric models are essential tasks in fields such as engineering, architecture, and medicine. Specifically, in medical education or surgical planning, anatomy data or medical data may consist of numerous model subsystems each of which may be composed of many individual curving anatomical structures that are often closely adjoined, intertwined, or enclosed. Therefore, in order to actively explore the data, objects and their constituent parts must be quickly and repeatedly rearranged relative to one another to reveal obscured relationships, and hidden objects must be revealed by creating contextual cutaway views.

This paper presents Aperio[1] (Fig. 1), an interactive system for exploring spatial relationships between geometric models (i.e. surface meshes) that uses an interaction metaphor based on a familiar mechanical tool analogy. Aperio utilizes a small collection of virtual metal tools, such as rods, rings, "cookie cutters", a scalpel and spreader (Fig. 1 left), to support a coherent set of intuitive, simple

---

[1] A video of Aperio can be viewed at www.youtube.com/watch?v=KcGfVjDlnpU.

© Springer International Publishing Switzerland 2015
G. Bebis et al. (Eds.): ISVC 2015, Part I, LNCS 9474, pp. 797–808, 2015.
DOI: 10.1007/978-3-319-27857-5_71

Fig. 1. Left: Aperio tools - "cookie" cutter, knife, ring, rod. Right: Using a combination of virtual mechanical tools to reveal spatial relationships between model parts.

and fast model rearrangements and model cutaways. The familiar shapes and metallic appearance of the tools provide strong visual cues to the user allowing tools to be quickly positioned and oriented unambiguously while also providing the user with a clear choice of which tool to use. Tools can be smoothly moved around the surface of the object models and dynamically oriented, scaled, and shaped. Model rearrangement, including exploded views, is performed by *sliding* models along the tools, akin to a beads on a wire. In the remainder of this paper, we describe Aperio, its interaction model, a description of each tool and its implementation. We demonstrate Aperio using a human anatomy data set and also present user studies to provide supporting evidence of Aperio's interaction simplicity, controllability and effectiveness for visual analysis of this data type.

Aperio presents several contributions to occlusion management of curving, twisting and intertwining organic shape models created by digital artists or derived from 3D medical images. Firstly, the system combines model cutaway and real-time cutaway previewing, constrained rigid transformation control of individual models, model parts, and model subsystems, and exploded view capabilities, all under a single, coherent interaction model based on a mechanical/surgical tool analogy. The highly-controllable tools can be flexibly combined and repeatedly applied to model parts cut away from the original. Secondly, Aperio supports user-defined dynamically configurable curving explosion paths, enabling clear views of the spatial relationship between closely adjoined curving objects by not only moving them apart but also "opening" them up relative to each other. Thirdly, unlike many other systems, we maintain a rendering of the tools/tool outlines to visually reinforce the tool operations and to aid in the overall perception of model part relationships. A user study suggests this type of visual cue aids in visual analysis and also helps the user maintain an understanding of what operations have been performed, thereby allowing users to more easily undo, extend, and modify the operations. Finally, the single, compact underlying mathematical formulation of the tools enables specialized, dynamically-configurable cutaways views such as ribbons.

2   Related Work

A wide range of novel visualization techniques have been proposed for managing 3D scene occlusion [1], for both volume data and surface mesh data (which is the focus here). The majority of these techniques fall into one (or more) of the following categories: transparency, cutaway, explosion and deformation. Aperio supports cutaway, transparency and explosions, as well as model rearrangement via rigid transformations. In this section we review related work in the cutaway, explosion and deformation categories only. We also briefly compare and contrast these works with Aperio to elucidate the contributions.

Cutaway views have a long history and are a widely used technique for revealing hidden objects [2–7]. In Knodel et al. [2] users generate cutaways using simple sketching actions and then refine the shape of the cut using widgets. Li et al. [3] present a system for experts/artists to interactively author cutaway illustrations of complex 3D models. Users can then use a pre-authored viewer application for exploring the data. To minimize the loss of contextual information, McInerney and Crawford [4] remove only part of the occluding geometry and retain polygonal strips called "ribbons" or solid thick "slices". Pindat et al. [5] use a moveable lens that combines a cut-away technique with multiple detail-in-context views of the data. Trapp and Döllner [6] generalize *clip-planes* to *clip-surfaces* that support the generation of curving cut surfaces with user definable contours. Burns and Finkelstein [7] uses a depth texture along with a depth parameter to generate view dependent cutaways. The cutter tool in Aperio is most similar to Pindat et al. [5], and Li et al. [3]. Pindat et al. [5] use a cone shaped cutter "lens" whereas Aperio uses a more flexible superellipsoid. Unlike the pre-authored system in Li et al. [3], which restricts the user's control over view generation, Aperio's real-time previewing cutter "lens" tool is controlled by the end user. However, Li et al. are currently able to generate a wider shape range of cutaways. Aperio also supports "ribbons" but unlike the statically generated ribbons of McInerney and Crawford [4], they are generated in real-time and are interactively configurable.

Exploded views attempt to reveal hidden surfaces or spatial relationships by interactively spreading objects apart along a path [8–11]. Radial explosion paths are common [8] but this simple strategy does not provide the user with much control and may result in visual clutter. Tatzgern et al. [9] introduced an automatic approach where only subsets of model part assemblies are exploded in an attempt to reduce visual clutter. Li et al. [10] generate exploded views by automatically creating an explosion graph that encodes how parts are moved. Many of the above techniques may be better suited to machine part models than the curving, twisting and closely adjoining or intertwined object mesh models found in anatomy/medical data. That is, the simple radial paths or the pre-computed/automatically generated explosion graphs/paths may make assumptions about the shape and spatial arrangement of the models. For example, Li et al. [10] assume that parts can be separated via linear translations, that models are two-sided and that parts fit together without interference. These assumptions may not apply to many human anatomy structures and subsystems. Furthermore, pre-computed/automated explosion graphs/paths restrict the

Fig. 2. Ring tool sliding along a model surface. The ring partially penetrates the model surface and automatically aligns itself with the model surface normal vector.

ability to explore individual/group relationships in model subsystems. Aperio was expressly designed to be less automatic and more user-controllable and flexible, so as to better handle systems of curving/twisting models. User-configurable path orientation, curving paths, and model sliding control (with/without "explosion") of individual models or model groups can be dynamically created with rods/rings.

Unlike exploded views, deformation techniques employ nonrigid transformations [12–14], such as peeling, bending, and retracting objects. Correa et al. [13] coin the term "Illustrative Deformation" to describe an approach to volume and surface data visualization that uses 3D displacement maps to perform a wide range of deformations. Birkeland and Viola [12] present a view-dependent peel-away technique for volume data. McGuffin et al. [14] propose an interactive system for browsing pre-labeled iso-surfaces in volume data where users can perform deformations to cut into and open up, spread apart, or peel away parts of the volume in real time. When dealing with anatomy mesh data, the complex spatial interrelationships of curving/twisting models may complicate deformation operations and result in unrealistic model interpenetration. Geometric-based deformation techniques may result in model transformations that are physically realistic (such as peeling) but typically do not consider real tissue deformation properties and may therefore generate un-intuitive and unexpected deformations of some structures if used improperly. Furthermore, for multiple adjoining subsystems, such as arteries, muscles and bones, it may be difficult to quickly and simply apply geometric-based deformations. For these reasons, we decided to support constrained *rigid* transformations combined with cutaways. Rigid transformations are efficient, familiar and easily understandable, enable simple, precise controllability, are flexible and easily combined, and result in predictable *geometric* behavior that is not restricted by expectations of physical realism.

3   Aperio Tool Interaction

Currently Aperio uses a mouse and modifier keys to position tools. A tool is instantly instantiated and made active by clicking on a tool icon from a control panel. The user can then smoothly slide the active tool along the surface of a model or (optionally) smoothly slide it (partially) off of the model surface by switching to a mode that uses a pre-computed oriented bounding box (OBB) of

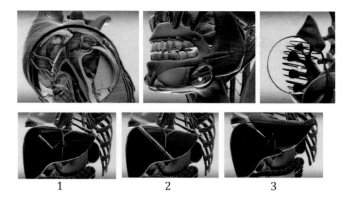

Fig. 3. Top row: cutter tool cutting into several objects in real-time: (1) the heart, (2) head with facial nerve is exposed, (3) kidney cutaway using "ribbons". Bottom row: knife and spreader tools are used cut a liver model into two pieces. A ring tool has been added to "open up" the cut liver (Color figure online).

the model. The tool will automatically orient itself to match the normal vector of the current model surface (Fig. 2) or OBB. The active tool can also be optionally "planted" at any time to establish a view, and "picked up" at a later time to modify the view. We use modifier keys and the mouse wheel to quickly and fluidly change the active tool size, orientation and depth without disturbing its current position. For tool parameters changed less often, GUI sliders are provided.

3.1   Cutting Tools

The ("cookie") cutter tool (Fig. 3) can interactively slide along (multiple) selected model surfaces and cutaway the parts of models inside the cutter boundaries, to a user-defined depth, in real-time. The selected model's cut-surface is automatically "capped" by the renderer so that the object model always appears solid. Furthermore, the user can double click as they move the mouse, selecting deeper objects visible inside the cutter. These objects are instantly cut away providing the ability to dynamically "drill-down" into the data and reveal inner layers. A cutter can optionally cut away a pattern of the occluding model, forming surface "ribbons" [4] contained within the superellipsoid cutter boundaries (Fig. 3 top row, right). The user can dynamically control ribbon orientation by spinning the cutter with the mouse wheel, and properties such as ribbon width, frequency and "tilt" can also be dynamically modified using GUI sliders.

The knife tool (Fig. 3 bottom row) is analogous to a surgical scalpel and is designed to automatically cut an object model into two separate pieces. The knife is rendered as a tapered, flattened superquadratic cylinder that resembles a knife blade. The user simply draws the knife across a selected object surface[2] and a shallow, narrow "surgical incision" cutaway region is rendered in real-time

---

[2] It is not necessary to draw completely across the selected object.

Fig. 4. Top left: rod tool used to create exploded view of heart model parts. Top right: ring tool used for "opening" models like pages in a ringed notebook or exploding models along a curving path. Bottom: two rings and two rods are used to open up and explode several brain parts.

to provide visual reinforcement. Once the cut action is finished, an automatically generated narrow rectangular cutter, known as a "spreader" tool, is instantiated and rendered (Fig. 3 bottom row center) inside the incision. A cutter scaling algorithm is executed based on the model's OBB, as is a highly optimized computational solid geometry (CSG) difference operation using the Carve library that typically completes in under a second. The result is the model is divided into two pieces, with a slight separation between them, along the "incision" path. The spreader tool can be used to interactively widen the incision cut to any desired degree. The user can treat the two pieces of the model in the same manner as any other object model. For example, rods and rings may be used to translate/rotate each piece and each piece can be divided with knife cuts.

3.2   Rearrangement Tools

The rod tool is a user-extendable cylinder-shaped superquadric (Fig. 4). Upon selection, it is oriented along the model surface normal at the current mouse point and partially penetrates the model surface. The simple cylindrical shape, pose, metallic appearance and partial model surface penetration suggests to the user that a model can "slide" along the rod. The user positions a rod along a model surface until it penetrates all target selected models[3]. GUI sliders allow the user to control model sliding (and, optionally, model fanning) back and forth along the rod, and to restore the models to their original position. Models can be individually translated along the rod or translated in groups. Furthermore, an additional "Spread" GUI button switches to an exploded view. In this mode,

---

[3] If a selected model is not penetrated, a point on the rod closest to the center of the model's OBB is used.

the back and forth movement of the GUI slider will cause the penetrated objects to automatically slide apart and together relative to each other.

The ring tool is represented using a supertoroid and is primarily suggestive of rotation and secondarily of translation (Fig. 4 upper right). Like the rod tool, each model that is selected by the user and penetrated by the ring can be individually, or in combination, slid back and forth along the ring with a GUI slider, and each model is automatically aligned with the normal vector of the ring at the current ring-model intersection point. Ring shape is controlled using a separate GUI slider. If the ring is set to a circular shape, models slide and rotate in a manner similar to turning pages on a ringed notebook. Similar to the rod, the "Spread" button can be used to "explode" models along the ring, "opening" them up with respect to each other. If a selected model does not intersect the ring, we use the center of the model's OBB and compute the closest point on the ring from this center point.

In the bottom row of Fig. 4 we show two stages of iterative model rearrangement (i.e. sliding apart and back together) when exploring a multi-part brain model. Two rings and two rods are used to open up the brain. If the user selects all penetrated models then a single GUI slider will slide all models back and forth along their respective rods/rings. In this example, establishing the initial exploded view takes approximately 30 s for an experienced user.

4   Implementation

This section will provide a brief, high-level description of Aperio's implementation. A detailed description can be found in [15]. In addition, Aperio software is open source and is available at github.com/eternallite/Aperio. Aperio is written in C++ and is constructed using the Visualization Toolkit (VTK) [16]. It uses OpenGL and GLSL shaders for rendering and Carve CSG, a constructive solid geometry library for performing cutting and splitting mesh models. Aperio uses superquadrics [17] to represent all tools - compactly defined geometric shapes that resemble ellipsoids and toroids but with a more expressive shape range. VTK contains classes for creating, transforming, and rendering superquadrics. A superquadric [15] has an implicit function formulation that provides a simple test to determine if a point is inside, outside, or on its surface, as well as a corresponding parametric function formulation. The vectors (**right, up, forward**) form the basis of the superellipsoid coordinate system (CS) (Fig. 5) and we convert points between this local CS and world coordinates using a transformation matrix constructed from these vectors.

Cutaway and Capping Algorithm. We use VTK's multi-pass rendering pipeline for "capping" the cut surface of the "hollow" mesh models to create the illusion of solid models. Our cut-surface capping algorithm requires two render passes: a pre-pass to render information into texture images of a Frame Buffer Object (FBO) and a subsequent main-pass that reads in texture data and renders the final scene. Each of these render passes triggers the execution of a special

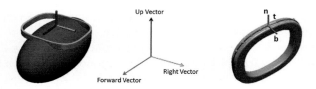

Fig. 5. Left, middle: A local superquadric coordinate system is constructed from the camera's "view up" vector and the data surface normal at the current mouse position. Right: A coordinate system is constructed at each sampled point along the middle of the ring's outer surface and is used to orient a model during sliding.

GPU fragment shader. In both passes, both fragment shaders first discard all selected mesh fragments that are inside the boundaries of the superellipsoid cutter, making use of the convenient superellipsoid implicit inside-outside function. The removal of these fragments will expose back-facing fragments of a cut model. We must replace these fragments with back-facing cutter fragments to achieve a solid cut (Fig. 3 upper left). In the pre-pass, the fragment shader outputs the depths and colors of all selected model front and back facing fragments, as well as the depth of all back facing cutter fragments into an FBO. Depths are encoded using an RGBA color vector. The pre-pass fragment shader also sets the color of all discarded/non-filled fragments in the texture to the RGBA color $(1, 1, 1, 1)$ (i.e. "infinite" depth or pure white) so that they are distinguishable from actual depths. In the main-pass we read in the FBO textures generated by the pre-pass and check if a fragment was discarded/non-filled. If so, it likely requires capping. For every back-facing fragment of the cutter, the main-pass fragment shader tests if the front-facing fragment of a selected model is discarded/non-filled, and the depth of the back-facing cutter fragment is less than (i.e. closer) the depth of the corresponding back-facing selected model fragment. If true then a selected model fragment color is output but using the back-facing cutter fragment depth. This algorithm will only render cap fragments that are within the bounds of a selected model.

Ring and Rod Path Generation. The parametric surface representation of a superquadric provides a basis for easily definable, unambiguously oriented sliding paths. To implement sliding along a ring, we use VTK's superquadric class to generate a position and normal vector $\mathbf{n}$ for points sampled along a path of the supertoroid that symmetrically divides the supertoroid in half (Fig. 5 right). We can then determine what orientation to give the models (at each point along the path) as they slide on the ring by constructing a local CS at each path point. The tangent vector $\mathbf{t}$ of a path-point CS is calculated by subtracting the current point from the next path point. The bi-tangent vector $\mathbf{b}$ is simply the cross-product of the normal and tangent vectors. The normal, tangent and bi-tangent vectors are used to construct a rotation matrix which is then multiplied by all mesh model points to orient it at the current path point. To determine the initial path point for a selected model, we use VTK to determine the intersection point between the ring and the model. If a selected model does not intersect the ring, we use the center of the model's OBB and compute the closest point on the ring from this center point. For the rod tool, the path generation is greatly simplified as the rod central axis is a line segment. Path points are generated evenly along this line segment and the rod central axis is used as the rotation axis to spin models. We again calculate the intersection point between the rod central axis and a selected mesh for the initial path point.

5   Validation

We performed a qualitative user study to gather some supporting evidence of the effectiveness of the visual cues provided by the tools, and the intuitiveness and controllability of user interactions. Eighteen participants each had a one hour session where they were asked to perform visualization tasks using Aperio. The 18 participants were aged 19–34 with an average age of roughly 26 years; there were 14 males and 4 females. The user study consisted of two trials and one practice trial. In each trial participants were shown several scene images depicting a subsystem of the anatomy data set in various "stages" of model rearrangement. Participants were asked to use Aperio to match each stage[4]. The matching task also included restoring the position and orientation of models to their original state. Participants were informed that their performance was not being measured. To begin the study, each participant was shown the functionality and interface of Aperio and were then allowed to play with the system using a demonstration data set. They were then asked to perform a practice trial followed by the matching task in two subsequent trials, with all trials using different anatomical model subsystems. Finally, once the scene matching tasks were completed, participants were asked to play with the cutter tool and ribbon view cutter on one of the trial data sets.

After the trials the participants filled out a questionnaire and indicated their level of agreement or disagreement (using a 7-point Likert scale) with statements pertaining to the perception of Aperio tools (Fig. 6 left) and to Aperio's user interface (Fig. 6 right). As is evident from the bar graphs (Fig. 6), the results were very positive in this qualitative study. In addition, with respect to the cutter tool versus the ribbon cutter, 15 users thought the ribbon cuts were an effective alternative to a full cutaway (for preserving the shape of the cut object), 2 did not think it was effective and 1 chose "OK". With respect to Aperio's user interface, of the 18 participants, 14 found the Aperio's tool interface intuitive, 4 did not. For tool preference, 9 users preferred the rod tool, 8 users preferred the ring tool and 1 user did not like either.

Finally, we performed two additional online perception studies using Survey Monkey [18]. In each study, 50 people participated and the two groups did

---

[4] We also initially used a separate system with a simplified user interface, similar to the interface of common object modeling packages, as a control for the study but it was found to be too simplistic and was rated poorly by almost all users. The results for this simplified system are not included in this paper but can be found in [15].

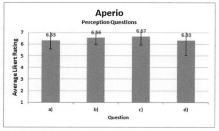
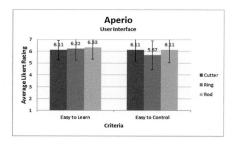

Aperio Tools help to...
a) perform actions (move and orient objects)
b) maintain understanding of what has been done to objects
c) restore original positioning and orientation of objects
d) understand layering and spatial relationships between objects

Fig. 6. Left: results of user study related to perceiving model spatial relationships using Aperio tools. Right: results of user study related to Aperio's user interface.

Fig. 7. Example of images used in the online user study using a visual cue of (a) metal rings, (b) gray curve, (c) bright green curve (Color figure online).

not overlap. The ages of the participants ranged from 18 to 60+ and all had at least four years of college level education. In these studies the goal was to determine if visual cues (such as Aperio tools) were useful for understanding the spatial interrelationships of models after the tools had been applied, as well as for understanding how the model parts had been moved apart or cut away. We created a series of static images of various model subsystems (e.g. the parts of the brain) with our tools to move apart or cut models (Fig. 7). We asked several multiple choice questions related to the understanding of the model rearrangements. In the first study, we presented multiple sets of four images to the participants. The first image showed a pre-transformed view and three images showed the result of the tool action with rings/rods/cutters as visual aids, with simple gray lines/curves as visual aids with each having the same shape and dimension as a corresponding rod/ring/cutter, and with no visual aids. We chose simple line/curve visual cues as a comparison to determine if the realistic shiny metal surface renderings of the Aperio tools were advantageous in these visual analysis scenarios. The second study was identical to the first except the gray lines/curves were replaced with bright green lines/curves that visually stood out more strongly against the data models.

In both studies the results showed that for rod/ring tool operations, on average, the participants preferred some visual aid (56.18 %) to no visual aid (28.64 %), while some had no preference (15.18 %). For cutter tool operations, on average the participants preferred some visual aid to help them determine what model region had been cut (35.85 %) to no visual aid (21.36 %). However, the results also showed that, on average, while more participants preferred the metal tools (34.88 %) to gray lines/curves (18.42 %), more participants preferred the bright green lines/curves (43.85 %) to the metal tools (24.42 %). We conjecture that when performing a visual analysis on images of 3D models that have already been transformed, as long as the visual cues are clearly visible and their shape/curvature discernable, then the least distracting visual cue (i.e. with the smallest visual impact with respect to the data models) is preferable.

6   Conclusion

User interaction models are an integral part of effective interactive visual exploration of 3D data. If the interaction model is too complex, for example one based on a widget with many handles and controls and artificial appearance, the user may spend too much time learning and/or applying the model. On the other hand, if the interaction model is too simple, then tool/option proliferation may result. This complicates the interface, reduces interaction consistency and forces the user to perform too much work. Furthermore, automated techniques or pre-computed views may overly restrict the user's control over view generation. Aperio lets the user decide which data models to rearrange and how to rearrange them. This approach may result in a heavier cognitive load with respect to which tools to use and how to apply and combine them to obtain the desired view. However, by adhering to this simpler approach of highly controllable model rearrangements and cutaways, users may discover/view relationships between complex organic-shaped multi-part models in ways not predicted or not possible by more automatic techniques. To help offset the increased cognitive and interaction load on the user, Aperio uses a small set of easily-applied tools with familiar shapes, appearance and affordance, to guide and visually reinforce user actions. For expert users who may want to create more complex views such as free-form shaped cutaways or custom curving explosion paths, it may be desirable to allow tool constraints to be "loosened". For example, superquads can be interactively deformed using global deformation functions, such as bending and tapering [17], expanding their range of possible tools, tool paths, and cutaway shapes. In addition, multiple superquads can be blended to form complex shapes while still providing a well-defined inside-outside function. These features are the subject of future research.

References

1. Elmqvist, N., Tsigas, P.: A taxonomy of 3d occlusion management for visualization. IEEE Trans. Visual Comput. Graphics **14**, 1095–1109 (2008)
2. Knödel, S., Hachet, M., Guitton, P.: Interactive generation and modification of cutaway illustrations for polygonal models. In: Butz, A., Fisher, B., Christie, M., Krüger, A., Olivier, P., Therón, R. (eds.) SG 2009. LNCS, vol. 5531, pp. 140–151. Springer, Heidelberg (2009)

3. Li, W., Ritter, L., Agrawala, M., Curless, B., Salesin, D.: Interactive cutaway illustrations of complex 3d models. ACM Trans. Graph. **26**, 31:1–31:11 (2007)

4. McInerney, T., Crawford, P.: RibbonView: interactive context-preserving cutaways of anatomical surface meshes. In: Bebis, G., et al. (eds.) ISVC 2010, Part II. LNCS, vol. 6454, pp. 533–544. Springer, Heidelberg (2010)

5. Pindat, C., Pietriga, E., Chapuis, O., Puech, C.: Drilling into complex 3d models with gimlenses. In: VRST 2013, pp. 223–230 (2013)

6. Trapp, M., Döllner, J.: 2.5d clip-surfaces for technical visualization. J. WSCG **21**, 89–96 (2013)

7. Burns, M., Finkelstein, A.: Adaptive cutaways for comprehensible rendering of polygonal scenes. ACM Trans. Graph. **27**, 154:1–154:7 (2008)

8. Sonnet, H., Carpendale, S., Strothotte, T.: Integrating expanding annotations with a 3d explosion probe. In: Proceedings of the Working Conference on Advanced Visual Interfaces, AVI 2004, pp. 63–70. ACM, New York (2004)

9. Tatzgern, M., Kalkofen, D., Schmalstieg, D.: Compact explosion diagrams. In: Proceedings of the 8th International Symposium on Non-photorealistic Animation and Rendering, NPAR 2010. ACM, New York, pp. 17–26 (2010)

10. Li, W., Agrawala, M., Curless, B., Salesin, D.: Automated generation of interactive 3d exploded view diagrams. ACM Trans. Graph. **27**, 101:1–101:7 (2008)

11. Bruckner, S., Gröller, M.E.: Exploded views for volume data. IEEE Trans. Visual Comput. Graphics **12**, 1077–1084 (2006)

12. Birkeland, S., Viola, I.: View-dependent peel-away visualization for volumetric data. In: Hauser, H., Spencer, S.N. (eds.) Spring Conference on Computer Graphics SCCG, pp. 121–128. ACM (2009)

13. Correa, C.D., Silver, D., Chen, M.: Feature aligned volume manipulation for illustration and visualization. IEEE Trans. Vis. Comput. Graph. **12**, 1069–1076 (2006)

14. McGuffin, M.J., Tancau, L., Balakrishnan, R.: Using deformations for browsing volumetric data. In: Proceedings of the 14th IEEE Visualization 2003 (VIS 2003), p. 53. IEEE Computer Society, Washington, D.C. (2003)

15. Tran, D.: Aperio: Managing 3d scene occlusion using a mechanical analogy for visualizing multi-part mesh data. Master's thesis, Dept. of Computer Science, Ryerson University, Toronto, ON, Canada (2015)

16. Schroeder, W.J., Martin, K., Lorensen, W.: The Visualization Toolkit: An Object-Oriented Approach to 3d Graphics, 4th edn. Kitware Inc., Clifton (2006)

17. Barr, A.H.: Superquadrics and angle-preserving transformations. IEEE Comput. Graph. Appl. **1**, 11–23 (1981)

18. SurveyMonkey Inc.: SurveyMonkey. www.surveymonkey.com. Accessed 08 October 2015

Quasi-Conformal Hybrid Multi-modality Image Registration and its Application to Medical Image Fusion

Ka Chun Lam[(✉)] and Lok Ming Lui

Department of Mathematics, The Chinese University of Hong Kong,
Room 220, Lady Shaw Building, Shatin N.T., Hong Kong, China
jefferykclam@gmail.com, lmlui@math.cuhk.edu.hk

Abstract. Fusion of images with same or different modalities has been conquering medical imaging field more rapidly due to the presence of highly accessible patients' information in recent years. For example, cross platform non-rigid registration of CT with MRI images has found a significant role in different clinical application. In some instances labelling of anatomical features by medical experts are also involved to further improve the accuracy and authenticity of the registration. Being motivated by these, we propose a new algorithm to compute diffeomorphic hybrid multi-modality registration with large deformations. Our iterative scheme consists of mainly two steps. First, we obtain the optimal Beltrami coefficient corresponding to the diffeomorphic mapping that exactly superimposes the feature points. The second step detects the intensity difference in the framework of mutual information. A non-rigid deformation which minimizes the intensity difference is then obtained. Experiments have been carried out on both synthetic and real data. Results demonstrate the stability and efficacy of the proposed algorithm to obtain diffeomorphic image registration.

1 Introduction

Image registration is one of the important steps in various fields which aims to align images [1–3]. Existing methods for image registration can be classified into three categories: landmark-based, intensity-based and hybrid methods. For landmark-based method, optimal deformation is obtained by aligning the sparse geometric feature points on both source and target domain. For example, the thin-plate spline registration method proposed by Bookstein et al. [4] aligns landmarks by using the biharmonic regularizer. Later, Joshi et al. [5] proposed to obtain the landmark matching diffeomorphism through the construction of vector field governed by the Navier-Stokes equation. A diffeomorphism with exact landmark alignment can be computed even with large deformation. One main advantages of landmark-based method is the straightforward incorporation of medical expert during the registration process [2]. This provides a reliable deformation approximation once the features alignment is accurate. In addition, landmark-based method is usually computationally efficient when

© Springer International Publishing Switzerland 2015
G. Bebis et al. (Eds.): ISVC 2015, Part I, LNCS 9474, pp. 809–818, 2015.
DOI: 10.1007/978-3-319-27857-5_72

modelling large deformations. Intensity-based registration method aims to match the intensity information without the help of sparse geometric knowledge. By taking into account more image information during the registration process, delineation of feature landmarks is not required. To list a few, the Diffeomorphic Demons Registration method in [6], which is proposed based on Thirion's demons algorithm [7], obtains the registration result in the space of diffeomorphic transformations. Glocker proposed the DROP algorithm [8] to register images by making use of the Markov random field formulation. But without human supervision, inaccurate registration with underlying large deformation may be resulted. Hybrid approaches make use of both sparse geometric and intensity information to guide the registration. By aligning landmark and intensity, we integrate both merits of matching approaches: reliability of meaningful feature alignment provided by medical practitioners and accurate registration of local intensity information. As a consequence, this type of approach can usually provide better registration results. For instance, Christensen et al. [9] proposed to apply the unidirectional landmark thin-plate-spline (UL-TPS) registration technique with the minimization of intensity difference to register images with inverse consistency property. Chanwimaluang et al. [10] proposed a hybird retina image registration by the combination of area-based and feature-based alignment techniques.

A large amount of work has also been published on medical image fusion recently [11] due to the increasing use of medical diagnostic devices and the improved accessibility of medical data. There are mainly two stages: (1) image registration and (2) fusion of image information from the registered images [11]. In the fusion part, various methods have been proposed. To name a few, Li et al. [12] combine images under the space of wavelet coefficients. The integrated images is obtained by taking the inverse wavelet transform of the fused wavelet coefficients. Naidu et al. [13] proposed to use principal component analysis (PCA) to obtain a weighted average for the registered images to be fused.

In this paper, we extend the landmark matching algorithm in [14] to obtain diffeomorphic hybrid image registration which can handle different modalities. The main idea of the algorithm is to find the optimizer of an energy functional involving the Beltrami coefficients term, which is effective in controlling the bijectivity and the conformality distortion of the mapping. Diffeomorphism associated to the optimized Beltrami coefficient will satisfy the landmark constraints and maximize the mutual information between the source and target images. In addition, our proposed algorithm can also control the conformality distortion of the transformation. The obtained transform thus preserves as much local geometric information as possible. Noted that an accurate diffeomorphic alignment is also a key issue as severe artefact in the fused image may be produced due to misalignment or loss of image information in folding regions. We therefore propose to restrict the class of registration transformation to Quasi-Conformal mapping for image fusion problem. To validate the scheme, we have tested it on different synthetic examples and real medical images. Results show that our proposed algorithm can successfully align images according to the prescribed landmark constraints and the similarity of image intensity. The use of our registration result can improve the image fusion quality and place a significant amount of trust under the inclusion of experts' labelling.

2 Mathematical Background

Our algorithm obtains the optimal transformation in the class of Quasi-Conformal (QC) mapping. QC maps are generalization of conformal maps which are orientation preserving homeomorphisms between Riemann surfaces with bounded conformality distortion. Mathematically, let $z = x + \sqrt{-1}y$ with $x, y \in \mathbb{R}$, $f : \mathbb{C} \to \mathbb{C}$ is a QC map if it satisfies the following Beltrami equation:

$$\frac{\partial f}{\partial \bar{z}} = \mu(z)\frac{\partial f}{\partial z} \tag{1}$$

for some complex-valued function μ satisfying $\|\mu\|_\infty < 1$. The function μ is called the Beltrami coefficient, which measures the nonconformality of the mapping f. By the first order Taylor expansion $f(z) \approx f(p) + f_z(p)(z - p + \mu(p)(\bar{z} - \bar{p}))$, the Beltrami coefficient provides us all the information about the conformality of the mapping. In addition, the following theorem describes the relation between the set of Beltrami coefficients and the set of orientation preserving homeomorphisms (See [15] for details).

Theorem 1. *Suppose $\mu : \mathbb{D} \to \mathbb{D}$ is Lebesgue measurable satisfying $\|\mu\|_\infty < 1$, then there is a Quasi-Conformal homeomorphism ϕ from the unit disk to itself, which is in the Sobolev space $W^{1,2}(\Omega)$ and satisfies the Beltrami equation (1) in the distribution sense. Furthermore, by fixing 0 and 1, the associated Quasi-Conformal homeomorphism ϕ is uniquely determined.*

This theorem motivates us to transform the problem of finding the optimal deformation into the problem of finding the corresponding Beltrami coefficient. The following theorem helps us to understand the relationship between the regularity of the Beltrami coefficient $\mu(f)$ and the associated mapping f.

Theorem 2. *For any smooth μ with $\|\mu\|_\infty < 1$, the corresponding Quasi-Conformal homeomorphism f is a C^∞ diffeomorphism.*

To measure the similarity of two images with different modalities, we apply the mutual transform proposed by Kroon [16]. Let M and S be the moving and the static image. The mutual transform M_T of M to S is defined to be

$$M_T(x) = \arg\max_K H_{G_x^r}(M, S, K) \tag{2}$$

where x is the pixel position in the image; K is the intensity level; G_x^r is the Guassian windows with center x and radius r and $H_{G_x^r}(M, S, K)$ is the number of pixel in G_x^r of intensity matrix $M(x)$, which is linked to intensity level K in the static image S. With mutual transform, we define the similarity measurement to be:

$$\text{Similar}(M, S) = \frac{1}{2}\int_\Omega (S_T - M)^2 + \frac{1}{2}\int_\Omega (S - M_T)^2 \tag{3}$$

where Ω is the domain of the image M.

3   Methodology

We now formulate the hybrid multi-modality registration problem as the following mathematical model. Let M and S to be the moving image and the static image respectively. Denote $\{p_i\}_{i=1,\ldots,m} \in M$ and $\{q_i\}_{i=1,\ldots,m} \in S$ to be the prescribed landmark correspondences. We also let $\mu(f) = \frac{\partial f}{\partial \bar{z}} / \frac{\partial f}{\partial z}$ and f_μ to be the solution f of the Beltrami equation (1) with Beltrami coefficient μ. The registration problem can be modelled as follows:

$$f = \arg\min_{g} \text{Similar}(M \circ g), \quad g : M \to S; \tag{4}$$

subject to:

– f is diffeomorphic;
– f satisfies the landmark constraints: $f(p_i) = q_i$ for $i = 1, \ldots, m$;

It is well-known that restricting the transformation to be bijective is difficult. However, by the Quasi-Conformal theory, there is a one-one correspondence between the set of Beltrami coefficients and the set of Quasi-conformal homeomorphisms. Therefore, we avoid to find the optimal deformation f by optimizing the associated Beltrami coefficient instead. In other words, we have the following energy-based variational framework for solving the hybrid multi-modality registration problem:

$$(\bar{\mu}, f) = \arg\min_{\nu, g} \int_\Omega |\nabla \nu|^2 + \alpha \int_\Omega |\nu|^p +$$
$$\frac{1}{2} \left[\int_\Omega (S_T - M \circ g)^2 + \int_\Omega (S - M_T \circ g)^2 \right]$$

subject to:

– $\|\bar{\mu}\|_\infty < 1$;
– $f(p_i) = q_i \; \forall i = 1, 2, \ldots, m$;
– $\mu(f) = \bar{\mu}$.

To solve this minimization problem, we propose to use the penalty splitting method. We consider to minimize:

$$(\bar{\mu}, \bar{\nu}) = \arg\min_{\nu, \mu} \int_\Omega |\nabla \nu|^2 + \alpha \int_\Omega |\nu|^p + \sigma \int_\Omega |\nu - \mu|^2$$
$$+ \frac{1}{2}\beta \left[\int_\Omega (S_T - M \circ g_\mu)^2 + \int_\Omega (S - M_T \circ g_\mu)^2 \right]$$

and subject to $\|\bar{\mu}\|_\infty < 1$ and $g_{\bar{\mu}}(p_i) = q_i \; \forall i = 1, 2, \ldots, m$.

Different from the ordinary penalty method, we fix σ to be a large enough constant to improve the efficiency of the algorithm. We have also set $p = 2$ to ensure $\|\bar{\mu}\|_\infty < 1$ in practice. Experiments show that this simplification can give satisfactory result even with large deformation.

μ-Subproblem

We first discuss the minimization problem over μ fixing ν_n:

$$\mu_{n+1} = \arg\min_\mu \left\{ \frac{1}{2}\beta \left[\int_\Omega (S_T - M \circ g_\mu)^2 + \int_\Omega (S - M_T \circ g_\mu)^2 \right] + \sigma \int_\Omega |\nu_n - \mu|^2 \right\} \quad (5)$$

To solve the minimization problem, we applied the modified Demon's algorithm proposed by Kroon [16] to obtain the descent direction dg_μ which minimizes $\int_\Omega (S_T - M \circ g_\mu)^2 + \int_\Omega (S - M_T \circ g_\mu)^2$:

$$dg_\mu = (M_T \circ g_\mu - S) \left(\frac{\nabla M_T}{|\nabla S|^2 + \phi^2 (M_T \circ g_\mu - S)^2} \right)$$

$$+ (M \circ g_\mu - S_T) \left(\frac{\nabla M}{|\nabla M|^2 + \phi^2 (M \circ g_\mu - S_T)} \right)$$

The modified Demon's direction provide us the adjustment of the mapping dg_μ which minimizes the energy functional. Theoretically, we know that g_μ is perturbed by $g(t) = g_\mu + t dg_\mu + o(|t|)$, in which

$$dg_\mu(p) = -\frac{g_\mu(g_\mu(p) - 1)}{\pi} \left(\int_\Omega \frac{d\mu_1(z)((g_\mu)_z(z))^2}{g_\mu(z)(g_\mu(z) - 1)(g_\mu(z) - g_\mu(p))} dxdy \right.$$

$$\left. + \int_\Omega \frac{\overline{d\mu_1(z)}(\overline{(g_\mu)_z(z)})^2}{\overline{g_\mu(z)}(1 - \overline{g_\mu(z)})(1 - \overline{g_\mu(z)}g_\mu(p))} dxdy \right)$$

when μ is perturbed by $\mu = \mu + td\mu_1 + t\epsilon(t)$, where $\|\epsilon(t)\|_\infty \to 0$ as $t \to 0$. However, it is inefficient to obtain $d\mu_1$ from the above equality. Instead, we consider the first order approximation:

$$\frac{\partial(g_\mu + dg_\mu)}{\partial \bar{z}} = (\mu + d\mu_1)\frac{\partial(g_\mu + dg_\mu)}{\partial z} \quad (6)$$

By further substituting the Beltrami equation in (1), we have

$$d\mu_1 = \left(\frac{\partial dg_\mu}{\partial \bar{z}} - \mu \frac{\partial dg_\mu}{\partial z} \right) \Big/ \frac{\partial(g_\mu + dg_\mu)}{\partial z} \quad (7)$$

For the second term, the descent direction is simply

$$d\mu_2 = -2(\mu - \nu_n) \quad (8)$$

Therefore, the overall descent direction for the μ-subproblem is given by

$$d\mu = \frac{1}{2}\beta d\mu_1 + \sigma d\mu_2 \quad (9)$$

With the updated Beltrami coefficient $\tilde{\mu}_{n+1} = \mu_n + td\mu$ for some step size t, we then solve the Beltrami equation for $f_{n+\frac{1}{2}}$ in least square sense with given Beltrami coefficients $\tilde{\mu}_{n+1}$ and landmark constraints to ensure that feature points can be superimposed exactly after the registration. The detail for solving the Beltrami equation will be discussed in Sect. 4. Once $f_{n+\frac{1}{2}}$ is calculated, we obtain the local minimum $\mu_{n+1} = \mu(f_{n+\frac{1}{2}})$ of the sub-problem.

ν-Subproblem

After updating μ, we optimize the energy function over ν fixing μ_{n+1}:

$$\tilde{\nu}_{n+1} = \arg\min_{\nu} \int_{\Omega} |\nabla\nu|^2 + \alpha \int_{\Omega} |\nu|^2 + \sigma \int_{\Omega} |\nu - \mu_{n+1}|^2 \tag{10}$$

A straightforward calculation shows that the Euler-Lagrange equation of the above energy function is

$$(-\Delta + 2\alpha I + 2\sigma I)\tilde{\nu}_{n+1} = \mu_{n+1} \tag{11}$$

Similar to the case of μ-subproblem, the Beltrami coefficient $\tilde{\nu}_{n+1}$ obtain from solving (11) is used to solve the Beltrami equation together with landmark constraints. This updates the $\nu_{n+1} = \mu(f_{n+1})$ in which the associated deformation f_{n+1} will match the prescribed feature points. We then keep the iteration going to obtain a sequence of pairs $\{(\mu_n, \nu_n)\}_n$. Iteration stops when $\|\nu_{n+1} - \nu_n\| \le \epsilon$ for some threshold ϵ.

4   Implementation

We now describe the numerical implementation of our proposed algorithm.

Solving the Beltrami Equation

Given the Beltrami coefficient μ and the landmark constraints, we need to solve (1) for f_{μ} which closely resembles to the given μ and satisfies $f(p_i) = q_i$. We follow the idea in [17,18] to transform the Beltrami equation into an elliptic partial differential equation and discretize it by using finite element method:

$$\nabla \cdot \left(A \begin{pmatrix} u_x \\ u_y \end{pmatrix} \right) = 0; \quad \nabla \cdot \left(A \begin{pmatrix} v_x \\ v_y \end{pmatrix} \right) = 0, \quad \text{where } A = \begin{pmatrix} \frac{(\rho-1)^2 + \tau^2}{1 - \rho^2 - \tau^2} & -\frac{2\tau}{1 - \rho^2 - \tau^2} \\ -\frac{2\tau}{1 - \rho^2 - \tau^2} & \frac{1 + 2\rho + \rho^2 + \tau^2}{1 - \rho^2 - \tau^2} \end{pmatrix} \tag{12}$$

A linear system with symmetric positive definite matrix can be formulated after discretization. We can then impose the landmark constraints and solve for a least square solution.

Choice of Parameters

For the step size appears in the μ-subproblem, we adopted an approximation of the Barzilai and Borwein approach [19] to set the step size t:

$$t = \frac{(dg_{\mu_{n+1}} - dg_{\mu_n})^T (\mu_{n+1} - \mu_n)}{(dg_{\mu_{n+1}} - dg_{\mu_n})^T (dg_{\mu_{n+1}} - dg_{\mu_n})} \tag{13}$$

The parameter α controls the conformality distortion of the deformation and we set $\alpha = 0.1$. The parameter β is responsible for the matching for the intensity similarity. We set $\beta = 1$. The penalty parameter σ is set to 10 which is large enough for all experiments we reported in the next session. The threshold ϵ is set to 0.05 for optimal solution.

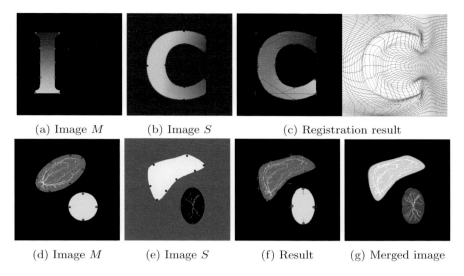

(a) Image M	(b) Image S	(c) Registration result

(d) Image M	(e) Image S	(f) Result	(g) Merged image

Fig. 1. Image registration of Example 1((a)–(c)) and Example 2((d)–(g)) respectively (Color figure online).

Image Fusion

Once we have the registered image $f(M)$, we can integrate the two images by using different well-established method. In this work, we adopted the Discrete wavelet transform (DWT) fusion method [12] to integrate the registered images. The main idea of this method is to merge the wavelet decomposition of the two images in the transformed domain. Since a larger absolute value of the coefficients correspond to some salient features in the images, important features on images can be integrated by the fusion of wavelet coefficients effectively.

5   Experimental Result

To validate the efficacy of our algorithm, we have tested it on both synthetic data together with the real medical data.

Example 1: In the first example, we validate our proposed algorithm by registering an image with English character "I" to an image with English character "C" with different modalities. Figure 1(a) and (b) show the moving image with character "I" and the static image with character "C" respectively. The red and blue dots on the images illustrate the prescribed landmark criteria $f(p_i) \rightarrow q_i$, $i = 1, \ldots, m$ in the registration. The green cross in (c) shows the original landmark position in (a), which indicates that a large displacement is present in this registration problem. By using our proposed algorithm, a large diffeomorphic transformation f superimposing the feature points and maximizing the intensity similarity is obtained. The registered result is shown in Fig. 1(c).

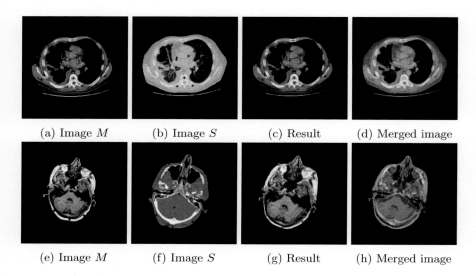

(a) Image M (b) Image S (c) Result   (d) Merged image

(e) Image M (f) Image S (g) Result   (h) Merged image

Fig. 2. Medical image registration and fusion of Example 3((a)–(d)) and Example 4((e)–(h)) respectively (Color figure online).

Example 2: Figure 1(d) and (e) shows the moving image M and S respectively. Synthetic blood vessels are planted in both images. The registered result is shown in (f), which is constructed by deforming the moving image M to superimpose S with maximum intensity similarity. The green cross indicates the original landmarks position in (d). The registered image and the static image S are then combined to have the fusion image (g). Note that the blood vessels from both images are integrated. In other words, local information from both images are merged together.

Example 3: We have also tested our algorithm with real medical images. Figure 2(a)–(d) shows an example of image fusion of a lung MRI image and a lung CT image. Figure 2(a) and (b) show the moving and static image M and S respectively. By the proposed algorithm, we obtained the registration result, which is shown in (c). By using the fusion technique discussed in Sect. 4, we have the integrated image (d). Note that the detail of blood vessels in image S and the detail of the heart in image M are both included in the integrated image. This shows that our proposed algorithm can effectively align the images with different modalities.

Example 4: In this example, we validate our proposed algorithm on the image registration between brain MRI and CT images. Figure 2(e) and (f) show the MRI and CT brain images respectively. The red and blue dots represent the artificial landmark constraints for testing purpose. We have set the parameter $\alpha = 1$ due to the highly dissimilarity of the intensity distribution. This prevents the perturbation of local deformation created by mismatching of intensity. In other words, conformality distortion contributes a larger portion in the energy

Table 1. Summary of the registration results using different registration methods

	$LM_{\max}$	$LM_{\mean}$	Landmark [14]		MI [16]		Proposed	
			$e_{\max}$ $\|\mu\|_\infty$	$e_{\mean}$ MI	$e_{\max}$ $\|\mu\|_\infty$	$e_{\mean}$ MI	$e_{\max}$ $\|\mu\|_\infty$	$e_{\mean}$ MI
Example 1	0.3309	0.1293	0 0.9950	0 0.5324	0.3078 0.9993	0.1140 0.1781	0 0.9775	0 1.0736
Example 2	0.0196	0.0057	0 0.7295	0 0.8088	0.0176 0.9061	0.0062 0.8311	0 0.8807	0 0.9712
Example 3	0.0023	0.0013	0 0.4126	0 1.3806	$2.42e^{-3}$ 0.4126	$1.73e^{-3}$ 1.4787	0 0.2220	0 1.4758
Example 4	0.0087	0.0027	0 0.4650	0 1.1929	0.0094 0.5243	0.0037 0.8670	0 0.4126	0 1.2025

functional under this setting and so the registration is less sensitive to the intensity difference. Figure 2(g) shows the registration result. It demonstrates that our proposed algorithm can align images with different modalities well, which is also illustrated by the image fusion result as shown in (h).

To validate our proposed algorithm, we compare the results obtained with landmark-based image registration method proposed in [14] and the multi-modality image registration in [16]. Quantitative measures are reported in Table 1. We first normalize the image domain into the $[0, 1] \times [0, 1]$ domain. $LM_{\max}$ and $LM_{\mean}$ represent the maximum and mean landmark displacement in the example respectively. $e_{\max}$ and $e_{\mean}$ measure the maximum and the mean error of the landmark alignment. Note that method in [16] is intensity-based and so landmark mismatching errors always exist. $\|\mu\|_\infty$ measures the maximum value of the conformality distortion. $\|\mu\|_\infty < 1$ indicates that the mapping is diffeomorphic. MI measures the mutual information between the target image and the deformed image. A larger MI value implies a more accurate image superimposition is achieved. The table shows that our proposed algorithm can accurately align landmarks exactly and maximize the mutual information during the registration.

6 Conclusion

In this work, an iterative scheme is proposed to register images with different modalities under prescribed landmark constraints. The main idea of our method is to find an optimized Beltrami coefficients for which the associated diffeomorphism satisfies the landmark constraints and superimposes the intensities in the sense of similarity of the images. We have applied our algorithm to register medical images with different modalities for image fusion. Experimental results show that our proposed algorithm can align images and match landmarks well, which is important for obtaining accurate image fusion.

Acknowledgements. This project is supported by HKRGC GRF (Project ID: 2130363 Reference: 402413)

References

1. Heckbert, P.S.: Survey of texture mapping. IEEE Comput. Graphics Appl. **6**, 56–67 (1986)
2. Sotiras, A., Davatzikos, C., Paragios, N.: Deformable medical image registration: a survey. IEEE Trans. Med. Imaging **32**, 1153–1190 (2013)
3. Zitova, B., Flusser, J.: Image registration methods: a survey. Image Vis. Comput. **21**, 977–1000 (2003)
4. Bookstein, F.L.: Principal warps: thin-plate splines and the decomposition of deformations. IEEE Trans. Pattern Anal. Mach. Intell. **11**, 567–585 (1989)
5. Joshi, S.C., Miller, M.I.: Landmark matching via large deformation diffeomorphisms. IEEE Trans. Image Process. **9**, 1357–1370 (2000)
6. Vercauteren, T., Pennec, X., Perchant, A., Ayache, N.: Diffeomorphic demons: efficient non-parametric image registration. NeuroImage **45**, S61–S72 (2009)
7. Thirion, J.P.: Image matching as a diffusion process: an analogy with Maxwell's demons. Med. Image Anal. **2**, 243–260 (1998)
8. Glocker, B., Sotiras, A., Komodakis, N., Paragios, N.: Deformable medical image registration: setting the state of the art with discrete methods. Annu. Rev. Biomed. Eng. **13**, 219–244 (2011)
9. Christensen, G.E., Johnson, H.J.: Consistent image registration. IEEE Trans. Med. Imaging **20**, 568–582 (2001)
10. Chanwimaluang, T., Fan, G., Fransen, S.R.: Hybrid retinal image registration. IEEE Trans. Inf. Technol. Biomed. **10**, 129–142 (2006)
11. James, A., Dasarathy, B.: Medical image fusion: a survey of the state of the art. Inf. Fusion **19**, 4–19 (2014)
12. Li, H., Manjunath, B., Mitra, S.: Multisensor image fusion using the wavelet transform. Graph. Models Image Process. **57**, 235–245 (1995)
13. Naidu, V., Raol, J.: Pixel-level image fusion using wavelets and principal component analysis. Def. Sci. J. **58**, 338–352 (2008)
14. Lam, K.C., Lui, L.M.: Landmark and intensity based registration with large deformations via quasi-conformal maps. SIAM J. Imaging Sci. **7**, 2364–2392 (2014)
15. Gardiner, F.P., Lakic, N.: Quasiconformal Teichmüller Theory. Mathematical Surveys and Monographs. American Mathematical Society, Providence (2000)
16. Kroon, D.: Multimodality non-rigid demon algorithm image registration. Robust Non-rigid Point Matching **14**, 120–126 (2008)
17. Astala, K., Iwaniec, T., Martin, G.: Elliptic Partial Differential Equations and Quasiconformal Mappings in the Plane. Oxford Graduate Texts in Mathematics. Princeton University Press, Princeton (2008)
18. Lui, L.M., Lam, K.C., Wong, T.W., Gu, X.F.: Texture map and video compression using Beltrami representation. SIAM J. Imaging Sci. **6**, 1880–1902 (2013)
19. Barzilai, J., Borwein, J.: Two-point step size gradient methods. IMA J. Numer. Anal. **8**, 141–148 (1988)

CINAPACT-Splines: A Family of Infinitely Smooth, Accurate and Compactly Supported Splines

Bita Akram, Usman R. Alim$^{(\boxtimes)}$, and Faramarz F. Samavati

Department of Computer Science, University of Calgary, Calgary AB, Canada
{bakram,ualim,samavati}@ucalgary.ca

Abstract. We introduce CINAPACT-splines, a class of C^∞, accurate and compactly supported splines. The integer translates of a CINAPACT-spline form a reconstruction space that can be tuned to achieve any order of accuracy. CINAPACT-splines resemble traditional B-splines in that higher orders of accuracy are achieved by successive convolutions with a B-spline of degree zero. Unlike B-splines however, the starting point for CINAPACT-splines is an infinitely smooth and compactly supported bump function that has been properly normalized so that it fulfills the partition of unity criterion. We use our construction to design two CINAPACT-splines, and explore their properties in the context of rendering volumetric data sampled on Cartesian grids. Our results show that CINAPACT-splines, while being infinitely smooth, are capable of providing similar reconstruction accuracy compared to some well-established filters of similar cost.

1 Introduction

Volume visualization is now a common way to explore three-dimensional images arising from modalities such as Computational Tomography (CT) or Magnetic Resonance Imaging (MRI). These images typically consist of arrays of scalar values that represent information gathered from regularly sampled volumes. The reconstruction of a continuous approximation from this discrete data is a fundamental operation in volume rendering. A careful choice of this reconstruction method can benefit the rendering process by providing a smooth, precise and efficient continuous representation of the data. As the primary goal of rendering is to provide a comprehensible image, accuracy is vital. Smoothness is beneficial for many rendering applications such as shading and feature line extraction for which higher order derivatives are required. Finally, rendering is a multi-step process dealing with large amounts of data. The cost of reconstruction therefore plays an important role in the efficiency of the visualization procedure.

In visual computing, kernel-based reconstruction methods — such as linear or cubic interpolation — are well-known. These methods place a modulated kernel at each of the sample locations to reconstruct values at arbitrary locations. The accuracy and smoothness of the reconstruction is therefore directly influenced

© Springer International Publishing Switzerland 2015
G. Bebis et al. (Eds.): ISVC 2015, Part I, LNCS 9474, pp. 819–829, 2015.
DOI: 10.1007/978-3-319-27857-5_73

by the kernel. Compactly supported piecewise polynomial splines are a popular choice due to their efficiency. The accuracy and support-size of these kernels are tightly knit in that more accurate reconstructions require kernels with wider support. Smoothness is achieved as a by-product, more accurate splines are usually composed of higher-degree polynomials. This has important implications for volume rendering where higher order derivatives are sometimes needed. When the reconstruction kernel does not possess sufficient smoothness, a digital derivative filter is employed to approximate the derivative at the sample locations, and then combined with the reconstruction kernel to approximate the derivative at arbitrary locations [1]. While this method ensures that the approximated derivative has the same level of smoothness as the reconstruction kernel, it adds a filtering overhead that adversely affects rendering performance.

In this paper, we introduce a novel family that consists of kernels that are explicitly constructed to be infinitely smooth (C^∞), but can be tuned to achieve any order of accuracy while maintaining compact support. We call these kernels CINAPACT-splines; they are a generalization of the recently proposed CIN-PACT (C^∞ and compactly supported) splines of Runions and Samavati [2] that are based on the Partition of Unity Parametrics (PUPs) framework [3]. CINAPACT-splines satisfy the partition of unity criterion and therefore guarantee first-order accuracy. CINAPACT-splines improve upon the accuracy using a convolution procedure that is similar to the procedure used in the construction of uniform B-splines. Like the B-splines, CINAPACT-splines also exhibit a trade-off between higher levels of accuracy and efficiency. Unlike the B-splines however, CINAPACT splines are infinitely smooth irrespective of the order of accuracy. Despite this advantage, our tests show that CINAPACT-splines possess approximation characteristics that are comparable to some of the best known kernels. This makes them ideal candidates for applications like volume rendering.

The remainder of the paper is organized as follows. Before reviewing some important prior art (Sect. 2) that has inspired the design of CINAPACT-splines, we summarize key concepts from signal processing that are needed in our construction scheme (Sect. 1.1). The details of our construction procedure and some concrete examples that of practical significance are presented in Sect. 3. Finally, Sect. 4 presents some rendering tests that show that CINAPACT-splines, while being infinitely smooth, yield highly accurate results as compared to some well-known polynomial-spline kernels.

1.1   Preliminaries

In signal processing, interpolation is a common way to reconstruct an approximation of a function from its samples. According to Shannon's sampling theorem, a band-limited univariate function can be perfectly recovered using an ideal *sinc* kernel, if and only if the sampling rate $T < \frac{\pi}{w_{max}}$, where w_{max} is the highest frequency in the band-limited function [4].

Aliasing: As most real-world functions are not band-limited, and applying ideal reconstruction kernels is not practical owing to their infinite support, aliasing issues occur. Pre-aliasing happens when the function is not band-limited or the sampling rate is not adequately high. Post-aliasing on the other hand, can occur when the reconstruction kernel is not ideal.

Generalized Sampling Theory: Developed by Blu and Unser [5], this theory extends Shannon's sampling theory by allowing a wider range of non-ideal kernels, ψ, while trying to minimize aliasing errors [6]. These kernels constitute an approximation space that is spanned by uniform shifts of a kernel. Approximated functions then belong to the following shift-invariant space:

$$V(\psi, T) = \left\{ \tilde{f}(x) = \sum_{n \in \mathbb{Z}} c_n \psi(x/T - n) : [c] \in l_2 \right\}. \tag{1}$$

Here, c_n is the nth component of the set of coefficients $[c]$, and the kernel $\psi \in L_2$.

Approximation Order and the Strang-Fix Conditions: The overall quality of reconstruction is affected by various factors such as the choice of the kernel ψ, and how the set of coefficients $[c]$ is determined from the available data. Approximation order is an effective tool for quantifying the quality of a reconstruction scheme. A reconstruction scheme has approximation order k if there exists a constant C such that [7]:

$$\|f - \tilde{f}\|_2 = CT^k \|f^{(k)}\|_2, \quad \text{as } T \to 0. \tag{2}$$

Here, $f^{(k)}$ is the kth derivative of the function to be approximated, and $\|\cdot\|_2$ denotes the L_2 norm. If a kernel ψ provides a kth order reconstruction, it is referred to as a kth order kernel. A kth order kernel ψ satisfies the following Fourier domain conditions (*Strang-Fix conditions*) [7]:

$$\hat{\psi}(0) = 1 \quad \text{and} \quad \hat{\psi}^{(\alpha)}(\beta) = 0 \text{ for } \alpha < k \text{ and } \beta \in \{2\pi l \mid l \in \mathbb{Z}\setminus\{0\}\}, \tag{3}$$

where $\hat{\psi}(\cdot)$ denotes the Fourier transform of ψ. In other words, ψ is a kth order kernel if and only if its Fourier transform and the derivatives of its Fourier transform up to order k vanish at integer multiples of 2π. The Fourier transform of the ideal kernel, i.e. the *sinc* function, is a rectangular pulse (or box function) which is completely localized; the ideal kernel therefore trivially satisfies the Strang-Fix conditions. Due to the uncertainty principle, the Fourier transform of a compactly supported kernel is not localized. The Strang-Fix conditions quantify how the transform decays to zero; the higher the approximation order, the faster the decay and the more accurate the approximation.

Partition of Unity: In order to ensure that a reconstruction scheme can approximate a given function with arbitrary accuracy, k must at least be one. Using the Poisson summation formula, it can be shown that the first approximation order ($k = 1$) implies that the integer translates of ψ form a partition of unity [5], i.e.

$$\sum_{n \in \mathbb{Z}} \psi(x - n) = 1, \quad \forall x \in \mathbb{R}. \tag{4}$$

Interpolation: Once a kth order kernel ψ has been chosen, one needs to ensure that the coefficients $[c]$ are determined in a way that respects the approximation order. In applications such as volume rendering, the function f is known through its idealized samples $[f]$ only. In this case, it is sufficient to ensure that the approximation $\tilde{f}$ interpolates the sample values, i.e. $\tilde{f}(Tn) = f_n$ [5]. As explained in Sect. 3.1, this is achieved through the application of a discrete interpolation pre-filter that is completely determined from the integer samples of ψ.

2   Related Work

Reconstruction kernels play an important role in the fields of data visualization and graphics. In the literature, a variety of kernels that satisfy different desirable properties such as simplicity, interpolation, smoothness, compact support, efficiency, and accuracy, is in use. There is usually a trade-off between different properties, and the nature of the application dictates the choice of the kernel. In this section, we look through some of the most common and useful designs of reconstruction kernels. We focus on univariate kernels with the understanding that they can be extended to higher dimensional integer grids via a simple tensor product extension. It should be noted that the design methodology behind CINAPACT splines is general and can also be applied to non-separable multivariate kernels such as radial basis functions [8] and box splines [9].

Piecewise polynomial splines are a good alternative to the ideal *sinc* function. They are built through junctions of polynomials pieced together at points called knots [10]. Uniform B-splines, introduced by Schoenberg [11], are well-known examples of this family. They have good reconstruction characteristics such as partition of unity, approximation order and compact support of $k+1$, and C^{k-1} continuity for polynomial degree k.

In 2001, Blu *et al.* introduced a family of compactly supported kernels with maximal order and minimal support (MOMS) [12]. Using their error kernel [13], they proved that the minimally-supported kernel of approximation order $k+1$ is piecewise-polynomial with degree k and support $k+1$. Uniform B-splines are a well-known member of the MOMS family and have the highest continuity among their peers [12]. Though all members of the MOMS family have maximal order for a specified support, the optimal MOMS (O-MOMS) subcategory optimizes the constant C in Eq. 2. However, these kernels have C^0 continuity and thus are not differentiable. While O-MOMS can provide us with high accuracy, we cannot neglect the importance of continuity in the quality of reconstruction [14].

Partition of Unity Parametrics (PUPs) is a flexible framework for meta-modeling introduced by Runions and Samavati [3]. In this framework, any function $R(x)$ can be employed after proper normalization to satisfy partition of unity. The general form of PUPs is as follows:

$$\psi(x) = \frac{R(x)}{\sum_{j\in\mathbb{Z}} R(x-j)}, \text{ where } \sum_{j\in\mathbb{Z}} R(x-j) \neq 0. \tag{5}$$

Partition of unity also ensures a first-order approximation which, according to the Strang-Fix theory, implies a linear relationship between error reduction and sampling rate increase. Satisfying partition of unity is therefore a necessary criterion for a kernel. Runions and Samavati suggested B-spline based PUPs and employed it in the reconstruction of curves and surfaces, and also demonstrated their advantages in feature sketching and converting planar meshes into parametric surfaces [3]. In a follow-up work, they used the PUPs framework to design a new family of C^∞ and compactly supported kernels called CINPACT-splines [2]. These kernels are obtained by truncating an exponential function to preserve its smoothness while providing compact support. The resulting *bump* function is then normalized to satisfy partition of unity. Moreover, this kernel can be designed to interpolate the tangents of curves and surfaces as well as their sample points. The bump function used in the design of CINPACT-splines is defined as:

$$R(x) := \begin{cases} \exp(\frac{-kx^2}{c^2-x^2}), & x \in (-c,c), \\ 0, & \text{otherwise.} \end{cases} \qquad (6)$$

The support of $R(x)$ is $2c$, and k is a continuous parameter that behaves like the polynomial degree of a B-spline if chosen correctly [2]. We therefore refer to k as the degree of the CINPACT-spline. Thus, CINPACT-splines can be considered to be a potential replacement for B-splines.

Though PUPs can be designed to serve as non-separable filters, their application in providing a framework for designing separable filters applicable to regular grids is the focus of our research. As demonstrated by some of the works described above, several properties that were believed to be strongly related, are in fact independent. For instance, Runions and Samavati have shown that smoothness and support can vary independently [2,3]. The interrelation between approximation order and level of continuity was also relaxed by the introduction of O-MOMS [12]. Drawing inspiration from these studies, the question is if we can design a filter that is infinitely smooth and has arbitrary approximation order, while preserving other beneficial properties such as compact support.

3 C^∞, Accurate and Compact Splines

CINPACT-splines provide two of the most important features for rendering applications: infinite smoothness and compact support. However, partition of unity only guarantees a first-order approximation. We need a technique to increase the approximation order while preserving infinite smoothness and compact support.

Using the Fourier convolution theorem and the Strang-Fix conditions, it can be readily shown that the convolution of two kernels having approximation orders a and b, yields a kernel with approximation order $a + b$. Hence, convolving a CINPACT-spline with the box-function (B-spline of degree zero and order one) increases its approximation order by one, and yields a *CINAPACT* spline with a minimum guaranteed order of two. Thus, we have

$$\psi^2(x) := \frac{R(x)}{\sum_{j \in \mathbb{Z}} R(x-j)} * \beta^0(x), \tag{7}$$

where $R(x)$ is the bump function introduced in Eq. 6, $\beta^0(x)$ denotes the B-spline of degree zero, and the symbol '*' indicates the continuous convolution operation. In order to further improve the approximation order, the initial CINPACT-spline is successively convolved with the box-function. The CINAPACT-spline with a minimum guaranteed order of L is constructed through the convolution of the normalized bump-function with the B-spline of degree $L-2$:

$$\psi^L(x) := \frac{R(x)}{\sum_{j \in \mathbb{Z}} R(x-j)} * \beta^{L-2}(x), \text{ where } L \geq 2. \tag{8}$$

Each convolution adds one unit to the support and increases the approximation order by one. Although the accuracy of CINAPACT-splines can increase arbitrarily, the cost of reconstruction increases exponentially in higher dimensions as the support size grows.

The infinite smoothness of CINPACT-splines is inherited by CINAPACT-splines through the convolution. The mth derivative of the CINAPACT spline $\psi^L(x)$ is given by

$$\psi^{L(m)}(x) = \left(\frac{R(x)}{\sum_j R(x-j)} * \beta^{L-2}(x) \right)^{(m)} = \left(\frac{R(x)}{\sum_j R(x-j)} \right)^{(m)} * \beta^{L-2}(x). \tag{9}$$

One difficulty that arises is due to the nature of the convolution integral in Eq. 8. We are currently unaware of an analytical closed form for this integral. We therefore use a quadrature method to table the values of CINAPACT splines at closely sampled locations within their support. The derivatives are also computed in a similar manner using Eq. 9. Figure 1 shows a second-order CINAPACT spline of support size four along with its first and second derivatives.

3.1   Interpolation

Similar to B-splines, the kernels within the CINAPACT-spline family are not interpolative. Recall that a kernel ψ is interpolative if and only if $\psi(x)$ vanishes at all non-zero integers. For an interpolative kernel, one can simply use the sample values $[f]$ as coefficients in Eq. 1. When a kernel is not interpolative, a discrete pre-filtering operation is necessary to ensure that the approximation $\tilde{f}(x)$ exactly reproduces the sample values. The resulting approximation (for $T = 1$) is given by

$$\tilde{f}(x) = \sum_{n \in \mathbb{Z}} c_n \psi(x-n), \text{ where } [c] = [\psi]^{-1} \otimes [f]. \tag{10}$$

Here, $[\psi]$ is a discrete filter that consists of the integer samples of ψ, i.e. $\psi_n = \psi(n)$, and $[\psi]^{-1}$ is its inverse. The symbol '$\otimes$' denotes the discrete convolution operation. In practice, for finite dimensional data, the discrete convolution in the above equation can be applied in the Fourier domain using the discrete Fourier transform which imposes periodic boundary conditions. Other types of boundary conditions can also be imposed by suitably padding the data vector $[f]$.

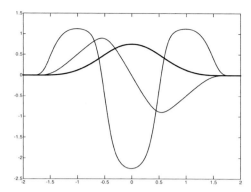

Fig. 1. The CINAPACT-spline $\psi^2(x)$ ($c = 1.5$, $k = 4$) and its first two derivatives.

3.2 Cardinal Kernel Optimization

The interpolative pre-filter and the reconstruction kernel in Eq. 10 can be combined into an interpolative *cardinal kernel* ψ_{int} so that

$$\tilde{f}(x) = \sum_{n \in \mathbb{Z}} f_n \psi_{\text{int}}(x - n), \text{ where } \psi_{\text{int}}(x) := \sum_{j \in \mathbb{Z}} \psi_j^{-1} \psi(x - j). \tag{11}$$

This interpolative scheme respects the overall approximation order of ψ. By inspecting the behaviour of the Fourier transform of the cardinal kernel ψ_{int}, one can also reason about the overall quality of a reconstruction scheme. Using Eq. 11, we infer that the Fourier transform of the cardinal kernel is

$$\hat{\psi}_{\text{int}}(\omega) = \hat{\psi}(\omega)/\widehat{[\psi]}(\omega), \tag{12}$$

where $\widehat{[\psi]}(\omega)$ is the discrete time Fourier transform of the filter $[\psi]$. We use Eq. 12 to tune the free parameters of CINAPACT-splines as explained below. Since we are unaware of a closed form Fourier transform of the CINAPACT-splines, we use the fast Fourier transform to approximate it.

Examples: We present two concrete examples of CINAPACT-splines that can be tuned to yield cardinal spectra that closely match the spectra of the cubic B-spline and the cubic O-MOMS. In order to ensure that the kernels are comparable, we fix the support size to four using the following two procedures.

1. $\psi_k^2(x)$: We set $c = 1.5$ and $L = 2$ in Eq. 8. The resulting CINAPACT-spline has a minimum approximation order of two.
2. $\psi_k^3(x)$: We set $c = 1$ and $L = 3$ in Eq. 8. The resulting CINAPACT-spline has a minimum approximation order of three.

In the above constructions, the degree $k \in \mathbb{R}$ is a free parameter which allows us to adjust the cardinal kernels' spectra. Figure 2a and b show the cardinal spectra

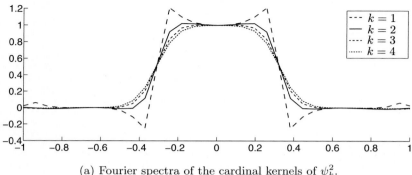

(a) Fourier spectra of the cardinal kernels of ψ_k^2.

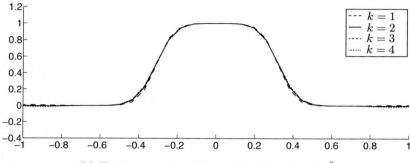

(b) Fourier spectra of the cardinal kernels of ψ_k^3.

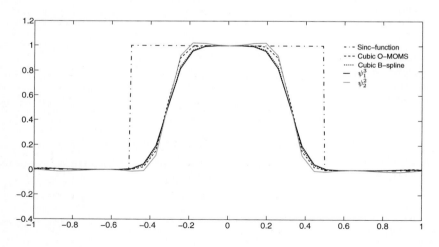

(c) Comparison of cardinal kernels in the Fourier domain.

Fig. 2. Cardinal kernels of the CINAPACT-splines ψ_2^2 and ψ_1^3 as compared to the cubic B-spline and the cubic O-MOMS.

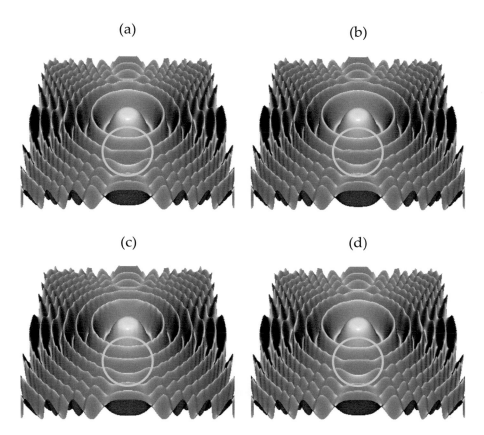

Fig. 3. Isosurface renderings of the sampled ML function using (a) cubic B-spline, (b) cubic O-MOMS, (c) ψ_2^2, and (d) ψ_1^3 (Color figure online).

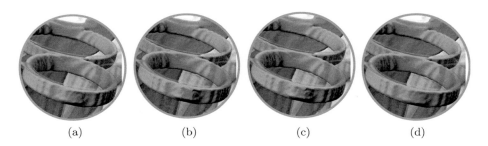

Fig. 4. Direct volume rendering results for the engine dataset using (a) cubic B-spline, (b) cubic O-MOMS, (c) ψ_2^2, and (d) ψ_1^3.

of ψ_k^2 and ψ_k^3 for various values of k. Observe that the CINAPACT-splines ψ_k^2 are much more sensitive to the degree parameter as compared to ψ_k^3 due to the lower approximation order. Among the values tested for ψ_k^2, $k = 2$ yields the best spectrum which is compared with the cardinal spectra of ψ_1^3, the cubic B-spline and the cubic O-MOMS in Fig. 2c. Observe that the cardinal spectrum of ψ_1^3 has the same behaviour as the cubic B-spline even though it has a lower guaranteed approximation order. More interestingly, the cardinal spectrum of ψ_2^2 — despite the lower approximation order — follows the O-MOMS' cardinal spectrum with some overshooting. This suggests that, with careful tuning, CINAPACT-splines can exhibit higher approximation orders than what is guaranteed; both ψ_2^2 and ψ_3^3 resemble fourth-order kernels despite their lower orders.

4   Volume Rendering Tests

We evaluated the accuracy of our proposed kernels ψ_2^2 and ψ_1^3 by rendering the 0.5 isosurface of the synthetic function suggested by Marschner and Lobb (ML) [15], and compared the results to the cubic O-MOMS and B-spline. All of these kernels have the same support, and hence the same reconstruction cost. In order to ensure that smoothness related artifacts are clear, we used the analytic first derivatives of the kernels when computing the gradient for shading. The results are shown in Fig. 3. At first glance, all the renditions look pretty similar. This is due to the fact that all four schemes behave similarly in the low-pass regime as is evident from the plot in Fig. 2c. On closer inspection, we can see subtle differences between the renditions; the cubic O-MOMS (Fig. 3b) seems to be slightly better at reconstructing the inner rings of the isosurface as compared to the cubic B-spline (Fig. 3a). However, it is also noticeably less smooth as indicated by the red circle. The rendition provided by ψ_2^2 (Fig. 3c) exhibits slightly higher undulations in the reproduction of the rings. This is owing to the overshoot of the cardinal Fourier spectrum (Fig. 2c). As predicted, ψ_1^3 (Fig. 3d) yields a rendition that is very close to the cubic B-spline. Observing the encircled regions, we see that the ψ_1^3 rendition is also slightly smoother.

When testing these four schemes in the context of direct volume rendering (DVR), we observed a similar trend. Figure 4 shows DVR images obtained from the engine CT dataset. As predicted by our Fourier analysis, the cubic B-spline and ψ_1^3 renditions (Fig. 4a and d) resemble each other, and the cubic O-MOMS are ψ_2^2 renditions (Fig. 4b and c) are very close to each other.

These results corroborate the fact that both ψ_2^2 and ψ_1^3 have higher approximation orders than anticipated, and practically behave like fourth-order B-spline kernels. It appears that the choice of the parameters c and k may be more influential than the B-spline degree (L) in determining the exact approximation order. This topic warrants further investigation.

5   Conclusion

We proposed a construction scheme that generalizes CINPACT-splines [2] to CINAPACT-splines, kernels that can be tuned to achieve any order of accuracy while maintaining infinite smoothness. We presented two examples of CINAPACT-splines: ψ_2^2 and ψ_1^3, that behave like fourth-order kernels. We also presented some preliminary results that suggest that, in the context of volume visualization, CINAPACT-splines may be a good alternative to B-splines due to their infinite smoothness. In future, we plan to investigate the smoothness advantage of CINAPACT-splines in more detail; potential applications include shading and feature-line extraction in a real-time volume rendering environment.

References

1. Hossain, Z., Alim, U.R., Möller, T.: Toward high quality gradient estimation on regular lattices. IEEE Trans. Visual. Comput. Graph. **17**, 426–439 (2011)
2. Runions, A., Samavati, F.: CINPACT-splines: a class of C-infinity curves with compact support. In: Boissonnat, J.-D., Cohen, A., Gibaru, O., Gout, C., Lyche, T., Mazure, M.-L., Schumaker, L.L. (eds.) Curves and Surfaces. LNCS, vol. 9213, pp. 384–398. Springer, Heidelberg (2015)
3. Runions, A., Samavati, F.F.: Partition of unity parametrics: a framework for meta-modeling. Visual Comput. **27**, 495–505 (2011)
4. Shannon, C.E.: Communication in the presence of noise. Proc. IRE **37**, 10–21 (1949)
5. Unser, M.: Sampling-50 years after shannon. Proc. IEEE **88**, 569–587 (2000)
6. Nehab, D., Hoppe, H.: A fresh look at generalized sampling. Found. Trends Comput. Graph. Vis. **8**, 1–84 (2014)
7. Strang, W., Fix, G.: An Analysis of the Finite Element Method. Prentice-Hall, Englewood Cliffs (1973). Prentice-Hall series in automatic computation
8. Buhmann, M.: Radial Basis Functions: Theory and Implementations. Cambridge University Press, New York (2003). Cambridge Monographs on Applied and Computational Mathematics
9. de Boor, C., Höllig, K., Riemenschneider, S.D.: Box splines, vol. 98. Springer, New York (1993)
10. Unser, M.: Splines: a perfect fit for signal and image processing. IEEE Signal Process. Mag. **16**, 22–38 (1999)
11. Schönberg, I.J.: Contributions to the problem of approximation of equidistant data by analytic functions. Quart. Appl. Math **4**, 45–99 (1946)
12. Blu, T., Thévenaz, P., Unser, M.: Moms: maximal-order interpolation of minimal support. IEEE Trans. Image Process. **10**, 1069–1080 (2001)
13. Blu, T., Unser, M.: Quantitative fourier analysis of approximation techniques. i. interpolators and projectors. IEEE Trans. Signal Process. **47**, 2783–2795 (1999)
14. Kindlmann, G., Whitaker, R., Tasdizen, T., Möller, T.: Curvature-based transfer functions for direct volume rendering: methods and applications. In: Visualization 2003, pp. 513–520. IEEE Computer Society Press (2003)
15. Marschner, S.R., Lobb, R.J.: An evaluation of reconstruction filters for volume rendering. In: Visualization 1994, pp. 100–107. IEEE Computer Society Press (1994)

Vis3D+: An Integrated System for GPU-Accelerated Volume Image Processing and Rendering

I. Nisar and T. McInerney(✉)

Department of Computer Science, Ryerson University,
Toronto, ON M5B 2K3, Canada
tmcinern@ryerson.ca

Abstract. This paper presents a prototype 3D image processing and rendering system that extends an existing interactive 3D image visualization system. The extensions consist of software modules implemented using graphics processing unit (GPU) programs known as "compute shaders". Compute shaders are able to utilize the massively parallel, general-purpose computing capabilities provided by modern GPUs and can also be tightly integrated as new stages in a GPU-based volume and surface rendering pipeline. The compute shaders in this paper are designed to support the execution of volume image processing algorithms, as well as to support the interactive editing of the algorithms' output. An example volume image processing algorithm known as level set segmentation is implemented and demonstrated. A new editing module is developed that enables user modification of the segmentation algorithm's output by extending a pre-existing volume "painting" interface.

1 Introduction

The general purpose computing capability of GPUs, known as GPGPU, is well-suited for efficient processing of medical volume images. Volume images are typically represented as a 3D grid of *voxels* (i.e. volume elements), and many volume image processing algorithms are data-parallel in nature, requiring repeated operations on individual voxels or on a small local neighborhood of voxels. Modern volume visualization systems provide real-time volume rendering typically by programming the fragment shader stage of the GPU-based rendering pipeline to accommodate the 3D grid structure of a volume image. However, few of these systems are able to integrate general volume image processing algorithms, such as image filtering and segmentation, into this GPU pipeline in a flexible and modular manner. Recently, a new programmable stage of the OpenGL [1] pipeline has been made available on graphics hardware called a "compute shader". Compute shaders can execute general purpose numerical calculations and can be inserted into various stages of the rendering pipeline.

In this paper we describe extensions to our existing interactive volume visualization system [2]. The extensions consist of volume image processing capabilities which are added to the system in a tightly integrated yet modular fashion.

© Springer International Publishing Switzerland 2015
G. Bebis et al. (Eds.): ISVC 2015, Part I, LNCS 9474, pp. 830–841, 2015.
DOI: 10.1007/978-3-319-27857-5_74

Our existing system is based on a well-known open source visualization framework called *ImageVis3D* [3], and combines both volume rendering of 3D medical images with surface rendering of polygonal meshes, and allows users to define 3D regions within the volume, delineated by surface envelopes, using an intuitive "volume painting" style interface. The volume image processing extensions are implemented using general compute shader modules. In this paper, we describe and demonstrate example volume image processing compute shaders for performing image filtering and segmentation, both necessary components for visualizing noisy volume images and for performing image analysis. We have also extended the existing system's front-end volume painting interface. This extension supports the editing of labelled 3D regions outputted by the segmentation algorithm (or other volume image processing modules). Finally, we also add a user-controllable 3D image slice that is rendered together with the volume, allowing the user to edit 3D regions in a slice-by-slice manner and providing precise painting and editing control in noisy volume images.

2   Related Work

Graphics hardware is typically structured such that the rendering process is executed in a staged pipeline fashion and many of the stages are now programmable using a high-level programming language. These programs are commonly referred to as "shaders" and examples are vertex, geometry and fragment shaders. The graphics hardware has quickly evolved resulting in a "unified" shader architecture that provides one large grid of general data-parallel floating-point processors that can be used by the various stages. This hardware advance coincided with the emergence of general purpose computing (GPGPU) on the graphics card, along with API's to create GPGPU programs such as CUDA [4]. It is possible to mix CUDA programs and OpenGL programs in several steps including mapping and unmapping of a buffer into formats understood by CUDA and by OpenGL. As mentioned, compute shaders are a recently released stage of the OpenGL graphics pipeline that not only provide similar general-purpose computation functionality as that of CUDA, but also can be more tightly and seamlessly integrated into the pipeline.

As volume images continue to grow in size due to advances in scanning technology, highly efficient image processing algorithms that can filter and label a 3D image are becomingly increasingly important. Examples of processing algorithms that have been implemented on the GPU include median filtering [5], an implementation of Canny edge detection [6], nonlinear anisotropic diffusion-based 3D image denoising using CUDA [7], and level set segmentation [8]. For a recent and thorough survey of medical volume image processing on the GPU, the reader is referred to Eklund et al. [9].

3   Vis3D+

In this section we describe the various GPU-based modules that we integrated to extend the existing interactive volume visualization system. The new modules consist of the following: compute shaders providing basic volume image filtering in the form of Gaussian smoothing and edge detection, a compute shader that implements a variant of the level set segmentation algorithm [10], and compute shaders and modifications to an existing volume rendering fragment shader to extend the existing system's 3D ROI painting mechanism for use in ROI editing.

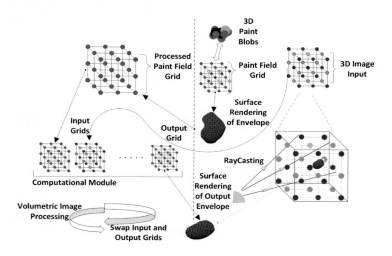

Fig. 1. High-level overview of the extended framework.

Volume processing algorithms, such as filtering, can be inserted into the processing pipeline of the extended system as long as they are parallelizable at the 3D grid point level. Furthermore, the filters can be cascaded - the output of one filtering stage can be used as input to the next. In this paper we implemented Gaussian Filtering to smooth a volume image and edge detection using a simple image gradient magnitude calculation. The output of these cascaded image filtering stages are used as input to a level set segmentation algorithm. We have chosen level set segmentation [10] to showcase our GPU-based volume image-processing support because it can segment topologically complex objects, is a highly parallelizable algorithm and fits well with the existing 3D painting interface. We have used Shader Storage Buffer Objects [1] as buffers, using their binding points as input or output hooks, allowing different shaders to pick up the same buffer and process them as they see fit, avoiding copying or moving data around on the GPU.

Figure 1 shows the extended system, with the left side of the vertical dotted line showing the extensions added in this paper. A compute shader module (Fig. 1 lower left) accepts a volume image grid and algorithm global parameters as input, iteratively executes the algorithm, and generates an output grid which is then used for volume rendering, if desired. In the extended system, the 3D painting interface can be used to define 3D regions of interest as an optional input to any volume image-processing algorithm. The upper middle and upper left part of Fig. 1 shows an optional input grid generated from the result of painting a 3D ROI. More complex compute shader programs, such as the level set segmentation compute shader, can make use of this input. In the case of the level set segmentation algorithm for example, the algorithm refines the 3D region and labels this region in an output grid (Fig. 1 lower left). The output grid can then be optionally input to a geometry shader where a Marching Cubes algorithm will generate a boundary surface representation of the labeled 3D region.

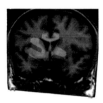

Fig. 2. Left: paint brush "tip" blob, rendered opaque here, can slide along a 3D slice clip plane. Right: applying paint strokes to a slice plane (blob envelope is transparent).

We have extended the system's original 3D painting interface to support painting a 3D ROI on a 3D user-orientable image slice plane. We alter the volume ray casting algorithm in the fragment shader to generate a 3D slice plane that is rendered along with the volumetric data during the volume rendering stage (Fig. 2), achieving the effect of a clipped volume rendering of the data. The volume ray casting is altered by computing the start of each ray from where it intersects the clip plane. These starting ray sample points ensure that everything in front of the plane is clipped away. The starting ray sample points are then used to look up the corresponding image intensity value in the volume image via interpolation. These volume image samples are mapped, using a simple transfer function, to a color and an opacity value and each sample corresponds to a fragment which will appear as a screen pixel. The fragments are shaded using the normal vector of the clip plane rather than a normal vector computed from the volume image. If the fragments are mapped to an opacity equal to 1.0, the ray casting algorithm is terminated for this ray; otherwise, the ray casting algorithm continues as usual.

It is often not possible to use a TF to isolate and volume render a target structure in noisy volumes, preventing the direct painting on the surface of the structure to create a surrounding envelope. This situation also occurs when the target structure is adjacent or connected to neighboring structures with similar intensity characteristics. In these cases, the user can use the image slice-plane painting approach, along with "flattened" superellipsoid paint blobs (Fig. 2 left), to define a 3D ROI that envelopes a cross-section of a target anatomical structure. The thickness of the flattened paint blobs can be precisely controlled by the user to range from a single slice thickness to many slices thick. The user can paint thick envelopes (i.e. several image slices thick) on several cross-sectional slices of the target structure such that these slice-painted envelopes overlap. The slice-painted envelopes are automatically blended to form a single envelope tightly bounding the entire target structure.

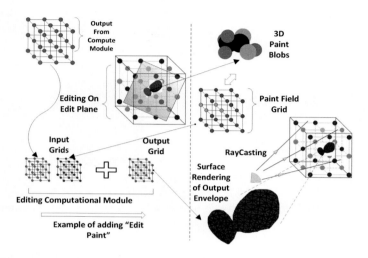

Fig. 3. High-level overview of editing a labelled 3D region via 3D slice paint painting.

Some image processing algorithms, including the level set segmentation algorithm, generate 3D grids with labeled regions. In our extended system, these labeled regions can be interactively edited by using the painting interface to erase parts of the region or to add "edit paint" to them. An illustration of this process is shown in Fig. 3. The left side of the dotted line depicts the slice plane painting and editing extensions. The volume image-processing algorithm for erasing and adding of labels to the voxels is described in Sect. 4.4. The appearance of the voxels within the labeled region can be controlled with a separate transfer function, allowing these voxels to be volume rendered using a distinctive "highlight" color. The altered volume ray-casting algorithm shades the color of voxels that are on the slice plane using the normal vector of the slice plane. The color of the labeled voxels is a blend of the highlight color and the color of the voxel intensity value assigned via the transfer function. The result is the labeled region voxels appear as semi-transparent and highlighted 2D "paint" (Fig. 2 right). Furthermore, the user can dynamically set the flattened superellipsoid paint blob thickness to be just thicker than a single slice plane and the surface of the paint blob can be made completely transparent. Any voxels on the slice plane that are inside the blob can therefore also be made to appear as semi-transparent and highlighted 2D paint (Fig. 2 right). Thus, this special slice-plane rendering capability gives the user the illusion of erasing and adding 2D paint to the labeled 3D region. The user visually discerns the boundaries of the target structure underneath the semi-transparent 2D paint of both the labeled region and the paint blob. Corrections can be made to the labeled region on the current slice plane. The user can then continue to another oriented slice plane to make further corrections.

4   Implementation

This section provides an overview of the extended system implementation. Details can be found in [11].

4.1   Initial Level Set Construction

In the level set segmentation method, a surface (in 3D) is defined to be the zero level set of a continuous function, $\phi(t, x, y, z)$. The movement of the level set surface is governed by an evolution equation of this function. On a computer, the level set function ϕ is sampled at points on a regular 3D grid and is referred to as the ϕ-grid. Typically the ϕ-grid dimensions are set equal to the input volume image dimensions. The level set method is initialized by creating an initial surface envelope that either loosely surrounds the target structure or is contained inside it [10]. We use the user-painted surface envelope to construct the initial level set function ϕ_0. Specifically, to initialize the ϕ_0-grid, at each grid point we determine if it is inside, outside, or on the painted envelope boundary using the blended paint blobs' defining implicit function [12]. From Li et al. [10] the initial level set function, ϕ_0, is defined at each grid point of a 3D grid as:

$$\phi_0(x, y, z) = \begin{Bmatrix} -\rho & (x, y, z) \in \Omega_0 - \partial\Omega_0 \\ 0 & (x, y, z) \in \partial\Omega_0 \\ \rho & (x, y, z) \in \Omega - \Omega_0, \end{Bmatrix}, \qquad (1)$$

where Ω is the volumetric domain, Ω_0 is a subset of the volumetric domain containing all points inside the painted envelope and $\partial\Omega_0$ is the set of all points exactly on the boundary of Ω_0 (i.e. the painted envelope boundary surface). Values inside the painted envelope are marked as $-\rho$ and outside are marked as $+\rho$, where ρ is a constant [10]. This initial level set function construction process is implemented in a separate compute shader, which accepts the array of paint blobs defining the painted envelope as input and writes out a 3D ϕ-grid.

4.2   Level Set Implementation

The level set segmentation algorithm uses edge image features to determine if the evolving level set surface has reached the boundary of the target structure. Input volume images are commonly convolved with a smoothing filter to remove noise before performing edge detection. We currently use two cascaded compute shaders - a Gaussian filter shader and an edge detection shader - to compute the edge detected image. The level set evolution is computed inside another compute shader, *updatephi.cs*. It takes an input ϕ-grid, along with the edge detected intensity grid. It then outputs a grid containing updated values of the function ϕ referred to as ϕ-gridOut. As mentioned previously, we use buffer binding indices as hooks to interchange the input and output ϕ-grid buffers, which avoids copying or moving the grids. The final output of the level set segmentation algorithm, ϕ-gridOut, contains scalar values. This output grid is sent to a geometry shader that executes the Marching Cubes algorithm and generates a mesh of triangles representing the zero level set surface.

4.3   Compute Shader Implementation

Shader Buffer Objects containing 3D grids of scalar values are stored as a contiguous one-dimensional array on the GPU. In a GPU thread, we are able to use the thread identifier to compute a 3D grid point position. Individual threads are grouped in a local work-group or block. The local work-groups together make up the larger global work-group. Shader Buffer Objects are stored in the GPU's $L2$ Cache and each thread looks up its required data from textures that are backed by these buffer objects.

In Kuo et al. [13], the authors mention several criteria for stopping the level set evolution, which is typically evaluated after each iteration of the segmentation algorithm. We use a simple but often effective stopping condition. The volume of the segmented 3D region is denoted V and the difference in the volume between the previous and current iteration is denoted ΔV. The evolution is stopped when $\Delta V/V$ falls below a small threshold value (e.g. 0.005). We have utilized atomic operations in the compute shader threads, supported as of OpenGL 4.3, to implement the simple stopping condition. Atomic operations write to (or read from) shared memory uninterrupted; if multiple threads attempt to access the same location simultaneously, they will be serialized. We use atomic operations that operate on a special stopping condition buffer that stores the previous and current computed volume of the segmented region, as well as the number of grid points that have been processed. When a thread begins executing at a current time step t (i.e. iteration), it checks the stopping condition ($\Delta V/V < 0.005$). If the condition is met, the thread returns. Otherwise the thread uses an atomic add operation to add 1 to the number of processed grid points. The thread then executes the level set evolution equation for its assigned grid point. If the ϕ function field value for this grid point is less than 0, we use the atomic add operation to add 1 to the current segmented region volume; that is, the number of voxels (i.e. grid points) inside the segmented region is used as a measure of the region's volume. When all grid points have been processed, we set the previous segmented region volume equal to the current volume and then reset the current volume and number of processed grid points to 0.

4.4   Editing a Labeled 3D Region

The user uses the painting interface to erase or add labels to the labeled region outputted from the segmentation algorithm in the form of the ϕ-grid. Edit compute shaders accept the ϕ-grid as input as well as the array of paint blobs defining the edit region. In Museth et al. [14], the authors define editing operators based on level sets, which provide advantages such as avoiding boundary surface self-intersection and coping with topological genus changes. In this paper, we use a simple version of the Constructive Solid Geometry (CSG) operations mentioned in Museth et al. [14]. A remove or erase operation is analogous to the difference operation in CSG and an add operation is equivalent to a union operation. Both the erase and add operations are parallelized on the ϕ-grid points. For the add operation, at each ϕ-grid point, we read its field value and store it at a corresponding output grid point. Using the array of painted blobs and the blended blobs' inside-outside function, we then check if this output grid point is inside the painted region. If so, we overwrite the field value from the ϕ-grid, with a $-\rho$ value, making it a part of the labeled region. For the erase operation, at each ϕ-grid point, we read its field value and set the corresponding output grid point to the same value. We then check if this output grid point is inside the painted region and if its value is less than 0 (i.e. indicating it is currently part of the labeled region). If so, it is overwritten with a $+\rho$ value, removing it from the labeled region.

5   Experimental Results

The paper is concerned with the modular integration of volume image processing algorithms into an existing volume image rendering pipeline, and with the algorithms' compute shader implementation. Consequently, in this section we present experiments to demonstrate a working compute shader implementation of the level set segmentation algorithm and some rough measurement of its performance, rather than on a formal analysis of segmentation accuracy and efficiency. Level set segmentation has been heavily researched over the years and its accuracy measured numerous times. The reader is referred to [15] as a representative example. In this paper, we currently use the system's 3D slice plane capability and visually inspect slices containing the target anatomical structures as well as the segmentation "paint" to assess segmentation accuracy. In addition, our level set segmentation implementation currently utilizes simple Gaussian smoothed gradient magnitude edges to stop the level set evolution. More accurate edges may lead to improved segmentation performance. We are currently investigating GPU-based median filtering [5] combined with more sophisticated edge detectors [6] and we plan to collect quantitative data of segmentation accuracy in the immediate future. Finally, all experiments were performed on a machine with an NVIDIA GTX 570M graphics card containing 1.5 GB GDDR5 of random access memory (RAM) and 7 streaming multiprocessors, each with 48 cores.

5.1   Segmenting Synthetic Data Sets

We use two synthetic "cloverleaf" volume images, with voxel values inside the cloverleaf smoothly graded and background voxels set to 0, and with dimensions $128 \times 128 \times 128$ and $256 \times 256 \times 256$, respectively. We painted an initial envelope

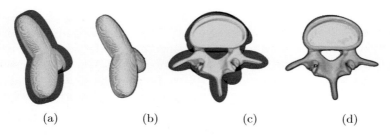

(a)   (b)   (c)   (d)

Fig. 4. Segmenting a $256 \times 256 \times 256$ synthetic cloverleaf data set. (a) Initial painted level set surface. (b) Final segmentation result. (c)(d) Segmenting a $168 \times 160 \times 92$ CT image of a human vertebra phantom. (c) Initial painted level set surface. (d) Final segmentation result.

surrounding the cloverleaf directly in 3D using three paint brush strokes (Fig. 4a). Figure 4b shows the segmentation after 100 iterations. The time required for the segmentation was 39 s. The result is visually very accurate. For the smaller data set (i.e. $128 \times 128 \times 128$ voxels) only 50 iterations (2.4 s) were required to generate an accurate result. The parameter settings for all tests, real and synthetic were $\mu = 0.02$, $\gamma = 5$, $\lambda = 5$, $\epsilon = 1.5$, $\tau = 5$. See Li et al. [10] for parameter descriptions.

In a second test, we use a $168 \times 160 \times 92$ CT volume image of a topologically complex human vertebra phantom (Fig. 4c and d) to further validate a working compute shader implementation of the segmentation algorithm. We painted an initial envelope without holes, directly in 3D, that surrounds the vertebra. The level set segmentation algorithm correctly captures the topology of the vertebra. The segmentation ran for 200 iterations. The approximate time required for the segmentation was 11 s and the result again is visually very accurate.

5.2   Segmenting a Real Data Set

We also ran our segmentation algorithm on a $240 \times 240 \times 192$ MRI brain volume to segment the lateral ventricle. Segmentation of structures in MRI scans is challenging due to noise, voxel intensity inhomogeneity and the similar voxel intensities of neighboring structures. We used the system's 3D slice painting facility and painted an envelope on several slices containing the lateral ventricle. Figure 5a–c shows the initial level set envelope, the segmentation result and an expert manually segmented guide. The 3D slice painting facility allows the user to quickly paint a highly accurate initial level set (Fig. 5a) - an important factor for segmentation algorithm robustness. The segmentation ran for 75 iterations and required approximately 14 s. Figure 5d–f shows several 3D slice views to demonstrate segmentation accuracy.

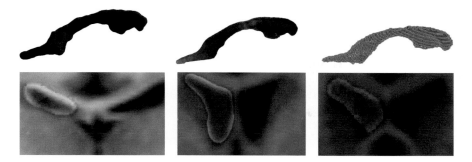

Fig. 5. Segmenting the lateral ventricle in a $240 \times 240 \times 192$ MRI brain image. Top row left: initial 3D slice painted level set surface, middle: final segmentation result, right: volume rendering of manually segmented ventricle. Bottom row: example image slices showing segmentation result.

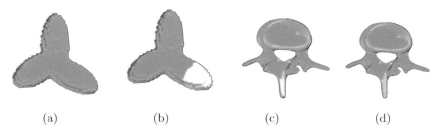

(a)   (b)   (c)   (d)

Fig. 6. (a)(b) Example of erasing paint on a synthetic clover leaf and, (c)(d) adding paint to the vertebrae segmentation.

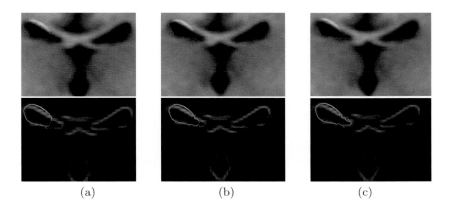

(a)   (b)   (c)

Fig. 7. Editing a 3D image slice of an MR brain volume to correct the lateral ventricle segmentation by adding paint. The bottom row shows corresponding edge detected images.

5.3   Editing

In this section we demonstrate the editing capability. Figure 6 shows a simple example of direct 3D editing of the cloverleaf (erasing the segmentation) and the vertebra phantom (adding edit paint). In Fig. 7 we demonstrate editing of the lateral ventricle segmentation result on a 3D image slice of the MR brain volume. The ventricle is under-segmented on this slice and is corrected by adding paint using a small paint brush tip. In this example, a circular brush tip is used. However, the superquadric brush tip can be shaped to a flattened ellipsoid or rectangular solid to better match a desired boundary curvature.

6   Conclusions

The paper demonstrates the potential of compute shaders for creating a flexible and modular software system that tightly integrates interactive 3D medical image visualization and processing. The prototype system described in this paper uses a single intuitive painting interface for all selection, initialization and editing interactions with the data, and supports the processing of noisy volume images by integrating a 3D slice plane view directly with the volume rendered view. Further improvements can be made to the framework. The GPU level set implementation can be further optimized by implementing the narrow band algorithm using stream compaction as in Roberts et al. [8] to obtain a subset of the current narrow band as it evolves. Slice-by-slice editing can be labor intensive if many slices require editing. We intend to experiment with interactively painted "barriers" that reinforce target structure boundaries in regions with no edge features. The idea is to allow the user to paint thin 3D regions that are then used to modify the edge detected image used by the level set segmentation algorithm.

References

1. Shreiner, D., Sellers, G., Kessenich, J.M., Licea-Kane, B.M.: OpenGL Programming Guide, Version 4.3, 8th edn. Addison-Wesley Professional (2013)
2. Faynshteyn, L., McInerney, T.: Context-preserving volumetric data set exploration using a 3D painting metaphor. In: Bebis, G., Boyle, R., Parvin, B., Koracin, D., Fowlkes, C., Wang, S., Choi, M.-H., Mantler, S., Schulze, J., Acevedo, D., Mueller, K., Papka, M. (eds.) ISVC 2012, Part I. LNCS, vol. 7431, pp. 336–347. Springer, Heidelberg (2012)
3. Fogal, T., Kruger, J.: Tuvok, an architecture for large scale volume rendering. In: Proceedings of the 15th International Workshop on Vision, Modeling, and Visualization (2010)
4. Nickolls, J., Buck, I., Garland, M., Skadron, K.: Scalable parallel programming with CUDA. Queue **6**, 40–53 (2008)
5. Perrot, G., Domas, S., Couturier, R.: Fine-tuned high-speed implementation of a gpu-based median filter. J. Signal Process. Syst. **75**, 185–190 (2014)
6. Luo, Y., Duraiswami, R.: Canny edge detection on nvidia cuda. In: Computer Vision and Pattern Recognition Workshops, CVPRW 2008, pp. 1–8 (2008)

7. Schwarzkopf, A., Kalbe, T., Bajaj, C., Kuijper, A., Goesele, M.: Volumetric nonlinear anisotropic diffusion on GPUs. In: Bruckstein, A.M., ter Haar Romeny, B.M., Bronstein, A.M., Bronstein, M.M. (eds.) SSVM 2011. LNCS, vol. 6667, pp. 62–73. Springer, Heidelberg (2012)
8. Roberts, M., Packer, J., Sousa, M.C., Mitchell, J.R.: A work-efficient GPU algorithm for level set segmentation. In: Proceedings of the Conference on High Performance Graphics, HPG 2010, pp. 123–132 (2010)
9. Eklund, A., Dufort, P., Forsberg, D., LaConte, S.M.: Medical image processing on the GPU - past, present and future. MIA **17**, 1073–1094 (2013)
10. Li, C., Xu, C., Gui, C., Fox, M.D.: Level set evolution without re-initialization: a new variational formulation. In: Computer Vision and Pattern Recognition, CVPR 2005, pp. 430–436 (2005)
11. Nisar, I.: Vis3D+: a tightly integrated GPU-accelerated computation and rendering framework for interactive 3D image visualization. Master's thesis, Department of Computer Science, Ryerson University, Toronto, ON, Canada (2015)
12. Faynshteyn, L.: Context-preserving volumetric data set exploration using a 3D painting metaphor. Master's thesis, Department of Computer Science, Ryerson University, Toronto, ON, Canada (2012)
13. Kuo, H.C., Giger, M.L., Reiser, I.S., Boone, J.M., Lindfors, K.K., Yang, K., Edwards, A.: Level set segmentation of breast masses in contrast-enhanced dedicated breast ct and evaluation of stopping criteria. J. Digital Imaging **27**, 237–247 (2014)
14. Museth, K., Breen, D.E., Whitaker, R.T., Barr, A.H.: Level set surface editing operators. ACM Trans. Graph. **21**, 330–338 (2002)
15. Li, C., Huang, R., Ding, Z., Gatenby, C., Metaxas, D.N., Gore, J.C.: A level set method for image segmentation in the presence of intensity inhomogeneities with application to MRI. IEEE Trans. Image Process. **20**, 2007–2016 (2011)

Ontology-Based Visual Query Formulation: An Industry Experience

Ahmet Soylu[1,2]($\boxtimes$), Evgeny Kharlamov[3], Dmitriy Zheleznyakov[3], Ernesto Jimenez-Ruiz[3], Martin Giese[1], and Ian Horrocks[3]

[1] Department of Informatics, University of Oslo, Oslo, Norway
{ahmets,martingi}@ifi.uio.no, ahmet.soylu@hig.no
[2] Faculty of Informatics and Media Technology,
GjøVik University College, GjøVik, Norway
[3] Department of Computer Science, University of Oxford, Oxford, UK
{Evgeny.Kharlamov,Dmitriy.Zheleznyakov,
Ernesto.Jimenez-Ruiz,Ian.Horrocks}@cs.ox.ac.uk

Abstract. Querying is an essential instrument for meeting ad hoc information needs; however, current approaches for querying semantic data sources mostly target technologically versed users. Hence, there is a need for methods that make it possible for users with limited technological skills to express relatively complex ad hoc information needs in an easy and intuitive way. Visual methods for query formulation undertake the challenge of making querying independent of users' technical skills and the knowledge of the underlying textual query language and the structure of data. In this paper, we present an ontology-based visual query system, OptiqueVQS, and report user experiments in two industrial settings.

Keywords: Visual query formulation · Ontology · Usability · SPARQL

1 Introduction

In the semantic web community, various visualisations and user interfaces have been developed to aid the understanding of different domains – often represented by large and complex ontologies – and value creation out of vast data sources (cf. [1, 2]). Amongst these, query interfaces are important as they enable users to express ad hoc information needs, which could not be addressed by predefined visualisations or queries embedded into applications. However, current approaches for querying semantic data sources mostly target technology-experienced users, although semantic data consumers come from different backgrounds, and have varying levels of expertise. Hence, there is a need for providing semantic data consumers who are not technology-experienced users with the flexibility to pose relatively complex ad hoc queries in an easy and intuitive way.

Formal textual languages, keyword search, natural language interfaces, visual query languages (VQL), and visual query systems (VQS) are known approaches

G. Bebis et al. (Eds.): ISVC 2015, Part I, LNCS 9474, pp. 842–854, 2015.
DOI: 10.1007/978-3-319-27857-5_75

for querying semantic data sources (cf. [3]). Formal textual languages are inaccessible to end users, since they demand a sound technical background. Keyword search and natural language interfaces remain insufficient for querying structured data, due to low expressiveness and ambiguities respectively. VQLs are based on a formal visual syntax and notation, and are comparable to formal textual languages from an end-user perspective, as users need to understand the semantics of visual syntax and notation. A VQS [4] differs from a VQL, since it is primarily a system of interactions formed by an arbitrary set of user actions that effectively capture a set of syntactic rules specifying a (query) language. A VQS might use a VQL for query representation; however, VQSs built on non-formal visualisations are expected to offer a good usability-expressiveness balance.

In this respect, VQSs primarily undertake the challenge of making querying independent of users' technical skills and the knowledge of the underlying textual query language and the structure of data. To this end, we have been developing an ontology-based visual query system, namely OptiqueVQS [3], within a large industrial project, called Optique[1] [5], for end users, i.e., domain experts. OptiqueVQS distinguishes itself from other query interfaces as it (*a*) does not use a formal notation and syntax for query representation, but still conforms to the underlying formalism; (*b*) employs a formal approach projecting the underlying ontology into a graph for navigation, which constitutes the backbone of the query formulation process; (*c*) possesses a set of important quality attributes such as adaptivity, modularity, and multi-paradigm design; and (*d*) has been evaluated with different sets of end users in different contexts, and found to be promising.

In this paper, we introduce OptiqueVQS from an end-user perspective, present its quality attributes, describe the underlying formal approach, and then present the results of usability experiments with domain experts.

2   OptiqueVQS

OptiqueVQS is meant for end users who have no or very limited technical skills and knowledge, such as on programming, databases, query languages, and have low/no tolerance, intention, nor time to use and learn formal textual query languages. As such, they often use computers in their daily life and work, such as for web browsing, e-mail, and office and entertainment applications. OptiqueVQS is a visual query system and it is not our concern to reflect the underlying formality (i.e., query language and ontology) per se. However, user behaviour is constrained so as to enforce the generation of valid queries, and ontologies are formally projected into graphs in order to provide simpler representation and interaction styles for end users. We are also not interested in providing full expressivity, as simpler interfaces will suffice for majority of end user queries [6]. End users make a very little use of advanced functionalities and are likely to drop their own requirements for the sake of having simpler ways for basic tasks [4].

---

[1] http://www.optique-project.eu.

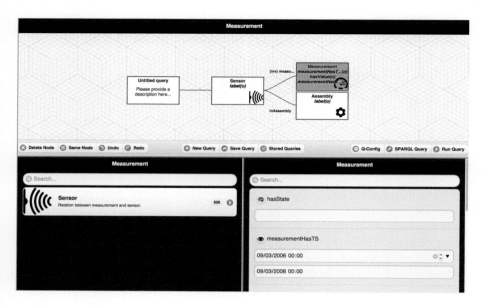

Fig. 1. An example query in visual mode is depicted.

2.1   User Interface

The interface of OptiqueVQS is designed as a widget-based user-interface mashup (UI mashup). Apart from flexibility and extensibility, such a modular approach provides us with the ability to combine multiple representations, interaction, and query formulation paradigms, and distribute functionality appropriately.

In Fig. 1, a query is shown as a tree in the upper widget (W1), representing typed variables as nodes and object properties as arcs. New typed variables can be added to the query by using the list in the bottom-left widget (W2). If a query node is selected, the faceted widget (W3) at the bottom-right shows controls for refining the corresponding typed variable, e.g. setting a value for a data property or switching to a more specific concept. Once a restriction is set on a data property or a data property is selected for output (i.e., using the eye icon), it is reflected in the label of the corresponding node in the query graph. The user has to follow the same steps to involve new concepts in the query and can always jump to a specific part of the query by clicking on the corresponding variable-node in W1. These three widgets are orchestrated by the system, through harvesting event notifications generated by each widget as the user interacts. At each step of the query formulation process W2 and W3 provide automatically generated ranked suggestions to guide users in constructing the query (see [7]).

The user can delete nodes, save/load queries, access query catalogue, and undo/redo actions by using the buttons at the bottom part of W1. The user can

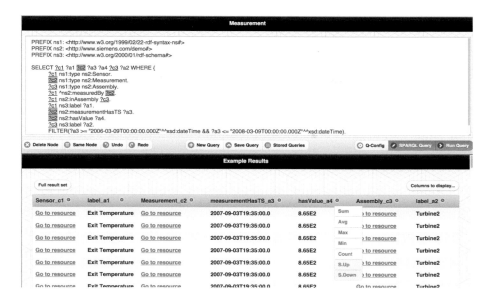

Fig. 2. An example query in textual mode and result view are depicted.

also switch to editable textual SPARQL mode by clicking on "SPARQL Query" button at the bottom-right part of the W1 as depicted in Fig. 2. The availability of a textual mode synchronised with the visual representation enables collaboration between end users and technology-experienced users. Note that SPARQL mode is compliant, in terms of expressiveness, to what can be represented in the visual mode.

Finally, we recently extended OptiqueVQS with two new widgets, which provide evidence on how a widget-based architecture allows us to hide complex functionality behind layers and combine different paradigms. The first widget is tabular result widget (W4 – see Fig. 2). It provides an example result list and also means for aggregation and sequencing operations. Aggregation and sequencing operations fit naturally to a tabular view, since it is a related and familiar metaphor. The user can also view the full result list, inspect the individuals, and export data. The second widget is a map widget (W5 – see Fig. 3), which is a domain-specific (i.e., geospatial) component allowing the user to constrain attributes by selecting an input value from a map. For this purpose, a button with a pin icon is placed next to every appropriate attribute.

There are limits to the supported expressiveness, e.g. no union, negation etc., however it is possible to construct rather complex queries with a number of classes, restrictions, and branches. As far as the design rationale is considered, OptiqueVQS combines multiple familiar representation and interaction paradigms into a single view. This way the user can have a constant and global overview of the query, while working with the list and faceted widgets to manipulate and extend it (i.e., view/overview). OptiqueVQS also provides simple

Fig. 3. An example query with the map widget is depicted.

three-shaped query representation, which is free of any SPARQL or ontology jargon. A unidirectional tree-shaped query representation is employed to avoid a graph representation for simplicity. For more details on the design and implementation of OptiqueVQS, we refer interested readers to our earlier work [3].

2.2 Quality Attributes

Quality attributes are non-functional requirements that affect run-time behaviour, design, and user experience. OptiqeVQS possesses the following interrelated quality attributes, which effectively increase the benefits gained and decrease the cost of adoption for end users. Usability is the primary quality attribute, and all other quality attributes directly or indirectly affect the usability. The attributes are derived from our conceptual review [8] and discussed in another work [9].

(A1) **Usability:** The design of OptiqueVQS emphasises harmonies between view and overview, and exploration and construction. It combines multiple representation, interaction, and query formulation paradigms to address different set of users and tasks. The functionality is distributed among widgets with respect to their suitability. Users formulate queries iteratively and could collaborate with different types of users.

(A2) **Modularity:** OptiqueVQS employs a widget-based mashup approach, which provides us with the flexibility to add/remove components easily. This could include alternative/complementary components for query formulation, exploration, visualisation, etc. with respect to context.

(A3) **Scalability:** OptiqueVQS provides gradual and on-demand access to the relevant parts of the underlying ontology to cope with large ontologies, while the employed ranking approach filters down the amount of ontological knowledge to be presented at each step.

(A4) **Adaptivity & adaptability:** The modular architecture of OptiqueVQS, availability of multiple representation, interaction, and query formulation paradigms, and ranked suggestions enable OptiqueVQS to provide diverse user experiences by altering presentation, content, and behaviour automatically or manually with respect to context.

(A5) **Extensibility:** OptiqeVQS provides flexibility against changing requirements both from architectural and design perspectives for sustainable evolution. The modular architecture allows new components to be easily introduced and combined, while new functionalities could be added easily without overloading the interface due to the multi-paradigm design.

(A6) **Interoperability:** The ability of OptiqueVQS to export data in different formats ensures that it fits into organisational contexts and broader user experiences, as the extracted data could be utilised by other applications in the workflow or the digital ecosystem.

(A7) **Portability:** OptiqueVQS relies on a domain-agnostic backend, which projects the underlying ontology into a graph for exploration and query construction. This provides ability to query other domains, rather than only a specific domain, without high installation and configuration costs.

(A8) **Reusability:** OptiqueVQS allows users to store, load, and modify queries. Queries are stored in a query catalogue with descriptive texts to facilitate their search and retrieval. End users can reuse existing queries or modify them to formulate more complex queries.

2.3 Navigation Graph

In OptiqueVQS, user queries have a graph-like structure and interaction with the ontology happens through graph navigation. However, OWL 2 axioms are not well-suited for a graph-based navigation. Indeed, note that OWL 2 axioms do not have a natural correspondence to a graph. Therefore, we need a technique to extract a suitable graph-like structure from a set of OWL 2 axioms.

Intuitively, OptiqueVQS allows users to construct tree-shaped conjunctive queries where each path is of the form: $Person(x), livesIn(x, y), City(y), \ldots$ Each such path essentially 'connects' classes like $Person$ and $City$ via properties like $livesIn$. At each query construction step OptiqueVQS suggests the user classes and properties that are semantically relevant to the already constructed partial query. We determine this relevance by exploiting the input OWL 2 ontology: we project the input ontology onto a graph structure that is called *navigation graph* [10] and use this graph at query construction time. More precisely, for each class in the partial query OptiqueVQS suggests only those properties and classes, which are reachable in the navigation graph in one step. Note that OWL 2 ontologies are essentially sets of first-order logic axioms and thus there is no immediate relationship between them and a graph. This makes projection of OWL 2 ontologies onto a navigation graph a non-trivial task.

In the remaining part of this section we will formally introduce navigation graph, define when a query is meaningful with respect to it, and finally we define the grammar of queries that users can construct with the help of OptiqueVQS.

The nodes of a navigation graph are unary predicates and constants, and edges are labelled with possible relations between such elements, that is, binary predicates or a special symbol type. The key property of a navigation graph is that every X-labelled edge (v, w) is justified by a rule or fact entailed by $\mathcal{O} \cup D$ which "semantically relates" v to w via X. We distinguish three kinds of semantic relations: *(i) existential*, where X is a binary predicate and (each element of) v must be X-related to (an element of) w in the models of $\mathcal{O} \cup D$; *(ii) universal*, where (each instance of) v is X-related only to (instances of) w in the models of $\mathcal{O} \cup D$; and *(iii) typing*, where $X = $ type, and (the constant) v is entailed to be an instance of (the unary predicate) w. Formally:

Definition 1. *Let $\mathcal{O}$ be an OWL 2 ontology and D a knowledge graph. A navigation graph for $\mathcal{O}$ and D is a directed labelled multigraph G having as nodes unary predicates or constants from $\mathcal{O}$ and D and s.t. each edge is labelled with a binary predicate from $\mathcal{O}$ or type. Each edge e is justified by a fact or rule α_e s.t. $\mathcal{O} \cup \mathcal{C} \models \alpha_e$ and α_e is of the form given next, where c, d are constants, A, B unary predicates, and R a binary predicate:*

(i) if e is $c \xrightarrow{R} d$, then α_e is of the form $R(c, d)$ or $\forall y.[R(c, y) \rightarrow y \approx d]$;

(ii) if e is $c \xrightarrow{R} A$, then α_e is a rule of the form $\top(c) \rightarrow \exists y.[R(c, y) \wedge A(y)]$ or $\forall y.[R(c, y) \rightarrow A(y)]$;

(iii) if e is $A \xrightarrow{R} B$, then α_e is a rule of the form
    $\forall x.[A(x) \rightarrow \exists y.[R(x, y) \wedge B(y)]]$ or $\forall x, y.[A(x) \wedge R(x, y) \rightarrow B(y)]$;

(iv) if e is $A \xrightarrow{R} c$, then α_e is a rule of the form $\forall x.[A(x) \rightarrow R(x, c)]$ or $\top(c) \rightarrow$
    $\exists y.[R(y, c) \wedge A(y)]$ or $\forall x, y.[A(x) \wedge R(x, y) \rightarrow y \approx c]$;

(v) if e is $c \xrightarrow{\text{type}} A$, then $\alpha_e = A(c)$.

The first (resp., second) option for each α_e in *(i)–(iii)* encodes the existential (resp., universal) R-relation between nodes in e; the first and second (resp., third) options for each α_e in *(iv)* encode the existential (resp., universal) R-relation between nodes in e; and *(v)* encodes typing. A graph may not contain all justifiable edges, but rather those that are deemed relevant to the given application.

To realise the idea of ontology and data guided navigation, we require that interfaces *conform to* the navigation graph. We assume that all the following definitions are parametrised with a fixed ontology $\mathcal{O}$ and a knowledge graph D.

Definition 2. *Let Q be a conjunctive query. The graph of Q is the smallest multi-labelled directed graph G_Q with a node for each term in Q and a directed edge (x, y) for each atom $R(x, y)$ occurring in Q, where R is different from $\approx$. We say that Q is tree-shaped if G_Q is a tree. Moreover, a variable node x is* *labelled with a unary predicate A if the atom $A(x)$ occurs in Q, and an edge (t_1, t_2) is labelled with a binary predicate R if the atom $R(t_1, t_2)$ occurs in O.*

Finally, we are ready to define the notion of conformation.

Definition 3. *Let Q be a conjunctive query and G a navigation graph. We say that Q conforms to G if for each edge (t_1, t_2) in the graph G_Q of Q the following holds:*

- *If t_1 and t_2 are variables, then for each label B of t_2 there is a label A of t_1 and a label R of (t_1, t_2) such that $A \xrightarrow{R} B$ is an edge in G.*
- *If t_1 is a variable and t_2 is a constant, then there is a label A of t_1 and a label R of (t_1, t_2) such that $A \xrightarrow{R} t_1$ is an edge in G.*
- *If t_1 is a constant and t_2 is a variable, then for each label B of t_2 there is a label R of (t_1, t_2) such that $t_1 \xrightarrow{R} t_2$ is an edge in G.*
- *If t_1 and t_2 are constants, then a label R of (t_1, t_2) such that $t_1 \xrightarrow{R} t_2$ is an edge in G.*

OptiqueVQS allows constructing conjunctive tree-shaped queries. The generation is done via reasoning over the navigation graph, which contain edges of types (iii)–(v) (see Definition 1).

Now we describe the class of queries that can be generated using OptiqueVQS and show that they conform to the navigation graph underlying the system. First, observe that the OptiqueVQS queries follow the following grammar:

$$\texttt{query}:: = A(x)(\wedge \texttt{ constr}(x))^*(\wedge \texttt{ expr}(x))^*$$
$$\texttt{expr}(x):: = \texttt{sug}(x, y)(\wedge \texttt{ constr}(x))^*(\wedge \texttt{ expr}(y))^*$$
$$\texttt{constr}(x):: = \exists y R(x, y) \mid R(x, y) \mid R(x, c)$$
$$\texttt{sug}(x, y):: = Q(x, y) \wedge A(y)$$

where A is an atomic class, R is an atomic data property, Q is an object property, and c is a data value. The expression of the form $A(\wedge B)^*$ designates that B-expressions can appear in the formula 0, 1, and so on, times. An OptiqueVQS `query` is constructed using suggestions `sug` and constraints `constr`, that are combined in expressions `expr`. Such queries are conjunctive and tree-shaped. All the variables that occur in classes and object properties are output variables and some variables occurring in data properties can also be output variables.

3   Evaluation

Three user experiments are conducted with different types of users. The first experiment is reported in our previous work [3] and involved a movie ontology and 15 casual users, who are bachelor students in different social science disciplines. The second and third experiments are conducted with our industrial partners, which are Statoil ASA and Siemens AG. In the Statoil experiment,

Table 1. Information needs used in the experiments (T1-9 Statoil and T10-14 Siemens).

T	Information need
1	List all *fields*.
2	What is the water depth of the "Snorre A" *platform* (facility)?
3	List all *fields* operated by "Statoil Petroleum AS" *company*.
4	List all *exploration wellbores* with the *field* they belong to and the *geochronological era(s)* with which they are recorded.
5	List the *fields* that are currently operated by the *company* that operates the "Alta" *field*.
6	List the *companies* that are *licensees* in *production licenses* that own *fields* with a recoverable oil equivalent over more than "300" in the *field reserve*.
7	List all *production licenses* that have a *field* with a *wellbore* completed between "1970" and "1980" and recoverable oil equivalent greater than "100" in the *company reserve*.
8	List the *blocks* that contain *wellbores* that are drilled by a *company* that is a *field operator*.
9	List all *producing fields* operated by "Statoil Petroleum AS" *company* that has a *wellbore* containing "gas" and a *wellbore* containing "oil".
10	Find all *assemblies* that exist in system.
11	Show all *messages* that *tribune* "NA0101/01" generated from "01.12.2009" to "02.12.2009".
12	Show all *turbines* that sent a *message* containing the text "Trip"' between "01.12.2009" and "02.12.2009".
13	Show all *event* categories known to the system.
14	Show all *turbines* that sent a *message* category "Shutdown" between "01.12.2009" and "02.12.2009".

an oil &gas ontology, which in total includes 253 concepts, 208 relationships (including inverse properties), and 233 attributes, is used. In the Siemens experiment, a diagnostic ontology, which in total includes five concepts, five relationships (excluding inverse properties), and nine attributes, is used. In both cases neither ontologies nor data sets are public. The tasks used in Statoil and Siemens experiments are all conjunctive, see Table 1. A total of seven domain experts are engaged in the experiments, see Table 2 (Likert scale 1 for "not familiar at all" and 5 for "very familiar").

The experiments are designed as a think-aloud study and only a 5 min. introduction is given to participants. Each participant performs the experiment in a single session, while being watched by an observer. Formulating the query, executing it, and inspecting the result set equals to one attempt. Participants have a maximum of three attempts per task. A task is ended, when the participant acknowledges completion or exhausts his/her three attempts.

Table 2. Participant profiles (P1-3 Statoil and P4-7 Siemens).

P	Age	Occupation	Education	Technical skills	Similar tools
1	39	Geologist	Master	3	3
2	40	Biostrat	Master	2	1
3	49	IT advisor	Master	5	4
4	33	Software engineer	Bachelor	5	2
5	27	Diagnostic Engineer	Bachelor	5	5
6	60	Mechanical Engineer	Master	3	1
7	45	Mechanical Engineer	Bachelor	1	2

In the Statoil experiment, participants overall have 84 percent correct completion rate and 69 percent first-attempt correct completion rate (i.e., percentage of correctly formulated queries in the first attempt), while in the Siemens experiment, correct completion rate is 88 percent and first-attempt correct completion rate is 72 percent (see Fig. 4). In our earlier experiment with casual users, there is a full correct completion rate and 80 percent first-attempt correct completion rate. The results are comparatively better, and this could be attributed to genericness of the ontology. Statoil users have lower scores and often commented that the ontology does not match to their understanding of the domain. This is because the ontology used in the Statoil experiment is automatically generated (i.e., bootstrapped), while the others are manually created. We acknowledge the situation and believe that the usability of an ontology is as crucial as the usability of a query formulation tool and is an overlooked issue in the research community.

Overall, the results indicate high effectiveness and efficiency suggesting that OptiqueVQS is a viable tool for users without any technical background to construct considerably complex queries. OptiqueVQS also offers a good learnability as users can solve complex tasks without any training. The participants praised the capability of OptiqueVQS for formulating complex information needs into queries. A common statement was that such a solution will not only improve their current practices, but also augment their value creation potential.

4   Related Work

We distinguish existing visual methods for querying semantic data sources into two categories. The first category includes approaches that are primarily built on a VQL, which has a formal visual syntax and notation. The second category of approaches mainly employs a system of interactions, i.e., VQSs, which generates queries in target linguistic form.

The notable examples of the fist category are LUPOSDATE [11], RDF-GL [12], GQL [13], and QueryVOWL [14]. LUPOSDATE and RDF-GL follow RDF syntax at a very low level through node-link diagrams representing the subject-predicate-object notation, while GQL and QueryVOWL represent queries at

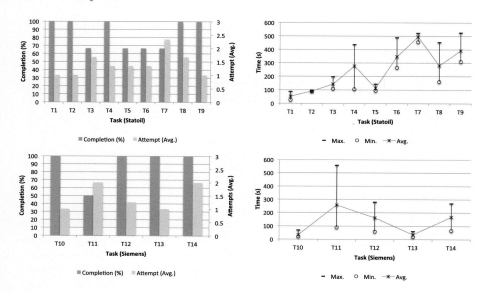

Fig. 4. The results of Statoil and Siemens experiments.

comparatively higher level, such as with UML-based diagrams. Each of these languages are managed by a VQS providing means for construction and manipulation of queries in a visual form. Albeit VQL-based approaches with higher level of abstraction are closer to end users, they still need to posses a higher level of knowledge and skills to understand the semantics of visual notation and syntax and to use it.

The prominent examples of the second category are gFacet [15], SparqlFilter-Flow [16], Konduit VQB [17], and Rhizomer [18]. gFacet and SparqlFilterFlow employ a diagram-based approach; however, diagrams representing the queries are rather informal. Konduit VQB and Rhizomer employ a form-based paradigm. Diagram-based approaches are good in providing a global overview; however, they remain insufficient for view (i.e., zooming into a specific concept for filtering and projection). This is because the visual space as a whole is mostly occupied for query overview. Form-based approaches provide a good view; however, they provide a poor overview, since the visual space as a whole is mostly occupied with the properties of the focus concept. Approaches combining multiple representation and interaction paradigms are known to be better as they could combine view and overview. Finally, gFacet and Rhizomer are originally meant for data browsing, that is they operate on data level rather than schema level and every user interaction generates and sends SPARQL queries in the background. Yet they are highly data-intensive, which is often impractical for large data sources.

5   Conclusion

OptiqueVQS enables end users to easily formulate comparatively complex queries at a conceptual level. It employs a formal ontology to graph projection method to support query formulation and ontology navigation, while the user interface remains rather informal and possesses important quality attributes.

OptiqueVQS is intentionally limited in expressiveness to achieve a usability - expressiveness balance, for instance unions, cardinality restrictions, intersection, and individuals are not supported. However the future work includes implementation of more features without compromising the usability, such as optionals and inequality relationships for data properties.

Acknowledgements. This research is funded by "Optique" (EC FP7 318338), as well as the EPSRC projects Score!, DBOnto, and MaSI³.

References

1. Dadzie, A.S., Rowe, M.: Approaches to visualising linked data: a survey. Seman. Web **2**(2), 89–124 (2011)
2. Katifori, A., et al.: Ontology visualization methods - a survey. ACM Comput. Surv. **39**(4), 1–43 (2007)
3. Soylu, A., et al.: Experiencing OptiqueVQS: a multi-paradigm and ontology-based visual query system for end users. Univ. Access Inf. Soc. (in press) 1–24 (2014)
4. Catarci, T., et al.: Visual query systems for databases: a survey. J. Vis. Lang. Comput. **8**(2), 215–260 (1997)
5. Giese, M., et al.: Optique: zooming in on big data. IEEE Comput. Mag. **48**(3), 60–67 (2015)
6. Leone, S., et al.: Exploiting tag clouds for database browsing and querying. In: CAiSE 2010 (2011)
7. Soylu, A., Giese, M., Jimenez-Ruiz, E., Kharlamov, E., Zheleznyakov, D., Horrocks, I.: Towards exploiting query history for adaptive ontology-based visual query formulation. In: Closs, S., Studer, R., Garoufallou, E., Sicilia, M.-A. (eds.) MTSR 2014. CCIS, vol. 478, pp. 107–119. Springer, Heidelberg (2014)
8. Soylu, A., et al.: Ontology-based end-user visual query formulation: why, what, who, how, and which? Universal Access in the Information Society (submitted)
9. Soylu, A., Martin, G.: Qualifying ontology-based visual query formulation. In: FQAS 2015 (2015)
10. Arenas, M., et al.: Faceted search over ontology-enhanced RDF data. In: CIKM 2014 (2014)
11. Ambrus, O., et al.: Visual query system for analyzing social semantic web. In: WWW 2011 (2011)
12. Hogenboom, F., et al.: RDF-GL: A SPARQL-based graphical query language for RDF. In: Chbeir, R., Badr, Y., Abraham, A., Hassanien, A.-E. (eds.) Emergent Web Intelligence, pp. 87–116. Springer London (2010)
13. Barzdins, G., et al.: Graphical query language as SPARQL frontend. In: ADBIS 2009 (2009)
14. Haag, F., et al.: Visual querying of linked data with QueryVOWL. In: SumPre 2015 and HSWI 2014–2015 (2015)

15. Heim, P., Ziegler, J.: Faceted visual exploration of semantic data. In: HCIV 2009 (2011)
16. Haag, F., et al.: Visual SPARQL querying based on extended filter/flow graphs. In: AVI 2014 (2014)
17. Ambrus, O., et al.: Konduit VQB: a visual query builder for SPARQL on the social semantic desktop. In: VISSW 2010 (2010)
18. Brunetti, J.M., et al.: From overview to facets and pivoting for interactive exploration of semantic web data. Int. J. Seman. Web Inf. Syst. 9(1), 1–20 (2013)

ST: Visual Perception
and Robotic Systems

Dynamic Target Tracking and Obstacle Avoidance using a Drone

Alexander C. Woods and Hung M. La$^{(\boxtimes)}$

Advanced Robotics and Automation Laboratory,
Department of Computer Science and Engineering,
University of Nevada, Reno, NV 89557, USA
hla@unr.edu

Abstract. This paper focuses on tracking dynamic targets using a low cost, commercially available drone. The approach presented utilizes a computationally simple potential field controller expanded to operate not only on relative positions, but also relative velocities. A brief background on potential field methods is given, and the design and implementation of the proposed controller is presented. Experimental results using an external motion capture system for localization demonstrate the ability of the drone to track a dynamic target in real time as well as avoid obstacles in its way.

1 Introduction

This paper focuses on dynamic target tracking and obstacle avoidance on a quadrotor drone. Drones are useful in many practical applications such as disaster relief efforts, infrastructure inspections, search and rescue, mapping, and military or law enforcement surveillance. However, in order to perform any of these tasks, a drone must be able to perform both localization as well as effective navigation and obstacle avoidance. The contribution of this paper is the development of a drone control system which allows the drone to track a dynamic target while simultaneously avoiding obstacles using an expanded potential field control method. A demonstration of the capabilities is performed utilizing the ARDrone 2.0 shown in Fig. 1 and an external motion capture system [1].

Currently, two main challenges prevent drones from being used regularly for application such as those mentioned. The first is localization. GPS is a great tool for localization in environments which allow communication with satellites, and there has been a lot of research activity in the field of localization in GPS-denied environments. For example, light detection and ranging (LIDAR) is a promising option [2–4] which has been used in simultaneous localization and mapping (SLAM) [5]. Another method is computer vision, utilizing both forward and downward looking stereo-cameras [6]. Additionally, tools such as Microsoft's Kinect sensor have been used which provide depth information much like LIDAR, but with a much larger sensing area [7]. All of these options, and more, provide drones with location information about themselves and their environment, including obstacles.

© Springer International Publishing Switzerland 2015
G. Bebis et al. (Eds.): ISVC 2015, Part I, LNCS 9474, pp. 857–866, 2015.
DOI: 10.1007/978-3-319-27857-5_76

The second challenge is accurate navigation and crash avoidance which enable the drone to perform any given task safely and efficiently. Similar to localization, this is also a field which has attracted much research. Many use multi-drone flocking strategies in order to track and maintain contact with a target for information gathering purposes [8,9]. Several methods have been developed for fixed-wing aircrafts, utilizing numerous methods ranging from Lyapunov-LaSalle to seventh order Bézier curves [10–12]. For use on quadrotor drones, a PID controller has been used for position control in a hover state coupled with a three dimensional PD controller for trajectory tracking [13,14] with

Fig. 1. A low cost, commercially available quadrotor drone is used as the experimental platform for testing and demonstrating the effectiveness of the proposed control method.

good results. In this paper, a novel potential field controller is presented which has been expanded to include not only relative position but also relative velocity in order to meet the unique challenges presented by drones. By expanding the method to include more than just two dimensional position data, greater performance can be achieved and allow the drone to perform in fast moving environments.

This paper is organized as follows. Section 2 describes the problem definition, assumptions, constrains and evaluation criteria for the paper. Section 3 provides a brief background on potential field methods, and their limitations. Section 4 discusses the design of the controller used in this paper. Section 5 presents the experimental results of the paper, and an evaluation of the performance of the controller. Finally, Sect. 6 provides a brief conclusion, with recommendations for future work.

2   Problem Definition

The goals of this paper are to develop a control system for a drone which allows it to safely navigate to, and follow a dynamic human target. The drone should never run into the target or any obstacles in its vicinity. For the purposes of this paper, we assume that the position of the drone and the target are known. In this case, we utilize an external motion tracking system [1] to provide position feedback, which is this paper's main constraint. It is important to note that the approach presented maybe be adapted to work with onboard sensing in future work.

To evaluate the success or failure of this paper, the drone should be able to navigate successfully from a starting location to a target location, while avoiding obstacles along the way. Additionally, the drone should be able to track the target while it moves around the testing environment.

3   Potential Field Navigation

Because of their simplicity and elegance, potential field methods are often used for navigation of ground robots [15–17]. Potential fields are aptly named, because they use attractive and repulsive potential equations to draw the drone toward a goal (attractive potential) or push it away from an obstacle (repulsive potential). For example, imagine a stretched spring which connects a drone and a target. Naturally, the spring will draw the drone to the target location.

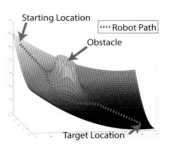

Fig. 2. An example of a traditional potential field which can be used for navigating toward a target while avoiding obstacles.

Conveniently, potential fields for both attractive and repulsive forces can be summed together, to produce a field such as that shown in Fig. 2. The example shown includes only one repulsive field, but multiple fields representing multiple obstacles could easily be added as they are sensed.

Traditionally, potential forces work in the x and y spatial dimensions, and are defined by a quadratic function given by

$$U_{att}(q_d) = \frac{1}{2}\lambda_1 d^2(q_d, q_t),\tag{1}$$

where λ_1 is positive scale factor, and $d(q_d, q_t)$ is the relative distance between the drone and the target. In order to achieve the target location in this scenario, the potential function should be minimized, thus the velocity of the drone should be

$$\begin{aligned}p_d^{att}(q_d) = \nabla U_{att}(q_d) &= \nabla\left(\frac{1}{2}\lambda_1 d^2(q_d, q_t)\right),\\ &= \frac{1}{2}\lambda_1 \nabla d^2(q_d, q_t),\\ &= \lambda_1(q_t - q_d).\end{aligned}\tag{2}$$

The velocity in (2) is appropriate for a stationary target but for a moving target, the velocity of the target should be also taken into account. If we consider that the drone should move with the same velocity as the target, in addition to the motion dictated by the attractive potential, then (2) becomes

$$p_d^{att}(q_d) = \lambda_1(q_t - q_d) + p_t,\tag{3}$$

where p_t is the velocity of the target. Through simulation and mathematical proofs not presented here, it can be shown that the relative position between the drone and target converges to zero using the method described. Thus, this method forms the foundation for the controller developed in this paper.

To account for obstacles, a potential function is generated which grows increasingly repulsive as the distance to the obstacle decreases to zero. Thus, the repulsive potential function is given by

$$U_{rep}(q_d) = \frac{1}{2}\eta_1 \frac{1}{d^2(q_d, q_o)}, \tag{4}$$

for which the resulting velocity of the drone is in the direction of the gradient, and is given by

$$p_d^{rep}(q_d) = \nabla U_{rep}(q_d) = -\eta_1 \frac{1}{(q_o - q_d)^3}, \tag{5}$$

where η_1 is again a scaling factor, and $d(q_d, q_o)$ is the relative distance between the drone and the obstacle. Because the drone should never actually come in contact with the obstacle, the relative distance should be computed between the drone and the a minimum distance it can be from the obstacle.

4   Controller Design

Because the potential field methods presented above are developed for ground robots, they do not address many of the factors that must be accounted for when designing a controller for aerial systems. For example, drones move very quickly and are inherently unstable which means they cannot simply move to a particular location and stop moving. They are consistently making fine adjustments to their position and velocity.

In order to account for factors unique to aerial platforms, not only is the relative position between the drone and the target taken into account, but also the relative velocity between the two. Using this new information, a potential function based on the relative velocity is given by

$$U_{att}(p_d) = \frac{1}{2}\lambda_2 v^2(p_d, p_t), \tag{6}$$

where λ_2 is a positive scale factor, and $v(p_d, p_t)$ is the relative velocity between the drone and the target. Again, this function should be minimized by moving in the direction of the gradient. Thus the speed of the drone due to attractive potential from the target should be the sum of the gradient of this potential function and (3) which yields

$$p_d^{att}(q_d, p_d) = \lambda_1(q_t - q_d) + p_t + \lambda_2(p_d - p_t), \tag{7}$$

Similarly, the potential function for obstacle avoidance derived in Sect. 3 should be augmented to include the a potential function for relative velocity between the drone and the obstacle. This potential function should react when the drone is moving toward the obstacle, but have no effect if the drone is moving away. Thus a piecewise potential function is generated and is described by

$$U_{rep}(p_d) = \begin{cases} -\frac{1}{2}\eta_2 v^2(p_d, p_o) : (p_o - p_d) < 0 \\ 0 \qquad\qquad\quad : (p_o - p_d) \geq 0 \end{cases} \tag{8}$$

and the resulting total velocity from repulsive forces is

$$p_d^{rep}(q_d, p_d) = -\eta_1 \frac{1}{(q_o - q_d)^3} - \eta_2(p_o - p_d). \tag{9}$$

Now that we have the velocity contributions from all of the attractive and repulsive potentials, we simply add them together to get the final velocity of the drone.

$$p_d(q_d, p_d) = p_d^{att} + p_d^{rep} \tag{10}$$

This velocity is calculated in the world frame, so must be transformed to the body frame of the drone. Given the heading of the drone, θ_d, the speed in the x_{body} and y_{body} is calculated using

$$\begin{aligned} v_{x,body} &= p_d cos(\theta_d) + p_d sin(\theta_d) \\ v_{y,body} &= -p_d sin(\theta_d) + p_d cos(\theta_d) \end{aligned} \tag{11}$$

This controller will seek out a moving target, and will also avoid obstacles that are in close proximity.

5 Experimental Results

The experimental platform used for this paper was the commercially available Parrot ARDrone 2.0 shown in Fig. 1 which has built in wifi connectivity which allows an easy way to control it through a computer interface, such as Matlab. In addition to the ARDrone, an external motion capture system was used to provide localization capabilities for the drone, target, and obstacle. The lab environment with drone, obstacle, target, and external tracking system is shown in Fig. 3.

Fig. 3. The experimental environment with a drone, target, obstacle, and external motion tracking system is shown.

In order to control the drone and display its location along with the target and any obstacles present, the Matlab GUI shown in Fig. 4 was created. It allows the user to determine when the drone takes off, lands, or tracks the target. This GUI is critical in efficient testing of the drone. Additionally, for the safety of the drone, if the controller does not behave as expected the user can request that the drone simply hover in place to avoid fly-aways.

In order to test the performance of the controller, several experiments were performed. First, the ARDrone is placed across the room from a stationary target, and is commanded to take off and reach the goal location. The performance of the controller here is evaluated by the overshoot and settling time of the system. Second, the first experiment is repeated but with an obstacle introduced. The performance of the controller in this case is evaluated by the success or failure of reaching the target without running into the obstacle.

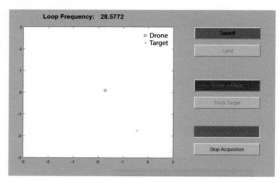

Fig. 4. A Matlab GUI was created to show the positions of the drone, target, and any obstacles present. It also allows the user to control when the drone takes off, lands, tracks the target, or simply hovers in place.

Finally, the ARDrone is commanded to follow the target as it moves about in the lab environment. This task is harder to evaluate, and a demonstration of the drone performing the requested action is used to verify that this functionality works. Future works should address how to evaluate the performance of this method quantitatively.

During the first experiment, the drone was placed approximately 4.2 meters from the target's location. The drone was requested to fly to 1 m away from the target location, since the purpose of this is to eventually track a human. The drone's response shown in Fig. 5 demonstrates the capability of the drone to achieve a goal position effectively. Starting at approximately 7.75 seconds, the drone enters an autonomous mode, and achieves stable hover 1 m away from the target at approximately 12.75 seconds. It is important to note that while the drone did overshoot it's goal location, it did not overshoot enough to get close to hitting the target. The closest that the drone got to the target was just under 0.75 meters.

In the second experiment, the drone was placed approximately 5.2 meters away from the target, and an obstacle was located in the path lying directly between the drone and the target. Similar to the first test, the drone's mission was to fly to within 1 m of the target, this time while avoiding the obstacle and still achieving the task. As the drone begins moving towards the target, it also moves towards the obstacle. Because of the repulsive forces generated by the

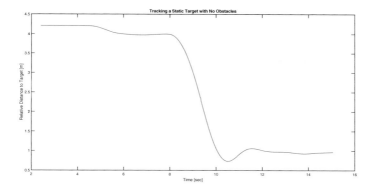

Fig. 5. The results of tracking a static target with no obstacles are very good. With an initial condition of approximately 3.2 m, the drone achieves position in under 5 seconds.

drones relative location and velocity with respect to the obstacle, the drone is elegantly "pushed" around the obstacle and still makes it to the target location. Figure 6 shows the results of this test, demonstrating that the drone maintains a safe distance from the obstacle (1 meter minimum) and also achieves the goal.

In the third experiment the drone is directed to follow the target around, while maintaining a safe distance at all times. In addition, the drone should always face the target. This task was performed several times to evaluate the performance. In each of the tests the drone successfully completes the task. Even under extreme circumstances (e.g., very fast maneuvers) the drone is able to recover and maintain the desired behavior. Figure 7 shows frames from a video [18] taken of the drone performing this task. In the video it can clearly

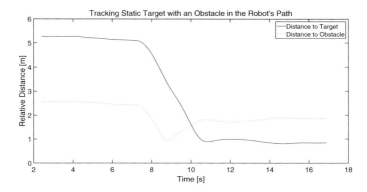

Fig. 6. The results of tracking a static target with an obstacle in the way demonstrates the controllers effectiveness at avoiding collisions.

Fig. 7. The third experiment demonstrates the drone tracking a dynamic target, while maintaining the proper heading to always face the target.

be seen that the drone follows the target around while always maintaining the proper heading to face the target.

In addition to tracking a dynamic target moving in an arbitrary pattern, the drone was instructed to follow the target as it moved in an approximate rectangle around the lab. The results shown in Fig. 8 illustrate the path of the drone as it follows the target through the pattern. As shown, the drone does in fact track the rectangle as instructed, with slight deviations. Because the path of the target was moved manually by a person as in previous experiments, the target trajectory is not a perfect rectangle either. In future work, a set trajectory will be established without human error which would allow repeatability to be addressed as well as a much better analysis of the tracking error.

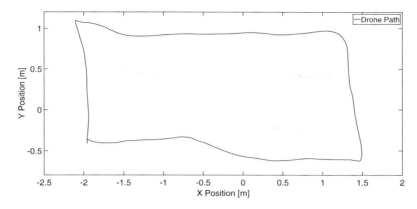

Fig. 8. The final experiment has the drone track a target which moves in an approximate rectangle. As shown, the drone does track, but because the desired trajectory is human-controlled the reference is not perfect. Future work will include using a predetermined and set trajectory.

6   Conclusion

This paper presented a novel, extended potential field controller designed to enable a drone to track a moving target in a dynamic environment. An experiment setup comprising of an ARDrone 2.0 and an external motion tracking system was used to implement the controller and evaluate the performance. The drone was able to successfully track stationary targets, with and without obstacles in it's path, as well as dynamic targets moving about in the lab.

Future work to build upon this paper will include determining a quantifiable evaluation method for the performance of dynamic target tracking. More importantly, because this system relies upon an external tracking system future work should include a method for estimating the state of both the drone and the target using onboard sensing and systems. Potential solutions might be LIDAR, optic flow sensing, and more.

Acknowledgements. The authors would like to specially thank the Motion Analysis Corporation for their support of the Motion Tracking System (MTS) setup at the Advanced Robotics and Automation (ARA) Lab at the University of Nevada, Reno. This project is partially supported by University of Nevada, Reno and NSF-NRI grant number 1426828.

References

1. Motion analysis systems. http://www.motionanalysis.com
2. Shaohua, M., Jinwu, X., Zhangping, L.: Navigation of micro aerial vehicle in unknown environments. In: 2013 25th Chinese Control and Decision Conference (CCDC), pp. 322–327 (2013)

3. Sa, I., Corke, P.: System identification, estimation and control for a cost effective open-source quadcopter. In: 2012 IEEE International Conference on Robotics and Automation (ICRA), pp. 2202–2209, May 2012

4. Shen, S., Michael, N., Kumar, V.: Autonomous multi-floor indoor navigation with a computationally constrained mav. In: 2011 IEEE International Conference on Robotics and Automation (ICRA), pp. 20–25, May 2011

5. Grzonka, S., Grisetti, G., Burgard, W.: A fully autonomous indoor quadrotor. IEEE Trans. Robot. **28**(1), 90–100 (2012)

6. Meier, L., Tanskanen, P., Heng, L., Lee, G., Fraundorfer, F., Pollefeys, M.: Pixhawk: a micro aerial vehicle design for autonomous flight using onboard computer vision. Auton. Robot. **33**(1–2), 21–39 (2012)

7. Shen, S., Michael, N., Kumar, V.: Autonomous indoor 3d exploration with a micro-aerial vehicle. In: 2012 IEEE International Conference on Robotics and Automation (ICRA), pp. 9–15 (2012)

8. Shanmugavel, M., Tsourdos, A., White, B., Zbikowski, R.: Co-operative path planning of multiple uavs using dubins paths with clothoid arcs. Control Eng. Pract. **18**(9), 1084–1092 (2010)

9. Wallar, A., Plaku, E., Sofge, D.: Reactive motion planning for unmanned aerial surveillance of risk-sensitive areas. IEEE Trans. Autom. Sci. Eng. **12**(3), 969–980 (2015)

10. Neto, A.A., Macharet, D., Campos, M.: Feasible path planning for fixed-wing uavs using seventh order bezier curves. J. Braz. Comput. Soc. **19**(2), 193–203 (2013)

11. Altmann, A., Niendorf, M., Bednar, M., Reichel, R.: Improved 3d interpolation-based path planning for a fixed-wing unmanned aircraft. J. Intell. Robot. Syst. **76**(1), 185–197 (2014)

12. Dogancay, K.: Uav path planning for passive emitter localization. IEEE Trans. Aerosp. Electron. Syst. **48**(2), 1150–1166 (2012)

13. Michael, N., Mellinger, D., Lindsey, Q., Kumar, V.: The grasp multiple micro-uav testbed. IEEE Robot. Autom. Mag. **17**(3), 56–65 (2010)

14. Mellinger, D., Kumar, V.: Minimum snap trajectory generation and control for quadrotors. In: 2011 IEEE International Conference on Robotics and Automation (ICRA), pp. 2520–2525, May 2011

15. La, H.M., Lim, R.S., Sheng, W., Chen, J.: Cooperative flocking and learning in multi-robot systems for predator avoidance. In: 2013 IEEE 3rd Annual International Conference on Cyber Technology in Automation, Control and Intelligent Systems (CYBER), pp. 337–342, May 2013

16. La, H.M., Sheng, W.: Multi-agent motion control in cluttered and noisy environments. J. Commun. **8**(1), 32 (2013)

17. La, H.M., Sheng, W.: Dynamic target tracking and observing in a mobile sensor network. Robot. Auton. Syst. **60**(7), 996 (2012)

18. Woods, A.C.: Dynamic target tracking with ardrone. https://youtu.be/v85hs8-uc1s

An Interactive Node-Link Visualization of Convolutional Neural Networks

Adam W. Harley$^{(\boxtimes)}$

Department of Computer Science, Ryerson University,
Toronto, ON M5B 2K3, Canada
`aharley@scs.ryerson.ca`

Abstract. Convolutional neural networks are at the core of state-of-the-art approaches to a variety of computer vision tasks. Visualizations of neural networks typically take the form of static diagrams, or interactive toy-sized networks, which fail to illustrate the networks' scale and complexity, and furthermore do not enable meaningful experimentation. Motivated by this observation, this paper presents a new interactive visualization of neural networks trained on handwritten digit recognition, with the intent of showing the actual behavior of the network given user-provided input. The user can interact with the network through a drawing pad, and watch the activation patterns of the network respond in real-time. The visualization is available at http://scs.ryerson.ca/~aharley/vis/.

1 Introduction

Convolutional neural networks (CNNs) are at the core of state-of-the-art approaches to a variety of computer vision tasks, including image classification [1] and object detection [2]. Despite this prevalence, interactive neural network visualization is still a relatively unexplored topic. Interactive simulations of toy networks have long existed [3], and visualizations of individual learned filters and features have emerged [4], but few visualizations illustrate how large-scale CNNs abstract from input to output. Motivated by this observation, this paper presents a new interactive visualization of a CNN trained on a specific task, with the intent of showing not only what it has learned, but how it behaves given new user-provided input.

Interactively visualizing the behavior of a neural network has a number of practical applications. First, having a detailed understanding of how neural networks behave is important for making practical use of them, so it is useful to have a visualization to support this understanding. Second, an interactive visualization of a neural network gives users the power to easily experiment with the input, and observe the effects of their experimentation immediately. This type of experimentation facilitates the process of exploring the strengths and weaknesses of a network, which is a critical part of designing better networks [4]. A third benefit of visualizing the behavior of a network is that it allows users to

© Springer International Publishing Switzerland 2015
G. Bebis et al. (Eds.): ISVC 2015, Part I, LNCS 9474, pp. 867–877, 2015.
DOI: 10.1007/978-3-319-27857-5_77

Fig. 1. The proposed visualization: an interactive node-link diagram of a convolutional neural network trained to recognize handwritten digits. On the left is a drawing pad, where the user can draw numbers for the network to classify. The activation level of each node is encoded in hue and brightness. Studying the architecture of the network, and experimenting with its input-output process, can enlighten students of machine learning as to how the network performs its abstraction from images to digits.

explore the layer-by-layer output of a network, and build intuitions about how neural networks perform hierarchical abstraction from input to output [5].

This visualization is targeted towards students of machine learning, who are learning how to design, code, and train new neural networks. To make the visualization useful to that audience, the visualization is based on the well-established node-link diagram representation of fully-connected neural networks. The visualization is supported by an actual neural network, designed and trained to recognize handwritten digits with high (99 %) accuracy. In the neural network literature, handwritten digit recognition is a well-known solved problem, and often serves as an example of an appropriate application of neural networks [6]. Users are able to interact with this network through a "drawing pad", on which they can write new numbers for the network to recognize. A screenshot of the visualization is shown in Fig. 1.

This paper begins by reviewing prior work on visualizing neural networks, with a special emphasis on identifying why the classic node-link diagram representation has endured the test of time. The paper then proceeds to describe the approach to developing the current visualization, considering the challenges of revealing inner detail, the use of color, and the elements of interaction. Finally, the effectiveness of the visualization is discussed, and future work is proposed.

Contributions: The proposed visualization is the first to accurately and interactively illustrate the structure, scale, and low-level inner workings of a CNN applied to a practical computer vision problem. Prior work on this topic was limited to simpler architectures, smaller problems, or static visualizations. The new visualization can be explored at http://scs.ryerson.ca/~aharley/vis/.

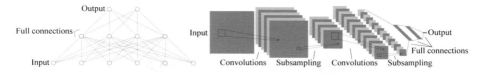

Fig. 2. Typical illustrations of fully-connected (left) and convolutional (right) neural networks (adapted from [12]).

2 Background and Related Work

This section establishes the context of the current work, by: (i) defining neural networks, and exploring why node-link diagrams are typically used to represent them, (ii) examining the challenges of complexity and scale that arise when using node-link diagrams for large neural networks, and (iii) considering the effective use of interaction.

2.1 Neurons as Nodes in a Graph

Neural networks compose many small functions in a network-like architecture, creating a larger function capable of pattern recognition [7]. In biological neural networks, the unitary function is a neuron, and neurons are connected together in extremely complex arrangements [8]. Artificial neural networks can be arbitrarily simple. One of the simplest arrangements is as a feed-forward graph with stacked "layers" of nodes, where every pair of neighboring layers is fully connected [9,10] (see Fig. 2, left). This arrangement is called a *fully-connected neural network*. A node-link visualization of this type of network is a straightforward outcome of (i) treating all neurons as identical processor units, and (ii) choosing an arrangement of neurons defined by a simple graph architecture. For these reasons, node-link diagrams of fully-connected networks are a mainstay in the inventory of visualizations for machine learning educators and researchers (e.g., see [6,11]).

Modern implementations of neural networks often use more complex architectures, but the variations typically appear underneath a traditional fully-connected network. That is, these implementations process the raw input (such as an image) with an alternative network, and then use the output of that network as input to a fully-connected network. Convolutional networks interact with fully-connected networks in this way. Figure 2 (right) shows a typical diagram for illustrating a CNN. The algorithm for CNNs relies heavily on the convolution operation, which is used to sequentially apply a series of (learned) filters to the input. A CNN can also be interpreted as a graph, which although is different in appearance from the fully-connected network graph, uses all the same mathematics for learning [12]. Figure 3 illustrates how convolutions can be interpreted as graph-like connections. For students learning about CNNs for the first time, it can be difficult to mentally assimilate the relationship between node-link diagrams and convolutional nets. One of the goals of the current visualization is to make this relationship easier to understand.

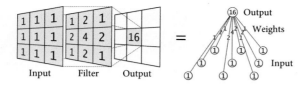

Fig. 3. A 3×3 convolution filter is equivalent to a node with nine weighted connections, where the filter values correspond to the weights. A convolution layer applies this filter to every location in the input image, producing a new (filtered) image.

2.2 Challenges of Complexity and Scale

The visualizations of neural networks in machine learning literature have changed only slightly over the years. The most significant trend is that the node-link diagrams have become smaller and less detailed over time. This is consistent with fully-connected networks' declining novelty, and reflects an intent of emphasizing things other than the structure of the networks. The most common simplified representation replaces each layer of nodes with a solid block, and replaces the dense connections between layers with either a single arrow pointing from one layer to the next [1], or with edges connecting the outskirts of the layers [13]. Simplifying the node-link diagram into a block diagram also addresses a problem of scale. To solve problems of practical value, the required network is often enormously large. A node-link diagram of such a network would have enough edges that the space in between layers would be opaque with lines. A block diagram is therefore an effective way of overcoming this problem of scale. The fully-connected portion of the CNN in Fig. 2 shows a simplification along these lines.

The main issue with grouping the nodes together by layer is that it eliminates the possibility for interaction and detailed analysis. In other domains, where analysis is often the primary focus, other solutions have been introduced. For example, it is sometimes possible to reveal a great deal of information about a set of connected nodes by bundling related edges together [14]. An alternative is to display only a representative subset of the nodes, or perhaps allow users to hover over a node to see edges or stubs leading into it [15,16]. A closely related strategy to these is the use of multi-scale navigation to "search, show context, expand on demand" [17]. In all, visualization research indicates that focus-plus-context techniques [18] could serve as reasonable alternatives to simply not showing the edges. The current work makes use of these ideas.

2.3 Prior Interactive Visualizations

Many interactive neural network visualizations exist. An early and popular visualization is the Stuttgart Neural Network Simulator (SNNS) [19], which shows neural nets as 2D and 3D node-link diagrams, in which the nodes and edges are colored in a scheme that maps negative and positive values to different ends of a palette. However, SNNS is targeted toward researchers, and accordingly its interface and basic usage demand some expertise. A similar tool is N^2VIS [20], which additionally makes use of a compact "matrix-like" visualization of a neuron's weights, in which each cell of the matrix represents a weight, and the cell is coloured according to the weight's magnitude. This is very similar to visualizing a CNN's parameters as filters, although the networks and visualizations in N^2VIS do not actually scale to convolutional networks and vision tasks. Interaction in these and similar research-targeted applications (e.g., [21,22]) typically centers around designing new networks, and making minute adjustments to trained networks, to see how these changes affect performance.

Visualizations designed for a tutorial context are different. For example, *Neural Java* [3] provides an extensive set of web-based exercises and demos, allowing students to experiment with and learn about a variety of neural network designs. In one application, the user can make design choices on a network tasked with solving a toy version of the handwritten digit recognition problem. Once the network is trained according to the user's settings, the user can use a cursor to draw new numbers for the network to classify, and view the network's classification output. Despite there being no depiction of the actual network being trained, this type of application enables students to empirically determine reasonable answers to a variety of challenging questions, concerning (for example) convergence, the optimal number of nodes and layers, translation invariance, and more. These are the types of benefits the current visualization aims to deliver.

Most interactive visualizations only depict fully-connected networks, and furthermore only visualize networks that are too small to be effective at computer vision problems. A recent exception to this is *ConvNetJS* [23], which is a JavaScript library for training neural networks. The website for the project features a set of visualizations, which have interactive examples of neural networks. As in *Neural Java*, one example features a neural network solving a handwritten digit recognition task, although in this case using a standard dataset (MNIST [12]). The visualization shows the network's layer-by-layer activation patterns in response to example inputs, which gives the user an in-depth look at how the network arrives at its final classification. A weakness of the visualization is that the activation patterns are not organized in a way that shows the architecture of the network (e.g., in a node-link diagram); instead these are simply shown in order, from the zeroth layer to the final layer. Also, unlike the example in *Neural Java*, this visualization does not allow users to interactively create new inputs for the network to classify. Nonetheless, *ConvNetJS* sets a high benchmark for scale and practical realism, which the current visualization aims to match.

3   Technical Approach

This section describes the technical approach to creating the visualization framed in the previous section. The discussion begins with an analysis of the activities that users of the visualization are expected to be interested in, then proceeds to describe the major visual elements employed to meet the requirements of those tasks. The section concludes with a discussion of the various interactive elements implemented.

3.1   Task Analysis

The visualization should meet the following goals, which summarize the unique properties of convolutional networks in computer vision, and reflect lessons learned from prior work. First, the visualization should handle networks large enough to solve practical vision tasks, such as handwritten digit recognition. Second, the visualization should depict the entire network architecture with a node-link diagram. Third, the visualization should allow users to easily experiment with the input-output process of the network, allowing them to judge the network's robustness to translational variance, rotational variance, and ambiguous input. Fourth, the visualization should allow users to view details on individual nodes, such as the activation level, the calculation being performed, the learned parameters (*i.e.*, the node's weights), and the numerical inputs and outputs.

3.2   Visual Elements

This section describes the major visual elements of the visualization, emphasizing their justification through the points established in the task analysis, as well as through theoretical principles of visualization. Screenshots are shown in Fig. 4.

Node-link Diagram: The only issue with a straightforward implementation of a node-link diagram is that large networks yield a dense mass of edges between layers. This was addressed by only showing edges for one node at a time, and only on request. This strategy is uniquely possible in domains with simple network architectures: unlike networks describing natural phenomena (*e.g.*, [15,24]), the edges of CNNs can be implied by their regular pattern. This strategy achieves a compromise between the block diagram's simplicity, and the node-link diagram's potential for detail.

Camera: Users can zoom in and out by scrolling, and translate the network by dragging the right mouse button. While exploring the visualization through these controls, the users can focus their attention on particular layers or features, gain familiarity with the architecture of the network, and also build an appreciation for the network's scale.

Node Activations and Edge Strengths: Displaying the activation levels of individual nodes, as well as the strength of edges between nodes, is crucial for creating an informative visualization of a neural network's behavior. At the zeroth layer of the network, the individual nodes correspond to pixels in the input image, so the activation level of each node simply corresponds to the brightness of the respective pixel. At every other layer of the network, the activation level corresponds to how closely the output from the layer below matches with the node's learned "ideal input". The image of the ideal input is represented in the strengths of the node's edges. In this visualization, edge strengths and activation

Fig. 4. Screenshots from the visualization. From left to right, the images show: edges revealed by hovering on a fully-connected node, edges revealed by hovering on a convolutional node, details revealed by clicking on a fully-connected node, details revealed by clicking on a convolutional node, and finally the top layer activations of a network given ambiguous input. This figure is best viewed in a zoomable PDF.

levels were encoded in a (shared) colormap, which mapped low and high values to different ends of a palette. The colormap is a variation on the rainbow palette, making use of a wide range of hues and saturations to ensure that the colors "pop out" from the background; it also uses a linear value curve, to effectively differentiate low and high levels of activation or edge strength. The colormap is consistent with colour-encoding theory, which encourages a mapping across value for ordered elements [25], and discourages the use of a full rainbow pallette [26].

Background: Preliminary user testing revealed that if the background has a solid color, and that color is drawn from the same palette as the activation patterns, then many of the nodes will inevitably blend in with the background. If the background has a color that none of the nodes have, the color schemes may clash, and attention will be drawn away from the nodes and toward the background. The solution was to give the background a large blurred ellipse, which faded in brightness toward the edges, creating the illusion of a soft light emanating from (or being projected into) the center of the visualization. Since the ellipse has a smooth color gradient, the solid-colored nodes cannot blend into it. The solution thus makes it unnecessary to alter the color palette, and also adds an aesthetically pleasing soft lighting effect.

Drawing Pad: The drawing pad is positioned at the top-left corner of the visualization, and includes three buttons just below it: a pen, an eraser, and a "clear" button. The eraser encourages the user to make small adjustments to their input. One of the greatest learning experiences to be derived from interacting with the network is in seeing how the network responds to slight changes in input. If the user gradually morphs a "1" into a "2", the network will respond in its top layer by gradually losing confidence (*i.e.*, activation strength) in the "1" node and gaining confidence in the "2" node. It is worth noting that the drawing pad has more pixels than the actual input layer of the network. The drawn input is thus downsampled before it is used by the network. An alternative strategy (pursued in *Neural Java* [3]) is to give the drawing pad "large pixels". However, the high-resolution strategy is more in line with a current issue of neural networks in computer vision: high-resolution input is available, but a high-resolution network would be computationally intractable. Showing this downsampling stage in the visualization presents another learning opportunity for the user.

Node Information on Focus: Two major focus-plus-context tools were implemented in the visualization. First, by hovering on a node, the user is able to see all of the edges leading toward that node. Second, by clicking on a node, the user can access detailed information about that node, including node's position and role in the network, the numerical input and output of the node, as well as the calculation being performed. For convolutional nodes, since each node's action can be interpreted as applying a filter to a small location in the image, clicking such a node reveals the corresponding image and filter, in addition to the numerical data.

4   Implementation Details

To support the visualization, two networks were trained on an augmented version of the MNIST dataset [12] to classify 28×28 pixel images of handwritten digits. The dataset was augmented with translated, rotated, and skewed versions of the original training data, to improve the network's robustness to these types of variations.

First, a convolutional network was trained, with 1024 nodes on the bottom layer (corresponding to pixels), 6 5×5 (stride 1) convolutional filters in the first hidden layer, followed by 16 5×5 (stride 1) convolutional filters in the second hidden layer, then three fully-connected layers, with 120 nodes in the first, 100 nodes in the second, and 10 nodes in the third (corresponding to the 10 digits). The convolutional layers are each followed by downsampling layer that does 2×2 max pooling (with stride 2). This network achieved a 99 % classification accuracy on the same dataset, which is consistent with the related work [12].

Second, a fully-connected network was trained, with 784 nodes on the bottom layer (corresponding to pixels), 300 nodes in the first hidden layer, 100 nodes in the second hidden layer, and 10 nodes in the output layer (corresponding to the 10 digits). In earlier work, this network architecture was shown to be capable of achieving approximately 97 % classification accuracy on the MNIST test set [12]. The newly trained network achieved approximately 96 % classification accuracy on the same dataset, but showed slightly better invariance to translation, rotation, and skewing, as expected.

This visualization was implemented as a WebGL application, written in JavaScript. As a WebGL application, the visualization is compatible with any device that has an HTML5-ready browser and a GPU. This includes mobile devices like the iPad 2 and higher, iPhone 4 and higher, and more. The drawing pad works especially well with touch-screen devices, since the user can draw with his or her finger, rather than with a mouse. The open source libraries *jquery*, *three.js*, and *math.js* were instrumental in achieving the final result. MATLAB was used to train the neural network, and also to export parameter matrices, colormaps, and 3D network layouts for JavaScript.

5 Empirical Evaluation

A thorough evaluation of the visualization's efficacy as an educational tool would ideally involve a two-arm randomized controlled trial, with machine learning students as participants. Future work may develop an evaluation along these lines. The present evaluation explores instead an important factor of usability known as responsiveness. This factor represents the ability of the system to provide quick and meaningful interaction, and it is tested by measuring the system's response time to user inputs [27]. Studies on human perception have identified three approximate limits for a computer application's response time [28, 29]: 0.1 seconds is the limit for providing an impression of continuous feedback (*e.g.*, the manipulation of objects in the interface should meet this criteria); 1.0 seconds is the limit for providing an impression of an immediate response (*e.g.*, a command given to the application should be carried out within this limit); 10 seconds is the limit for keeping the user engaged (*e.g.*, a delay exceeding this limit may cause the user to switch away from the task).

The current visualization has two main types of interaction: manipulation of the camera, which is carried out on the GPU, and drawing pad interaction which is carried out on the CPU. On a PC with a mid-range graphics card (ATI Radeon HD 4850), the visualization runs at a consistent 60 frames per second while the camera is being manipulated, which is well within the 0.1-second criterion. With a modern processor (Intel Core i7-4770), drawing pad interaction is processed at an average of 36 milliseconds, which is again within the "continuous feedback" 0.1-second criterion. On mobile devices with GPUs (such as the iPad 2), the frame rate of the visualization stays at 60, while the drawing pad interaction slows to approximately 250ms, which places the interaction in the "immediate response" limit. These results suggest that the application is responsive enough to support engaging user interaction on a variety of devices.

In preliminary user testing, observing usage of the drawing pad led to some changes in its functionality: in an early iteration of the visualization, the drawing pad had a "Go" button, which caused the drawing to be input into the bottom layer of the network. This two-stage process allowed the user to prepare a detailed drawing before seeing the network's interpretation of it. However, users did not immediately know what to do with the button, and were confused as to why the network did not respond automatically once a number was drawn. Based on these observations, the button was removed, and the drawing pad was configured to send its data through the network after every stroke.

6 Conclusion

This paper presented a new interactive visualization of convolutional neural networks. Users can interact with the visualization by drawing new digits for the network to classify. The visualization shows the nodes of the network arranged in a node-link diagram, which is novel for visualizations of convolutional networks. Given an input image, the visualization shows the activation level of every node in the network, by setting the brightness and color of each node according to the magnitude of the activation. A variety of details-on-demand techniques were employed to incorporate additional information into the visualization, including node labels, weight images, edges, and a summary of the mathematics being computed at every node. The visualization also enables responsive experimentation with the network's input-output process, allowing users to learn about the network's strengths and weaknesses through interaction. Whereas prior visualizations have only depicted fully-connected, non-interactive, or toy-sized neural networks, the current work demonstrates that practical vision-trained convolutional nets can be effectively depicted in an interactive node-link diagram. The visualization, and its source code, are available at http://scs.ryerson.ca/~aharley/vis/.

Acknowledgements. The author gratefully thanks Tim McInerney and Kosta Derpanis for insightful discussions, and for helping improve the manuscript.

References

1. Krizhevsky, A., Sutskever, I., Hinton, G.E.: Imagenet classification with deep convolutional neural networks. In: NIPS, pp. 1106–1114 (2012)
2. Girshick, R., Donahue, J., Darrell, T., Malik, J.: Rich feature hierarchies for accurate object detection and semantic segmentation. In: CVPR (2014)
3. Corbett, F.D., Card, H.C.: Neural Java: Neural networks tutorial with Java applets (2000). http://lcn.epfl.ch/tutorial/english/. Accessed on 31 Nov 2014
4. Zeiler, Matthew D., Fergus, Rob: Visualizing and understanding convolutional networks. In: Fleet, David, Pajdla, Tomas, Schiele, Bernt, Tuytelaars, Tinne (eds.) ECCV 2014, Part I. LNCS, vol. 8689, pp. 818–833. Springer, Heidelberg (2014)
5. Craven, M.W., Shavlik, J.W.: Visualizing learning and computation in artificial neural networks. Int. J. Artif. Intell. Tools **1**, 399–425 (1992)
6. Bishop, C.M.: Pattern Recognition and Machine Learning. Springer-Verlag New York Inc, Secaucus, NJ, USA (2006)
7. Rosenblatt, F.: The perceptron: a probabilistic model for information storage and organization in the brain. Psychol. Rev. **65**, 386–408 (1958)
8. Lodish, H., Berk, A., Zipursky, S.L., Matsudaira, P., Baltimore, D., Darnell, J.: Molecular Cell Biology, 4th edn. W.H. Freeman, New York (2001)
9. Werbos, P.J.: Beyond Regression: New Tools for Prediction and Analysis in the Behavioral Sciences. Ph.D thesis, Harvard University (1974)
10. Rumelhart, D.E., Hinton, G.E., Williams, R.J.: Learning representations by back-propagating errors. Nature **323**, 533–536 (1986)
11. Jarrett, K., Kavukcuoglu, K., Ranzato, M., LeCun, Y.: What is the best multi-stage architecture for object recognition? In: ICCV (2009)
12. LeCun, Y., Bottou, L., Bengio, Y., Haffner, P.: Gradient-based learning applied to document recognition. Proc. IEEE **86**, 2278–2324 (1998)
13. LeCun, Y., Kavukvuoglu, K., Farabet, C.: Convolutional networks and applications in vision. In: ISCAS, pp. 253–256 (2010)
14. Holten, D.: Hierarchical edge bundles: visualization of adjacency relations in hierarchical data. TVCG **12**, 741–748 (2006)

15. Al-Awami, A., Beyer, J., Strobelt, H., Kasthuri, N., Lichtman, J., Pfister, H., Hadwiger, M.: NeuroLines: a subway map metaphor for visualizing nanoscale neuronal connectivity. TVCG **20**, 2369–2378 (2014)
16. Lex, A., Partl, C., Kalkofen, D., Streit, M., Gratzl, S., Wassermann, A.M., Schmalstieg, D., Pfister, H.: Entourage: visualizing relationships between biological pathways using contextual subsets. TVCG **19**, 2536–2545 (2013)
17. Van Ham, F., Perer, A.: Search, show context, expand on demand: supporting large graph exploration with degree-of-interest. TVCG **15**, 953–960 (2009)
18. Shneiderman, B.: The eyes have it: a task by data type taxonomy for information visualizations. In: Visual Languages, pp. 336–343 (1996)
19. Zell, A., Mache, N., Hbner, R., Mamier, G., Vogt, M., Schmalzl, M., Herrmann, K.U.: SNNS (Stuttgart Neural Network Simulator). In: Skrzypek, J. (ed.) Neural Network Simulation Environments: The Kluwer International Series in Engineering and Computer Science, vol. 254, pp. 165–186. Springer, US (1994)
20. Streeter, M.J., Ward, M.O., Alvarez, S.A.: N2Vis: an interactive visualization tool for neural networks. In: Visual Data Exploration and Analysis VII, pp. 234–241 (2001)
21. Tzeng, F.Y., Ma, K.L.: Opening the black box - Data driven visualization of neural networks. In: Visualization, pp. 383–390 (2005)
22. Srp, J., Stehlík, L., Suda, M., Šašek, P., Zvánovcová, K.: Interactive neural network simulator (2007). http://sourceforge.net/projects/isns/. Accessed on 31 Nov 2014
23. Karpathy, A.: ConvNetJS: Deep learning in your browser (2014). http://cs.stanford.edu/people/karpathy/convnetjs/. Accessed on 31 Nov 2014
24. Henry, N., Fekete, J.D., McGuffin, M.J.: NodeTrix: a hybrid visualization of social networks. TVCG **13**, 1302–1309 (2007)
25. Ware, C.: Information Visualization: Perception for Design. Elsevier, Amsterdam (2012)
26. Rogowitz, B.E., Treinish, L.A.: How not to lie with visualization. Comput. Phys. **10**, 268–273 (1996)
27. Nielsen, J.: Usability Engineering. Elsevier, Boston (1993)
28. Miller, R.B.: Response time in man-computer conversational transactions. In: Proceedings of AFIPS Fall Joint Computer Conference, vol. 33, pp. 267–277 (1968)
29. Card, S.K., Robertson, G.G., Mackinlay, J.D.: The information visualizer: an information workspace. In: ACM CHI, pp. 181–188 (1991)

DPN-LRF: A Local Reference Frame for Robustly Handling Density Differences and Partial Occlusions

Shuichi Akizuki$^{(\boxtimes)}$ and Manabu Hashimoto

Graduate School of Science and Technology, Chukyo University,
101-2, Yagoto-Honmachi, Showa-ku, Nagoya, Aichi, Japan
{akizuki,mana}@isl.sist.chukyo-u.ac.jp

Abstract. For the purpose of 3D keypoint matching, a Local Reference Frame (LRF), a local coordinate system of the keypoint, is one important information source for achieving repeatable feature descriptions and accurate pose estimations. We propose a robust LRF for two main point cloud disturbances: density differences and partial occlusions. To generate LRFs that are robust to such disturbances, we employ two strategies: normalizing the effects of point cloud density by approximating the surface area in the local region and using the dominant orientation of a normal vector around the keypoint. Experiments confirm that the proposed method has higher repeatability than state-of-the-art methods with respect to density differences and partial occlusions. It was also confirmed that the method enhances the reliability of keypoint matching.

1 Introduction

3D object detection and localization using point clouds are basic tasks for robots working in human living environments. However, the performance of vision systems for achieving such tasks is sometimes disturbed by two types of conditions. One is the difference in point cloud density between an object model and a target object in the input scene, because the distance between a range sensor and the target object is unknown. The other is the occurrence of missing point clouds, because in this case objects frequently occlude each other in the input scene.

Model-based methods are among the types of methods typically used for 3D object recognition. Such methods generally have main four modules, one each for:

1. Extracting keypoints from the object model and the input scene
2. Describing 3D features
3. Finding corresponding points between the object model and the input scene
4. Estimating position and pose of the object model in the input scene.

After the module 1, a Local Reference Frame (LRF), a local coordinate system of the keypoint, is calculated. The LRF consists of three orthogonal unit 3D vectors. A rigid transformation that aligns corresponding LRFs can be determined as the pose parameter of the object model in the input scene. In addition,

© Springer International Publishing Switzerland 2015
G. Bebis et al. (Eds.): ISVC 2015, Part I, LNCS 9474, pp. 878–887, 2015.
DOI: 10.1007/978-3-319-27857-5_78

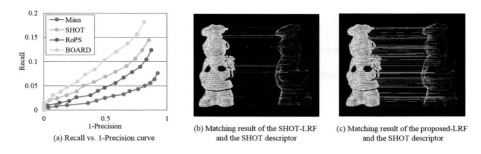

(a) Recall vs. 1-Precision curve

(b) Matching result of the SHOT-LRF and the SHOT descriptor

(c) Matching result of the proposed-LRF and the SHOT descriptor

Fig. 1. 3D keypoint matching performance. (a) Recall vs. 1-precision curve. Keypoints were matched to each other by using a SHOT descriptor and four types of LRFs. (b), (c) Example matching results for SHOT-LRF and the proposed LRF with the SHOT descriptor. Green/red lines show correct/incorrect correspondences.

most of the 3D features [1,2] are calculated by dividing support regions into multiple cells according to each axis direction of the LRF, and the geometric relations of point clouds in the cells are converted to a feature vector. In this way, a repeatable LRF can enhance not only the accuracy of pose estimation, but also the reliability of keypoint matching. Figure 1 (a) shows recall vs. 1-precision curves we obtained in a keypoint matching experiment. In this experiment, we used a pair of point clouds with different densities and missing region.

We used a SHOT descriptor and four types of LRFs: Mian [3], SHOT [4], RoPS [2], and BOARD [5]. Note that the only difference among the curves is due to the methods used to calculate the LRF. Replacing the LRF confirmed the keypoint matching performance changed. Figure 1 (b) and (c) show keypoint matching results; (b) shows those obtained with SHOT-LRF and (c) shows those obtained with the proposed LRF. Many more corresponding points were obtained for the (c) results than for the (b) results.

These results led us to conclude that we should select a suitable LRF by considering the disturbances that occur in the application environment. The purpose of our research is to achieve a repeatable LRF for changing point cloud density and occlusions.

2 Related Work

There are two types of LRFs. One uses the Eigenvector of a covariance matrix generated by a point cloud within the support region centered in the keypoint. The methods described in [3,6] are the simplest of these. With these methods, Eigenvectors are assigned as each of the LRF axes. However, the directions of Eigenvectors have sign ambiguities that decrease the uniqueness of the LRF. In order to solve this problem, the SHOT [4] determines the major direction by counting the number of 3D points existing on both sides of the support regions.

RoPS [2] and DosSants [7] have been used to tackle the problem of point cloud density. With these methods, a normalized scatter matrix is used for generating the LRF. This matrix is normalized by the total area of the local surface, which is approximated as the summation of the area of each mesh. However, the repeatability of the direction of each axis will decrease when point clouds within the support region are partially missed due to occlusion. This is because in such cases the point cloud distribution will be changed. The problem with these Eigenvector-based methods is that all points within the local region contribute to the calculation of the axis direction.

The other LRFs calculate three axes one by one [1,5,8–12]. These methods commonly assign the normal vector of a local surface as the z-axis of the LRF because the direction of it has high repeatability. Therefore, the method used to calculate the x-axis characterizes each LRF. In the methods proposed in [9,10], an arbitrary point within the support region is projected onto the tangent plane of the z-axis. A vector that indicates the projected point from the origin of the z-axis is assigned as the x-axis of the LRF. However, the uniqueness of the x-axis direction is not so high because it is difficult to obtain the repeatable projected point.

The BOARD [5] has solved this problem by detecting the point that has the most inclined normal vector respective to the z-axis, and it is projected onto the tangent plane. This method is robust to occlusions because it estimates whether the most inclined normal vector exists or not in the missing region. The Mesh HoG [12] determines the dominant orientation of the polar angle on the tangent plane of the z-axis by using the polar histogram of the surface gradient. This LRF is also robust to occlusions because "dominant" information has been used for generating the x-axis. However, it has been reported in [2] that the repeatability of LRF decreases when point clouds have different densities. Therefore, there is no method that is robust to density differences and occlusions at the same time.

3   DPN-LRF: Dominant Projected Normal LRF

3.1   Overview

To deal with the effects of differences in point cloud density and those of occlusions, the proposed LRF consists of two strategies. The first is to take into account the weighting factor for the area of the local surface mesh as reported in [2]. This is one good solution for solving the problem of point cloud density. It also uses information about the area of the local surface. We also propose a method for quickly calculating the weighting factor for the area of the local surface mesh. The other strategy is to determine an x-axis that is robust to partial occlusions. Since the proposed method assigns the Dominant Projected Normal vector as the x-axis of the LRF, we named this method DPN-LRF for short.

3.2   Method for Generating the DPN-LRF

The DPN-LRF at the keypoint $\mathbf{p}$ and its normal vector $\mathbf{n}$ is calculated by using point cloud $\{\mathbf{p}_0, \mathbf{p}_2, ..., \mathbf{p}_N\}$ and its normal vectors $\{\mathbf{n}_0, \mathbf{n}_1, ..., \mathbf{n}_N\}$ within the support region centered in the $\mathbf{p}$. Figure 2 shows an overview of the method used to generate the DPN-LRF.

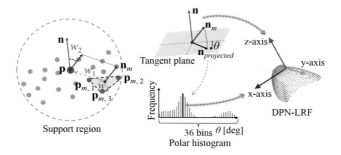

Fig. 2. Overview of the method used to generate DPN-LRF. At left is a point cloud within the support region centered on the keypoint (red point). At center top is the tangent plane of the z-axis of the DPN-LRF; θ shows orientation of the projected normal vector. At center bottom is the polar histogram. At right is the DPN-LRF on the point cloud within the support region.

z-axis: This axis is defined as the Eigenvector corresponding to the smallest Eigenvalue of covariance matrix generated from point cloud around keypoints. This vector has sign ambiguity, but this problem can easily be solved by determining that the positive direction is the direction towards the viewpoint. This vector corresponds to the normal vector $\mathbf{n}$ of each point. We used point cloud within 10 mr ("mr"=mesh resolution) p for calculating covariance matrix.

x-axis: This axis is generated as the dominant orientation of the polar direction on the tangent plane of the z-axis centered in the $\mathbf{p}$. Here, the orientation θ of the projected normal vector $\mathbf{n}_{projected}$ onto the tangent plane is voted to a polar histogram that has 36 bins (covering 360 degrees). To deal with aliasing, each vote is smoothed by using Gaussian distribution. In this histogram, a direction that has a high voted value is assigned as the x-axis direction. To generate a highly accurate axis, Bilinear interpolation is applied by using neighboring bins. Because this process uses histogram-formed data, the generated x-axis becomes stable even if point clouds are missed due to occlusions. The voted value is calculated by multiplication of the following three weighting factors.

w_1 is the weighting factor for normalizing density differences in point clouds. In the corresponding points of the object model and the input scene, the local surface areas within the support regions are the same even if the point densities differ. The voting value is normalized by using the local surface area.

First of all, point cloud is converted to triangle meshes $\{t_0, t_1, ..., t_M\}$. Here, the m-th mesh consists of $\mathbf{p}_{m,1}$, $\mathbf{p}_{m,2}$, and $\mathbf{p}_{m,3}$, and the normal vector $\mathbf{n}_m$ of t_m is calculated as $(\mathbf{n}_{m,1} + \mathbf{n}_{m,2} + \mathbf{n}_{m,3})/3$. The voting process is performed for each triangle. The weighting value w_1 of the m-th mesh is described by

$$w_{m,1} = \frac{(\mathbf{p}_{m,2} - \mathbf{p}_{m,1}) \times (\mathbf{p}_{m,3} - \mathbf{p}_{m,1})}{\sum_{m=0}^{M}(\mathbf{p}_{m,2} - \mathbf{p}_{m,1}) \times (\mathbf{p}_{m,3} - \mathbf{p}_{m,1})} . \tag{1}$$

Since a larger mesh represents the local shape more strongly, a large mesh gets a large weighting value. w_2 represents the distance between the keypoint and the mesh and is described by

$$w_{m,2} = r - \left\| \frac{(\mathbf{p}_{m,1} + \mathbf{p}_{m,2} + \mathbf{p}_{m,3})}{3} - \mathbf{p} \right\|. \tag{2}$$

In general, a mesh that is distant from the keypoint is sometimes affected by the presence of clutter. Therefore, by decreasing the weighing value of such normal vectors, the effects of noise can be decreased. w_3 represents the stability of the projected normal vector and is defined by

$$w_{m,3} = 1 - \mathbf{n} \cdot \mathbf{n}_m. \tag{3}$$

Here, $\cdot$ represents a dot product. When $\mathbf{n}_m$ is steeply inclined towards $\mathbf{n}$, the polar direction of the projected vector becomes stable. Therefore, the weighing value of the normal vector should become large.

y-axis: This is the perpendicular axis of the z-axis and the x-axis. Therefore, it is calculated by z-axis $\times$ x-axis.

3.3   Fast DPN-LRF Generation from Unorganized Point Cloud

By taking the area of each triangle mesh into account, the LRF can obtain robustness for differences in point cloud density. Unfortunately, triangle meshes are not always available, because unorganized point clouds are generally captured by the range sensor. Therefore, for generating the DPN-LRF, it is necessary to generate triangle meshes as an additional processing. Moreover, many 3D features do not require triangle meshes for computing. In practical use, it is preferable to generate the LRF without triangle meshes.

In this subsection, we propose a method to calculate the DPN-LRF without triangle meshes. In particular, the area of a triangle mesh is approximated as the distance between $\mathbf{p}_n$ and its nearest neighbor. (1) is replaced with

$$w'_{n,1} = \frac{\|\mathbf{p}_{nearest} - \mathbf{p}_n\|}{\sum_{n=0}^{N} \|\mathbf{p}_{nearest} - \mathbf{p}_n\|}. \tag{4}$$

Here, $\mathbf{p}_{nearest}$ represents the nearest point of $\mathbf{p}_n$. Also, (2) and (3) are respectively replaced

$$w'_{n,2} = r - \|\mathbf{p}_n - \mathbf{p}\|, \tag{5}$$

$$w'_{n,3} = 1 - \mathbf{n} \cdot \mathbf{n}_n. \tag{6}$$

The computational cost of the above modified equations is lower than that of the equations described in Sect. 3.2. Therefore, we named the LRF calculated using (4) – (6) "Fast DPN-LRF".

4   Experiments

4.1   Performance for Point Density

To evaluate the various methods' performance for density differences in point clouds, we performed an experiment in which LRF repeatability was calculated on corresponding points. In this experiment, we used the object models provided by [3]. For simulating a point cloud captured by a range sensor, we applied the Hidden Point Removal (HPR) operator [13] to each model. We define these models as M and present them in Fig. 3.

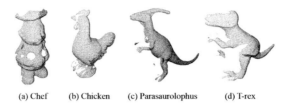

(a) Chef   (b) Chicken   (c) Parasaurolophus   (d) T-rex

Fig. 3. Overview of the object models M.

Sparse point clouds S were generated by adding Gaussian noise with a standard deviation of 0.1 mr and then downsampled to $1.0 - 5.0$ mr. We randomly extracted 1,000 keypoints from each model. The corresponding point in the S is determined as the nearest point to the keypoint of the M.

Figure 4 shows the percentage of LRF that has error within 10 [deg]. In this experiment, we compared our methods with the Mian, the SHOT, the RoPS,the PS-LRF and BOARD methods. In order to estimate the effect of the proposed weighting factors w_1, w_2, and w_3, we also compared the performance of the DPN (b). This method do not use weighting factors when the DPN-LRF is generated. All methods are implemented in C++, and we used the Point Cloud Library [14]. In this experiment, we tested different values of support radius r (5, 10, 15, and 20 mr). $r = 20$ mr is best for all method except for z-axis of the PS. Oprimal radius of this method was 5 mr.

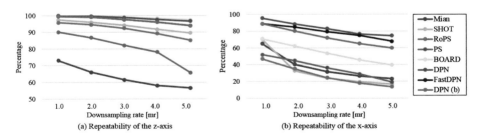

(a) Repeatability of the z-axis   (b) Repeatability of the x-axis

Fig. 4. The relationship between the angular error of z, the x-axis of LRF, and the point cloud density. The reliability of the z-axis and x-axis are shown in (a) and (b). The vertical axis represents the percentage of LRF that has error within 10 degrees.

In terms of z-axis repeatability, all methods achieved highly accurate estimation except for the Mian, with which some of the z-axes generated were directed toward the opposite side. When the downsampling rate was changed from 1.0 to 5.0 *mr*, the performance of the BOARD decreased by 2.7%, while that of DPN and Fast DPN decreased by 5.7%. This shows the z-axis repeatability is almost equal for BOARD-LRF and the proposed methods.

In terms of x-axis repeatability, the performance of the previous methods significantly decreased. The decrease was 54.0% for SHOT and 31.6% for BOARD, but was only 21.1% for DPN and the Fast DPN. Moreover, for a 5.0 *mr* downsampling rate, the repeatability of DPN was 8.1% higher than that of DPN (b). This confirmed that the proposed weighting factors helped to increase the repeatability of the axis direction when the point cloud density was different.

4.2   Performance for Partial Occlusions

To evaluate the methods' performance for partially missed point clouds within the support region, we calculated the relationship between the number of missing points and the LRF repeatability. In this experiment, we used M as the model data. The scene data was one-sided point clouds generated by using rotated object models and the HPR operator. Gaussian noise was added to each target. The range of rotation was 0 – 90 [deg] and the pitch was 10 [deg]. We randomly extracted keypoints $\mathbf{p}_m$ and $\mathbf{p}_s$ from the model and the scene, respectively. Corresponding points were determined as those that satisfied $|\mathbf{p}_m - \mathbf{R}\mathbf{p}_s| < 2.5[mr]$, where $\mathbf{R}$ is the rotation matrix. By using this process, we were able to generate partially missing point clouds in the support region of corresponding points. Figure 5 shows the relationship between the percentage of angular error of LRF within 10 [deg] and the missing rate of point clouds calculated by *missing rate* $= |N_M - N_S|/max(N_M, N_S)$. Here, N_M and N_S respectively represent the number of points within the support regions of $\mathbf{p}_m$ and $\mathbf{p}_s$. A higher value means that a lot of points have been missed. In this experiment, we calculated the angular error e of LRF by

$$e = \arccos\left(\frac{trace(\mathbf{L}_S\mathbf{L}_M^{-1}) - 1}{2}\right)\frac{180}{\pi} . \tag{7}$$

This equation proposed by [2]. $\mathbf{L}_S$ and $\mathbf{L}_M$ represent corresponding LRFs and are formed as 3×3 matrix. If the corresponding LRFs have no error, $trace(\mathbf{L}_S\mathbf{L}_M^{-1})$ will become the identity matrix. In this case, the value e takes zero.

In this experiment, we used over 20k corresponding points and tested different values of support radius r (5, 10, 15, and 20 *mr*). The results for the highest (optimal) radius for each method are 20, 20, 15, 20, 20, 20, and 20 in order from 'Mian' to 'Fast DPN'.

The stability of the Eigenvector direction is not so high when point clouds within the support region are partially missed. As a result, the repeatability of the Eigenvector-based methods Mian, SHOT, and RoPS significantly decreased when the missing rate was high. In contrast, PS and BOARD has a function to deal with the missed regions and so its repeatability was higher than that of

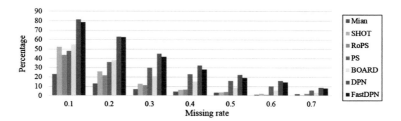

Fig. 5. Relationship between the angular error of LRF and the missing rate. Vertical axis represents percentage of LRF has error within 10 degree.

Eigen vector-based method. For every missing rate, it was confirmed that the repeatability of the proposed method is higher than that of the other methods.

4.3 Performance Evaluation for Keypoint Matching

The results obtained from the aforementioned experiments confirmed that the proposed LRF has higher repeatability than any of the other LRFs to which it was compared. Therefore, in the application of keypoint matching for object localization, it can be expected that the reliability of matching will be enhanced by replacing any of the previous LRF with the proposed one. In this subsection, we report the performance obtained for each method by recall vs. 1-precision curves as recommended in [15].

We used point clouds with four types of conditions: (a) Gaussian noise with a standard deviation of 0.1 mr, (b) Gaussian noise and 3.0 mr downsampling, (c) Gaussian noise and occlusion, and (d) Gaussian noise, downsampling and occlusion. In order to eliminate keypoint detector errors, we used 1,000 randomly extracted corresponding points. Occluded corresponding points are generated by method used in Sect. 4.2. The results obtained are shown in Fig. 6.

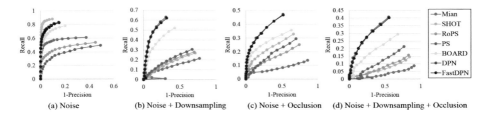

Fig. 6. Recall vs. 1-precision curves for four controlled conditions.

Under condition (a), the best results were obtained with SHOT and the results obtained with BOARD, DPN, and Fast DPN were also good. This confirmed that these LRFs are robust to noise. Under condition (b), the SHOT

Table 1. Computational time of each LRF.

Method	Mian [3]	SHOT [4]	RoPS [2]	PS [9]	BOARD [5]	DPN	Fast DPN
T [msec]	0.12	0.19	0.60	0.24	0.18	0.25	0.27

curve dropped significantly because when the downsampling rate was 3.0 mr, the x-axis repeatability of SHOT was less than 25 % (see Fig. 4 (b)). In contrast, the BOARD, DPN, and Fast DPN performances were relatively higher than those of other LRFs. Under conditions (c) and (d), all methods showed lower performance than they did under conditions (a) and (b). The DPN and Fast DPN performances were also relatively higher than those of the other LRFs.

In the all conditions, curves of the DPN-LRF and Fast DPN-LRF are almost equal. Therefore, it was confirmed that Eq. (4) is a practical solution for approximation of the area of meshes.

4.4 Processing Time Comparison

We calculated the processing time in an experiment for generating LRF. In this experiment, we measured the average processing time of the LRF at a keypoint for all methods. The results are shown in Table 1. This experiment is performed on a desktop with an Intel CORE^{TM}i7 860 CPU and 12GB RAM.

Since RoPS and DPN require mesh generation and the mesh generation computation time depends on algorithms, we did not include these methods in the experiment. Although we were able to improve the Fast DPN processing time by approximating the area of meshes, it was still slightly slower than that of the other methods. However, this is not a problem because it requires more LRFs to get the same number of correct corresponding points by using previous LRFs.

5 Conclusion

We have proposed a novel Local Reference Frame (LRF) for robustly handling two main point cloud disturbances: density differences and partial occlusions. The method comprises two types, Dominant Projected Normal (DPN)-LRF and Fast DPN-LRF. Experiment results confirmed that the method is more robust to differences in point cloud density and partially missing clouds than other methods to which it was compared. In a keypoint matching experiment, we also confirmed that the reliability of matching was enhanced by replacing previous LRF with the proposed one. This suggests that a good combination of LRF and 3D features will provide even more reliable recognition results. Accordingly, in future work, we will consider ways of achieving such a combination.

Acknowledgements. This work was partially supported by Grant-in-Aid for Scientific Research (C) 26420398.

References

1. Frome, A., Huber, D., Kolluri, R., Bülow, T., Malik, J.: Recognizing objects in range data using regional point descriptors. In: Pajdla, T., Matas, J.G. (eds.) ECCV 2004. LNCS, vol. 3023, pp. 224–237. Springer, Heidelberg (2004)
2. Guo, Y., Sohel, F.A., Bennamoun, M., Lu, M., Wan, J.: Rotational projection statistics for 3D local surface description and object recognition. Int. J. Comput. Vis. **105**, 63–86 (2013)
3. Mian, A.S., Bennamoun, M., Owens, R.A.: On the repeatability and quality of keypoints for local feature-based 3d object retrieval from cluttered scenes. Int. J. Comput. Vis. **89**, 348–361 (2010)
4. Tombari, F., Salti, S., di Stefano, L.: Unique signatures of histograms for local surface description. In: European Conference on Computer Vision, pp. 356–369 (2010)
5. Petrelli, A., di Stefano, L.: On the repeatability of the local reference frame for partial shape matching. In: IEEE International Conference on Computer Vision, pp. 2244–2251 (2011)
6. Zhong, Y.: Intrinsic shape signatures: a shape descriptor for 3d object recognition. In: Proceedings of the International Conference on Computer Vision Workshops, pp. 689–696 (2009)
7. dos Santos, T.R., Franz, A.M., Meinzer, H., Maier-Hein, L.: Robust multi-modal surface matching for intra-operative registration. In: IEEE International Symposium on Computer-Based@Medical Systems, pp. 1–6 (2011)
8. Stein, F., Medioni, G.G.: Structural indexing: Efficient 3-d object recognition. IEEE Trans. Pattern Anal. Mach. Intell. **14**, 125–145 (1992)
9. Chua, C.S., Jarvis, R.: Point signatures: a new representation for 3D object recognition. Int. J. Comput. Vis. **25**, 63–85 (1997)
10. Sun, Y., Abidi, M.A.: Surface matching by 3D point's fingerprint. In: ICCV, pp. 263–269 (2001)
11. Novatnack, J., Nishino, K.: Scale-Dependent/Invariant local 3D shape descriptors for fully automatic registration of multiple sets of range images. In: Forsyth, D., Torr, P., Zisserman, A. (eds.) ECCV 2008, Part III. LNCS, vol. 5304, pp. 440–453. Springer, Heidelberg (2008)
12. Zaharescu, A., Boyer, E., Varansi, K., Horaud, R.: Surface feature detection and description with applications to mesh matching. In: Proceedings of the Computer Vision and Pattern Recognition (CVPR), pp. 373–380 (2009)
13. Katz, S., Tal, A., Basri, R.: Direct visibility of point sets. ACM Trans. Graph. **26**, 24 (2007)
14. Rusu, R.B., Cousins, S.: 3d is here: Point cloud library (PCL). In: IEEE International Conference on Robotics and Automation, ICRA, IEEE (2011)
15. Ke, Y., Sukthankar, R.: PCA-SIFT: a more distinctive representation for local image descriptors. In: CVPR (2), pp. 506–513 (2004)

3D Perception for Autonomous Robot Exploration

Jiejun Xu$^{(\boxtimes)}$, Kyungnam Kim, Lei Zhang, and Deepak Khosla

HRL Laboratories, LLC, Malibu, USA
{jxu,kkim,lzhang,dkhosla}@hrl.com

Abstract. We propose an online 3D sensor-based algorithm for autonomous robot exploration in an indoor setting. Our algorithm consists of two modules, a proactive open space detection module, and a reactive obstacle avoidance module. The former, which is the primary contribution of the paper, is responsible for guiding the robot towards meaningful open spaces based on high level navigation goals. This generally translates to identifying open doors or corridor vanishing points in a typical indoor setting. The latter is a necessary component that enables safe autonomous exploration by preventing the robot from colliding with objects along the moving path. Assuming a 3D range sensor is mounted on the robot, it continues to scan and acquire signal from its surroundings as it explores in an unknown environment. From each 3D scan, the two modules function cooperatively to identify any open spaces and obstacles within the generated point cloud using robust geometric estimation methods. Combination of the two modules provides the basic capability of a autonomous robot to explore an unknown environment freely. Experimental results with the proposed algorithm on both real world and simulated data are promising.

1 Introduction

The ability to explore unknown and dynamic environments is an essential component of an autonomous mobile robot system. The basic problem is the following: after being deployed into an unknown environment, a robot must continuously perform sensing operations to decide the next exploration location, and possibly maintain an internal representation of the environment. A well-studied and related problem is the SLAM (Simultaneous Localization and Mapping) problem, whose goal is to integrate the information collected during navigation into an accurate map [1]. However, SLAM does not address the question of where the robot should go next. Another related and well-known problem is path planning [2]. Unlike the original problem, where complete or partial knowledge of the environment is given, exploring an unknown environment is usually performed in a step-by-step greedy fashion. Instead of planning the entire trajectory of the mobile robot in advance, we emphasize an exploration strategy which plans one step (or a few steps) ahead based on the information about the environment acquired by the onboard sensors. In other words, the main problem we address

© Springer International Publishing Switzerland 2015
G. Bebis et al. (Eds.): ISVC 2015, Part I, LNCS 9474, pp. 888–900, 2015.
DOI: 10.1007/978-3-319-27857-5_79

in this work is determining where the robot will move next. Specifically we focus our attention on an indoor environment, such as corridors or warehouses. In such a setting, the next "good" candidate positions to explore are typically meaningful open spaces based on high-level navigation goals (e.g., open doors and vanishing point direction).

Another necessary component for autonomous robot exploration is the "sense and avoid" capability [3,4]. To ensure safe navigation, a robot must be able to achieve self-separation and collision avoidance with other objects (e.g., pedestrians and other robots) on its moving path. To this end, we propose a robust online algorithm consisting of two modules: a proactive open space detection module and a reactive obstacle avoidance module. We believe an effective combination of the two modules provides the basic capability to an autonomous robot to freely explore an unknown environment.

Many 2D-based vision algorithms have been proposed for robot exploration in both indoor [5,6] and outdoor spaces [7,8]. However, 2D visual information could be difficult to use in a 3D world, especially when the 3D structure needs to be inferred for the specific application [9]. The advent of inexpensive 3D (or RGB-D) sensing hardware such as the Microsoft Kinect, Asus Xtion and PrimeSense Capri sensors brings new opportunities to capture 3D environment in a more straightforward manner [10–12]. In this paper, we focus on 3D perception for the indoor exploration task. Our proposed algorithm functions by continuously acquiring and processing 3D data as it moves in an unknown environment. From each 3D scan, the two modules in the proposed algorithm work cooperatively to identify any open spaces and obstacles within the 3D point cloud using robust geometric estimation methods. The next moving direction can be determined by using results from both of these modules. A key benefit of our algorithm is that it is platform-agnostic and can be applied to both ground-based and aerial (e.g., Micro Air Vehicles) robots. We believe our work is applicable to other related areas, such as search-and-rescue, mapping, and 3D modeling.

2 Related Work

In this section, we provide a brief survey of existing literature related to our work on indoor exploration, specifically focusing on door detection and obstacle avoidance. Several visual door detection algorithms have been proposed since it is a topic of much relevance in both robot navigation and manipulation tasks. Most of the earlier works focused on extracting 2D appearance and shape features to recognize doors. For example, Murillo et al. [13] proposed a probabilistic approach by defining the likelihood of generating the door configurations with various features. Tian et al. [14] further proposed a more generic geometric-based approach to detect doors based on stable features, such as edges and corners. In [15], a dedicated real-time edge-based tracker is used to locate and track both open and closed doors within a corridor. A similar approach is also seen in [16]. However, as in most 2D vision algorithms, these approaches are sensitive to common factors such as illumination changes, occlusion, clutter, and perspective distortion. On the other hand, recent advances in sensor technology have enabled capture of 3D data with inexpensive range sensors (e.g., Kinect, Xtion). The additional depth information is especially helpful in detecting open spaces in indoor environments. In addition, the quality of 3D point cloud data is less sensitive to the aforementioned environmental factors. Recently, promising results have been shown by Rusu et al. [17] in detecting doors with only 3D point clouds. In their work, the main focus was to open and close doors through precise arm manipulation based on a pre-learned door template. Similarly Derry et al. proposed a 3D based algorithm to detect doors for assistive wheelchairs control [18]. Their algorithm imposes a prior assumption on the geometric orientation of the point cloud. For instance, the corners and boundaries of the doors are defined based on the assumption that the robot and doors reside on the same surface; thus it is not applicable to aerial robots. Furthermore, their algorithm only detects doors from a wall that is mostly front-facing. In contrast, our proposed algorithm explicitly computes the geometry of the 3D point cloud to detect open doors independent of robot position and is also capable of detecting doors from multiple walls at different angles with respect to the robot.

Another research area relevant to this paper is obstacle detection and avoidance. Due to the growing interest in small robots navigating in indoor and other GPS denied environments, quite a few research works have been carried out to utilize visual sensors to ensure safe maneuver in these conditions. Techniques focusing on monocular (2D) sensors have long been the main studied area. This includes an optical flow-based approach by Zingg et al. [5] and an imitation learning-based approach by Ross et al. [19]. In the recent years, there has been much interest in developing similar types of methods for 3D range sensor-based obstacle avoidance. However, regardless of 2D or 3D vision, a common theme across most works on obstacle avoidance is the reactive action control for robots. This means that the control system will react and prevent the robot from colliding with an impending obstacle. However, it does not plan ahead for exploring far away spaces. Our proposed algorithm overcomes this weakness by coupling obstacle avoidance with active open space detection in autonomous exploration.

3   Methodology

This section presents the details of our algorithm. We assume a 3D range sensor is mounted on the robot. In our work, we utilize the Xtion sensor. The device consists of two cameras and an Infrared (IR) laser light source. The IR source projects a pattern of dark and bright spots onto the environment. This pattern is then received by one of the two cameras which are designed to be sensitive to IR light. Depth measurements can be obtained from the IR pattern by triangulation. Subsequently, a point cloud is constructed internally in the sensor and is then requested by our algorithm directly [10]. Note that the returned point cloud is in real-world scale, and is described in a common metric system. Given the 3D point cloud, the outputs of the algorithm are the direction of the open space (e.g., open doors or the vanishing point) and the direction to avoid the obstacle (if applicable) with respect to the sensor coordinates. Both of these directions

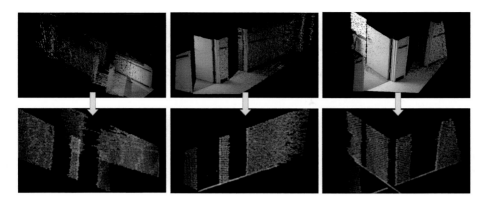

Fig. 1. Examples of vertical plane extraction. Top: input point clouds from an Xtion sensor. Note that viewing angle w.r.t sensor decreases from top to bottom. Bottom: extracted vertical planes after the filtering and downsamping step (color indicates distance from the sensor) (Color figure online).

can be combined to guide the robot towards the next exploration position. We now describe the two modules of the proposed algorithm.

3.1 Proactive Open Space Detection

Filtering and Downsampling: 3D range sensors typically generate voluminous data that cannot be processed in its entirety for many real-time applications. Thus the first step of our algorithm is to reduce the number of points in the point cloud by filtering and downsampling for efficient subsequent computations. Filtering consists of removing points which lie outside of a specific range. Typically an Xtion sensor has an effective range of 5 meters. Thus in our work, we remove all points beyond 5 meters.

In terms of downsampling the point cloud, a typical voxelized grid approach is taken. A 3D voxel grid is essentially a set of fixed width 3D boxes in space over the input point cloud data. In each voxel, all the points will be approximated by their centroid. A 3D voxel grid can be created efficiently with a hierarchical Octree [20] data structure. Each Octree node has either eight children or no children. The root node describes a cubic bounding box which contains all points. At every tree level, this space is further subdivided by a fixed factor, which results in an increased voxel resolution. In this work, we utilize the VoxelGrid functionality implemented in the Point Cloud Library (PCL). The size of each voxel is fixed at 0.05 meter. A significant portion of the points are removed by the end of this step.

Plane Extraction: In this step, we extract ground surface and walls (e.g., vertical planes which reside on the ground surface) from the previously processed point cloud. The ground surface is important as it serves as the key reference for geometrical estimates. The walls are the candidate search space for open doors in the context of our application. This step is essentially done by fitting a planar model to the point cloud and finding the ones with sufficient number of points. To speed up the search process, Random Sample Consensus (RANSAC) [21] algorithms is used to generate plane model hypotheses. The Point Cloud Library provides a convenient implementation to extract the planes in their parameter form $ax + by + cz + d = 0$. Planes are extracted according to their size in a sequential order. At each iteration, the set of points (inliers) aligned with the model hypotheses is selected as the support for the planar model, and they are archived and removed from the original point cloud. The remaining points will be used to identify the next best plane. This process continues to detect planes and remove points until the original point cloud is exhausted or its size reaches a certain threshold. Note that, for each detected plane, an additional step is taken to project all inlier points to the plane such that they lie in a perfect plane model. This makes subsequent computation more efficient and less error prone.

With the extracted planes, the next step is to determine the one that corresponds to the ground surface and the ones that correspond to the walls. Recall that the normal of a plane can be computed using the planar model coefficient directly. Given the plane $(y = 0)$ with normal $\mathbf{n_0} = <0, 1, 0>$, the angle θ between an arbitrary plane $\mathbf{n_1} = <a_1, b_1, c_1>$ and $\mathbf{n_0}$ is computed as $\theta = \arccos\left(\frac{b_1}{\sqrt{a_1^2+b_1^2+c_1^2}}\right) 180/\pi$. Thus, the ground plane can be identified by computing the angles between all planes with respect to $\mathbf{n_0}$ and keeping the one pair which is in parallel. If more than one pair is found, the lower plane will be kept. In our implementation, we allow a $\pm 15°$ to compensate for possible sensor movements. Similarly we can identify walls by keeping planes which are vertical to the identified ground surface. A similar strategy has been applied in [22] to identify walls, ceiling and floor to help maintain a Micro Air Vehicle (MAV) in the center of a corridor. Naturally this is also applicable to ground-based robots. Examples of extracted vertical planes from various input point clouds are shown in Fig. 1. Subsequent steps will operate on these planes for open door extraction.

Open Door Detection: At a high level, our algorithm scans horizontally at a certain height of the vertical plane (wall) for gaps which are within the width of a typical door. When a gap of the appropriate width is identified ($th_g^{min} \leq g_s \leq th_g^{max}$), a potential open door is detected. In order to detect open gaps, we first define a specific height h_s for scanning. Note that instead of defining the height with respect to sensor as in [18], we define the height with respect to the ground surface in the point cloud. After that, we collect a set of points which are nearby with respect to the reference height. A small distance threshold th_s (0.05 meter in our work) is used in our implementation to define the nearby radius. In actual implementation, our algorithm first goes through each point in the point cloud and computes its distance to the ground surface. The computation of the distance between a point and a plane in 3D is as follow:

Let $a_0x + b_0y + c_0z + d_0 = 0$ be the parameter form of the ground plane, and $\mathbf{x_q} = (x_q, y_q, z_q)$ be a point in 3D space. A vector from the plane to the point $\mathbf{x_q}$ is given by

$$\mathbf{w} = - \begin{bmatrix} x - x_q \\ y - y_q \\ z - z_q \end{bmatrix}. \tag{1}$$

Projecting $\mathbf{w}$ onto the normal vector of the plane $\mathbf{n_0} = <a_0, b_0, c_0>$ gives the distance D from the point to the plane as

$$D = \frac{|\mathbf{n_0} \cdot \mathbf{w}|}{|\mathbf{n_0}|} \\ = \frac{a_0 x_q + b_0 y_q + c_0 z_q + d_0}{\sqrt{a_0^2 + b_0^2 + c_0^2}}. \tag{2}$$

Since the point cloud has been previously filtered and downsampled, conducting the above operation on the point cloud is very efficient. A point whose distance to the ground surface is approximately equal to the defined height (i.e., $h_s \pm th_s$) will be included in the point set. Once the points are collected, the algorithm computes the dominant direction of the point set (either in X-axis or Z-axis), and then sorts all the points in the set according to the computed direction. After that, the algorithm goes through each point sequentially in the sorted order. All the gaps can be efficiently identified in just one pass by simply checking the distances between each consecutive pair of points. Whenever the distance between two consecutive points (p_s^1, p_s^2) falls within the pre-defined width range (0.8 to 1.6 meters in this work), a candidate open space is reported to be detected. Although it is usually enough to determine open doors in the indoor setting with one scan, our algorithm scans a wall at two different heights. An open door is only reported if the two detected gaps are sufficiently close. In the case when precise open space is required, we can then connect the boundary points based on the open gaps. In the rare event of a false positive, the obstacle avoidance module (see Sect. 3.2) will be activated to prevent collisions.

Vanishing Point Estimation: In the case where doors are not available, the robot should continue to explore the indoor environment via wall following or moving to the end of the corridor. This can be usually achieved by computing vanishing points from the sensed data. Typically they are estimated by finding the intersecting point of major line segments in 2D images [6, 15]. In the 3D case, we take a more direct approach to approximate vanishing points based on the wall-floor intersecting line. This line has been shown to provide important geometrical heuristics in indoor navigation [15, 23].

To detect the wall-floor intersecting line, we utilize the identified ground plane and vertical walls. The computation is carried out with a standard Lagrange multiplier approach suggested in [24]. Basically the intersecting line from the two planes is represented by a point on the line $\mathbf{x} = (x, y, z)$ and a directional vector $\mathbf{n} = <a, b, c>$ originating from the point. Thus the goal is to identify the two. Assuming the two planes (the wall and the ground surface) are given by the normal vectors, $\mathbf{n_1} = <a_1, b_1, c_1>$ and $\mathbf{n_2} = <a_2, b_2, c_2>$, and arbitrary points on the two planes are $\mathbf{x_1} = (x_1, y_1, z_1)$ and $\mathbf{x_2} = (x_2, y_2, z_2)$. The directional vector of the intersecting line is easily computed as the cross product of the two normal vectors, i.e., $\mathbf{n} = \mathbf{n_1} \times \mathbf{n_2}$.

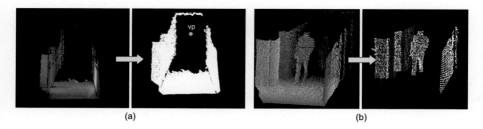

Fig. 2. (a) Left: a corridor snapshot. Right: detected wall-floor intersecting lines and the derived vanishing point. (b) Example of obstacle extraction. (Second pedestrian is not kept as he falls out of the predefined 5 m range).

To compute the point $\mathbf{x}$, we introduce another arbitrarily chosen point $\mathbf{x_0} = (x_0, y_0, z_0)$, which could simply be the origin or any other point. The point $\mathbf{x}$ on the intersecting line should be as close as possible to $\mathbf{x_0}$. This is a standard problem for Lagrange multipliers, with one objective function and two constraints: $(\mathbf{x} - \mathbf{x_1}) \cdot \mathbf{n_1} = 0$ and $(\mathbf{x} - \mathbf{x_2}) \cdot \mathbf{n_2} = 0$ (i.e., $\mathbf{x}$ must be on both planes). The cost function W containing both factors is

$$
\begin{aligned}
W &= \|\mathbf{x} - \mathbf{x_0}\|^2 + \lambda(\mathbf{x} - \mathbf{x_1}) \cdot \mathbf{n_1} + \mu(\mathbf{x} - \mathbf{x_2}) \cdot \mathbf{n_2} \\
&= (x - x_0)^2 + (y - y_0)^2 + (z - z_0)^2 + \lambda x a_1 + \lambda y b_1 \\
&\quad + \lambda z c_1 - \lambda \mathbf{x_1} \cdot \mathbf{n_1} + \mu x a_2 + \mu y b_2 + \mu z c_2 - \mu \mathbf{x_2} \cdot \mathbf{n_2}
\end{aligned}
\tag{3}
$$

where λ and μ are the two Lagrange multipliers. In order to solve for $\mathbf{x}$, we compute the partial derivatives $\left(\frac{\partial W}{\partial x}, \frac{\partial W}{\partial y}, \frac{\partial W}{\partial z}, \frac{\partial W}{\partial \lambda}, \frac{\partial W}{\partial \mu} \right)$ and set them to zero. This can be rewritten as five linear equations stacking in matrix form

$$
\begin{pmatrix}
2 & 0 & 0 & a_1 & a_2 \\
0 & 2 & 0 & b_1 & b_2 \\
0 & 0 & 2 & c_1 & c_2 \\
a_1 & b_1 & c_1 & 0 & 0 \\
a_2 & b_2 & c_2 & 0 & 0
\end{pmatrix}
\begin{pmatrix}
x \\ y \\ z \\ \lambda \\ \mu
\end{pmatrix}
=
\begin{pmatrix}
2x_0 \\ 2y_0 \\ 2z_0 \\ \mathbf{x_1} \cdot \mathbf{n_1} \\ \mathbf{x_2} \cdot \mathbf{n_2}
\end{pmatrix}.
\tag{4}
$$

Solving the linear system for (x, y, z, λ, μ) gives the point $\mathbf{x}$ on the intersecting line. Together with the directional vector $\mathbf{n}$, the wall-floor intersecting line is then determined. Finally, the vanishing point is obtained as the furthest point on the line parallel to the wall-floor intersecting line staring from the origin (i.e., sensor location). Figure 2(a) shows an example of detected wall-floor intersecting lines and the vanishing point for the input point cloud.

3.2   Reactive Obstacle Avoidance

In order for the robot to navigate autonomously, an important step is to enable it to avoid obstacles reliably in its environment. We have developed a variant of the navigation strategy proposed in [4] to determine an alternate moving direction for the robot when obstacles are present. The basic idea is to compute and compare open path lengths available to the robot for different angular directions. Prior to this step, the obstacles must be first identified. Recall that the ground plane has been identified and segmented out from the input point cloud for open space detection as described in Sect. 3.1. The rest of the point cloud contains points that may correspond to obstacles residing on the current moving path of the robot. Figure 2(b) shows the remaining points after the ground plane has been removed from the input point cloud. As can be seen, multiple clusters are clearly formed based on these points. Following this observation, we simply perform obstacle checks using these remaining 3D points. Given a point cloud O consisting of obstacles, a robot with radius r, and the current desired direction of travel θ_d, we first define the open path length $d(\theta)$ as a function of the direction of travel θ as: $d(\theta) = \min_{p \in \widehat{O}} \Big(\max(0, \big\| p \cdot \widehat{\theta} \big\| - r) \Big)$, assuming the origin of the coordinate system coincides with the sensor in the center of the robot. p is a 3D point in vector form, $\widehat{\theta}$ is a unit vector in the direction of the angle θ. Projecting p on $\widehat{\theta}$ through the dot product gives the open path length in the direction of θ. Subtracting r compensates for the size of the robot and excludes any open paths which are shorter than the robot radius. Note that we did not go through every point in the input point cloud as suggested in [4]. Instead, we modified the original strategy by only considering obstacle points $\widehat{O}$ within a certain range with respect to the desired direction of travel θ_d. In this work, only points within $\pm 45°$ with respect to θ_d are considered. Subsequently the chosen obstacle avoidance direction θ^* is calculated as $\theta^* = \text{argmax}_\theta(d(\theta)cos(\theta - \theta_d))$. Essentially it trades off between: maximizing the open path length and minimizing the deviating angles.

Algorithm 1 summarizes the proposed autonomous robot exploration algorithm, which includes both proactive open space detection and reactive obstacle avoidance. Figure 3(a) shows a graphical example of the overall 3D point cloud processing by the proposed algorithm. Please refer to Figs. 4, 5 and 6 for more details.

4   Experiments and Results

We have fully implemented the proposed algorithm in C++ under the ROS (Robot Operating System) framework. Specifically, the proactive open space detection module and the reactive obstacle avoidance module are implemented in separate ROS packages complied with the general design principle. Experiments are carried out with a laptop computer with Intel Core i7 CPU and 8 GB of RAM. The computer is configured with ROS Hydro, OpenNi driver v1.5.4, and PCL (Point Cloud Library) 1.6.

We primarily focus on two indoor scenarios, a corridor and a warehouse. In the first scenario, data is captured live by an Asus Xtion Pro RGB-D sensor mimicking the behavior of a ground-based robot. In the second scenario, data

---

Algorithm 1. Autonomous Robot Exploration

Input: A point could P captured by a range sensor, and current desired direction θ_d
Output: Direction θ_o for nearest open door, direction θ_v for vanishing point, direction
　　　θ_a to avoid obstacles if applicable.

1: $P_f \leftarrow \{\}, O \leftarrow \{\}, D \leftarrow \{\}, V \leftarrow \{\}, A \leftarrow \{\}$
2: $P_f \leftarrow DistanceFilter(P)$
3: $P_f \leftarrow VoxelGridDownsample(P_f)$
4: $N_p \leftarrow \alpha \cdot |P_f|$
5: Find dominant ground plane within P_f; project nearby points onto the plane ($\rightarrow$
　　P_{ground})
6: $O \leftarrow P_f \backslash P_{ground}$
7: **while** $N_p < |P_f|$ **do**
8:　　$G \leftarrow \{\}$
9:　　Extract dominant plane, project nearby points onto plane ($\rightarrow P_{curr}$)
10:　　**if** P_{curr} is a vertical plane **then**
11:　　　　Collect a set of points L_s at predefined height $h_s \pm th_s$ w.r.t. ground plane
12:　　　　Scan points in L_s sequentially according dominant direction
13:　　　　**if** $\|p_s^1 - p_s^2\| \in \left[th_g^{min}, th_g^{max}\right]$ for two consecutive points **then**
14:　　　　　　$g_s \leftarrow (p_s^1, p_s^2)$
15:　　　　　　$G \leftarrow \{G \cup g_s\}$
16:　　　　**end if**
17:　　　　(Optional) Repeat scanning at different height
18:　　　　**if** $g_s \in G$ are consistent **then**
19:　　　　　　$D \leftarrow \{D \cup mean(G)\}$
20:　　　　**end if**
21:　　　　Compute vanishing point v_{curr} based on wall-floor intersecting line
22:　　　　$V \leftarrow \{V \cup v_{curr}\}$
23:　　**end if**
24:　　$P_f \leftarrow P_f \backslash P_{curr}$
25: **end while**
26: Compute θ_o based on coordinates of nearest door in D
27: Compute θ_v based on coordinates of VPs in V
28: Compute θ_a based on O and θ_d
29: **return** θ_o, θ_v, and θ_a
Note: $\alpha = 0.2$, $h_s = 0.8$, $th_s = 0.05$, $th_g^{min} = 0.8$, $th_g^{max} = 1.6$

---

is generated in simulation through Gazebo[1] mimicking the behavior of a aerial-based robot. In particular, we utilize the ROS Hector Quadrotor package[2] to model the dynamics and control of a MAV. Experimental parameters are consistent in both real world and simulated data with minor exceptions. Specifically, we increase the radius threshold th_s from 0.05 to 0.15 when collecting points for wall-scanning (see Sect. 3.1). This is because simulated point cloud appears to have different density (i.e.,sparser) compared to real world collected data. In

---

[1] http://wiki.ros.org/gazebo.
[2] http://wiki.ros.org/hector_quadrotor.

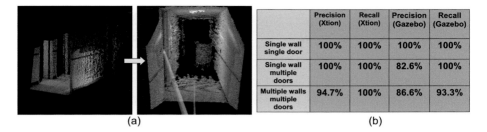

	Precision (Xtion)	Recall (Xtion)	Precision (Gazebo)	Recall (Gazebo)
Single wall single door	100%	100%	100%	100%
Single wall multiple doors	100%	100%	82.6%	100%
Multiple walls multiple doors	94.7%	100%	86.6%	93.3%

(a)   (b)

Fig. 3. (a) An integrated view showing various detections from 3D processing: open doors are marked with text "Door" (in red), yellow arrow indicates the nearest door, "VP" (in pink) indicates the detected vanishing point, green arrow indicates evading direction if obstacle is detected. (b) Quantitative results of open doorway detection on real world data captured with an Xtion sensor and simulated data generated in Gazebo (Color figure online).

addition, we adjust the distance filter to keep points within 8 meters from the sensor along the Z-axis, as simulated depth data has larger effective range.

In the first experiment, we focus on evaluating the effectiveness of the proactive open door detection module. Figure 3(b) summarizes the quantitative results of the proposed algorithm under different test configurations: single wall single door, single wall multiple doors, and multiple walls multiple doors. For each configuration, a total of fifteen test cases (at different angles and orientations) are randomly selected from experimental data. This is essentially done by taking point cloud snapshots from the 3D data. The first two configurations effectively cover open doors with angles ranging from 0° to 75° offset with respect to the sensor. The third configuration cover angle offset at about 90°. Overall, detection precision and recall are very high for real-world data (100 % in most cases). Detection results are slightly lower for simulation data, as the scenes are more diverse. We observe that most of false positives come from (non-door) open spaces that coincide with typical door width. Currently, the proposed algorithm does not provide a means to distinguish the two cases. However, this could be solved by introducing additional geometrical templates as done in [17]. Figures 4 and 5 show visual examples of detecting open doors in both real world and simulated data.

We also evaluate the obstacle avoidance module under different indoor settings. In total, we test the algorithm against six types of obstacles (e.g., pedestrian, box, and four types of chairs) in three different locations (e.g., corridor, office, and large conference room). In all cases, the reactive obstacle avoidance module is able to provide effective evading directions. Figure 6 presents a number of examples of the algorithm in action. The current implementation allows both modules to run simultaneously at about 10 Hz on the laptop computer, which makes it suitable for online applications.

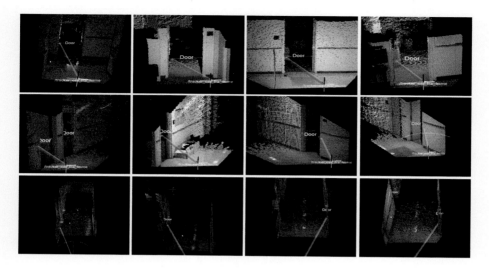

Fig. 4. Correctly detected doors from a ground-based robot platform using real world corridor data. Examples include doors detected at different angles (rows correspond to roughly 0°, 45°, 90°) and orientations with respect to the range sensor. Detected doors are indicated by green lines and marked with red text "Door". Yellow arrows indicate nearest doors to the robot based on the current scanned point cloud (Color figure online).

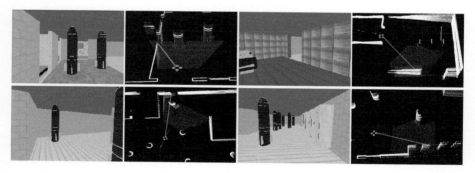

Fig. 5. Correctly detected doors from an aerial platform (i.e., MAV) using simulated warehouse data. Views from the MAV are shown as well.

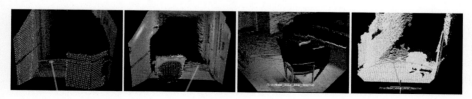

Fig. 6. Examples of obstacle avoidance. Green arrow indicates computed optimal moving direction to avoid collision. (Point cloud on the right is captured with a Capri sensor, thus lacks the color information) (Color figure online).

5   Discussion and Conclusions

This work presents an online algorithm for autonomous robot exploration in indoor settings using a 3D range sensor. Our system combines a proactive open space detection module with a reactive obstacle avoidance module to provide the basic autonomous functionality for a robot to navigate and explore an unknown environment. A key advantage of our proposed proactive algorithm's door detection capability is that it can detect doors accurately (1) irrespective of robot position and, (2) simultaneously from multiple walls at different angles with respect to the robot. By incorporating this accurate detection capability into a high level proactive controller and combining it with a reactive controller, we provide the robot with an autonomous exploration strategy. Promising results with this exploration strategy have been shown for a robot motion in both real and simulated worlds.

3D sensors are gradually becoming mainstream and available in small form factor at low cost. This makes them an attractive choice for mobile power-constrained robotic platforms such as MAV. This paper is a step towards enabling low-cost MAVs that can explore and map unknown indoor environments equipped with 3D range sensors. By combining the proposed algorithms with more traditional 2D sensors (e.g., camera) and algorithms, we can further improve the exploration and navigation capabilities and also extend navigation to outdoor settings. A future research direction is to integrate the proposed algorithms with long-term path planning and control schemes.

Acknowledgements. This material is based upon work supported by Defense Advanced Research Projects Agency under contract numbers W31P4Q-08-C-0264 and HR0011-09-C-0001. Any opinions, findings and conclusion or recommendations expressed in this material are those of the author and do not necessarily reflect the view of the Defense Advanced Research Projects Agency. The views and conclusions contained in this document are those of the authors and should not be interpreted as representing the official policies, either expressly or implied, of the Defense Advanced Research Projects Agency or the U.S. Government.

References

1. González-Baños, H.H., Latombe, J.C.: Navigation strategies for exploring indoor environments. I. J. Robotic Res. **21**, 829–848 (2002)
2. Dudek, G., Jenkin, M.R.M.: Computational principles of mobile robotics. Cambridge University Press, New York (2000)
3. Lai, J., Mejías, L., Ford, J.J.: Airborne vision-based collision-detection system. J. Field Robot. **28**, 137–157 (2011)
4. Biswas, J., Veloso, M.M.: Depth camera based localization and navigation for indoor mobile robots. In: RGB-D Workshop in Robotics: Science and Systems (RSS) (2011)
5. Zingg, S., Scaramuzza, D., Weiss, S., Siegwart, R.: MAV navigation through indoor corridors using optical flow. In: ICRA. IEEE (2010)

6. Bills, C., Chen, J., Saxena, A.: Autonomous MAV flight in indoor environments using single image perspective cues. In: ICRA, pp. 5776–5783. IEEE (2011)
7. Fraundorfer, F., Heng, L., Honegger, D., Lee, G.H., Meier, L., Tanskanen, P., Pollefeys, M.: Vision-based autonomous mapping and exploration using a quadrotor mav. In: IROS. (2012)
8. Meier, L., Tanskanen, P., Heng, L., Lee, G.H., Fraundorfer, F., Pollefeys, M.: Pixhawk: a micro aerial vehicle design for autonomous flight using onboard computer vision. Auton. Robots **33**, 21–39 (2012)
9. Celik, K., Chung, S.J., Clausman, M., Somani, A.K.: Monocular vision slam for indoor aerial vehicles. In: IROS. IEEE (2009)
10. Cruz, L., Lucio, D., Velho, L.: Kinect and RGBD images: challenges and applications. In: SIBGRAPI Tutorials (2012)
11. Herbst, E., Ren, X., Fox, D.: RGB-D flow: dense 3-D motion estimation using color and depth. In: ICRA, pp. 2276–2282 (2013)
12. Shen, S., Michael, N., Kumar, V.: Autonomous indoor 3D exploration with a micro-aerial vehicle. In: ICRA, pp. 9–15. IEEE (2012)
13. Murillo, A.C., Kosecká, J., Guerrero, J.J., Sagüés, C.: Visual door detection integrating appearance and shape cues. Robot. Auton. Syst. **56**, 512–521 (2008)
14. Tian, Y., Yang, X., Arditi, A.: Computer vision-based door detection for accessibility of unfamiliar environments to blind persons. In: Miesenberger, K., Klaus, J., Zagler, W., Karshmer, A. (eds.) ICCHP 2010, Part II. LNCS, vol. 6180, pp. 263–270. Springer, Heidelberg (2010)
15. Sekkal, R., Pasteau, F., Babel, M., Brun, B., Leplumey, I.: Simple monocular door detection and tracking. In: IEEE International Conference on Image Processing, ICIP 2013, Melbourne, Australie (2013)
16. Fernández-Caramés, C., Moreno, V., Curto, B., Rodríguez-Aragón, J., Serrano, F.: A real-time door detection system for domestic robotic navigation. J. Intell. Robot. Syst. **76**(1), 119–136 (2013)
17. Rusu, R.B., Meeussen, W., Chitta, S., Beetz, M.: Laser-based perception for door and handle identification. In: International Conference on Advanced Robotics (ICAR) (2009)
18. Derry, M., Argall, B.: Automated doorway detection for assistive shared-control wheelchairs. In: ICRA, pp. 1254–1259 (2013)
19. Ross, S., Melik-Barkhudarov, N., Shankar, K.S., Wendel, A., Dey, D., Bagnell, J.A., Hebert, M.: Learning monocular reactive uav control in cluttered natural environments. In: CoRR (2012)
20. Meagher, D.: Geometric modeling using octree encoding. Comput. Graph. Image Process. **19**, 129–147 (1982)
21. Fischler, M.A., Bolles, R.C.: Random sample consensus: a paradigm for model fitting with applications to image analysis and automated cartography. Commun. ACM **24**, 381–395 (1981)
22. Lange, S., Sünderhauf, N., Neubert, P., Drews, S., Protzel, P.: Autonomous corridor flight of a UAV using a low-cost and light-weight RGB-D camera. In: Rueckert, U., Joaquin, S., Felix, W. (eds.) Advances in Autonomous Mini Robots. Non-series, vol. 101, pp. 183–192. Springer, Heidelberg (2012)
23. Pasteau, F., Babel, M., Sekkal, R.: Corridor following wheelchair by visual servoing. In: IEEE/RSJ International Conference on Intelligent Robots and Systems, IROS 2013, Tokyo, Japon (2013)
24. Krumm, J.: Intersection of two planes (2000)

Group Based Asymmetry–A Fast Saliency Algorithm

Puneet Sharma[✉] and Oddmar Eiksund

Department of Engineering and Saftey(IIS),
UiT-The Arctic University of Norway, Tromso, Norway
puneet.sharma@uit.no

Abstract. In this paper, we propose a saliency model that makes two major changes in a latest state-of-the-art model known as group based asymmetry. First, based on the properties of the dihedral group D_4 we simplify the asymmetry calculations associated with the measurement of saliency. This results is an algorithm which reduces the number of calculations by at-least half that makes it the fastest among the six best algorithms used in this paper. Second, in order to maximize the information across different chromatic and multi-resolution features the color image space is de-correlated. We evaluate our algorithm against 10 state-of-the-art saliency models. Our results show that by using optimal parameters for a given data-set our proposed model can outperform the best saliency algorithm in the literature. However, as the differences among the (few) best saliency models are small we would like to suggest that our proposed fast GBA model is among the best and the fastest among the best.

1 Introduction

In literature, visual attention has been mainly classified as: top-down, and bottom-up [16]. Top-down, is voluntary, goal-driven, and slow, i.e., usually in the range between 100 milliseconds to several seconds [16]. It is assumed that the top-down attention is closely linked with cognitive aspects such as memory, thought, and reasoning. For instance, by using top-down mechanisms we can read this text one word at a time, while neglecting other aspects of the scene such as, words in other lines. In contrast, bottom-up attention (also known as visual saliency) is associated with attributes of a scene that draw our attention to a particular location. These attributes include: motion, contrast, orientation, brightness, and color [13]. Bottom-up mechanisms are involuntary, and faster as compared to top-down [16]. For instance, a red object among green objects, and an object placed horizontally among vertical objects are some stimuli that would automatically capture our attention in the environment.

In a recent study by Alsam et al. [1,2] it was proposed that asymmetry can be used as a measure of saliency. In order to calculate asymmetry of an image region the authors used dihedral group D_4, which is the symmetry group of the square. D_4 consists of 8 group elements namely, rotation by 0, 90, 180 and 270

© Springer International Publishing Switzerland 2015
G. Bebis et al. (Eds.): ISVC 2015, Part I, LNCS 9474, pp. 901–910, 2015.
DOI: 10.1007/978-3-319-27857-5_80

degrees and reflection about the horizontal, vertical and two diagonal axes. The saliency maps obtained from their algorithm show good correspondence with the saliency maps calculated from the classic visual saliency model by Itti et al. [11].

Inspired by the fact that bottom-up calculations are fast, in this paper, we use the symmetries present in the dihedral group D_4 to make the calculations associated with the D_4 group elements simpler and faster to implement. In doing so, we modify the saliency model proposed by Alsam et al. [1,2]. For details, please see Sect. 2.

Next, we are motivated from the study by Garcia-Diaz et al. [8] which implies that in order to quantify distinct information in a scene, our visual system de-correlates its chromatic and multi-resolution features. Based on this, we perform the de-correlation of input color image by calculating its principal components (details in Sect. 2.3).

2   Method

2.1   Background

Alsam et al. [1,2] proposed a saliency model that uses asymmetry as a measure of saliency. In order to calculate saliency, the input image is decomposed into square blocks, and for each block the absolute difference between the block itself and the result of the D_4 group elements acting on the block is calculated. The sum of the absolute differences (also known as L_1 norm) for each block is used as a measure of asymmetry for the block. The asymmetry values for all the blocks are then collected in an image matrix and scaled up to the size of original image using bilinear-interpolation. In order to capture both the local and the global salient details in an image three different image resolutions are used. All maps are combined linearly to get a single saliency map.

In their algorithm, asymmetry of a square region is calculated as follows: M (i.e., the square block) is defined as an $n \times n$-matrix and σ_i as one of the eight group elements of D_4. The eight elements are the rotations along $0°$, $90°$, $180°$ and $270°$, and the reflections along horizontal, vertical and two diagonal axis of the square. Asymmetry of M by σ_i is denoted by $A(M)$ to be,

$$A(M) = \sum_{i=1}^{8} ||M - \sigma_i M||_1, \tag{1}$$

where $||_1$ represents L_1 norm. Instead of calculating asymmetry value associated with each group element and followed by their sum, we believe that the algorithm can run faster if the calculations in Eq. 1 are made simpler. For this we propose a fast implementation of these operations pertaining to the D_4 group elements.

2.2   Fast Implementation of the Group Operations

Let us assume M as 4 by 4 matrix,

$$M = \begin{bmatrix} \boxed{\alpha_1} & a & b & \boxed{\beta_1} \\ c & \boxed{\alpha_2} & \boxed{\beta_2} & d \\ e & \boxed{\gamma_2} & \boxed{\delta_2} & f \\ \boxed{\gamma_1} & g & h & \boxed{\delta_1} \end{bmatrix}$$

The asymmetry $A(M)$ of the matrix M is measured as the sum of absolute differences of the different permutations of the matrix entries pertaining to the D_4 group elements and the original. The total number of such differences are determined to be 40. As the calculations associated with absolute differences are repeated for the rotation and reflection elements of the dihedral group D_4, our objective is to find the factors associated with these repeated differences.

For our calculations we divide the set of matrix entries into two computational categories: the diagonal entries (highlighted in yellow) and the rest of the entries of M. Please note that these calculations can be generalized to any matrix of size n by n, given that n is even.

For the *rest* of the entries, first, we can look at $|a - b|$. This element will only be possible if we flip the matrix about the vertical axis. This will result in two parts in the sum, $|a - b|$ and $|b - a|$, giving a factor 2. Here a and b represents a reflection symmetric pair, and all other reflection symmetric pairs will behave in the same way. Now let's focus on $|a - d|$. This represents a rotational symmetric pair. Rotating the matrix counterclockwise will move d onto the position of a giving a part $|a - d|$ in the sum. Rotating clockwise gives us, $|d - a|$. As these differences are not plausible in any other way, this gives us a factor of 2. All other rotational symmetric pairs will behave in the same way. This means that the *asymmetry* for the *rest* of the entries can be calculated as follows:

$$2|a - b| + 2|a - c| + 2|a - d| + \cdots + 2|g - h|. \tag{2}$$

For the *diagonal* entries, we can see that they exhibit both rotation and reflection symmetries. For instance, we can move β to the place of α and α to β with one reflection and two rotations. This gives us a factor of 4. The *asymmetry* of one set of *diagonal* entries can be calculated as follows:

$$4|\alpha - \beta| + 4|\alpha - \gamma| + 4|\alpha - \delta| + 4|\beta - \gamma| + 4|\beta - \delta| + 4|\gamma - \delta|. \tag{3}$$

The *asymmetry* for both the diagonal entries and the rest is represented as,

$$\begin{aligned} A(M) = \quad & 4|\alpha_1 - \beta_1| + 4|\alpha_1 - \gamma_1| + \cdots + 4|\gamma_1 - \delta_1| \\ & + 4|\alpha_2 - \beta_2| + 4|\alpha_2 - \gamma_2| + \cdots + 4|\gamma_2 - \delta_2| \\ & + 2|a - b| + 2|a - c| + \cdots + 2|g - h|. \end{aligned} \tag{4}$$

As shown in Eq. 4, the asymmetry calculations associated with the matrix M are reduced to a quarter for the diagonal entries and one-half for the rest of the entries. This makes the proposed algorithm at least twice as fast.

2.3 De-correlation of Color Image Channels

De-correlation of color image channels is done as follows: First, using bilinear interpolation we create three resolutions (original, half and one-quarter) of the RGB color image. In order to collect all the information in a matrix the (half and one-quarter) resolutions are rescaled to the size of original. This gives us a matrix I of size w by h by n, where w is the width of the original, h is the height and n is the number of channels ($3 \times 3 = 9$).

Second, by rearranging the matrix entries of I we create a two dimensional matrix A of size $w \times h$ by n. We do normalization of A around the the mean as,

$$B = A - \mu, \tag{5}$$

where μ is the mean for each of the channels, and B is $w \times h$ by n.

Third, we calculate correlation matrix of B as,

$$C = B^T B, \tag{6}$$

where the size of C is n by n.

Four, the Eigen decomposition of a symmetric matrix is represented as,

$$C = VDV^T, \tag{7}$$

where V is a square matrix whose columns are Eigen-vectors of C and D is the diagonal matrix whose diagonal entries are the corresponding Eigen-values.

Finally, the image channels are transformed into Eigenvector space (also known as principal components) as:

$$E = V^T(A - \mu), \tag{8}$$

where E is the transformed space matrix which is rearranged to get back the de-correlated channels.

2.4 Implementation of the Algorithm

First, the input color image is rescaled to half the original resolution. Second, by using the de-correlation procedure described in Sect. 2.3 on resulting image we get 9 de-correlated multi-resolution and chromatic channels. Third, a fixed block size (e.g.,12) is selected– as discussed later in Sect. 3.6, this choice is governed by the data-set. If the rows and columns of the de-correlated channels are not divisible by the block size then they are padded with neighboring information along the right and bottom corners. Finally, the saliency map is generated by using the procedure outlined in Sect. 2.2. The code is open source and will be available at Matlab Central for the research community.

3 Comparing Different Saliency Models

The performance of visual saliency algorithms is usually judged by how well the two-dimensional saliency maps can predict the human eye fixations for a given image. Center-bias is a key factor that can influence the evaluation of saliency algorithms [15].

3.1   Center-Bias

While viewing images, observers tend to look at the center regions more as compared to peripheral regions. As a result of that a majority of fixations fall at the image center. This effect is known as center-bias and is well documented in vision studies [17,18]. The two main reasons for this are: first, the tendency of photographers to place the objects at the center of the image. Second, the viewing strategy employed by observers, i.e., to look at center locations more in order to acquire the most information about a scene [19]. The presence of center bias in fixations makes it difficult to analyze the correspondence between the fixated regions and the salient image regions.

3.2   Shuffled AUC Metric

Shuffled AUC metric was proposed by Tatler et al. [18] and later used by Zhang et al. [20] to mitigate the effect of center-bias in fixations. The shuffled AUC metric is a variant of AUC [7] which is known as area under the receiver operating characteristic curve. For a detailed description of AUC, please see the study by Fawcett [7].

To calculate the shuffled AUC metric for a given image and one observer, the locations fixated by the observer are associated with the positive class (in a manner similar to the regular AUC metric), however, the locations for the negative class are selected randomly from the fixated locations of other unrelated images, such that they do not coincide with the locations from the positive class. Similar to the regular AUC, the shuffled AUC metric gives us a scalar value in the interval [0,1]. If the value is 1 then it indicates that the saliency model is perfect in predicting fixations. If Shuffled $AUC <= 0.5$ then it implies that the performance of the saliency model is not better than a random classifier or chance prediction.

3.3   Dataset

For the analysis, we used the eye tracking database from the study by Judd et al. [12]. The database consists of 1003 images selected randomly from different categories and different geographical locations. In the eye tracking experiment [12], these images were shown to fifteen different users under free viewing conditions for a period of 3 seconds each. In the database, a majority of the images are 1024 pixels in width and 768 pixels in height. These landscape images were specifically used in the evaluation.

3.4   Saliency Models

For our comparison, eleven state-of-the-art saliency models, namely, **AIM** by Bruce & Tsotsos [5], **AWS** by Garcia-Diaz et al. [8], **Erdem** by Erdem & Erdem [6], **Hou** by Hou & Zhang [10], **Spec** by Schauerte & Stiefelhagen [14], **GBA** by Alsam et al. [1,2], **Fast GBA** proposed in this paper, **GBVS** by

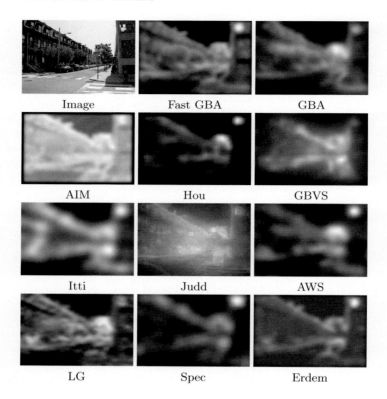

Fig. 1. Figure shows a test image (from database [12]) and the associated the saliency maps from different saliency algorithms used in the paper.

Harel et al. [9], **Itti** by Itti et al. [11], **Judd** by Judd et al. [12], and **LG** by Borji & Itti [3] are used. In line with the study by Borji et al. [4], two models are selected to provide a baseline for the evaluation. **Gauss** is defined as a two-dimensional Gaussian blob at the center of the image. Different radii of the Gaussian blob are tested and the radius that corresponds best with human eye fixations is selected. Figure 1 shows a test image and the associated saliency maps from different saliency algorithms.

3.5   Ranking Among the Saliency Models

We compare the ranking of saliency models using the shuffled AUC metric. From the results in Fig. 2, we note that, first, the **Gauss** model is ranked the worst indicating that the shuffled AUC metric counters the effects associated with the center-bias. Second the **AWS** model is ranked the best followed by the proposed **Fast GBA** model. It is important to note that a majority of the state-of-the-art saliency models such as: **Itti, Hou, Spec, GBA, Fast GBA LG, Erdem, AIM**, and **AWS** are quite close to each other in terms of their performance.

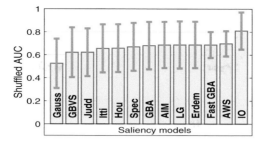

Fig. 2. Ranking of different saliency models using the Shuffled AUC metric. The results are obtained from the fixations data of 463 landscape images and fifteen observers.

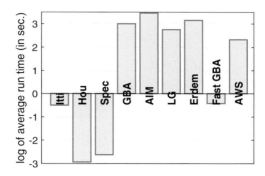

Fig. 3. Average run time across 463 landscape images for different saliency models, **Itti** = 0.60, **Hou** = 0.05, **Spec** = 0.07, **GBA** = 20.13, **AIM** = 31.75, **LG** = 15.70, **Erdem** = 23.35, **Fast GBA** = 0.65, **AWS** = 10.27. All run times are in seconds. For a better visualization we use the natural logarithm of the average run times.

Next, we compare the average run times (for 463 landscape images) of the saliency models that rank at the same or better than **Itti** i.e., the classic saliency model. For a better visualization we use the natural logarithm of the average run times. For this, we used Matlab R2015 on a 64 bit windows PC with a 3.16 Ghz Intel processor and 4 GB RAM. From the Fig. 3, we observe that the algorithms, **Hou**, and **Spec** are the fastest. However, among the top six algorithms, the proposed **Fast GBA** model is the fastest. Furthermore, it shows that **Fast GBA** is nearly 31 times faster than the original **GBA** algorithm.

3.6 Optimizing the Proposed Fast GBA model

The performance of the proposed model is influenced by the choice of parameters such as, block size, which depends on the size of an average image in the database used for testing. To find the optimal parameters for our algorithm we use three variables: image scaling factor S_f (which rescales the original image in order to reduce the number of calculations), block size b, number of resolutions N_r

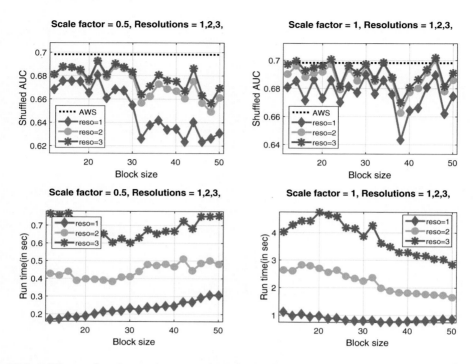

Fig. 4. The results obtained by using the Shuffled AUC metric for the three variables are shown in the first row. The figure on the top-left shows the Shuffled AUC values for $S_f = 0.5$, with the red, green, and blue lines depicting the N_r as 1, 2, and 3 respectively, while, the figure on the top-right shows the Shuffled AUC values for the $S_f = 1$. In the second row, we show the average run time of the algorithm for the different values of S_f, b, and N_r (Color figure online).

(different resolutions to capture local and global details). For this analysis, we use $S_f = 0.5$ (half size) and $S_f = 1$, b in the range [12, 50], and $N_r = 1, 2$, and, 3. The results obtained by using the Shuffled AUC metric for the three variables are shown in the first row of the Fig. 4. The figure on the top-left shows the Shuffled AUC values for $S_f = 0.5$, with the red, green, and blue lines depicting the N_r as 1, 2, and 3 respectively, while, the figure on the top-right shows the Shuffled AUC values for the $S_f = 1$. In the second row of the Fig. 4, we depict the average run time of the algorithm for the different values of S_f, b, and N_r. The results indicate that: first, increasing the number of resolutions improves the performance of the proposed model. Second, based on the figures in the second row we note that using $S_f = 0.5$ (i.e., working with an image of half the original resolution) reduces the run time to less than one second. Third, we observe (in the figure on top-right) that the Shuffled AUC values for our algorithm exceed that of the values obtained from the **AWS** model (i.e., the best saliency model–represented by the black dashed line) for the following parameters: $S_f = 1$, $N_r = 3$, $b = 14, 22, 34, 46$ and $S_f = 1$, $N_r = 2$ and $b = 46$. In other words, using the optimal parameters (mentioned above) our proposed model outranks the best saliency model in literature, however, we believe that the difference between the top 5 algorithms (**AIM**, **LG**, **Erdem**, **Fast GBA**, and **AWS**) are too small to rank one as the best over the rest. Four, from the figure on the bottom-right, we note that using the optimal parameters increases the run time to a few seconds (minimum of 1.7 to maximum of 4.7 s) which are still faster than the run time of **AWS** model (i.e., 10.2 s). Please note that in order to highlight the intrinsic nature of the **Fast GBA** model no GPU computing was employed.

4   Conclusion

In this paper, we improve a state-of-the-art saliency model called group based asymmetry as follows: first, based on the properties of the Dihedral group D_4 we simplify the asymmetry calculations associated with the measurement of saliency. This results is an algorithm which reduces the number of calculations by at-least half that makes it the fastest among the six best algorithms used in this paper. Second, in order to maximize the differences across the different image features we de-correlated the color image space.

We compare our algorithm with 10 state-of-the-art saliency models. Our results clearly show that by using optimal parameters for a given data-set our proposed model can outperform the best saliency algorithm in the literature. However, as the differences among the (few) best saliency models are small we would like to suggest that our proposed model is among the best and the fastest among the best. We believe that our proposed model can be used for calculating saliency in real-time.

References

1. Alsam, A., Sharma, P., Wrålsen, A.: Asymmetry as a measure of visual saliency. In: Kämäräinen, J.-K., Koskela, M. (eds.) SCIA 2013. LNCS, vol. 7944, pp. 591–600. Springer, Heidelberg (2013)
2. Alsam, A., Sharma, P., Wrålsen, A.: Calculating saliency using the dihedral group d4. J. Imaging Sci. Technol. **58**(1), 10504:1–10504:12 (2014)
3. Borji, A., Itti, L.: Exploiting local and global patch rarities for saliency detection. In: Proceedings IEEE Conference on Computer Vision and Pattern Recognition (CVPR), Providence, Rhode Island, pp. 1–8 (2012)
4. Borji, A., Sihite, D.N., Itti, L.: Quantitative analysis of human-model agreement in visual saliency modeling: a comparative study. IEEE Trans. Image Process. **22**(1), 55–69 (2013)
5. Bruce, N., Tsotsos, J.: Saliency based on information maximization. In: The Proceedings of the Neural Information Processing Systems Conference (NIPS 2005), pp. 155–162, Vancouver, British Columbia, Canada (2005)
6. Erdem, E., Erdem, A.: Visual saliency estimation by nonlinearly integrating features using region covariances. J. Vis. **13**(4:11), 1–20 (2013)
7. Fawcett, T.: Roc graphs with instance-varying costs. Pattern Recogn. Lett. **27**(8), 882–891 (2004)

8. Garcia-Diaz, A., Fdez-Vidal, X.R., Pardo, X.M., Dosil, R.: Saliency from hierarchical adaptation through decorrelation and variance normalization. Image Vis. Comput. **30**(1), 51–64 (2012)

9. Harel, J., Koch, C., Perona, P.: Graph-based visual saliency. In: Proceedings of Neural Information Processing Systems (NIPS), pp. 545–552. MIT Press (2006)

10. Hou, X., Zhang, L.: Computer vision and pattern recognition. In: IEEE Conference on Saliency Detection: A Spectral Residual Approach, CVPR 2007, pp. 1–8 (2007)

11. Itti, L., Koch, C., Niebur, E.: A model of saliency-based visual attention for rapid scene analysis. IEEE Trans. Pattern Anal. Mach. Intell. **20**(11), 1254–1259 (1998)

12. Judd, T., Ehinger, K., Durand, F., Torralba, A.: Learning to predict where humans look. In: The Proceedings of the 2009 IEEE International Conference on Computer Vision (ICCV), pp. 2106–2113, Kyoto, Japan. IEEE (2009)

13. Koch, C., Ullman, S.: Shifts in selective visual attention: towards the underlying neural circuitry. Human Neurobiol. **4**, 219–227 (1985)

14. Schauerte, B., Stiefelhagen, R.: Predicting human gaze using quaternion dct image signature saliency and face detection. In: Proceedings of the IEEE Workshop on the Applications of Computer Vision (WACV), Breckenridge, CO, USA. IEEE 9–11 January 2012

15. Sharma, P.: Evaluating visual saliency algorithms: past, present and future. J. Imaging Sci. Technol. **59**(5), 50501-1 (2015)

16. Suder, K., Worgotter, F.: The control of low-level information flow in the visual system. Rev. Neurosci. **11**, 127–146 (2000)

17. Tatler, B.W.: The central fixation bias in scene viewing: selecting an optimal viewing position independently of motor biases and image feature distributions. J. Vis. **7**, 1–17 (2007)

18. Tatler, B.W., Baddeley, R.J., Gilchrist, I.D.: Visual correlates of fixation selection: effects of scale and time. Vis. Res. **45**(5), 643–659 (2005)

19. Tseng, P.-H., Carmi, R., Cameron, I.G., Munoz, D.P., Itti, L.: Quantifying center bias of observers in free viewing of dynamic natural scenes. J. Vis. **9**(7), 1–16 (2009)

20. Zhang, L., Tong, M.H., Marks, T.K., Shan, H., Cottrell, G.W.: Sun: a bayesian framework for saliency using natural statistics. J. Vis. **8**(7), 1–20 (2008)

Prototype of Super-Resolution Camera Array System

Daiki Hirao and Hitoshi Iyatomi$^{(\boxtimes)}$

Applied Informatics, Graduate School of Science and Engineering,
Hosei University, Tokyo, Japan
iyatomi@hosei.ac.jp

Abstract. We present a prototype of a super-resolution camera array system. Since the proposed system consists of a number of low-cost camera devices, all of which operate synchronously, it is a low-cost, high quality imaging system, and capable of handling moving targets. However, when the targets are located near the system, parallax and differences in photographic conditions among the cameras become pronounced. In addition, conventional super-resolution techniques frequently emphasize noise, as well as edges, contours, and so on, when the number of the observed (i.e., low resolution) images is limited. Therefore, we also propose the following procedures for our camera-array system: (1) color calibration among cameras, (2) automated region of the interest (ROI) detection under large parallax, and (3) effective noise reduction with effective edge preservation. We developed a camera array system comprising 12 low-cost Web camera devices. We confirm that the proposed system in general reduces the drawbacks of the array system and achieves approximately a 2 dB higher S/N ratio, i.e., equivalent to the effect of two additional images.

1 Introduction

Super-resolution is a technique for generating a high-resolution (HR) image using observed low-resolution (LR) image (s), and many studies on these techniques have been reported [1]. The super-resolution techniques addressed in these research studies researches can be divided into two broad types. The first is the so-called "learning type" super-resolution technique. This type usually utilizes only one base image from the observed images and predicts unknown details using a pre-trained database and/or estimators [2] or an interpolation approach based on signal processing techniques [3]. These methods generate better quality images by estimating unknown high-frequency components in many cases; however, there is no guarantee that the estimations of the components are true.

The second type is the so-called "registration type" super-resolution technique. This registration-based super-resolution technique utilizes a large number of images in order to increase the pixel density of the image. Accordingly, the true high-frequency components of the image can be estimated by using an appropriate reconstruction algorithm. Typical registration type super-resolution techniques utilize a maximum likelihood (ML) method [4], maximum a

© Springer International Publishing Switzerland 2015
G. Bebis et al. (Eds.): ISVC 2015, Part I, LNCS 9474, pp. 911–920, 2015.
DOI: 10.1007/978-3-319-27857-5_81

posteriori (MAP) method [5,6], iterative backward projection (IBP) method [7], frequency domain approach [8], or projection onto convex sets (POCS) method [9,10] as a reconstruction algorithm. However, because these methods require not a few observed images, precise correction of the positional deviations among the images, namely sub-pixel image registration, is necessary for the successive reconstruction process. It should be noted that the positional deviation among the images is not only translation but also constitutes complex deformation, such as rotation and scaling.

Various image registration techniques, including region-based matching [11], local-feature-based matching based on robust image features such as SIFT [12] and SURF [13], and so on, have been widely used. In most cases, the super-resolution process is performed based on images taken by a single camera device and therefore when moving objects are present in the target image frame, sophisticated image processing techniques [14] are necessary.

In this paper, we propose a synchronously controlled super-resolution camera array system. This system is a high quality imaging system that can also handle moving objects and does not require expensive optical equipment or complex image processing. However, the camera array system may give rise to the following issues. (1) A mismatch of color and brightness in the images captured by different cameras, (2) parallax among cameras, and (3) limited image quality due to the restriction of the number of cameras. Thus, in this paper, we also propose three effective measures to address these issues.

2   Registration-Type Super-Resolution

Our super-resolution camera-array system is composed of a large number of low-cost cameras and a registration type super-resolution algorithm. The process of the super-resolution consists of two processes:

1. Image alignment process.
2. Reconstruction process.

The details are described in the following.

2.1   Image Alignment Process

The registration type super-resolution technique requires sub-pixel registration accuracy. A flow chart of the image alignment process is shown in Fig. 1. Now, let us consider the image alignment between two images, namely, the HR image estimated at the time of the imaging and the up-sampled observed LR image, for instance.

When an arbitrary point object is captured by the proposed camera array system, we assume it should be observed at a similar position in each image. Based on this assumption, we compared the coordinates of each matching point of the images and, if their coordinates were significantly different, we considered this matching point inappropriate, and therefore, eliminated it. According to the results of preliminary experiments, we eliminated 50 % of matching points. Then, we selected four

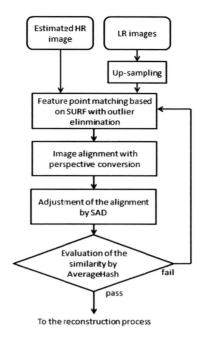

Fig. 1. Flow of the image alignment process.

matching points between the images for performing image alignment using perspective projection transformation. In order to achieve accurate image alignment, it is more effective to select four points having a bounded area that is as large as possible. Accordingly, we selected the matching point from each image quadrant that was the shortest distance from the corresponding corner. In order to achieve accurate and reliable image alignment, we conducted further image registration using the sum of absolute difference (SAD) measure. We generated several shifted images from one of the images and sought the location on the second image that had the lowest SAD. Thus, we realized a sub-pixel precise image alignment, i.e., alignment is performed in up-sampled images. Furthermore, in order to ensure accurate alignment, we evaluated the accuracy of the image alignment process. When we determined that the image alignment was inappropriate, we re-performed the alignment process using a different selection of SURF matching points. Here, we applied the AverageHash algorithm [15] for the evaluation of the alignment by calculating the similarity of the images. This algorithm first defines the region of interest (ROI) in the image and down-samples it to an 8×8 gray scale image. Then, the binarization of the ROI is performed using its average intensity as the threshold. Finally, the similarity is calculated by comparing the bit sequences. In this study, we defined arbitrary 30×30 square region extracted from the aligned images as the ROI and calculated the similarity.

2.2   Reconstruction Process

In the reconstruction process, we reconstruct an HR image from the observed LR images from the camera array system based on Farsiu's method [16]. According to Farsiu's model, the original HR image was degraded for several reasons and accordingly only the LR image was observed. This image degradation model is expressed as

$$Y_k = D_k H_k F_k X. \quad k = 1, \cdots, N. \tag{1}$$

Here, Y_k is the observed k-th LR image, D_k is the down-sampling process, H_k is the blurring function, F_k is the moving effect, X is the original (estimated) HR image, and N is the number of the input LR images.

We reconstructed the HR image based on the frequently used ML estimation method and a steepest descent method to solve the minimization problem of the differences between the original and estimated HR image. According to this, Eq. (1) can be rewritten as the following iterative formula.

$$X_{n+1} = X_n - \left\{ \sum_{k=1}^{N} F_k^T H_k^T D_k^T \text{sign} \left(D_k H_k F_k X_n - Y_k \right) \right\}. \tag{2}$$

Here, X_n is the n-th estimated HR image produced by steepest descent method. In the reconstruction process, we used the Butterworth low-pass filter as the blur kernel. Because of space limitation, for details, please refer the original article [16].

3   Super-Resolution Camera Array System

The prototype of our super-resolution camera-array system is shown in Fig. 2. This prototype is composed of 12 Axis M1011-W cameras. The resolution of each camera is 640×480 pixel. The horizontal and vertical viewing angle of each camera is $47°$ and $35°$, respectively. The size of the overall system is $18\,\text{cm} \times 24\,\text{cm}$.

Since multiple cameras are arranged in our system, parallax and the differences in photographic conditions among the cameras cannot be ignored. We therefore propose three approaches to address these issues and improved the image quality.

Fig. 2. Camera array system.

Fig. 3. Observed image.

Fig. 4. Determined ROI image.

3.1 Color Calibration Among Cameras

We performed a color calibration among the observed images. We selected an arbitrary ROI from an HR image estimated as the basis for the calibration. We searched the corresponding areas among LR images in a manner similar to the procedure described in Sect. 2.1 and calibrated them using a histogram equalization strategy.

3.2 Automated ROI Detection-Strategy for Parallax

In a camera array system, parallax among equipped cameras occurs and it becomes particularly pronounced when the targets are located near the system. It causes the appearance of the objects to differ, in particular, in terms of depth. It is therefore quite important that the issue of parallax be treated appropriately. We propose a semi-automated detection method for the areas required for performing high resolution processing, i.e., the ROIs. First, we select the arbitrarily observed i-th image $(1, 2, \cdots, i, \cdots, N)$ and manually determine the area ROI_i that requires HR processing. Based on ROI_i, the corresponding areas among the observed images ROI_j $(1, 2, \cdots, j, \cdots, N, j \neq i)$ are automatically estimated using the image alignment procedure based on the method described in Sect. 2.1. An example of the observed image and the determined ROI_i is shown in Figs. 3 and 4, respectively.

3.3 Effective Noise Reduction with Effective Edge Preservation

Usually, a low-cost camera device comprises a small optical sensor (measuring "1/4", "1/2.33," similar). In situations where the illumination is insufficient, including even in normal indoor environments, gain noise is frequently incurred in the observed image produced by these sensor devices. Unfortunately, in some cases, this cannot be ignored in the super-resolution technique process. As

Fig. 5. Result of bi-linear method.

Fig. 6. Super-resolution result of conventional method from 12 LR images.

Fig. 7. Super-resolution result of proposed method from 12 LR images.

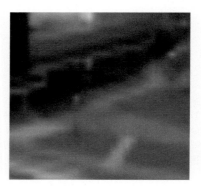

Fig. 8. Magnified view of result of bilinear interpolation.

mentioned earlier, conventional super-resolution techniques frequently emphasize noise, as well as the requisite edges, contours, and so on, when the number of the observed images is limited. Since one of our objectives is to develop a low-cost super-resolution system, effective noise reduction is crucial.

At the beginning of our super-resolution process, the up-sampled observed LR image is treated as the initial estimation of the HR image. The initial HR image frequently includes noise originating from the base LR image. In addition, ML estimation during the iterative updating process sometimes amplifies noise, in particular when the given LR images have relatively large noise or the number of given LR images is limited. In order to address each of the above issues, we introduce the idea of using two filters; (1) a median filter applied to the initial HR image to reduce the initial noise, and (2) a bi-lateral filter applied to the sum of the residual between the HR image estimated and the i-th up-sampled LR image $(1, 2, \cdots, i, \cdots, N)$. It can be expected that introducing these two procedures will reduce noise, while preserving native edges, contours, and other important components. These filtering processes improved the image quality according to the qualitative visual assessment in our preliminary experiments.

Fig. 9. Magnified view of super-resolution result of conventional method from 12 LR images.

Fig. 10. Magnified view of super-resolution of proposed method from 12 LR images.

Fig. 11. Result of bi-linear method.

Fig. 12. Super-resolution result of proposed method from 12 LR images.

4   Results

First, we tested our system using a distant view. It should be noted that the magnification ratio in this experiment was 3×3. Figures 5, 6 and 7 show the results of bi-linear interpolation (Fig. 5), conventional super-resolution techniques (Fig. 6), and the proposed method (Fig. 7). Figures 8, 9 and 10 show enlarged areas of the images in Figs. 5, 6 and 7, respectively, for visual assessment.

Second, we tested our system using a short distance view. As explained previously, Fig. 4 illustrates the sample of the ROI in the LR image (ROIs were appropriately determined also in other LR images) and Figs. 11 and 12 show the results of bi-linear interpolation and the proposed system reconstructed from 12 LR images, respectively. In order to investigate the effectiveness of the proposed method, we compared the conventional super-resolution method and the proposed method based on only 4 LR images. The results are shown in Figs. 13 and 14, respectively. Figures 15, 16, 17 and 18 show enlarged areas of these images

Fig. 13. Super-resolution result of conventional method from 4 LR images.

Fig. 14. Super-resolution result of proposed method from 4 LR images.

Fig. 15. Magnified view of bi-linear method.

Fig. 16. Magnified view of super-resolution result of conventional method from 12 LR images.

Fig. 17. Magnified view of super-resolution result pf conventional method from 4 LR images.

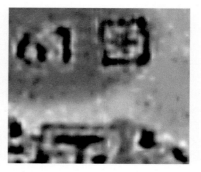

Fig. 18. Magnified view of super-resolution result of proposed method from 4 LR images.

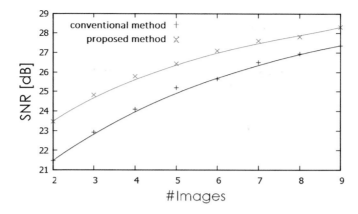

Fig. 19. Relationship between # number of images and the S/N ratio.

(Fig. 15: bi-linear, Fig. 16: proposed from 12 LR, Fig. 17: proposed method from 4 LR, and Fig. 18: conventional method from 4 LR).

As for the quantitative evaluation, we defined the result of the super-resolution using 40 LR images as the gold standard in this study and compared the proposed method and the conventional algorithm. Figure 19 illustrates the relationship between the number of base LR images and the S/N ratio. Note that, the S/N ratio of the bi-linear method was 20.1 dB.

It can be seen that the proposed method achieved superior results.

5   Discussion

We confirmed the advantages of the super-resolution technique in all cases, although in the cases of long distance imaging (see Figs. 6 and 7), the visual advantage of the proposed method over the conventional one was limited. We consider that the reasons are that the outdoor imaging with sufficient light did not produce a large gain noise in the original LR images and the number of the LR images used for reconstruction was sufficient.

In the short distance example, we confirmed that our proposed method successfully improved the resolution by resolving the parallax and color divergence issues (see Figs. 15 and 16). Now, we consider the cases where both the amount of the light and the number of reconstruction images were limited. It can be seen in Figs. 17 and 18 that the conventional method emphasizes the edges as well as noise, while the proposed method emphasizes the edges to the same degree as the conventional method, but successfully suppresses noise. In addition, it can be seen in Fig. 19 that the proposed method shows an S/N ratio superior to that of the conventional method by 2 dB. This is almost equivalent to the effect of two additional observed images.

The above results confirm that the proposed method is robust against noise and produces a high quality high-resolution image, even when the number of observed LR images is insufficient.

6   Summary

In this study, we developed a prototype of a super-resolution camera array system. The camera array system has several merits, on the other hand; however, on the other hand, it also suffers from several problem that need to be addressed to allow its practical usage. We proposed three processes to reduce the drawbacks of the array system and achieved an approximately 2 dB higher S/N ratio (i.e., equivalent to the effect of two additional images) than the conventional method.

The miniaturization of the system reduces the drawbacks of the array system. We will develop the next version of the model in the near future.

References

1. Park, S.C., Park, M.K., Kang, M.G.: Super-resolution image reconstruction: a technical overview. IEEE Signal Process. Mag. **20**, 21–36 (2003)
2. Freeman, W.T., Jones, T.R., Pasztor, E.C.: Example-based super-resolution. IEEE Comput. Graphics Appl. **22**, 56–65 (2002)
3. Takashi, K., Takahiro, S.: Super-resolution decoding of the JPEG coded image data using total-variation regularization. In: Picture Coding Symposium, pp. 114–117 (2010)
4. Tom, B., Katsaggelos, A.: Reconstruction of a high-resolution image by simultaneous registration, restoration, and interpolation of low-resolution images. In: IEEE International Conference on Image Processing, vol. 2, pp. 539–542 (1995)
5. Schultz, R., Stevenson, R.: A bayesian approach to image expansion for improved definition. IEEE Trans. Image Process. **3**, 233–242 (1994)
6. Schultz, R., Stevenson, R.: Extraction of high-resolution frames from video sequences. IEEE Trans. Image Process. **5**, 996–1011 (1996)
7. Irani, M., Peleg, S.: Improving resolution by image registration. Graph. Models Image Process. **53**, 231–239 (1991)
8. Tsai, R., Huang, T.: Multiframe image restoration and registration. In: Advances in Computer Vision and Image, vol. 1, pp. 317–339 (1984)
9. Eren, P., Sezan, M., Tekalp, A.: Robust, object-based high resolution image reconstruction from low-resolution video. IEEE Trans. Image Process. **6**, 1446–1451 (1997)
10. Patti, A.J., Altunbasak, Y.: Artifact reduction for set theoretic super resolution image reconstruction with edge adaptive constraints and higher-order interpolants. IEEE Trans. Image Process. **10**, 179–186 (2001)
11. Zitova, B., Flusser, J.: Image registration methods: a survey. Image Vis. Comput. **21**, 977–1000 (2003)
12. Lowe, D.G.: Distinctive image features from scaleinva-invariant keypoints. Proc. Int. J. Comput. Vis. (IJCV) **60**, 91–110 (2004)
13. Bay, H., Ess, A., Tuytelaars, T., Gool, L.V.: Speeded-up robust features (surf). Comput. Vis. Image Underst. **110**, 346–359 (2008)
14. Tung, T., Nobuhara, S., Matsuyama, T.: Simultaneous super-resolution and 3d video using graph-cuts. In: IEEE Conference Computer Vision Pattern Recognition, Anchorage (2008)
15. Aghav, S., Kumar, A., Gadakar, G., Mehta, A., Mhaisane, A.: Mitigation of rotational constraints in image based plagiarism detection using perceptual hash. Int. J. Comput. Sci. Trends Technol. **2**, 28–32 (2014)
16. Farsiu, S., Robinson, M.D., Elad, M., Milanfar, P.: Fast and robust multi-frame super-resolution. IEEE Trans. Image Process. **13**, 1327–1344 (2004)

Author Index

Agamennoni, Gabriel   I-432
Agu, Emmanuel   II-150
Akhoury, Sharat   II-237
Akizuki, Shuichi   I-878
Akram, Bita   I-819
Al Delail, Buti   II-618
Alba, Alfonso   I-151
Aldhahab, Ahmed   II-672
Alexis, Kostas   I-444, I-455
AlHalawani, Sawsan   II-702
Al-Hamadi, Ayoub   II-337
Alim, Usman R.   I-819
Al-Mualla, Mohammed   II-618
Álvarez, Mauricio A.   I-349, I-491, II-499,
    II-682, II-692
Alves, W.A.L.   II-45
Amditis, Aggelos   I-162
Amorim, Paulo   I-45
Andaluz, Víctor H.   I-476
Anselmo, K.P.   II-45
Aoki, Yoshimitsu   I-635
Arango, Ramiro   I-349
Araújo, S.A.   II-45
Arbab-Zavar, Banafshe   I-162
Arce-Santana, Edgar   I-151
Atia, George   II-672
Auttila, Miina   II-227

Bae, Sujung   I-396
Baghaie, Ahmadreza   I-406
Bahnsen, Chris   I-741, I-774
Baldrich, Ramon   II-463
Balsa-Barreiro, José   I-361
Balta, Haris   I-466
Banic, Amy Ulinski   II-443, II-745
Barnada, Marc   I-115
Bebis, George   II-213
Belan, P.A.   II-45
Belkhouche, Yassine   II-655
Benavides, Angela   II-443
Benes, Bedrich   II-126
Benligiray, Burak   II-35
Bhagavatula, Sravan   I-243
Bhaskar, Harish   II-618

Biswas, K.K.   II-195
Blache, Ludovic   I-92
Bouakaz, Saida   II-304
Bousetouane, Fouad   II-379
Braffort, Annelies   II-293
Branch, John W.   II-102
Brandão, Iara   II-586
Brun, Anders   I-231, I-297
Brutch, Tasneem   II-160
Buckles, Bill   II-655
Burch, Michael   II-733
Bush, Mark   II-347

Cabezas, Ivan   II-102
Cardona, Hernán Darío Vargas   I-349,
    II-499, II-692
Cardoso, Jaime S.   II-268, II-509
Carillo, Fabio Martinez   II-463
Carneiro, Allan   II-410
Celebi, M. Emre   II-712
Chan, Chee Seng   II-776
Chen, Chen   I-613
Chen, Cheng   I-14
Chen, Hsiao-Tzu   I-127
Chen, Xianming   II-248
Chen, Xiaoping   I-253, II-389
Chendeb, Safwan   II-797
Cheng, Irene   II-489
Chicaiza, Fernando   I-476
Choi, Min-Hyung   II-474
Cobb, Jorge A.   II-755
Correndo, Gianluca   I-162

D'Souza, Roshan M.   I-406
Dai, Shuanglu   I-657
Dam, Nhan   II-420
Davatzikos, Christos   I-521
Daza, Juan M.   II-173
De Cubber, Geert   I-466
De Meyst, Olivier   I-466
Dekkiche, Djamila   I-762
Demer, Joseph L.   I-501
Desai, Alok   II-819
Desbrun, Mathieu   I-92

Díez, German   II-173
Dings, Laslo   II-337
Do, Minh N.   II-420
Dodds, Z.   I-373
Dong, Chun   I-511
Doulamis, Anastasios   I-693, I-717
Doulamis, Nikolaos   I-162, I-706
Duan, Xiaodong   II-258
Duong, Anh-Duc   II-420
Duraisamy, Prakash   II-655

Eerola, Tuomas   I-187, II-227
Egges, Arjan   I-57
Eichner, Christian   II-722
Eiksund, Oddmar   I-901
Einakian, Sussan   I-318
El-etriby, Sherif   II-337
ElTantawy, Agwad   I-275
Elzobi, Moftah   II-337
Emami, Ebrahim   II-213
Esteves, Rui   II-268

Fanani, Nolang   I-115
Feng, Chaolu   I-531
Fiadeiro, Paulo Torrão   II-268
Flores-Barranco, Martha Magali   II-489
Fong, Terry   II-213
Forrest, Peter M.   I-265
Fritsch, Dieter   I-361
Fujisaki, Masahiro   II-137

Gao, Xiaoli   II-664
Garcia Bravo, Esteban   II-767
García, Hernán F.   I-349, I-491, II-682
Garg, Sourav   I-138
Gaumer, M.   I-373
Gaüzère, Benoit   II-574
Ge, Shuzi Sam   II-842
Geist, Robert   I-82
Giese, Martin   I-842
Giraldo, Jhony   II-173
Goethals, Thijs   I-466
Gomes, Helder T.   I-25
Gómez, Alexander   II-173
Gong, Huajun   II-93
Gong, Zhaoxuan   I-521, I-531
Gouiffès, Michèle   II-293, II-463
Grammenos, D.   I-551
Granger, Eric   II-509

Greco, Claudia   II-574
Grimm, Cindy   I-35, I-307
Guamushig, Franklin   I-476
Guha, Prithwijit   I-138
Guijarro, Enrique   I-349
Guo, Shuxu   II-205
Guo, Yanan   II-664

Haario, Heikki   I-187
Haelterman, Rob   I-466
Hanmandlu, M.   II-195
Harley, Adam W.   I-867
Hashimoto, Manabu   I-878
Hassan, Ehtesham   I-138
Hassannia, Hamid   II-542
He, Guotian   I-561
Hentschel, Christian   II-400
Hinkenjann, André   II-115
Hirao, Daiki   I-911
Holcomb, Jeffrey W.   II-755
Holguín, Germán A.   II-682
Höllerer, Tobias   I-599
Holloway, Michelle   I-307
Hollowood, Kathryn   II-347
Horrocks, Ian   I-842
Hou, Yuqing   I-71
Hou, Zhenjie   I-613
Hsu, Leighanne   II-24
Hu, Fei   II-586
Hu, Woong   II-532
Huber, David J.   I-419

Ibarra-Mazano, Mario-Alberto   II-489
Iwata, Kenji   I-635
Iyatomi, Hitoshi   I-911, II-16, II-638, II-712

Jacquemin, Christian   II-463
Jafari, Mohammad   II-628
Jamal Zemerly, M.   II-618
Jensen, Morten B.   I-774
Jeong, Moonsik   I-396
Jiang, Junjun   I-613
Jimenez-Ruiz, Ernesto   I-842
Jin, Dakai   I-14

Kagiwada, Satoshi   II-638
Kälviäinen, Heikki   I-187, II-227
Kán, Peter   I-574
Kang, Byeongkeun   I-586

Kapoor, Aditi   II-195
Karantzalos, Konstantinos   I-717
Karatzas, Dimosthenis   II-327
Kataoka, Hirokatsu   I-635
Kaufmann, Hannes   I-574
Kawasaki, Yusuke   II-638
Kenwright, Ben   II-787
Khan, Rizwan Ahmed   II-304
Khan, Waqar   II-57
Kharlamov, Evgeny   I-842
Khoshrou, Samaneh   II-509
Khosla, Deepak   I-419, I-888
Kim, Byung-Gyu   II-532
Kim, Kyungnam   I-419, I-888
Klette, Reinhard   II-57
Koivuniemi, Meeri   II-227
Koutlemanis, P.   I-551
Krigger, Amali   I-307
Kular, Dalwinderjeet   I-647, II-347
Kumar, Swagat   I-138
Kunnasranta, Mervi   II-227

La, Hung Manh   I-857, II-628
Lam, Ka Chun   I-809
Le, Justin H.   I-682
Lee, Dah-Jye   II-646, II-819
Lee, Seungkyu   I-396
Lee, Yeejin   I-586
Leondar, A.   I-373
Leroy, Laure   II-797
Levänen, Riikka   II-227
Levine, Joshua A.   I-82
Li, Binglin   II-355
Li, Chunming   I-521, I-531, II-205
Li, Robert (Bo)   II-160
Li, Yifan   I-501
Li, Yongmin   II-115
Liao, Wentong   I-198
Liebeskind, David S.   I-538
Lien, Kuo-Chin   I-599
Lima, Rui   I-25
Lin, Jung-Hsuan   II-521
Lin, Yuanchang   I-561
Liu, Chao   I-671
Liu, Yongchun   II-93
Liu, Yu   II-205
Lladós, Josep   II-327
Lobo, Joana   II-268

López, Edison   I-476
López-Lopera, Andrés F.   II-692
Loscos, Céline   I-92
Loupos, Konstantinos   I-693, I-717
Lucas, Laurent   I-92
Lui, Lok Ming   I-809
Luo, Heng   I-561
Luo, Zhiping   I-57

Ma, Yide   II-664
Maiero, Jens   II-115
Makantasis, Konstantinos   I-693, I-717
Makrogiannis, Sokratis   I-221
Malpica, Norberto   II-692
Man, Hong   I-657
Manceau, Jérôme   I-386
Manobanda, David   I-476
Manzanera, Antoine   II-293
Martinello, Manuel   I-671
Martínez, Fabio   II-293
McGraw, Tim   I-3, I-102, II-767
McInerney, T.   I-797, I-830
Mecocci, Alessandro   II-81
Medeiros, Fatima   II-410
Medina, J.   I-373
Mérigot, Alain   I-762
Mester, Rudolf   I-115
Meyer, Alexandre   II-304
Micheli, Francesco   II-81
Middleton, Lee   I-162
Mikhael, Wasfy B.   II-672
Milkosky, Nathaniel   II-24
Miller, Corey A.   I-728
Misawa, Hiroki   II-16
Mitra, Niloy J.   II-702
Mody, Shreeya   II-819
Moeslund, Thomas B.   I-741, I-774
Møgelmose, Andreas   I-774
Monteiro, João C.   II-268
Moolla, Yaseen   II-237
Moraes, Thiago   I-45
Morales, Vicente   I-476
Morishima, Shigeo   II-137
Morris, Brendan   I-752, II-379, II-809
Mostafa, Ahmed E.   II-453
Müller, Oliver   II-365
Muthukumar, Venkatesan   I-682, II-554

Nagai, Kazuhiro   I-328
Nakada, Masaki   I-339
Nam, Jae-Hyun   II-532
Nefian, Ara   II-213
Newman, Timothy S.   I-318, I-511
Nguyen, Truong Q.   I-586
Nguyen, Vinh-Tiep   II-420
Niemi, Marja   II-227
Nisar, I.   I-830
Nixon, Mark S.   I-265
Ntelidakis, A.   I-551
Nyolt, Martin   II-722

Obara, Takashi   II-16
Ogura, Yohei   II-433
Ohki, Keiichi   II-712
Olson, Clark F.   II-597
Ommojaro, Olatide   II-347
Orozco, Álvaro A.   I-349, I-491, II-499,
    II-682, II-692
Ostermann, Jörn   II-365

Padilla, José Bestier   I-349
Palma, Carlos   I-174
Papachristos, Christos   I-444, I-455
Patil, Yogendra   II-586
Payandeh, Shahram   II-314, II-563
Pedrini, Helio   I-45
Pérez, Fabricio   I-476
Phan, Ly   I-35
Philipsen, Mark P.   I-774
Piastowski, Patrick   I-784
Piziak, Dee   II-542
Pourashraf, Payam   II-609
Pradhan, Neera   II-443
Protopapadakis, Eftychios   I-706

Qin, Yalun   I-599
Quachtran, Benjamin   I-538
Quint, Franz   II-70

Regentova, Emma E.   I-682, II-554
Ren, Hongliang   II-842
Reso, Matthias   I-198
Ribeiro, Eraldo   I-647, II-347
Ridene, Taha   II-797
Ritrovato, Pierluigi   II-574
Rivas, David   I-476
Rivera, V.   I-373

Rodrigues, Pedro J.S.   I-25
Rodrigues, Raquel O.   I-25
Rong, Guodong   II-160
Rosen, Paul   I-328
Rosenhahn, Bodo   I-198
Roth, Thorsten   II-115
Rueben, Matthew   II-830
Rugonyi, Sandra   I-35
Rusiñol, Marçal   II-327

Sabeur, Zoheir   I-162
Sack, Harald   II-400
Saggese, Alessia   II-574
Saha, Punam K.   I-14
Saith, Ebrahim   II-237
Saito, Hideo   II-433
Sakou, Lionel B.   II-745
Salazar, Augusto   I-174, II-173
Salgian, Andrea   II-24
Samavati, Faramarz F.   I-819
Sampo, Jouni   I-187
Sanandaji, Anahita   I-307
Sánchez, Carlos   I-476
Sánchez, Jorge S.   I-476
Santos, Gil   II-268
Sapra, Yash   I-82
Satoh, Yutaka   I-634
Saver, Jeffrey L.   I-538
Scalzo, Fabien   I-538
Schaefer, Gerald   II-712
Schumann, Heidrun   II-722
Schwenk, Andreas   II-283
Séguier, Renaud   I-386
Sengupta, Shamik   II-628
Sephus, Nashlie   I-243
Setkov, Aleksandr   II-463
Shahraki, Farideh Foroozandeh   II-554
Shalunts, Gayane   I-285, II-185
Sharlin, Ehud   II-453
Sharma, Puneet   I-901
She, Bohao   II-597
Shehata, Mohamed S.   I-275
Shen, Yi   II-160
Sheth, Sunil   I-538
Shirazi, Mohammad Shokrolah   I-752, II-809
Shu, Chang   II-160
Siegwart, Roland   I-432
Silva, Adrián M.T.   I-25
Silva, Jorge   I-45
Singh, Ann   II-237

Siqueira, Guilherme   II-586
Skaff, Sandra   I-671
Smart, Katrina   II-347
Smart, William D.   II-830
Soladié, Catherine   I-386
Song, Yihua   I-531
Sousa, Mario Costa   II-453
Souza, Marcelo   II-410
Sowell, Ross   I-307
Soylu, Ahmet   I-842
Stastny, Thomas   I-432
Stilla, Uwe   II-70
Sun, Che   II-150

Tafti, Ahmad P.   I-561, II-542
Takashima, Kazuki   II-453
Tamames, Juan Antonio Hernández   II-692
Tan, Wenjun   I-521
Tan, Zheng-Hua   II-258
Tang, Chih-Wei   I-127
Tang, Jun   I-198
Tang, Keke   I-253, II-389
Tavakkoli, Alireza   I-210
Tavares, João Manuel R.S.   I-25
Tavares, Pedro B.   I-25
Teixeira, Luís F.   II-509
Tenorio, D.   I-373
Terzopoulos, Demetri   I-339
Thomas, Gabriel   II-355
Tian, Yingli   II-3
Tomuro, Noriko   II-609
Topal, Cihan   II-35
Torres-Valencia, Cristian A.   II-682
Tran, D.   I-797
Tran, Minh-Triet   II-420
Transue, Shane   II-474
Trivedi, Mohan M.   I-774
Turk, Matthew   I-599
Tzes, Anthony   I-444, I-455
Tzoumanikas, Dimos   I-444

Uga, Hiroyuki   II-638
Unterguggenberger, Johannes   I-574
Urhahne, Joseph A.   I-784
Ushizima, Daniela   II-410

van der Voort, Mascha C.   I-784
Van, Mien   II-842
Vanrell, Maria   II-463
Vargas, Camilo   II-102

Vargas, Francisco   I-174
Vasko, Ross   II-70
Veltkamp, Remco C.   I-57
Vempati, Anurag Sai   I-432
Vento, Mario   II-574
Vickerman, David   II-24
Vincke, Bastien   I-762
Vogt, Karsten   II-365

Walls, Thomas J.   I-728
Wang, Jing-wei   II-521
Wang, Jingya   II-314
Wang, Keju   II-664
Wang, Rong-Sheng   II-521
Watcharopas, Chakrit   I-82
Wei, Qi   I-501
Weier, Martin   II-115
West, Ruth   I-307
Whittinghill, David   II-126
Wiggins, Brendan   I-501
Wilches, Daniel   II-745
Wilkinson, Tomas   I-231, I-297
Williams, Dexter   II-355
Wilson, Brandon   I-210
Wiradarma, Timur Pratama   II-400
Wong, Jessica Soo Mee   II-776
Woods, Alexander C.   I-857
Wu, Denglu   II-842

Xian, Yang   II-3
Xing, Qi   I-501
Xu, George   II-563
Xu, Jiejun   I-419, I-888

Yang, Michael Ying   I-198
Yang, Seung-Hoon   II-532
Yang, Xiaodong   II-3
Yang, Yun   I-613
Yang, Zhen   II-664
Yates, Deniece   I-307
Yazdanpanah, Ali Pour   I-682, II-554
Ye, Cang   I-624
Yi, Jaeho   I-396
Yu, Zeyun   I-406, II-542
Yuan, Chunrong   II-283

Zabulis, X.   I-551
Zafari, Sahar   I-187
Zeller, Niclas   II-70

Zhang, Baochang   I-613
Zhang, Chao   I-221
Zhang, Chaoyang   II-248
Zhang, He   I-624
Zhang, Lei   I-888
Zhang, Meng   II-646
Zhang, Shaoxiang   II-205
Zhang, Ting-hao   I-127
Zhang, Wenyin   II-248
Zhang, Zhaoxing   II-93
Zhao, Dazhe   I-521, I-531

Zhao, Mengyuan   I-561
Zhao, Yue   II-205
Zhao, Zhe   I-253, II-389
Zhelezniakov, Artem   II-227
Zheleznyakov, Dmitriy   I-842
Zhou, Shengchuan   II-126
Zhou, Zhaoxian   II-248
Zhu, Qin   II-443
Zhuo, Huilong   II-126
Zikou, Lida   I-455
Zoppetti, Claudia   II-81